Advanced Java Programming

Advanced Java Programming

B Prasanalakshmi MCE, PhD

Associate Professor
Department of Computer Science and Engineering
Professional Group of Institutions
Palladam, Tirupur, Tamil Nadu, India

CBS Publishers & Distributors Pvt Ltd

New Delhi • Bengaluru • Chennai • Kochi • Mumbai • Pune
Hyderabad • Kolkata • Nagpur • Patna • Vijayawada

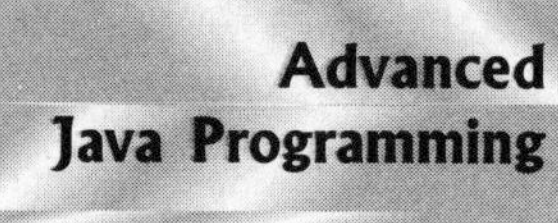

ISBN: 978-81-239-2383-3

First Edition: 2015

Published by Satish Kumar Jain for
CBS Publishers & Distributors Pvt Ltd
4819/XI Prahlad Street, 24 Ansari Road, Daryaganj, New Delhi 110 002, India.
Ph: 23289259, 23266861, 23266867   Fax: 011-23243014   Website: www.cbspd.com
e-mail: delhi@cbspd.com; cbspubs@airtelmail.in.

Corporate Office: 204 FIE, Industrial Area, Patparganj, Delhi 110 092
Ph: 4934 4934   Fax: 4934 4935   e-mail: publishing@cbspd.com; publicity@cbspd.com

Branches

- **Bengaluru:** Seema House 2975, 17th Cross, K.R. Road, Banasankari 2nd Stage, Bengaluru 560 070, Karnataka
  Ph: +91-80-26771678/79   Fax: +91-80-26771680   e-mail: bangalore@cbspd.com
- **Chennai:** 20, West Park Road, Shenoy Nagar, Chennai 600 030, Tamil Nadu
  Ph: +91-44-26260666, 26208620   Fax: +91-44-42032115   e-mail: chennai@cbspd.com
- **Kochi:** 36/14 Kalluvilakam, Lissie Hospital Road, Kochi 682 018, Kerala
  Ph: +91-484-4059061-65   Fax: +91-484-4059065   e-mail: kochi@cbspd.com
- **Mumbai:** 83-C, Dr E Moses Road, Worli, Mumbai-400018, Maharashtra
  Ph: +91-22-24902340/41   Fax: +91-22-24902342   e-mail: mumbai@cbspd.com
- **Pune:** Bhuruk Prestige, Sr. No. 52/12/2+1+3/2 Narhe, Haveli (Near Katraj-Dehu Road Bypass), Pune 411 041, Maharashtra
  Ph: +91-20-64704058/59, 32392277   Fax: +91-20-24300160   e-mail: pune@cbspd.com

Representatives

- **Hyderabad** 0-9885175004
- **Nagpur** 0-9021734563
- **Vijayawada** 0-9000660880
- **Kolkata** 0-9831437309, 0-9051152362
- **Patna** 0-9334159340

Printed at India Binding House, Noida

to

Mr S Balaji, my hubby
Master B Nandha Kumar and
Master B Vijay Sriram
and my sweet kids

Preface

The importance of programming is well known nowadays for development of any logic functionality. Overwhelming response to my previous books as recognition from universities inspired me to write this book. The book is structured to cover all the object-oriented programming aspects. This book uses plain and lucid language to explain the concepts of programming right from the basic to various complicated and important topics. Each topic is well supported with sufficient illustrations and output for the worked out programs. All chapters in the book are arranged in a proper sequence so that the idea may develop for the user himself to write a program on his own. Interview-based questions are included to help the student community to prepare themselves for their future and also all methods have been provided as a ready reckoner including syntax and examples.

I wish to extend my profound thanks to all those who helped me in making this book a reality. Thanks to the moral support provided by my supervisor Dr A Kannammal, Professor, Department of CA, Coimbatore Institute of Technology, India, and technical support provided by Mr A Selvakumar, Assistant Professor, Department of MCA, Thirumalai Engineering College, Kanchipuram, India. Also my sincere thanks to the publisher and the entire team for their immense pain taken to get this book in time with quality printing. My sincere thanks to Mr YN Arjuna, CBS Publishers & Distributors, New Delhi, for his tremendous support and guidance throughout the publication of this book.

B Prasanalakshmi

Contents

Chapter 3

EVENT-DRIVEN PROGRAMMING

Chapter 4

GENERIC PROGRAMMING

Chapter 5

CONCURRENT PROGRAMMING

Chapter 6

JAVA LIBRARY

Chapter 7

NETWORK PROGRAMMING IN JAVA

Chapter 8

MULTI-TIER APPLICATION DEVELOPMENT

Chapter 9

APPLICATIONS IN DISTRIBUTED ENVIRONMENT

1
Programming Paradigms and Object-Oriented Programming

1. PROGRAMMING PARADIGMS

A programming paradigm is a general approach to programming or to the solution of problems using a programming language. Thus programming languages that have similar characteristics are grouped together into a single paradigm. A programming language is a way to express the solution to a problem. They are different from the system programs or hardware enhancement programs.

A programming paradigm is both a method of problem solving and an approach to programming language design. A distinction can be made between symbolic and non-symbolic (or numerical) programming.

The major paradigms are:

- Procedural programming
- Object-oriented programming
- Functional programming
- Logic programming
- Rule-based programming

1.1 Procedural Programming

Procedural programming is by far the most common form of programming. A program is a series of instructions which operate on variables, sometimes known as *imperative programming*. Procedures also known as routines, subroutines, methods or functions contain a series of computational steps to be carried out.

Advantages of procedural programming include its relative simplicity, and ease of implementation of compilers and interpreters. Disadvantages of procedural programming include the difficulties of reasoning about programs and to some degree of difficulty in parallelisation. Procedural programming tends to be relatively low level compared to some other paradigms, and as a result can be very much less productive.

1.2 Object-Oriented Programming

Object-oriented programming (OOP) was introduced by Xerox with the language Smalltalk. It is a programming paradigm using objects, which are instances of a class consisting of data fields and methods. It is generally viewed as a collection of interacting objects. The object-oriented approach encourages the programmer to place the data where it is not directly accessed by the rest of the program. This hidden data is accessed by calling specially written functions, commonly called methods, which are either bundled in with the data or inherited from "class objects." These act as the intermediaries for retrieving or modifying the data they control. The programming construct that combines data with a set of methods for accessing and managing those data is called an *object*.

Object-oriented programming is characterized by the defining of classes of objects, and their properties. Inheritance of properties is one way of reducing the amount of programming, and provision of class libraries in the programming environment can also reduce the effort required. Object-oriented programming has proved to be particularly successful in the design of user interfaces.

1.3 Functional Programming

Functional programming is based upon the notion of a program as a function in a similar sense to its usage in mathematics. Programs are designed by the composition of functions. In functional programming, programs are executed by evaluating *expressions,* in contrast with imperative programming where programs are composed of *statements* which change global *state* when executed. Functional programming typically avoids using mutable state. Functional programming requires that programs are treated like any other values and can be passed as arguments to other functions or be returned as a result of a function. It is also possible to define and manipulate functions from within other functions. Special attention needs to be given to functions that reference local variables from their scope. If such a function escapes their block after being returned from it, the local variables must be retained in memory, as they might be needed later when the function is called. Often it is difficult to determine when those resources can be released, so it is necessary to use automatic memory management. Some of the functional programs are Ocaml, Erlang, Haskell, etc.

The major advantages of functional programming are that programs are easy to understand and to formally reason about, and that functions are very reusable. It is possible to develop and maintain a very large programs consisting of thousands of functions, because functions have no side effects and it is therefore straightforward to fully test functions and the resulting systems.

The major disadvantage of functional programming is the difficulty of doing input–output since this is inherently nonfunctional. There are also other aspects of problem solving that cannot easily or sensibly be performed in a functional manner. Nevertheless, large programs can be developed with about 80% of the code being designed in purely functional manner.

1.4 Logic Programming

Logic programming is characterized by programming with relations and inference. **Logic programming** is based on the idea of using logical sentences in the causal form: G if G_1 and ... and G_n , to represent programs and to perform computation.

Logic programming is usually equated with PROLOG programming, although in fact there are other logic programming languages available. Nevertheless, due to the widespread usage of PROLOG, it is natural that it has become almost interchangeable with logic programming.

Logic programming adopts a different approach to problem solving to both procedural programming and functional programming. In essence, logic programming requires a logical declarative description of the nature of the problem. The designer need not worry about the execution of the program, and this is left to the PROLOG inference engine to deal with. As a result, this raises the level of abstraction considerably as the program contains no explicit control information. A PROLOG program consists of facts and rules. Running a program consists of asking a query about a fact.

Advantages of logic programming include the development of concise solutions to problems, and the ability to reason about the programs.

Disadvantages of PROLOG programming include slowness of execution, the difficulties encountered in learning complex PROLOG programming. Some people would add the difficulties in understanding complex PROLOG programs and the problems in debugging large PROLOG programs. However, with careful design it should be possible to design PROLOG programs so that they can be understood purely declaratively. It should not therefore be necessary for the reader to attempt to work out how the execution of the program will proceed, which is known as the procedural semantics. Nevertheless, many PROLOG programs cannot be understood without consideration of the procedural semantics.

1.5 Rule-based Programming

In computer science, **rule-based systems** are used as a way to store and manipulate knowledge to interpret information in a useful way. They are often used in artificial intelligence applications and research.

A typical rule-based system has four basic components:

- A list of rules or **rule base**, which is a specific type of knowledge base.
- An inference engine or semantic reasoner, which infers information or takes action based on the interaction of input and the rule base. The interpreter executes a production system program by performing the following match-resolve-act cycle:
    - ✓ ***Match***: In this first phase, the left-hand sides of all productions are matched against the contents of working memory. As a result a conflict set is obtained, which consists of instantiations of all satisfied productions. An instantiation of a production is an ordered list of working memory elements that satisfies the left-hand side of the production.
    - ✓ ***Conflict-resolution***: In this second phase, one of the production instantiations in the conflict set is chosen for execution. If no productions are satisfied, the interpreter halts.
    - ✓ ***Act***: In this third phase, the actions of the production selected in the conflict-resolution phase are executed. These actions may change the contents of working memory. At the end of this phase, execution returns to the first phase.
- Temporary working memory.
- A user interface or other connection to the outside world through which input and output signals are received and sent.

Rule-based programming is also known as IF-THEN programming and as expert system design. It shares many similarities with logic-based programming, and in principle is simply a variant of it. Nevertheless, practical rule-based systems are very different in their design and usage to PROLOG.

Rule-based programming has its origins in the very first electronic computer, the British Colossus developed at Bletchley park by Alan Turing and his research team to decode the German submarine messages. Rule-based systems are also sometimes known as Post production systems and Post was probably the first developer of the idea.

A major advantage of rule-based programming is that prototypes can often be developed very quickly. Some rule-based languages are also easy to learn.

The main disadvantages of rule-based systems include slowness of execution and the difficulties of debugging. More advanced rule-based systems include the ability to automatically generate procedural code (typically in C) which performs the same function as the expert system. This usually solves the execution speed problem.

There are also some paradigms left over without discussion, since they are of least importance to the subject concerned.

2. OBJECT-ORIENTED PROGRAMMING—FUNDAMENTALS

Object-oriented approach removes the flaws encountered in the procedural approach. OOP treats data as a critical element in the program development and does not allow it to flow freely around the system. It ties data more closely to the functions that operate on it, and protects it from accidental modification from outside functions. OOP allows decomposition of a problem into a number of entities called objects and then builds data and functions around these objects.The data of an object can be accessed only by the functions associated with that object. However, functions of one object can access the functions of other objects.

Characteristics of Object-Oriented Programming

- Emphasis is on data rather than procedure.
- Programs are organized into a number of objects.
- Data structures are designed such that they characterize the objects and functions that operate on the data of an object are tied together in the data structure.
- Data is hidden and cannot be accessed by external functions.
- Objects may communicate with each other through functions.
- New data and functions can be easily added whenever necessary.
- Follows bottom-up approach in program design.

Definition: "Object-oriented programming is an approach that provides a way of modularizing programs by creating partitioned memory area for both data and functions that can be used as templates for creating copies of such modules on demand." Thus, an object is considered to be a partitioned area of computer memory that stores data and a set of operations that can access that data. Since the memory partitions are independent, the objects can be used in a variety of different programs without modifications.

Review of OOP

It is necessary to understand some of the concepts used extensively in object-oriented programming.

These are:

- Objects and classes
- Inheritance
- Polymorphism
- Data abstraction and encapsulation
- Interfaces

Objects and classes and interfaces are discussed in the consecutive topics. Inheritance and polymorphism accompanied by data abstraction and encapsulation is discussed in Part 2.

3. OBJECTS AND CLASSES IN JAVA

A *class* defines the structure and behavior of an object or a set of objects. For example, peacock, sparrow, and kingfisher are categorized as birds because all of them share some common characteristics, such as they are covered with feathers, have hollow bone structures, and have the ability to fly. Similarly, the various entities that have similar attributes comprise a class. For example, in a university, all the students have similar attributes. They have a student ID, name, date of birth, gender, and address. Therefore, to automate the student management system of the university, you can create a Students class that consists of the attributes needed to store the student details. The general form of a class definition is:

```
class classname
{
        Type instance-variable;
        Type methodname (parameter-list)
        {
        ____
        ____
        }
}
```

Objects are the basic building blocks of object-oriented programming (OOP). The concept of the object-oriented methodology is based on objects. You create independent entities that can be reused across various programs. Objects display the following characteristics:

- State: The state of an object is indicated by a set of attributes and the values to these attributes. For example, when you design an online shopping site for books, each book, which is an object in the program, has its size, number of pages, type of binding, and its shipping details.
- Behavior: The behavior of an object refers to a change of its state over a period of time. For example, in the online shopping site, when you place an order for a book, the status of the book might change from available to sold out. The changed state of a book is its behavior.
- Identity: Each object has a unique identity, just as each person has a unique identity. For example, multiple books consist of the same size, number of pages, type of binding, and shipping details. However, each book has its distinct identity, for example, the book number. The book number identifies a book among other books.

An object is an instance of a class. A class in object-oriented methodology is a collection of various attributes, such as data and methods. You can access the data of a class by using its methods.

Consider a pay phone where you put in a coin and then dial the number that you want to call. The pay phone exhibits a certain kind of behaviour, that is, when you drop a coin and dial a number, it connects you to the person you want to talk to at the other end of the line. The order in which the operations are carried out is important. The pay phone behaves in this manner because of the existence of a state. One state of the pay phone is that a caller inserts a coin but the number is not dialled and in the other state a coin is inserted as well as the number is dialled.

The state and behavior together comprise the properties of an object. For example, one property of the object pay phone could be the amount of money or the number of coins to be inserted.

Grady Booch, the famous author, has written many books on OOP has defined a property as:

"An inherent or distinct characteristic trait, quality, or feature that contributes to making an object uniquely that object."

Objects do not exist in isolation. They interact with other objects and react to these interactions. These interactions take place through messages. Grady Booch has defined behaviour as:

"Behaviour is how an object acts and reacts, in terms of its state changes and message passing."

The identity of an object distinguishes it from all other objects. In the case of the pay phone, its identity will be the serial number assigned to it by the manufacturer. Two objects may have the same behavior, may or may not have the same state but will never have the same identity. The identity of an object never changes in its lifetime.

Several people have described techniques for identifying classes. According to one of these techniques, you first write the English description of the problem. Then you underline the nouns. The nouns represent candidate classes. For example, in an outlet, there can be several counters, each one managed by a single salesperson selling a specific product. A customer approaches any counter, depending on the product the customer wishes to purchase. The salesperson hands over the product and accepts the payment from the customer. The different classes that can be identified are:

- o Counter
- o Salesperson
- o Product
- o Customer
- o Payment

An object is declared as follows:

```
classname objectname=new classname();
```

This statement can also be rewritten as:

```
className objectName;
objectName=new className( );
```

A class defines user-defined objects and their characteristics. Any concept that you need to implement in a Java program is encapsulated within a class. A class defines the attributes and methods of objects of the same type sharing common characteristics. For example, when you create an online shopping site for books, each book is an instance of the Books class. All books have attributes, such as the number of pages, type of binding, and title. In addition, the books have various common methods, such as being selected, being bought, and being sold. The state of each book is independent of the state of another book.

The main components of a class are:

- Data members (attributes)
- Methods

Classes contain statements that include the declaration of data members, which specify the type of data to be stored. Methods of a class contain a set of executable statements that gives the desired output. Methods define the action to be carried out on the data members of the class. The class block or the class body is included within a set of braces, { }, which indicate the start and end of the class.

3.1 Creating Classes in Java

Various data members and methods of a class are defined inside a class. The statements written in a Java class must end with a semicolon, ;. The following syntax shows how to declare a class:

```
Class ClassName
{
// Declaration of data members
//Declaration of methods
}
```

In the preceding syntax, the world *class* is a keyword in Java used to declare a class and *ClassName* is the name given to the class. This name is required to create objects of the class.

The preceding syntax consists of a few statements that are prefixed with double slash, //. These statements are considered as comment entries by the compiler. The compiler ignores these statements and these are not executed at run time. It is the best practice to include comment entries to describe the code.

You can also declare an empty class that does not contain any data members and methods. For example, consider a class, Organization, which does not have data members and methods. The following syntax shows the declaration of an empty class, Organization, without any data members and methods:

```
Class Organization
{
    // No data members and methods
}
```

You can use the following code snippet to declare the Employee class that defines various data members, such as employeeName, employeeId, and employeeDesignation.

```
Class Employee
{
    String employeeName;
    int employeeID;
    string employeeDesignation;
}
```

3.2 Creating Objects of Classes

An object is an instance of a class and has a unique identity. The identity of an object distinguishes it from other objects. Classes and objects are closely linked to each other. While an object has a unique identity, a class is an abstraction of the common properties of various objects. To create an object, you need to perform the following steps:

1. Declaration: Declares a variable that holds the reference to the object. The following syntax shows how to declare an object of the class:

```
Class_name object_name;
```

2. Instantiation or creation: Creates an object of the specified class. When you declare an object, memory is not allocated to it. Therefore, you cannot store data in the data members of the object. To allocate memory to the object, you need to use the new operator. The new operator allocates memory to an object and returns a reference to that memory location in the object variable. The following syntax shows how to create an object:

```
Class_name object_name= new class_name( );
```

You can use the following statement to declare and instantiate an object, e1, of the Employee class in a single statement:

```
Employee e1= new Employee ( );
```

You can use the following code snippet to create objects for four employees of an organization:

```
Employee e1= new Employee ( );
Employee e2= new Employee ( );
Employee e3= new Employee ( );
Employee e4= new Employee ( );
```

3.3 Accessing Data Members of a Class

You need to assign values to data members of the object before using them. You can access the data members of a class outside the class by specifying the object name followed by the dot operator and the data member name. The following syntax shows how to access the data members of a class outside the class:

```
Object_name.data_member_name
```

In the preceding syntax, *object_name* refers to the name of the object and *data_member_name* refers to the name of the data variable inside the object that you want to access.

You can use the following code snippet to access the data members of the Employee class such as employeeName, employeeID, and employeeDesignation and assign values to them through objects, e1 and e2:

```
e1.employeeName="John";
e2.employeeName="Andy";
e1.employeeID=1;
e2.employeeID=2;
e1.employeeDesignation= "Manager";
e2.employeeDesignation= "Director";
```

3.4 Methods

A Java method is a collection of statements that are grouped together to perform an operation. When you call the System.out. println method, for example, the system actually executes several statements in order to display a message on the console.

Now you will learn how to create your own methods with or without return values, invoke a method with or without parameters, overload methods using the same names, and apply method abstraction in the program design.

I. Creating a Method

In general, a method has the following syntax:

```
modifier returnValueType methodName (list of parameters) {
  // Method body;
}
```

A method definition consists of a method header and a method body. Here are all the parts of a method:

- **Modifiers:** The modifier, which is optional, tells the compiler how to call the method. This defines the access type of the method.
- **Return Type:** A method may return a value. The returnValueType is the data type of the value the method returns. Some methods perform the desired operations without returning a value. In this case, the returnValueType is the keyword **void.**
- **Method Name:** This is the actual name of the method. The method name and the parameter list together constitute the method signature.
- **Parameters:** A parameter is like a placeholder. When a method is invoked, you pass a value to the parameter. This value is referred to as actual parameter or argument. The parameter list refers to the type, order, and number of the parameters of a method. Parameters are optional; that is, a method may contain no parameters.
- **Method Body:** The method body contains a collection of statements that define what the method does.

```
(1) public static int sum(int first,int second)
(2)   {
(3)   int sum;
(4)   sum= first + second;
(5)   return sum;
(6)   }
```

Where,

Line 1 – method header
 public, static – modifiers
 int – return value type
 sum –method name
 int first, int second – formal parameters
line 2–6 – method body
line 5 – return value

> **Note :** In certain other languages, methods are referred to as procedures and functions. A method with a nonvoid return value type is called a function; a method with a void return value type is called a procedure.

Example: Here is the source code of the above defined method called max(). This method takes two parameters num1 and num2 and returns the maximum between the two:

Routine to find maximum of two numbers

```
/** Return the max between two numbers */
public static int max(int num1, int num2) {
  int result;
  if (num1 > num2)
    result = num1;
  else
    result = num2;
  return result;
}
```

II. Calling a Method

In creating a method, you give a definition of what the method is to do. To use a method, you have to call or invoke it. There are two ways to call a method; the choice is based on whether the method returns a value or not.

When a program calls a method, program control is transferred to the called method. A called method returns control to the caller when its return statement is executed or when its method-ending closing brace is reached.

If the method returns a value, a call to the method is usually treated as a value. For example:

```
int larger = max(30, 40);
```

If the method returns void, a call to the method must be a statement. For example, the method println returns void. The following call is a statement:

```
system.out.println("Welcome to Java!");
```

Defining and using methods—maximum of 2 numbers

```
public class TestMax {
  /** Main method */
  public static void main(String[] args) {
    int i = 5;
    int j = 2;
    int k = max(i, j);
    System.out.println("The maximum between " + i + " and " + j + " is " + k);
}
/** Return the max between two numbers */
public static int max(int num1, int num2) {
  int result;
  if (num1 > num2)
    result = num1;
  else
    result = num2;
  return result;
}
```

Output

The maximum between 5 and 2 is 5

This program contains the main method and the max method. The main method is just like any other method except that it is invoked by the JVM.

The main method's header is always the same, like the one in this example, with the modifiers public and static, return value type void, method name main, and a parameter of the String[] type. String[] indicates that the parameter is an array of String.

III. The Void Keyword

This section shows how to declare and invoke a void method. Following example gives a program that declares a method named printGrade and invokes it to print the grade for a given score.

Finding grade using if-else ladder

```
public class TestVoidMethod {
  public static void main(String[] args) {
    printGrade(78.5);
  }
  public static void printGrade(double score) {
    if (score >= 90.0) {
      System.out.println('A');
    }
    else if (score >= 80.0) {
      System.out.println('B');
    }
    else if (score >= 70.0) {
      System.out.println('C');
    }
    else if (score >= 60.0) {
      System.out.println('D');
    }
    else {
      System.out.println('F');
    }
  }
}
```

This would produce following result:

C

Here the printGrade method is a void method. It does not return any value. A call to a void method must be a statement. So, it is invoked as a statement in line 3 in the main method. This statement is like any Java statement terminated with a semicolon.

IV. Passing Parameters by Values

When calling a method, you need to provide arguments, which must be given in the same order as their respective parameters in the method specification. This is known as parameter order association.

Method to print a message n times

```
public static void nPrintln(String message, int n) {
  for (int i = 0; i < n; i++)
    System.out.println(message);
}
```

Here, you can use nPrintln("Hello", 3) to print "Hello" three times. The nPrintln("Hello", 3) statement passes the actual string parameter, "Hello", to the parameter, message; passes 3 to n; and prints "Hello" three times. However, the statement nPrintln(3, "Hello") would be wrong.

When you invoke a method with a parameter, the value of the argument is passed to the parameter. This is referred to as pass-by-value. If the argument is a variable rather than a literal value, the value of the variable is passed to the parameter. The variable is not affected, regardless of the changes made to the parameter inside the method.

For simplicity, Java programmers often say passing an argument *x* to a parameter *y*, which actually means passing the value of *x* to *y*.

The following code demonstrates the effect of passing by value. The program creates a method for swapping two variables. The swap method is invoked by passing two arguments.

Swapping numbers

```
public class TestPassByValue {
  public static void main(String[] args) {
    int num1 = 1;
    int num2 = 2;
    System.out.println("Before swap method, num1 is " +num1 + " and num2 is "
+ num2);
    // Invoke the swap method
    swap(num1, num2);
    System.out.println("After swap method, num1 is " +num1 + " and num2 is " +
num2);
  }
  /** Method to swap two variables */
  public static void swap(int n1, int n2) {
    System.out.println("\t\tBefore swapping n1 is " + n1+ " n2 is " + n2);
    // Swap n1 with n2
```

```
    int temp = n1;
    n1 = n2;
    n2 = temp;
    System.out.println("\t\tAfter swapping n1 is " + n1+ " n2 is " + n2);
  }
}
```

This would produce the following result:

```
Before swap method, num1 is 1 and num2 is 2
    Before swapping n1 is 1 and n2 is 2
    After swapping n1 is 2 and n2 is 1
After swap method, num1 is 1 and num2 is 2
```

V. Overloading Methods

The max method that was used earlier works only with the int data type. But what if you need to find which of two floating-point numbers has the maximum value? The solution is to create another method with the same name but different parameters, as shown in the following code:

```
public static double max(double num1, double num2) {
  if (num1 > num2)
    return num1;
  else
    return num2;
}
```

If you call max with int parameters, the max method that expects int parameters will be invoked; if you call max with double parameters, the max method that expects double parameters will be invoked. This is referred to as **method overloading**; that is, two methods have the same name but different parameter lists within one class.

The Java compiler determines which method is used based on the method signature. Overloading methods can make programs clearer and more readable. Methods that perform closely related tasks should be given the same name.

Overloaded methods must have different parameter lists. You cannot overload methods based on different modifiers or return types. Sometimes there are two or more possible matches for an invocation of a method due to similar method signature, so the compiler cannot determine the most specific match. This is referred to as ambiguous invocation.

VI. The Scope of Variables

The scope of a variable is the part of the program where the variable can be referenced. A variable defined inside a method is referred to as a local variable. The scope of a local variable starts from its declaration and continues to the end of the block that contains the variable. A local variable must be declared before it can be used. A parameter is actually a local variable. The scope of a method parameter covers the entire method. Variable declared in the initial action part of a for loop header has its scope in the entire loop. But a variable declared inside a for loop body has its scope limited in the loop body from its declaration to the end of the block. You can declare a local variable with the same name multiple times in different non-nesting blocks in a method, but you cannot declare a local variable twice in the nested blocks.

VII. Using Command-Line Arguments

Sometimes you want to pass information into a program when you run it. This is accomplished by passing command-line arguments to main( ).

A command-line argument is the information that directly follows the program's name on the command line when it is executed. To access the command-line arguments inside a Java program is quite easy. They are stored as strings in the String array passed to main( ).

Displays all of the command-line arguments

```
class CommandLine {
  public static void main(String args[]){
    for(int i=0; i<args.length; i++){
      System.out.println("args[" + i + "]: " +args[i]);
    }
  }
}
```

java CommandLine this is a command line 200 -100

Output

```
args[0]: this
args[1]: is
args[2]: a
args[3]: command
args[4]: line
args[5]: 200
args[6]: -100
```

3.5 Adding Methods to a Class

In a program, referring to multiple data members of a class can be a tedious task. In addition, accessing data members directly overrules the concept of encapsulation. You can create a method that can be used to access the data members. Using methods in a Java program provides the following advantage on reusability. It enables the user to reuse the whole code or a part of it. A function or a task is encapsulated in a method that can be accessed from anywhere in the program. The practice of reusing a method is also called "write once, use many".

The following syntax shows how to define a method:

```
void methodName ( )
{
  // Method body
}
```

In the preceding syntax, the void keyword specifies that the function does not return any value. The *methodName* specifies the name of the method.

3.6 Declaring the Main ( ) Method

A Java program consists of the *main ( )* method that calls the methods defined in a class. You can create a number of classes in a Java program. The Java compiler compiles all the classes in an application but to execute a program, you need to include a *main ( )* method in the program. The following syntax shows how to declare the *main ( )* method:

```
public static void main (String [ ] args)
{
// code for main ( ) method
}
```

In the preceding syntax, the method header contains three words—public, static, and void. The implications of these words are:

- public: The public keyword indicates that the method can be accessed from any object in a Java program.
- static: The static keyword is used with the *main ( )* method that associates the method with its class. You need not create an object of the class to call the *main ( )* method.
- void: The void keyword signifies that the *main ( )* method returns no value.

The *main ( )* method can be declared in any class, but the name of the file and the class name in which the *main ( )* method is declared should be the same. The file must have the .java extension. For example, if the *main ( )* method is declared in class Employee, the name of the file should be *Employee.java.*

The *main ( )* method accepts a single argument in the form of an array of elements of type string as shown in the following command:

```
public static void main (String arg [ ])
```

In the preceding statement, the method header shows one argument in parentheses. The argument implies that when the class Employee is executed, an array of strings is sent to the program that helps in program initialization. Strings are a sequence of characters. arr[ ] is the array containing the string values. An argument is a variable that you need to pass to a method for providing an input value to the method.

4. DEFINING CONSTRUCTORS

To initialize the data members of a class, values should be assigned to each data member. However, it might turn out to be a tedious task if you need to initialize a large number of data members. In Java, you can create constructors of the classes that automatically initialize the data members of the class when you create an object.

A *constructor* is a method with the same class name. A constructor of a class is automatically invoked every time an instance of a class is created. Constructors do not have a return type.

For example, to calculate the interest on the balance amount in the account of a customer you can create an *Account* class. You can use the following code snippet to add a constructor, *Account ( )* of the class to initialize the values of the data members of the Account class.

```
Class Account
{
int principal;          // Data members of the class
int interestRate;
int time;
Account ( )             //constructor method
{
Principal=10000;            //Initialization of data members
interestRate=5;
time=2;
}
void interest( )  //method declaration
{
  //method accessing the data members of the class
float amount= (principal*interestRate*time)/100;
}
}
```

In the preceding code snippet, each time an object of the Account class is created, the constructor is invoked and the object data members are initialized to the default values as specified in the constructor. Constructors have the following characteristics:

- A constructor has the same name as the class itself.
- There is no return type for a constructor. A constructor returns the instance of the class instead of a value.
- A constructor is used to assign values to the data members of each object created from a class.

5. GARBAGE COLLECTION AND FINALIZE METHOD

Garbage collection is the feature of Java that helps to automatically destroy the objects created and release their memory for future reallocation. When no reference to an object exists and the object is assumed to be no longer required, the memory occupied by that object can be released and used by another object. The various activities involved in garbage collection are:

- Monitoring the objects used by a program and determining when they are not in use.
- Destroying objects that are no more in use and reclaiming their resources, such as memory space.

The Java Virtual Machine (JVM) acts as a garbage collector that keeps a track of the memory allocated to various objects and the objects being referenced. The JVM performs garbage collection when it needs more memory to continue execution of programs. You cannot explicitly reclaim the memory of a specific de-referenced object. The Runtime class in Java encapsulates the Java runtime environment. The various methods of the Runtime class used in memory management are:

- *static Runtime.getRuntime ( ):* Returns the reference of current runtime object.
- *void gc ( ):* Invokes garbage collection.

- *long totalMemory ( ):* Returns the total number of bytes of memory available in JVM.

Implementing Garbage Collection

You can explicitly run the garbage collector in Java to collect information regarding how large the object heap is or to determine the number of objects of a certain type that can be instantiated.

To obtain these values, you can use the *totalMemory ( )* and *freeMemory ( )* methods. The *totalMemory ( )* method returns the total memory in the JVM, and the *freeMemory ( )* method returns the amount of memory free in the JVM. You can run the garbage collector by calling the *gc ( )* method. The *gc ( )* method enables you to run the garbage collector.

To run the garbage collector explicitly, you need to perform two steps:

1. Create an object of the Java Runtime class. The following syntax shows how to create an object of the Java Runtime class:
   Runtime r = Runtime.getRuntime ( );
2. Invoke the *gc ( )* method of the *Runtime ( )* class to request the garbage collection. The following syntax shows how to invoke the garbage collector:
   r.gc ( );

It is possible to define a method that will be called just before an object's final destruction by the garbage collector. This method is called **finalize( )**, and it can be used to ensure that an object terminates cleanly.

For example, you might use finalize( ) to make sure that an open file owned by that object is closed.

Code to invoke the garbage collector

```
public class GarbageCollDemo
{
    int mem = 30000;
    int [ ] ArrayA = new int[mem];
    void occupyMemory ( )
    {
        for (int i=0; i<mem; i++)
        {
            Array[i] = i;
    }
    }
    void DiscardArray ( )
    {
        for (int i=0; i<mem; i++)
        {
            Array[i] = 0;
        }
}
public static void main (String args [ ])
```

```
    {
        GarbageCollDemo gc = new GarbageCollDemo( );
        Runtime r = Runtime.getRuntime ( );
        long freemem = r.freeMemory ( );
// Determine the current amount of free memory.
        System.out.println ("Initial free memory before creating array:" +freemem);
        r.gc ( );
        freemem = r.freeMemory ( );
        System.out.println ("Free memory after garbage collection:" + freemem);
        // Consume some memory
        gc.occupyMemory ( );
 // Determine amount of memory left after consumption.
        freeMem = r.freeMemory ( );
        System.out.println ("free memory after creating array:" +freeMem);
        gc.DiscardArray ( ); //  Discard the array.
        r.gc ( ); // Run the garbage collector
        freeMem = r.freeMemory ( );
        System.out.println ("free memory after running gc ( ):" +freemem);
    }
}
```

It is possible to define a method that will be called just before an object's final destruction by the garbage collector. This method is called **finalize( )**, and it can be used to ensure that an object terminates cleanly. The finalize( ) method may be used to make sure that an open file owned by that object is closed.

Inside the finalize( ) method you will specify those actions that must be performed before an object is destroyed.

The finalize( ) method has the following general form:

```
protected void finalize( )
{
  // finalization code here
}
```

Here, the keyword protected is a specifier that prevents access to finalize( ) by code defined outside its class.

This means that you cannot know when.or even if.finalize( ) will be executed. For example, if your program ends before garbage collection occurs, finalize( ) will not execute.

6. ACCESS SPECIFIERS AND MODIFIERS

Classes enable an object to access data variables or methods of another class. For example, in a banking application you might need to hide information, such as customer balance, from unauthorized access by other classes of the application. Within the Account class, the methods access the information, but outside the class you need to restrict access to this information.

Java provides access specifiers and modifiers to decide which part of the class, such as data members and methods will be accessible to other classes or objects and how the data members are used in other classes and objects.

6.1 Access Specifiers

An access specifier controls the access of class members and variables by other objects. The various types of access specifiers in Java are:

- public
- private
- protected
- friendly or package

The Public Access Specifier

Class members with *public* specifier can be accessed anywhere in the same class, package in which the class is created, or a package other than the one in which the class is declared. You can use a public class, data member, or a method from any object in a Java program. *A package is a collection of classes.*

Sample code to define a class, methods and variables

```
public class Account
{
public int account_no;        //Data members are accessible outside the class
public String name;
  public void show ( )        // Method declaration
  {
    System.out.println ("Name ="+ name); //Statement of the method
    System.out.println("Account number of this customer is= "+ account_no);
  }
}
```

The *public* keyword is used to declare a member as public. The following statement shows how to declare a data member of a class as public:

public <data type> <variable name>;

The class Account defines the *show ( )* method and various data members, such as name and account number. All the classes in the program can access the various details of a customer, such as name and account number. Therefore, these data members and the method are declared public. The *show ( )* method is used to display the account number and customer name of a customer.

The Private Access Specifier

The *private* access specifier provides most restricted level access. A data member of a class declared *private* is accessible at the class level only in which it is defined. You can use the *private* access specifier to declare members that should be available to the class within which they are declared. The *private* keyword is used to declare a member as private.

The following syntax shows how to declare a data member of a class as *private:*

```
private float <variableName>;      //Private data member of float type
private methodName ( );            // Private method
```

The *Account* class defines the *show ( )* method and the various data members, such as balance and age. These members are to be accessed only by the objects of the same class. Therefore, these methods are declared private. You can use the following code snippet that shows the *Acccount* class with private data variables, such as *age* and *balance*. The following code snippet, the objects of the Account class can call the *show ( )* method, but objects of other classes cannot access or invoke the private members of *Account* class.

```
class Account
{
private int account_no;   // Data members converted to private to encapsulate data
private String name ;
private int age;
private float balance;
public void show ( ) //Method can be called from outside the class to access the data members
   {
      System.out.println("Age ="+ age);
      System,out.println("Balance of this customer is="+ balance);
   }
}
```

The Protected Access Specifier

The variables and methods that are declared *protected* are accessible only to the subclasses of the class in which they are declared. The *protected* keyword is used to declare a member as protected.

The following statement shows how to declare a member as protected:

```
protected <data type> <name of the variable>;
```

In the airline reservation application, you can create the Ticket class that consists of various data members, such as *flightNumber, date, time,* and *destination.* You can derive the *ConfirmedTicket* subclass from the *Ticket* class that consists of an additional data member, *seatNumber*. You can declare the data members of the *Ticket* class as protected, which can be accessed by the *ConfirmedTicket* subclass.

You can use the following code snippet to define the *Ticket* class that has protected data variables:

```
public class Ticket
{
      protected int flightNumber; //protected data members accessible to derived classes
      protected String date;
      protected String time;
      protected String destination;
      protected void showData ( )
```

```
{
  // code body
}
}
```

In the preceding code snippet various data members and methods are declared protected.

The Friendly or Package Access Specifier

If you do not specify any access specifier, the scope of data members and methods is friendly. Java provides a large number of classes, which are organized into groups in a package. A class, variable, or method that has friendly access is accessible only to the classes of a package. The data members, such as *pageNumbers* and *price,* and the *showData ( )* method of the *Books* class are not given access specifiers. The following code snippet shows the *Books* class that has friendly access specifier:

```
class Books
{
int pageNumbers;     //The default friendly access is provided to the data members
float price;
void showData ( )
  {
    // code body
  }
}
```

6.2 Types of Permitted Modifiers

Modifiers determine or define how the data members and methods are used in other classes and objects. The main difference between access specifiers and modifiers is that access specifiers define the accessibility of the data members in a class and the modifiers determine how these methods are used and modified by other classes. The various modifiers permitted in Java are:

- static
- final
- abstract
- native
- synchronized

Static

The static keyword is used with methods, variables and inner classes. The *static* keyword is used to define class variables and methods that belong to a class and not to any particular instance of the class.

A static method associates the data members with a class and not the objects of the class. Therefore, all the objects of a class share the same static data members and methods. Non-static methods are those, which are associated with objects, and as a result the values of the data members differ for different objects. You cannot access non-static data members and methods from a static method.

For example, in an online shopping application, you can keep track of the number of books sold by keeping static counter data member in the Books class that increment each time a book is sold.

Final

The *final* keyword is used with methods, variables and classes. The *final* modifier indicates that the data member cannot be modified. For example, consider a variable that has been assigned a value. If the variable has been declared *final,* you cannot modify the value of the variable and if you try to do so, it will cause runtime errors. A variable declared *final* is initialized at the time it is declared. *A class can be declared as final if you do not want the class to be subclassed.*

The *final* modifier does not allow the class to be inherited. It is used to create classes that serve as a standard, and you do not want anybody to modify the methods in a subclass and use them in a different manner. The *final* modifier has the following characteristics:

- A *final* method cannot be modified in the subclass.
- A *final* class cannot be inherited.
- All the methods and data members in a *final* class are implicitly final.

Abstract

The *abstract* keyword is used to declare classes that only define common properties and behavior of other classes. An *abstract* class is used as a base class to derive specific classes of the same type. For example, you can create an abstract Books class that contains the common data members, such as title, page numbers, and type of binding for all the books.

Native

The *native* modifier is used only with methods. It is used to inform the compiler that the method has been coded in a programming language other than Java, such as C or C++. The *native* keyword with a method indicates that the method lies outside the Java Runtime Environment (JRE). *The native method makes a program platform-dependent. In addition, writing **native** methods must be avoided. They are used when you have an existing code in another language and do not want to rewrite the code in Java.*

The following syntax shows how to declare a native modifier:

```
public native void nativeMethod (var1, var2,..);
```

Synchronized

The *synchronized* modifier is used for methods. The *synchronized* modifier controls the access to a block of code in a multithreaded programming environment. A thread is a unit of execution within a process. Java supports multithreaded programming and each thread defines a separate path of execution.

In a multithreaded program, you need to synchronize various threads. As a result of synchronization, only one thread can access a shared resource when two or more threads need access to the resource at the same time. For example, if multiple threads need to print a document, only one thread can access the printer as a result of synchronization.

7. ARRAYS

Java provides a data structure, the **array**, which stores a fixed-size sequential collection of elements of the same type. An array is used to store a collection of data, but it is often more useful to think of an array as a collection of variables of the same type.

Instead of declaring individual variables, such as number0, number1, …, and number99, you declare one array variable such as numbers and use numbers[0], numbers[1], and …, numbers[99] to represent individual variables.

This tutorial introduces how to declare array variables, create arrays, and process arrays using indexed variables.

Declaring Array Variables

To use an array in a program, you must declare a variable to reference the array, and you must specify the type of array the variable can reference. Here is the syntax for declaring an array variable:

```
dataType[] arrayRefVar;   // preferred way.
or
dataType arrayRefVar[];  //  works but not preferred way.
```

The following code snippets are examples of this syntax:

```
double[] myList;         // preferred way.
or
double myList[];         //  works but not preferred way.
```

Creating Arrays

The length of an array is fixed at the time of its creation. An array represents related entities having the same data type in contiguous or adjacent memory locations. The related data items form a group and are referred to by the same name. For example, the automated employee management system might need an array of integers to hold the Ids of the employees. The following command shows an array named employee:

employee [5];

The complete set of values is known as array and the individual entities are called as elements of the array. A specific value in an array is indicated by writing the array name and placing the index of the desired element in square brackets. The index is also known as subscript and the array may be called a collection of subscripted elements of the same data type. The advantage of using array lies in the fact that you can refer to a large number of elements by just specifying the subscript preceded by the array name. Arrays make it easy to do calculations in a loop, such as determining the total number of values stored in an array.

You can create an array by using the new operator with the following syntax:

```
arrayRefVar = new dataType[arraySize];
```

The above statement does two things:

- It creates an array using new dataType[arraySize];
- It assigns the reference of the newly created array to the variable arrayRefVar.

Declaring an array variable, creating an array, and assigning the reference of the array to the variable can be combined in one statement, as shown below:

```
dataType[] arrayRefVar = new dataType[arraySize];
```

Alternatively, you can create arrays as follows:

dataType[] arrayRefVar = {value0, value1,..., valuek};

The array elements are accessed through the **index**. Array indices are 0-based; that is, they start from 0 to **arrayRefVar.length-1**.

Example: The following statement declares an array variable, myList, creates an array of 10 elements of double type, and assigns its reference to myList.:

double[] myList = new double[10];

The following picture represents array myList. Here myList holds ten double values and the indices are from 0 to 9.

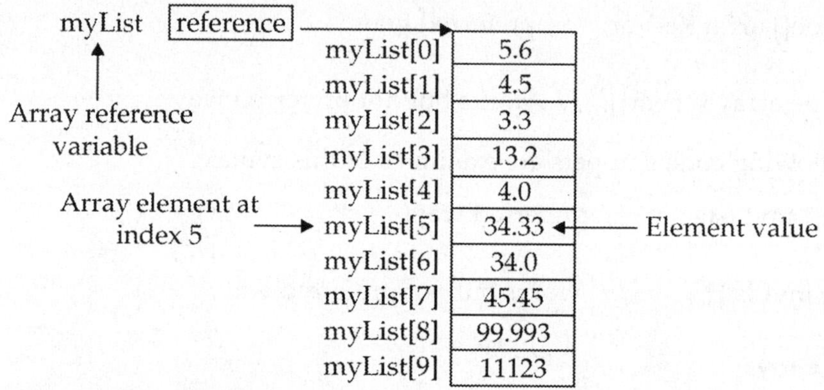

The various types of arrays in Java are:

- One-dimensional arrays
- Multi-dimensional arrays

One-dimensional Arrays

One-dimensional array is a list of variables of the same data type. The following syntax shows how to declare a one-dimensional array:

Type arr [ ]

In the preceding syntax, type is the data type of the array. All the elements of the array contain variables of the same data type as the array. For example, you have an array that contains the names of the various designations offered by SimpleSystemsSolutions Inc. for its employees. The following syntax shows how to declare a one-dimensional string array:

String designation_types [10];

In the preceding syntax, the designation_types array is declared that can store 10 designation types in it.

Allocating Memory to Arrays

You use the new operator to allocate memory to an array. The following syntax shows how to allocate memory to a one-dimensional array:

arr = new type [size];

You can use the following statement to allocate memory to an array having 3 elements:

Emp_ID = new string[3];

The SimpleSolutionsSystem Inc. needs to store the employee Ids in an array, emp_ID[ ].

Multi-dimensional Arrays

In addition to one-dimensional arrays, you can create multi-dimensional arrays. To declare multi-dimensional arrays, you need to specify multiple square brackets after the array name. For example, the following syntax displays the declaration of a two-dimensional array:

```
int multiDim [ ]=new int[3] [ ];
```

In multi-dimensional array, you need to allocate memory for only the first dimension, as shown in the preceding syntax. You can allocate the remaining dimensions separately. The following code snippet manually allocates memory to the second dimension:

```
multiDim [0] = new int[4];

multiDim [1] = new int[4];

multiDim [2] = new int[4];
```

In addition, when you allocate memory to the second dimension of a multi-dimensional array, you need not allocate the same number to each dimension. You can use the following code snippet to create a multi-dimensional array in which the sizes of the second dimension are not equal:

```
int multiDim [ ]= new int [3] [ ];

multiDim [0] = new int[1];

multiDim [1] = new int[2];

multiDim [2] = new int[4];
```

Assigning Values to the Elements of an Array

To access a specific array, you need to specify the name of the array and the index number of the element. The index position of the first element in the array is 0. For example, the following statement assigns the value, *designation_types [ ]:*

```
designation_types [0] ="General Manager";
```

You can use the following code snippet to assign values to different elements of the given array:

```
string designation_types [ ];
designation_types= new string[3];
designation_types [0] = "General Manager";
designation_types [1] = "Assistant Manager";
designation_types [2] = "Managing Director";
```

You can declare and allocate memory to a user-defined array in a single statement. The following syntax shows how to declare an array and allocate the array in a single statement:

```
type arr[ ] = new type [size];
```

You can use the following code snippet to declare and initialize arrays in the same statement:

```
String Designation_types [ ]= { "General Manager", "Assistant Manager",
                                "Managing Director"};
```

Accessing Values from Various Elements of An Array

Similar to assigning values to array elements, you can access values from elements in the array by referring to the element by its index number. The following statement is used to access and display an element stored at a specific index position:

```
Designation_types [0] =designation_types [2];
```

In the preceding statement, the value of the third element of the array is assigned to the first element of the array.

Processing Arrays

When processing array elements, we often use either for loop or foreach loop because all the elements in an array are of the same type and the size of the array is known.

Here is a complete example of showing how to create, initialize and process arrays:

```
public class TestArray {
  public static void main(String[] args) {
    double[] myList = {1.9, 2.9, 3.4, 3.5};
    // Print all the array elements
    for (int i = 0; i < myList.length; i++) {
      System.out.println(myList[i] + " ");
    }
    // Summing all elements
    double total = 0;
    for (int i = 0; i < myList.length; i++) {
      total += myList[i];
    }
    System.out.println("Total is " + total);
    // Finding the largest element
    double max = myList[0];
    for (int i = 1; i < myList.length; i++) {
      if (myList[i] > max) max = myList[i];
    }
    System.out.println("Max is " + max);
  }
}
```

This would produce the following result:

```
1.9
2.9
3.4
3.5
Total is 11.7
Max is 3.5
```

The Foreach Loops

JDK 1.5 introduced a new for loop, known as foreach loop or enhanced for loop, which enables you to traverse the complete array sequentially without using an index variable.

Example: The following code displays all the elements in the array myList:

```
public class TestArray {
  public static void main(String[] args) {
    double[] myList = {1.9, 2.9, 3.4, 3.5};

    // Print all the array elements
    for (double element: myList) {
      System.out.println(element);
    }
  }
}
```

This would produce the following result:

```
1.9
2.9
3.4
3.5
```

Passing Arrays to Methods

Just as you can pass primitive type values to methods, you can also pass arrays to methods. For example, the following method displays the elements in an int array:

```
public static void printArray(int[] array) {
  for (int i = 0; i < array.length; i++) {
    System.out.print(array[i] + " ");
  }
}
```

You can invoke it by passing an array. For example, the following statement invokes the printArray method to display 3, 1, 2, 6, 4, and 2:

```
printArray(new int[]{3, 1, 2, 6, 4, 2});
```

Returning an Array from a Method

A method may also return an array. For example, the method shown below returns an array that is the reversal of another array:

```
public static int[] reverse(int[] list) {
  int[] result = new int[list.length];

  for (int i = 0; i = result.length - 1;
                 i < list.length; i++, j—) {
    result[j] = list[i];
  }
  result result;
}
```

The Arrays Class

The java.util.Arrays class contains various static methods for sorting and searching arrays, comparing arrays, and filling array elements. These methods are overloaded for all primitive types.

Methods of Array Class

Method	Usage	Use
binarySearch	Public static int binarySearch(Object[] a, Object key)	Searches the specified array of Object for the specified value using the binary search algorithm. The array must be sorted prior to making this call. This returns index of the search key, if it is contained in the list.
equals	Public static boolean equals(long[] a, long[] a2)	Returns true if the two specified arrays of longs are equal to one another. Two arrays are considered equal if both arrays contain the same number of elements, and all corresponding pairs of elements in the two arrays are equal.
fill	Public static void fill(int[] a, int val)	Assigns the specified int value to each element of the specified array of ints. Same method could be used by all other premitive data types (Byte, short, Int, etc.)
sort	Public static void sort(Object[] a)	Sorts the specified array of objects into ascending order, according to the natural ordering of its elements. Same method could be used by all other premitive data types (Byte, short, Int, etc.)

8. STRINGS

Strings, which are widely used in Java programming, are a sequence of characters. In the Java programming language, strings are objects.

The Java platform provides the String class to create and manipulate strings.

Creating Strings

The most direct way to create a string is to write:

String greeting = "Hello world!";

Whenever it encounters a string literal in your code, the compiler creates a String object with its value in this case, "Hello world!'.

As with any other object, you can create String objects by using the new keyword and a constructor. The String class has eleven constructors that allow you to provide the initial value of the string using different sources, such as an array of characters:

```
public class StringDemo{
  public static void main(String args[]){
    char[] helloArray = { 'h', 'e', 'l', 'l', 'o', '.'};
    String helloString = new String(helloArray);
    System.out.println( helloString );
  }
}
```

This would produce the following result:

hello

Note: The String class is immutable, so that once it is created a String object cannot be changed. If there is a necessity to make a lot of modifications to Strings of characters, then you should use String Buffer & String Builder Classes.

String Length

Methods used to obtain information about an object are known as accessor methods. One accessor method that you can use with strings is the length() method, which returns the number of characters contained in the string object.

After the following two lines of code have been executed, len equals 17:

```
public class StringDemo{
  public static void main(String args[]){
    String palindrome = "Dot saw I was Tod";
    int len = palindrome.length();
    System.out.println("String Length is : " + len);
  }
}
```

Concatenating Strings

The String class includes a method for concatenating two strings:

string1.concat(string2);

This returns a new string that is string1 with string2 added to it at the end. You can also use the concat() method with string literals, as in:

"My name is ".concat("Jamal");

Strings are more commonly concatenated with the + operator, as in:

"Hello," + " world" + "!"

which results in:

"Hello, world!"

Let us look at the following example:

```
public class StringDemo{
  public static void main(String args[]){
                              String string1 = "saw I was ";

    System.out.println("Dot " + string1 + "Tod");
  }
}
```

This would produce the following result:

Dot saw I was Tod

Creating Format Strings

You have printf() and format() methods to print output with formatted numbers. The String class has an equivalent class method, format(), that returns a String object rather than a PrintStream object.

Use of String's static format() method allows you to create a formatted string that you can reuse, as opposed to a one-time print statement. For example, instead of:

```
System.out.printf("The value of the float variable is " +
      "%f, while the value of the integer " +
      "variable is %d, and the string " +
      "is %s", floatVar, intVar, stringVar);
```

you can write:

```
String fs;
fs = String.format("The value of the float variable is " +
```

```
    "%f, while the value of the integer " +
    "variable is %d, and the string " +
    "is %s", floatVar, intVar, stringVar);
System.out.println(fs);
```

String Methods

Here is the list methods supported by String class:

String Methods

S.No.	Methods usage	Use
1.	char charAt(int index)	Returns the character at the specified index.
2.	int compareTo(Object o)	Compares this string to another object.
3.	int compareTo(String anotherString)	Compares two strings lexicographically.
4.	int compareToIgnoreCase(String str)	Compares two strings lexicographically, ignoring case differences.
5.	String concat(String str)	Concatenates the specified string to the end of this string.
6.	boolean contentEquals(StringBuffer sb)_	Returns true if and only if this string represents the same sequence of characters as the specified string buffer.
7.	static String copyValueOf(char[] data)_	Returns a String that represents the character sequence in the array specified.
8.	static String copyValueOf(char[] data, int offset, int count)	Returns a String that represents the character sequence in the array specified.
9.	boolean endsWith(String suffix)	Tests if this string ends with the specified suffix.
10.	boolean equals(Object anObject)	Compares this string to the specified object.
11.	boolean equalsIgnoreCase(String anotherString)	Compares this String to another String, ignoring case considerations.
12.	byte getBytes()	Encodes this String into a sequence of bytes using the platform's default charset, storing the result into a new byte array.

(contd.)

(*contd.*)

13.	byte[] getBytes (String charsetName)	Encodes this String into a sequence of bytes using the named charset, storing the result into a new byte array.
14.	void getChars(int srcBegin, int srcEnd, char[] dst, int dstBegin)	Copies characters from this string into the destination character array.
15.	int hashCode()	Returns a hash code for this string.
16.	int indexOf (int ch)_	Returns the index within this string of the first occurrence of the specified character.
17.	int indexOf(int ch, int fromIndex)	Returns the index within this string of the first occurrence of the specified character, starting the search at the specified index.
18.	int indexOf(String str)	Returns the index within this string of the first occurrence of the specified substring.
19.	int indexOf (String str, int fromIndex)	Returns the index within this string of the first occurrence of the specified substring, starting at the specified index.
20.	String intern ()	Returns a canonical representation for the string object.
21.	int lastIndexOf(int ch)	Returns the index within this string of the last occurrence of the specified character.
22.	int lastIndexOf(int ch, int fromIndex)	Returns the index within this string of the last occurrence of the specified character, searching backward starting at the specified index.
23.	int lastIndexOf(String str)	Returns the index within this string of the rightmost occurrence of the specified substring.
24.	int lastIndexOf(String str, int fromIndex)	Returns the index within this string of the last occurrence of the specifiedsubstring, searching backward starting at the specified index.
25.	int length()	Returns the length of this string.

(*contd.*)

(contd.)

26.	boolean matches(String regex)	Tells whether or not this string matches the given regular expression.
27.	boolean regionMatches(boolean ignoreCase, int toffset, String other, int ooffset, int len)	Tests if two string regions are equal.
28.	boolean regionMatches(int toffset, String other, int ooffset, int len)	Tests if two string regions are equal.
29.	String replace(char oldChar, char new char)	Returns a new string resulting from replacing all occurrences of old char in this string with new char.
30.	String replaceAll(String regex, String replacement)	Replaces each substring of this string that matches the given regular expression with the given replacement.
31.	String replaceFirst(String regex, String replacement)	Replaces the first substring of this string that matches the given regular expression with the given replacement.
32.	String[] split(String regex)	Splits this string around matches of the given regular expression.
33.	String[] split(String regex, int limit)	Splits this string around matches of the given regular expression.
34.	boolean startsWith(String prefix)	Tests if this string starts with the specified prefix.
35.	boolean startsWith(String prefix, int toffset) specified prefix	Tests if this string starts with the beginning of a specified index.
36.	CharSequence subSequence(int beginIndex, int endIndex)	Returns a new character sequence that is a subsequence of this sequence.
37.	String substring(int beginIndex)	Returns a new string that is a substring of this string.
38.	String substring(int beginIndex, int endIndex)	Returns a new string that is a substring of this string.
39.	char[] toCharArray()	Converts this string to a new character array.

(contd.)

(contd.)

40.	String toLowerCase()	Converts all of the characters in this string to lower case using the rules of the default locale.
41.	String toLowerCase(Locale locale)	Converts all of the characters in this string to lower case using the rules of the given locale.
42.	String toString()	This object (which is already a string!) is itself returned.
43.	String toUpperCase()	Converts all of the characters in this string to upper case using the rules of the default locale.
44.	String toUpperCase(Locale locale)	Converts all of the characters in this string to upper case using the rules of the given Locale.
45.	String trim()	Returns a copy of the string, with leading and trailing whitespace omitted.
46.	static String valueOf(primitive data type x)	Returns the string representation of the passed data type argument.

The description of the methods and their usage with example are discussed in Part 6.

9. PACKAGES AND INTERFACES

A. Java Package

In simple, it is a way of categorizing the classes and interfaces. When developing applications in Java, hundreds of classes and interfaces will be written, therefore categorizing these classes is a must as well as makes life much easier.

B. Import Statements

In Java if a fully qualified name, which includes the package and the class name, is given, then the compiler can easily locate the source code or classes. Import statement is a way of giving the proper location for the compiler to find that particular class.

For example, the following line would ask compiler to load all the classes available in directory java_installation/java/io:

```
import java.io.*;
```

A *package* is a collection of classes and interfaces. Each package has its own name and organizes its top-level (that is, nonnested) classes and interfaces into a separate *namespace,* or name collection. Although same-named classes and interfaces cannot appear in the same package, they can appear in different packages because a separate namespace assigns to each package.

From an implementation perspective, equating a package with a directory proves helpful, as does equating a package's classes and interfaces with a directory's classfiles. Keep in mind other approaches—such as the use of databases—to implement packages, so do not get into the habit of always equating packages with directories. But because many JVMs use directories to implement packages, this article equates packages with directories. The Java 2 SDK organizes its vast collection of classes and interfaces into a tree-like hierarchy of packages within packages, which is equivalent to directories within directories. That hierarchy allows Sun Microsystems to easily distribute (and you to easily work with) those classes and interfaces. Examples of Java's packages include:

- **java.lang**: A collection of language-related classes, such as Object and String, organized in the java package's lang subpackage.
- **java.lang.ref**: A collection of reference-related language classes, such as SoftReference and ReferenceQueue, organized in there sub-subpackage of the java package's lang subpackage.
- **javax.swing**: A collection of Swing-related component classes, such as JButton, and interfaces, such as ButtonModel, organized in the javax package's swing subpackage

Period characters separate package names. For example, in javax.swing, a period character separates package name javax from subpackage name swing. A period character is the platform-independent equivalent of forward slash characters (/), backslash characters (\), or other characters to separate directory names in a directory-based package implementation, database branches in a hierarchical database-based package implementation, and so on.

Create a Package of Classes and Interfaces

Every source file's classes and interfaces organize into a package. In the package directive's absence, those classes and interfaces belong to the unnamed package (the directory the JVM regards as the current directory—the directory where a Java program begins its execution via the Windows java.exe, or OS-equivalent, program—and contains no subpackages). But if the package directive appears in a source file, that directive names the package for those classes and interfaces. Use the following syntax to specify a package directive in source code:

'package' *packageName* ['.' *subpackageName* ...] ';'

A package directive begins with the package keyword. An identifier that names a package, *packageName,* immediately follows. If classes and interfaces are to appear in a subpackage (at some level) within *packageName,* one or more period-separated *subpackageName* identifiers appear after *packageName*. The following code fragment presents a pair of package directives:

```
package game;
package game.devices;
```

The first package directive identifies a package named game. All classes and interfaces appearing in that directive's source file organize in the game package. The second package directive identifies a subpackage named devices, which resides in a package named game. All classes and interfaces appearing in that directive's source file organize in the game package's devices subpackage. If a JVM implementation maps package names to directory names, game.devices maps to a game\devices directory hierarchy under Windows and a game/devices directory hierarchy under Linux or Solaris.

Only one package directive can appear in a source file. Furthermore, the package directive must be the first code (apart from comments) in that file. Violating either rule causes Java's compiler to report an error.

```
// A.java
package testpkg;
public class A
{
  int x = 1;
  public int y = 2;
  protected int z = 3;
  int returnx ()
  {
    return x;
  }
  public int returny ()
  {
    return y;
  }
  protected int returnz ()
  {
    return z;
  }
  public interface StartStop
  {
    void start ();
    void stop ();
  }
}
class B
{
  public static void hello ()
  {
    System.out.println ("hello");
  }
}
```

To help you get comfortable with packages, I have prepared an example that spans all topics in this article. In this section, you learn how to create the example's package. In later sections, you will learn how to import a class and an interface from this package, how to move this package to another location on your hard drive and still access the package from a program, and how to store the package in a jar file. The package's source code introduces the source code to your first named package. The package testpkg; directive names that package testpkg. Within testpkg, there are classes A and B. Within A, there are three field declarations, three method declarations, and an inner interface declaration. Within B, there is a single method declaration. The entire source code stores in A.java because A is a public class. Our task: Turn this source code into a package that consists of two classes and an inner interface (or a directory that contains three classfiles). The following Windows-specific steps accomplish that task:

1. Open a Windows command window and ensure you are in the c:drive's root directory (the main directory—represented by an initial backslash (\) character). To do that, type the c: command followed by the cd \ command. (If you use a different drive, replace c: with your chosen drive. Also, do not forget to press the Enter key after typing a command.)
2. Create a testpkg directory by typing md testpkg.
   Note: When following this article's steps, do not type periods after the commands.
3. Make testpkg the current directory by typing cd testpkg.
4. Use an editor to enter Listing 1's source code and save that code to an A.java file in testpkg.
5. Compile A.java by typing javac A.java. You should see classfiles A$StartStop.class, A.class, and B.class appear in the testpkg directory.

Figure 1.1 illustrates steps 3 to 5.

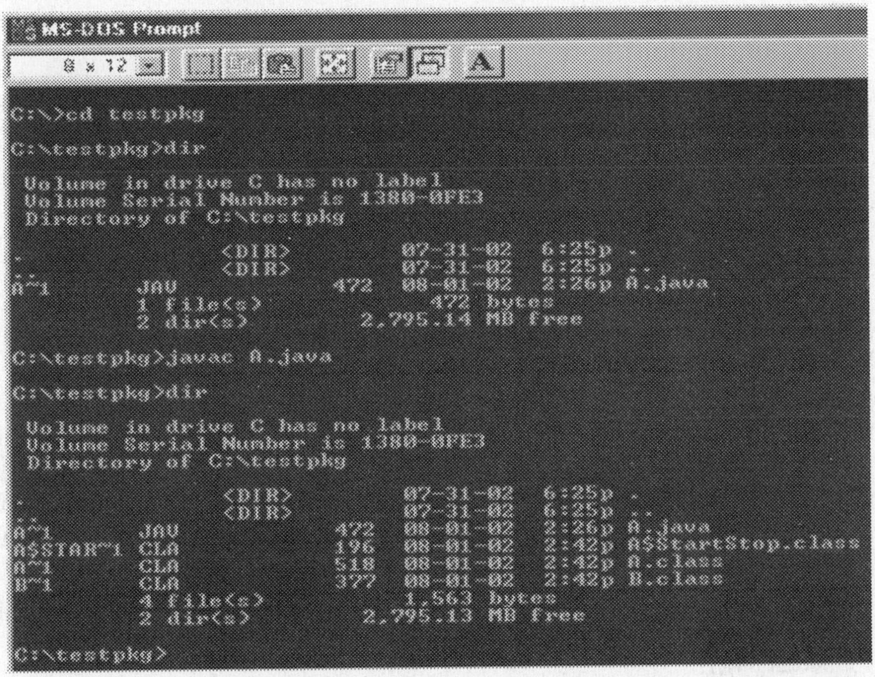

```
MS-DOS Prompt

C:\>cd testpkg

C:\testpkg>dir

 Volume in drive C has no label
 Volume Serial Number is 1380-0FE3
 Directory of C:\testpkg

.              <DIR>        07-31-02  6:25p .
..             <DIR>        07-31-02  6:25p ..
A~1      JAV           472  08-01-02  2:26p A.java
         1 file(s)              472 bytes
         2 dir(s)        2,795.14 MB free

C:\testpkg>javac A.java

C:\testpkg>dir

 Volume in drive C has no label
 Volume Serial Number is 1380-0FE3
 Directory of C:\testpkg

.              <DIR>        07-31-02  6:25p .
..             <DIR>        07-31-02  6:25p ..
A~1      JAV           472  08-01-02  2:26p A.java
A$STAR~1 CLA           196  08-01-02  2:42p A$StartStop.class
A~1      CLA           518  08-01-02  2:42p A.class
B~1      CLA           377  08-01-02  2:42p B.class
         4 file(s)            1,563 bytes
         2 dir(s)        2,795.13 MB free

C:\testpkg>
```

Fig. 1.1: The testpkg directory with its three classfiles equates to a testpkg package with classes A and B, and A's inner interface StartStop.

You have just created your first package. Think of this package as containing two classes (A and B) and A's single inner interface (StartStop). You can also think of this package as a directory containing three classfiles:A$StartStop.class, A.class, and B.class.

To minimize package name conflicts (especially among commercial packages), Sun has established a convention in which a company's Internet domain name reverses and prefixes a package name. For example, a company with x.com as its Internet domain name and a.b as a package name (a) followed by a subpackage name (b) prefixes com.x to a.b, resulting in com.x.a.b. My article does not follow this convention because the testpkg package is a throw-away designed for teaching purposes only.

```
// Usetestpkg1.java
class Usetestpkg1 implements testpkg.A.StartStop
{
  public static void main (String [] args)
  {
    testpkg.A a = new testpkg.A ();
    System.out.println (a.y);
    System.out.println (a.returny ());
    Usetestpkg1 utp = new Usetestpkg1 ();
    utp.start ();
    utp.stop ();
  }
  public void start ()
  {
    System.out.println ("Start");
  }
  public void stop ()
  {
    System.out.println ("Stop");
  }
}
```

Import a Package's Classes and Interfaces

Once you have a package, you will want to import classes and/or interfaces—actually, class and/or interface names—from that package to your program, so it can use those classes and/or interfaces. One way to accomplish that task is to supply the fully qualified package name (the package name and all subpackage names) in each place where the reference type name (the class or interface name) appears, as Listing 2 demonstrates.

By prefixing testpkg. to A, Usetestpkg1 accesses testpkg's class A in two places and A's inner interface StartStop in one place. Complete the following steps to compile and run Usetestpkg1:

1. Open a Windows command window and make sure you are in the c: drive's root directory.

2. Ensure the classpath environment variable does not exist by executing set classpath=. (I discuss classpath later in this article.)
3. Use an editor to enter Listing 2's source code and save that code to a Usetestpkg1.java file in the root directory.
4. Compile Usetestpkg1.java by typing javac Usetestpkg1.java. You should see classfileUsetestpkg1.class appear in the root directory.
5. Type java Usetestpkg1 to run this program.

Figure 1.2 illustrates steps 3 to 5 and shows the program's output.

According to Usetestpkg1's output, the main()method's thread successfully accesses testpkg. A's yfield and calls the returny() method. Furthermore, the output shows a successful implementation of the testpkg.A.StartStop inner interface.

For Usetestpkg1, prefixing testpkg. to A in three places does not seem a big deal. But who wants to specify a fully qualified package name prefix in a hundred places? Fortunately, Java supplies the import directive to import a package's public reference type name(s), so you do not have to enter fully qualified package name prefixes. Express an import directive in source code via the following syntax:

'import' *packageName* ['.' *subpackageName* ...] '.' (*referencetypeName* | '*') ';'

```
MS-DOS Prompt
8 x 12

C:\>dir Usetestpkg1.java

 Volume in drive C has no label
 Volume Serial Number is 1380-0FE3
 Directory of C:\

USETES~1 JAV           490  08-01-02  4:23p Usetestpkg1.java
         1 file(s)            490 bytes
         0 dir(s)       2,791.14 MB free

C:\>javac Usetestpkg1.java

C:\>dir *.class

 Volume in drive C has no label
 Volume Serial Number is 1380-0FE3
 Directory of C:\

USETES~1 CLA           780  08-01-02  4:25p Usetestpkg1.class
         1 file(s)            780 bytes
         0 dir(s)       2,791.13 MB free

C:\>dir testpkg

 Volume in drive C has no label
 Volume Serial Number is 1380-0FE3
 Directory of C:\testpkg

.            <DIR>        07-31-02  6:25p .
..           <DIR>        07-31-02  6:25p ..
A~1      JAV           472  08-01-02  2:26p A.java
A$STAR~1 CLA           196  08-01-02  4:25p A$StartStop.class
A~1      CLA           518  08-01-02  4:25p A.class
B~1      CLA           377  08-01-02  4:25p B.class
         4 file(s)          1,563 bytes
         2 dir(s)       2,791.13 MB free

C:\>java Usetestpkg1
2
2
Start
Stop

C:\>
```

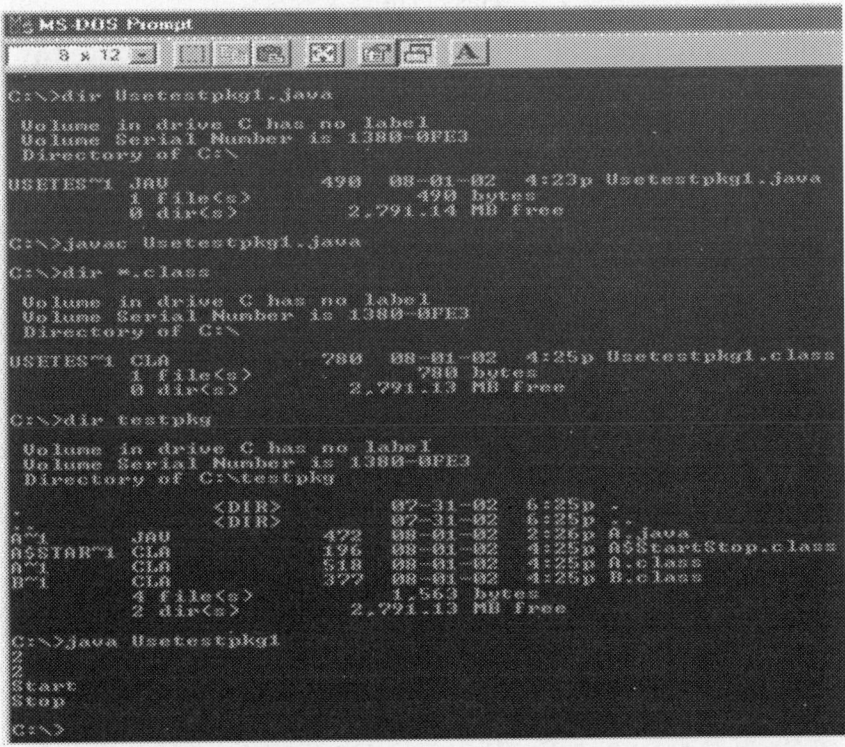

Fig. 1.2: The root directory contains Usetestpkg1.class and the testpkg directory that corresponds to the testpkg package.

An import directive consists of the import keyword immediately followed by an identifier that names a package, *packageName*. An optional list of subpackageName identifiers follows to identify the appropriate subpackage (if necessary). The directive concludes with either a *referencetypeName* identifier that identifies a specific class or interface from the package, or an asterisk (*) character. If *referencetypeName* appears, the directive is a *single-type import* directive. If an asterisk character appears, the directive is a *type-on-demand import* directive.

As with the package directive, import directives must appear before any other code, with three exceptions: a package directive, other import directives, or comments. The single-type import directive imports the name of a single public reference type from a package, as the following code fragment demonstrates:

```
import java.util.Date;
```

The previous single-type import directive imports class name Date into source code. As a result, you specify Date instead of java.util.Date in each place that class name appears in source code. For example, when creating a Date object, specifyDate d = new Date (); instead of java.util.Date d = new java.util.Date ();.

Exercise care with single-type import directives. If the compiler detects a single-type import directive that specifies a reference type name also declared in a source file, the compiler reports an error, as the following code fragment demonstrates:

```
import java.util.Date;

class Date {}
```

The compiler regards the code fragment as an attempt to introduce two reference types with the same Date name:

1. The imported Date reference type name from java's util subpackage via the single-type import directive
2. The Date class declaration

If the compiler permitted the above code fragment to compile, which Date reference type would the compiler refer to upon encountering Date d = new Date ();?

In contrast to a single-type import directive, a type-on-demand import directive imports public reference type names from a package on an as-needed basis, as the following code fragment demonstrates:

```
import java.net.*;
```

According to the above code fragment, if Socket or another java.net reference type name appears in a source file, that name can appear without a java.net prefix. During compilation, when the compiler detects Socket, it searches java.net to verify that Socket is a member of java's net subpackage. Once verified, the compiler stores java.net with Socket in the classfile.

As with the single-type import directive, the type-on-demand import directive has problems. Consider the following code fragment:

Listing 3. Usetestpkg2.java

```
// Usetestpkg2.java
import testpkg.*;
class Usetestpkg2 implements A.StartStop
{
  public static void main (String [] args)
  {
    A a = new A ();
```

```
   // Following code does not compile because x is not public outside testpkg.A
   // System.out.println (a.x);
   // Following code does not compile because returnx() is not public outside
testpkg.A
   // System.out.println (a.returnx ());
   System.out.println (a.y);
   System.out.println (a.returny ());
  // Following code does not compile because z has protected access in testpkg.A
   // System.out.println (a.z);
   // Following code does not compile because returnz() has protected access in
testpkg.A
   // System.out.println (a.returnz ());
   SubA sa = new SubA ();
   System.out.println (sa.returnZ ());
   // Following code does not compile because testpkg.B is not public in testpkg
and //hello() is not defined in a public class
   // B.hello ();
   Usetestpkg2 utp = new Usetestpkg2 ();
   utp.start ();
   utp.stop ();
  }
  public void start ()
  {
   System.out.println ("Start");
  }
  public void stop ()
  {
   System.out.println ("Stop");
  }
}
class SubA extends A
{
  int returnZ ()
  {
   // It is legal (regardless of package) for a subclass method to
   // call a superclass's protected method
   return super.returnz ();
  }}
import a.*;
import b.*;
X x = new X ();
```

Suppose packages a and b each contain an X class. When the compiler determines that X is present in a and b, the compiler reports an error because each package might have a different X implementation, and the compiler has no way of knowing which X to import. To solve the problem, the developer must prefix each occurrence of X with the fully qualified package name. For the code fragment above, this leads to either

```
a.X x = new a.X ();
```

or

```
b.X x = new b.X ();
```

You can use a type-on-demand import directive to save yourself from the trouble of specifying testpkg., wherever A appears in Usetestpkg 1, as shown in Listing 3:

Usetestpkg 2's type-on-demand import directive imports reference type names A and A.StartStop from testpkg. What about class B? We cannot import that class name because B is not a public class within testpkg. As a result, only code within testpkg's class A (and any other class you subsequently add to that package) can access B.

Usetestpkg 2 demonstrates that you can subclass a package's public class (A) and access that class's protected members (such as int returnz()). To prove that, compile Usetestpkg 2 (which you place in the same directory as Usetestpkg1) and run the program. If successful, you see the following output:

```
2
2
3
Start
Stop
3 is the value of testpkg.
```

Unlike returnz(), you cannot call int returnx() because the absence of an access modifier keyword (private, public, or protected) implies package access—and a field or method with package access is only accessible or callable from code within its class's (or interface's) package.

A Simple Case Study

```
import java.io.*;
public class Student{
  String name;
  int age;
  String designation;
  double Mark;

  // This is the constructor of the class Student
  public Student(String name){
    this.name = name;
  }
  // Assign the age of the Student  to the variable age.
  public void empAge(int empAge){
```

```
    age =  empAge;
  }
  /* Assign the designation to the variable designation.*/
  public void empDesignation(String empDesig){
    designation = empDesig;
  }
  /* Assign the Mark to the variable Mark.*/
  public void empMark(double empMark){
    Mark = empMark;
  }
  /* Print the Student details */
  public void printStudent(){
    System.out.println("Name:"+ name );
    System.out.println("Age:" + age );
    System.out.println("Designation:" + designation );
    System.out.println("Mark:" + Mark);
  }
}
```

For our case study we will be creating two classes. They are Student and StudentTest.

First open notepad and add the following code. Remember this is the Student class and the class is a public class. Now save this source file with the name Student.java.

The Student class has four class variables name, age, designation and Mark. The class has one explicitly defined constructor which takes a parameter.

As mentioned previously in this tutorial processing starts from the main method. Therefore in order to run this Student class, there should be main method and objects should be created. We will be creating a separate class for these tasks.

Given below is the *StudentTest* class which creates two instances of the class Student and invokes the methods for each object to assign values for each variable.

Save the following code in StudentTest.java file

```
import java.io.*;
public class StudentTest{
public static void main(String args[]){
/* Create two objects using constructor */
Student empOne = new Student("James Smith");
Student empTwo = new Student("Mary Anne");
// Invoking methods for each object created
empOne.empAge(26);
empOne.empDesignation("Senior Software Engineer");
empOne.empMark(1000);
empOne.printStudent();
```

```
empTwo.empAge(21);
empTwo.empDesignation("Software Engineer");
empTwo.empMark(500);
empTwo.printStudent();
  }
}
```

Now compile both the classes and then run *StudentTest* to see the result as follows:

```
C :> javac Student.java
C :> vi StudentTest.java
C :> javac  StudentTest.java
C :> java StudentTest
Name:James Smith
Age:26
Designation:Senior Software Engineer
Mark:1000.0
Name:Mary Anne
Age:21
Designation:Software Engineer
Mark:500.0
```

INTERFACE

An interface is a collection of abstract methods. A class implements an interface, thereby inheriting the abstract methods of the interface.

An interface is not a class. Writing an interface is similar to writing a class, but they are two different concepts. A class describes the attributes and behaviors of an object. An interface contains behaviors that a class implements.

Unless the class that implements the interface is abstract, all the methods of the interface need to be defined in the class.

An interface is similar to a class in the following ways:

- An interface can contain any number of methods.
- An interface is written in a file with a **.java** extension, with the name of the interface matching the name of the file.

  The bytecode of an interface appears in a **.class** file.
- Interfaces appear in packages, and their corresponding bytecode file must be in a directory structure that matches the package name.

However, an interface is different from a class in several ways, including:

- You cannot instantiate an interface.
- An interface does not contain any constructors.
- All of the methods in an interface are abstract.
- An interface cannot contain instance fields. The only fields that can appear in an interface must be declared both static and final.
- An interface is not extended by a class; it is implemented by a class.
- An interface can extend multiple interfaces.

Declaring Interfaces

The **interface** keyword is used to declare an interface. Here is a simple example to declare an interface:

Example : Let us look at an example that depicts encapsulation:

```
/* File name : NameOfInterface.java */import java.lang.*;
//Any number of import statements

public interface NameOfInterface{
//Any number of final, static fields
//Any number of abstract method declarations\
}
```

Interfaces have the following properties:

- An interface is implicitly abstract. There is no need to use the **abstract** keyword when declaring an interface.
- Each method in an interface is also implicitly abstract, so the abstract keyword is not needed.
- Methods in an interface are implicitly public.

Implementing Interfaces

When a class implements an interface, one can think of the class as signing a contract, agreeing to perform the specific behaviors of the interface. If a class does not perform all the behaviors of the interface, the class must declare itself as abstract.

A class uses the **implements** keyword to implement an interface. The implements keyword appears in the class declaration following the extends portion of the declaration.

When overriding methods defined in interfaces, there are several rules to be followed:

- Checked exceptions should not be declared on implementation methods other than the ones declared by the interface method or subclasses of those declared by the interface method.
- The signature of the interface method and the same return type or subtype should be maintained when overriding the methods.
- An implementation class itself can be abstract and if so, interface methods need not be implemented.

When implementation interfaces, there are several rules:

- A class can implement more than one interface at a time.
- A class can extend only one class, but implement many interfaces.
- An interface itself can extend another interface. An interface cannot extend another interface.

Extending Interfaces

An interface can extend another interface, similarly to the way that a class can extend another class. The **extends** keyword is used to extend an interface, and the child interface inherits the methods of the parent interface.

```
public class Main {
   public static void main(String[] args) {
       shape circleshape=new circle();
          circleshape.Draw();
   }
}
interface shape
{
    public  String baseclass="shape";
       public void Draw();
}
 class circle implements shape
{
   public void Draw() {
       System.out.println("Drawing Circle here");
   }
 }
```

Extending Multiple Interfaces

A Java class can only extend one parent class. Multiple inheritance is not allowed. Interfaces are not classes, however, and an interface can extend more than one parent interface.

The extends keyword is used once, and the parent interfaces are declared in a comma-separated list.

Tagging Interfaces

The most common use of extending interfaces occurs when the parent interface does not contain any methods. For example, the MouseListener interface in the java.awt.event package extended java.util.EventListener, which is defined as:

```
package java.util;
public interface EventListener
{}
```

An interface with no methods in it is referred to as a **tagging** interface. There are two basic design purposes of tagging interfaces:

Creates a common parent: As with the EventListener interface, which is extended by dozens of other interfaces in the Java API, you can use a tagging interface to create a common parent among a group of interfaces. For example, when an interface extends EventListener, the JVM knows that this particular interface is going to be used in an event delegation scenario.

Adds a data type to a class: This situation is where the term tagging comes from. A class that implements a tagging interface does not need to define any methods (since the interface does not have any), but the class becomes an interface type through polymorphism.

10. JAVADOC COMMENTS

Java supports three types of comments. The first two are the // and the /* */. The third type is called a documentation comment. It begins with the character sequence /** and it ends with */.

Documentation comments allow you to embed information about your program into the program itself. You can then use the javadoc utility program to extract the information and put it into an HTML file.

Documentation comments make it convenient to document your programs.

The Javadoc Tags

The javadoc utility recognizes the following tags:

Javadoc Utility Tags

Tag	Description	Example
@author	Identifies the author of a class.	@author description
@deprecated	Specifies that a class or member description is deprecated.	@deprecated
{@docRoot}	Specifies the path to the root directory of the current documentation	Directory Path
@exception	Identifies an exception thrown by a method.	@exception exception-name explanation
{@inheritDoc}	Inherits a comment from the immediate superclass.	Inherits a comment from the immediate surperclass.
{@link}	Inserts an in-line link to another topic.	{@link name text}
{@linkplain}	Inserts an in-line link to another topic, but the link is displayed in a plain-text font.	Inserts an in-line link to another topic.
@param	Documents a method's parameter.	@paramparameter-name explanation
@return	Documents a method's return value.	@return explanation
@see	Specifies a link to another topic.	@see anchor
@serial	Documents a default serializable field.	@serial description
@serialData	Documents the data written by the write object ( ) or writeExternal ( ) methods	@serial data description
@serialField	Documents an object stream field description	@serial field name type component

(contd.)

(*contd.*)

@since	States the release when a specific change was introduced.	@since release
@throws	Same as @exception.	The @throws tag has the same meaning as the @exception tag.
{@value}	Displays the value of a constant, which must be a static field.	Displays the value of a constant, which must be a static field.
@version	Specifies the version of a class.	@version info

Documentation Comment

After the beginning /**, the first line or lines become the main description of your class, variable, or method.

After that, you can include one or more of the various @ tags. Each @ tag must start at the beginning of a new line or follow an asterisk (*) that is at the start of a line.

Multiple tags of the same type should be grouped together. For example, if you have three @see tags, put them one after the other.

Here is an example of a documentation comment for a class:

```
/**
This class draws a documentation example.
* @BPL- AJP*
@version 1.1
*/
```

javadoc Output

The javadoc program takes as input your Java program's source file and outputs several HTML files that contain the program's documentation.

Information about each class will be in its own HTML file. Java utility **javadoc** will also output an index and a hierarchy tree. Other HTML files can be generated.

Since different implementations of javadoc may work differently, you will need to check the instructions that accompany your Java development system for details specific to your version.

Example: Following is a sample program that uses documentation comments. Notice the way each comment immediately precedes the item that it describes.

After being processed by javadoc, the documentation about the SquareNum class will be found in SquareNum.html.

```
import java.io.*;
/*** This class demonstrates documentation comments.
* @BPL- AJP
* @version 1.1
*/
```

```
public class SquareNum {
/**
* This method returns the square of num. This is a multiline description. You can
use  * as many lines as you like. @param num—the value to be squared.
* @return num squared.
*/
public double square(double num) { return num * num;
}  /**
* This method inputs a number from the user.
* @return The value input as a double.
* @exception IOException On input error.   * @see IOException
*/
public double getNumber() throws IOException {
InputStreamReader isr = new InputStreamReader(System.in);
BufferedReader inData = new BufferedReader(isr);
String str;
str = inData.readLine();
return (new Double(str)).doubleValue();
}
/**
* This method demonstrates square().
* @param args Unused.
* @return Nothing.
* @exception IOException On input error.
* @see IOException
*/
public static void main(String args[]) throws IOException
{
SquareNum ob = new SquareNum();
double val;
System.out.println("Enter value to be squared: ");
val = ob.getNumber();
val = ob.square(val);
System.out.println("Squared value is " + val);
}
}
```

Now process above SquareNum.java file using javadoc utility as follows:

```
$ javadoc SquareNum.java
Loading source file SquareNum.java...
Constructing Javadoc information...
```

```
Standard Doclet version 1.5.0_13
Building tree for all the packages and classes...
Generating SquareNum.html...
SquareNum.java:39: warning - @return tag cannot be used in method with void
return type.
Generating package-frame.html...
Generating package-summary.html...
Generating package-tree.html...
Generating constant-values.html...
Building index for all the packages and classes...
Generating overview-tree.html...
Generating index-all.html...
Generating deprecated-list.html...
Building index for all classes...
Generating allclasses-frame.html...
Generating allclasses-noframe.html...
Generating index.html...
Generating help-doc.html...
Generating stylesheet.css...
1 warning
$
```

2

Object-Oriented Programming and Java Basics

1. INHERITANCE

In object-oriented methodology, *inheritance* enables the user to extend the functionality of an existing class. The user creates a class that inherits the attributes and behavior of another class. In addition, the new class can consist of a few new attributes and behaviors that are specific to the class. In terms of classes and objects, attributes refer to the data and behavior refers to the methods.

For example, you might have the student class that consists of the basic characteristics of students, such as name, date of birth, and gender. You can extend the functions of the student class to a day scholar, who in addition to characteristics of the student class contains its own features, such as locker number and residential address.

Inheritance enables you to add new features and functionality to an existing class without modifying the existing class. Inheritance also enables you to share data and methods among multiple classes.

a. Superclass and Subclass

In inheritance, a *Superclass* or *parent* class is the one from which another class inherits attributes and behavior. A *subclass* or *child* class is a class that inherits attributes and behavior from a superclass. For example, an Air ticket can be of two types, Confirmed and Requested. It also has various attributes, such as flight number, date, time and destination. Both types of air tickets inherit these common attributes. The confirmed ticket will, however, have a seat number while a requested ticket will have a status attribute.

A subclass has the following characteristics:

- A subclass is a specialized form of the superclass.
- A subclass can inherit the properties and methods of its parent or superclass.
- A subclass may have some additional and unique features in addition to the features inherited from its superclass.

To implement inheritance in object-oriented programming, various relationships exist between the various classes of the program.

b. Relationships Between Classes

Classes and objects are related to interact with each other. In object-oriented programming, objects perform actions in response to messages from other objects defining the receiving object's behaviour. This approach specifies the relationships among classes based on the behavior of individual class.

The various relationships that can exist between the various classes of a program are:

- Kind-of
- Is-a
- Part-of
- Has-a

Kind-of Relationship

The *kind-of relationship* is used at the class level to describe the relationship between a superclass and a subclass. A subclass can inherit the attributes of a superclass. Therefore, a subclass is a type of superclass. In the ticketing example discussed earlier, the subclasses, such as confirmed ticket and requested tickets are the kind of tickets.

Is-A Relationship

The relationship between objects of superclass and subclass is referred to as an *is-a relationship*. For example, in the ticketing system, a ticket class is a superclass and confirmed ticket is a subclass. A confirmed ticket for Chennai is an object of the confirmed ticket class. The relationship between these three reads as the ticket for Chennai is a confirmed ticket, which in turn is a kind of ticket.

Part-of Relationship

When a class is an element of another class, it depicts the *part-of relationship*. For example, while developing a program, address of various entities, such as students, customers, and organization must be stored. Address consists of various sub-attributes, such as address line 1, address line 2, city, state, country, and zip code. To reuse the entire set of these attributes, a class is to be created that consists of the required attributes. This class can be used as an element of the classes where the address is to be stored. The address class exhibits the part-of relationship with respect to the class in which it is included.

Has-A Relationship

The *has-a relationship* is reverse of the part-of relationship. When a class consists of another class as its element, it depicts the has-a relationship. For example, when the address class is to be included in the student class, the relationship between the student and address class reads as the student class has an address. The has-a relationship is also known as *aggregation or composition*.

```
class subclass extends superclass
{
  .....
  ...
}
```

In a program, various classes that exhibit several relationships among them can be created. The types of relationships among classes result in different types of inheritance. Inheritance is the process of creating a new class with the characteristics of an existing class, along with additional characteristics unique to the new class. When a class is based on another class, it inherits all the properties of that class. The class, which is inheriting, is referred to as the subclass (child class) and the class providing the information to inherit is referred to as the superclass (parent class). The keyword used for inheritance extends.

c. Types of Inheritance

There are 4 types of inheritance that can be used in Java programs:

- Simple inheritance
- Multilevel inheritance
- Multiple inheritance
- Hybrid inheritance

i. Simple Inheritance

When a subclass is derived simply from its parent class, then this mechanism is known as simple inheritance. In case of simple inheritance there is only a subclass and its parent class. It is also called single inheritance or one level inheritance.

```
//creating a super class
class sup_class
{
static int i=5,j=10;
static void showij()
{
System.out.println("i and j:" + " "+ i + " ," +j);
}
}
//creating the subclass
class sub_class extends sup_class
{
static int k=15;
static void showk()
{
System.out.println("k"+k);
}
static void sum()
{
System.out.println("i+j+k:"+(i+j+k));
}
}
class main_prg
{
public static void main(string arg[ ])
{
sup_class superobj=new sup_class;
sub_class subobj=new sub_class;
//the super class can be called by itself
System.out.println("contents of super class");
superobj.showij
System.out.println();
//the subclass has access to tall public class of super class
System.out.println("Contents of sub");
```

```
Subobj.showk();
System.out.println( );
System.out.println( "sum in Super and Sub class is");
Subobj.sum();
}
}
```

Output

```
Contents of super
i and j : 5, 10
Contents of sub
k 15
sum in Super and Subclass is
i + j + k : 30
```

Another simple example:

```
class A {
int x;
int y;
int get(int p, int q){
x=p; y=q; return(0);
}
void Show(){
System.out.println(x);
}
}

class B extends A{
public static void main(String args[]){
A a = new A();
a.get(5,6);
a.Show();
}
void display(){
System.out.println("B");
}
}
```

ii. Multilevel Inheritance

It is the enhancement of the concept of inheritance. When a subclass is derived from a derived class, then this mechanism is known as the multilevel inheritance. The derived class is called the subclass or child class for its parent class and this parent class works as the child class for its just above (parent) class. Multilevel inheritance can go up to any number of level.

```
class A {
int x;
int y;
int get(int p, int q){
x=p; y=q; return(0);
}
void Show(){
System.out.println(x);
}
}
class B extends A{
void Showb(){
System.out.println("B");
}
}

class C extends B{
void display(){
System.out.println("C");
}
public static void main(String args[]){
A a = new A();
a.get(5,6);
a.Show();
}
}
```

iii. Multiple Inheritance

The mechanism of inheriting the features of more than one base class into a single class is known as multiple inheritance. Java does not support multiple inheritance but the multiple inheritance can be achieved by using the interface.

In Java Multiple Inheritance can be achieved through use of interfaces by implementing more than one interfaces in a class.

Super Keyword

The super is java keyword. As the name suggest super is used to access the members of the superclass. It is used for two purposes in java.

The first use of keyword super is to access the hidden data variables of the superclass hidden by the subclass.

For example, suppose class A is the superclass that has two instance variables as int a and float b. Class B is the subclass that also contains its own data members named a and b. Then we can access the superclass (class A) variables a and b inside the subclass class B just by calling the following command.

Super Member

Here member can either be an instance variable or a method. This form of super most is useful to handle situations where the local members of a subclass hides the members of a superclass having the same name. The following example clarify all the confusions.

```
class A{
int a;
float b;
void Show(){
System.out.println("b in super class:  " + b);
}

}

class B extends A{
int a;
float b;
B(int p, float q){
a = p;
super.b = q;
}
void Show(){
super.Show();
System.out.println("b in super class:  " + super.b);
System.out.println("a in sub class:  " + a);
}

public static void main(String[] args){
B subobj = new B(1, 5);
subobj.Show();
}
}
```

Use of super to call superclass constructor: The second use of the keyword super in java is to call superclass constructor in the subclass. This functionality can be achieved just by using the following command.

Super (param-list);

Here parameter list is the list of the parameter requires by the constructor in the superclass. Super must be the first statement executed inside a superclass constructor. If we want to call the default constructor then we pass the empty parameter list. The following program illustrates the use of the super keyword to call a superclass constructor.

```
class B extends A{
int d;
B(int l, int m, int n, int o){
super(l,m,n);
d=o;
}
```

```
void Show(){
 System.out.println("a = " + a);
 System.out.println("b = " + b);
 System.out.println("c = " + c);
 System.out.println("d = " + d);
 }

 public static void main(String args[]){
 B b = new B(4,3,8,7);
 b.Show();
 }
 }
```

iv. Hybrid Inheritance

A combination of one or many types of inheritance to be done in a single program is called Hybrid inheritance.

Class Hierarchy

A class in Java can only extend **one** other class; that is, it can only have one immediate superclass. However, a superclass can also be derived from another class.

Classes often are depicted graphically in a **class hierarchy** (also called an inheritance hierarchy).

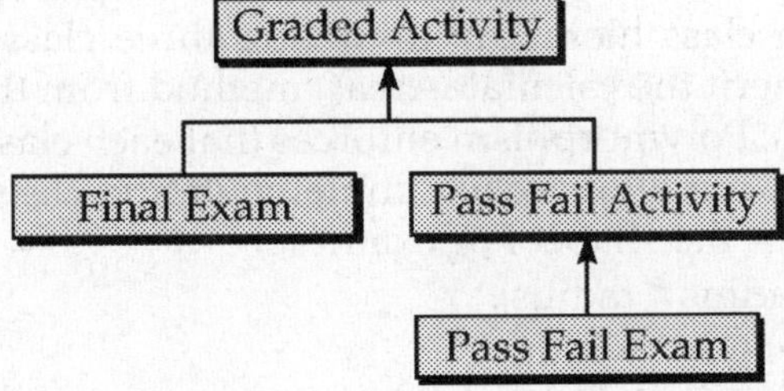

Classes higher up in the hierarchy are more generalized; classes lower down in the hierarchy are more specialized.

By placing general code as far up in the hierarchy as possible, we are able to use inherited methods in sub classes **without having to duplicate code**. This leads to smaller programs.

Should we ever need to change this code, we are able to **change it one place** (the superclass) and all the derived classes will use the corrected version through inheritance. This leads to maintainable programs.

When you design a more complicated system, your class hierarchy will have more levels. For example, if you were designing a library catalog, you might layers such as these:

- Library Object
- Loan Object vs Reference Object
- Periodical vs Book vs Audio Tape vs Video Tape vs DVD
- Daily Periodical vs Monthly Periodical
- Foreign Language Monthly Periodical vs English Monthly Periodical
- Asian Language Monthly Periodical vs European Language Monthly Periodical
- Japanese Monthly Periodical vs Chinese Monthly Periodical

Business rules are enforced by inheritance:

All the objects in the hierarchy inherit a characteristic from the LibraryObject, that is, belonging to the library.

- All the objects for loan, unlike the reference objects, can be "checked out" to customers.
- All the periodicals, unlike books and such, are published on a recurring basis.

Therefore, whatever object is a member of the Japanese Monthly Periodical class, will automatically inherit the logic common to Loan Object, that is, the ability to be "checked out". I do not have to rewrite the logic for the procedure of checking out a Japanese Monthly Periodical and make it different from the procedure for checking out a European Language Monthly Periodical. It is like inheriting money or kingship: I do not have to do work to get the money or the crown.

Inheritance as Specialization

However, it might WANT to override what an object inherits. For example, I might want the

- checkout period for Asian Language Monthly Periodical to be shorter or longer than for European Language Monthly Periodical
- the late Return Fine for faculty to be higher than the late Return Fine for students
- materials listed as critical To Current Course to be temporarily considered reference, with no checkout allowed this semester

In other words, I might want to have different styles or forms of checkout rules. The ability to have many (poly) forms (morphs) is called polymorphism.

Suppose you have a class hierarchy involving three classes, Circle, Square, and Rectangle, which all inherit the calculateArea() method from theShape class (or, as we shall see later, interface). Polymorphism enforces that each class implementation has a calculateArea() method, but allows each implementation class redefine the inherited or required method to fit the shape. For example,

- circleArea = pi * radius * radius;
- rectangleArea = length * width;
- triangleArea = base * (height/2);

```
class Circle implements Shape {
public double calculateArea() {
circleArea = pi * radius * radius;
return circleArea;
}
}
class Triangle implements Shape {
public double calculateArea() {
triangleArea = base * (height/2);
return triangleArea;
}
}
```

The basic packages available in java are:

java.lang — basic language functionality and fundamental types
java.util — collection data structure classes
java.io— file operations

java.math— multiprecision arithmetics
java.nio— the New I/O framework for Java
java.net — networking operations, sockets, DNS lookups, ...
java.security — key generation, encryption and decryption
java.sql — Java Database Connectivity (JDBC) to access databases
java.awt — basic hierarchy of packages for native GUI components
javax.swing — hierarchy of packages for platform-independent rich GUI components
java.applet — classes for creating an applet

2. POLYMORPHISM

Polymorphism is the ability of an object to take on many forms. The most common use of polymorphism in OOP occurs when a parent class reference is used to refer to a child class object. Any java object that can pass more than one IS-A test is considered to be polymorphic. In Java, all java objects are polymorphic since any object will pass the IS-A test for their own type and for the class Object. It is important to know that the only possible way to access an object is through a reference variable. A reference variable can be of only one type. Once declared the type of a reference variable cannot be changed. The reference variable can be reassigned to other objects provided that it is not declared final. The type of the reference variable would determine the methods that it can invoke on the object. The concept of polymorphism is often expressed by phrase "One interface, Multiple methods". This means that it is possible to design a generic interface to a group of related activities. Java implements the **polymorphisim** in one of the two ways, that is, **Method Overriding** and **Method Overloading.**

a. Method Overriding

Method overriding occurs when sub class declares a method that has the same signature (name, plus the number and the type of its parameters) and return type as a method declared by one of its superclass. The key benefit of overriding is the ability to define behavior that is specific to a particular subclass type. An overriding method can also return a subtype of the type returned by the overridden method. This is called a *covariant return type*. Which overridden version of the method to call is decided at runtime based on the object type.

The rules for overriding a method are as follows:

- The argument list must exactly match that of the overridden method. If does not match, we will get a overloaded method, not a overriding method.
- The return type must be same as, or a subtype of, the return type declared in the original overridden method in the superclass.
- We cannot override method marked final. Static methods cannot be overridden but they can be redefined.
- If a method cannot be inherited , you cannot override it. Overriding is nothing but reimplementing a method that we are inherited.
- Overriding method must not throw checked exceptions that are new and broader than those declared by the overridden method.
- The access level cannot be more restrictive than the overridden method. It can be less restrictive than that of overridden method.
- The overriding method can throw any unchecked or runtime exception, regardless of whether overridden method declares the exception or not.
- The overbidding method does not have to declare any exceptions that it will never throw, regardless of what the overridden method declares.

- A subclass within the same package as the instance's superclass can override any superclass method that is not declared private or final. A subclass in a different package can only override the non-final methods declared public or protected.
- Constructors cannot be overridden.

A program to explain the above rules of Java Method Overridding

```
public  class SuperClass{
  public void method1() {
    System.out.println("SuperClass.method1()");
  }
  private void method2() {
      System.out.println("SuperClass.method2()");
  }
  public final void method3(){
      System.out.println("SuperClass.method3()");
}
private final void method4(){
      System.out.println("SuperClass.method4()");
}
public static void method5() {
      System.out.println("SuperClass.method5()");
}
public void method6() throws Exception {
      System.out.println("SuperClass.method6()");
}
private void method7(){
      System.out.println("SuperClass.method7()");
}
private void method8(int x){
      System.out.println("SuperClass.method8()");
}
public static void method9() {
      System.out.println("SuperClass.method9()");
}
}
public class SubClass extends SuperClass {
  public void method1() {
        System.out.println("OverridingClass.method1()");
}
private void method2(){
        System.out.println("OverridingClass.method2()");
}
//We can't override a public final method
/*  public final void method3(){
        System.out.println("OverridingClass.method3()");
  }*/
```

```
 private final void method4(){
    System.out.println("OverridingClass.method4()");
}
public static void method5() {
    System.out.println("OverridingClass.method5()");
}
public void method6() throws IOException{
      System.out.println("OverridingClass.method6()");
}

public void method7(){
    System.out.println("OverridingClass.method7()");
}

public void method8(final int x){
     System.out.println("OverridingClass.method8()");
}

  //A static method cannot be overridden to be non-static instance method
  /*public void method9() {
     System.out.println("OverridingClass.method9()");
  }*/
}
public class MethodOverrdingSample{
   public static void main(String[] args) {
      SubClass sub = new SubClass();
      SuperClass sc = new SubClass();
      sub.method1();
      sub.method3();
      /*  Since its private, the below 2 methods are not visible
      sub.method2();
      sub.method4();*/

      sub.method5();
      try {
            sub.method6();
      } catch (IOException e) {
            e.printStackTrace();
}
sub.method7();
sub.method8(100);
sc3.method5();
SubClass subClass = new SubClass();
SuperClass supClass = (SuperClass)subClass;
supClass.method5();
  supClass.method1();
 }
}
```

To invoke a superclass version of overridden method we use a **super** keyword. For example

```
public class Superclass{
    public void method1(){
        System.out.println("Superclass Version");}
    }
}
public class SubClass{
   public method1(){
        super.method1();      //Invoke first superclass code
                              //then do the subclass specific code
        System.out.println("Subclass Version");
    }
}
```

b. Method Overloading

Method overloading means to have two or more methods with same name in the same class and its subclass with different arguments and optionally different return type. Which overloaded version of the method to be called is based on the reference type of the argument passed at compile time.

The rules for overloading a method are as follows:

- Overloaded methods **must** change the argument list.
- Overloaded methods can change the return type.
- Overloaded methods can declare new or broader checked exceptions
- Overloaded methods can change the access modifier.
- Constructors can be overloaded.

A program to explain the above rules of Java Method Overloading

```
public class A {
   protected A(){
      System.out.println("Hi A Constructor");
   }
   public A(int i,int j){
      System.out.println("Hi A Constructor A,B");
   }
   public void method1(int i){
      System.out.println("Hi A in method1(i)");
   }
}
public class B extends A{
   public B(){
      System.out.println("Hi B Constructor");
   }
   public B(int i, int j){
      super(i,j);
      System.out.println("Hi B Constructor with i,j");
   }
    public void method1(int j){
```

```
        System.out.println("Hi B in method1(i)");

    }
    public void method1(int i, int j){
        System.out.println("Hi B in method1(i,j)");
    }
}
public class C extends A{
    public C(){
        System.out.println("Hi C Constructor");
    }
    public C(int i, int j){
        super(i,j);
        System.out.println("Hi C Constructor with i,j");
    }
    public void method1(int j){
        System.out.println("Hi C in method1(i)");
    }

    public void method1(int i, int j){
        System.out.println("Hi C in method1(i,j)");
    }
}
public class D {
    public  static void main(String[] args) {
        // TODO Auto-generated method stub
        int k=0;
        int l=0;
        A a=new A();
        B b= new B();
        a= b;
        C c= new C();
        A ac=c;
        a.method1(k);
        b.method1(k,l);
        b.method1(k);
        c.method1(k);
    }
}
```

Output for the above program

Hi A Constructor

Hi A Constructor

Hi B Constructor

Hi A Constructor
Hi C Constructor
Hi B in method1(i)
Hi B in method1(i,j)
Hi B in method1(i)
Hi C in method1(i)

3. BINDING

Dynamic Binding *or* Late Binding

Dynamic Binding refers to the case where compiler is not able to resolve the call and the binding is done at runtime only. Suppose we have a class named 'SuperClass' and another class named 'SubClass' extends it. Now a 'SuperClass' reference can be assigned to an object of the type 'SubClass' as well. If we have a method (say 'someMethod()') in the 'SuperClass' which we override in the 'SubClass', then a call of that method on a 'SuperClass' reference can only be resolved at runtime as the compiler cannot be sure of what type of object this reference would be pointing to at runtime.

```
...
SuperClass superClass1 = new SuperClass();
SuperClass superClass2 = new SubClass();

superClass1.someMethod(); // SuperClass version is called
superClass2.someMethod(); // SubClass version is called
...
```

Here, we see that even though both the object references superClass1 andsuperClass2 are of type 'SuperClass' only, but at runtime they refer to the objects of types 'SuperClass' and 'SubClass' respectively. Hence, at compile time the compiler cannot be sure if the call to the method 'someMethod()' on these references actually refer to which version of the method—the superclass version or the subclass version. Thus, dynamic binding in Java simply binds the method calls (inherited methods only as they can be overriden in a subclass and hence compiler may not be sure of which version of the method to call) based on the actual object type and not on the declared type of the object reference.

```
class Shape{
public void draw() {
System.out.println("\n\tDrawing a shape.");
}
}
class Circle extends Shape{
public void draw()      {
System.out.println("\n\tDrawing a Circle.");
}
}
```

```
class Rectangle extends Shape{
public void draw()      {
System.out.println("\n\tDrawing a Rectangle.");
}
}
public class DynamicBindingDemo{
public static void main(String args[])      {
Shape obj;
obj = new Shape();
obj.draw();
obj = new Circle();
obj.draw();obj = new Rectangle();
obj.draw();
}
}
```

Result

```
Drawing a shape.
Drawing a Circle.
Drawing a Rectangle.
```

Static Binding *or* Early Binding

If the compiler can resolve the binding at the compile time only, then such a binding is called Static Binding or Early Binding. All the instance method calls are always resolved at runtime, but all the static method calls are resolved at compile time itself and hence we have static binding for static method calls. Because static methods are class methods and hence they can be accessed using the class name itself (in fact they are encouraged to be used by their corresponding class names only and not by using the object references) and therefore access to them is required to be resolved during compile time only using the compile time type information. That is the reason why static methods cannot actually be overriden. Similarly, access to all the member variables in Java follows static binding as Java does not support (in fact, it discourages) polymorphic behavior of member variables. For example:

```
class SuperClass{
...
public String someVariable = "Some Variable in SuperClass";
...
}
class SubClass extends SuperClass{
...
public String someVariable = "Some Variable in SubClass";
...
}
...
...
SuperClass superClass1 = new SuperClass();
SuperClass superClass2 = new SubClass();
System.out.println(superClass1.someVariable);
System.out.println(superClass2.someVariable);
...
```

Output

Some Variable in SuperClass
Some Variable in SuperClass

We can observe that in both the cases, the member variable is resolved based on the declared type of the object reference only, which the compiler is capable of finding as early as at the compile time only and hence a static binding in this case. Another example of static binding is that of 'private' methods as they are never inherited and the compile can resolve calls to any private method at compile time only.

Polymorphism and Dynamic Binding

```
/**
 * Polymorphism & dynamic binding Example
 * Overridden area(double) defintions for Circle & Square
**/

abstract class Shape{
protected final static double PI = 22.0/7.0;
protected double length;
public abstract double area();
}

class Square extends Shape{
Square(double side){
length=side;// initialises inherited length
}
public double area(){// overrides area()    of Shape
return length*length;// length inherited from Shape
}
}

class Circle extends Shape{
Circle(double radius){
length=radius;// initialises inherited length
}
public double area(){// overrides area()     of Shape
return PI*length*length;// PI & length inherited from Shape
}
}
/**
* Polymorphism & Dynamic binding test class
**/
public class l1PolyTest{
public static void main(String[] args){
Shape sh;// no object instance just variable declaration
```

```
Square sq = new Square(10.0);// sq is a Square object reference
Circle circ = new Circle(10.0);// circ is a Circle object reference

sh=sq;// sh dynamically bound to the Square object referenced by sq
System.out.println("Area of Square = " + sh.area());

sh=circ; // sh dynamically bound to the Circle object referenced by circ
System.out.println("Area of circle = " + sh.area());
}
}
/*********** Compile & Run *************
Area of Square = 100.0
Area of circle = 314.2857142857143
#
*****************************************/
```

4. FINAL KEYWORD AND METHODS

Like other **keywords in java**, *final* is also a keyword used in several different contexts to define an entity which cannot be changed later.

Final Classes

The Java Programming language permits us to apply the keyword *final* to classes. If the class is made *final*, then it cannot be subclassed. For example, *java.lang.String* is a *final* class. This is done for security reasons, because it ensures that if the method is referenced as *String*, then the method is a definite string of class *String* and not a string of class which has subclassed *String* class.

Final Methods

Like class we can also mark methods as *final*. Methods that are marked as *final* cannot be overridden in any case. For security reasons only you must make the method as *final* if the method has implementation which you do not want others to change.

Methods declared as *final* can be optimized. The compiler can generate a code that causes a direct call to the method, rather than invoking it the usual way, i.e. during the runtime.

Final Variables

If a variable is marked as *final*, then the value of that variable cannot be changed, i.e. *final* keyword when used with a variable makes it a **constant**. And if you try to change the value of that variable during the course of your program, the compiler will give you an error. If you mark variable of a reference type as *final*, that variable cannot refer to any other object. However, you can change the object's contents, because only the reference itself is *final*.

Blank Final Variables

A *blank final variable* is a variable that is not initialized during its declaration. The initialization is delayed. **A blank final instance variable must be assigned in a constructor, *but it can be set only once*.** A blank final variable that is local variable can be set at any time in the body of the method, but it can be set only once.

The following code fragment is an example of how a blank final variable can be used in class.

```
public class Customer{
    private final long customerID;

public Customer(){
    customerID = createID();
}
  public long getID(){
    return customerID;
}

  public long createID(){
    return ... // generates new ID
  }
  ... // more declarations
}
```

Discussion

When you use the keyword 'final' with a variable declaration, the value stored inside that variable cannot be changed.

So suppose you do this:

final int i =10;

After this, nowhere in your code can you change the value of 'i', i.e. 'i' will always have a value '102'.

So i = 11; //will give you a compile time error.

Now, suppose you did not initialize the final integer 'i' while declaring it.

final int i;

Value may be assigned to this variable, but only one.

final int i;

i = 10; //This is fine

i = 11; //Compiler error

Now that was the core stuff about 'final'.

The 'final' keyword can be used for a class declaration or method declaration.

When the keyword 'final' is used for a class declaration, no subclass of a class can be declared final, i.e. no other class can extend this class. If a method is declared final, then the method cannot be overridden. Making a class both 'private' and 'final' makes the 'final' keyword redundant as a private method cannot be accessed in its subclass. But you will be able to declare a method of the same name as in the base class if the method has been made private in the base class. But then that does not mean you are overriding the method. You are simply declaring a new method in the subclass.

Note : You cannot make an 'abstract' class or method as 'final' because an 'abstract' class needs to be extended which will not be possible if you mark it as 'final'!

5. DATA ABSTRACTION AND ENCAPSULATION

Grady Booch, a famous author who has written many books on OOPS, has defined the encapsulation feature as:

"Encapsulation is the process of hiding all of the details of an object that do not contribute to its essential characteristics."

Encapsulation implies that non-essential details of an object are hidden from the user and an access is provided to its essential details. Therefore, encapsulation is also called information hiding. For example, when you plug in the cord of a vacuum cleaner and switch it on, the vacuum cleaner starts functioning. An end-user need not know the working principle of a vacuum cleaner to convert electricity into suction power. The switch of the vacuum cleaner encapsulates the complex process of conversion of electricity into suction power. The complexity of an object is hidden as a result of encapsulation.

Computer games also use the feature of encapsulation. The user only needs to know how to play the game. However, the complex working of game is hidden from the user.

In Object-oriented methodology, need of encapsulation arises because the emphasis is on designing classes in such a manner that the classes share data and methods among themselves. Encapsulation is the feature that provides security to the data and methods of a class.

ATM machines that you use to withdraw money also has an encapsulated function. You insert an ATM card in the machine and provide the pin code. The only access that you have to the ATM is to provide the pin code of your account and not to the process of validating the password. If the pin code is correct, you can make the desired transaction. The backend transaction process is encapsulated. In addition, if you try to access another account, access is denied, as you are not authorized to access any account other than your account.

Abstraction

Grady Booch defined the abstraction feature as:

"An Abstraction denotes the essential characteristics of an object that distinguishes it from all other kinds of objects and thus provides crisply defined conceptual boundaries, relative to the perspective of the viewer."

Abstraction refers to the attributes of an object that clearly demarcates it from other objects. For example, while developing an online shopping site for books, you use objects of various items, such as books, compact disks (CDs), and cassettes. All these objects have a well-defined set of attributes that distinguish these objects from each other. For example, a book has page numbers, type of binding, and subject. However, a CD has type of media, such as audio or visual, duration, and storage capacity of the CD.

The concept of abstraction is implemented in object-oriented programming by creating classes. All the attributes of the objects of the classes are defined in the class. However, you cannot store any data in a class because creating a class does not allocate any memory space to the class. To store data, you need to create objects of the class, which have memory allocated as soon as it is created. Classes form the templates for creating objects.

In addition to the well-defined characteristics of an object, abstraction enables you to provide a restricted access to data. You come across hundreds of advertising messages every day through magazines or newspapers. You do not read, understand,

and respond to all of these because these are not of your interest. Instead, you concentrate only on messages that are of your specific interest. For example, if you want to buy a refrigerator, you will concentrate on the advertisements featuring refrigerators. You will not concentrate on the advertisements of other products, such as vaccum cleaner, washing machine, and air conditioner.

In object-oriented programming, abstraction means ignoring the non-essential details of an object and concentrating on its essential details. As discussed earlier, in an ATM, the user is not required to know the entire process of a transaction and how the data is stored. However, the program provides a restricted access to the user's account.

Similarly, when you want to send e-mail messages, you should know the processes of writing e-mail messages and sending it to the receiver. However, it is not necessary for you to know the entire process of sending the e-mail messages across the network.

To implement abstraction, you also use the encapsulation feature. Encapsulation hides the irrelevant details of an object and abstraction makes only the relevant details of an object visible. For example, the operation of a washing machine is hidden or encapsulated from the user. The details, like switching on and off the washing machine are the only details relevant for the user to know. This is implemented by abstraction. Rest of the working of the washing machine is encapsulated from the user.

6. ABSTRACT CLASS

Use the **abstract** keyword to declare a class abstract. The keyword appears in the class declaration somewhere before the class keyword.

```
/* File name : Student.java */
public abstract class Student
{
private String name;
private String address;
private int number;
public Student(String name, String address, int number)
{
System.out.println("Constructing an Student");
this.name = name;
this.address = address;
this.number = number;
}
public double computeMark()
{
System.out.println("Inside Student computeMark");
return 0.0;
}
public void mailResult()
{
System.out.println("Mailing the result to " + this.name + " " + this.address);
}
public String toString()
{
```

```
return name + " " + address + " " + number;
}
public String getName()
{
return name;
}
public String getAddress()
{
return address;
}
public void setAddress(String newAddress)
{
address = newAddress;
}
public int getNumber()
{
return number;
}
}
```

Notice that nothing is different in this Student class. The class is now abstract, but it still has three fields, seven methods, and one constructor.

Now if you would try as follows:

```
/* File name : AbstractDemo.java */
public class AbstractDemo
{
public static void main(String [] args)
  {

  /* Following is not allowed and would raise error */
  Student e = new Student("Santosh", "Street1, TN", 43);

  System.out.println("\n Call mailResult using Student reference—");
  e.mailResult();
  }
}
```

When you would compile above class, then you would get the following error:

```
Student.java:46: Student is abstract; cannot be instantiated
Student e = new Student("Santosh", "Street1, TN", 43);
                ^

1error
```

Extending Abstract Class

We can extend Student class in normal way as follows:

```
/* File name : Mark.java */
public class Mark extends Student
```

```
{
private double Mark; //Annual Mark
public Mark(String name, String address, int number, double Mark)
{
    super(name, address, number);
    setMark(Mark);
  }
  public void mailResult()
  {
    System.out.println("Within mailResult of Mark class ");
    System.out.println("Mailing check to " + getName() + " with Mark " +
Mark);
  }
  public double getMark()
  {
    return Mark;
  }
  public void setMark(double newMark)
  {
    if(newMark >= 0.0)
    {
      Mark = newMark;
    }
  }
  public double computeMark()
  {
    System.out.println("Computing Mark for " + getName());
    return Mark/52;
  }
}
```

Here we cannot instantiate a new Student, but if we instantiate a new Mark object, the Mark object will inherit the three fields and seven methods from Student.

```
/* File name : AbstractDemo.java */
public class AbstractDemo
{
  public static void main(String [] args)
  {
Mark s = new Mark("Ajay", "Lane1, TN", 3, 600.00);
Mark e = new Mark("John ", "Lane108, MA", 2, 400.00);
System.out.println("Call mailResult using  Mark reference —");
s.mailResult();
System.out.println("\n Call mailResult using  Student reference—");
e.mailResult();
  }
}
```

This would produce the following result:
Constructing an Student
Constructing an Student
Call mailResult using Mark reference —
Within mailResult of Mark class
Mailing check to Ajay with Mark 600.0
Call mailResult using Student reference—
Within mailResult of Mark class
Mailing check to John with Mark 400.

Abstract Methods

If you want a class to contain a particular method but you want the actual implementation of that method to be determined by child classes, you can declare the method in the parent class as abstract.

The abstract keyword is also used to declare a method as abstract. An abstract method consists of a method signature, but no method body.

Abstract method would have no definition, and its signature is followed by a semicolon, not curly braces as follows:

```
public abstract class Student
{
private String name;
private String address;
private int number;
public abstract double computeMark();
//Remainder of class definition}
```

Declaring a method as abstract has two results:

- The class must also be declared abstract. If a class contains an abstract method, the class must be abstract as well.
- Any child class must either override the abstract method or declare itself abstract.
- A child class that inherits an abstract method must override it. If they do not, they must be abstract, and any of their children must override it.
  Eventually, a descendant class has to implement the abstract method; otherwise, you would have a hierarchy of abstract classes that cannot be instantiated.

If Mark is extending Student class, then it is required to implement computeMark() method as follows:

```
/* File name : Mark.java */
public class Mark extends Student
{
private double Mark; //Annual Mark
public double computeMark()
{
System.out.println("Computing Mark pay for " + getName());
return Mark/52;
}
//Remainder of class definition
}
```

Encapsulation is one of the four fundamental OOP concepts. The other three are: Inheritance, polymorphism, and abstraction.

Encapsulation is the technique of making the fields in a class private and providing access to the fields via public methods. If a field is declared private, it cannot be accessed by anyone outside the class, thereby hiding the fields within the class. For this reason, encapsulation is also referred to as data hiding.

Encapsulation can be described as a protective barrier that prevents the code and data being randomly accessed by other code defined outside the class. Access to the data and code is tightly controlled by an interface.

The main benefit of encapsulation is the ability to modify our implemented code without breaking the code of others who use our code. With this feature encapsulation gives maintainability, flexibility and extensibility to our code.

Example : Let us look at an example that depicts encapsulation:

```
/* File name : EncapTest.java */
public class EncapTest{
private String name;
private String idNum;
private int age;
public int getAge(){
  return age;
}
public String getName(){
  return name;
}  public String getIdNum(){
  return idNum;
}
public void setAge( int newAge){
  age = newAge;
}
public void setName(String newName){
  name = newName;
}
public void setIdNum( String newId){
  idNum = newId;
}
}
```

The public methods are the access points to this class field from the outside java world. Normally these methods are referred to as getters and setters. Therefore any class that wants to access the variables should access them through these getters and setters.

The variables of the EncapTest class can be accessed as follows:

```
/* File name : RunEncap.java */
public class RunEncap{
public static void main(String args[]){
  EncapTest encap = new EncapTest();
  encap.setName("James");
  encap.setAge(20);
  encap.setIdNum("12343ms");
  System.out.print("Name : " + encap.getName()+  " Age : "+ encap.getAge());
  }
}
```

This would produce the following result:
Name: James Age: 20

Benefits of Encapsulation

- The fields of a class can be made read-only or write-only.
- A class can have total control over what is stored in its fields.
- The users of a class do not know how the class stores its data. A class can change the data type of a field, and users of the class do not need to change any of their code.

7. THE OBJECT CLASS

The object class sits at the top of the class hierarchy tree in the Java development environment. Every class in the Java system is a descendent (direct or indirect) of the object class. The object class defines the basic state and behavior that all objects must have the ability, such as to compare oneself to another object, to convert to a string, to wait on a condition variable, to notify other objects that a condition variable has changed, and to return the object's class.

The Equals Method

Use the equals to compare two objects for equality. This method returns true if the objects are equal, false otherwise. Note that equality does not mean that the objects are the same object. Consider this code that tests two Integers, one and another One, for equality:

```
Integer one = new Integer(1), another One = new Integer(1);
if (one. equals (another One))
  System.out.println("objects are equal");
```

This code will display objects which are equal even though one and another One reference two different and distinct objects. They are considered equal because they contain the same integer value.

Your classes should override this method to provide an appropriate equality test. Your equals method should compare the contents of the objects to see if they are functionally equal and return true if they are.

The Get Class Method

The get Class method is a final method (cannot be overridden) that returns a runtime representation of the class of this object. This method returns a Class object. You can query the Class object for a variety of information about the class, such as its name, its superclass, and the names of the interfaces that it implements. The following method gets and displays the class name of an object:

```
void Print Class Name (Object obj) {
    System.out.println("The Object's class is " + obj.get Class(). get Name());
}
```

One handy use of the get Class method is to create a new instance of a class without knowing what the class is at compile time. This sample method creates a new instance of the same class as object which can be any class that inherits from object (which means that it could be any class):

```
Object createNew Instance Of (Object obj) {
    return obj.getClass().newInstance();
}
```

The to String Method

Object's to string method returns a string representation of the object. You can use toString to display an object. For example, you could display a string representation of the current thread like this:

System.out.println(Thread.current Thread().toString());

The string representation for an object is entirely dependent on the object. The string representation of an Integer object is the integer value displayed as text. The string representation of a Thread object contains various attributes about the thread, such as its name and priority. For example, the previous code above display the following: Thread[main,5,main]

8. REFLECTION

Reflection API is a powerful technique (that provides the facility) to find-out its environment as well as to inspect the class itself. Reflection API was included in Java 1.1. The classes of reflection API are the part of the package java.lang.reflect and the methods of reflection API are the parts of the package java.lang.class. It allows the user to get the complete information about interfaces, classes, constructors, fields and various methods being used. It also provides an easy way to create a Java Application that was not possible before Java 1.1. You can create methods like event handlers, hash code, etc and also find-out the objects and classes.

With the help of reflection API you can get the information about any class of the java. lang package. There are some useful methods like get Name () and getInterfaces(), which allows us to retrieve the name of the class and the interfaces of the package respectively.

Avoid using reflection API in those applications wherever it affects the application's performance, security related code of the application such as in Applet programming. Reflection API also affects the application if the private fields and methods are there.

Reflection, also known as introspection, is a somewhat lofty term to describe the ability to "look inside" a class or an object and get information about that object's variables and methods as well as actually set and get the values of those variables and to call methods. Object reflection is useful for tools such as class browsers or debuggers, where getting at the information of an object on-the-fly allows you to explore what that object can do, or for component-based programs such as Java Beans, where the ability for one object to query another object about what it can do (and then ask it to do something) is useful to building larger applications.

The classes that support reflection of Java classes and objects will be part of the core Java 1.1 API (they are not available in the 1.0.2 version of the JDK). A new package, java.lang.reflect, will contain new classes to support reflection, which include the following:

- Field, for managing and finding out information about class and instance variables
- Method, for managing class and instance methods
- Constructor, for managing the special methods for creating new instances of classes.
- Array, for managing arrays
- Modifier, for decoding modifier information about classes, variables and methods

In addition, there will be a number of new methods available in the class to help tie together the various reflection classes.

Uses of Reflection

Reflection is commonly used by programs which require the ability to examine or modify the runtime behavior of applications running in the Java virtual machine. This is a relatively advanced feature and should be used only by developers who have a strong grasp of the fundamentals of the language. With that caveat in mind, reflection is a powerful technique and can enable applications to perform operations which would otherwise be impossible.

a. Extensibility Features

An application may make use of external, user-defined classes by creating instances of extensibility objects using their fully-qualified names.

b. Class Browsers and Visual Development Environments

A class browser needs to be able to enumerate the members of classes. Visual development environments can benefit from making use of type information available in reflection to aid the developer in writing correct code.

c. Debuggers and Test Tools

Debuggers need to be able to examine private members on classes. Test harnesses can make use of reflection to systematically call a discoverable set APIs defined on a class, to insure a high level of code coverage in a test suite.

Drawbacks of Reflection

Reflection is powerful, but should not be used indiscriminately. If it is possible to perform an operation without using reflection, then it is preferable to avoid using it. The following concerns should be kept in mind when accessing code via reflection.

a. Performance Overhead

Because reflection involves types that are dynamically resolved, certain Java virtual machine optimizations cannot be performed. Consequently, reflective operations have slower performance than their non-reflective counterparts, and should be avoided in sections of code which are called frequently in performance-sensitive applications.

b. Security Restrictions

Reflection requires a runtime permission which may not be present when running under a security manager. This is in an important consideration for code which has to run in a restricted security context, such as in an Applet.

c. Exposure of Internals

Since reflection allows code to perform operations that would be illegal in non-reflective code, such as accessing private fields and methods, the use of reflection can result in unexpected side-effects, which may render code dysfunctional and may destroy porta bility. Reflective code breaks abstractions and therefore may change behavior with upgrades of the platform.

The reflection API represents, or reflects the classes, interfaces, and objects in the current Java Virtual Machine. You will want to use the reflection API if you are writing development tools such as debuggers, class browsers, and GUI builders. With the reflection API you can:

- Determine the class of an object.
- Get information about a class's modifiers, fields, methods, constructors, and super-classes.

- Find out what constants and method declarations belong to an interface.
- Create an instance of a class whose name is not known until runtime.
- Get and set the value of an object's field, even if the field name is unknown to your program until runtime.
- Invoke a method on an object, even if the method is not known until runtime.
- Create a new array, whose size and component type are not known until runtime, and then modify the array's components.

First, a note of caution. Do not use the reflection API when other tools more natural to the Java programming language would suffice. For example, if you are in the habit of using function pointers in another language, you might be tempted to use the Method objects of the reflection API in the same way. Resist the temptation! Your program will be easier to debug and maintain if you do not use Method objects. Instead, you should define an interface, and then implement it in the classes that perform the needed action. Other trails use the term "member variable" instead of "field." The two terms are synonymous. Because the Field class is part of the reflection API, this trail uses the term "field."

The following topics are to be discussed:

- **Examining Classes** explain how to determine the class of an object, and how to get information about classes and interfaces.
- **Manipulating Objects** show you how to instantiate classes, get or set field values, and invoke methods. With the reflection API, you can perform these tasks even if the names of the classes, fields, and methods are unknown until runtime.
- **Working with Arrays** describes the APIs used to create and to modify arrays whose names are not known until runtime.

8.1 Examining Classes

8.1.1 Retrieving Class Objects

You can retrieve a Class object in several ways:

If an instance of the class is available, you can invoke Object.getClass. The getClass method is useful when you want to examine an object but you do not know its class. The following line of code gets the Class object for an object named mystery:

```
Class c = mystery.getClass();
```

If you want to retrieve the Class object for the superclass that another Class object reflects, invoke the getSuperclass method. In the following example, getSuperclass returns the Class object associated with the the TextComponent class, because TextComponent is the superclass of TextField:

```
TextField t = new TextField();
Class c = t.getClass();
Class s = c.getSuperclass();
```

If you know the name of the class at compile time, you can retrieve its Class object by appending .class to its name. In the next example, the Class object that represents the Button class is retrieved:

```
Class c = java.awt.Button.class;
```

If the class name is unknown at compile time, but available at runtime, you can use the forName method. In the following example, if the String named strg is set to "java.awt.Button", then forName returns the Class object associated with the Button class:

```
Class c = Class.forName(strg);
```

8.1.2 Examining Interfaces

Class objects represent interfaces as well as classes. To make sure whether a Class object represents an interface or a class, the isInterface() method is called. Class methods which are invoked to get information about an interface. To find the public constants of an interface, invoke the getFields () method upon the Class object that represents the interface. getMethods() are used to get information about an interface's methods. To find out about an interface's modifiers, invoke the getModifier's method.

By calling isInterface, the following program reveals that Observer is an interface and that Observable is a class:

```
import java.lang.reflect.*;
import java.util.*;
class SampleCheckInterface {
public static void main(String[] args) {
Class observer = Observer.class;
Class observable = Observable.class;
verifyInterface(observer);
verifyInterface(observable);
}
static void verifyInterface(Class c) {
String name = c.getName();
if (c.isInterface()) {
System.out.println(name + " is an interface.");
} else {
System.out.println(name + " is a class.");
}
}
}
```

The output of the preceding program is:
java.util.Observer is an interface.
java.util.Observable is a class.

8.1.3 Getting the Class Name

Every class in the Java programming language has a name. When you declare a class, the name immediately follows the class keyword. In the following class declaration, the class name is Point:

```
public class Point {int x, y;}
```

At runtime, you can determine the name of a Class object by invoking the getName method. The String returned by getName is the fully-qualified name of the class. The following program gets the class name of an object. First, it retrieves the corresponding Class object, and then it invokes the getName method on that Class object.

```
import java.lang.reflect.*;
import java.awt.*;
class SampleName {
```

```
public static void main(String[] args) {
Button b = new Button();
printName(b);
}
static void printName(Object o) {
Class c = o.getClass();
String s = c.getName();
System.out.println(s);
}
}
```

The sample program prints the following line:

java.awt.Button

8.1.4 Discovering Class Modifiers

A class declaration may include the following modifiers: Public, abstract, or final. The class modifiers precede the class keyword in the class definition. In the following example, the class modifiers are public and final:

public final Coordinate {int x, int y, int z}

To identify the modifiers of a class at runtime these steps are performed

1. Invoke getModifiers on a Class object to retrieve a set of modifiers.
2. Check the modifiers by calling isPublic, isAbstract, and isFinal.

The following program identifies the modifiers of the String class.

```
import java.lang.reflect.*;
import java.awt.*;
class SampleModifier {
public static void main(String[] args) {
String s = new String();
printModifiers(s);}
public static void printModifiers(Object o) {
Class c = o.getClass();
int m = c.getModifiers();
if (Modifier.isPublic(m))
System.out.println("public");
if (Modifier.isAbstract(m))
System.out.println("abstract");
if (Modifier.isFinal(m))
System.out.println("final");
}
}
```

The output of the sample program reveals that the modifiers of the String class are public and final:

public
final

8.1.5 Finding Superclasses

Because the Java programming language supports inheritance, an application such as a class browser must be able to identify superclasses. To determine the superclass of a class, the get Superclass method is invoked. This method returns a Class object representing the superclass, or returns null if the class has no superclass. To identify all ancestors of a class, call getSuperclass iteratively until it returns null. The program that follows finds the names of the Button class's ancestors by calling getSuperclass iteratively.

```
import java.lang.reflect.*;
import java.awt.*;
class SampleSuper {
public static void main(String[] args) {
Button b = new Button();
printSuperclasses(b);
}
static void printSuperclasses(Object o) {
Class subclass = o.getClass();
Class superclass = subclass.getSuperclass();
while (superclass != null) {
String className = superclass.getName();
System.out.println(className);
subclass = superclass;
superclass = subclass.getSuperclass();
}
}
}
```

The output of the sample program verifies that the parent of Button is Component, and that the parent of Component is Object:

```
java.awt.Component
java.lang.Object
```

8.1.6 Identifying the Interfaces Implemented by a Class

The type of an object is determined not only by its class and superclass, but also by its interfaces. In a class declaration, the interfaces are listed after the implements keyword. For example, the RandomAccessFile class implements the DataOutput and DataInput interfaces:

```
public class RandomAccessFile implements DataOutput, DataInput
```

The getInterfaces () is invoked to determine which interfaces a class implements. The getInterfaces method returns an array of Class objects. The reflection API represents interfaces with Class objects. Each Class object in the array returned by getInterfaces represents one of the interfaces implemented by the class. You can invoke the getName method on the Class objects in the array returned by getInterfaces to retrieve the interface names.

The program that follows prints the interfaces implemented by the RandomAccessFile class.

```
import java.lang.reflect.*;
import java.io.*;
class SampleInterface {
```

```
public static void main(String[] args) {
try {
RandomAccessFile r = new RandomAccessFile("myfile", "r");
printInterfaceNames(r);
} catch (IOException e) {
System.out.println(e);
}
}
static void printInterfaceNames(Object o) {
Class c = o.getClass();
Class[] theInterfaces = c.getInterfaces();
for (int i = 0; i < theInterfaces.length; i++) {
String interfaceName = theInterfaces[i].getName();
System.out.println(interfaceName);
}
}
}
```

Note that the interface names printed by the sample program are fully qualified:
java.io.DataOutput
java.io.DataInput

8.1.7 Identifying Class Fields

If you are writing an application such as a class browser, you might want to find out what fields belong to a particular class. You can identify a class's fields by invoking the getFields method on a Class object. The getFields method returns an array of Field objects containing one object per accessible public field.

A public field is accessible if it is a member of either:

- this class
- a superclass of this class
- an interface implemented by this class
- an interface extended from an interface implemented by this class

The methods provided by the Field class allow to retrieve the field's name, type, and set of modifiers.

The following program prints the names and types of fields belonging to the GridBagConstraints class. Note that the program first retrieves the Field objects for the class by calling getFields, and then invokes the getName and getType methods on each of these Field objects.

```
import java.lang.reflect.*;
import java.awt.*;
class SampleField {
public static void main(String[] args) {
GridBagConstraints g = new GridBagConstraints();
printFieldNames(g);
}
static void printFieldNames(Object o) {
```

```
Class c = o.getClass();
Field[] publicFields = c.getFields();
for (int i = 0; i < publicFields.length; i++) {
String fieldName = publicFields[i].getName();
Class typeClass = publicFields[i].getType();
String fieldType = typeClass.getName();
System.out.println("Name: " + fieldName + ", Type: " + fieldType);
}
}
}
```

A truncated listing of the output generated by the preceding program follows:

```
Name: RELATIVE, Type: int
Name: REMAINDER, Type: int
Name: NONE, Type: int
Name: BOTH, Type: int
Name: HORIZONTAL, Type: int
Name: VERTICAL, Type: int
```

8.1.8 Discovering Class Constructors

To create an instance of a class, you invoke a special method called a constructor. Like methods, constructors can be overloaded and are distinguished from one another by their signatures. You can get information about a class's constructors by invoking the getConstructors method, which returns an array of Constructor objects. You can use the methods provided by the Constructor class to determine the constructor's name, set of modifiers, parameter types, and set of throwable exceptions. You can also create a new instance of the Constructor object's class with the Constructor.newInstance method.

The sample program that follows print out the parameter types for each constructor in the Rectangle class. The program performs the following steps:

1. It retrieves an array of Constructor objects from the Class object by calling getConstructors.
2. For every element in the Constructor array, it creates an array of Class objects by invoking getParameterTypes. The Class objects in the array represent the parameters of the constructor.
3. The program calls getName to fetch the class name for every parameter in the Class array created in the preceding step.

It is not as complicated as it sounds. Here is the source code for the sample program:

```
import java.lang.reflect.*;
import java.awt.*;
class SampleConstructor {
public static void main(String[] args) {Rectangle r = new Rectangle();
showConstructors(r);
}
static void showConstructors(Object o) {
Class c = o.getClass();
Constructor[] theConstructors = c.getConstructors();
```

```
for (int i = 0; i < theConstructors.length; i++) {
System.out.print("( ");
Class[] parameterTypes =
theConstructors[i].getParameterTypes();
for (int k = 0; k < parameterTypes.length; k ++) {
String parameterString = parameterTypes[k].getName();
System.out.print(parameterString + " ");
}
System.out.println(")");
}
}
}
```

In the first line of output generated by the sample program, no parameter types appear because that particular Constructor object represents a no-argument constructor. In subsequent lines, the parameters listed are either int types or fully qualified object names. The output of the sample program is:

```
( )
(int int)
(int int int int)
(java.awt.Dimension)
(java.awt.Point)
(java.awt.Point java.awt.Dimension)
(java.awt.Rectangle)
```

8.1.9 Obtaining Method Information

To find out what public methods belong to a class, invoke the method named get Methods. The array returned by getMethod() contains Method objects. You can use a Method object to uncover a method's name, return type, parameter types, set of modifiers, and set of throwable exceptions. All of this information would be useful if you were writing a class browser or a debugger. With Method.invoke, you can even call the method itself.

The following sample program prints the name, return type, and parameter types of every public method in the Polygon class. The program performs the following tasks:

1. It retrieves an array of Method objects from the Class object by calling getMethods.
2. For every element in the Method array, the program:
   a. retrieves the method name by calling getName
   b. gets the return type by invoking getReturnType
   c. creates an array of Class objects by invoking getParameterTypes
3. The array of Class objects created in the preceding step represents the parameters of the method. To retrieve the class name for every one of these parameters, the program invokes getName against each Class object in the array.

Not many lines of source code are required to accomplish these tasks:

```
import java.lang.reflect.*;
import java.awt.*;
class SampleMethod {
```

```
public static void main(String[] args) {
Polygon p = new Polygon();
showMethods(p);
}
static void showMethods(Object o) {
Class c = o.getClass();
Method[] theMethods = c.getMethods();
for (int i = 0; i < theMethods.length; i++) {
String methodString = theMethods[i].getName();
System.out.println("Name: " + methodString);
String returnString =
theMethods[i].getReturnType().getName();
System.out.println(" Return Type: " + returnString);
Class[] parameterTypes = theMethods[i].getParameterTypes();
System.out.print(" Parameter Types:");
for (int k = 0; k < parameterTypes.length; k ++) {
String parameterString = parameterTypes[k].getName();
System.out.print(" " + parameterString);
}
System.out.println();
}
}
}
```

An abbreviated version of the output generated by the sample program is as follows:

```
Name: getBoundingBox
Return Type: java.awt.Rectangle
Parameter Types:
Name: contains
Return Type: boolean
Parameter Types: java.awt.geom.Point2D
Name: contains
Return Type: boolean
Parameter Types: double double
Name: contains
Return Type: boolean
Parameter Types: double double double double
Name: contains
Return Type: boolean
Parameter Types: java.awt.geom. Rectangle2D
Name: contains
Return Type: boolean
Parameter Types: java.awt.Point
Name: contains
Return Type: boolean
Parameter Types: int int
Name: reset
```

Return Type: void
Parameter Types:
Name: intersects
Return Type: boolean
Parameter Types: java.awt.geom. Rectangle2D
Name: intersects
Return Type: boolean
Parameter Types: double double double double
Name: getBounds
Return Type: java.awt.Rectangle
Parameter Types:
Name: inside
Return Type: boolean
Parameter Types: int int
Name: invalidate
Return Type: void
Parameter Types:
Name: translate
Return Type: void
Parameter Types: int int
Name: getPathIterator
Return Type: java.awt.geom.PathIterator
Parameter Types: java.awt.geom. AffineTransform double
Name: getPathIterator
Return Type: java.awt.geom.PathIterator
Parameter Types: java.awt.geom. AffineTransform
Name: addPoint
Return Type: void
Parameter Types: int int
Name: getBounds2D
Return Type: java.awt.geom.Rectangle2D
Parameter Types:
Name: hashCode
Return Type: int
Parameter Types:
Name: getClass
Return Type: java.lang.Class
Parameter Types:
Name: wait
Return Type: void
Parameter Types:
Name: wait
Return Type: void
Parameter Types: long int
Name: wait
Return Type: void
Parameter Types: long
Name: equals

Return Type: boolean
Parameter Types: java.lang.Object
Name: notify
Return Type: void
Parameter Types:
Name: notifyAll
Return Type: void
Parameter Types:
Name: toString
Return Type: java.lang.String
Parameter Types:

8.2 Manipulating Objects

8.2.1 Getting Field Values

If you are writing a development tool such as a debugger, you must be able to obtain field values. This is a three-step process:

1. Create a Class object.
2. Create a Field object by invoking getField on the Class object.
3. Invoke one of the get methods on the Field object.

The Field class has specialized methods for getting the values of primitive types. For example, the getInt() method returns the contents as an int value, get Float returns a float, and so forth. If the field stores an object instead of a primitive, then use the get method to retrieve the object.

The following sample program demonstrates the three steps listed previously. This program gets the value of the height field from a Rectangle object. Because the height is a primitive type (int), the object returned by the get method is a wrapper object (Integer).

In the sample program, the name of the height field is known at compile time. However, in a development tool such as a GUI builder, the field name might not be known until runtime.

Here is the source code for the sample program:

```
import java.lang.reflect.*;
import java.awt.*;
class SampleGet {
public static void main(String[] args) {
Rectangle r = new Rectangle(100, 325);
printHeight(r);
}
static void printHeight(Rectangle r) {
Field heightField;
Integer heightValue;
 Class c = r.getClass();
try {
heightField = c.getField("height");
heightValue = (Integer) heightField.get(r);
System.out.println("Height: " + heightValue.toString());
```

```
} catch (NoSuchFieldException e) {
System.out.println(e);
} catch (SecurityException e) {
System.out.println(e);
} catch (IllegalAccessException e) {
System.out.println(e);
}
}
}
```

The output of the sample program verifies the value of the height field:
Height: 325

8.2.2 Setting Field Values

Some debuggers allow users to change field values during a debugging session. If you are writing a tool that has this capability, you must call one of the Field class's set methods. To modify the value of a field, perform the following steps:

1. Create a Class object.
2. Create a Field object by invoking getField on the Class object.
3. Invoke the appropriate set method on the Field object.

The Field class provides several set methods. Specialized methods, such as setBoolean and setInt, are for modifying primitive types. If the field you want to change is an object, invoke the set method. You can call set to modify a primitive type, but you must use the appropriate wrapper object for the value parameter. The sample program that follows modifies the width field of a Rectangle object by invoking the set method. Since the width is a primitive type, an int, the value passed by set is an Integer, which is an object wrapper.

```
import java.lang.reflect.*;
import java.awt.*;
class SampleSet {
public static void main(String[] args) {
Rectangle r = new Rectangle(100, 20);
System.out.println("original: " + r.toString());
modifyWidth(r, new Integer(300));
System.out.println("modified: " + r.toString());
}
static void modifyWidth(Rectangle r, Integer widthParam) {
Field widthField;
Integer widthValue;
Class c = r.getClass();
try {
widthField = c.getField("width");
widthField.set(r, widthParam);
} catch (NoSuchFieldException e) {
System.out.println(e);
} catch (IllegalAccessException e) {
```

```
System.out.println(e);
}
}
}
```

The output of the sample program verifies that the width changed from 100 to 300:
original: java.awt.Rectangle [x=0,y=0,width =100,height=20]
modified: java.awt.Rectangle[x=0,y=0, width =300,height=20]

8.2.3 Invoking Methods

Suppose that you are writing a debugger that allows the user to select and then invoke methods during a debugging session. Since you do not know at compile time which methods the user will invoke, you cannot hardcode the method name in your source code. Instead, you must follow these steps:

1. Create a Class object that corresponds to the object whose method you want to invoke.
2. Create a Method object by invoking getMethod on the Class object. The getMethod() has two arguments: A String containing the method name, and an array of Class objects. Each element in the array corresponds to a parameter of the method you want to invoke.
3. Invoke the method by calling invoke. The invoke method has two arguments: An array of argument values to be passed to the invoked method, and an object whose class declares or inherits the method.

The sample program that follows shows you how to invoke a method dynamically. The program retrieves the Method object for the String.concat method and then uses invoke to concatenate two String objects.

```
import java.lang.reflect.*;
class SampleInvoke {
public static void main(String[] args) {
String firstWord = "Hello ";
String secondWord = "everybody.";
String bothWords = append(firstWord, secondWord);
System.out.println(bothWords);
}
public static String append(String firstWord, String secondWord) {
String result = null;
Class c = String.class;
Class[] parameterTypes = new Class[] {String.class};
Method concatMethod;
Object[] arguments = new Object[] {secondWord};
try {
concatMethod = c.getMethod("concat", parameterTypes);
result = (String) concatMethod.invoke(firstWord, arguments);
} catch (NoSuchMethodException e) {
System.out.println(e);
} catch (IllegalAccessException e) {
```

```
System.out.println(e);
} catch (InvocationTargetException e) {
System.out.println(e);
}
return result;
}
}
```

The output of the preceding program is:
Hello everybody.

Creating Objects

The simplest way to create an object in the Java programming language is to use the new operator:

```
Rectangle r = new Rectangle();
```

This technique is adequate for nearly all applications, because usually you know the class of the object at compile time. However, if you are writing development tools, you may not know the class of an object until runtime. For example, a GUI builder might allow the user to drag and drop a variety of GUI components onto the page being designed. In this situation, you may be tempted to create the GUI components as follows:

```
String className;
// . . . load className from the user interface
Object o = new (className); // WRONG!
```

The preceding statement is invalid because the new operator does not accept arguments. Fortunately, with the reflection API you can create an object whose class is unknown until runtime. The method you invoke to create an object dynamically depends on whether or not the constructor you want to use has arguments. This section discusses these topics:

- Using no-argument constructors
- Using constructors that have arguments

(i) Using No-argument Constructors

If you need to create an object with the no-argument constructor, you can invoke the newInstance method on a Class object. The newInstance method throws a NoSuchMethodException if the class does not have a no-argument constructor. The following sample program creates an instance of the Rectangle class using the no-argument constructor by calling the newInstance method:

```
import java.lang.reflect.*;
import java.awt.*;
class SampleNoArg
{
public static void main(String [] args)
{
 Rectangle r=(Rectangle) createObject("java.awt.Rectangle");
 System.out.println(r.toString());
}
static Object createObject(String className)
{
```

```
  Object object=null;
 try
 {
  Class classDefinition=Class.forName(className);
  object=classDefinition.newInstance();
 }
 catch(InstantiationException e)
 {
  System.out.println(e);
 }
 catch(IllegalAccessException e)
 {
  System.out.println(e);
 }
 catch(ClassNotFoundException e)
 {
  System.out.println(e);
 }
 return object;
 }
 }
```

The output of the preceding program is:

java.awt.Rectangle[x=0,y=0,width=0, height=0]

(ii) Using Constructors that Have Arguments

To create an object with a constructor that has arguments, you invoke the newInstance method on a Constructor object, not a Class object. This technique involves several steps:

1. Create a Class object for the object you want to create.
2. Create a Constructor object by invoking getConstructor on the Class object. The getConstructor method has one parameter: An array of Class objects that correspond to the constructor's parameters.
3. Create the object by invoking newInstance on the Constructor object. The newInstance method has one parameter: An Object array whose elements are the argument values being passed to the constructor.

The sample program that follows creates a Rectangle with the constructor that accepts two integers as parameters. Invoking newInstance on this constructor is analogous to this statement:

Rectangle rectangle = new Rectangle(12, 34);

This constructor's arguments are primitive types, but the argument values passed to newInstance must be objects. Therefore, each of the primitive int types is wrapped in an Integer object.

The sample program hardcodes the argument passed to the getConstructor method. In a reallife application such as a debugger, you would probably let the user select the constructor.

The source code for the sample program follows:

```
import java.lang.reflect.*;
```

```
import java.awt.*;
class SampleInstance {
public static void main(String[] args) {
Rectangle rectangle;
Class rectangleDefinition;
Class[] intArgsClass = new Class[] {int.class, int.class};

Integer height = new Integer(12);
Integer width = new Integer(34);
Object[] intArgs = new Object[] {height, width};
Constructor intArgsConstructor;
try {
rectangleDefinition = Class.forName("java.awt.Rectangle");
intArgsConstructor =rectangleDefinition.getConstructor(intArgsClass);
rectangle =(Rectangle) createObject(intArgsConstructor, intArgs);}
catch (ClassNotFoundException e) {
System.out.println(e);
} catch (NoSuchMethodException e) {
System.out.println(e);
}
}
public static Object createObject(Constructor constructor, Object[] arguments) {
System.out.println ("Constructor: " + constructor.toString());
Object object = null;
try {
object = constructor.newInstance(arguments);
System.out.println ("Object: " + object.toString());
return object;
} catch (InstantiationException e) {
System.out.println(e);
} catch (IllegalAccessException e) {
System.out.println(e);
} catch (IllegalArgumentException e) {
System.out.println(e);
} catch (InvocationTargetException e) {
System.out.println(e);
}
return object;
}
}
```

The sample program prints a description of the constructor and the object that it creates:

```
Constructor: public java.awt.Rectangle(int,int)
Object: java.awt.Rectangle[x=0, y=0, width=12, height=34]
```

8.2.5 Working with Arrays

(i) Identifying Arrays

If you are not certain that a particular object is an array, you can check it with the Class.isArray method. Let us take a look at an example.

The sample program that follows prints the names of the arrays that are encapsulated in an object. The program performs these steps:

1. It retrieves the Class object that represents the target object.
2. It gets the Field objects for the Class object retrieved in step 1.
3. For each Field object, the program gets a corresponding Class object by invoking the getType method.
4. To verify that the Class object retrieved in the preceding step represents an array, the program invokes the isArray() method.

Here is the source code for the sample program:

```
import java.lang.reflect.*;
import java.awt.*;
class SampleArray {
public static void main(String[] args) {
KeyPad target = new KeyPad();
printArrayNames(target);
}
static void printArrayNames(Object target) {
Class targetClass = target.getClass();
Field[] publicFields = targetClass.getFields();
for (int i = 0; i < publicFields.length; i++) {
String fieldName = publicFields[i].getName();
Class typeClass = publicFields[i].getType();
String fieldType = typeClass.getName();
if (typeClass.isArray()) {
System.out.println("Name: " + fieldName +", Type: " + fieldType);
}
}
}
}
class KeyPad {
public boolean alive;
public Button power;
public Button[] letters;
public int[] codes;
public TextField[] rows;
public boolean[] states;
}
```

The output of the sample program is as follows. Note that the left bracket indicates that the object is an array.

```
Name: letters, Type: [Ljava.awt.Button;
Name: codes, Type: [I
```

Name: rows, Type: [Ljava.awt.TextField;
Name: states, Type: [Z

(ii) Retrieving Component Types

The component type is the type of an array's elements. For example, the component type of the arrowKeys array in the following line of code is Button:

```
Button[] arrowKeys = new Button[4];
```

The component type of a multidimensional array is an array. In the next line of code, the component type of the array named matrix is int[]:

```
int[][] matrix = new int[100][100];
```

By invoking the getComponentType method against the Class object that represents an array, you can retrieve the component type of the array's elements.

The sample program that follows invokes the getComponentType method and prints out the class name of each array's component type.

```
import java.lang.reflect.*;
import java.awt.*;
class SampleComponent {
public static void main(String[] args) {
int[] ints = new int[2];
Button[] buttons = new Button[6];
String[][] twoDim = new String[4][5];
printComponentType(ints);
printComponentType(buttons);
printComponentType(twoDim);
}
static void printComponentType(Object array) {
Class arrayClass = array.getClass();
String arrayName = arrayClass.getName();
Class componentClass = arrayClass.getComponentType();
String componentName = componentClass.getName();
System.out.println("Array: " + arrayName +
", Component: " + componentName);
}
}
```

The output of the sample program is:

```
Array: [I, Component: int
Array: [Ljava.awt.Button;, Component: java.awt.Button
Array: [[Ljava.lang.String;, Component: [Ljava.lang.String;
```

(iii) Creating Arrays

If you are writing a development tool such as an application builder, you may want to allow the end user to create arrays at runtime. Your program can provide this capability by invoking the Array.newInstance method.

The following sample program uses the newInstance method to create a copy of an array that is twice the size of the original array. The newInstance method accepts as arguments the length and component type of the new array. The source code is as follows:

```
import java.lang.reflect.*;
class SampleCreateArray {
public static void main(String[] args) {
int[] originalArray = {55, 66};
int[] biggerArray = (int[]) doubleArray(originalArray);
System.out.println("originalArray:");
for (int k = 0; k < Array.getLength(originalArray); k++)
System.out.println(originalArray[k]);
System.out.println("biggerArray:");
for (int k = 0; k < Array.getLength(biggerArray); k++)
System.out.println(biggerArray[k]);
}
static Object doubleArray(Object source) {
int sourceLength = Array.getLength (source);
Class arrayClass = source.getClass();
Class componentClass = arrayClass.getComponentType();
Object result = Array.newInstance(componentClass,
sourceLength * 2);
System.arraycopy(source, 0, result, 0, sourceLength);
return result;
}
}
```

The output of the preceding program is:

```
originalArray:
55
66
biggerArray:
55
66
0
0
```

You can also use the newInstance method to create multidimensional arrays. In this case, the parameters of the method are the component type and an array of int types representing the dimensions of the new array.

The next sample program shows how to use newInstance to create multidimensional arrays:

```
import java.lang.reflect.*;
class SampleMultiArray {
public static void main(String[] args) {
// The oneDimA and oneDimB objects are one
// dimensional int arrays with 5 elements.
int[] dim1 = {5};
int[] oneDimA = (int[]) Array.newInstance(int.class, dim1);
int[] oneDimB = (int[]) Array.newInstance(int.class, 5);
// The twoDimStr object is a 5 X 10 array of String objects.
```

```
int[] dimStr = {5, 10};
String[][] twoDimStr =
(String[][]) Array.newInstance(String.class,dimStr);
// The twoDimA object is an array of 12 int arrays. The tail
// dimension is not defined. It is equivalent to the array
// created as follows:
// int[][] ints = new int[12][];
int[] dimA = {12};
int[][] twoDimA = (int[][]) Array.newInstance(int[].class, dimA);
}
}
```

(iv) Getting and Setting Element Values

In most programs, to access array elements you merely use an assignment expression as follows:

```
int[10] codes;
codes[3] = 22;
aValue = codes[3];
```

This technique will not work if you do not know the name of the array until runtime. Fortunately, you can use the Array class set and get methods to access array elements when the name of the array is unknown at compile time. In addition to get and set, the Array class has specialized methods that work with specific primitive types. For example, the value parameter of setInt is an int, and the object returned by getBoolean is a wrapper for a boolean type.

The sample program that follows uses the set and get methods to copy the contents of one array to another.

```
import java.lang.reflect.*;
class SampleGetArray {
public static void main(String[] args) {
int[] sourceInts = {12, 78};
int[] destInts = new int[2];
copyArray(sourceInts, destInts);
String[] sourceStrgs = {"Hello ", "there ", "everybody"};
String[] destStrgs = new String[3];
copyArray(sourceStrgs, destStrgs);
}
public static void copyArray(Object source, Object dest) {
for (int i = 0; i < Array.getLength(source); i++) {
Array.set(dest, i, Array.get(source, i));
System.out.println(Array.get(dest, i));
}
}
}
```

The output of the sample program is:
12
78
Hello
there
everybody

8.3 Summary of Classes

The following table summarizes the classes that compose the reflection API. The Class and Object classes are in the java.lang package. The other classes are contained in the java.lang.reflect package.

Class	Description
Array	Provides static methods to dynamically create and access arrays.
Class	Represents, or reflects, classes and interfaces.
Constructor	Provides information about, and access to, a constructor for a class. Allows you to instantiate a class dynamically.
Field	Provides information about, and dynamic access to, a field of a class or an interface.
Method	Provides information about, and access to, a single method on a class or interface. Allows you to invoke the method dynamically.
Modifier	Provides static methods and constants that allow you to get information about the access modifiers of a class and its members.
Object	Provides the getClass method.

9. INNER CLASSES

An inner class is a class that is defined inside another class. Inner classes let you make one class a member of another class. Just as classes have member variables and methods, a class can also have member classes.

Regular Inner Class

You define an inner class within the curly braces of the outer class, as follows:

```
class MyOuter {
        class MyInner { }
}
```

And if you compile it, **%javac MyOuter.java** , you will end up with two class files:
My Outer. class
MyOuter$MyInner.class

The inner class is still, in the end, a separate class, so a class file is generated. But the inner class file is not accessible to you in the usual way. The only way you can access the inner class is through a live instance of the outer class.

Instantiating an Inner Class

To instantiate an instance of an inner class, you must have an instance of the outer class. An inner class instance can never stand alone without a direct relationship with a specific instance of the outer class.

Instantiating an Inner Class from Within Code in the Outer Class

From inside the outer class instance code, use the inner class name in the normal way:

```
class MyOuter {
private int x = 7;
MyInner mi = new MyInner();
class MyInner {
public void seeOuter() { System.out.println("Outer x is " + x);
}
}
public static void main(String arg[]){
MyOuter mo=new MyOuter();
mo.mi.seeOuter();
}
```

Output

Outer x is 7

Method-Local Inner Classes

A method-local inner class is defined within a method of the enclosing class.

```
class MyOuter {
    void inner()
    {
    final int c=9;
    class MyInner {
        int x=5;
        public void display() { System.out.println ("Inner x is " + x);
            System.out.println("Inner c is " + c);
        }
        }
        MyInner mi = new MyInner();
        mi.display();
    }
    public static void main(String arg[]){ MyOuter mo = new MyOuter();
    mo.inner();
    }
}
```

Output

x is 5 c is 9

Anonymous Inner Classes

- Anonymous inner classes have no name, and their type must be either a subclass of the named type or an implementer of the named interface.
- An anonymous inner class is always created as part of a statement, so the syntax will end the class definition with a curly brace, followed by a closing parenthesis to end the method call, followed by a semicolon to end the statement: });

- An anonymous inner class can extend one subclass, or implement one interface. It cannot extend a class and implement an interface, nor can it implement more than one interface.

```
public class Test{
public static void main(String arg[]){
B b=new B();
b.ob.display();
}
}
class A{ }
class B {
void display(){System.out.println("Hai");
}
A ob=new A(){
void display(){ System.out.println ("Hello");
}

};
}
```

Output

Hello

And if you compile it, **%javac Test.java**, you will end up with two class files: A.class B.class B$1.class Test.class

Static Nested Classes

Static nested classes are inner classes marked with the static modifier.

- Technically, a static nested class is not an inner class, but instead is considered a top-level nested class.
- Because the nested class is static, it does not share any special relationship with an instance of the outer class. In fact, you do not need an instance of the outer class to instantiate a static nested class.
- Instantiating a static nested class requires using both the outer and nested class names as follows:

```
BigOuter.Nested n = new BigOuter.Nested();
```

A static nested class cannot access nonstatic members of the outer class, since it does not have an implicit reference to any outer instance (in other words, the nested class instance does not get an outer this reference).

```
public class Test{
  public static void main(String arg[]){
    A.B b=new A.B();
    b.display();
  }}
```

```
class A {
static class B {
  int m=5;
  void display(){ System.out.println ("m=" +m);
  }
}
}
```

Output: *m=5*

10. LANGUAGE BASICS

Java programming language was originally developed by Sun Microsystems, which was initiated by James Gosling and released in 1995 as core component of Sun Microsystems, Java platform (Java 1.0 [J2SE]). As of December 08, the latest release of the Java Standard Edition is 6 (J2SE). With the advancement of Java and its widespread popularity, multiple configurations were built to suite various types of platforms. For example, J2EE for Enterprise Applications, J2ME for Mobile Applications. Sun Microsystems has renamed the new J2 versions as Java SE, Java EE and Java ME respectively. Java is guaranteed to be **Write Once, Run Anywhere**.

Java is:

- **Object oriented:** In Java everything is an Object. Java can be easily extended since it is based on the Object model.
- **Platform independent:** Unlike many other programming languages including C and C++ when Java, is compiled, it is not compiled into platform specific machine, rather into platform independent byte code. This byte code is distributed over the web and interpreted by virtual machine (JVM) on whichever platform it is being run.
- **Simple:** Java is designed to be easy to learn. If you understand the basic concept of OOP Java would be easy to master.
- **Secure:** With Java.s secure feature it enables to develop virus-free, tamper-free systems. Authentication techniques are based on public-key encryption.
- **Architectural-neutral:** Java compiler generates an architecture-neutral object file format which makes the compiled code to be executable on many processors, with the presence of Java runtime system.
- **Portable:** Being architectural neutral and having no implementation dependent aspects of the specification makes Java portable. Compiler and Java is written in ANSI C with a clean portability boundary which is a POSIX subset.
- **Robust:** Java makes an effort to eliminate error prone situations by emphasizing mainly on compile time error checking and runtime checking.
- **Multi-threaded:** With Java, multi-threaded feature it is possible to write programs that can do many tasks simultaneously. This design feature allows developers to construct smoothly running interactive applications.
- **Interpreted:** Java byte code is translated on the fly to native machine instructions and is not stored anywhere. The development process is more rapid and analytical since the linking is an incremental and light weight process.
- **High performance:** With the use of Just-In-Time compilers, Java enables high performance.
- **Distributed:** Java is designed for the distributed environment of the internet.

- **Dynamic:** Java is considered to be more dynamic than C or C++ since it is designed to adapt to an evolving environment. Java programs can carry extensive amount of runtime information that can be used to verify and resolve accesses to objects on runtime.

10.1 Environment Setup

Before we proceed further it is important that we set up the java environment correctly. This section guides you on how to download and set up Java on your machine. Please follow the following steps to set up the environment.

Java SE is freely available from the link

http://www.oracle.com/technetwork/java/javase/downloads/index-jdk5-jsp-142662.html. So you download a version based on your operating system.

Follow the instructions to download java and run the **.exe** to install Java on your machine. Once you installed Java on your machine, you would need to set environment variables to point to correct installation directories:

1. Setting up the Path for Windows 2000/XP

Assuming you have installed Java in *c:\Program Files\java\jdk* directory:

- Right-click on 'My Computer' and select 'Properties'.
- Click on the 'Environment variables' button under the 'Advanced' tab.
- Now alter the 'Path' variable so that it also contains the path to the Java executable. For example, if the path is currently set to 'C:\WINDOWS\SYSTEM32', then change your path to read 'C:\WINDOWS\SYSTEM32;c:\Program Files\java\jdk\bin'.

2. Setting up the Path for Windows 95/98/ME

Assuming you have installed Java in *c:\Program Files\java\jdk* directory:

Edit the 'C:\autoexec.bat' file and add the following line at the end: 'SET PATH=%PATH%;C:\Program Files\java\jdk\bin'

3. Setting up the Path for Linux, UNIX, Solaris, FreeBSD

Environment variable PATH should be set to point to where the java binaries have been installed. Refer to your shell documentation if you have trouble doing this.

For example, if you use *bash* as your shell, then you would add the following line to the end of your '.bashrc: export PATH=/path/to/java:$PATH'

Popular Java Editors

To write your java programs you will need a text editor. There are even more sophisticated IDE available in the market. But for now, you can consider one of the following:

- **Notepad:** On Windows machine you can use any simple text editor like Notepad, TextPad.
- **Netbeans:** It is a Java IDE that is open source and free which can be downloaded from http://www.netbeans.org/index.html.
- **Eclipse :** It is also a Java IDE developed by the eclipse open source community and can be downloaded from http://www.eclipse.org/.

When we consider a Java program, it can be defined as a collection of objects that communicate via invoking each others methods. Let us now briefly look into what do class, object, methods and instant variables mean.

- **Object:** Objects have states and behaviors. For example, a dog has states—color, name, breed as well as behaviors—wagging, barking, eating. An object is an instance of a class.

- **Class:** A class can be defined as a template/ blue print that describe the behaviors/ states that object of its type support.
- **Methods:** A method is basically a behavior. A class can contain many methods. In methods the logics are written, data is manipulated and all the actions are executed.
- **Instant variables:** Each object has its unique set of instant variables. An object's state is created by the values assigned to these instant variables.

First Java Program

Let us look at a simple code that would print the words *Hello World.*

```
public class MyFirstJavaProgram{

/* This is my first java program.
* This will print 'Hello World' as the output
*/

public static void main(String []args){
System.out.println("Hello World"); // prints Hello World
}
}
```

Let us look at how to save the file, compile and run the program. Please follow the steps given below:

1. Open notepad and add the code as above.
2. Save the file as : MyFirstJavaProgram.java.
3. Open a command prompt window and go to the directory where you saved the class. Assume its C:\.
4. Type ' javac MyFirstJavaProgram.java ' and press enter to compile your code. If there are no errors in your code, the command prompt will take you to the next line.(Assumption : The path variable is set).
5. Now type ' java MyFirstJavaProgram ' to run your program.
6. You will be able to see ' Hello World ' printed on the window.

```
C : > javac MyFirstJavaProgram.java
C : > java MyFirstJavaProgram
Hello World
```

10.2 Java Primer

Having discussed some of the ideas and core concepts on Java, this prerequisite knowledge is necessary to recall or implement the further topics.

About Java programs, it is very important to keep in mind the following points.

- **Case sensitivity:** Java is case sensitive which means identifier **Hello** and **hello** would have different meaning in Java.
- **Class names:** For all class names the first letter should be in upper case. If several words are used to form a name of the class, each inner word's first letter should be in upper case.

  For example, *class MyFirstJavaClass*

- **Method names:** All method names should start with a lower case letter. If several words are used to form the name of the method, then each inner word's first letter should be in upper case.
  For example, *public void myMethodName()*
- **Program file name:** Name of the program file should exactly match the class name. When saving the file you should save it using the class name (Remember java is case sensitive) and append '.java' to the end of the name. (If the file name and the class name do not match, your program will not compile).

For example, assume 'MyFirstJavaProgram' is the class name. Then the file should be saved as *'MyFirstJavaProgram.java'*

- **Public static void main(String args[]):** Java program processing starts from the main() method which is a mandatory part of every java program.

a. Java Identifiers

All java components require names. Names used for classes, variables and methods are called identifiers.

In java there are several points to remember about identifiers. They are as follows:

- All identifiers should begin with a letter (A to Z or a to z), currency character ($) or an underscore (–).
- After the first character identifiers can have any combination of characters.
- A key word cannot be used as an identifier.
- Most importantly identifiers are case sensitive.
- Examples of legal identifiers: Age, $Mark, _value, __1_value
- Examples of illegal identifiers: 123abc, -Mark

b. Java Modifiers

Like other languages it is possible to modify classes, methods, etc. by using modifiers. There are two categories of modifiers.

- **Access modifiers:** Public, protected, private, default.
- **Non-access modifiers:** Final, abstract, strictfp

c. Java Variables

We would see the following type of variables in Java:

- Local variables
- Class variables (static variables)
- Instance variables (non-static variables)

d. Java Arrays

Arrays are objects that store multiple variables of the same type. However, an array itself is an object on the heap. Ways to declare, construct and initialize arrays are discussed in upcoming chapters.

e. Java Enums

Enums were introduced in java 5.0. Enums restrict a variable to have one of only a few predefined values. The values in this enumerated list are called enums.

With the use of enums it is possible to reduce the number of bugs in your code.

For example, if we consider an application for a fresh juice shop it would be possible to restrict the glass size to small, medium and Large. This would make sure that it would not allow anyone to order any size other than the small, medium or large.

Example:

```
class Month{
enum Month_names{ JANUARY, FEBRUARY, MARCH, APRIL, MAY, JUNE,
JULY, AUGUST, SEPTEMBER, OCTOBER, NOVEMBER, DECEMBER}}
Month_names nameofmonth;
}
public class Monthtest{
  public static void main(String args[]){
    Monthtest test = new Monthtest();
      test.nameofmonth = Month.Month_names.MAY;
}
}
```

Note: Enums can be declared as their own or inside a class. Methods, variables, constructors can be defined inside enums as well.

f. Java Keywords

The following list shows the reserved words in Java. These reserved words may not be used as constant or variable or any other identifier names.

abstract	assert	boolean	break
byte	case	catch	char
class	const	continue	efault
do	double	else	enum
extends	final	finally	float
for	goto	if	implements
import	instanceof	int	interface
long	native	new	package
private	protected	public	return
short	static	strictfp	super
switch	synchronized	this	throw
throws	transient	try	void
volatile	while		

g. Comments in Java

Java supports single line and multi-line comments very similar to C and C++. All characters available inside any comment are ignored by Java compiler.

```
public class MyFirstJavaProgram{

/* This is my first java program.
* This will print 'Hello World' as the output
* This is an example of multi-line comments.
*/
```

```
public static void main(String []args){
// This is an example of single line comment
/* This is also an example of single line comment. */
System.out.println("Hello World");
}
        }
```

h. Using Blank Lines

A line containing only whitespace, possibly with a comment, is known as a blank line, and Java totally ignores it.

i. Source file Declaration Rules

These rules are essential when declaring classes, *import* statements and *package* statements in a source file.

- There can be only one public class per source file.
- A source file can have multiple non-public classes.
- The public class name should be the name of the source file as well which should be appended by **.java** at the end. For example, the class name is . *public class Student{}* Then the source file should be as Student.java.
- If the class is defined inside a package, then the package statement should be the first statement in the source file.
- If import statements are present, then they must be written between the package statement and the class declaration. If there are no package statements, then the import statement should be the first line in the source file.
- Import and package statements will imply to all the classes present in the source file. It is not possible to declare different import and/or package statements to different classes in the source file.

Classes have several access levels and there are different types of classes; abstract classes, final classes, etc.

10.3 Data Types, Literals and Variables

Variables are nothing but reserved memory locations to store values. This means that when you create a variable, you reserve some space in memory.

Based on the data type of a variable, the operating system allocates memory and decides what can be stored in the reserved memory. Therefore, by assigning different data types to variables, you can store integers, decimals, or characters in these variables.

There are two data types available in Java:

1. Primitive Data Types
2. Reference/Object Data Types

10.3.1 Primitive Data Types

There are eight primitive data types supported by Java. Primitive data types are predefined by the language and named by a key word. Let us now look into detail about the eight primitive data types.

Byte

- Byte data type is a 8-bit signed two.s complement integer.
- Minimum value is –128 (–2^7)
- Maximum value is 127 (inclusive)(2^7 –1)
- Default value is 0

- Byte data type is used to save space in large arrays, mainly in place of integers, since a byte is four times smaller than an int.
- For example, byte a = 100, byte b = –50

Short

- Short data type is a 16-bit signed two's complement integer.
- Minimum value is –32,768 (–2^15)
- Maximum value is 32,767(inclusive) (2^15 –1)
- Short data type can also be used to save memory as byte data type. A short is 2 times smaller than an int
- Default value is 0.
- For example, short s = 10000 , short r = –20000

Int

- Int data type is a 32-bit signed two's complement integer.
- Minimum value is –2, 147, 483, 648. (–2^31)
- Maximum value is 2, 147, 483, 647(inclusive).(2^31 –1)
- Int is generally used as the default data type for integral values unless there is a concern about memory.
- The default value is 0.
- For example, int a = 100000, int b = –200000

Long

- Long data type is a 64-bit signed two's complement integer.
- Minimum value is –9, 223, 372, 036, 854, 775, 808. (–2^63)
- Maximum value is 9, 223, 372, 036, 854, 775, 807 (inclusive). (2^63 –1)
- This type is used when a wider range than int is needed. Default value is 0L.
- For example, int a = 100000L, int b = –200000L

Float

- Float data type is a single-precision 32-bit IEEE 754 floating point.
- Float is mainly used to save memory in large arrays of floating point numbers.
- Default value is 0.0f.
- Float data type is never used for precise values such as currency.
- For example, float f1 = 234.5f

Double

- Double data type is a double-precision 64-bit IEEE 754 floating point.
- This data type is generally used as the default data type for decimal values, generally the default choice.
- Double data type should never be used for precise values such as currency.
- Default value is 0.0d.
- For example, double d1 = 123.4

Boolean

- Boolean data type represents one bit of information.
- There are only two possible values: True and false.
- This data type is used for simple flags that track true/false conditions.
- Default value is false.
- For example, boolean one = true

Char

- char data type is a single 16-bit Unicode character.
- Minimum value is '\u0000' (or 0).
- Maximum value is '\uffff' (or 65,535 inclusive).

- Char data type is used to store any character.
- For example, char letterA ='A'

10.3.2 Reference Data Types

- Reference variables are created using defined constructors of the classes. They are used to access objects. These variables are declared to be of a specific type that cannot be changed. For example, student, puppy, etc.
- Class objects, and various type of array variables come under reference data type.
- Default value of any reference variable is null.
- A reference variable can be used to refer to any object of the declared type or any compatible type.
- For example, Animal animal = new Animal("giraffe");

10.3.3 Java Literals

A literal is a source code representation of a fixed value. They are represented directly in the code without any computation.

Literals can be assigned to any primitive type variable. For example:

byte a = 68;

char a = 'A' byte, int, long, and short can be expressed in decimal(base 10), hexadecimal(base 16) or octal(base 8) number systems as well.

Prefix 0 is used to indicate octal and prefix 0x indicates hexadecimal when using these number systems for literals. For example:

int decimal = 100;

int octal = 0144;

int hexa = 0x64; String literals in Java are specified like they are in most other languages by enclosing a sequence of characters between a pair of double quotes. Examples of string literals are:

"Hello World"

"two\nlines"

"\"This is in quotes\"" String and char types of literals can contain any Unicode characters.

For example:

char a = '\u0001';

String a = "\u0001"; Java language supports a few special escape sequences for String and char literals as well. They are:

Notation	Character represented
\n	Newline (0x0a)
\r	Carriage return (0x0d)
\f	Formfeed (0x0c)
\b	Backspace (0x08)
\s	Space (0x20)
\t	tab
\"	Double quote
\'	Single quote
\\	backslash
\ddd	Octal character (ddd)
\uxxxx	Hexadecimal UNICODE character (xxxx)

10.3.4 Variable

A variable is the basic unit of storage in a java program. A variable is defined by the combination of an identifier, a type, and an optional initializer.

Variable types

In Java, all variables must be declared before they can be used. The basic form of a variable declaration is shown here:

type identifier [= value][, identifier [= value] ...] ;

The *type* is one of Java's datatypes. The *identifier* is the name of the variable. To declare more than one variable of the specified type, use a comma-separated list. There are several examples of variable declarations of various types. Note that some include an initialization.

```
int a, b, c;          // declares three ints, a, b, and c.
int d = 3, e, f = 5; // declares three more ints, initializing d and f.
byte z = 22;          // initializes z.
double pi = 3.14159; // declares an approximation of pi.
char x = 'x';         // the variable x has the value 'x'.
```

This chapter will explain various variable types available in Java Language. There are three kinds of variables in Java:

1. Local variables
2. Instance variables
3. Class/static variables

Local Variables

- Local variables are declared in methods, constructors, or blocks.
- Local variables are created when the method, constructor or block is entered and the variable will be destroyed once it exits the method, constructor or block.
- Access modifiers cannot be used for local variables.
- Local variables are visible only within the declared method, constructor or block.
- Local variables are implemented at stack level internally.
- There is no default value for local variables so local variables should be declared and an initial value should be assigned before the first use.

Example: Here *age* is a local variable. This is defined inside *pupAge()* method and its scope is limited to this method only.

```
public class Test{
public void pupAge(){
int age = 0;
age = age + 7;    System.out.println("Puppy age is : " + age)
}
public static void main(String args[]){
Test test = new Test();
Test.pupAge();
}
}
```

This would produce the following result:
Puppy age is: 7

The following example uses *age* without initializing it, so it would give an error at the time of compilation.

```
public class Test{
public void pupAge(){
int age;
age = age + 7;
System.out.println("Puppy age is : " + age)
}

public static void main(String args[]){
Test test = new Test();
Test.pupAge();
}
}
```

This would produce the following error while compiling it:

```
Test.java:4:variable number might not have been initialized
age = age + 7;
      ^
1 error
```

Instance Variables

- Instance variables are declared in a class, but outside a method, constructor or any block.
- When a space is allocated for an object in the heap, a slot for each instance variable value is created.
- Instance variables are created when an object is created with the use of the key word 'new' and destroyed when the object is destroyed.
- Instance variables hold values that must be referenced by more than one method, constructor or block, or essential parts of an object's state that must be present throughout the class.
- Instance variables can be declared in class level before or after use.
- Access modifiers can be given for instance variables.
- The instance variables are visible for all methods, constructors and block in the class. Normally it is recommended to make these variables private (access level). However, visibility for subclasses can be given for these variables with the use of access modifiers.
- Instance variables have default values. For numbers the default value is 0, for booleans it is false and for object references it is null. Values can be assigned during the declaration or within the constructor.
- Instance variables can be accessed directly by calling the variable name inside the class. However, within static methods and different classes (when instance variables are given accessibility) this should be called using the fully qualified name. *Object Reference. Variable Name.*

Example:

```
import java.io.*;
class Student{
// this instance variable is visible for any child class.
public String name;

// Mark  variable is visible in Student class only.
private double Mark;

// The name variable is assigned in the constructor.
public Student (String empName){
name = empName;
}
// The Mark variable is assigned a value.
public void set Mark(double empSal){
Mark = empSal;
}
// This method prints the Student details.
public void printEmp(){
System.out.println("name  : " + name );
System.out.println("Mark :" + Mark);
}
public static void main(String args[]){
Student empOne = new Student("Ransika");
empOne.setMark(1000);
empOne.printEmp();
}
}
```

This would produce the following result:

```
name  : Ransika
Mark :1000.0
```

Class/static Variables

- Class variables also known as static variables are declared with the *static* keyword in a class, but outside a method, constructor or a block.
- There would only be one copy of each class variable per class, regardless of how many objects are created from it.
- Static variables are rarely used other than being declared as constants. Constants are variables that are declared as public/private, final and static. Constant variables never change from their initial value.
- Static variables are stored in static memory. It is rare to use static variables other than declared final and used as either public or private constants.
- Static variables are created when the program starts and destroyed when the program stops.

- Visibility is similar to instance variables. However, most static variables are declared public since they must be available for users of the class.
- Default values are same as instance variables. For numbers the default value is 0, for Booleans it is false and for object references it is null. Values can be assigned during the declaration or within the constructor. Additionally values can be assigned in special static initializer blocks.
- Static variables can be accessed by calling with the class name . *Class Name. Variable Name.*
- When declaring class variables as public static final, then variables names (constants) are all in upper case. If the static variables are not public and final, the naming syntax is the same as instance and local variables.

Example:

```
import java.io.*;
class Student{
// Mark  variable is a private static variable
private static double Mark;
// DEPARTMENT is a constant
public static final String DEPARTMENT = "Development";
public static void main(String args[]){      Mark = 1000;
System.out.println(DEPARTMENT+"average Mark:"+Mark);
}
}
```

This would produce the following result:
Development average Mark:1000

Note: If the variables are access from an outside class, the constant should be accessed as Student.DEPARTMENT

10.4 Basic Operators

Java provides a rich set of operators to manipulate variables. We can divide all the Java operators into the following groups:

- Arithmetic operators
- Relational operators
- Bitwise operators
- Logical operators
- Assignment operators
- Misc operators

10.4.1 Arithmetic Operators

Arithmetic operators are used in mathematical expressions in the same way that they are used in algebra. The following table lists the arithmetic operators: Assume integer variable A holds 10 and variable B holds 20 then:

Operator	Description	Example
+	Addition - Adds values on either side of the operator	A + B will give 30
–	Subtraction - Subtracts right hand operand from left hand operand	A – B will give –10
*	Multiplication - Multiplies values on either side of the operator	A * B will give 200
/	Division - Divides left hand operand by right hand operand	B / A will give 2
%	Modulus - Divides left hand operand by right hand operand and returns remainder	B % A will give 0
++	Increment - Increase the value of operand by 1	B++ gives 21
--	Decrement - Decrease the value of operand by 1	B-- gives 19

```
class Test {
public static void main(String args[]) {
  int a = 10;
  int b = 20;
  int c = 25;
  int d = 25;
  System.out.println("a + b = " + (a + b) );
  System.out.println("a – b = " + (a – b) );
  System.out.println("a * b = " + (a * b) );
  System.out.println("b / a = " + (b / a) );
  System.out.println("b % a = " + (b % a) );
  System.out.println("c % a = " + (c % a) );
  System.out.println("a++   = " + (a++) );
  System.out.println("b—   = " + (a—) );
  // Check the difference in d++ and ++d
  System.out.println("d++   = " + (d++) );
  System.out.println("++d   = " + (++d) );
}
}
```

Output

a + b = 30

a – b = –10

a * b = 200

b / a = 2

b % a = 0

c % a = 5

a++ = 10

b-- = 11

d++ = 25

++d = 27

10.4.2 Relational Operators

There are the following relational operators supported by Java language.

Assume variable A holds 10 and variable B holds 20 then:

Operator	Description	Example
==	Checks if the value of two operands are equal or not, if yes, then condition becomes true.	(A == B) is not true.
!=	Checks if the value of two operands are equal or not, if values are not equal, then condition becomes true.	(A != B) is true.
>	Checks if the value of left operand is greater than the value of right operand, if yes, then condition becomes true.	(A > B) is not true.
<	Checks if the value of left operand is less than the value of right operand, if yes, then condition becomes true.	(A < B) is true.
>=	Checks if the value of left operand is greater than or equal to the value of right operand, if yes, then condition becomes true.	(A >= B) is not true.
<=	Checks if the value of left operand is less than or equal to the value of right operand, if yes, then condition becomes true.	(A <= B) is true.

```
class Test {
public static void main(String args[]) {
  int a = 10;
  int b = 20;
  System.out.println("a == b = " + (a == b) );
  System.out.println("a != b = " + (a != b) );
  System.out.println("a > b = " + (a > b) );
  System.out.println("a < b = " + (a < b) );
  System.out.println("b >= a = " + (b >= a) );
  System.out.println("b <= a = " + (b <= a) );
}
}
```

Output

```
a == b = false
a != b = true
a > b = false
a < b = true
b >= a = true
b <= a = false
```

10.4.3 Bitwise Operators

Java defines several bitwise operators which can be applied to the integer types, long, int, short, char, and byte. Bitwise operators work on bits and perform bit by bit operation. Assume if a = 60; and b = 13; Now in binary format they will be as follows:

a = 0011 1100

b = 0000 1101

a&b = 0000 1100

a | b = 0011 1101

a^b = 0011 0001

~a = 1100 0011

The following table lists the bitwise operators:

Assume integer variable A holds 60 and variable B holds 13, then:

Operator	Description	Example
&	Binary AND Operator copies a bit to the result if it exists in both operands.	(A & B) will give 12 which is 0000 1100
|	Binary OR Operator copies a bit if it exists in either operand.	(A | B) will give 61 which is 0011 1101
^	Binary XOR Operator copies the bit if it is set in one operand but not both.	(A ^ B) will give 49 which is 0011 0001
~	Binary Ones Complement Operator is unary and has the effect of 'flipping' bits.	(~A) will give -60 which is 1100 0011
<<	Binary Left Shift Operator. The left operands value is moved left by the number of bits specified by the right operand.	A << 2 will give 240 which is 1111 0000
>>	Binary Right Shift Operator. The left operands value is moved right by the number of bits specified by the right operand.	A >> 2 will give 15 which is 1111
>>>	Shift right zero fill operator. The left operands value is moved right by the number of bits specified by the right operand and shifted values are filled up with zeros.	A >>>2 will give 15 which is 0000 1111

```
class Test {
public static void main(String args[]) {
  int a = 60; /* 60 = 0011 1100 */
  int b = 13; /* 13 = 0000 1101 */
  int c = 0;  c = a & b;      /* 12 = 0000 1100 */
  System.out.println("a & b = " + c );
  c = a | b;      /* 61 = 0011 1101 */
  System.out.println("a | b = " + c );
  c = a ^ b;      /* 49 = 0011 0001 */
  System.out.println("a ^ b = " + c );
  c = ~a;         /*-61 = 1100 0011 */
  System.out.println("~a = " + c );
  c = a << 2;     /* 240 = 1111 0000 */
  System.out.println("a << 2 = " + c );
  c = a >> 2;     /* 215 = 1111 */
  System.out.println("a >> 2  = " + c );
  c = a >>> 2;    /* 215 = 0000 1111 */
  System.out.println("a >>> 2 = " + c );
}
}
```

Output

```
a & b = 12
a | b = 61
a ^ b = 49
~a = -61
a << 2 = 240
a >> 15
a >>> 15
```

10.4.4 Logical Operators

The following table lists the logical operators:

Assume boolean variable A holds true and variable B holds false then:

Operator	Description	Example
&&	Called Logical AND operator. If both the operands are non-zero, then condition becomes true.	(A && B) is false.
||	Called Logical OR Operator. If any of the two operands are non-zero, then condition becomes true.	(A || B) is true.
!	Called Logical NOT Operator. Use to reverses the logical state of its operand. If a condition is true, then Logical NOT operator will make false.	!(A && B) is true.

```
class Test {
public static void main(String args[]) {
  int a = true;
  int b = false;
  System.out.println("a && b = " + (a&&b) );
  System.out.println("a || b = " + (a||b) );
  System.out.println("!(a && b) = " + !(a && b) );
}
}
```

Output

```
a && b = false
a || b = true
!(a && b) = true
```

10.4.5 Assignment Operators

These are the following assignment operators supported by Java language:

Operator	Description	Example
=	Simple assignment operator: Assigns values from right side operands to left side operand	C = A + B will assign value of A + B into C
+=	Add AND assignment operator: It adds right operand to the left operand and assign the result to left operand	C += A is equivalent to C = C + A
-=	Subtract AND assignment operator: It subtracts right operand from the left operand and assign the result to left operand	C –= A is equivalent to C = C - A
*=	Multiply AND assignment operator: It multiplies right operand with the left operand and assign the result to left operand	C *= A is equivalent to C = C * A
/=	Divide AND assignment operator: It divides left operand with the right operand and assign the result to left operand	C /= A is equivalent to C = C / A
%=	Modulus AND assignment operator: It takes modulus using two operands and assign the result to left operand	C %= A is equivalent to C = C % A
<<=	Left shift AND assignment operator	C <<= 2 is same as C = C << 2
>>=	Right shift AND assignment operator	C >>= 2 is same as C = C >> 2
&= C	Bitwise AND assignment operator	C &= 2 is same as C = & 2
^=	Bitwise exclusive OR and assignment operator	C ^= 2 is same as C = C ^ 2
|=	Bitwise inclusive OR and assignment operator	C |= 2 is same as C = C | 2

```
class Test {
public static void main(String args[]) {
int a = 10;
int b = 20;
int c = 0;
c = a + b;
System.out.println(“c = a + b = “ + c );
c += a ;
System.out.println(“c += a  = “ + c );
c –= a ;
System.out.println(“c –= a = “ + c );
```

```
c *= a ;
System.out.println("c *= a = " + c );
a = 10;
c = 15;
c /= a ;
System.out.println("c /= a = " + c );
a = 10;
c = 15;
c %= a ;
System.out.println("c %= a  = " + c );
c <<= 2 ;
System.out.println("c <<= 2 = " + c );
c >>= 2 ;
System.out.println("c >>= 2 = " + c );
c >>= 2 ;
System.out.println("c >>= a = " + c );
c &= a ;
System.out.println("c &= 2  = " + c );
c ^= a ;
System.out.println("c ^= a  = " + c );
c |= a ;
System.out.println("c |= a  = " + c );
}
}
```

Output

```
c = a + b = 30
c += a  = 40
c -= a = 30
c *= a = 300
c /= a = 1
c %= a  = 5
c <<= 2 = 20
c >>= 2 = 5
c >>= 2 = 1
c &= a  = 0
c ^= a   = 10
c |= a   = 10
```

10.4.6 Misc Operators

There are a few other operators supported by Java Language.

Conditional Operator (? :)

Conditional operator is also known as the ternary operator. This operator consists of three operands and is used to evaluate boolean expressions. The goal of the operator is to decide which value should be assigned to the variable. The operator is written as:

variable x = (expression) ? value if true : value if false

Following is the example:

```
public class Test {
public static void main(String args[]){
int a , b;
a = 10;
b = (a == 1) ? 20: 30;
System.out.println( "Value of b is : " +  b );

b = (a == 10) ? 20: 30;
System.out.println( "Value of b is : " + b );
}}
```

This would produce the following result:

```
Value of b is : 30
Value of b is : 20
```

Instance of Operator

This operator is used only for object reference variables. The operator checks whether the object is of a particular type(class type or interface type). instanceOf operator is written as: (Object reference variable) instance of (class/interface type)

If the object referred by the variable on the left side of the operator passes the IS-A check for the class/interface type on the right side, then the result will be true. Following is the example:

```
String name = = 'James';
boolean result = s instanceOf String;
// This will return true since name is type of String
```

This operator will still return true if the object being compared is the assignment compatible with the type on the right. Following is one more example:

```
class Vehicle {}
public class Car extends Vehicle {
  public static void main(String args[]){
  Vehicle a = new Car();
  boolean result =  a instanceof Car;
  System.out.println( result);
}}
```

This would produce the following result:

```
true
```

Precedence of Java Operators

Operator precedence determines the grouping of terms in an expression. This affects how an expression is evaluated. Certain operators have higher precedence than others; for example, the multiplication operator has higher precedence than the addition operator:

For example x = 7 + 3 * 2; Here x is assigned 13, not 20 because operator * has higher precedence than + so it first get multiplied with 3*2 and then adds into 7.

Here operators with the highest precedence appear at the top of the table, those with the lowest appear at the bottom. Within an expression, higher precedence operators will be evaluated first.

<table>
<tr><th>Category</th><th>Operator</th><th>Associativity</th></tr>
<tr><td>Postfix</td><td>() [] . (dot operator)</td><td>Left to right</td></tr>
<tr><td>Unary</td><td>++ – – ! ~</td><td>Right to left</td></tr>
<tr><td>Multiplicative</td><td>* / %</td><td>Left to right</td></tr>
<tr><td>Additive</td><td>+ –</td><td>Left to right</td></tr>
<tr><td>Shift</td><td>>> >>> <<</td><td>Left to right</td></tr>
<tr><td>Relational</td><td>> >= < <=</td><td>Left to right</td></tr>
<tr><td>Equality</td><td>== !=</td><td>Left to right</td></tr>
<tr><td>Bitwise AND</td><td>&</td><td>Left to right</td></tr>
<tr><td>Bitwise XOR</td><td>^</td><td>Left to right</td></tr>
<tr><td>Bitwise OR</td><td>|</td><td>Left to right</td></tr>
<tr><td>Logical AND</td><td>&&</td><td>Left to right</td></tr>
<tr><td>Logical OR</td><td>| |</td><td>Left to right</td></tr>
<tr><td>Conditional</td><td>?:</td><td>Right to left</td></tr>
<tr><td>Assignment</td><td>= += –= *= /= %= >>= <<=
& = ^= | =</td><td>Right to left</td></tr>
<tr><td>Comma</td><td>,</td><td>Left to right</td></tr>
</table>

10.5 Control and Decision Making Statements

10.5.1 Loop Control /Iteration Statements

There may be a situtation when we need to execute a block of code several number of times, and is often referred to as a loop.

Java has very flexible three looping mechanisms. You can use one of the following three loops:

- While Loop
- Do ... while Loop
- For Loop

As of java 5 the *enhanced for loop* was introduced. This is mainly used for Arrays.

a. While Loop

A while loop is a control structure that allows you to repeat a task a certain number of times.

Syntax:

The syntax of a while loop is:

```
while(Boolean_expression)
{
  //Statements
}
```

When executing, if the *boolean_expression* result is true, then the actions inside the loop will be executed. This will continue as long as the expression result is true.

Here key point of the *while* loop is that the loop might not ever run. When the expression is tested and the result is false, the loop body will be skipped and the first statement after the while loop will be executed.

Example:

```
public class Test {
  public static void main(String args[]){
   int x= 10;
     while( x < 20 ){
       System.out.print("value of x : " + x );
        x++;
        System.out.print("\n");
     }
  }
}
```

This would produce the following result:

```
value of x : 10
value of x : 11
value of x : 12
value of x : 13
value of x : 14
value of x : 15
value of x : 16
value of x : 17
value of x : 18
value of x : 19
```

The following code shows the use of while loop by generating the Fibonacci series:

```
class Test While Loop
{
        public static void main ( String args [ ] )
        {
                int num1 = 1, num2 = 1;
                System.out.println("The Fibonacci series between 1 and 100:");
                System.out.println(num1);
                while (num2<100)
                {
                        System.out.println(num2);
                        num2 += num1;  //  adding the value of num1 to num2
                        num1 = num2 – num1;  // reassigning num1 to the difference
                                                        between num2 and num1
                }
        }
}
```

In the preceding code, the Fibonacci series between 1 and 100 is generated. In this series, each number is the sum of its two preceding numbers. The series starts with 1. The *num2* variable has value 1 outside the loop. The *while* loop checks the value of *num2* and finds it less than 100. Therefore, the condition in *while* loop comes out as true and control enters the body of the loop. In each loop iteration, the *num2* variable gets incremented by adding the value of *num1* variable with the value of *num2*. When the value of *num2* exceeds 100, the control comes out of the *while* loop.

b. Do...while Loop

A do...while loop is similar to a while loop, except that a do...while loop is guaranteed to execute at least one time.

Syntax

The syntax of a do...while loop is:

```
do
{
  //Statements
}while(Boolean_expression);
```

Notice that the Boolean expression appears at the end of the loop, so the statements in the loop execute once before the Boolean is tested. If the Boolean expression is true, the flow of control jumps back up to do, and the statements in the loop execute again. This process repeats until the Boolean expression is false.

Example:

```
public class Test {
 public static void main(String args[]){
  int x= 10;

  do{
   System.out.print("value of x : " + x );
   x++;
  System.out.print("\n");
  }while( x < 20 );
 }}
```

This would produce the following result:

```
value of x : 10
value of x : 11
value of x : 12
value of x : 13
value of x : 14
value of x : 15
value of x : 16
value of x : 17
value of x : 18
value of x : 19
```

Another example to implement a do-while statement is as follows:
The following code shows the use of the *do-while* loop:

```
class TestDoWhile
{
        public static void main ( String args [ ] )
        {
                int num=100;
                do
                {
                        System.out.println("Inside the do-while loop");
                        System.out.println("The value of the variable num is:" +num);
                }
                while (num<100);
        }
}
```

In the preceding code, the value assigned to *num* variable is 100. The control enters into the *do-while* construct without testing the value of the *num* variable and executes the statement inside the *do-while* loop. The *while* construct then checks the value of *num* and since the value is not less than 100, the control comes out of the *do-while* loop.

c. For Loop

A for loop is a repetition control structure that allows you to efficiently write a loop that needs to execute a specific number of times.

A for loop is useful when you know how many times a task is to be repeated.

Syntax

The syntax of a for loop is:

```
for(initialization; Boolean_expression; update)
{
   //Statements
}
```

Here is the flow of control in a for loop:

1. The initialization step is executed first, and only once. This step allows you to declare and initialize any loop control variables. You are not required to put a statement here, as long as a semicolon appears.
2. Next, the Boolean expression is evaluated. If it is true, the body of the loop is executed. If it is false, the body of the loop does not execute and flow of control jumps to the next statement past the for loop.
3. After the body of the for loop executes, the flow of control jumps back up to the update statement. This statement allows you to update any loop control variables. This statement can be left blank, as long as a semicolon appears after the Boolean expression.
4. The Boolean expression is now evaluated again. If it is true, the loop executes and the process repeats itself (body of loop, then update step, then Boolean expression). After the Boolean expression is false, the for loop terminates.

Example:

```
public class Test {
  public static void main(String args[]){
    for(int x = 10; x < 20; x = x+1){
      System.out.print("value of x : " + x );
       System.out.print("\n");
       }
  }
}
```

This would produce the following result:

```
value of x : 12
value of x : 13
value of x : 14
value of x : 15
value of x : 16
value of x : 17
value of x : 18
value of x : 19
```

Enhanced for Loop in Java

As of Java 5 the enhanced for loop was introduced. This is mainly used for Arrays.

Syntax

The syntax of enhanced for loop is:

```
for(declaration : expression)
{
  //Statements
}
```

- **Declaration:** The newly declared block variable, which is of a type compatible with the elements of the array you are accessing. The variable will be available within the for block and its value would be the same as the current array element.
- **Expression:** This evaluate to the array you need to loop through. The expression can be an array variable or method call that returns an array.

Example:

```
public class Test {
  public static void main(String args[]){
   int [] numbers = {10, 20, 30, 40, 50};

   for(int x : numbers ){
     System.out.print( x );
     System.out.print(",");
   }
   System.out.print("\n");
   String [] names ={"James", "Larry", "Tom", "Lacy"};
   for( String name : names ) {
     System.out.print( name );
```

```
        System.out.print(",");
      }
    }
  }
```

Example:

```
public class Test {
    public static void main(String args[]){
      int [] numbers = {10, 20, 30, 40, 50};
       for(int x : numbers){
          System.out.print(x);
          System.out.print(",");
       }
       System.out.print("\n");
       String [] names ={"James", "Larry", "Tom", "Lacy"};
       for( String name : names ) {
          System.out.print( name );
          System.out.print(",");
          }
    }
}
```

This would produce the following result:
10, 20, 30, 40, 50,
James,Larry,Tom,Lacy

10.5.2 Jump Statements

Java supports three jump statements as break, continue and switch.

a. The break Keyword

The *break* keyword is used to stop the entire loop. The break keyword must be used inside any loop or a switch statement.

The break keyword will stop the execution of the innermost loop and start executing the next line of code after the block.

Syntax

The syntax of a break is a single statement inside any loop:

```
break;
```

Example:

```
public class Test {
  public static void main(String args[]){
   int [] numbers = {10, 20, 30, 40, 50};

   for(int x : numbers){
    if(x == 30){
    continue;
   }
```

```
      System.out.print( x );
      System.out.print("\n");
    }
  }
}
```

This would produce the following result:
10
20

b. The continue Keyword

The *continue* keyword can be used in any of the loop control structures. It causes the loop to immediately jump to the next iteration of the loop.

- In a for loop, the continue keyword causes flow of control to immediately jump to the update statement.
- In a while loop or do/while loop, flow of control immediately jumps to the Boolean expression.

Syntax

The syntax of an if statement is:

```
if(Boolean_expression)

{
  //Statements will execute if the Boolean expression is true
}
```

If the boolean expression evaluates to true, then the block of code inside the if statement will be executed. If not, the first set of code after the end of the if statement(after the closing curly brace) will be executed.

Example:

```
public class Test {
  public static void main(String args[]){
    int x = 10;

    if( x < 20 ){
      System.out.print("This is if statement");
    }
  }
}
```

This would produce the following result:
This is if statement

The if...else Statement

An if statement can be followed by an optional *else* statement, which executes when the Boolean expression is false.

Syntax

The syntax of if...else is:

```
if(Boolean_expression 1){
  //Executes when the Boolean expression 1 is true
```

```
}else if(Boolean_expression 2){
  //Executes when the Boolean expression 2 is true
}else if(Boolean_expression 3){
  //Executes when the Boolean expression 3 is true
}else {
  //Executes when the none of the above condition is true.
}
```

Example:

The syntax of a if...else is:

```
if(Boolean_expression){
  //Executes when the Boolean expression is true
}else{
  //Executes when the Boolean expression is false
}
```

Example:

```
public class Test {
 public static void main(String args[]){
  int x = 30;

  if( x < 20 ){
    System.out.print("This is if statement");
  }else{
    System.out.print("This is else statement");
  }
 }
}
```

This would produce the following result:

This is else statement

The if...else if...else Statement:

An if statement can be followed by an optional *else if...else* statement, which is very useful to test various conditions using single if...else if statement.

When using if, else if, else statements, there are a few points to keep in mind.

- An if can have zero or one else's and it must come after any else if's.
- An if can have zero to many else if's and they must come before the else.
- Once an else if succeeds, none of the remaining else if's or else's will be tested.

Syntax

The syntax of if...else is:

```
if(Boolean_expression 1){
  //Executes when the Boolean expression 1 is true
}else if(Boolean_expression 2){
  //Executes when the Boolean expression 2 is true
}else if(Boolean_expression 3){
  //Executes when the Boolean expression 3 is true
}else {
  //Executes when the none of the above condition is true.
}
```

Example:

```
public class Test {
  public static void main(String args[]){
    int x = 30;

    if( x == 10 ){
      System.out.print("Value of X is 10");
    }else if( x == 20 ){
      System.out.print("Value of X is 20");
    }else if( x == 30 ){
      System.out.print("Value of X is 30");
    }else{
      System.out.print("This is else statement");
    }
  }
}
```

This would produce the following result:
Value of X is 30

Nested if...else Statement

It is always legal to nest if-else statements, which means you can use one if or else if statement inside another if or else if statement.

Syntax

The syntax for a nested if...else is as follows:

```
if(Boolean_expression 1){
  //Executes when the Boolean expression 1 is true
  if(Boolean_expression 2){
    //Executes when the Boolean expression 2 is true
  }
}
```

You can nest *else if...else* in the similar way as we have nested *if* statement.

Example:

```
public class Test {
  public static void main(String args[]){
    int x = 30;
      int y = 10;
    if( x == 30 ){
      if( y == 10 ){
      System.out.print("X = 30 and Y = 10");
    }
  }
}
```

This would produce the following result:
X = 30 and Y = 10

c. The Switch Statement

A *switch* statement allows a variable to be tested for equality against a list of values. Each value is called a case, and the variable being switched on is checked for each case.

Syntax

The syntax of enhanced for loop is:

```
switch(expression){
   case value:
      //Statements
      break; //optional
   case value:
      //Statements
      break; //optional
   //You can have any number of case statements.
   default: //Optional
      //Statements
}
```

The following rules apply to a switch statement:

- The variable used in a switch statement can only be a byte, short, int, or char.
- You can have any number of case statements within a switch. Each case is followed by the value to be compared to and a colon.
- The value for a case must be the same data type as the variable in the switch, and it must be a constant or a literal.
- When the variable being switched on is equal to a case, the statements following that case will execute until a *break* statement is reached.
- When a *break* statement is reached, the switch terminates, and the flow of control jumps to the next line following the switch statement.
- Not every case needs to contain a break. If no break appears, the flow of control will *fall through* to subsequent cases until a break is reached.
- A *switch* statement can have an optional default case, which must appear at the end of the switch. The default case can be used for performing a task when none of the cases is true. No break is needed in the default case.

Example:

```
public class Test {
   public static void main(String args[]){
      char grade = args[0].charAt(0);

      switch(grade)
      {
      case 'A' :
         System.out.println("Excellent!");
         break;
      case 'B' :
      case 'C' :
         System.out.println("Well done");
```

```
      break;
    case 'D' :
      System.out.println("You passed");
    case 'F' :
      System.out.println("Better try again");
      break;
    default :
      System.out.println("Invalid grade");
    }
    System.out.println("Your grade is " + grade);
  }
}
```

Compile and run the above program using various command line arguments. This would produce the following result:

```
$ java Test a
Invalid grade
Your grade is a a
$ java Test A
Excellent!
Your grade is a A
$ java Test C
Well done
Your grade is a C
$
```

3

Event-Driven Programming

1. GRAPHICS PROGRAMMING

A graphics context provides the capabilities of drawing on the screen. The graphics context maintains states such as the color and font used in drawing, as well as interacting with the underlying operating system to perform the drawing. In Java, custom painting is done via the java.awt. graphics class, which manages a graphics context, and provides a set of device-independent methods for drawing texts, figures and images on the screen on different platforms.

The java.awt.Graphics is an abstract class, as the actual act of drawing is system-dependent and device-dependent. Each operating platform will provide a subclass of graphics to perform the actual drawing under the platform, but conform to the specification defined in graphics.

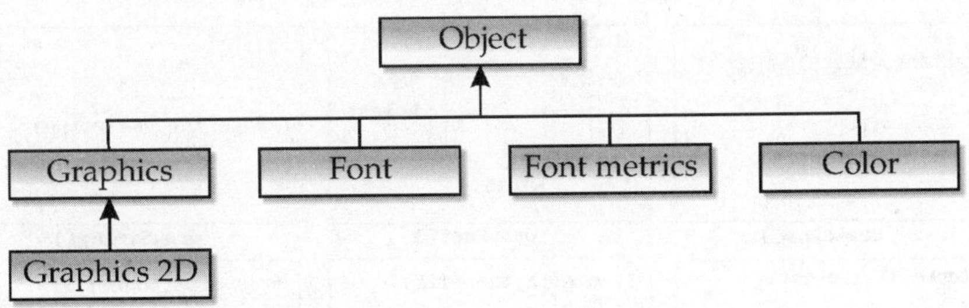

Graphics' Drawing Methods—2D Shapes

The graphics class provides methods for drawing three types of graphical objects:

Text strings: To note that System.out.println () prints to the system console, not to the graphics screen. The drawString () is used in this case.

Vector-graphic primitives and shapes: Methods like drawXxx () and fill Xxx (), are used where Xxx could be Line, Rect, Oval, Arc, Poly Line, Round Rect, or 3D Rect.

Bitmap images: The drawImage () method is used.

Drawing (or printing) Texts on the Graphics Screen

drawString (String str, int xBaselineLeft, int yBaselineLeft);

Drawing Lines

drawLine (int x1, int y1, int x2, int y2);
drawPolyline (int [] x Points, int [] y Points, int num Point);

Drawing Primitive Shapes

drawRect (int x Top Left, int y Top Left, int width, int height);

drawOval (int x Top Left, int y Top Left, int width, int height);

drawArc(int x Top Left, int y Top Left, int width, int height, int start Angle, int arc Angle);

draw 3 D Rect (int x Top Left, int, y Top Left, int width, int height, boolean raised);

draw Round Rect (int x Top Left, int yTop Left, int width, int height, int arcWidth, int arc Height)

draw Polygon (int [] x Points, int [] y Points, int num Point);

Filling Primitive Shapes

fillRect(int x Top Left, int y TopLeft, int width, int height);

fillOval(int xTopLeft, int yTopLeft, int width, int height);

fillArc(int xTopLeft, int yTopLeft, int width, int height, int startAngle, int arcAngle);

fill3DRect(int xTopLeft, int, yTopLeft, int width, int height, boolean raised);

fillRoundRect(int xTopLeft, int yTopLeft, int width, int height, int arcWidth, int arcHeight)

fillPolygon(int[] xPoints, int[] yPoints, int numPoint);

Drawing (or Displaying) Images

drawImage(Image img, int xTopLeft, int yTopLeft, ImageObserver obs);

drawImage(Image img, int xTopLeft, int yTopLeft, int width, int height, ImageObserver o);

The drawXxx() methods draw the outlines; while fillXxx() methods fill the internal with the graphics context current color. Shapes with negative width and height will not be painted.

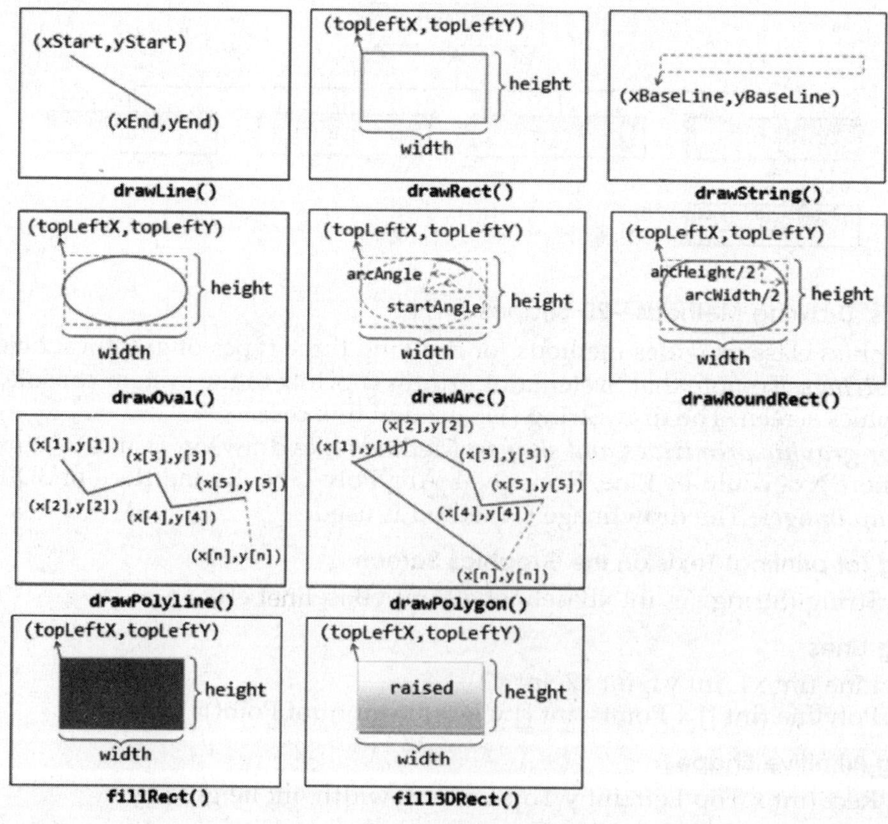

Graphics' Methods for Maintaining the Graphics Context

The graphic context maintains states (or attributes) such as the current painting color, the current font for drawing text strings, and the current painting rectangular area (called clip). The following methods are used to manipulate or retrieve the properties of graphics context.

Graphics context's current color.

```
void setColor(Color c)
Color getColor()
```

Graphics context's current font.

```
void setFont(Font f)
Font getFont()
```

Set/Get the current clip area.

Clip area shall be rectangular and no rendering is performed outside the clip area.

```
void setClip(int xTopLeft, int yTopLeft, int width, int height)
void setClip(Shape rect)
public abstract void clipRect(int x, int y, int width, int height) // intersects
the current clip with the given rectangle
Rectangle getClipBounds() // returns an Rectangle
Shape getClip() // returns an object (typically Rectangle) implements Shape
```

The paintComponent (Graphics g) methods that we override take an argument of a java.awt.Graphics object. Whenever a component (e.g. a button, a panel) is displayed, the Java Windowing Subsystem create a graphics object for the component. This graphics object may be retrieved using thegetGraphics() method of the component.

Other Methods

```
void clearRect(int x, int y, int width, int height)
  // Clear the rectangular area to background
void copyArea(int x, int y, int width, int height, int dx, int dy)
  // Copy the rectangular area to offset (dx, dy).
void translate(int x, int y)
  // Translate the origin of the graphics context to (x, y). Subsequent drawing
  // uses the new origin.
FontMetrics getFontMetrics()
FontMetrics getFontMetrics(Font f)
  // Get the FontMetrics of the current font / the specified font
```

Graphics Coordinate System

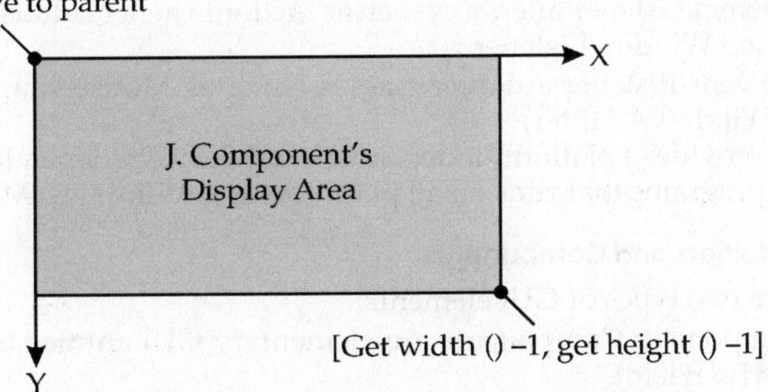

In Java Windowing Subsystem (like most of the 2D Graphics systems), the origin (0, 0) is located at the top-left corner.

Each component/container has its own coordinate system, ranging for (0, 0) to (width-1, height-1) as illustrated.

You can use method get Width () and get Height () to retrieve the width and height of a component/container. You can use get X () or get Y () to get the (x, y) of this component's origin relative to its parent.

There are two sets of Java APIs for graphics programming: AWT (Abstract Windowing Toolkit) and Swing.

1. AWT API was introduced in JDK 1.0. Most of the AWT components have become obsolete and should be replaced by newer Swing components.
2. Swing API, a much more comprehensive set of graphics libraries that enhances the AWT, was introduced as part of Java Foundation Classes (JFC) after the release of JDK 1.1. JFC, which consists of Swing, Java 2D, Accessibility API, Internationalization, and Pluggable Look-and-Feel Support, was an add-on to JDK 1.1 but has been integrated into core Java since JDK 1.2.

Other than AWT/Swing Graphics APIs provided in JDK, others have also provided Graphics APIs that work with Java, such as Eclipse's Standard Widget Toolkit (SWT), Google Web Toolkit (GWT), 3D Graphics API such as Java bindings for Open GL (JOGL) and Java 3D.

2. PROGRAMMING WITH AWT

Java Graphics APIs—AWT and Swing—provide a huge set of reusable GUI components, such as button, text field, label, choice, panel and frame for building GUI applications. These classes may be reuseds. AWT classes are discussed in detail before moving on to swing. AWT is now used only in exceptional circumstances when the JRE supports only JDK 1.1.

2.1 AWT Packages

AWT consists of 12 packages (Swing is even bigger, with 18 packages as of JDK 1.7!). Fortunately, only 2 packages—java.awt and java.awt.event—are commonly used.

1. The java.awt package contains the *core* AWT graphics classes:
   - GUI Component classes (such as Button, TextField, and Label),
   - GUI Container classes (such as Frame, Panel, Dialog and ScrollPane),
   - Layout managers (such as FlowLayout, BorderLayout and GridLayout),
   - Custom graphics classes (such as Graphics, Color and Font).
2. The java.awt.event package supports event handling:
   - Event classes (such as ActionEvent, MouseEvent, KeyEvent and WindowEvent)
   - Event Listener Interfaces such as ActionListener, MouseListener, KeyListener and WindowListener
   - Event Listener Adapter classes (such as MouseAdapter, KeyAdapter, and WindowAdapter).

AWT provides a platform-independent and device-independent interface to develop graphic programs that runs on all platforms, e.g. Windows, Mac, Unix, etc.

2.2 Containers and Components

There are two types of GUI elements:

1. Component: Components are elementary GUI entities (such as Button, Label, andTextField).

2. *Container*: Containers (such as Frame, Panel and Applet) are used to *hold components in a specific layout*. A container can also hold sub-containers.
   GUI components are also called *controls* (Microsoft Active X Control), *widgets* (Eclipse's Standard Widget Toolkit, Google Web Toolkit), which allow users to interact with the application via mouse, keyboard, and other forms of inputs such as voice.
   A Frame is the *top-level container* of an AWT GUI program. A Frame has a title bar (containing an icon, a title, and the minimize/maximize(restore-down)/close buttons), an optional menu bar and the content display area. A Panel is a *rectangular area* (or partition) used to group related GUI components. In a GUI program, a component must be kept in a container. A container should be identified to hold the components. Every container has a method called add(Component c). A container (says *a Container*) can invoke *a Container*. add (*a Component*) to add *a Component* into itself. For example,

```
Panel panel = new Panel(); // Panel is a Container
Button btn = new Button(); // Button is a Component
panel.add(btn); // The Panel Container adds a Button Component
```

2.2.1 AWT Container Class

Top-Level Containers: Frame, Dialog and Applet

Each GUI program has a *top-level container*. The commonly-used top-level containers in AWT are Frame, Dialog and Applet:

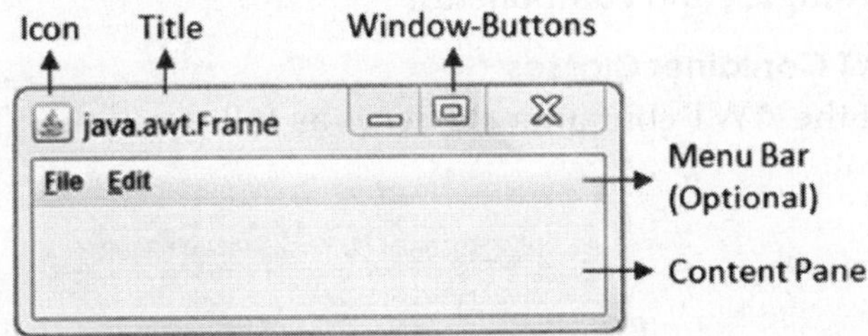

A Frame provides the "main window" for the GUI application, which has a title bar (containing an icon, a title, the minimize, maximize/restore-down and close buttons), an optional menu bar, and the content display area. To write a GUI program, we typically start with a subclass extending from java.awt.Frame to inherit the main window as follows:

```
import java.awt.Frame; // Using Frame class in package java.awt
// A GUI program is written as a subclass of Frame—the top-level container
// This subclass inherits all properties from Frame, e.g. title, icon, buttons,
content-pane
public class MyGUIProgram extends Frame {
// Constructor to setup the GUI components
public MyGUIProgram() { ...... }
......
......

// The entry main() method
public static void main(String[] args) {
```

```
        // Invoke the constructor (to setup the GUI) by allocating an instance
        MyGUIProgram m = new MyGUIProgram();
    }
}
```

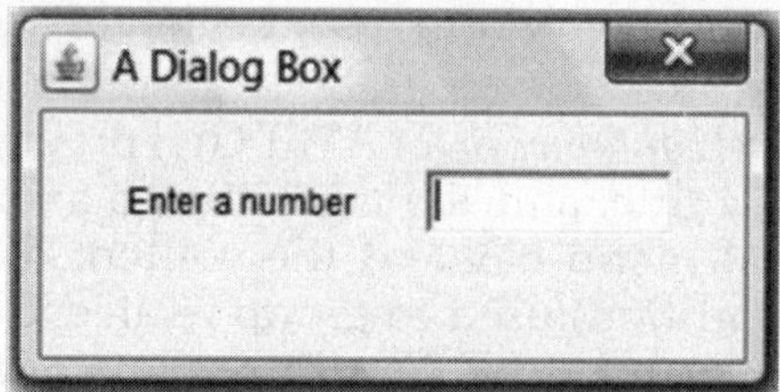

An AWT Dialog is a *"pop-up window"* used for interacting with the users. A Dialog has a title-bar (containing an icon, a title and a close button) and a content display area, as illustrated. An AWT Applet (in package java.applet) is the top-level container for an applet, which is a Java program running inside a browser.

Secondary Containers: Panel and Scroll Pane

Secondary containers are placed inside a top-level container or another secondary container. AWT also provides secondary containers such as Panel and Scroll Pane.

- A Panel is a rectangular box (partition) under a higher-level container, used to *layout* a set of related GUI components. See the above examples for illustration.
- Others, such as Scroll Pane (which provides automatic horizontal and/or vertical scrolling for a single child component).

Hierarchy of the AWT Container Classes

The hierarchy of the AWT container classes is as follows:

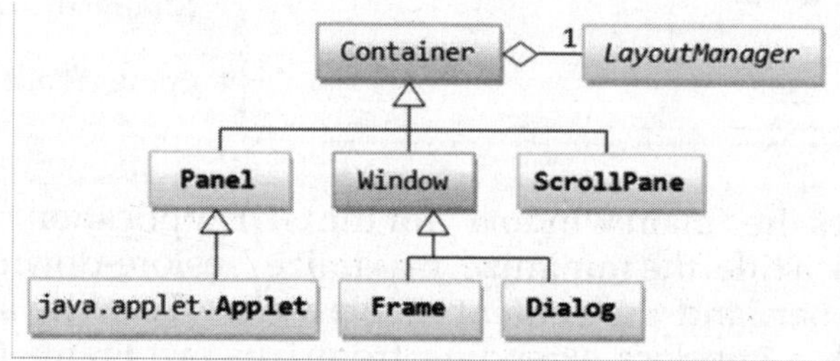

2.2.2 AWT Component Class

AWT provides many ready-made and reusable GUI components. The frequently used are: Button, Text Field, Label, Check box, Check box Group(radio buttons), List, and Choice, as given below.

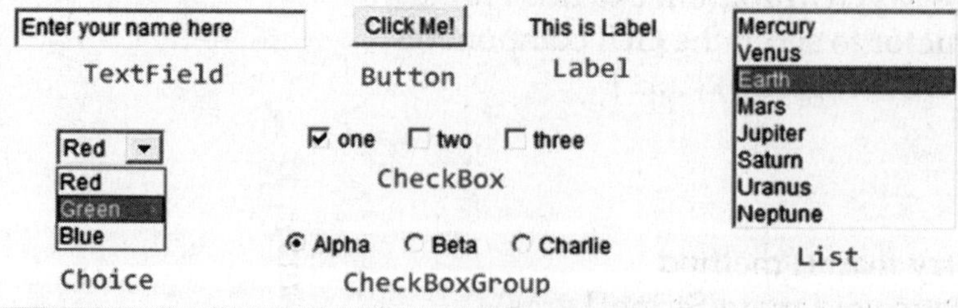

(i) Label

A java.awt.Label provides a text description message. Take note that System.out.println() prints to the system console, not to the graphics screen. You could use a Label to label another component (such as text field) or provide a text description.

Constructors

```
//Construct a Label with the given text String, of the Text alignment
public Label(String strLabel, int alignment);
//Construct a Label with the given text String
public Label(String strLabel);
//Construct an initially empty label
public Label();
```

The Label class has three constructors:

1. The first constructor constructs a Label object with the given text string in the given alignment. Note that three static constants Label.LEFT,Label.RIGHT, and Label.CENTER are defined in the class for you to specify the alignment (rather than asking you to memorize arbitrary integer values).
2. The second constructor constructs a Label object with the given text string in default of left-aligned.
3. The third constructor constructs a Label object with an initially empty string. You could set the label text via the setText() method later.

Constants

```
public static final LEFT;   // Label.LEFT
public static final RIGHT;  // Label.RIGHT
public static final CENTER; // Label.CENTER
```

These three constants are defined for specifying the alignment of the Label's text.

Public Methods

```
// Examples
public String getText();
public void setText(String strLabel);
public int getAlignment();
public void setAlignment(int alignment);
```

The getText() and setText() methods can be used to read and modify the Label's text. Similarly, the getAlignment() and setAlignment()methods can be used to retrieve and modify the alignment of the text.

Constructing a Component and Adding the Component into a Container

Three steps are necessary to create and place a GUI component:

1. Declare the component with an *identifier*;
2. Construct the component by invoking an appropriate constructor via the new operator;

3. Identify the container (such as Frame or Panel) designed to hold this component. The container can then add this component onto itself via *a Container*.add (*aComponent*) method. Every container has a add(Component) method. Take note that it is the container that actively and explicitly adds a component onto itself, instead of the other way.

```
/*<applet code=bpl_ajp_label width=300 height=300>
    </applet>*/

import java.awt.*;
import java.applet.*;
public class bpl_ajp_label extends Applet
{
Label lab=new Label("LEFT",Label.LEFT);
Label lab1=new Label("CENTER",Label.CENTER);
Label lab2=new Label("RIGHT",Label.RIGHT);
public void init()
{
    add(lab);
    add(lab1);
    add(lab2);
}
public void paint(Graphics g)
{}
}
```

An Anonymous Instance

A Label can be created without specifying an identifier, called *anonymous instance*. In the case, the Java compiler will assign an *anonymous identifier* for the allocated object. The user will not be able to reference an anonymous instance in your program after it is created. This is usually alright for a Label instance as there is often no need to reference a Label after it is constructed.

```
// Allocate an anonymous Label instance. "this" container adds the instance into itself.
// You CANNOT reference an anonymous instance to carry out further operations.
add(new Label("Enter Name: ", Label.RIGHT));
// Same as
Label lblXxx = new Label("Enter Name: ", Label.RIGHT); // lblXxx assigned by compiler
add(lblXxx);
```

(ii) Button

A java.awt.Button is a GUI component that triggers a certain programmed *action* upon clicking.

Constructors

//construct a Button with the givenlabel
public Button(String buttonLabel);
//Construct a Button with empty label
public Button();

Public Methods

public String get Label (); // Get the label of this Button instance
public void set Label (String buttonLabel); // Set the label of this Button instance
public void set Enable (boolean enable); // Enable or disable this Button. Disabled Button cannot be clicked.

The get Label () and set Label () methods can be used to read the current label and modify the label of a button, respectively.

Note: The latest Swing's J Button replaces get Label ()/set Label () with get Text () / set Text () to be consistent with all the components. We will describe Swing later.

Event

Clicking a button fires a so-called ActionEvent and triggers a certain programmed action.

Button btnColor = new Button ("Red"); // Declare and allocate a Button instance called btnColor
add(btnColor); // "this" Container adds the Button
...
btnColor.setLabel("green"); // Change the button's label
btnColor.getLabel(); // Read the button's label
...
add(Button("Blue")); //Create an anonymous Button. It CANNOT be referenced later

(iii) Text Field

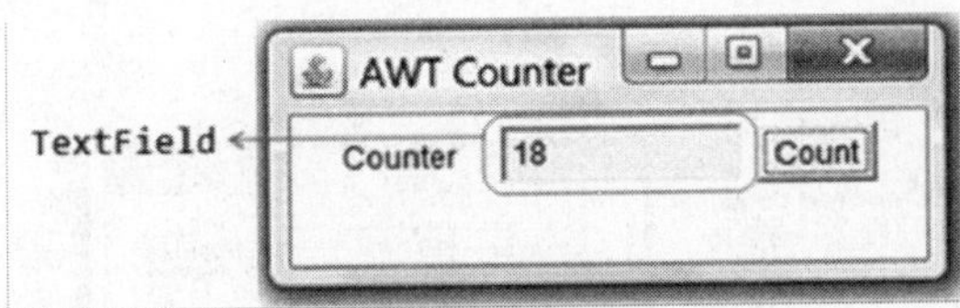

A java. awt. Text Field is single-line text box for users to enter texts. (There is a multiple-line text box called TextArea.) Hitting the "ENTER" key on a Text Field object triggers an action-event.

Constructors

public TextField(String strInitialText, int columns);
//construct a Textfield instance with the given
// initial text string with number of columns
public TextField(String strInitialText, int columns);
//Construct a TextField instance with the given initial
//textstring
public TextField(int columns);

Public Methods

```
public String getText();
  // Get the current text on this TextField instance
public void setText(String strText);
  // Set the display text on this TextField instance
public void setEditable(boolean editable);
  // Set this TextField to editable (read/write) or non-editable (read-only)
```

Event

Hitting the "ENTER" key on a TextField fires a ActionEvent, and triggers a certain programmed action.

```
TextField tfInput = new TextField(30); // Declare and allocate an TextField
instance called tfInput
add(tfInput);                    // "this" Container adds the TextField
TextField tfResult = new TextField(); // Declare and allocate an TextField
instance called tfResult
tfResult.setEditable(false) ;       // Set to read-only
add(tfResult);                   // "this" Container adds the TextField
......
// Read an int from TextField "tfInput", square it, and display on "tfResult".
// getText() returns a String, need to convert to int
int number = Integer.parseInt(tfInput.getText());
number *= number;
// setText() requires a String, need to convert the int number to String.
tfResult.setText(number + "");
```

Take note that get Text ()/ Set Text () operates on String. You can convert a String to a primitive, as int or double via static method integer. parse Int () or Double. parse Double (). To convert a primitive to a String, simply concatenate the primitive with an empty string.

AWT Counter

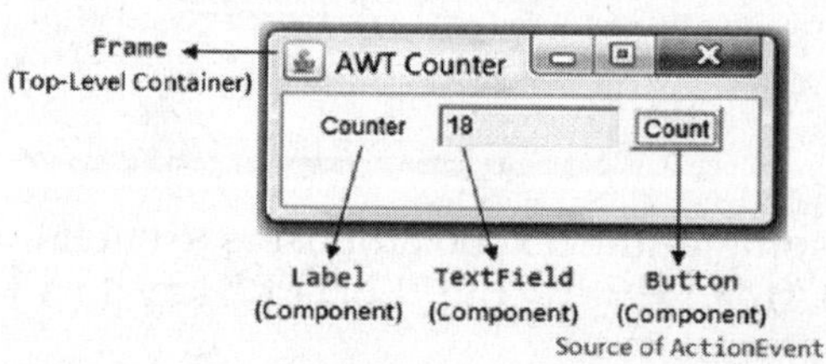

Let us assemble some components together into a simple GUI counter program, as illustrated. It has a top-level container frame, which contains three components—a label "Counter", a non-editable Text Field to display the current count, and a "Count" button. The Text Field displays "0" initially.

Each time you click the button, the counter's value increases by 1.

```
import java.awt.*;       // using AWT containers and components
import java.awt.event.*; // using AWT events and listener interfaces
// An AWT GUI program inherits the top-level container java.awt.Frame
```

```
public class AWTCounter extends Frame implements ActionListener {
private Label lblCount;    // declare component Label
private TextField tfCount; // declare component TextField
private Button btnCount;   // declare component Button
private int count = 0;     // counter's value
/** Constructor to setup GUI components */
public AWTCounter () {
setLayout(new FlowLayout());
// "this" Frame sets its layout to FlowLayout, which arranges the components
// from left-to-right, and flow to next row from top-to-bottom.

lblCount = new Label("Counter"); // construct Label
add(lblCount);                   // "this" Frame adds Label

tfCount = new TextField("0", 10); // construct TextField
tfCount.setEditable(false);       // set to read-only
add(tfCount);                     // "this" Frame adds tfCount

btnCount = new Button("Count"); // construct Button
add(btnCount);                  // "this" Frame adds Button

btnCount.addActionListener(this); // for event-handling

setTitle("AWT Counter"); // "this" Frame sets title
setSize(250, 100);       // "this" Frame sets initial window size
setVisible(true);        // "this" Frame shows
}

/** The entry main() method */
public static void main(String[] args) {
// Invoke the constructor to setup the GUI, by allocating an instance
AWTCounter app = new AWTCounter();
}

/** ActionEvent handler - Called back when user clicks the button. */
@Override
public void actionPerformed(ActionEvent evt) {
count++; // increase the counter value
// Display the counter value on the TextField tfCount
tfCount.setText(count + ""); // convert int to String
}}
```

To exit this program, you have to close the CMD-shell (or press "control-c"); or push the red-square close button in Eclipse's Application Console. This is because we have yet to write the handler for the window's close button. We shall do that in the later example.

Dissecting the AWTCounter.java

- The import statements (Lines 1-2) are needed, as AWT container and component classes, such as frame, Button, Text Field, and label, are kept in the java. awt package; while AWT events and event-listener interfaces, such as action event and Action Listener are kept in the java awt event package.
- A GUI program needs a top-level container, and is often written as a subclass of Frame (Line 5). In other words, this class AWT Counter *is a* Frame, and inherits all the attributes and behaviors of a Frame, such as the title bar and content pane.
- Lines 12 to 31 define a constructor, which is used to setup and initialize the GUI components.
- The set layout () method (in Line 13) is invoked without an object and the dot operator. Hence, defaulted to "this" object, i.e. this set layout (). The set layout () is inherited from the superclass Frame and is used to set the layout of the components of the containerFrame. Flow layout is used in this example, which arranges the GUI components in left-to-right and flows into next row in a top-to-bottom manner.
- A Label, Text Field (non-editable), and button are constructed. "this" object (frame container) adds these components into it via this add () inherited from the superclass frame.
- The set size () and the set title () (Line 28–29) are used to set the initial size and the title of "this" Frame. The set visible (true) method (Line 30) is then invoked to show the display.
- The statement btn count add action listener (this) (Line 26) is used to setup the event-handling mechanism, which will be discussed in length later. In brief, whenever the button is clicked, the action performed () will be called. In the action performed () (Lines 34–39), the counter value increases by 1 and displayed on the Text Field.
- In the entry main () method (Lines 43–45), an instance of AWT counter is constructed. The constructor is executed to initialize the GUI components and setup the event-handling mechanism. The GUI program then waits for the user input.

toString()

It is interesting to inspect the GUI objects via the toString(), to gain an insight to these classes. (Alternatively, study the source code!) For example, if we insert the following code before and after the setvisible():

```
System.out.println(this);
System.out.println(lblCount);
System.out.println(tfCount);
System.out.println(btnCount);
setVisible(true);        // "this" Frame shows
System.out.println(this);
System.out.println(lblCount);
System.out.println(tfCount);
System.out.println(btnCount);
```

The output (with my comments) are as follows. You could have an insight of the variables defined in the class.

```
// Before setVisible()
AWTCounter[frame0,0,0,250×100,invalid,hidden,layout=java.awt.FlowLayout,title=AWT
Counter,resizable,normal]
   // name (assigned by compiler) is "frame0"; top-left (x,y) at (0,0); width/
   // height is 250x100 (via setSize());
java.awt.Label[label0,0,0,0 × 0,invalid,align=left,text=Counter]
   // name is "Label0"; align is "Label.LEFT" (default); text is "Counter"
   // (assigned in constructor)
java.awt.TextField[textfield0,0,0,0x0,invalid,text=0,selection=0-0]
   // name is "Textfield0"; text is "0" (assigned in constructor)
java.awt.Button[button0,0,0,0x0,invalid,label=Count]
   // name is "button0"; label text is "Count" (assigned in constructor)
   // Before setVisible(), all components are invalid (top-left (x,y), width/height
are invalid)

   // After setVisible(), all components are valid
AWTCounter[frame0,0,0,250x100,layout=java.awt.FlowLayout,title=AWT
Counter,resizable,normal]
   // valid and visible (not hidden)
java.awt.Label[label0,20,41,58x23,align=left,text=Counter]
   // Top-left (x,y) at (20,41) relative to the parent Frame; width/height = 58 × 23
java.awt.TextField[textfield0,83,41,94x23,text=0,selection=0-0]
   // Top-left (x,y) at (83,41) relative to the parent Frame; width/height = 94 × 23;
no text selected (0-0)
java.awt.Button[button0,182,41,47 × 23,label=Count]
   // Top-left (x,y) at (182,41) relative to the parent Frame; width/height = 47 × 23
```

2.2.3 Frame

Frame is a top-level window that has a title bar, menu bar, borders, and resizing corners. By default, a frame has a size of 0×0 pixels and it is not visible. The default layout for a frame is BorderLayout.Frames are examples of containers. It can contain other user interface components such as buttons and text fields.

Class Hierarchy for Frame

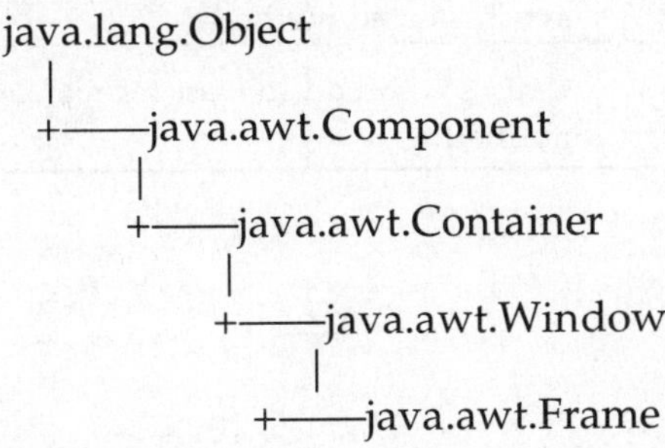

Java Lang Object

It is the root of the class hierarchy. Every class in the system has Object as its ultimate parent. Every variable and method defined here is available in every Object.

The general usage is given as

public class ***Object***

Constructor

public Object()

Methods

Methods	Usage	Use
getClass	public final Class getClass()	Returns the Class of this Object.
hashCode	public int hashCode()	Returns a hashcode for this Object.
equals	public boolean equals(Object obj)	Compares two Objects for equality. Returns a boolean that indicates whether this Object is equivalent to the specified Object.
copy	protected void copy(Object src)	Copies the contents of the specified Object into this Object.
clone	protected Object clone()	Creates a clone of this Object. A new instance is allocated and the copy() method is called to copy the contents of this Object into the clone.
toString	public String toString()	Returns a String that represents the value of this Object.
notify	public final void notify()	Notifies a single waiting thread on a change in condition of another thread. The thread effecting the change notifies the waiting thread using notify().
notifyAll	public final void notifyAll()	Notifies all of the threads waiting for a condition to change.
wait	public final void wait(long timeout)	Causes a thread to wait until it is notified or the specified timeout expires.
wait	public final void wait(long timeout, int nanos)	More accurate wait. The method wait() can only be called from within a synchronized method.
wait	public final void wait()	Causes a thread to wait forever until it is notified.

Class Java Awt Component

```
java.lang.Object
  |
  +——java.awt.Component
```

A generic Abstract Window Toolkit component.

It is generally used as

public class Component extends object implements image observer

S.No.	Usage	Use
1.	public Container getParent()	Gets the parent of the component.
2.	public ComponentPeer getPeer()	Gets the peer of the component.
3.	public Toolkit getToolkit()	Gets the toolkit of the component. This toolkit is used to create the peer for this component.
4.	public boolean isValid()	Checks if this Component is valid. Components are invalidated when the are first shown on the screen.
5.	public boolean isVisible()	Checks if this Component is visible. Components are initially visible (with the exception of top level components such as Frame).
6.	public boolean isShowing()	Checks if this Component is showing on screen. This means that the component must be visible, and it must be in a container that is visible and showing.
7.	public boolean isEnabled()	Checks if this Component is enabled. Components are initially enabled.
8.	public Point location()	Returns the current location of this component. The location will be in the parent's coordinate space.
9.	public Dimension size()	Returns the current size of this component.
10.	public Rectangle bounds()	Returns the current bounds of this compo-nent.
11.	public synchronized void enable()	Enables a component.
12.	public void enable(boolean cond)	Conditionally enables a component. Parameters: cond—if true, enables component; disables otherwise.
13.	public synchronized void disable()	Disables a component.
14.	public synchronized void show()	Shows the component.
15.	public void show(boolean cond)	Conditionally shows the component. Parameters: cond—if true, it shows the component; hides otherwise.
16.	public synchronized void hide()	Hides the component.
17.	public Color getForeground()	Gets the foreground color. If the component does not have a foreground color, the foreground color of its parent is returned.
18.	public synchronized void setForeground(Color c)	Sets the foreground color.

19.	public Color getBackground()	Gets the background color. If the component does not have a background color, the background color of its parent is returned.
20.	public synchronized void setBackground(Color c)	Sets the background color. Parameters: c—the Color
21.	public Font getFont()	Gets the font of the component. If the component does not have a font, the font of its parent is returned.
22.	public synchronized void setFont(Font f)	Sets the font of the component. Parameters: f—the font
23.	public synchronized ColorModel getColorModel()	Gets the ColorModel used to display the component on the output device.
24.	public void move(int x, int y)	Moves the Component to a new location. The x and y coordinates are in the parent's co-ordinate space.
25.	public void resize(int width, int height)	Resizes the Component to the specified width and height.
26.	public void resize(Dimension d)	Resizes the Component to the specified dimension.
27.	public synchronized void reshape(int x, int y, int width, int height)	Reshapes the Component to the specified bounding box.
28.	public Dimension preferredSize()	Returns the preferred size of this component.
29.	public Dimension minimumSize()	Returns the minimum size of this component.
30.	public void layout()	Lays out the component. This is usually called when the component is validated.
31.	public void validate()	Validates a component.
32.	public void invalidate()	Invalidates a component.
33.	public Graphics getGraphics()	Gets a Graphics context for this component. This method will return null if the component is currently not on the screen.
34.	public FontMetrics getFontMetrics(Font font)	Gets the font metrics for this component. This will return null if the component is currently not on the screen.
35.	public void paint(Graphics g)	Paints the component.
36.	public void update(Graphics g)	Updates the component. This method is called in response to a call to repaint. You can assume that the background is not cleared. Component G refers to the specified Graphics window.
37.	public void paintAll(Graphics g)	Paints the component and its subcomponents.

38.	public void repaint()	Repaints the component. This will result in a call to update as soon as possible.
39.	public void repaint(long tm)	Repaints the component. This will result in a call to update within *tm* milliseconds.
40.	public void repaint(int x, int y, int w, int h)	Repaints part of the component. This will result in a call to update as soon as possible.
41.	public void repaint(long tm, int x, int y, int w, int h)	Repaints part of the component. This will result in a call to update width *tm* millseconds.
42.	public void print(Graphics g)	Prints this component. The default implementation of this method calls paint.
43.	public void printAll(Graphics g)	Prints the component and its subcomponents.
44.	public boolean imageUpdate(Image img, int flags, int x, int y, int w, int h)	Repaints the component when the image has changed.
45.	public Image createImage(ImageProducer producer)	Creates an image from the specified image producer.
46.	public Image createImage(int width, int height)	Creates an off-screen drawable image to be used for double buffering.
47.	public boolean prepareImage(Image image, ImageObserver observer)	Prepares an image for rendering on this component. The image data is downloaded asynchronously in another thread and the appropriate screen representation of the image is generated.
48.	public boolean prepareImage(Image image, int width, int height, ImageObserver observer)	Prepares an image for rendering on this component at the specified width and height. The image data is downloaded asynchronously in another thread and an appropriately scaled screen representation of the image is generated.
49.	public int checkImage(Image image, ImageObserver observer)	Returns the status of the construction of a screen representation of the specified image. This method does not cause the image to begin loading, use the prepareImage method to force the loading of an image.
50.	public int checkImage(Image image, int width, int height, ImageObserver observer)	Returns the status of the construction of a scaled screen representation of the specified image. This method does not cause the image to begin loading, use the prepareImage method to force the loading of an image.
51.	public synchronized boolean inside(int x, int y)	Checks whether a specified x,y location is "inside" this component. By default, x and y are inside a component if they fall within the bounding box of that component.

52.	public Component locate(int x, int y)	Returns the component or subcomponent that contains the x,y location.
53.	public void deliverEvent(Event e)	Delivers an event to this component or one of its subcomponents.
54.	public void postEvent(Event e)	Posts an event to this component. This will result in a call to handleEvent. If handleEvent returns false, the event is passed on to the parent of this component.
55.	public boolean handleEvent(Event evt)	Handles the event. Returns true if the event is handled and should not be passed to the parent of this component. The default event handler calls some helper methods to make life easier on the programmer.
56.	public boolean mouseDown(Event evt, int x, int y)	Called if the mouse is down.
57.	public boolean mouseDrag(Event evt, int x, int y)	Called if the mouse is dragged (the mouse button is down).
58.	public boolean mouseUp(Event evt,int x, int y)	Called if the mouse is up. Parameters: evt—the event, x—the x coordinate, y—the y coordinate
59.	public boolean mouseMove(Event evt,int x, int y)	Called if the mouse moves (the mouse button is up). Parameters: evt—the event, x—the x coordinate, y—the y coordinate
60.	public boolean mouseEnter(Event evt,int x, int y)	Called when the mouse enters the component. Parameters: evt—the event, x—the x coordinate, y—the y coordinate
61.	public boolean mouseExit(Event evt, int x, int y)	Called when the mouse exits the component. Parameters: evt—the event, x—the x coordinate, y—the y coordinate
62.	public boolean keyDown(Event evt, int key)	Called if a character is pressed. Parameters: evt—the event key—the key that's pressed
63.	public boolean keyUp(Event evt, int key)	Called if a character is released. Parameters: evt—the event, key—the key that's released
64.	public boolean action(Event evt, Object what)	Called if an action occurs in the component. Parameters: evt—the event, what—the action that's occuring
65.	public void addNotify()	Notifies the component to create a peer.
66.	public synchronized void removeNotify()	Notifies the component to destroy the peer.
67.	public boolean gotFocus(Event evt, Object what)	Indicates that this component has received the input focus.

68.	public boolean lostFocus(Event evt, Object what)	Indicates that this component has lost the input focus.
69.	public void requestFocus()	Requests the input focus. The gotFocus() method will be called if this method is successful.
70.	public void nextFocus()	Moves the focus to the next component.
71.	protected String paramString()	Returns the parameter String of this component.
72.	public String toString()	Returns the String representation of this component's values.
73.	public void list()	Prints a listing to a print stream.
74.	public void list(PrintStream out)	Prints a listing to the specified print out stream.
75.	public void list(PrintStream out, int indent)	Prints out a list, starting at the specified idention, to the specified print stream. Parameters: out—the Stream name, indent—the start of the list

Class Java awt Component Container

```
java.lang.Object
   |
   +——java.awt.Component
        |
        +——java.awt.Container
```

A generic Abstract Window Toolkit(AWT) container object is a component that can contain other AWT components. It is generally used as

public class **Container** extends Component

Methods

Class Java Awt Component Container

Usage	Use
public int countComponents()	Returns the number of components in this panel.
public synchronized Component getComponent(int n)	Gets the nth component in this container.
public synchronized Component[] getComponents()	Gets all the components in this container.
public Insets insets()	Returns the insets of the container. The insets indicate the size of the border of the container. A Frame, for example, will have a top inset that corresponds with the height of the Frame's title bar.

public synchronized Component add(Component comp)	Adds the specified component to this container.
public synchronized Component add(String name, Component comp)	Adds the specified component to this container. The component is also added to the layout manager of this container using the name.
public synchronized void remove(Component comp)	Removes the specified component from this container.
public synchronized void removeAll()	Removes all the components from this container.
public LayoutManager getLayout()	Gets the layout manager for this container.
public void setLayout(LayoutManager mgr)	Sets the layout manager for this container.
public synchronized void layout()	Does a layout on this container.
public synchronized void validate()	Validates this container and all of the components contained within it.
public synchronized Dimension preferredSize()	Returns the preferred size of this container.
public synchronized Dimension minimumSize()	Returns the minimum size of this container.
public void paintComponents(Graphics g)	Paints the components in this container. Parameters: g—the specified Graphics window
public void printComponents(Graphics g)	Prints the components in this container.
public void deliverEvent(Event e)	Delivers an event. The appropriate component is located and the event is delivered to it.
public Component locate(int x, int y)	Locates the component that contains the x,y position. RETURNS null if the component is not within the x and y coordinates; returns the component otherwise.
public synchronized void addNotify()	Notifies the container to create a peer. It will also notify the components contained in this container.
public synchronized void removeNotify()	Notifies the container to remove its peer. It will also notify the components contained in this container.
protected String paramString()	Returns the parameter String of this container.
public void list(PrintStream out, int indent)	Prints out a list, starting at the specified indention, to the specified out stream. Parameters: out—the Stream name, indent—the start of the list

Class java.awt. Component Container Window

```
java.lang.Object
   |
   +——java.awt.Component
```

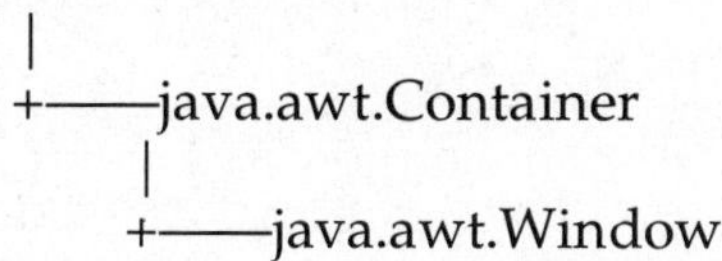

A Window is a top-level window with no borders and no menubar. It could be used to implement a pop-up menu. The default layout for a window is BorderLayout. It is generally used as:

public class Window extends Container

Constructor

Window (Frame): Constructs a new Window initialized to an invisible state.
USAGE: public Window(Frame parent)
USE: Constructs a new Window initialized to an invisible state. It behaves as a modal dialog in that it will block input to other windows when shown.

Methods

Usage	Use
public synchronized void addNotify()	Creates the Window's peer. The peer allows us to modify the appearance of the Window without changing its functionality.
public synchronized void pack()	Packs the components of the Window.
public synchronized void show()	Shows the Window. This will bring the window to the front if the window is already visible.
public synchronized void dispose()	Disposes of the Window. This method must be called to release the resources that are used for the window.
public void toFront()	Brings the frame to the front of the Window.
public void toBack()	Sends the frame to the back of the Window.
public Toolkit getToolkit()	Returns the toolkit of this frame.
public final String getWarningString()	Gets the warning string for this window. This is a string that will be displayed somewhere in the visible are of windows that are not secure.

Class java.awt. Component Container Window Frame

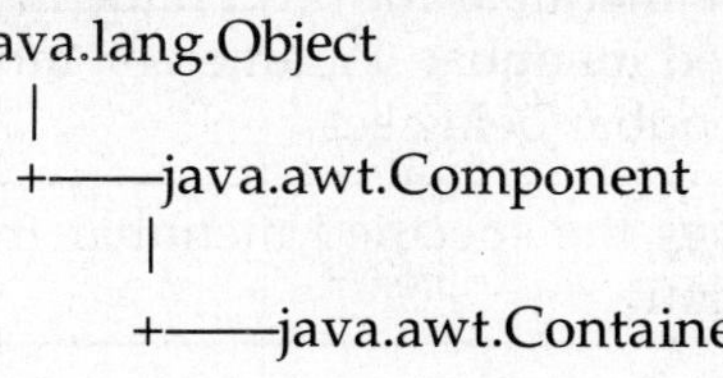

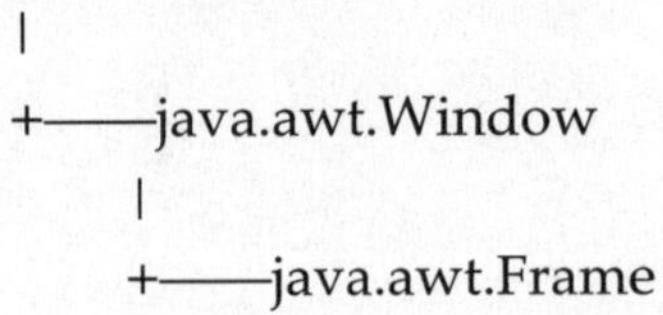

A Frame is a top-level window with a title. The default layout for a frame is Border Layout. It is generally represented as:

public class **Frame** extends Window implements Menu Container

Constructors

1. Frame

USAGE: public Frame()

USE: Constructs a new Frame that is initially invisible.

2. Frame

USAGE: public Frame(String title)

USE: Constructs a new, initially invisible Frame with the specified title.

Parameters: title—the specified title

Methods

Usage	Use
public synchronized void addNotify()	Creates the Frame's peer. The peer allows us to change the look of the Frame without changing its functionality.
public String getTitle()	Gets the title of the Frame.
public void setTitle(String title)	Sets the title for this Frame to the specified title. Parameter: title—the specified title of this Frame
public Image getIconImage()	Returns the icon image for this Frame.
public void setIconImage(Image image)	Sets the image to display when this Frame is iconized. Note that not all platforms support the concept of iconizing a window. Parameter: image— the icon image to be displayed
public MenuBar getMenuBar()	Gets the menubar for this Frame.
public synchronized void setMenuBar(MenuBar mb)	Sets the menubar for this Frame to the specified menubar. Parameters: mb—the menubar being set
public synchronized void remove(MenuComponent m)	Removes the specified menubar from this Frame.

public synchronized void dispose()	Disposes of the Frame. This method must be called to release the resources that are used for the frame.
public boolean is Resizable()	Returns true if the user can resize the Frame.
public void setResizable(boolean resizable)	Sets the resizable flag. Paramter: resizable—true if resizable; false otherwise.
public void setCursor(Image img)	Set the cursor image.
protected String paramString()	Returns the parameter String of this Frame.

Steps in Creating a Frame Window

Method 1: In main () method

- Create an instance of a Frame class.
  Frame f=new Frame(frame_name);
- Set the frame size
  f.setSize(500,500);
- Make the frame visible
  f.setVisible(true);

The following java program creates a frame with the dimension as 600 × 400 and makes it visible in the screen. Import java.awt package because Frame class is available in that package.

```
import java.awt.*;
public class Demo{
public static void main(String arg[]){
Frame f=new Frame("Demo");
f.setSize(600,400);
f.setVisible(true);
}
}
```

Method 2:

- Create a subclass of **Frame.**
- In the subclass constructor change the frame title by calling superclass [Frame] constructor using ***super (String)*** method call.

- Set the size of the window explicitly by calling the ***set Size ( )*** method.
- Make the frame visible by calling ***set Visible ()*** method.
- In main () method Create an instance of subclass.

```
import java.awt.*;
public class Demo extends Frame
{
Demo(String s)
{
super(s);
setSize(600,400);
setVisible(true);
        }
}
public static void main(String arg[])
{
        Demo ob=new Demo("Demo");
}
```

3. WORKING WITH 2D SHAPES

The Graphics class is part of the ***java awt*** package. The **Graphics** class defines a number of drawing functions. Each shape can be drawn edge-only or filled. Objects are drawn and filled in the currently selected graphics color, which is black by default. When a graphics object is drawn that exceeds the dimensions of the window, output is automatically clipped.

Lines

To draw straight lines, use the drawLine method. drawLine takes four arguments: The x and y coordinates of the starting point and the x and y coordinates of the ending point.

void **drawLine**(int startX, int startY, int endX, int endY)

displays a line in the current drawing color that begins at startX, startY and ends at endX, endY.

lineDemo.java

```
import java.awt.*;
public class LineDemo extends Frame{ Demo(String s){
    super(s);
    setSize(100,100);
    setVisible(true);
```

```
}
public void paint(Graphics g) {
    g.drawLine(10,10,60,60);
}
public static void main(String arg[]){
LineDemo ob=new LineDemo("Line Demo");
}}
```

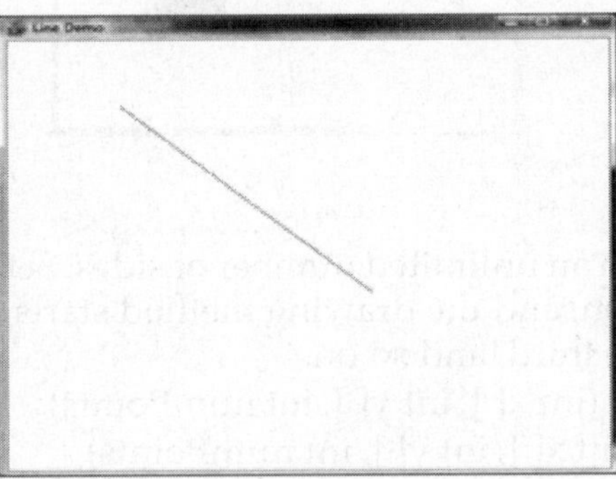

Rectangles

The Java graphics primitives provide two kinds of rectangles: Plain rectangles and Rounded rectangles (which are rectangles with rounded corners).

- void **draw Rect** (int x, int y, int width, int height)
- void **fill Rect** (int x, int y, int width, int height)
- void **draw Round Rect** (int x, int y, int width, int height, int xDiam, int yDiam)
- void **fill Round Rect** (int x, int y, int width, int height, int xDiam, int yDiam)

A rounded rectangle has rounded corners. The upper-left corner of the rectangle is at *x, y*. The dimensions of the rectangle are specified by *width* and *height*. The diameter of the rounding arc along the X axis is specified by *xDiam*. The diameter of the rounding arc along the Y axis is specified by *yDiam*.

rectDemo.java

```
import java.awt.*;
public class rectDemo extends Frame{ Demo(String s){
        super(s);
        setSize(500,500);
        setVisible(true);
}}
public void paint(Graphics g)
{ g.drawRect(100,100,60,60);
g.fillRect(250,100,60,60);
g.drawRoundRect(100,250,60,60,10,10);
g.fillRoundRect(250,250,60,60,20,20);
}
public static void main(String arg[]){
rectDemo ob=new rectDemo("Rectangle Demo");
}
```

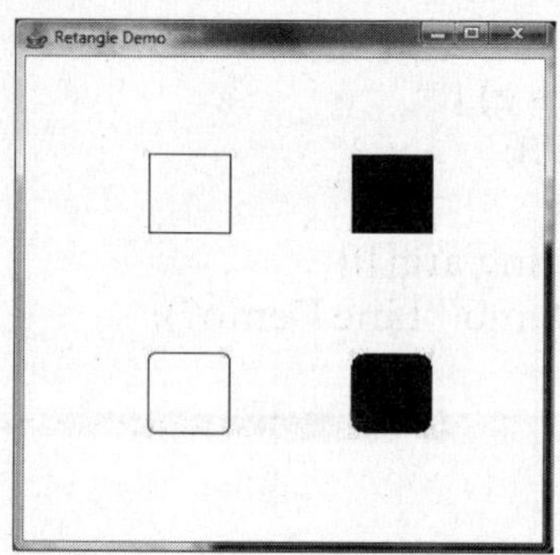

Polygons

Polygons are shapes with an unlimited number of sides. Set of x and y coordinates are needed to draw a polygon, and the drawing method starts at one, draws a line to the second, then a line to the third, and so on.

- void **draw Polygon** (int x[ ], int y[ ], int numPoints)
- void **fill Polygon** (int x[ ], int y[ ], int numPoints)

x[]- An array of integers representing x coordinates
y[]- An array of integers representing y coordinates
numPoints- An integer for the total number of points

polyDemo.java

```
import java.awt.*;
public class polyDemo extends Frame{
public void paint(Graphics g) {
      int x1[] = { 39,94,97,112,53,58,26 };
      int y1[] = { 133,174,136,170,208,180,206 };
      g.drawPolygon(x1,y1,7);
      int x2[] = { 139,194,197,212,153,126 };
      int y2[] = { 133,174,136,170,208,180,206 };
      g.fillPolygon(x2,y2,7);
}
}
public static void main(String arg[]){
      polyDemo ob=new polyDemo("Polygon Demo");
}
```

Ovals

Use ovals to draw ellipses or circles.

- void **drawOval**(int top, int left, int width, int height)
- void **fillOval**(int top, int left, int width, int height)

ovalDemo.java

```
import java.awt.*;
public class ovalDemo extends Frame{
}Demo(String s){
    super(s);
    setSize(300, 300);
    setVisible(true);
}
public void paint(Graphics g) { g.drawOval(20,120,70,70);
g.fillOval(140,120,100,70);
}public static void main(String arg[]){
        ovalDemo ob=new ovalDemo("Oval Demo");
}
}
```

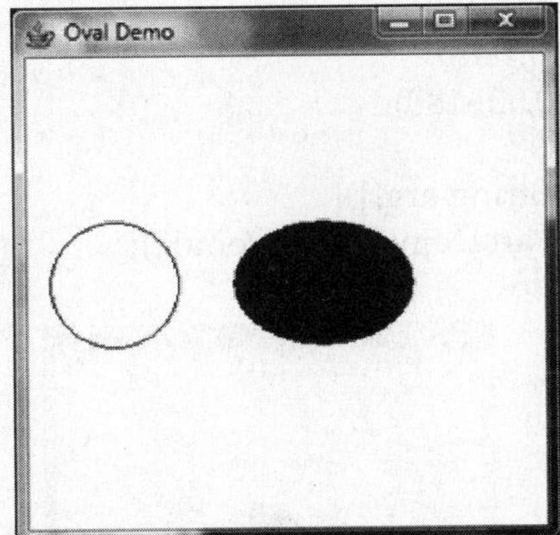

Arc

The arc is drawn from *start angle* through the angular distance specified by *ark angle.* Angles are specified in degrees. The arc is drawn counterclockwise if *sweep angle* is positive, and clockwise if *ark angle* is negative.

- void **draw arc** (int *x*, int *y*, int *width*, int *height*, int *start angle*, int *arc angle*)
- void **fill arc** (int *x*, int *y*, int *width*, int *height*, int *start angle*, int *arc angle*)

The arc is drawn from *start angle* through the angular distance specified by *ark angle.* Angles are specified in degrees. The arc is drawn counterclockwise if *sweep angle* is positive, and clockwise if *ark angle* is negative.

- void **draw arc** (int *x*, int *y*, int *width*, int *height*, int *start angle*, int *arc angle*)
- void **fill arc** (int *x*, int *y*, int *width*, int *height*, int *start angle*, int *arc angle*)

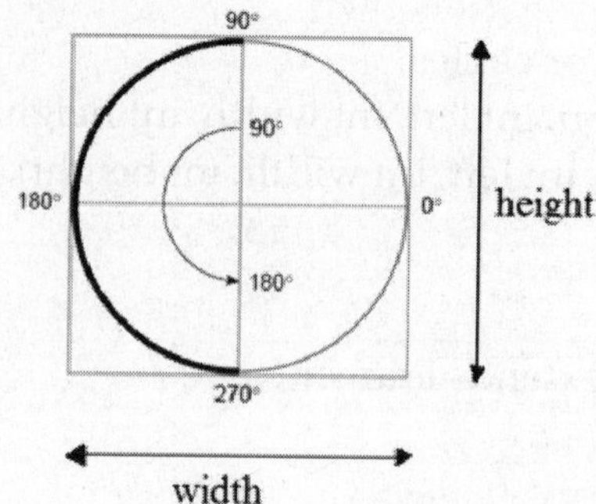

x=0, y=0, width=100, height=100, Startangle=90, arc angle=180

arcDemo.java

```
import java.awt.*;
public class arcDemo extends Frame{ Demo(String s){
        super(s);
        setSize(300,300);
        setVisible(true);
}
}
public void paint(Graphics g) {
g.drawArc(20,120,90,90,90,180);
g.fillArc(120,120,90,90,90,180);
g.drawArc(170,120,90,90,90,-180);
}
public static void main(String arg[]){
        arcDemo ob=new arcDemo("Arc Demo");
}
```

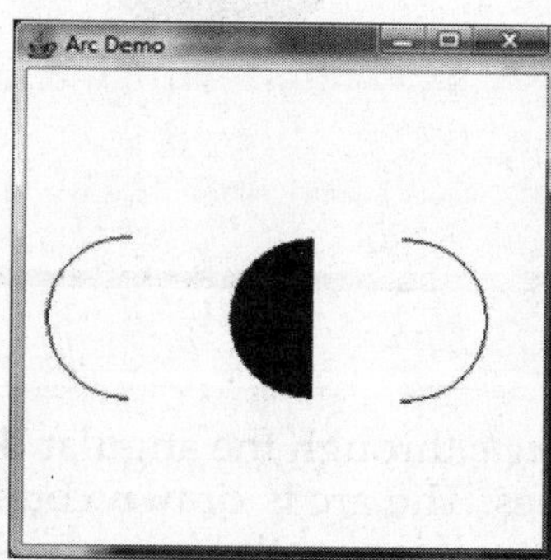

Drawing Text

Draw text on the screen using the method drawstring().

- void **drawstring**(String text, int x, int y)

draws a string in the current font and color.

```
import java.awt.*;
public class Demo extends Frame{ Demo(String s){
        super(s);
```

```
        setSize(200,200);
        setVisible(true);
}}
public void paint(Graphics g) {
g.drawString("welcome", 75, 100);
}
public static void main(String arg[]){
        Demo ob=new Demo("Text Demo");
}
```

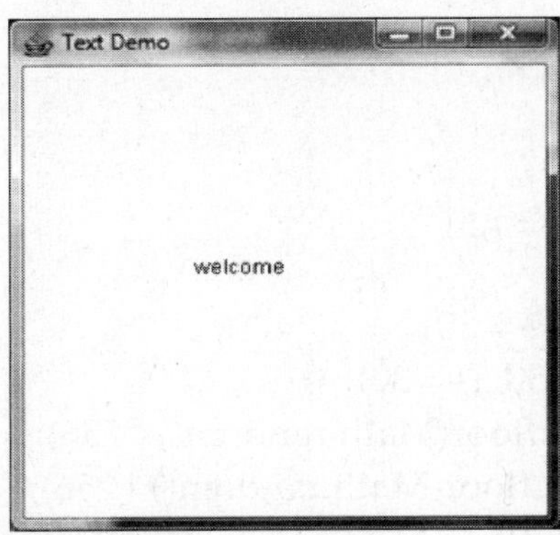

4. USING COLOR, FONTS, AND IMAGES

Java supports color in a portable, device independent fashion. The AWT color system allows the user to specify any color needed.

Color class creates a solid RGB color with the specified red, green, blue value in the range of 0–255. Java's abstract color model uses 24-bit color. The values of each color must be in the range of 0–255.

Constructor

Color*(int red, int green, int blue)*—create new color object for any combination of red, green, and blue.

Java.awt.Color class used to create new color object for any combination of red, green, and blue, and it predefines a few color constants. They are stored in class variables,

Color white (255, 255, 255)
Color black (0, 0, 0)
Color light gray (192, 192, 192)
Color gray (128, 128, 128)
Color dark gray (64, 64, 64)
Color red (255, 0, 0)
Color green (0, 255, 0)
Color blue (0, 0, 255)
Color yellow (255, 255, 0)
Color magenta (255, 0, 255)
Color cyan(0, 255, 255)
Color pink (255, 175, 175)
Color orange (255, 200, 0)

Component class defines set background () and set foreground () methods for setting background and foreground colors to the window. Since Frame is a subclass of Component class, it can use these methods.

- void **set background** (java awt color)—set background color to the window.
- void **set foreground** (java awt color)—set foreground color to the window.

The following program for designing a frame with square boxes, each of which has a randomly chosen color in it.

```
import java.awt.*;
public class colorDemo extends Frame{ Demo(String s){
super(s);
setSize(200,200);
setVisible(true);
setBackground(Color.black);
}
}
public void paint(Graphics g) {
int rval, gval, bval;
for (int j = 30; j < 200-30; j += 30)
     for (int i = 30; i < 200-30; i+= 30){
          rval = (int)Math.floor(Math.random() * 256);
          gval = (int)Math.floor(Math.random() * 256);
          bval = (int)Math.floor(Math.random() * 256);
          g.setColor(new Color(rval,gval,bval));
          g.fillRect(i,j,25,25);
}
}
public static void main(String arg[]){
colorDemo ob=new colorDemo("Color Demo");
}
```

Consolidated list describing the Constructors and methods of java.awt.Color are given below.

Constructors Color

1. public color (int r, int g, int b)

Creates a color with the specified red, green, and blue values in the range of 0 – 255. The actual color used in rendering will depend on finding the best match given the color space available for a given output device.

Parameters

r - the red component

g - the green component

b - the blue component

2. public color (int rgb)

Creates a color with the specified combined RGB value consisting of the red component in bits 16–23, the green component in bits 8–15, and the blue component in bits 0–7. The actual color used in rendering will depend on finding the best match given the color space available for a given output device.

Parameters

rgb - the combined RGB components

3. public color (float r, float g, float b)

Creates a color with the specified red, green, and blue values in the range of 0.0–1.0. The actual color used in rendering will depend on finding the best match given the color space available for a given output device.

Usage	Use
public int getRed()	Gets the red component.
public int getGreen()	Gets the green component.
public int getBlue()	Gets the blue component.
public int getRGB()	Gets the RGB value representing the color in the default RGB ColorModel.
public Color brighter()	Returns a brighter version of this color.
public Color darker()	Returns a darker version of this color.
public int hashCode()	Computes the hash code.
public boolean equals(Object obj)	Compares this object against the specified object.
public String toString()	Returns the String representation of this Color's values.
public static Color getColor(String nm)	Gets the specified Color property in argument nm.
public static Color getColor(String nm, Color v)	Gets the specified Color property of the specified Color and returns the new color.
public static Color getColor(String nm, int v)	Gets the specified Color property of the color value and returns the new color.
public static int HSBtoRGB(float hue, float saturation, float brightness)	Returns the RGB value defined by the default RGB ColorModel, of the color corresponding to the given HSB color components.

public static float[] RGBtoHSB(int r, int g, int b, float hsbvals[])	Returns the HSB values as array denoted as hsbvals[ ] corresponding to the color defined by the red, green, and blue components.
public static Color getHSBColor(float h, float s, float b)	A static Color factory for generating a Color object from HSB values. Returns the Color object for the corresponding RGB color.

Font

Usage	Use
public String getFamily()	Gets the platform specific family name of the font. Use getName to get the logical name of the font.
public String getName()	Gets the logical name of the font.
public int getStyle()	Gets the style of the font.
public int getSize()	Gets the point size of the font.
public boolean isPlain()	Returns true if the font is plain.
public boolean isBold()	Returns true if the font is bold.
public boolean isItalic()	Returns true if the font is italic.
public static Font getFont(String nm)	Gets a font from the system properties list. Argument nm represents name of the property.
public static Font getFont(String nm,Font font)	Gets the specified font from the system properties list.parameter font represents a default font to return if property 'nm' is not defined
public int hashCode()	Returns a hashcode for this font.
public boolean equals(Object obj)	Compares this object to the specified object.
public String toString()	Converts this object to a String representation.

Fontmetrics

```
java.lang.Object
   |
   +——java.awt.FontMetrics
```

A font metrics object. Note that the implementations of these methods are inefficient, they are usually overridden with more efficient toolkit specific implementations.

It is generally used as:
public class **Font metrics** extends Object

Constructor

Protected Font Metrics (Font font)
Creates a new FontMetrics object with the specified font.

Methods	Use
public Font getFont()	Gets the font.
public int getLeading()	Gets the standard leading, or line spacing, for the font. This is the logical amount of space to be reserved between the descent of one line of text and the ascent of the next line. The height metric is calculated to include this extra space.
public int getAscent()	Gets the font ascent. The font ascent is the distance from the base line to the top of the characters.
public int getDescent()	Gets the font descent. The font descent is the distance from the base line to the bottom of the characters.
public int getHeight()	Gets the total height of the font. This is the distance between the baseline of adjacent lines of text. It is the sum of the leading + ascent + descent.
public int getMaxAscent()	Gets the maximum ascent of all characters in this Font. No character will extend further above the baseline than this metric.
public int getMaxDecent()	Gets the maximum descent of all characters. No character will descend futher below the baseline than this metric.
public int getMaxAdvance()	Gets the maximum advance width of any character in this Font. Returns -1 if the max advance is not known.
public int charWidth(int ch)	Returns the width of the specified character in this Font.
public int charWidth(char ch)	Returns the width of the specified character in this Font.
public int stringWidth(String str)	Returns the width of the specified String in this Font.
public int charsWidth(char data[], off, int len)	Returns the width of the specified character int array in this Font.

public int bytesWidth(byte data[], int off, int len)	Returns the width of the specified array of bytes in this Font. The parameters data—the data to be checked, off—the start offset of the data, len—the maximum number of bytes checked
public int[] getWidths()	Gets the widths of the first 256 characters in the Font.
public String toString()	Returns the String representation of this FontMetric's values.

Images

```
java.lang.Object
  |
  +——java.awt.Image
```

The image class is an abstract class. The image must be obtained in a platform specific way. It is generally used as *public class **Image** extends Object.*

Constructor

Public Image () – Default constructor

Methods	Use
public abstract int getWidth(ImageObserver observer)	Gets the actual width of the image. If the width is not known yet, then the ImageObserver will be notified later and -1 will be returned.
public abstract int getHeight(ImageObserver observer)	Gets the actual height of the image. If the height is not known yet, then the ImageObserver will be notified later and -1 will be returned.
public abstract ImageProducer getSource()	Gets the object that produces the pixels for the image. This is used by the Image filtering classes and by the image conversion and scaling code.
public abstract Graphics getGraphics()	Gets a graphics object to draw into this image. This will only work for off-screen images.
public abstract Object getProperty(String name, ImageObserver observer)	Gets a property of the image by name. Individual property names are defined by the various image formats. If a property is not defined for a particular image, then this method will return the UndefinedProperty object. If the properties for this image are not yet known, then this method

will return null and the ImageObserver object will be notified later. The property name "comment" should be used to store an optional comment which can be presented to the user as a description of the image, its source, or its author.

5. BASICS OF EVENT HANDLING

A window based program is called event driven program. The signals that a program receives from the OS as a result of your actions are called events. Unlike the old rigid sequential programs, it puts user in charge, user control the sequence of program. Application waits for the user action. This approach is called event driven programming.

Event Delegation Model

The Delegation Event Model used by Java to handle user interaction with GUI components. It describes how your program can respond to user interaction. It has three important players.

- Event Source
- Event Listener/Handler
- Event Object

Event Source

- GUI component that generates the event.
- Example: Button

Event Listener/Handler

- It receives and handles events. It contains business logic.
- Example: Displaying information useful to the user, computing a value.

Event Object

- Created when an event occurs (i.e. user interacts with a GUI component). It contains all necessary information. It is represented by an Event class.

Event Basics

All Events are objects of *Event Classes*. All Event Classes derive from *EventObject*. When an Event occurs, Java sends a message to all registered Event Listeners from the Event source. The *EventObject class* is found in the *java.util package.* The *AWTEvent class* is an immediate subclass of *EventObject*, defined in*java.awt package*—root of all AWT-based events. Three steps for event Handler:

1. Either **implements** a listener interface or extends **a** class that implements a listener interface
2. Register your listener
   some components add action listener (instance of my class);
3. Implement user action
   public void actionPerformed(ActionEvent e)
   { ...//code that reacts to the action... }

```
public class TryWindow extends JFrame implements ActionListener{
static JTextField tf=new JTextField("    no    ");
static int count=0;
```

```
        public void actionPerformed(ActionEvent e){
        count++;
        tf.setText("hiiii "+count);  }
    Public static void main(String arg[]){
        JButton b = new JButton("one");
        TryWindow frame=new TryWindow();
        b.addActionListener(frame);
            panel1.setLayout(new FlowLayout());
            panel1.add(b);
            panel1.add(tf);
            //other code for frame
    }
```

Event Class	Listener Interface	Listener Methods
ItemEvent	ItemListener	itemStateChanged()
KeyEvent	KeyListener	keyPressed(), keyReleased(), keyTyped()
MouseEvent	MouseListener, Mouse MotionListener	mouseClicked(),mouseEntered(), mouseExited(),mousePressed(), mouseReleased(),mouseDragged(), mouseMoved()
TextEvent	TextListener	TextValueChanged()
WindowEvent	WindowListener	windowActivated(),windowClosed(), windowClosing(),windowDeactivated(), windowDeiconified(),window iconified(), windowOpened()

Mouse Listener Example

```
import java.awt.*;
import java.awt.event.*;
public class MouseDemo extends Frame
        implements MouseListener, MouseMotionListener
{
    TextField tf;
    public MouseDemo(String title){
    super(title);
    tf = new TextField(60);
    addMouseListener(this);    // Register event listener to the event source
}
public void launchFrame() {
  add(tf, BorderLayout.SOUTH); // Add components to the frame
  setSize(300,300);
  setVisible(true);
}
```

```
public void mouseClicked(MouseEvent me) {
String msg = "Mouse clicked.";
tf.setText(msg);
}
public void mouseEntered(MouseEvent me) {
String msg = "Mouse entered component.";
tf.setText(msg);
}
public void mouseExited(MouseEvent me) {
String msg = "Mouse exited component.";
tf.setText(msg);
}
public void mousePressed(MouseEvent me) {
String msg = "Mouse pressed.";
tf.setText(msg);
}
public void mouseReleased(MouseEvent me) {
   String msg = "Mouse released.";
   tf.setText(msg);
} public void mouseDragged(MouseEvent me) {
   String msg = "Mouse dragged at " + me.getX()    + "," + me.getY();
   tf.setText(msg);}
public void mouseMoved(MouseEvent me) {
   String msg = "Mouse moved at " + me.getX() + "," + me.getY();
   tf.setText(msg);}
   public static void main(String args[]) {
   MouseDemo med =new MouseDemo("Mouse Events Demo");
   med.launchFrame();
}}
```

Window Listener Example

```
import java.awt.*;
import java.awt.event.*;
class CloseFrame extends Frame implements WindowListener {

CloseFrame(String title) {
super(title);
this.addWindowListener(this);
}
void launchFrame() {
   setSize(300,300);
   setVisible(true);
  }
  //Implement methods of listener interface
  public void windowActivated(WindowEvent e) {
  System.out.println("Window is activated");
```

```
}
public void windowClosed(WindowEvent e) {}
public void windowClosing(WindowEvent e) {
  setVisible(false);
  System.exit(0);
}
public void windowDeactivated(WindowEvent e) {}
public void windowDeiconified(WindowEvent e) {
System.out.println("Window is Deiconified");
}
public void windowIconified(WindowEvent e) {}
public void windowOpened(WindowEvent e) {}
public static void main(String args[ ]) {
CloseFrame cf = new CloseFrame("Close Window Example");
cf.launchFrame();
}}
```

Adapter Classes (Alternate for EventListener)

Often we only really want to implement one or two of the event handlers. But interface implementation requires us to implement all of them. Java adapter classes allow us to solve this problem. Event-handling interfaces are implemented in a trivial way, just to satisfy the interface. We extend the class and override the methods that we need. Some adapter classes are:

- KeyAdapter (instead of KeyListener)
- WindowAdapter
- MouseAdapter (MouseListener)
- MouseMotionAdapter (MouseMotionListener), etc.

```
import java.awt.*;
import java.awt.event.*;
class CloseFrame extends Frame{
Label label;
CFListener w = new CFListener(this);
CloseFrame(String title) {
super(title);
label = new Label("Close the frame.");
this.addWindowListener(w); }
void launchFrame() {
setSize(300,300);
  setVisible(true); }
  public static void main(String args[]) {
  CloseFrame cf = new CloseFrame("Close Window Example");
  cf.launchFrame();
  }
  class CFListener extends WindowAdapter {
```

```
    CloseFrame ref;
    CFListener( CloseFrame ref ){
    this.ref = ref;
    }

    public void windowClosing(WindowEvent e) {
    ref.dispose();
    System.exit(1);
    }
}
```

6. ADAPTER CLASSES

An adapter class provides an empty implementation of all methods in an event listener interface. It is useful when you want to receive and process only some of the events that are handled by an event listener interface. Define a new class that extends the adapter class and overrides the desired methods only. Extends adapter class to create Listener and override the desired methods. If you implement the interface, you have to define all of the methods in it. This adapter class defines null methods for all events, so you can only redefine the methods that you needed. Some adapter classes are

- Key adapter
- Mouse adapter
- Mouse motion adapter

Example: If the user is only interested in mouse pressed event, then simply extends MouseAdapter class and redefine the mousePressed event. The following program listening MouseEvent whenever the user press the mouse, it will print "mouse pressed" text on command prompt.

```
import java.awt.*;
import java.awt.event.*;
public class Demo extends MouseAdapter
{
    public static void main(String arg[])
    {
        Frame f=new Frame("Mouse Adapter Demo");
        f.setSize(200,200);
        f.setVisible(true);
        Demo ob=new Demo();
        f.addMouseListener(ob);
    }
    public void mousePressed(MouseEvent me)
    {
        System.out.println("Mouse pressed..");
    }
}
```

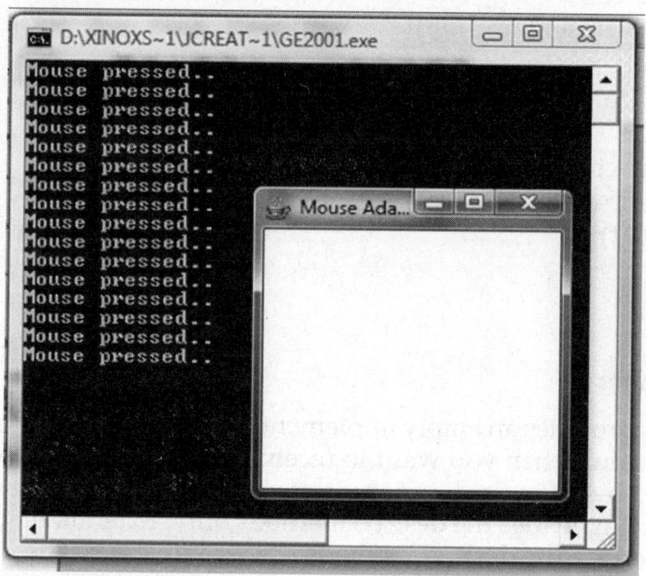

7. ACTIONS AND MOUSE EVENTS

Action Event

An object of this class represents a high-level action event generated by an AWT component. Instead of representing a direct user event, such as a mouse or keyboard event, Action event represents some sort of action performed by the user on an AWT component.

The `get ID ()` method returns the type of action that has occurred. For AWT-generated action events, this type is always `action event ACTION_PERFORMED`; custom components can generate action events of other types.

The `get action command ()` method returns a `string` that serves as a kind of name for the action that the event represents. The `Button` and `Menu Item` components have a`set action command ()` method that allows the programmer to specify an action command string to be included with any action event generated by those components. It is this value that is returned by the `get action command ()` method. When more than one `Button` or other component notifies the same `Action Listener`, you can useget `Action Command ()` to help determine the appropriate response to the event. This is generally a better technique than using the source object returned by `Get Source ()`. If no action command string is explicitly set, `get Action Command ()` returns the label of the `Button` or `Menu Item`. Internationalized programs should not rely on these labels being constant.

`Get Modifiers ()` returns a value that indicates the keyboard modifiers that were in effect when the action event was triggered. Use the various `_MASK` constants, along with the & operator, to decode this value.

public class **ActionEvent** extends AWTEvent {
// Public Constructors
public **ActionEvent** (Object *source*, int *id*, String *command*);
public **ActionEvent** (Object *source*, int *id*, String *command*, int *modifiers*);
// Public Constants

```
    public static final int ACTION_FIRST ;=1001
    public static final int ACTION_LAST ; =1001
    public static final int ACTION_PERFORMED ;    =1001
    public static final int ALT_MASK ; =8
    public static final int CTRL_MASK ;    =2
    public static final int META_MASK ;    =4
    public static final int SHIFT_MASK ;    =1
  // Public Instance Methods
    public String getActionCommand ();
    public int getModifiers ();
  // Public Methods Overriding AWTEvent
    public String paramString ();
  }
```

Hierarchy

`Object java.util.EventObject(Serializable)->AWTEvent->ActionEvent`

Passed To: Too many methods to list.

ActionListener

This interface defines the method that an object must implement to listen for action events on AWT components. When an Action Event occurs, an AWT component notifies its registered Action Listener objects by invoking their action Performed () methods.

```
public abstract interface ActionListener extends java.util.EventListener {
// Public Instance Methods
  public abstract void actionPerformed (ActionEvent e);
}
```

Hierarchy: `(Action Listener (java.util.Event Listener))`
Implementations: `AWT event multicaster, java.awt.dnd.Drop Target.DropTargetAutoScroller, javax.swing.Action, javax.swing.Default Cell Editor. Editor Delegate, javax.swing.JCombo Box, javax.swing.Tool Tip Manager.inside Timer Action, javax.swing.Tool TipManager.outside Timer Action, javax.swing.Tool Tip Manager.still Inside Timer Action, javax.swing.text. html.FormView, javax.swing.tree. Default Tree Cell Editor`
Passed To: Too many methods to list.
Returned By: AWTEvent Multicaster.{add(), remove()}, `javax.swing.AbstractButton.createActionListener(), javax.swing.JComponent.getActionForKeyStroke()`
Type of: `javax.swing.AbstractButton.actionListener`

Adjustment Event

An event of this type indicates that an adjustment has been made to an `Adjustable` object—usually, this means that the user has interacted with a `Scrollbar` component.

The getValue() method returns the new value of the `Adjustable` object. This is usually the most important piece of information stored in the event. get `Adjustable ()` returns the `Adjustable` object that was the source of the event. It is a convenient alternative to the inherited getSource().

The getID() method returns the type of an `Adjustment Event`. The standard AWT components only generate adjustment events of the `type Adjustment Event.ADJUSTMENT_VALUE_CHANGED`. There are several types of adjustments that can be made to an `Adjustable` object, however, and the `get AdjustmentType()` method returns one of five constants to indicate which type has occurred. `UNIT_INCREMENT` indicates that the `Adjustable` value has been incremented by one unit, as in a scroll-line-down operation. `UNIT_DECREMENT` indicates the opposite: scroll-line-up. `BLOCK_INCREMENT` and `BLOCK_DECREMENT` indicate that the `Adjustable` object has been incremented or decremented by multiple units, as in a scroll-page-down or scroll-page-up operation. Finally, the `TRACK` constant indicates that the `Adjustable` value has been set to an absolute value unrelated to its previous value, as when the user drags a scrollbar to a new position.

```
public class AdjustmentEvent extends AWTEvent {
// Public Constructors
  public AdjustmentEvent (Adjustable source, int id, int type, int value);
// Public Constants
  public static final int ADJUSTMENT_FIRST ; =601
  public static final int ADJUSTMENT_LAST ; =601
  public static final int ADJUSTMENT_VALUE_CHANGED ; =601
  public static final int BLOCK_DECREMENT ; =3
  public static final int BLOCK_INCREMENT ; =4
  public static final int TRACK ; =5
  public static final int UNIT_DECREMENT ; =2
  public static final int UNIT_INCREMENT ; =1
// Public Instance Methods
  public Adjustable getAdjustable ();
  public int getAdjustmentType ();
  public int getValue ();
// Public Methods Overriding AWTEvent
  public String paramString ();
}
```

Hierarchy: `Objectjava->util.EventObject(Serializable)->AWTEvent->AdjustmentEvent`

Passed To: `AWTEventMulticaster.adjustmentValueChanged(), Scrollbar.processAdjustmentEvent(), AdjustmentListener.adjustmentValueChanged()`

Adjustment Listener

This interface defines the method that an object must implement to listen for adjustment events on AWT components. When an `Adjustment Event` occurs, an AWT component notifies its registered `Adjustment Listener` objects by invoking their `adjustment Value Changed ()` methods.

```
public abstract interface AdjustmentListener extends java.util.EventListener {
//   Public Instance Methods
     public abstract void adjustmentValueChanged (AdjustmentEvent e);

}
```

Hierarchy: `(Adjustment Listener(java.util.Event Listener))`
Implementations: `AWT Event Multicaster`
Passed To:
`Adjustable.{addAdjustmentListener(),removeAdjustmentListener()}, AWTEventMulticaster.{add(),remove()},Scrollbar.{add AdjustmentListener(), removeAdjustmentListener()},javax.swing.JScrollBar.{addAdjustment Listener(), removeAdjustmentListener()}`
Returned By: AWTEventMulticaster.{add(), remove()}

AWT Event Listener

This interface is implemented by objects, such as GUI macro recorders, that want to be notified of all AWT events that occur on `Component` and `Menu Component` objects throughout the system. Register an `AWT Event Listener` with the `add AWT Event Listener()` method of `Toolkit.`

```
public abstract interface AWTEventListener extends java.util.EventListener {
// Public Instance Methods
  public abstract void eventDispatched (AWTEvent event);
}
```

Hierarchy: `(AWTEventListener(java.util.EventListener))`
Passed to: `Toolkit.{addAWTEventListener(), removeAWTEventListener()}`

Component Adapter

This class is a trivial implementation of `Component Listener;` it provides empty bodies for each of the methods of that interface. When you are not interested in all of these methods, it is often easier to subclass `Component Adapter` than it is to implement `Component Listener` from scratch.

```
public abstract class ComponentAdapter implements ComponentListener {
// Public Constructors
  public ComponentAdapter ();
// Methods Implementing ComponentListener
  public void componentHidden (ComponentEvent e);          empty
  public void componentMoved (ComponentEvent e);           empty
  public void componentResized (ComponentEvent e);         empty
  public void componentShown (ComponentEvent e);           empty
}
```

Hierarchy: `Object->Component Adapter(Component Listener(java. util. Event Listener))`
Subclasses: `javax.swing.JViewport.View Listener`

Component Event

An event of this type serves as notification that the source `Component` has been moved, resized, shown, or hidden. Note that this event is a notification only: the AWT handles these `Component` operations internally, and the recipient of the event needs to take no action itself. `get Component()` returns the component that was moved, resized, shown, or hidden. It is simply a convenient alternative to `get Source()`. `get ID()` returns one of four `COMPONENT_` constants to indicate what operation was performed on the `Component`.

```
public class Component Event extends AWTEvent {
//  Public Constructors
    public Component Event (Component source, int id);
//  Public Constants
    public static final int COMPONENT_FIRST ;          =100
    public static final int COMPONENT_HIDDEN ;         =103
    public static final int COMPONENT_LAST ;           =103
    public static final int COMPONENT_MOVED ;          =100
    public static final int COMPONENT_RESIZED ;        =101
    public static final int COMPONENT_SHOWN ;          =102
//  Public Instance Methods
    public Component get Component ();
//  Public Methods Overriding AWTEvent
    public String param String ();
}
```

Hierarchy: `Object—>java.util.EventObject(Serializable)—>AWTEvent—>ComponentEvent`

Subclasses: `ContainerEvent, FocusEvent, InputEvent, PaintEvent, WindowEvent`

Passed To: `AWTEventMulticaster.{componentHidden(), componentMoved(), componentResized(), componentShown()}, Component. processComponentEvent(), ComponentAdapter. {componentHidden(), componentMoved(), componentResized(), componentShown()}, ComponentListener. {componentHidden(), componentMoved(), componentResized(), componentShown()}, javax. swing. JViewport. ViewListener. componentResized()`

Component Listener

This interface defines the methods that an object must implement to listen for component events on AWT components. When a Component Event occurs, an AWT component notifies its registered Component Listener objects by invoking one of their methods. An easy way to implement this interface is by subclassing the Component Adapter class.

```
public abstract interface ComponentListener extends java.util.EventListener {
// Public Instance Methods
  public abstract void componentHidden (ComponentEvent e);
  public abstract void componentMoved (ComponentEvent e);
  public abstract void componentResized (ComponentEvent e);
  public abstract void componentShown (ComponentEvent e);
}
```

Hierarchy: (ComponentListener(java.util.EventListener))
Implementations: AWT Event Multicaster, Component Adapter
Passed To: AWT Event Multicaster.{add(), remove()}, Component.{add Component Listener(), remove Component Listener()}
Returned By: AWT Event Multicaster.{add(), remove()}

Container Adapter

This class is a trivial implementation of Container Listener; it provides empty bodies for each of the methods of that interface. When you are not interested in all of these methods, it is often easier to subclass Container Adapter than it is to implement Container Listener from scratch.

```
public abstract class ContainerAdapter implements ContainerListener {
// Public Constructors
  public ContainerAdapter();
// Methods Implementing ContainerListener
  public void componentAdded (ContainerEvent e);        empty
  public void componentRemoved (ContainerEvent e);      empty
}
```

Hierarchy: Object->Container Adapter(Container Listener(java.util.Event Listener))

Container Event

An event of this type serves as notification that the source Container has had a child added to it or removed from it. Note that this event is a notification only; the AWT adds or removes the child internally, and the recipient of this event needs to take no action itself.

get Child() returns the child Component that was added or removed, and get Container() returns the Container to which it was added or from which it was removed. get Container() is simply a convenient alternative to get Source(). get ID () returns the constant COMPONENT_ADDED or COMPONENT_REMOVED to indicate whether the specified child was added or removed.

```
public class Container Event extends Component Event {
// Public Constructors
  public ContainerEvent (Component source, int id, Component child);
// Public Constants
  public static final int COMPONENT_ADDED;            =300
  public static final int COMPONENT_REMOVED;          =301
  public static final int CONTAINER_FIRST;            =300
  public static final int CONTAINER_LAST;             =301
// Public Instance Methods
  public Component getChild();
  public Container getContainer();
// Public Methods Overriding ComponentEvent
  public String paramString();
}
```

Hierarchy: `Object—>java.util.EventObject(Serializable)—>AWTEvent—>ComponentEvent—>ContainerEvent`

Passed To: `AWTEventMulticaster.{componentAdded(), componentRemoved()}, Container.processContainerEvent(), ContainerAdapter.{componentAdded(), componentRemoved()}, ContainerListener.{componentAdded(), componentRemoved()}, javax.swing.JComponent.AccessibleJComponent.Accessible Container Handler.{componentAdded(), componentRemoved()}`

Container Listener

This interface defines the methods that an object must implement to listen for container events on AWT components. When a `Container Event` occurs, an AWT component notifies its registered `Container Listener` objects by invoking one of their methods. An easy way to implement this interface is by subclassing the `Container Adapter` class.

```
public abstract interface ContainerListener extends java.util.EventListener {
// Public Instance Methods
  public abstract void componentAdded (ContainerEvent e);
  public abstract void componentRemoved (ContainerEvent e);
}
```

Hierarchy: `(ContainerListener(java.util.EventListener))`

Implementations: `AWT Event Multicaster, Container Adapter, javax.swing.J Component. Accessible J Component. Accessible Container Handler`

Passed To: `AWT Event Multicaster.{add(), remove()}, Container.{add Container Listener(), remove Container Listener()}`

Returned By: `AWT Event Multicaster.{add(), remove()}`

Type of: `javax. swing. J Component. Accessible J Component. accessible Container Handler`

Focus Adapter

This class is a trivial implementation of `Focus Listener`; it provides empty bodies for each of the methods of that interface. When you are not interested in all of these methods, it is often easier to subclass `Focus Adapter` than it is to implement `Focus Listener` from scratch.

```
public abstract class FocusAdapter implements FocusListener {
// Public Constructors
  public FocusAdapter ();
// Methods Implementing FocusListener
  public void focusGained (FocusEvent e);          empty
  public void focusLost (FocusEvent e);            empty
}
```

Hierarchy: `Object—>Focus Adapter(Focus Listener(java. util. Event Listener))`

Focus Event

An event of this type indicates that a `Component` has gained or lost focus on a temporary or permanent basis. Use the inherited `get Component ()` method to determine which component has gained or lost focus. Use `get ID ()` to determine the type of focus event; it returns `FOCUS_GAINED` or `FOCUS_LOST`.

When focus is lost, you can call `is Temporary()` to determine whether it is a temporary loss of focus. Temporary focus loss occurs when the window that contains the component loses focus, for example, or when focus is temporarily diverted to a popup menu or a scrollbar. Similarly, you can also use `is Temporary()` to determine whether focus is being granted to a component on a temporary basis.

```
public class FocusEvent extends ComponentEvent {
// Public Constructors
  public FocusEvent (Component source, int id);
  public FocusEvent (Component source, int id, boolean temporary);
// Public Constants
  public static final int FOCUS_FIRST ;                    =1004
  public static final int FOCUS_GAINED ;                   =1004
  public static final int FOCUS_LAST ;                     =1005
  public static final int FOCUS_LOST ;                     =1005
// Public Instance Methods
  public boolean isTemporary ();
// Public Methods Overriding ComponentEvent
  public String paramString ();
}
```

Hierarchy: `Object->java.util.Event Object(Serializable)->AWT Event->Component Event->Focus Event`

Passed To: `AWT Event Multicaster.{focusGained(), focusLost()}, Component.processFocusEvent(), FocusAdapter.{focusGained(), focusLost()}, FocusListener.{focusGained(), focusLost()}, javax.swing.JComponent.processFocusEvent(), javax.swing.text.DefaultCaret.{focusGained(), focusLost()}`

Focus Listener

This interface defines the methods that an object must implement to listen for focus events on AWT components. When a `Focus Event` occurs, an AWT component notifies its registered `Focus Listener` objects by invoking one of their methods. An easy way to implement this interface is by subclassing the `Focus Adapter` class.

```
public abstract interface FocusListener extends java.util.EventListener {
// Public Instance Methods
  public abstract void focusGained (FocusEvent e);
  public abstract void focusLost (FocusEvent e);
}
```

Hierarchy: `(Focus Listener (java. util. Event Listener))`

Implementations: `AWT Event Multicaster, Focus Adapter, javax.swing.text.Default Caret`

Passed To: Too many methods to list.
Returned By: AWT Event Multicaster. {add(), remove()}

Input Event

This abstract class serves as the superclass for the raw user input event types `Mouse Event` and `Key Event`. Use the inherited `get Component()` method to determine in which component the event occurred. Use `get When()` to obtain a timestamp for the event. Use `get Modifiers()` to determine which keyboard modifier keys or mouse buttons were down when the event occurred. You can decode the `get Modifiers()` return value using the various `_MASK` constants defined by this class. The class also defines four convenience methods for determining the state of keyboard modifiers.

As of Java 1.1, input events are delivered to the appropriate listener objects before they are delivered to the AWT components themselves. If a listener calls the `consume ()` method of the event, the event is not passed on to the component. For example, if a listener registered on a `Button` consumes a mouse click, it prevents the button itself from responding to that event. You can use `is Consumed()` to test whether some other listener object has already consumed the event.

```
public abstract class InputEvent extends ComponentEvent {
// No Constructor
// Public Constants
1.2 public static final int ALT_GRAPH_MASK;                 =32
    public static final int ALT_MASK;                       =8
    public static final int BUTTON1_MASK;                   =16
    public static final int BUTTON2_MASK;                   =8
    public static final int BUTTON3_MASK;                   =4
    public static final int CTRL_MASK;                      =2
    public static final int META_MASK;                      =4
    public static final int SHIFT_MASK;                     =1
// Property Accessor Methods (by property name)
    public boolean isAltDown ();
1.2 public boolean isAltGraphDown ();
    public boolean isConsumed (); Overrides:AWTEvent
    public boolean isControlDown ();
    public boolean isMetaDown ();
    public int getModifiers ();
    public boolean isShiftDown ();
    public long getWhen ();
// Public Methods Overriding AWTEvent
    public void consume ();
}
```

Hierarchy: `Object->java.util.EventObject(Serializable)->AWTEvent->ComponentEvent->InputEvent`
Subclasses: `KeyEvent, MouseEvent`
Passed To: `java.awt.dnd.DragGestureRecognizer.appendEvent ()`

Returned By: `java.awt.dnd.DragGestureEvent.getTriggerEvent ()`, `java.awt.dnd.DragGestureRecognizer.get.TriggerEvent ()`

Input Method Event

Events of this type are sent from an input method to the text input component or text composition window that is using the services of the input method. An `InputMethodEvent` is generated each time the user makes an edit to the text that is being composed. Input method details are usually hidden by text input components. Application-level code should never have to use this class.

The `getText ()` method returns a `java.text.AttributedCharacterIterator` that contains the current input method text. `getCommittedCharacterCount ()` specifies how many characters of that text have been fully composed and committed, so that they are ready to be integrated into the text input component. The input method does not send these committed characters again. The characters returned by the iterator beyond the specified number of committed characters are characters that are still undergoing composition and are not ready to be integrated into the text input component. These characters may be repeated in future `InputMethodEvent` objects, as the user continues to edit them.

```
public class InputMethodEvent extends AWTEvent {
// Public Constructors
  public InputMethodEvent (Component source, int id,
  java.awt.font.TextHitInfo caret,
  java.awt.font.TextHitInfo visiblePosition);
  public InputMethodEvent (Component source, int id,
  java.text.AttributedCharacterIterator text, int committedCharacterCount,
  java.awt.font.TextHitInfo caret,
  java.awt.font.TextHitInfo visiblePosition);
// Public Constants
  public static final int CARET_POSITION_CHANGED;                    =1101
  public static final int INPUT_METHOD_FIRST;                        =1100
  public static final int INPUT_METHOD_LAST;                         =1101
  public static final int INPUT_METHOD_TEXT_CHANGED;                 =1100
// Property Accessor Methods (by property name)
  public java.awt.font.TextHitInfo getCaret ();
  public int getCommittedCharacterCount ();
  public boolean isConsumed ();                          Overrides:AWTEvent
  public java.text.AttributedCharacterIterator getText ();
  public java.awt.font.TextHitInfo getVisiblePosition ();
// Public Methods Overriding AWTEvent
  public void consume ();
  public String paramString ();
}
```

Hierarchy: `Object->java.util.EventObject (Serializable)->AWTEvent->InputMethodEvent`

Passed To: `AWTEventMulticaster.{caretPositionChanged (),` inputMethodTextChanged()}, Component.processInputMethodEvent(), InputMethodListener.{caretPositionChanged(), inputMethodTextChanged()}, javax.swing.text.JTextComponent.processInputMethodEvent()

Input Method Listener

This interface defines the methods that a text input component must define in order to receive notifications from an input method. `caretPositionChanged()` is invoked when the user has moved the editing cursor. `inputMethodTextChanged()` is invoked when the user has edited text being composed by the input method.

Input method details are usually hidden by text input components. Application-level code should never have to use or implement this interface.

```
public abstract interface InputMethodListener extends java.util.EventListener {
// Public Instance Methods
  public abstract void caretPositionChanged (InputMethodEvent event);
  public abstract void inputMethodTextChanged (InputMethodEvent event);
}
```

Hierarchy: (InputMethodListener(java.util.EventListener))
Implementations: AWTEventMulticaster
Passed To: AWTEventMulticaster.{add(), remove()}, Component.{addInputMethodListener(), removeInputMethodListener()}, javax.swing.text.JTextComponent.addInputMethodListener()
Returned By: AWTEventMulticaster.{add(), remove()}

Invocation Event

An event of this type is not generated by an asynchronous external event, such as user input. Instead, an `InvocationEvent` is placed on the event queue by the `invokeLater()` and `invokeAndWait()` methods of `EventQueue`. `InvocationEvent` implements `java.awt.ActiveEvent`, which means that it is an event that knows how to dispatch itself, with its own `dispatch()` method. When an InvocationEvent reaches the front of the event queue, its `dispatch()` method is called, and this invokes the `run()` method of the `Runnable` object specified when the `InvocationEvent` was created. This technique provides a simple method for running arbitrary code from the event dispatch thread.

Applications need not be concerned with these details; they can simply use the `invokeLater()` and `invokeAndWait()` methods of `EventQueue`.

```
public class InvocationEvent extends AWTEvent implements ActiveEvent {
// Public Constructors
  public InvocationEvent (Object source, Runnable runnable);
  public InvocationEvent (Object source, Runnable runnable, Object notifier,
boolean catchExceptions);
// Protected Constructors
  protected InvocationEvent (Object source, int id, Runnable runnable, Object
notifier, boolean catchExceptions);
// Public Constants
```

```
  public static final int INVOCATION_DEFAULT ;                    =1200
  public static final int INVOCATION_FIRST ;                      =1200
  public static final int INVOCATION_LAST ;                       =1200
// Public Instance Methods
  public Exception getException ();
// Methods Implementing ActiveEvent
  public void dispatch ();
// Public Methods Overriding AWTEvent
  public String paramString ();
// Protected Instance Fields
  protected boolean catchExceptions ;
  protected Object notifier ;
  protected Runnable runnable ;
}
```

Hierarchy: `Object->java.util.EventObject(Serializable)->AWTEvent->InvocationEvent(ActiveEvent)`

Item Event

An event of this type indicates that an item within an `ItemSelectable` component has had its selection state changed. `getItemSelectable()` is a convenient alternative to `getSource()` that returns the `ItemSelectable` object that originated the event. `getItem()` returns an object that represents the item that was selected or deselected.

`getID()` returns the type of the `ItemEvent`. The standard AWT components always generate item events of type `ITEM_STATE_CHANGED`. The `getStateChange()` method returns the new selection state of the item: It returns one of the constants `SELECTED` or `DESELECTED`. (This value can be misleading for Checkbox components that are part of a `CheckboxGroup`. If the user attempts to deselect a selected component, a `DESELECTED` event is delivered, but the `CheckboxGroup` immediately reselects the component to enforce its requirement that at least one `Checkbox` be selected at all times.)

```
public class ItemEvent extends AWTEvent {
// Public Constructors
  public ItemEvent (ItemSelectable source, int id, Object item, int stateChange);
// Public Constants
  public static final int DESELECTED ;                            =2
  public static final int ITEM_FIRST ;                            =701
  public static final int ITEM_LAST ;                             =701
  public static final int ITEM_STATE_CHANGED ;                    =701
  public static final int SELECTED ;                              =1
// Public Instance Methods
  public Object getItem ();
  public ItemSelectable getItemSelectable ();
  public int getStateChange ();
// Public Methods Overriding AWTEvent
  public String paramString ();
}
```

Hierarchy: `Object->java.util.EventObject(Serializable)->AWTEvent->ItemEvent`

Passed To: `AWTEventMulticaster.itemStateChanged(), Checkbox.processItemEvent(), CheckboxMenuItem.processItemEvent(), Choice.processItemEvent(), java.awt.List.processItemEvent(), ItemListener.itemStateChanged(), javax.swing.AbstractButton.fireItemStateChanged(), javax.swing.DefaultButtonModel.fireItemStateChanged(), javax.swing.DefaultCellEditor.EditorDelegate.itemStateChanged(), javax.swing.JComboBox.fireItemStateChanged(), javax.swing.JToggleButton.AccessibleJToggleButton.itemStateChanged()`

Item Listener

This interface defines the method that an object must implement to listen for item events on AWT components. When an ItemEvent occurs, an AWT component notifies its registered `ItemListener` objects by invoking their `itemStateChanged()` methods.

```
public abstract interface ItemListener extends java.util.EventListener {
// Public Instance Methods
  public abstract void itemStateChanged (ItemEvent e);
}
```

Hierarchy: `(ItemListener(java.util.EventListener))`

Implementations: `AWTEventMulticaster, javax.swing.DefaultCellEditor.EditorDelegate, javax.swing.JToggleButton.AccessibleJToggleButton`

Passed To: Too many methods to list.

Returned By: `AWTEventMulticaster.{add(), remove()}, javax.swing.AbstractButton.createItemListener()`

Type of: `javax.swing.AbstractButton.itemListener`

Key Adapter

This class is a trivial implementation of `KeyListener`; it provides empty bodies for each of the methods of that interface. When you are not interested in all of these methods, it is often easier to subclass `KeyAdapter` than it is to implement `KeyListener` from scratch.

```
public abstract class KeyAdapter implements KeyListener {
// Public Constructors
  public KeyAdapter ();
// Methods Implementing KeyListener
  public void keyPressed (KeyEvent e);     empty
  public void keyReleased (KeyEvent e);    empty
  public void keyTyped (KeyEvent e); empty
}
```

Hierarchy: `Object->KeyAdapter (KeyListener (java. util. EventListener))`

Key Event

An event of this type indicates that the user has pressed or released a key on the keyboard. Call `getID()` to determine the particular type of key event that has occurred. The constant `KEY_PRESSED` indicates that a key has been pressed, while the constant `KEY_RELEASED` indicates that a key has been released. Not all keystrokes actually correspond to or generate Unicode characters. Modifier keys and function keys, for example, do not correspond to characters. Furthermore, for internationalized input, multiple keystrokes are sometimes required to generate a single character of input. Therefore, `getID()` returns a third constant, `KEY_TYPED`, to indicate a `KeyEvent` that actually contains a character value.

For `KEY_PRESSED` and `KEY_RELEASED` key events, use `getKeyCode()` to obtain the virtual keycode of the key that was pressed or released. `KeyEvent` defines a number of VK_ constants that represent these virtual keys. Note that not all keys on all keyboards have corresponding constants in the KeyEvent class, and not all keyboards can generate all of the virtual keycodes defined by this class. As of Java 1.1, the VK_ constants for letter keys, number keys, and some other keys have the same values as the ASCII encodings of the letters and numbers. You should not rely on this to always be the case, however. If the key that was pressed or released corresponds directly to a Unicode character, you can obtain that character by calling getKeyChar(). If there is not a corresponding Unicode character, this method returns the constantCHAR_UNDEFINED. The isActionKey() method returns true if the key that was pressed or released does not have a corresponding character.

For KEY_TYPED key events, use getKeyChar() to return the Unicode character that was typed. If you call getKeyCode() for this type of key event, it returnsVK_UNDEFINED.

See InputEvent for information on inherited methods you can use to obtain the keyboard modifiers that were down during the event and other important methods. UsegetComponent(), inherited from ComponentEvent, to determine over what component the event occurred. The static method getKeyText() returns a (possibly localized) textual name for a given keycode. The static method getKeyModifiersText() returns a (possibly localized) textual description for a set of modifiers.

`KeyEvent` has methods that allow you to change the keycode, key character, or modifiers of an event. These methods, along with the `consume()` method, allow a `KeyListener` to perform filtering of key events before they are passed to the underlying AWT component.

public class **KeyEvent** extends InputEvent {
// Public Constructors
public **KeyEvent** (Component *source*, int *id*, long *when*, int *modifiers*, int *keyCode*);
public **KeyEvent** (Component *source*, int *id*, long *when*, int *modifiers*, int *keyCode*, char *keyChar*);
// Public Constants
public static final char **CHAR_UNDEFINED** ; *='\uFFFF'*
public static final int **KEY_FIRST** ; *=400*
public static final int **KEY_LAST** ; *=402*
public static final int **KEY_PRESSED** ; *=401*
public static final int **KEY_RELEASED** ; *=402*

	public static final int **KEY_TYPED** ;	*=400*
	public static final int **VK_0** ;	*=48*
	public static final int **VK_1** ;	*=49*
	public static final int **VK_2** ;	*=50*
	public static final int **VK_3** ;	*=51*
	public static final int **VK_4** ;	*=52*
	public static final int **VK_5** ;	*=53*
	public static final int **VK_6** ;	*=54*
	public static final int **VK_7** ;	*=55*
	public static final int **VK_8** ;	*=56*
	public static final int **VK_9** ;	*=57*
	public static final int **VK_A** ;	*=65*
	public static final int **VK_ACCEPT** ;	*=30*
	public static final int **VK_ADD** ;	*=107*
1.2	public static final int **VK_AGAIN** ;	*=65481*
1.2	public static final int **VK_ALL_CANDIDATES** ;	*=256*
1.2	public static final int **VK_ALPHANUMERIC** ;	*=240*
	public static final int **VK_ALT** ;	*=18*
1.2	public static final int **VK_ALT_GRAPH** ;	*=65406*
1.2	public static final int **VK_AMPERSAND** ;	*=150*
1.2	public static final int **VK_ASTERISK** ;	*=151*
1.2	public static final int **VK_AT** ;	*=512*
	public static final int **VK_B** ;	*=66*
	public static final int **VK_BACK_QUOTE** ;	*=192*
	public static final int **VK_BACK_SLASH** ;	*=92*
	public static final int **VK_BACK_SPACE** ;	*=8*
1.2	public static final int **VK_BRACELEFT** ;	*=161*
1.2	public static final int **VK_BRACERIGHT** ;	*=162*
	public static final int **VK_C** ;	*=67*
	public static final int **VK_CANCEL** ;	*=3*
	public static final int **VK_CAPS_LOCK** ;	*=20*
1.2	public static final int **VK_CIRCUMFLEX** ;	*=514*
	public static final int **VK_CLEAR** ;	*=12*
	public static final int **VK_CLOSE_BRACKET** ;	*=93*
1.2	public static final int **VK_CODE_INPUT** ;	*=258*
1.2	public static final int **VK_COLON** ;	*=513*
	public static final int **VK_COMMA** ;	*=44*
1.2	public static final int **VK_COMPOSE** ;	*=65312*
	public static final int **VK_CONTROL** ;	*=17*
	public static final int **VK_CONVERT** ;	*=28*
1.2	public static final int **VK_COPY** ;	*=65485*
1.2	public static final int **VK_CUT** ;	*=65489*
	public static final int **VK_D** ;	*=68*
1.2	public static final int **VK_DEAD_ABOVEDOT** ;	*=134*
1.2	public static final int **VK_DEAD_ABOVERING** ;	*=136*
1.2	public static final int **VK_DEAD_ACUTE** ;	*=129*

1.2	public static final int **VK_DEAD_BREVE** ;	*=133*
1.2	public static final int **VK_DEAD_CARON** ;	*=138*
1.2	public static final int **VK_DEAD_CEDILLA** ;	*=139*
1.2	public static final int **VK_DEAD_CIRCUMFLEX** ;	*=130*
1.2	public static final int **VK_DEAD_DIAERESIS** ;	*=135*
1.2	public static final int **VK_DEAD_DOUBLEACUTE** ;	*=137*
1.2	public static final int **VK_DEAD_GRAVE** ;	*=128*
1.2	public static final int **VK_DEAD_IOTA** ;	*=141*
1.2	public static final int **VK_DEAD_MACRON** ;	*=132*
1.2	public static final int **VK_DEAD_OGONEK** ;	*=140*
1.2	public static final int **VK_DEAD_SEMIVOICED_SOUND** ;	*=143*
1.2	public static final int **VK_DEAD_TILDE** ;	*=131*
1.2	public static final int **VK_DEAD_VOICED_SOUND** ;	*=142*
	public static final int **VK_DECIMAL** ;	*=110*
	public static final int **VK_DELETE** ;	*=127*
	public static final int **VK_DIVIDE** ;	*=111*
1.2	public static final int **VK_DOLLAR** ;	*=515*
	public static final int **VK_DOWN** ;	*=40*
	public static final int **VK_E** ;	*=69*
	public static final int **VK_END** ;	*=35*
	public static final int **VK_ENTER** ;	*=10*
	public static final int **VK_EQUALS** ;	*=61*
	public static final int **VK_ESCAPE** ;	*=27*
1.2	public static final int **VK_EURO_SIGN** ;	*=516*
1.2	public static final int **VK_EXCLAMATION_MARK** ;	*=517*
	public static final int **VK_F** ;	*=70*
	public static final int **VK_F1** ;	*=112*
	public static final int **VK_F10** ;	*=121*
	public static final int **VK_F11** ;	*=122*
	public static final int **VK_F12** ;	*=123*
1.2	public static final int **VK_F13** ;	*=61440*
1.2	public static final int **VK_F14** ;	*=61441*
1.2	public static final int **VK_F15** ;	*=61442*
1.2	public static final int **VK_F16** ;	*=61443*
1.2	public static final int **VK_F17** ;	*=61444*
1.2	public static final int **VK_F18** ;	*=61445*
1.2	public static final int **VK_F19** ;	*=61446*
	public static final int **VK_F2** ;	*=113*
1.2	public static final int **VK_F20** ;	*=61447*
1.2	public static final int **VK_F21** ;	*=61448*
1.2	public static final int **VK_F22** ;	*=61449*
1.2	public static final int **VK_F23** ;	*=61450*
1.2	public static final int **VK_F24** ;	*=61451*
	public static final int **VK_F3** ;	*=114*
	public static final int **VK_F4** ;	*=115*
	public static final int **VK_F5** ;	*=116*

	public static final int **VK_F6** ;	*=117*
	public static final int **VK_F7** ;	*=118*
	public static final int **VK_F8** ;	*=119*
	public static final int **VK_F9** ;	*=120*
	public static final int **VK_FINAL** ;	*=24*
1.2	public static final int **VK_FIND** ;	*=65488*
1.2	public static final int **VK_FULL_WIDTH** ;	*=243*
	public static final int **VK_G** ;	*=71*
1.2	public static final int **VK_GREATER** ;	*=160*
	public static final int **VK_H** ;	*=72*
1.2	public static final int **VK_HALF_WIDTH** ;	*=244*
	public static final int **VK_HELP** ;	*=156*
1.2	public static final int **VK_HIRAGANA** ;	*=242*
	public static final int **VK_HOME** ;	*=36*
	public static final int **VK_I** ;	*=73*
	public static final int **VK_INSERT** ;	*=155*
1.2	public static final int **VK_INVERTED_EXCLAMATION_MARK** ;	*=518*
	public static final int **VK_J** ;	*=74*
1.2	public static final int **VK_JAPANESE_HIRAGANA** ;	*=260*
1.2	public static final int **VK_JAPANESE_KATAKANA** ;	*=259*
1.2	public static final int **VK_JAPANESE_ROMAN** ;	*=261*
	public static final int **VK_K** ;	*=75*
	public static final int **VK_KANA** ;	*=21*
	public static final int **VK_KANJI** ;	*=25*
1.2	public static final int **VK_KATAKANA** ;	*=241*
1.2	public static final int **VK_KP_DOWN** ;	*=225*
1.2	public static final int **VK_KP_LEFT** ;	*=226*
1.2	public static final int **VK_KP_RIGHT** ;	*=227*
1.2	public static final int **VK_KP_UP** ;	*=224*
	public static final int **VK_L** ;	*=76*
	public static final int **VK_LEFT** ;	*=37*
1.2	public static final int **VK_LEFT_PARENTHESIS** ;	*=519*
1.2	public static final int **VK_LESS** ;	*=153*
	public static final int **VK_M** ;	*=77*
	public static final int **VK_META** ;	*=157*
1.2	public static final int **VK_MINUS** ;	*=45*
	public static final int **VK_MODECHANGE** ;	*=31*
	public static final int **VK_MULTIPLY** ;	*=106*
	public static final int **VK_N** ;	*=78*
	public static final int **VK_NONCONVERT** ;	*=29*
	public static final int **VK_NUM_LOCK** ;	*=144*
1.2	public static final int **VK_NUMBER_SIGN** ;	*=520*
	public static final int **VK_NUMPAD0** ;	*=96*
	public static final int **VK_NUMPAD1** ;	*=97*
	public static final int **VK_NUMPAD2** ;	*=98*
	public static final int **VK_NUMPAD3** ;	*=99*

	public static final int **VK_NUMPAD4** ;	*=100*
	public static final int **VK_NUMPAD5** ;	*=101*
	public static final int **VK_NUMPAD6** ;	*=102*
	public static final int **VK_NUMPAD7** ;	*=103*
	public static final int **VK_NUMPAD8** ;	*=104*
	public static final int **VK_NUMPAD9** ;	*=105*
	public static final int **VK_O** ;	*=79*
	public static final int **VK_OPEN_BRACKET** ;	*=91*
	public static final int **VK_P** ;	*=80*
	public static final int **VK_PAGE_DOWN** ;	*=34*
	public static final int **VK_PAGE_UP** ;	*=33*
1.2	public static final int **VK_PASTE** ;	*=65487*
	public static final int **VK_PAUSE** ;	*=19*
	public static final int **VK_PERIOD** ;	*=46*
1.2	public static final int **VK_PLUS** ;	*=521*
1.2	public static final int **VK_PREVIOUS_CANDIDATE** ;	*=257*
	public static final int **VK_PRINTSCREEN** ;	*=154*
1.2	public static final int **VK_PROPS** ;	*=65482*
	public static final int **VK_Q** ;	*=81*
	public static final int **VK_QUOTE** ;	*=222*
1.2	public static final int **VK_QUOTEDBL** ;	*=152*
	public static final int **VK_R** ;	*=82*
	public static final int **VK_RIGHT** ;	*=39*
1.2	public static final int **VK_RIGHT_PARENTHESIS** ;	*=522*
1.2	public static final int **VK_ROMAN_CHARACTERS** ;	*=245*
	public static final int **VK_S** ;	*=83*
	public static final int **VK_SCROLL_LOCK** ;	*=145*
	public static final int **VK_SEMICOLON** ;	*=59*
	public static final int **VK_SEPARATER** ;	*=108*
	public static final int **VK_SHIFT** ;	*=16*
	public static final int **VK_SLASH** ;	*=47*
	public static final int **VK_SPACE** ;	*=32*
1.2	public static final int **VK_STOP** ;	*=65480*
	public static final int **VK_SUBTRACT** ;	*=109*
	public static final int **VK_T** ;	*=84*
	public static final int **VK_TAB** ;	*=9*
	public static final int **VK_U** ;	*=85*
	public static final int **VK_UNDEFINED** ;	*=0*
1.2	public static final int **VK_UNDERSCORE** ;	*=523*
1.2	public static final int **VK_UNDO** ;	*=65483*
	public static final int **VK_UP** ;	*=38*
	public static final int **VK_V** ;	*=86*
	public static final int **VK_W** ;	*=87*
	public static final int **VK_X** ;	*=88*
	public static final int **VK_Y** ;	*=89*
	public static final int **VK_Z** ;	*=90*

```
// Public Class Methods
   public static String getKeyModifiersText (int modifiers);
   public static String getKeyText (int keyCode);
// Property Accessor Methods (by property name)
   public boolean isActionKey ();
   public char getKeyChar ();
   public void setKeyChar (char keyChar);
   public int getKeyCode ();
   public void setKeyCode (int keyCode);
// Public Instance Methods
   public void setModifiers (int modifiers);
// Public Methods Overriding ComponentEvent
   public String paramString ();
}
```

Hierarchy: `Object->java.util.EventObject(Serializable)->AWTEvent->ComponentEvent->InputEvent->KeyEvent`

Subclasses: `javax.swing.event.MenuKeyEvent`

Passed To: Too many methods to list.

Key Listener

This interface defines the methods that an object must implement to listen for key events on AWT components. When a `KeyEvent` occurs, an AWT component notifies its registered `KeyListener` objects by invoking one of their methods. An easy way to implement this interface is by subclassing the `KeyAdapter` class.

```
public abstract interface KeyListener extends java.util.EventListener {
// Public Instance Methods
public abstract void keyPressed (KeyEvent e);
public abstract void keyReleased (KeyEvent e);
public abstract void keyTyped (KeyEvent e);
}
```

Hierarchy: `(KeyListener(java.util.EventListener))`

Implementations: `AWTEventMulticaster, KeyAdapter`

Passed To: `AWTEventMulticaster.{add(), remove()}, Component.{addKeyListener(), removeKeyListener()}`

Returned By: `AWTEventMulticaster.{add(), remove()}`

Mouse Adapter

This class is a trivial implementation of `MouseListener`; it provides empty bodies for each of the methods of that interface. When you are not interested in all of these methods, it is often easier to subclass `MouseAdapter` than it is to implement `MouseListener` from scratch.

```
public abstract class MouseAdapter implements MouseListener {
// Public Constructors
   public MouseAdapter ();
// Methods Implementing MouseListener
```

```
public void mouseClicked (MouseEvent e);        empty
public void mouseEntered (MouseEvent e);        empty
public void mouseExited (MouseEvent e);         empty
public void mousePressed (MouseEvent e);        empty
public void mouseReleased (MouseEvent e);       empty
}
```

Hierarchy: `Object->MouseAdapter(MouseListener(java. util. EventListener))`

Subclasses: `javax.swing.ToolTipManager, javax. swing. text. html. FormView.MouseEventListener, javax. swing. text. html. HTMLEditorKit. LinkController`

Mouse Event

An event of this type indicates that the user has moved the mouse or pressed one of the mouse buttons. Call `getID()` to determine the specific type of mouse event that has occurred. This method returns one of the following seven constants, which corresponds to a method in either the `MouseListener` or `MouseMotionListener` interface:

MOUSE_PRESSED

The user has pressed a mouse button.

MOUSE_RELEASED

The user has released a mouse button.

MOUSE_CLICKED

The user has pressed and released a mouse button without any intervening mouse drag.

MOUSE_DRAGGED

The user has moved the mouse while holding a button down.

MOUSE_MOVED

The user has moved the mouse without holding any button down.

MOUSE_ENTERED

The mouse pointer has entered the component.

MOUSE_EXITED

The mouse pointer has left the component.

Use `getX()` and `getY()` or getPoint() to obtain the coordinates of the mouse event. Use `translatePoint()` to modify these coordinates by a specified amount.

Use `getModifiers()` and other methods and constants inherited from `InputEvent` to determine the mouse button or keyboard modifiers that were down when the event occurred. See InputEvent for details. Note that mouse button modifiers are not reported for `MOUSE_RELEASED` events, since, technically, the mouse button in question is no longer pressed.

Use `getComponent()`, inherited from `ComponentEvent`, to determine over which component the event occurred. For mouse events of type `MOUSE_CLICKED`, `MOUSE_PRESSED`, or `MOUSE_RELEASED`, call `getClickCount()` to determine how many consecutive clicks have occurred. If you are using popup menus, use `is PopupTrigger()` to test whether the current event represents the standard platform-dependent popup menu trigger event.

```
public class MouseEvent extends InputEvent {
// Public Constructors
    public MouseEvent (Component source, int id, long when, int modifiers, int x, int
y, int clickCount, boolean popupTrigger);
// Public Constants
    public static final int MOUSE_CLICKED ;                    =500
    public static final int MOUSE_DRAGGED ;                    =506
    public static final int MOUSE_ENTERED ;                    =504
    public static final int MOUSE_EXITED ;                     =505
    public static final int MOUSE_FIRST ;                      =500
    public static final int MOUSE_LAST ;                       =506
    public static final int MOUSE_MOVED ;                      =503
    public static final int MOUSE_PRESSED ;                    =501
    public static final int MOUSE_RELEASED ;                   =502
// Property Accessor Methods (by property name)
    public int getClickCount ();
    public Point getPoint ();
    public boolean isPopupTrigger ();
    public int getX ();
    public int getY ();
// Public Instance Methods
    public void translatePoint (int x, int y);
// Public Methods Overriding ComponentEvent
    public String paramString ();
}
```

Hierarchy: `Object->java.util.EventObject(Serializable)->AWTEvent->ComponentEvent->InputEvent->MouseEvent`

Subclasses: `javax.swing.event.MenuDragMouseEvent`
Passed To: `Too many methods to list.`
Returned By: `javax.swing.SwingUtilities.convertMouseEvent()`

Mouse Listener

This interface defines the methods that an object must implement to listen for mouse events on AWT components. When a `MouseEvent` occurs, an AWT component notifies its registered `MouseListener` objects (or `MouseMotionListener` objects, if the event involves mouse motion) by invoking one of their methods. An easy way to implement this interface is by subclassing the `MouseAdapter` class.

```
public abstract interface MouseListener extends java.util.EventListener {
// Public Instance Methods
  public abstract void mouseClicked (MouseEvent e);
  public abstract void mouseEntered (MouseEvent e);
  public abstract void mouseExited (MouseEvent e);
  public abstract void mousePressed (MouseEvent e);
  public abstract void mouseReleased (MouseEvent e);
}
```

Hierarchy: (MouseListener(java.util.EventListener))

Implementations: AWTEventMulticaster, java. awt. dnd. MouseDragGestureRecognizer, MouseAdapter, javax. swing. event. MouseInputListener, javax.swing.text.DefaultCaret

Passed To: AWTEventMulticaster.{add(), remove()}, Component. {addMouseListener(), removeMouseListener()}

Returned By: AWTEventMulticaster.{add(), remove()}

Mouse Motion Adapter

This class is a trivial implementation of MouseMotionListener; it provides empty bodies for each of the methods of that interface. When you are not interested in all of these methods, it is often easier to subclass MouseMotionAdapter than it is to implement MouseMotionListener from scratch.

```
public abstract class MouseMotionAdapter implements MouseMotionListener {
// Public Constructors
    public MouseMotionAdapter ();
// Methods Implementing MouseMotionListener
    public void mouseDragged (MouseEvent e);
    public void mouseMoved (MouseEvent e);
}
```

Hierarchy: Object->MouseMotionAdapter(MouseMotionListener(java. util.EventListener))

Mouse Motion Listener

This interface defines the methods that an object must implement to listen for mouse motion events on AWT components. When a MouseEvent involving a mouse drag or mouse motion with no buttons down occurs, an AWT component notifies its registered MouseMotionListener objects by invoking one of their methods. An easy way to implement this is by subclassing the MouseMotionAdapter class.

```
public abstract interface MouseMotionListener extends java.util.EventListener {
// Public Instance Methods
  public abstract void mouseDragged (MouseEvent e);
  public abstract void mouseMoved (MouseEvent e);
}
```

Hierarchy: (MouseMotionListener(java.util.EventListener))

Implementations: AWTEventMulticaster, java. awt. dnd. MouseDragGestureRecognizer, MouseMotionAdapter, javax. swing. ToolTipManager, javax.swing.event.MouseInputListener, javax.swing.text.DefaultCaret

Passed To: AWTEventMulticaster.{add(), remove()}, Component.{addMouseMotionListener(), removeMouseMotionListener()}

Returned By: AWTEventMulticaster.{add(), remove()}

Paint Event

An event of this type indicates that a component should have its update() method invoked. (The update() method typically, by default, invokes the paint() method.)

`PaintEvent` differs from the other event types in `java.awt.event` in that it does not have a corresponding `EventListener` interface. `PaintEvent` is essentially for internal use by the AWT redisplay framework, so your programs should not try to handle it the way they handle other events. Instead, applets and custom components should simply override their `paint()` and/or `update()` methods to redraw themselves appropriately. AWT automatically invokes `update()` (which typically invokes `paint()`) when a `PaintEvent` arrives.

Although you do not typically use `PaintEvent`, redraw events are implemented through this class for simplicity, so that they are on equal footing with other event types and so that advanced programs can manipulate them through the `EventQueue`.

```
public class PaintEvent extends ComponentEvent {
// Public Constructors
    public PaintEvent (Component source, int id, Rectangle updateRect);
// Public Constants
    public static final int PAINT ;                                    =800
    public static final int PAINT_FIRST ;                              =800
    public static final int PAINT_LAST ;                               =801
    public static final int UPDATE ;                                   =801
// Public Instance Methods
    public Rectangle getUpdateRect ();
    public void setUpdateRect (Rectangle updateRect);
// Public Methods Overriding ComponentEvent
    public String paramString ();
}
```

Hierarchy: `Object->java.util.EventObject(Serializable)->AWTEvent->ComponentEvent->PaintEvent`

Text Event

An event of this type indicates that the user has edited the text value that appears in a `TextField`, `TextArea`, or other `TextComponent`. This event is triggered by any change to the displayed text. Note that this is not the same as the `ActionEvent` sent by the `TextField` object when the user edits the text and strikes the **Return** key.

Use the inherited `getSource()` to determine the object that was the source of this event. You have to cast that object to its TextComponent type. Call `getID()` to determine the type of a TextEvent. The standard AWT components always generate text events of type `TEXT_VALUE_CHANGED`.

```
public class TextEvent extends AWTEvent {
// Public Constructors
    public TextEvent (Object source, int id);
// Public Constants
    public static final int TEXT_FIRST ;                               =900
    public static final int TEXT_LAST ;                                =900
    public static final int TEXT_VALUE_CHANGED ;                       =900
// Public Methods Overriding AWTEvent
    public String paramString ();
    }
```

Hierarchy: `Object->java.util.EventObject(Serializable)->AWTEvent->TextEvent`
Passed To: `AWTEventMulticaster.textValueChanged()`, `TextComponent.processTextEvent()`, `TextListener.textValueChanged()`

Text Listener

This interface defines the method that an object must implement to listen for text events on AWT components. When a TextEvent occurs, an AWT component notifies its registered `TextListener` objects by invoking their `textValueChanged()` methods.

```
public abstract interface TextListener extends java.util.EventListener {
// Public Instance Methods
  public abstract void textValueChanged (TextEvent e);
}
```

Hierarchy: `(TextListener(java.util.EventListener))`
Implementations: `AWTEventMulticaster`
Passed To: `AWTEventMulticaster.{add(), remove()}`, `TextComponent.{addTextListener(), removeTextListener()}`
Returned By: `AWTEventMulticaster.{add(), remove()}`
Type of: `TextComponent.textListener`

WindowAdapter

This class is a trivial implementation of `WindowListener`; it provides empty bodies for each of the methods of that interface. When you are not interested in all of these methods, it is often easier to subclass `WindowAdapter` than it is to implement `WindowListener` from scratch.

```
public abstract class WindowAdapter implements WindowListener {
// Public Constructors
    public WindowAdapter ();
// Methods Implementing WindowListener
    public void windowActivated (WindowEvent e);
    public void windowClosed (WindowEvent e);
    public void windowClosing (WindowEvent e);
    public void windowDeactivated (WindowEvent e);
    public void windowDeiconified (WindowEvent e);
    public void windowIconified (WindowEvent e);
    public void windowOpened (WindowEvent e);
}
```

Hierarchy: `Object->WindowAdapter(WindowListener(java. util. EventListener))`
Subclasses: `javax.swing.JMenu.WinListener`

Window Event

An event of this type indicates that an important action has occurred for a Window object. Call `getWindow()` to determine the `Window` object that is the source of this event. Call `getID()` to determine the specific type of event that has occurred. Each of the following seven constants corresponds to one of the methods of the `WindowListener` interface:

WINDOW_OPENED
Indicates that the window has been created and opened; it is delivered only the first time that a window is opened.

WINDOW_CLOSING
Indicates that the user has requested that the window be closed through the system menu, through a close button on the window's border, or by invoking a platform-defined keystroke, such as **Alt-F4** in Windows. The application should respond to this event by calling `hide()` or `dispose()` on the Window object.

WINDOW_CLOSED
Delivered after a window is closed by a call to hide() or dispose().

WINDOW_ICONIFIED
Delivered when the user iconifies the window.

WINDOW_DEICONIFIED
Delivered when the user deiconifies the window.

WINDOW_ACTIVATED
Delivered when the window is activated—that is, when it is given the keyboard focus and becomes the active window.

WINDOW_DEACTIVATED
Delivered when the window ceases to be the active window, typically when the user activates some other window.

```
public class WindowEvent extends ComponentEvent {
// Public Constructors
    public WindowEvent (Window source, int id);
// Public Constants
    public static final int WINDOW_ACTIVATED ;        =205
    public static final int WINDOW_CLOSED ;           =202
    public static final int WINDOW_CLOSING ;          =201
    public static final int WINDOW_DEACTIVATED ;      =206
    public static final int WINDOW_DEICONIFIED ;      =204
    public static final int WINDOW_FIRST ;            =200
    public static final int WINDOW_ICONIFIED ;        =203
    public static final int WINDOW_LAST ;             =206
    public static final int WINDOW_OPENED ;           =200
// Public Instance Methods
    public Window getWindow ();
// Public Methods Overriding ComponentEvent
    public String paramString ();
}
```

Hierarchy: `Object->java.util.EventObject(Serializable)->AWTEvent ->ComponentEvent->WindowEvent`

Passed To: Too many methods to list.

Window Listener

This interface defines the methods that an object must implement to listen for window events on AWT components. When a `WindowEvent` occurs, an AWT component notifies its registered `WindowListener` objects by invoking one of their methods. An easy way to implement this interface is by subclassing the `WindowAdapter` class.

public abstract interface **WindowListener** extends java.util.EventListener {
// Public Instance Methods
  public abstract void **windowActivated** (WindowEvent *e*);
  public abstract void **windowClosed** (WindowEvent *e*);
  public abstract void **windowClosing** (WindowEvent *e*);
  public abstract void **windowDeactivated** (WindowEvent *e*);
  public abstract void **windowDeiconified** (WindowEvent *e*);
  public abstract void **windowIconified** (WindowEvent *e*);
  public abstract void **windowOpened** (WindowEvent *e*);
}

Hierarchy: `(WindowListener(java.util.EventListener))`
Implementations: `AWTEventMulticaster, WindowAdapter`
Passed To: `AWTEventMulticaster.{add(), remove()}, Window.{addWindowListener(), removeWindowListener()}`
Returned By: `AWTEventMulticaster.{add(), remove()}`

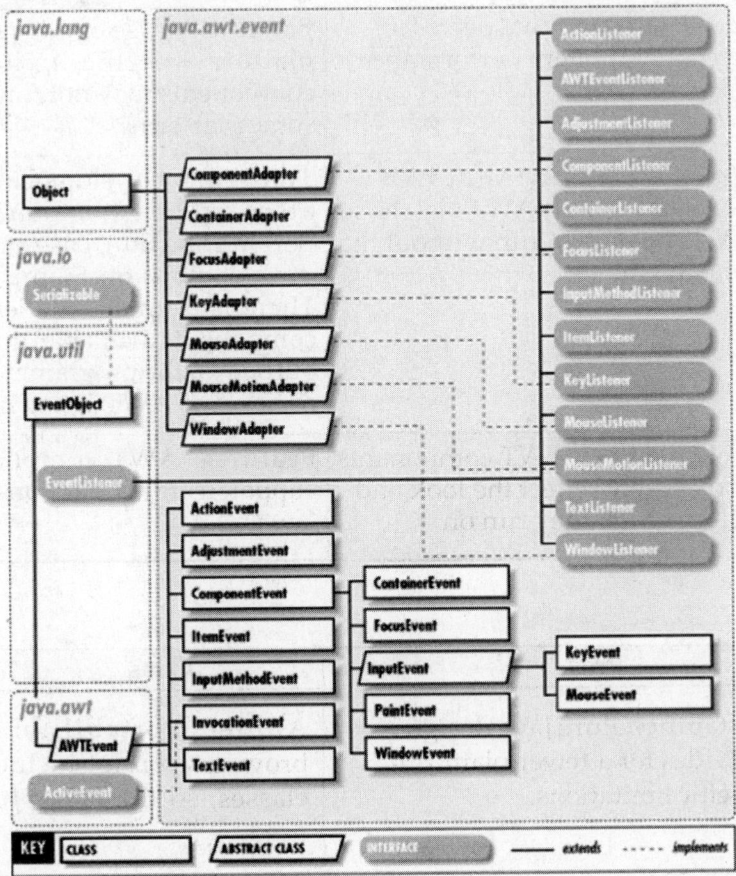

8. INTRODUCTION TO SWING

Swing is the primary Java GUI widget toolkit. It is a part of Oracle's Java Foundation Classes (JFC)—an API for providing a graphical user interface (GUI) for Java programs.

Swing was developed to provide a more sophisticated set of GUI components than the earlier **Abstract Window Toolkit (AWT)**. Swing provides a native look and feel that emulates, the look and feel of several platforms, and also supports a pluggable look and feel that allows applications to have a look and feel unrelated to the underlying platform. It has more powerful and flexible components than AWT. In addition to familiar components such as buttons, check box and labels, Swing provides several advanced components such as tabbed panel, scroll panes, trees, tables and lists.

The **Abstract Window Toolkit** (AWT) is Java's original platform-independent windowing, graphics, and user-interface widget toolkit. The AWT is now part of the Java Foundation Classes (JFC) — the standard API for providing a graphical user interface (GUI) for a Java program.

AWT is also the GUI toolkit for a number of Java ME profiles. For example, Connected Device Configuration profiles require Java runtimes on mobile telephones to support AWT.

AWT VS SWING

AWT

S.No.	Pros	Cons
1	**Speed:** Use of native peers speeds component performance	**Portability:** Use of native peers creates platform-specific limitations. Some components may not function at all on some platforms
2	**Applet Portability:** Most Web browsers support AWT classes so AWT applets can run without the Java plugin.	**Third Party Development:** The majority of component makers, including Borland and Sun, base new component development on Swing components. There is a much smaller set of AWT components available, thus placing the burden on the programmer to create his or her own AWT-based components.
3	**Look and Feel:** AWT components more closely reflect the look and feel of the OS they run on	**Features:** AWT components do not support features like icons and tool-tips.

SWING

S.No.	Pros	Cons
1	**Portability:** Pure Java design provides for a fewer platform-specific limitations.	**Applet Portability:** Most Web browsers do not include the Swing classes, so the Java plugin must be used.

2	**Behavior:** Pure Java design allows for a greater range of behavior for Swing components since they are not limited by the native peers that AWT uses.	**Performance:** Swing components are generally slower and bigger than AWT, due to both the fact that they are pure Java and to video issues on various platforms. Since Swing components handle their own painting (rather than using native API's like DirectX on Windows), you may run into graphical glitches.
3	**Features:** Swing supports a wider range of features like icons and pop-up tool-tips for components.	
4	**Vendor Support:** Swing development is more active. Sun puts much more energy into making Swing robust.	
5	**Look and Feel:** The pluggable look and feel lets you design a single set of GUI components that can automatically have the look and feel of any OS platform (Microsoft Windows, Solaris, Macintosh, etc.). It also makes it easier to make global changes to your Java programs that provide greater accessibility (like picking a hi-contrast color scheme or changing all the fonts in all dialogs, etc.).	**Look and Feel:** Even when Swing components are set to use the look and the OS they are run on, they may not look like their native counterparts.

Swing also has the ability to replace these objects on-the-fly.

- 100% Java implementation of components
- Pluggable look and feel
- Lightweight components
- **Uses MVC Architecture**
    - **Model** represents the data
    - **View** as a visual representation of the data
    - **Controller** takes input and translates it to changes in data
- **Three parts**
    - Component set (subclasses of JComponent)
    - Support classes
    - Interfaces

Swing Packages

The Swing API is powerful, flexible and immense. The Swing API has 18 public packages:

javax.accessibility	javax.swing.plaf	javax.swing.text
javax.swing	javax.swing.plaf.basic	javax.swing.text.html
javax.swing.border	javax.swing.plaf.metal	javax.swing.text.html.parser
javax.swing.colorchooser	javax.swing.plaf.multi	javax.swing.text.rtf
javax.swing.event	javax.swing.plaf.synth	javax.swing.tree
javax.swing.filechooser	javax.swing.table	javax.swing.undo

Fortunately, most programs use only a small subset of the API. This trail sorts out the API for you, giving you examples of common code and pointing you to methods and classes you are likely to need. Most of the code in this trail uses only one or two Swing packages:

- javax.swing
- javax.swing.event (not always required)

Java Swing Class Hierarchy

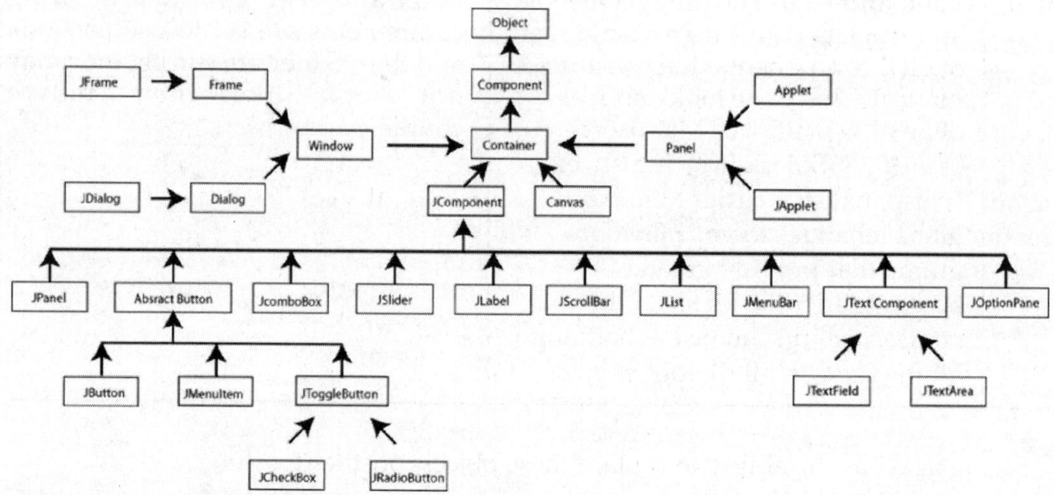


```
import javax.swing.JFrame;
import javax.swing.JLabel;
//import statements//
Check if window closes automatically. Otherwise add suitable code
public class HelloWorldFrame extends JFrame {
        public static void main(String args[]) {
                new HelloWorldFrame();
        } HelloWorldFrame() {
                JLabel jlbHelloWorld = new JLabel("Hello World");
                add(jlbHelloWorld);
                this.setSize(100, 100);
                // pack();
                setVisible(true);
        }
}
```

Compile

```
javac HelloWorldFrame.java
```

Run

```
java HelloWorldFrame
```

Output

9. MODEL-VIEW-CONTROLLER DESIGN PATTERN

Swing uses the *model-view-controller architecture* (MVC) as the fundamental design behind each of its components. Essentially, MVC breaks GUI components into three elements. Each of these elements plays a crucial role in how the component behaves. ***Model*** (includes state data for each component): The model encompasses the state data for each component. There are different models for different types of components. Model has no user interface. Model data always exists independent of the component's visual representation. For example, model of a Choice list might contain the information about list of items and currently selected item. This information remains the same no matter how the component is painted on the screen. ***View*** (to display component on screen): The view refers to how you see the component on the screen. It determines exactly where and how to draw the choice list by the information offered by the model. ***Controller*** (handles user Input): The controller decides the behavior of each component with respect to the events. Events come in many forms (a mouse click, a keyboard event). The controller decides how each component will react to the event—if it reacts at all.

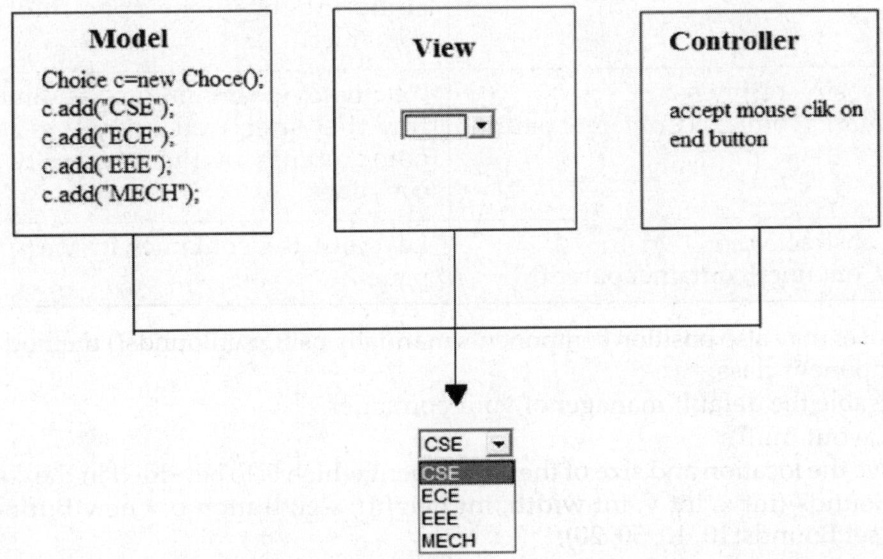

10. LAYOUT MANAGEMENT

Java AWT Layout Manager

Layout manager automatically arranges several components within a window. Each container object has a layout manager associated with it.

- Panel, Applet: Flow Layout
- Frame: Border layout

Whenever a container is resized, the layout manager is used to position each of the components within it. General syntax for setting layout to container is

Void set Layout(Layout Manager obj)

Layout managers are

- Flow layout
- Border layout
- Grid layout
- Gridbag layout
- Box layout

It defines the interface for classes that know how to layout containers. It is generally used as—public interface layout manager extends object.

The methods used in the layout manager are:

Usage	Use
public abstract void addLayoutComponent(String name,Component comp)	Adds the specified component with the specified name to the layout.
public abstract void removeLayoutComponent(String name,Component comp)	Removes the specified component with the specified name to the layout.
public abstract Dimension preferredLayoutSize(Container parent)	Calculates the preferred size dimensions for the specified panel given the components in the specified parent container.
public abstract Dimension minimumLayoutSize(Container parent)	Calculates the minimum size dimensions for the specified panel given the components in the specified parent container.
public abstract void layoutContainer(container parent)	Lays out the container in the specified panel.

The user may also position components manually using setBounds() method defined by Component class.

1. Disable the default manager of your container.

Set Layout (null);

2. Give the location and size of the component which is to be added in the container.

Set Bounds (int x, int y, int width, int height); e.g.: Button b = new Button(¯click me); b. set Bounds(10, 10, 50, 20);

Constructors and Methods

1. Flow Layout

Flow layout arranges the components in rows from left-to-right and top-to-bottom order based on the order in which they were added to the container. Flow Layout arranges components in rows, and the alignment specifies the alignment of the rows. For example, if you create a Flow Layout that is left aligned, the components in each row will appear next to the left edge of the container. The Flow Layout constructors allow you to specify the horizontal and vertical gaps that should appear between components, and if you use a constructor that does not accept these values, they both default to 5.

The flow layout manager lines the components horizontally until there is no more room and then starts a new row of components. When the user resizes the container, the layout manager automatically reflows the components to fill the available space. If you reduce the width of the frame further, then portions of the wider components begin to disappear. Similarly, if you reduce the frame's vertical size so that there's not enough vertical space to display all rows, some of the components will become partially or completely inaccessible .

The general method of implementation is

public class Flow Layout extends Object implements Layout Manager

Constructor

1. ***Flow Layout ():*** create default layout, which centers component and leaves 5 pixels spaces between each component.
2. ***Public Flow Layout (int align):*** Constructs a new Flow Layout with the specified alignment.
3. ***Public Flow Layout (int align, int hgap, int vgap):*** Constructs a new Flow Layout with the specified alignment and horizontal and vertical gap values.

Methods	Usage	Use
addLayout Component	public void addLayout Component(String name, Component comp)	Adds the specified component to the layout.
removeLayout Component	public void removeLayout Component(Component comp)	Removes the specified component from the layout. Does not apply.
preferredLayout Size	public Dimension preferred LayoutSize(Container target)	Returns the preferred dimensions for this layout given the components in the specified target container.
minimumLayout Size	public Dimension minimum LayoutSize(Container target)	Returns the minimum dimensions needed to layout the components contained in the specified target container.
layoutContainer	public void layoutContainer (Container target)	Lays out the container. This method will actually reshape

		the components in target in order to satisfy the constraints of the Border Layout object.
toString	public String toString()	Returns the String representation of this Flow Layout's values.

Sample program to demonstrate FlowLayout is given as below;

```
import java.awt.*;
public class FlowDemo extends Frame
{
        FlowLayout flow=new FlowLayout();
        Button b1=new Button("one");
        Button b2=new Button("Two");
        Button b3=new Button("Three");
        FlowDemo(String s)
        {
                super(s);
                setLayout(flow);
                add(b1);
                add(b2);
                add(b3);
                setSize(200,200);
                setVisible(true);
        }
        public static void main(String arg[])
        {
                FlowDemo ob=new FlowDemo("FlowLayout Demo");
        }
}
```

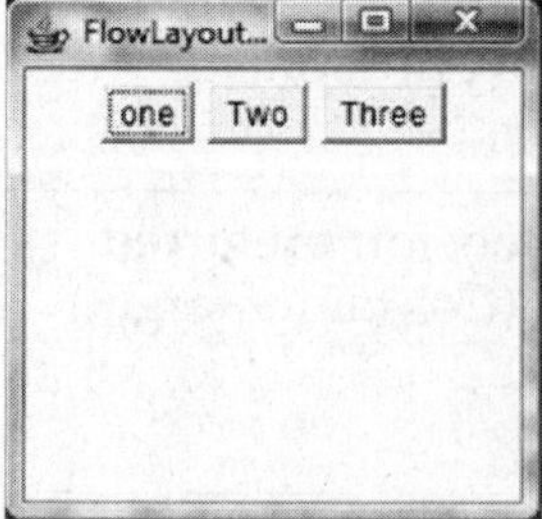

2. Border Layout

Border layout divides the container into five areas, and you can add a component to each area. The five regions correspond to the top, left, bottom, and right sides of the container, along with one in the center. Each of the five areas is associated with a constant value defined in BorderLayout: NORTH, SOUTH, EAST, WEST, and CENTER for the top, bottom, right, left, and center regions, respectively.

The general way to implement the border layout is
public class BorderLayout extends Object implements Layout Manager

Constructors

1. Border Layout () – Default constructor: Construct a new Border Layout
2. Border Layout (int hspace, int vspace) – Constructs a Border Layout with the specified horizontal and vertical gaps.

Methods	Usage	Use
addLayout Component	public void addLayout Component(String name, Component comp)	Adds the specified named component to the layout.
removeLayout Component	public void removeLayout Component(Component comp)	Removes the specified component from the layout.
minimumLayout Size	public Dimension minimum LayoutSize(Container target)	Returns the minimum dimensions needed to layout the components contained in the specified target container.
preferredLayout Size	public Dimension preferred LayoutSize(Container target)	Returns the preferred dimensions for this layout given the components in the specified target container.
layoutContainer	public void layoutContainer (Container target)	Lays out the specified container. This method will actually reshape the components in the specified target container in order to satisfy the constraints of the Border Layout object.
toString	public String toString()	Returns the String representation of this Border Layout's values.

Border Layout grows all components to fill the available space. You can add components by specifying a constraint—Border Layout. CENTER | NORTH | SOUTH | EAST | WEST
void add(Component obj, constraint)

```
import java.awt.*;
public class Demo extends Frame
{
        BorderLayout grid=new BorderLayout();
        Button b1=new Button("one");
        Button b2=new Button("Two");
        Button b3=new Button("Three");
        Button b4=new Button("four");
```

```
Button b5=new Button("five");
Demo(String s)
{
        super(s);
        setLayout(grid);
        add(b1,BorderLayout.NORTH);
        add(b2,BorderLayout.SOUTH);
        add(b3,BorderLayout.CENTER);
        add(b4,BorderLayout.EAST);
        add(b5,BorderLayout.WEST);
        setSize(200,200); setVisible(true);
}
public static void main(String arg[])
{
        Demo ob=new Demo("BorderLayout Demo");
}
}
```

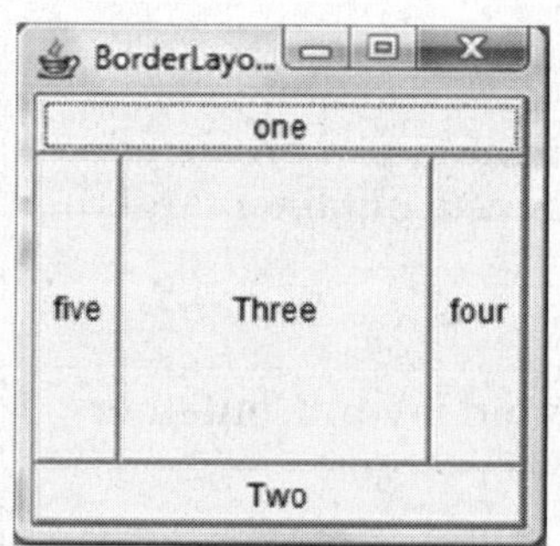

3. Grid Layout

The Grid Layout manager divides the available space into a grid of cells, evenly allocating the space among all the cells in the grid and placing one component in each cell. Cells are always the same size. When you resize the window, the cells grow and shrink, but all the cells have identical sizes. Constructor ? Grid Layout (int rows, int cols)—construct a grid with specified rows and cols.

? Grid Layout (int rows, int cols, int hspace, int vspace)—to specify the amount of horizontal and vertical space that should appear between adjacent components.

When you create a Grid Layout, you can specify a value of 0 for either the row count or the column count, but not both. If you set the number of rows to 0, Grid Layout creates as many rows as it needs to display all the components using the specified number of columns.

```
java.lang.Object
  |
  +——java.awt.GridLayout
```

It is generally represented as :

*public class **Grid Layout** extends Object implements Layout Manager*

A layout manager for a container that lays out grids.

Constructor

1. public GridLayout (int rows, int cols)
   Creates a grid layout with the specified rows and specified columns.
2. public GridLayout (int rows, int cols, int hgap, int vgap)
   Creates a grid layout with the specified rows, columns, horizontal gap, and vertical gap.

Throws: IllegalArgument Exception, if the rows and columns are invalid.

Methods	Usage	Use
addLayoutComponent	public void addLayout Component(String name, Component comp)	Adds the specified component with the specified name to the layout.
removeLayoutComponent	public void removeLayout Component(Component comp)	Removes the specified component from the layout. Does not apply.
preferredLayoutSize	public Dimension preferred LayoutSize(Container parent)	Returns the preferred dimensions for this layout given the components in the specified panel.
minimumLayoutSize	public Dimension minimum LayoutSize(Container parent)	Returns the minimum dimensions needed to layout the components contained in the specified panel.
layoutContainer	public void layoutContainer (Container parent)	Lays out the container in the specified panel.
toString	public String toString()	Returns the String representation of this GridLayout's values.

```
import java.awt.*;
public class Demo extends Frame
{
        GridLayout grid=new GridLayout(2,2);
        Button b1=new Button("one");
        Button b2=new Button("Two");
        Button b3=new Button("Three");
        Demo(String s)
        {
                super(s);
                setLayout(grid);
                add(b1);
                add(b2);
```

```
            add(b3);
            setSize(200,200);
            setVisible(true);
        }
        public static void main(String arg[])
        {
            Demo ob=new Demo("GridLayout Demo");
        }
}
```

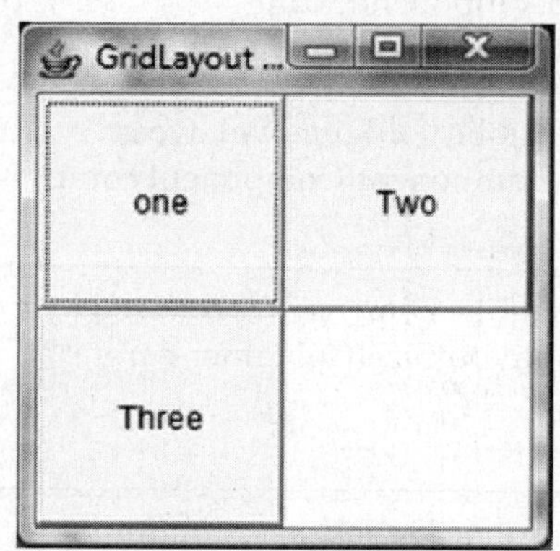

4. GridbagLayout

```
java.lang.Object
    |
    +——java.awt.GridBagLayout
```

It is generally represented as:

*public class **GridBagLayout** extends Object implements Layout Manager*

In Grid Bag Layout, the rows and columns have variable sizes. It is possible to merge two adjacent cells and make a space for placing larger components. To describe the layout to grid bag manager, you must follow the procedure.

1. Create an object of type Grid Bag Layout. No need to specify rows and column.
2. Set this Grid Bag Layout object to the container.
3. Create an object of type Grid Bag Constraints. This object will specify how the components are laid out within the grid bag.
4. For each components, fill in the Grid Bag Constraints object. Finally add the component with the constraint by using the call add (component, constraint);

Grid Bag Layout is a flexible layout manager that aligns components vertically and horizontally, without requiring that the components be the same size. Each Grid Bag Layout uses a rectangular grid of cells, with each component occupying one or more cells (called its *display area*). Each component managed by a GridBagLayout is associated with a Grid Bag Constraints instance that specifies how the component is laid out within its display area. How a Grid Bag Layout places a set of components depends on each component's Grid Bag Constraints and minimum size, as well as the preferred size of the components' container.

To use a Grid Bag Layout effectively, you must customize one or more of its components' GridBagConstraints. You customize a Grid Bag Constraints object by setting one or more of its instance variables:

gridx, gridy

Specifies the cell at the upper left of the component's display area, where the upper-left-most cell has address gridx=0, gridy=0. Use Grid Bag Constraints. RELATIVE (the default value) to specify that the component be just placed just to the right of (for gridx) or just below (for gridy) the component that was added to the container just before this component was added.

gridwidth, gridheight

Specifies the number of cells in a row (for gridwidth) or column (for gridheight) in the component's display area. The default value is 1. Use Grid Bag Constraints. REMAINDER to specify that the component be the last one in its row (for gridwidth) or column (for gridheight). Use Grid Bag Constraints. RELATIVE to specify that the component be the next to last one in its row (for gridwidth) or column (for gridheight).

fill

Used when the component's display area is larger than the component's requested size to determine whether (and how) to resize the component. Valid values are Grid Bag Constraint. NONE (the default), Grid Bag Constraint. HORIZONTAL (make the component wide enough to fill its display area horizontally, but do not change its height), Grid Bag Constraint. VERTICAL (make the component tall enough to fill its display area vertically, but do not change its width), and Grid Bag Constraint. BOTH (make the component fill its display area entirely).

ipadx, ipady

Specifies the internal padding: How much to add to the minimum size of the component. The width of the component will be at least its minimum width plus ipadx*2 pixels (since the padding applies to both sides of the component). Similarly, the height of the component will be at least the minimum height plus ipady*2 pixels.

insets

Specifies the external padding of the component — the minimum amount of space between the component and the edges of its display area.

anchor

Used when the component is smaller than its display area to determine where (within the area) to place the component. Valid values are Grid Bag Constraints. CENTER (the default), Grid Bag Constraints. NORTH, Grid Bag Constraints. NORTHEAST, Grid Bag Constraints. EAST, Grid Bag Constraints. SOUTHEAST, Grid Bag Constraints. SOUTH, Grid Bag Constraints. SOUTHWEST, Grid Bag Constraints. WEST, Grid Bag Constraints. NORTHWEST, Grid Bag Constraints.

weightx, weighty

Used to determine how to distribute space; this is important for specifying resizing behavior. Unless you specify a weight for at least one component in a row (weightx) and column (weighty), all the components clump together in the center of their container. This is because when the weight is zero (the default), the Grid Bag Layout puts any extra space between its grid of cells and the edges of the container.

The following figure shows ten components (all buttons) managed by a Grid Bag Layout:

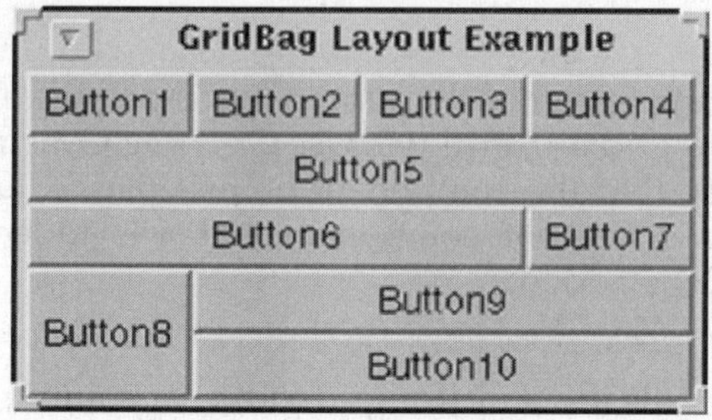

All the components have fill=Grid Bag Constraints.BOTH. In addition, the components have the following non-default constraints:

- Button1, Button2, Button3: weightx=1.0
- Button4: weightx=1.0, gridwidth=Grid Bag Constraints. REMAINDER
- Button5: gridwidth = Grid Bag Constraints. REMAINDER
- Button6: gridwidth = Grid Bag Constraints. RELATIVE
- Button7: gridwidth = Grid Bag Constraints. REMAINDER
- Button8: gridheight = 2, weighty = 1.0,
- Button9, Button 10: gridwidth = Grid Bag Constraints. REMAINDER

Here is the code that implements the example shown above:

```
import java.awt.*;
import java.util.*;
import java. applet. Applet;
public class Grid Bag Ex1 extends Applet {
 protected void make button (String name,
        Grid Bag Layout gridbag,
        Grid Bag Constraints c) {
 Button button = new Button (name);
 gridbag. set Constraints (button, c);
 add (button);
}
public void init() {
 GridBagLayout gridbag = new GridBagLayout();
 GridBagConstraints c = new GridBagConstraints();
 setFont(new Font("Helvetica", Font.PLAIN, 14));
 setLayout(gridbag);

 c.fill = GridBagConstraints.BOTH;
 c.weightx = 1.0;
 makebutton("Button1", gridbag, c);
 makebutton("Button2", gridbag, c);
 makebutton("Button3", gridbag, c);
    c.gridwidth = GridBagConstraints.REMAINDER; //end row
```

```
makebutton("Button4", gridbag, c);

    c.weightx = 0.0;            //reset to the default
makebutton("Button5", gridbag, c); //another row c.

    c.gridwidth = GridBagConstraints.RELATIVE; //next-to-last in row
makebutton("Button6", gridbag, c);

    c.gridwidth = GridBagConstraints.REMAINDER; //end row
makebutton("Button7", gridbag, c);

    c.gridwidth = 1;            //reset to the default
    c.gridheight = 2;
c.weighty = 1.0;
makebutton("Button8", gridbag, c);

c.weighty = 0.0;        //reset to the default
  c.gridwidth = GridBagConstraints.REMAINDER; //end row
  c.gridheight = 1;       //reset to the default
makebutton("Button9", gridbag, c);
makebutton("Button10", gridbag, c);

resize(300, 100);
}

public static void main(String args[]) {
    Frame f = new Frame("GridBag Layout Example");
    GridBagEx1 ex1 = new GridBagEx1();

    ex1.init();

    f.add("Center", ex1);
    f.pack();
    f.resize(f.preferredSize());
    f.show();
 }
}
```

Constructor

public GridBagLayout(): Creates a gridbag layout. It is a default constructor.

Methods

Methods	Usage	Use
setConstraints	public void setConstraints(Component comp, GridBagConstraints)	Sets the constraints for the specified component.
getConstraints	public GridBagConstraints getConstraints(Component comp)	Retrieves the constraints for the specified component. A copy of the constraints is returned.
lookupConstraints	protected GridBagConstraints lookupConstraints(Component comp)	Retrieves the constraints for the specified component. The return value is not a copy, but is the actual constraints class used by the layout mechanism.
addLayoutComponent	public void addLayout Component(String name, Component comp)	Adds the specified component with the specified name to the layout.
removeLayoutComponent	public void removeLayout Component(Component comp)	Removes the specified component from the layout. Does not apply.
preferredLayoutSize	public Dimension preferred LayoutSize(Container parent)	Returns the preferred dimensions for this layout given the components in the specified panel.
minimumLayoutSize	public Dimension minimum LayoutSize(Container parent)	Returns the minimum dimensions needed to layout the components contained in the specified panel.
layoutContainer	public void layoutContainer (Container parent)	Lays out the container in the specified panel.
toString	public String toString()	Returns the String representation of this GridLayout's values.
DumpLayoutInfo	protected void DumpLayoutInfo (GridBagLayoutInfo)	Print the layout information. Useful for debugging.
DumpConstraints	protected void DumpConstraints (GridBagConstraints)	Print the layout constraints. Useful for debugging.
GetLayoutInfo	protected GridBagLayoutInfo GetLayoutInfo(Container parent,int sizeflag)	Returns the layout information

```
import java.awt.*;
public class Demo extends Frame
{
        GridBagLayout gb=new GridBagLayout();
        GridBagConstraints gc1= new GridBagConstraints();
        GridBagConstraints gc2= new GridBagConstraints();
        GridBagConstraints gc3= new GridBagConstraints();
        Button b1=new Button("one");
        Button b2=new Button("Two");
        Button b3=new Button("Three");
        Demo(String s)
        {
                super(s);
                setLayout(gb);
                gc1.gridx=0;
                gc1.gridy=0;
                gc1.gridwidth=2;
                gc1.gridheight=1;
                gc2.gridx=0;
                gc2.gridy=1;
                gc2.gridwidth=1;
                gc2.gridheight=1;
                gc3.gridx=1;
                gc3.gridy=1;
                gc3.gridwidth=1;
                gc3.gridheight=1;
                add(b1,gc1);
                add(b2,gc2);
                add(b3,gc3);
                setSize(200,200);
                setVisible(true);
                }
                public static void main(String arg[])
                {
                        Demo ob=new Demo("GridBagLayout Demo");
                }
}
```

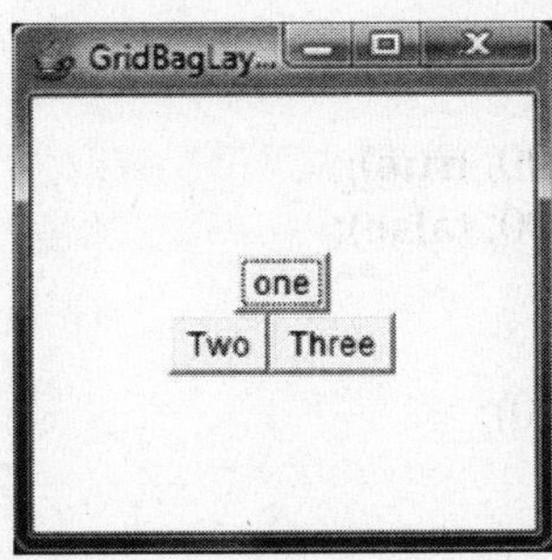

11. APPLET

The class hierarchy for applets is:

```
java.lang.Object
|
|___ java.awt.Component
        |
    |___java.awt.Container
              |
              |___java.awt.Panel
                      |
                      |___ java.applet.Applet
```

The Applet class inherits the paint method from the Container class. This method is called whenever the container needs to be redrawn onto the screen, and the Graphics object it is passed contains the object to draw on, its location, the default color to use, etc.

The Graphics class provides a wide variety of methods for drawing, which can be roughly broken up into two major groups: Draw and fill. For example,

public void drawRect (int x, int y, int width, int height)

draws the outline of a rectangle, whereas

public abstract void fillRect (int x, int y, int width, int height)

draws the outline of a rectangle and fills it with the current colour.

Here is the code for an applet that uses some of the Graphics class methods.

```
/* <applet code="GraphicsExample.class" width=500 height=500></applet>
*/

import java.applet.Applet;
import java.awt.Graphics;
import java.awt.Polygon;

public class GraphicsExample extends Applet
{
public void paint(Graphics g)
{
Polygon polygon = new Polygon();
int [] xPoints = {260, 280, 300, 320, 340, 360};
int [] yPoints = {60, 80, 60, 80, 60, 80 };
int numPoints = 6;

g.draw3DRect(20, 20, 400, 400, true);
g.draw3DRect(40, 40, 400, 400, false);

g.drawLine(20, 20, 40, 40);
g.drawLine(420, 420, 440, 440);
g.drawLine(20, 420, 40, 440);
g.drawLine(420, 20, 440, 40);
```

```
g.drawString("The Graphics class has many interesting methods.", 60, 210);

g.fillOval(80, 60, 50, 50);
g.drawOval(70, 50, 70, 70);

g.fillRoundRect(80, 300, 90, 90, 20, 20);
g.drawLine(170, 300, 200, 390);

g.fillArc( 300, 250, 40, 40, 0, 270);

polygon.addPoint(280, 350);
polygon.addPoint(350, 280);
polygon.addPoint(267, 235);
polygon.addPoint(243, 412);

g.drawPolygon(polygon);

g.drawPolyline(xPoints, yPoints, numPoints);

g.drawRect(200, 65, 200, 75);

}
}
```

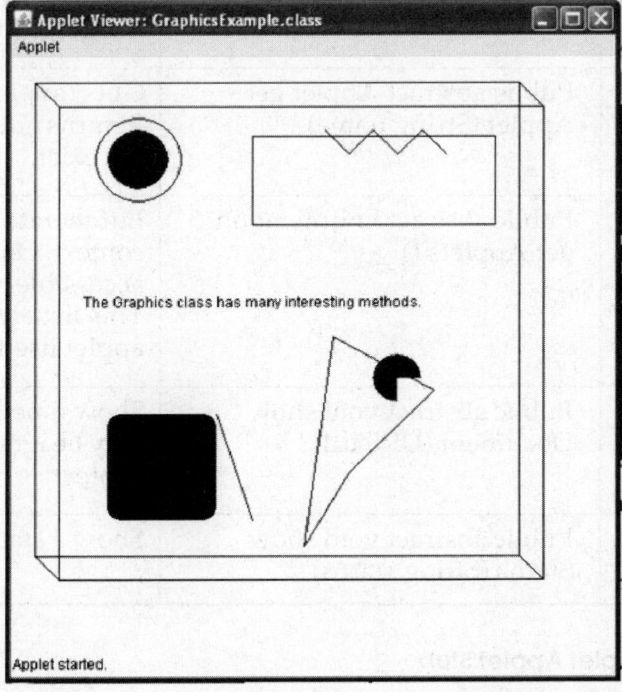

Applet is a small application that is accessed on an internet server, transported over the internet and run as part of web document. The java applet package has the Applet class , which contains several methods that provides detailed control over the execution of the applet.

The java applet package is detailed with its three interfaces: Applet Context, Applet Stub and Audio Clip and the class Applet as below:

- Interface java applet Applet Context

public interface **Applet Context** extends Object

Applet context is the interface that corresponds to an Applet's environment. It can be used by an applet to obtain information from the applet's environment, which is usually the browser or the applet viewer.

Applet Context Methods

Methods	Usage	Use
Get Audio Clip	Public abstract Audio Clip get Audio Clip (URL url)	Gets an audio clip.
Get Image	Public abstract Image get Image (URL url)	Gets an image. This usually involves downloading it over the net. However, the environment may decide to cache images. This methods takes an array of URLs, each of which will be tried until the images is found.
Get Applet	Public abstract Applet get Applet (String name)	Gets an applet by name. Returns null if the applet does not exist.
Get Applets	Public abstract Enumeration get Applets ()	Enumerate the applets in this context. Only applets that are accessible will be returned. This list always includes the applet itself.
Show Document	Public abstract void show Document (URL url)	Show a new document. This may be ignored by the applet context.
Show Status	Public abstract void show Status (String status)	Show a status string.

Interface Java Applet Applet Stub

Public interface **AppletStub** extends Object

Applet stub is the interface used to implement an applet viewer. It is not normally used by applet programmers.

Applet Stub-methods

Methods	Usage	Use
Is Active	Public abstract boolean is Active ()	Returns true if the applet is active.
Get Document Base	Public abstract URL get Document Base ()	Gets the document URL.
Get Code Base	Public abstract URL get Code Base ()	Gets the base URL.
Get Parameter	Public abstract String get Parameter (String name)	Gets a paramater of the applet.
Get Applet Context	Public abstract Applet Context get Applet Context ()	Gets a handler to the applet's context.
Applet Resize	Public abstract void applet Resize (int width, int height)	Is called when the applet wants to be resized.

- **Interface java applet Audio Clip**

Public interface **Audio Clip** extends Object
Audio Clip interface. A very high level abstraction of audio.

Methods	Usage	Use
Play	Public abstract void play ()	Start playing the clip. Each time this is called the clip is restarted from the beginning.
Loop	Public abstract void loop ()	Start playing the clip in a loop.
Stop	Public abstract void stop ()	Stop playing the clip.

The class hierarchy for the class java.applet. Applet is given as:

```
java.lang.Object
  |
  +——java.awt.Component
      |
      +——java.awt.Container
          |
          +——java.awt.Panel
              |
              +——java.applet.Applet
```

public class **Applet** extends Panel
This is the Base applet class.

- **java applet Applet – default Constructor**

public Applet()

Java applet Applet – Methods

Methods	Usage	Use
set Stub	Public final void setStub (Applet Stub stub)	Set the applet stub. This is done automatically by the system.
isActive	Public boolean isActive ()	Returns true if the applet is active. An applet is marked active just before the start method is called.
getDocument Base	Public URL getDocument Base ()	Gets the document URL. This is the URL of the document in which the applet is embedded.
getCodeBase	Public URL getCodeBase ()	Gets the base URL. This is the URL of the applet itself.
getParameter	Public String get Parameter (String name)	Gets a parameter of the applet.
getApplet Context	Public Applet Context getApplet Context ()	Gets a handle to the applet context. The applet context lets an applet control the applet's environment which is usually the browser or the applet viewer.
resize	Public void resize (int width, int height)	Request for the applet to be resized. Overrides resize () in class component
showStatus	Public void show Status (String msg)	Show a status message in the Applet's context.
getImage	Public Image getImage (URL url)	Gets an image given a URL. Note that this method always returns an image object immediately, even if the image does not exist. The actual image data is loaded when it is first needed.
getImage	Public Image get Image (URL url, String name)	Gets an image relative to a URL. This methods returns immediately, even if the image does not exist. The actual image data is loaded when it is first needed.

getAudioClip	Public Audio Clip getAudioClip (URL url)	Gets an audio clip.
getAudioClip	Public Audio Clip getAudioClip (URL url, String name)	Gets an audio clip.
getAppletInfo	Public String get Applet Info ()	Return a string containing information about the author, version and copyright of the applet.
getParameter Info	Public String[][] getParameterInfo ()	Returns an array of strings describing the parameters that are understood by this applet. The array consists of sets of 3 strings name/type/description.
play	Public void play (URL url)	Play an audio clip. Nothing happens if the audio clip could not be found.
play	Public void play (URL url, String name)	Play an audio clip. Nothing happens if the audio clip could not be found.
init	Public void init ()	Initializes the applet. You never need to call this directly, it is called automatically by the system once the applet is created.
start	Public void start ()	Called to start the applet. You never need to call this method directly, it is called when the applet's document is visited.
stop	Public void stop ()	Called to stop the applet. It is called when the applet's document is no longer on the screen. It is guaranteed to be called before destroy() is called. You never need to call this method directly.
destroy	Public void destroy ()	Cleans up whatever resources are being held. If the applet is active, it is first stopped.

APPLET BASICS

All applets are subclasses of Applet Class. So in order to load an Applet class java.applet package must be imported.

Sample code

```
import java.applet.*;
public class myapplet extends Applet
{
................
................
}
```

All applet classes are declared public, because it has to communicate with the browser and other classes. The applet class can also be declared without import statement as follows:

```
public class myapplet extends java.applet.Applet
{
...................
...................

}
Example
//demo of Applet
import java.applet.*;
//<applet code=sample width=300 height=300> < /applet>
//an empty applet will be created
public class sample extends Applet
{
public void init()
{
//this message will be displayed in the command prompt
System.out.println("This is my first applet");
}
}
```

SKELETON OF AN APPLET

Trivial applets overrides a set of methods that provide the basic mechanism by which the browser or applet viewer interfaces to the applet and controls its execution. Four of these methods—init(), start(), stop() and destroy()—are defined by the Applet class. Another method paint() is defined by the AWT component class. Default implementation for all these methods are provided. Applet do not need to override these methods that they do not use. These five methods can be assembled into the skeleton shown here.

```
import java.awt.*;
import java.applet.*;
public class classname extends Applet
//First method
public void init()
{
//Initialisation
}
//second method, this method is also called whenever the applet is restarted
public void start()
{
//start or resume execution
}
//third method, called when the applet is stopped
public void stop()
{
//suspends execution
}
//fourth method called when the applet is terminated- last method
{
public void destroy()
{
//Perform shutdown activities
}
//called when an applet's window must be restored.
public void paint(Graphics g)
{
//redisplay the contents of the window
}
}
```

Applet Initialisation and Termination Order

It is important to understand the order in which the various methods shown in the skeleton are called. When an applet begins, the AWT calls the following methods, in this sequence

Init()

When a web browser or an appletviewer runs a java applet, the applet's execution starts with the init method. The init() method is as similar as main() method in application programs, at which the program's execution starts. However, unlike the main method, when the init method ends, the applet does not end. Most applets use init to initialize the key variables. If the user does not define the init method, then java runs its own init method defined default in the applet class.

Start()

The start( ) method is called after init( ). It is also called to restart an applet after it has been stopped, whereas init( ) is called the first time an applet is loaded. Start( ) is called each time an applet's HTML document is displayed on the screen. So, id a user leaves a web page and comes back, the applet resumes execution at start( ) method.

Paint()

The paint method is called each time, when the applet's output must be redrawn. This situation can occur for several reasons. For example, the window in which the applet is running may be overwritten by another window and then uncovered, or the applet window may be minimized and then restored. Paint( ) is also called when the applet begins execution. Whatever the cause, whenever the applet must redraw its output, paint( ) is called. The paint( ) method has one parameter of type Graphics. This parameter will contain the graphics context, which describes the graphics environment in which the applet is running. This context is used whenever output to the applet is required.

Stop()

Stop( ) method is called when a web browser leaves the HTML document containing the applet, that is, whenever the user moves off the web page that contains the applet.

Destroy()

The destroy( ) method is called when the environment determines that the running applet needs to be removed completely from memory at this point, the resource used by the applet must be freed up, so stop( ) is always called before destroy( ).

EMBEDDING AN APPLET INTO HTML

After creating an applet , to run it from both an internet browser and an applet viewer, it has to be embedded in HTML. An HTML file uses keywords, known as tags <>, which defines the page's appearance to the browser. For example, the designer uses HTML entries to define the appearance of text and where and what images the browser displays on the web page when Java applet is added to web pages, HTML uses special <Applet> tag that provides the browser with information about the applet.

Using the <Applet> tag, the designer can specify a set of attributes that describe such items as the applet filename, the size and location of the applet window, an so on. The complete syntax of the <Applet> tag is as follows:

<APPLET

CODEBASE = *codebase URL*
ARCHIVE = *archive List*
CODE = *applet File* ...or... **OBJECT = *serialized Applet***
ALT = *alternate Text*
NAME = *applet Instance Name*
WIDTH = *pixels* HEIGHT = *pixels*
ALIGN = *alignment*
VSPACE = *pixels* HSPACE = *pixels*
>
<PARAM NAME = *applet Attribute 1* VALUE = *value*>
<PARAM NAME = *applet Attribute 2* VALUE = *value*>
. . .
alternate HTML

</APPLET>

CODE, CODEBASE, and so on are attributes of the applet tag; they give the browser information about the applet. The only mandatory attributes are CODE, WIDTH, and HEIGHT. Each attribute is described below.

CODEBASE = *codebaseURL*

This OPTIONAL attribute specifies the base URL of the applet—the directory that contains the applet's code. If this attribute is not specified, then the document's URL is used.

ARCHIVE = *archiveList*

This OPTIONAL attribute describes one or more archives containing classes and other resources that will be "preloaded". The classes are loaded using an instance of an Applet Class Loader with the given CODEBASE. The archives in *archiveList* are separated by ",". NB: in JDK1.1, multiple APPLET tags with the same CODEBASE share the same instance of a Class Loader. This is used by some client code to implement inter-applet communication. Future JDKs *may* provide other mechanisms for inter-applet communication. For security reasons, the applet's class loader can read only from the same codebase from which the applet was started. This means that archives in *archiveList* must be in the same directory as, or in a subdirectory of, the codebase. Entries in *archive List* of the form ../a/b.jar will not work unless explicitly allowed for in the security policy file (except in the case of an http codebase, where archives in *archiveList* must be from the same host as the codebase, but can have ".."'s in their paths.)

CODE = *appletFile*

This REQUIRED attribute gives the name of the file that contains the applet's compiled Applet subclass. This file is relative to the base URL of the applet. It cannot be absolute. One of CODE or OBJECT must be present. The value *appletFile* can be of the form *classname*.class or of the form*packagename.classname*.class.

OBJECT = *serializedApplet*

This attribute gives the name of the file that contains a serialized representation of an Applet. The Applet will be deserialized. The init() method will *not* be invoked; but its start() method will. Attributes valid when the original object was serialized are *not* restored. Any attributes passed to this APPLET instance will be available to the Applet; we advocate very strong restraint in using this feature. An applet should be stopped before it is serialized. One of CODE or OBJECT must be present.

ALT = *alternateText*

This OPTIONAL attribute specifies any text that should be displayed if the browser understands the APPLET tag but cannot run Java applets.

NAME = *appletInstanceName*

This OPTIONAL attribute specifies a name for the applet instance, which makes it possible for applets on the same page to find (and communicate with) each other.

WIDTH = *pixels* HEIGHT = *pixels*

These REQUIRED attributes give the initial width and height (in pixels) of the applet display area, not counting any windows or dialogs that the applet brings up.

ALIGN = *alignment*

This OPTIONAL attribute specifies the alignment of the applet. The possible values of this attribute are the same as those for the IMG tag: left, right, top, texttop, middle, absmiddle, baseline, bottom, absbottom.

VSPACE = *pixels* HSPACE = *pixels*

These OPTIONAL attributes specify the number of pixels above and below the applet (VSPACE) and on each side of the applet (HSPACE). They are treated in the same way as the IMG tag's VSPACE and HSPACE attributes.

<PARAM NAME = *appletAttribute1* VALUE = value>

<PARAM NAME = *appletAttribute2* VALUE = value> . . .

This tag is the only way to specify an applet-specific attribute. Applets access their attributes with the getParameter() method.

12. MESSAGE BOX AND SWING COMPONENTS

Message Box

Message dialog box is used to display informative messages to the user. J OptionPane class can be used to display the message Dialog box.

The J Option Pane class has three methods as follows:

- **showMessageDialog():** First is the **showMessageDialog()** method which is used to display a simple message.
- **showInputDialog():** Second is the **showInputDialog()** method which is used to display a prompt for inputting. This method returns a String value which is entered by you.
- **showConfirmDialog():** And the last or third method is the **showConfirmDialog()** which asks the user for confirmation (Yes/No) by displaying message. This method returns a numeric value either 0 or 1. If you click on the "Yes" button, then the method returns 1, otherwise 0.

Icons used by JOptionPane

Icon description	Java look and feel	Windows look and feel
question	?	?
information	i	i
warning	!	!
error	x	X

S. No.	Input Code	Output
1	JOptionPane showMessageDialog (frame, "Eggs are not supposed to be green.");	Message — Eggs aren't supposed to be green. [OK]
2	JOptionPane showMessageDialog (frame, "Eggs are not supposed to be be green.", "Inane warning", J Option Pane.WARNING_MESSAGE);	Inane warning — Eggs aren't supposed to be green [OK]
3	JOptionPane showMessageDialog (frame, "Eggs are not supposed to be green.", "Inane error", J Option Pane ERROR_MESSAGE);	Inane error — Eggs aren't supposed to be green. [OK]

4	JOptionPane show MessageDialog (frame, "Eggs are not supposed to be green.","A plain message", J OptionPane PLAIN_MESSAGE);	A plain message Eggs aren't supposed to be green. OK
5	JOptionPane showMessageDialog (frame, "Eggs are not supposed to be green.", "Inane custom dialog", J Option Pane INFORMATION_ MESSAGE, icon);	Inane custom dialog Eggs aren't supposed to be green. OK
6	int n = JOptionPane showConfirmDialog (frame, "Would you like green eggs and ham?", "An Inane Question", J Option Pane YES_NO_OPTION);	An Inane Question ? Would you like green eggs and ham? Yes No

Swing Components

JFrame

To add a component to a JFrame, we must use its content Pane instead. JFrame is a Window with border, title and buttons. When JFrame is set visible, an event dispatching thread is started. JFrame objects store several objects including a Container object known as the content pane. To add a component to a JFrame, add it to the content pane.

JFrame Features

- It is a window with title, border, (optional) menu bar and user-specified components.
- It can be moved, resized.
- It is not a subclass of J Component.
- Delegates responsibility of managing user-specified components to a content pane, an instance of JPanel.

Creating a JFrame Window

Step 1: Construct an object of the JFrame class.

Step 2: Set the size of the Jframe.

Step 3: Set the title of the Jframe to appear in the title bar (title bar will be blank if no title is set).

Step 4: Set the default close operation. When the user clicks the close button, the program stops running.

Step 5: Make the Jframe visible.

S.No.	Constructors	Purpose
1	**JFrame ()**	Constructs a new frame that is initially invisible.
2	**JFrame (Graphics Configuration gc)**	Creates a Frame in the Specified Graphics Configuration of a screen device and a blank title.

3	**JFrame (String title)**	Creates a new, initially invisible Frame with the specified title.
4	**JFrame (String title, Graphics Configuration gc)**	Creates a JFrame with the specified title and the specified Graphics Configuration of a screen device.
Methods		
1	void validate()	Check the layout of components in the window
2	void repaint()	Repaint the window

J Panel

Declare and create the J Panel
For example,
J Panel *p* = new J Panel();

Common Methods Used:

p.**setLayout**(*LayoutManager layout*);
p.**setLayout**(new BorderLayout());
p.**setPreferredSize**(new Dimension(200, 100));
p.**setBackground**(Color.blue);
p.**setPreferredSize**(new Dimension(*int width, int height*));
p.**setBackground**(*Color c*);
p.**add**(*some component*);
w = *p*.getWidth(); // Used for graphics, not components
h = *p*.getHeight(); // Used for graphics, not components

JLabel

A J Label object provides text instructions or information on a GUI — display a single line of read-only text, an image or both text and image.

S. No.	Constructors	Purpose
1	**JLabel()**	Creates a JLabel instance with no image and with an empty string for the title.
2	**JLabel(Icon image)**	Creates a JLabel instance with the specified image.
3	**JLabel(Icon image, int horizontalAlignment)**	Creates a JLabel instance with the specified image and horizontal alignment.
4	**JLabel(String text)**	Creates a JLabel instance with the specified text.

5	**JLabel (String text, Icon icon, int horizontal Alignment)**	Creates a JLabel instance with the specified text, image, and horizontal alignment.
6	**JLabel (String text, int horizontal Alignment)**	Creates a JLabel instance with the specified text and horizontal alignment.
Methods		
1	void set Horizontal Alignment (intalignment)	Sets the alignment of the label's contents along the X axis. Alignment should have one of the values in Swing Constants: RIGHT, LEFT, CENTER
2	void set Horizontal Text Position (inttext Position)	Sets the horizontal position of the label's text, relative to its image. Position has one of the values in Swing Constants: RIGHT, LEFT, CENTER
3	void set Text (String text)	Defines the single line of text this component will display.
4	void set Vertical Alignment (intalignment)	Sets the alignment of the label's contents along the Y axis. Alignment has one of the values in Swing Constants: TOP, BOTTOM, CENTER
5	void set Vertical Text Position (inttext Position)	Sets the vertical position of the label's text, relative to its image. Text Position has one of the values in Swing Constants: TOP, BOTTOM, CENTER

Ex:

```
import javax swing JFrame;
import javax swing JLabel;
public class Labeln Text {
public static void main (String[] args) {

      JFrame f = new JFrame ("Hello Swing"); //We create a new frame
      f.setSize(200, 100); //Setting size of width=200px and height=100px
      f.setDefaultCloseOperation(JFrame.EXIT_ON_CLOSE);
  //Frame must close when close button is pressed
      f.add(new JLabel("Diving into swing!")); //Adding a label
      f.setVisible(true);
    }
  }
```

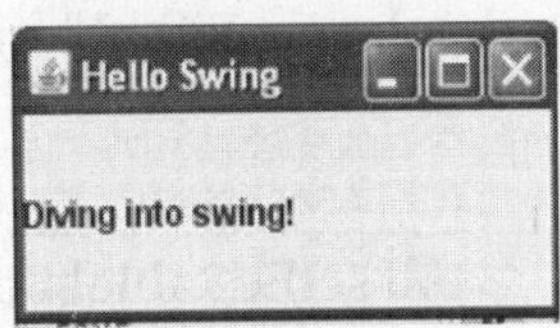

JTextField

JTextField allows editing/displaying of a single line of text. New features include the ability to justify the text left, right, or center, and to set the text's font. When the user types data into them and presses the Enter key, an action event occurs. If the program registers an event listener, the listener processes the event and can use the data in the text field at the time of the event in the program. J Text Field is an input area where the user can type in characters. If you want to let the user enter multiple lines of text, Jtextfield cannot be used unless you create several of them. The solution is to use J **Text Area**, which enables the user to enter multiple lines of text.

S. No.	Constructors	Purpose
1	**JTextField ()**	Constructs a new TextField.
2	**JTextField (Document doc, String text, int columns)**	Constructs a new JTextField that uses the given text storage model and the given number of columns.
3	**JText Field (int columns)**	Constructs a new empty Text Field with the specified number of columns.
4	**JText Field (String text)**	Constructs a new Text Field initialized with the specified text.
5	**JText Field (String text, int columns)**	Constructs a new Text Field initia-lized with the specified text and columns.
Methods		
1.	String getText()	Returns the text contained in this Text Component.
2.	boolean isEditable()	Returns the boolean indicating whether this Text Component is editable or not.
3.	void setEditable (boolean b)	Sets the specified boolean to indicate whether or not this Text Component should be editable.
4.	void setHorizontal Alignment (int alignment)	Sets the horizontal alignment of the text. Alignment is one of the values from Swing Constants: RIGHT, LEFT, CENTER
5.	void setText (String t) to the specified text.	Sets the text of this Text Component

Source Code

```
import java.awt.event.ActionEvent;
import javax.swing.JFrame;
import javax.swing.JTextField;
public class AddingActionListenerJTextField {
public static void main(String[] a) {
  JFrame frame = new JFrame();
  frame.setDefaultCloseOperation(JFrame.EXIT_ON_CLOSE);
  JTextField jTextField1 = new JTextField();
  jTextField1.setText("jTextField1");
  jTextField1.addActionListener(new java.awt.event.ActionListener() {
    public void actionPerformed(ActionEvent e) {
      System.out.println("action");
    }
  });
  frame.add(jTextField1);

  frame.setSize(300, 200);
  frame.setVisible(true);
  }

}
```

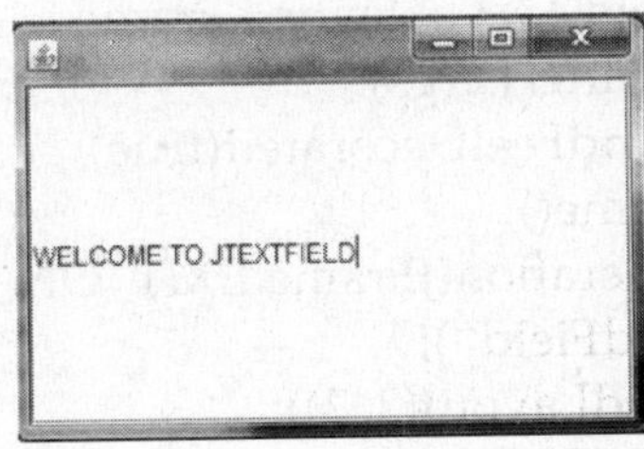

JPasswordField

JPasswordField (a direct subclass of JTextField)—you can suppress the display of input. Each character entered can be replaced by an echo character. This allows confidential input for passwords, for example. By default, the echo character is the asterisk, *. When the user types data into them and presses the Enter key, an action event occurs. If the program registers an event listener, the listener processes the event and can use the data in the text field at the time of the event in the program. If you need to provide an editable text field that does not show the characters the user types—use the JPasswordField class.

JPasswordField Constructor

S. No.	Constructors and Methods	Purpose
1	**JPasswordField()**	Constructs a new JPasswordField, with a default document, null starting text string, and 0 column width.

2	**JPasswordField (Document doc, String txt, int columns)**	Constructs a new JPasswordField that uses the given text storage model and the given number of columns.
3	**JPasswordField (int columns)**	Constructs a new empty JPassword-Field with the specified number of columns.
4	**JPasswordField (String text)**	Constructs a new JPasswordField initialized with the specified text.
5	**JPasswordField (String text, int columns)**	Constructs a new JPasswordField initialized with the specified text and columns.

Source Code

```
import java.awt.GridLayout;
import javax.swing.JFrame;
import javax.swing.JLabel;
import javax.swing.JPasswordField;
import javax.swing.JTextField;
import javax.swing.SwingConstants;
  public class J Password Field Test extends J Frame {
   public static void main(String[] args) {
    JFrame.setDefaultLookAndFeelDecorated(true);
    JFrame frame = new JFrame();
    frame.setDefaultCloseOperation(JFrame.EXIT_ON_CLOSE);
    frame.setTitle("JPasswordField");
    frame.setLayout(new GridLayout(2, 2));
    JLabel label = new JLabel("User Name:", SwingConstants.RIGHT);
    JLabel label2 = new JLabel("Password:", SwingConstants.RIGHT);
    JTextField userNameField = new JTextField(20);
    JPasswordField passwordField = new JPasswordField();
    frame.add(label);
    frame.add(userNameField);
    frame.add(label2);
    frame.add(passwordField);
    frame.setSize(200, 70);
    frame.setVisible(true);
   }
  }
```

JPasswordField
User Name: selva
Password: ••••

JTextArea

JTextArea allows editing of multiple lines of text. JTextArea can be used in conjunction with class JScrollPane to achieve scrolling. The underlying JScrollPane can be forced to always or never have either the vertical or horizontal scrollbar.

S. No.	Constructors	Purpose
1	**JTextArea()**	Constructs a new TextArea.
2	**JTextArea (Document doc)**	Constructs a new J Text Area with the given document model, and defaults for all of the other arguments (null, 0, 0).
3	**JTextArea (Document doc, String text, int rows, int columns)**	Constructs a new J Text Area with the specified number of rows and columns, and the given model.
4	**JTextArea (int rows, int columns)**	Constructs a new empty Text Area with the specified number of rows and columns.
5	**JTextArea (String text)**	Constructs a new Text Area with the specified text displayed.
6	**JTextArea (String text, int rows, int columns)**	Constructs a new Text Area with the specified text and number of rows and columns.
Methods		
1	void append (String str)	Appends the given text to the end of the document.
2	int getLineCount()	Determines the number of lines contained in the area.
3	String getText()	Returns the text contained in this Text Component.
4	Boolean isEditable()	Returns the boolean indicating whether this Text Component is editable or not.
5	void setEditable (boolean b)	Specifies whether or not this Text Component should be editable.
6	void setHorizontalAlignment (intalignment)	Sets the horizontal alignment of the text. Alignment is one of the values from Swing Constants: RIGHT, LEFT, CENTER
7	void setLineWrap (boolean wrap)	Sets the line-wrapping policy of the text area. If set to true the lines will be wrapped, if they are too long to within the allocated width.
8	void setText (String t)	Sets the text of this Text Component to the specified text.
9	void setWrapStyleWord (boolean word)	Sets the style of wrapping used if the text area is wrapping lines. If set to true, the lines will be wrapped at word boundaries (whitespace); if they are too long to within the allocated width.

Source Code

```
import java.awt.Dimension;
import java.awt.FlowLayout;
import javax.swing.JFrame;
import javax.swing.JScrollPane;
import javax.swing.JTextArea;

public class JTextAreaTest {
 public static void main(String[] args) {
  JFrame.setDefaultLookAndFeelDecorated(true);
  JFrame frame = new JFrame("JTextArea");
  frame.setLayout(new FlowLayout());
  frame.setDefaultCloseOperation(JFrame.EXIT_ON_CLOSE);
  String text = "A JTextArea object represents a multiline area for displaying
text. "
      + " When the user types data into them and presses the Enter key, an action
event occurs., "
      + " If the program registers an event listener, the listener processes the
event and can use the data in the text field at the time of the event in the
program. "
      + "JTextField allows editing/displaying of a single line of text. New
features include the ability to justify the text left, right, or center, and to set the
text's font.";
  JTextArea textAreal = new JTextArea(text, 5, 10);
  textAreal.setPreferredSize(new Dimension(100, 100));
  JTextArea textArea2 = new JTextArea(text, 5, 10);
  textArea2.setPreferredSize(new Dimension(100, 100));
  JScrollPane scrollPane = new JScrollPane(textArea2,
JScrollPane.VERTICAL_SCROLLBAR_ALWAYS,
     JScrollPane.HORIZONTAL_SCROLLBAR_ALWAYS);
  textAreal.setLineWrap(true);
  textArea2.setLineWrap(true);
  frame.add(textAreal);
  frame.add(scrollPane);
  frame.pack();
  frame.setVisible(true);
 }
}
```

J Button

The abstract class Abstract Button extends class J Component and provides a foundation for a family of button classes, including J Button. A button is a component the user clicks to trigger a specific action.

S. No.	Constructors	Purpose
1	**JButton()**	Creates a button with no set text or icon.
2	**JButton(Action a)**	Creates a button where properties are taken from the Action supplied.
3	**JButton(Icon icon)**	Creates a button with an icon.
4	**JButton(String text)**	Creates a button with text.
5	**JButton(String text, Icon icon)**	Creates a button with initial text and an icon.
Methods		
1	void setEnabled (boolean b)	Enables (or disables) the button. If disabled, then the button cannot be pressed.
2	void setHorizontal Alignment (int alignment)	Sets the horizontal alignment of the icon and text. Alignment is one of the values from Swing Constants: RIGHT, LEFT, CENTER
3	void setHorizontalText Position (int text Position)	Sets the horizontal position of the text relative to the icon. TextPosition is one of the values from Swing Constants: RIGHT, LEFT, CENTER
4	void setIconTextGap (int icon Text Gap)	If both the icon and text properties are set, this property defines the space between them. The default value of this property is 4 pixels.
5	void setText (String text)	Sets the button's text.
6	void setVerticalAlignment (int alignment)	Sets the vertical alignment of the icon and text. Alignment is one of the values from Swing Constants: TOP, BOTTOM, CENTER
7	void setVerticalTextPosition (int text Position)	Sets the vertical position of the text relative to the icon. Text Position is one of the values from Swing Constants: TOP, BOTTOM, CENTER

Source Code

```
import java.awt.*;
import java.awt.event.*;
import javax.swing.*;
public class JButtonDemo2 {
```

```
JFrame jtfMainFrame;
JButton jbnButton1, jbnButton2;
JTextField jtfInput;
JPanel jplPanel;
public JButtonDemo2() {
        jtfMainFrame = new JFrame("Which Button Demo");
        jtfMainFrame.setSize(50, 50);
        jbnButton1 = new JButton("Button 1");
        jbnButton2 = new JButton("Button 2");
        jtfInput = new JTextField(20);
        jplPanel = new JPanel();
        jbnButton1.setMnemonic(KeyEvent.VK_I); //Set ShortCut Keys
        jbnButton1.addActionListener(new ActionListener() {

                public void actionPerformed(ActionEvent e) {
                        jtfInput.setText("Button 1!");
                }
        });
        jbnButton2.setMnemonic(KeyEvent.VK_I);
        jbnButton2.addActionListener(new ActionListener() {

                public void actionPerformed(ActionEvent e) {
                        jtfInput.setText("Button 2!");
                }
        });
        jplPanel.setLayout(new FlowLayout());
        jplPanel.add(jtfInput);
        jplPanel.add(jbnButton1);
        jplPanel.add(jbnButton2);
        jtfMainFrame.getContentPane().add(jplPanel, BorderLayout.CENTER);
        jtfMainFrame.setDefaultCloseOperation(JFrame.EXIT_ON_CLOSE);
        jtfMainFrame.pack();
        jtfMainFrame.setVisible(true);
}
public static void main(String[] args) {
        // Set the look and feel to Java Swing Look
        try {
                UIManager.setLookAndFeel(UIManager
                                .getCrossPlatformLookAndFeelClassName());
        } catch (Exception e) {
        }
        JButtonDemo2 application = new JButtonDemo2();
}
}
```

J Radio Button

J Radio Button is similar to J Check box, except for the default icon for each class. A set of radio buttons can be associated as a group in which only one button at a time can be selected.

S. No.	Constructors	Purpose
1	**JRadio Button()**	Creates an initially unselected radio button with no set text.
2	**JRadioButton (Action a)**	Creates a radio button where properties are taken from the Action supplied.
3	**JRadioButton (Icon icon)**	Creates an initially unselected radio button with the specified image but no text.
4	**JRadioButton (Icon icon, boolean selected)**	Creates a radio button with the specified image and selection state, but no text.
5	**JRadioButton(String text)**	Creates an unselected radio button with the specified text.
6	**JRadioButton (String text, boolean selected)**	Creates a radio button with the specified text and selection state.
7	**JRadioButton (String text, Icon icon)**	Creates a radio button that has the specified text and image, and that is initially unselected.
8	**JRadioButton (String text, Icon icon, boolean selected)**	Creates a radio button that has the specified text, image, and selection state.
Methods		
1.	boolean isSelected()	Returns the state of the button.
2.	void setSelected (boolean b)	Sets the state of the button.

Source Code

```
import javax.swing.*;
import java.awt.*;
public class JRadioButtonExample
{
  public static void main(String [] args)
  {
    JFrame frame = new JFrame();
    JPanel panel = new JPanel();
    panel.setLayout(new GridLayout(1, 0, 5, 5));
    JRadioButton radio1 = new JRadioButton("Cut");
    JRadioButton radio2 = new JRadioButton("Copy");
    radio2.setSelected(true);
    JRadioButton radio3 = new JRadioButton("Paste");
```

```
        ButtonGroup group = new ButtonGroup();
        group.add(radio1);
        group.add(radio2);
        group.add(radio3);
    panel.add(radio1);
    panel.add(radio2);
    panel.add(radio3);
    frame.getContentPane().add(panel, BorderLayout.CENTER);
    frame.pack();
    frame.setVisible(true);
  }
}
```

JCheckbox

A **check box** represents an **option** to the user that can be *on* or *off* (or *selected* or *unselected*). It usually consists of two elements: A small box which contains a tick if the option is selected, and a piece of text that describes the option in question. The user clicks on the check box to select the option; clicking again deselects it.

S. No.	Constructors	Purpose
1	**JCheckBox** ()	Creates an initially unselected check box button with no text, no icon.
2	**JCheckBox** (Action a)	Creates a check box where properties are taken from the Action supplied.
3	**JCheckBox** (Icon icon)	Creates an initially unselected check box with an icon.
4	**JCheckBox** (Icon icon, boolean selected)	Creates a check box with an icon and specifies whether or not it is initially selected.
5	**JCheckBox** (String text)	Creates an initially unselected check box with text.
6	**JCheckBox** (String text, boolean selected)	Creates a check box with text and specifies whether or not it is initially selected.
7	**JCheckBox** (String text, Icon icon)	Creates an initially unselected check box with the specified text and icon.
8	**JCheckBox** (String text, Icon icon, boolean selected	Creates a check box with text and icon, and specifies whether or not it is initially selected.
Methods		
1.	boolean is Selected()	Returns the state of the button.
2.	void set Selected (boolean b)	Sets the state of the button.

Source Code

```
import javax.swing.*;
import java.awt.*;

public class JCheckBoxExample

{
  public static void main(String [] args)
  {
    JFrame frame = new JFrame();
    JPanel panel = new JPanel();
    panel.setLayout(new GridLayout(0, 1, 5, 5));

    JCheckBox check1 = new JCheckBox("Cut");
    JCheckBox check2 = new JCheckBox("Copy", true);
    JCheckBox check3 = new JCheckBox("Paste");
    JCheckBox check4 = new JCheckBox();
    check4.setText("Delete");
    check4.setSelected(true);

    panel.add(check1);
    panel.add(check2);
    panel.add(check3);
    panel.add(check4);

    frame.getContentPane().add(panel, BorderLayout.CENTER);
    frame.pack();
    frame.setVisible(true);
  }
}
```

Cut
Copy
Paste
Delete

JComboBox

JComboBox is like a drop down box — you can click a drop-down arrow and select an option from a list. It generates Item Event. For example, when the component has focus, pressing a key that corresponds to the first character in some entry's name selects that entry. A vertical scrollbar is used for longer lists.

S. No.	Constructors	Purpose
1	**JComboBox ()**	Creates a JComboBox with a default data model.
2	**JComboBox (Combo Box Model a Model)**	Creates a JComboBox that takes its items from an existing Combo Box Model.
3	**JComboBox (Object[] items)**	Creates a JComboBox that contains the elements in the specified array.
4	**JComboBox (Vector items)**	Creates a JComboBox that contains the elements in the specified Vector.

Source Code

```
import java.awt.event.ActionEvent;
import javax.swing.JButton;
import javax.swing.JFrame;
import javax.swing.JList;
public class GettingSettingSelectedItem {
 public static void main(String[] a) {
   JFrame frame = new JFrame();
   frame.setDefaultCloseOperation(JFrame.EXIT_ON_CLOSE);
   JButton jButton1 = new JButton("Button");
   String[] mystring = { "Java", "Swing", "JApplet", "Servlet" };
   final JList jList1 = new JList(mystring);
   jButton1.addActionListener(new java.awt.event.ActionListener() {
     public void actionPerformed(ActionEvent e) {
       Object contents = jList1.getSelectedValue();
       System.out.println(contents);
     }
   });

   frame.add(jList1, "Center");
   frame.add(jButton1,"South");
   frame.setSize(300, 200);
   frame.setVisible(true);
 }

}
```

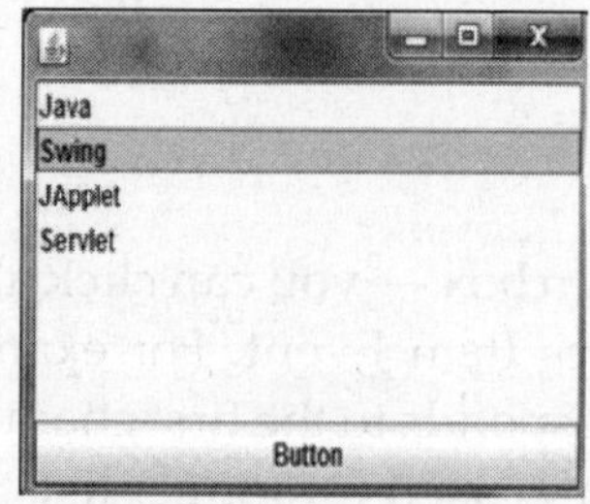

JList

JList provides a scrollable set of items from which one or more may be selected. JList can be populated from an Array or Vector. JList does not support scrolling directly—instead, the list must be associated with a scrollpane. The view port used by the scroll pane can also have a user-defined border. JList actions are handled using List Selection Listener.

S. No.	Constructors	Purpose
1	**JList()**	Constructs a JList with an empty model
2	**JList (List Model data Model)**	Constructs a JList that displays the elements in the specified, non-null model.
3	**JList (Object [] list Data)**	Constructs a JList that displays the elements in the specified array.
4	**JList (Vector list Data)**	Constructs a JList that displays the elements in the specified Vector.

Source Code

```
import java.awt.FlowLayout;
import javax.swing.JFrame;
import javax.swing.JList;
import javax.swing.JScrollPane;
public class JListTest {
 public static void main(String[] args) {
  JFrame.setDefaultLookAndFeelDecorated(true);
  JFrame frame = new JFrame("JList ");
  frame.setLayout(new FlowLayout());
  frame.setDefaultCloseOperation(JFrame.EXIT_ON_CLOSE);
  String[] selections = { "Lenovo", "HP", "Sony", "Dell" };
  JList list = new JList(selections);
  list.setSelectedIndex(1);
  System.out.println(list.getSelectedValue());
  frame.add(new JScrollPane(list));
  frame.pack();
  frame.setVisible(true);
 }
}
```

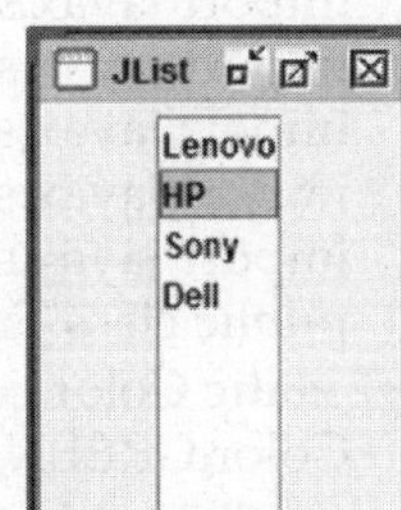

JTabbedPane

A JTabbedPane contains a tab that can have a tool tip and a mnemonic, and it can display both text and an image.

S. No.	Constructors	Purpose
1	**JTabbedPane()**	Creates an empty Tabbed Pane with a default tab placement of J Tabbed Pane.TOP.
2	**JTabbedPane (int tab Placement)**	Creates an empty Tabbed Pane with the specified tab placement of either: J Tabbed Pane.TOP, J Tabbed Pane.BOTTOM, JTab bed Pane.LEFT, or J Tabbed Pane.RIGHT.
3	**JTabbedPane (int tab Placement, int tab Layout Policy)**	Creates an empty Tabbed Pane with the specified tab placement and tab layout policy.

Source Code

```
import java.awt.BorderLayout;
import java.awt.Color;
import java.awt.event.KeyEvent;
import javax.swing.ImageIcon;
import javax.swing.JButton;
import javax.swing.JFrame;
import javax.swing.JTabbedPane;
import javax.swing.event.ChangeEvent;
import javax.swing.event.ChangeListener;
public class TabSample {
 static Color colors[] = { Color.RED, Color.ORANGE, Color.YELLOW,
Color.GREEN, Color.BLUE,
    Color.MAGENTA };

  static void add(JTabbedPane tabbedPane, String label) {
   int count = tabbedPane.getTabCount();
   JButton button = new JButton(label);
   button.setBackground(colors[count]);
   tabbedPane.addTab(label, new ImageIcon("yourFile.gif"), button, label);
  }

  public static void main(String args[]) {
   JFrame frame = new JFrame("Tabbed Pane Sample");
   frame.setDefaultCloseOperation(JFrame.EXIT_ON_CLOSE);

   JTabbedPane tabbedPane = new JTabbedPane();
   tabbedPane.setTabLayoutPolicy(JTabbedPane.SCROLL_TAB_LAYOUT);
   String titles[] = { "RED", "ORANGE", "YELLOW", "GREEN", "BLUE",
"MAGENTA" };
   for (int i = 0, n = titles.length; i < n; i++) {
    add(tabbedPane, titles[i]);
   }
```

```
  frame.add(tabbedPane, BorderLayout.CENTER);
  frame.setSize(400, 150);
  frame.setVisible(true);
 }

}
```

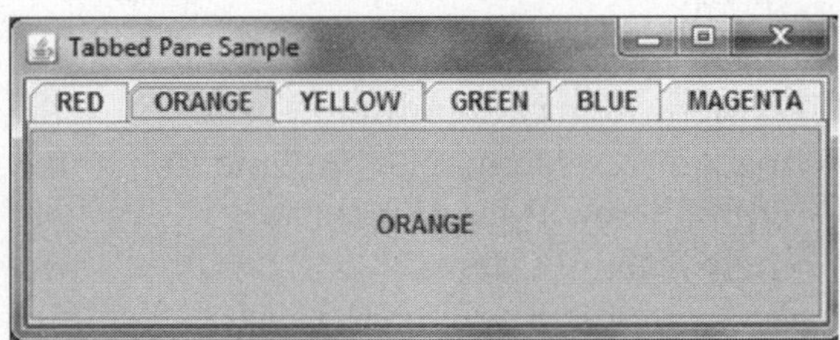

J Menu

Swing provides support for pull-down and popup menus. A J Menubar can contain several J Menu 's. Each of the JMenu 's can contain a series of J MenuItem 's that you can select.

How Menu's Are Created?

1. First, A JMenubar is created
2. Then, we attach all of the menus to this JMenubar.
3. Then we add JMenuItem 's to the J Menu
4. The JMenubar is then added to the frame. By default, each JMenuItem added to a JMenu is enabled—that is, it can be selected. In certain situations, we may need to disable a JMenuItem. This is done by calling setEnabled(). The setEnabled() method also allows components to be enabled.

Source Code

```
import java.awt.*;
import java.awt.event.*;
import javax.swing.JMenu;
import javax.swing.JMenuItem;
import javax.swing.JCheckBoxMenuItem;
import javax.swing.JRadioButtonMenuItem;
import javax.swing.ButtonGroup;
import javax.swing.JMenuBar;
import javax.swing.KeyStroke;
import javax.swing.ImageIcon;
import javax.swing.JPanel;
import javax.swing.JTextArea;
import javax.swing.JScrollPane;
import javax.swing.JFrame;

//Used Action Listner for JMenuItem & JRadioButtonMenuItem
//Used Item Listner for JCheckBoxMenuItem
public class JMenuDemo implements ActionListener, ItemListener {
```

```
JTextArea jtAreaOutput;
JScrollPane jspPane;
public JMenuBar createJMenuBar() {
    JMenuBar mainMenuBar;
    JMenu menu1, menu2, submenu;
    JMenuItem plainTextMenuItem, textIconMenuItem, iconMenuItem,
    subMenuItem;
    JRadioButtonMenuItem rbMenuItem;
    JCheckBoxMenuItem cbMenuItem;
    ImageIcon icon = createImageIcon("jmenu.jpg");
    mainMenuBar = new JMenuBar();
    menu1 = new JMenu("Menu 1");
    menu1.setMnemonic(KeyEvent.VK_M);
    mainMenuBar.add(menu1);
    // Creating the MenuItems
    plainTextMenuItem = new JMenuItem("Menu item with Plain Text",
            KeyEvent.VK_T);
    // can be done either way for assigning shortcuts
    // menuItem.setMnemonic(KeyEvent.VK_T);
    // Accelerators, offer keyboard shortcuts to bypass navigating the menu
    // hierarchy.
    plainTextMenuItem.setAccelerator(KeyStroke.getKeyStroke(
            KeyEvent.VK_1, ActionEvent.ALT_MASK));
    plainTextMenuItem.addActionListener(this);
    menu1.add(plainTextMenuItem);
    textIconMenuItem = new JMenuItem("Menu Item with Text & Image", icon);
    textIconMenuItem.setMnemonic(KeyEvent.VK_B);
    textIconMenuItem.addActionListener(this);
    menu1.add(textIconMenuItem);
    // Menu Item with just an Image
    iconMenuItem = new JMenuItem(icon);
    iconMenuItem.setMnemonic(KeyEvent.VK_D);
    iconMenuItem.addActionListener(this);
    menu1.add(iconMenuItem);
    menu1.addSeparator();
    // Radio Button Menu items follow a seperator
    ButtonGroup itemGroup = new ButtonGroup();
    rbMenuItem = new JRadioButtonMenuItem("Menu Item with Radio Button");
    rbMenuItem.setSelected(true);
    rbMenuItem.setMnemonic(KeyEvent.VK_R);
    itemGroup.add(rbMenuItem);
    rbMenuItem.addActionListener(this);
    menu1.add(rbMenuItem);
    rbMenuItem = new JRadioButtonMenuItem("Menu Item 2 with Radio Button");
    itemGroup.add(rbMenuItem);
```

```
            rbMenuItem.addActionListener(this);
            menu1.add(rbMenuItem);
            menu1.addSeparator();
            // Radio Button Menu items follow a seperator
            cbMenuItem = new JCheckBoxMenuItem("Menu Item with check box");
            cbMenuItem.setMnemonic(KeyEvent.VK_C);
            cbMenuItem.addItemListener(this);
            menu1.add(cbMenuItem);
            cbMenuItem = new JCheckBoxMenuItem("Menu Item 2 with check box");
            cbMenuItem.addItemListener(this);
            menu1.add(cbMenuItem);
            menu1.addSeparator();
            // Sub Menu follows a seperator
            submenu = new JMenu("Sub Menu");
            submenu.setMnemonic(KeyEvent.VK_S);
            subMenuItem = new JMenuItem("Sub MenuItem 1");
subMenuItem.setAccelerator(KeyStroke.getKeyStroke(KeyEvent.VK_2,
                                ActionEvent.CTRL_MASK));
            subMenuItem.addActionListener(this);
            submenu.add(subMenuItem);
            subMenuItem = new JMenuItem("Sub MenuItem 2");
            submenu.add(subMenuItem);
            subMenuItem.addActionListener(this);
            menu1.add(submenu);
            // Build second menu in the menu bar.
            menu2 = new JMenu("Menu 2");
            menu2.setMnemonic(KeyEvent.VK_N);
            mainMenuBar.add(menu2);
            return mainMenuBar;
    }
    public Container createContentPane() {
            // Create the content-pane-to-be.
            JPanel jplContentPane = new JPanel(new BorderLayout());
            jplContentPane.setLayout(new BorderLayout());// Can do it either way
            // to set layout
            jplContentPane.setOpaque(true);
            // Create a scrolled text area.
            jtAreaOutput = new JTextArea(5, 30);
            jtAreaOutput.setEditable(false);
            jspPane = new JScrollPane(jtAreaOutput);
            // Add the text area to the content pane.
            jplContentPane.add(jspPane, BorderLayout.CENTER);
            return jplContentPane;
    }
    /** Returns an ImageIcon, or null if the path was invalid. */
```

```
protected static ImageIcon createImageIcon(String path) {
        java.net.URL imgURL = JMenuDemo.class.getResource(path);
        if (imgURL != null) {
                return new ImageIcon(imgURL);
        } else {
                System.err.println("Couldn't find image file: " + path);
                return null;
        }
}
private static void createGUI() {
        JFrame.setDefaultLookAndFeelDecorated(true);
        // Create and set up the window.
        JFrame frame = new JFrame("JMenu Usage Demo");
        frame.setDefaultCloseOperation(JFrame.EXIT_ON_CLOSE);
        JMenuDemo app = new JMenuDemo();
        frame.setJMenuBar(app.createJMenuBar());
        frame.setContentPane(app.createContentPane());
        frame.setSize(500, 300);
        frame.setVisible(true);
}
public void actionPerformed(ActionEvent e) {
        JMenuItem source = (JMenuItem) (e.getSource());
        String s = "Menu Item source: " + source.getText()
                    + " (an instance of " + getClassName(source) + ")";
        jtAreaOutput.append(s + "\n");
        jtAreaOutput.setCaretPosition(jtAreaOutput.getDocument()
                    .getLength());
}
public void itemStateChanged(ItemEvent e) {
        JMenuItem source = (JMenuItem) (e.getSource());
        String s = "Menu Item source: "
                    + source.getText()
                    + " (an instance of "
                    + getClassName(source)
                    + ")"
                    + "\n"
                    + "    State of check Box: "
                    + ((e.getStateChange() == ItemEvent.SELECTED) ?
                      "selected"
                                : "unselected");
        jtAreaOutput.append(s + "\n");
        jtAreaOutput.setCaretPosition(jtAreaOutput.getDocument()
                    .getLength());
}
// Returns the class name, no package info
protected String getClassName(Object o) {
        String classString = o.getClass().getName();
        int dotIndex = classString.lastIndexOf(".");
        return classString.substring(dotIndex + 1); // Returns only Class name
```

```
    }
    public static void main(String[] args) {
        javax.swing.SwingUtilities.invokeLater(new Runnable() {

            public void run() {
                createGUI();
            }
        });
    }
}
```

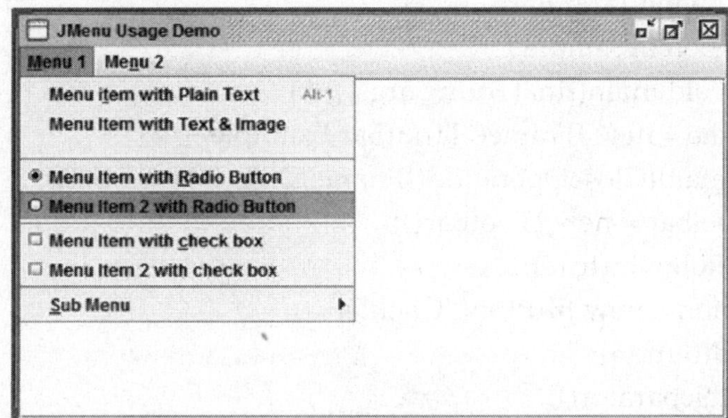

JTool Bar

A JToolbar contains a number of components whose type is usually some kind of button which can also include separators to group related components within the toolbar. The toolbar can be docked against any of the four edges of a container (panel or a frame). A toolbar can also be made to float.

Tool bars uses Box Layout, which arranges components in one horizontal row/ vertical column. This layout manager does not force each component to have the same height or width; instead, it uses their preferred height or width, and attempts to align them. You can adjust the resulting alignment by calling class Component's methods set Alignment X() and/or set Alignment Y() on each component.

S. No.	Constructors	Purpose
1.	**JToolBar ()**	Creates a new tool bar; orientation defaults to HORIZONTAL.
2.	**JToolBar (int orientation)**	Creates a new tool bar with the specified orientation.
3.	**JToolBar (String name)**	Creates a new tool bar with the specified name.
4.	**JToolBar (String name, int orientation)**	Creates a new tool bar with a specified name and orientation.

Source Code

```
import java.awt.BorderLayout;
import java.awt.Container;
import javax.swing.JButton;
import javax.swing.JComboBox;
import javax.swing.JFrame;
import javax.swing.JScrollPane;
import javax.swing.JTextArea;
import javax.swing.JToolBar;
public class ToolBarSample {
public static void main(final String args[]) {
    JFrame frame = new JFrame("JToolBar Example");
    frame.setDefaultCloseOperation(JFrame.EXIT_ON_CLOSE);
    JToolBar toolbar = new JToolBar();
    toolbar.setRollover(true);
    JButton button = new JButton("Click");
    toolbar.add(button);
    toolbar.addSeparator();
    toolbar.add(new JButton("Here"));
    toolbar.add(new JComboBox(new String[]{"java","swing","applet"}));
    Container contentPane = frame.getContentPane();
    contentPane.add(toolbar, BorderLayout.NORTH);
    JTextArea textArea = new JTextArea();
    JScrollPane pane = new JScrollPane(textArea);
    contentPane.add(pane, BorderLayout.CENTER);
    frame.setSize(350, 150);
    frame.setVisible(true);
  }
}
```

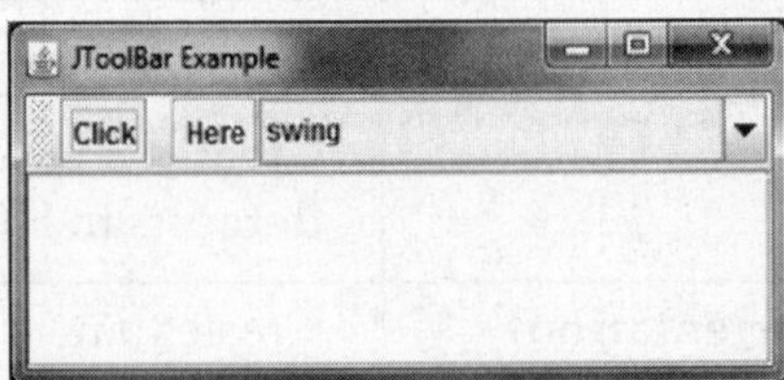

JTree

It can display hierarchical data. A JTree object does not actually contain your data; it simply provides a view of the data. Like any non-trivial Swing component, the tree gets data by querying its data model. JTree displays its data vertically. Each row displayed by the tree contains exactly one item of data, which is called a *node*. Every tree has a *root* node from which all nodes descend.

S. No.	Constructors	Purpose
1.	JTree (Tree Node) JTree (Tree Node, boolean) JTree (Tree Model) JTree () JTree (Hashtable) JTree (Object []) JTree (Vector)	Create a tree. The Tree Node argument specifies the root node, to be managed by the default tree model. The Tree Model argument specifies the model that provides the data to the table. The no-argument version of this constructor is for use in builders; it creates a tree that contains some sample data. If you specify a Hashtable, array of objects, or Vector as an argument, then the argument is treated as a list of nodes under the root node (which is not displayed), and a model and tree nodes are constructed accordingly.The boolean argument, if present, specifies how the tree should determine whether a node should be displayed as a leaf. If the argument is false (the default), any node without children is displayed as a leaf. If the argument is true, a node is a leaf only, if its get Allows Children method returns false.

Source Code

```
import java.awt.BorderLayout;
import java.util.Vector;
import javax.swing.JFrame;
import javax.swing.JScrollPane;
import javax.swing.JTree;
class TreeNodeVector<E> extends Vector<E> {
 String name;
 TreeNodeVector(String name) {
  this.name = name;
 }
 TreeNodeVector(String name, E elements[]) {
  this.name = name;
  for (int i = 0, n = elements.length; i < n; i++) {
   add(elements[i]);
  }
 }

 public String toString() {
  return "[" + name + "]";
 }
}
```

```
public class TreeArraySample {
  public static void main(final String args[]) {
   JFrame frame = new JFrame("JTreeSample");
   frame.setDefaultCloseOperation(JFrame.EXIT_ON_CLOSE);
   Vector<String> v1 = new TreeNodeVector<String>("Two", new String[] {
"Mercury", "Venus",
     "Mars" });
   Vector<Object> v2 = new TreeNodeVector<Object>("Three");
   v2.add(System.getProperties());
   v2.add(v1);
   Object rootNodes[] = {v1, v2 };
   Vector<Object> rootVector = new TreeNodeVector<Object>("Root",
rootNodes);
   JTree tree = new JTree(rootVector);
   frame.add(new JScrollPane(tree), BorderLayout.CENTER);
   frame.setSize(300, 300);
   frame.setVisible(true);
  }
}
```

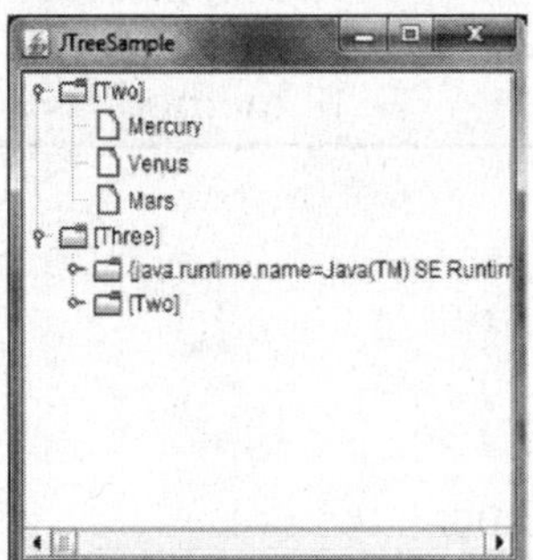

Tool Tips

Creating a tool tip for any J Component object is easy. Use the set Tool Tip Text method to set up a tool tip for the component. For example, to add tool tips to three buttons, you add only three lines of code:

```
b 1. setToolTipText ("Click this button to disable the middle button.");
b 2. setToolTipText ("This middle button does not react when you click it.");
b 3. setToolTipText ("Click this button to enable the middle button.");
```

S. No.	Constructors	Purpose
1.	**SetToolTipText (String)**	If the specified string is not null, then this method registers the component as having a tool tip and when displayed, gives the tool tip the specified text. If the argument is null, then this method turns off the tool tip for this component.
2.	**String getToolTipText ()**	Returns the string that was previously specified with set Tool Tip Text.

3.	**String getToolTipText (Mouse Event)**	By default, returns the same value returned by get Tool Tip Text (). Multi-part components such as J Tabbed Pane, J Table, and J Tree override this method to return a string associated with the mouse event location. For example, each tab in a tabbed pane can have different tool tip text.
4.	**Point getToolTip Location (Mouse Event)**	Returns the location (in the receiving component's coordinate system) where the upper left corner of the component's tool tip appears. The argument is the event that caused the tool tip to be shown. The default return value is null, which tells the Swing system to choose a location.

Source Code

```
import javax.swing.JButton;
import javax.swing.JFrame;
public class ToolTipDemo {
 public static void main(String args[]) {
  JFrame frame = new JFrame();
  frame.setDefaultCloseOperation(JFrame.EXIT_ON_CLOSE);
  JButton b = new JButton("Click Here");
  frame.add(b,"Center");
  b.setToolTipText("Press");
  frame.setSize(300, 200);
  frame.setVisible(true);
 }
}
```

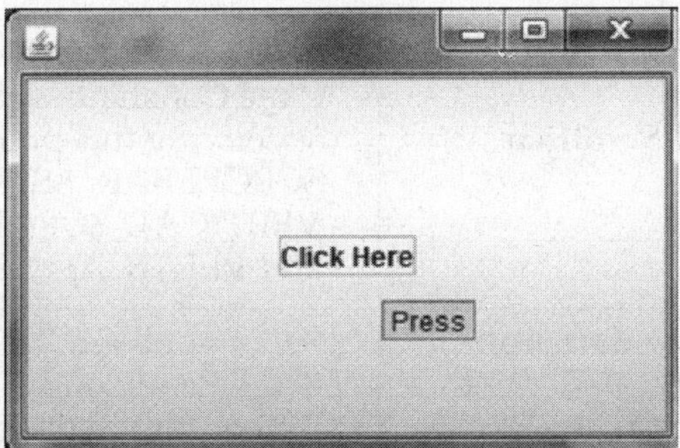

JScrollPane

A JScrollPane provides a scrollable view of a component. When screen real estate is limited, use a scroll pane to display a component that is large or one whose size can change dynamically.

J Scroll Pane scroll Pane = new JScrollPane (text Area)

S. No.	Policy	Description
1.	**VERTICAL_SCROLLBAR_AS_NEEDED**	The default. The scroll bar appears when the viewport is smaller than the client and disappears when the viewport is larger than the client.
2.	**HORIZONTAL_SCROLLBAR_AS_NEEDED**	
3.	**VERTICAL_SCROLLBAR_ALWAYS**	Always display the scroll bar. The knob disappears if the viewport is large enough to show the whole client.
4.	**HORIZONTAL_SCROLLBAR_ALWAYS**	
5.	**VERTICAL_SCROLLBAR_NEVER**	Never display the scroll bar. Use this option if you do not want the user to directly control what part of the client is shown, or if you want them to use only non-scroll-bar techniques (such as dragging).
6.	**HORIZONTAL_SCROLLBAR_NEVER**	

S. No.	Constructors	Purpose
1.	**JScrollPane ()** **JScrollPane (Component)** **JScrollPane (int, int)** **JScrollPane (Component, int, int)**	Create a scroll pane. The Component parameter, when present, sets the scroll pane's client. The two int parameters, when present, set the vertical and horizontal scroll bar policies (respectively).
2.	**Void setViewport View (Component)**	Set the scroll pane's client.
3.	**Void setverticalScrollbar Policy (int)** **int getVerticalScrollbar Policy ()**	Set or get the vertical scroll policy. Scroll Pane Constants defines three values for specifying this policy: CERTICAL_ SCROLLBAR_AS_NEEDED (the default), VERTICAL_SCROLLBAR_ALWAYS, and VERTICAL_SCROLLBAR_NEVER.
4.	**Void setHorizontalScroll BarPolicy (int)** **int getHorizontalScroll Bar Policy ()**	Set or get the horizontal scroll policy. Scroll Pane Constants defines three values for specifying this policy: HORIZONTAL_SCROLLBAR _AS_NEEDED (the default), HORIZONTAL_SCROLLBAR _ALWAYS, and HORIZONTAL _SCROLLBAR_NEVER.

5.	**Void setViewportBorder (Border) Border getView portBorder ()**	Set or get the border around the viewport. This is preferred over setting the border on the component.
6.	**Boolean isWheelScrolling Enabled ()**	Set or get whether scrolling occurs in response to the mouse wheel. Mouse-wheel scrolling is enabled by default.

Source code

```
import java.awt.BorderLayout;
import java.awt.Dimension;
import javax.swing.JFrame;
import javax.swing.JLabel;
import javax.swing.JScrollPane;
public class JScrollPaneViewport {
 public static void main(String args[]) {
  JFrame frame = new JFrame("Tabbed Pane Sample");
  frame.setDefaultCloseOperation(JFrame.EXIT_ON_CLOSE);
  JLabel label = new JLabel("Label");
  label.setPreferredSize(new Dimension(1000,1000));
  JScrollPane jScrollPane = new JScrollPane();
  jScrollPane.setViewportView(label);
  frame.add(jScrollPane, BorderLayout.CENTER);
  frame.setSize(400, 150);
  frame.setVisible(true);

 }
}
```

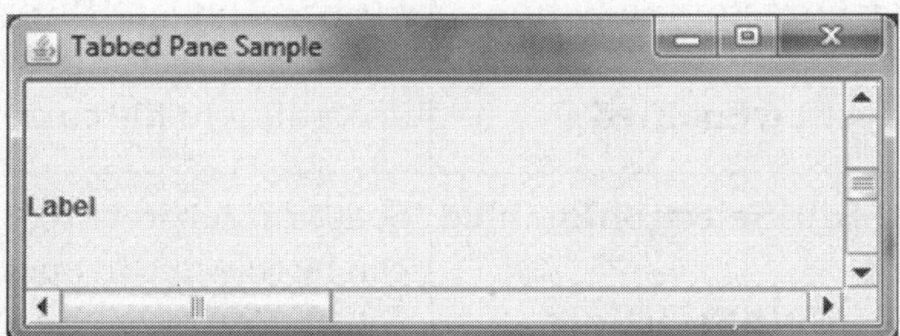

JFileChooser

File choosers provide a GUI for navigating the file system, and then either choosing a file or directory from a list, or entering the name of a file or directory. To display a file chooser, you usually use the J File Chooser API to show a modal dialog containing the file chooser. Another way to present a file chooser is to add an instance of J File Chooser to a container. The return value of the three methods is one of the following:

1. JFileChooser. CANCEL_OPTION, if the user clicks Cancel.
2. JFileChooser. APPROVE_OPTION, if the user click an OK/Open/Save button.
3. JFileChooser. ERROR_OPTION, if the user closes the dialog

JFileChooser fc = new JFileChooser ();

int return Val = fc. show Dialog (File Chooser Demo 2. this, "Attach");

S. No.	Constructors and Methods	Purpose
1.	**JFileChooser ()** **JFileChooser (File)** **JFileChooser (String)**	Creates a file chooser instance. The File and String arguments, when present, provide the initial directory.
2.	**int showOpenDialog (Component)** **int showSaveDialog (Component)** **int showDialog (Component, String)**	Shows a modal dialog containing the file chooser. These methods return APPROVE_OPTION if the user approved the operation and CANCEL_OPTION if the user cancelled it. Another possible return value is ERROR_OPTION, which means an unanticipated error occurred.
3.	**Void setSelectedFile (File)** **File getSelectedFile ()**	Sets or obtains the currently selected file or (if directory selection has been enabled) directory.
4.	**Void setSelectedFiles (File[])** **File[] getSelectedFiles ()**	Sets or obtains the currently selected files if the file chooser is set to allow multiple selection.
5.	**Void setFileSelectionMode (int)** **void GetFileSelectionMode ()** **boolean isDirectorySelection Enabled () boolean isFileSelection Enabled ()**	Sets or obtains the file selection mode. Acceptable values are FILES_ONLY (the default), DIRECTORIES_ONLY, and FILES_AND_DIRECTORIES.Interprets whether directories or files are selectable according to the current selection mode.
6.	**Void setMultiSelectionEnabled (boolean)** **boolean isMultiSelection-Enabled ()**	Sets or interprets whether multiple files can be selected at once. By default, a user can choose only one file.
7.	**Dialog createDialog (Component)**	Given a parent component, creates and returns a new dialog that contains this file chooser, is dependent on the parent's frame, and is centered over the parent.

Sample Code

```
import javax.swing.JFileChooser;
import javax.swing.JFrame;
```

```
public class MainClass extends JFrame {
 public MainClass() {
  JFileChooser fileChooser = new JFileChooser();
  fileChooser.setDialogTitle("Choose a file");
  this.getContentPane().add(fileChooser);
  fileChooser.setVisible(true);
 }

 public static void main(String[] args) {
  JFrame frame = new MainClass();
  frame.setDefaultCloseOperation(JFrame.EXIT_ON_CLOSE);

  frame.pack();
  frame.setVisible(true);
 }
}
```

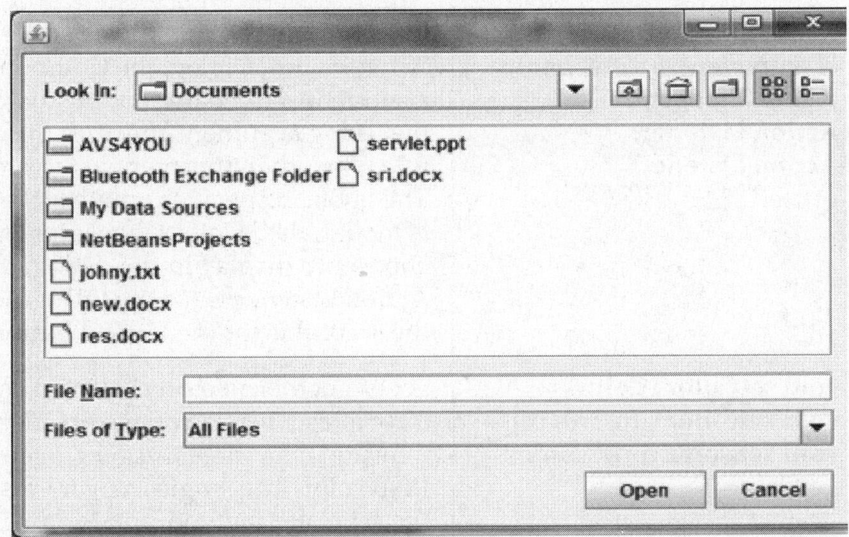

JColorChooser

JColorChooser is a component that is new to Swing; there is no AWT equivalent. It lets the user interactively select a Color. The default behavior is to present a dialog box containing a tabbed pane letting the user choose via swatches, HSB values, or RGB values.

The simplest use is very simple indeed; call JColorChooser to show Dialog, supplying three arguments. The first argument is the parent Component, the second is the title, and the third is the initially selected color. If the user selects "OK," the return value is the Color chosen. If the user cancels, the return value is null.

You can also allocate JColorChooser via a constructor. This is common if you want to display it somewhere other than in a popup dialog, or if you are likely to pop it up many times. In the latter case, pass the JColorChooser instance to JColorChooser create Dialog.

S. No.	Constructors and Methods	Purpose
1.	**JColorChooser ()** **JColorChooser (Color)** **JColorChooser (Color Selection Model)**	Create a color chooser. The default constructor creates a color chooser with an initial color of Color.white. Use the second constructor to specify a different initial color. The ColorSelectionModel argument, when present, provides the color chooser with a color selection model.
2.	**Color showDialog (Component, String, Color)**	Create and show a color chooser in a modal dialog. The Component argument is the parent of the dialog, the String argument specifies the dialog title, and the Color argument specifies the chooser's initial color.
3.	**JDialog createDialog (Component, String, boolean, JColorChooser, Action Listener, Action Listener)**	Create a dialog for the specified color chooser. As with showDialog, the Component argument is the parent of the dialog and the String argument specifies the dialog title. The other arguments are as follows: The boolean specifies whether the dialog is modal, the J Color Chooser is the color chooser to display in the dialog, the first Action Listener is for the **OK** button, and the second is for the **Cancel** button.
4.	**void setColor (Color)** **void setColor (int, int, int)** **void setColor (int) Color getColor ()**	Set or get the currently selected color. The three integer version of the set Color method interprets the three integers together as an RGB color. The single integer version of the set Color method divides the integer into four 8-bit bytes and interprets the integer as an RGB color as follows: | | R | G | B |
5.	**Void setSelectionModel (Color Selection Model) ColorSelectionModel getSelectionModel ()**	Set or get the selection model for the color chooser. This object contains the current selection and fires change events to registered listeners whenever the selection changes.

Sample Code

```
import java.awt.*;
import java.awt.event.*;
import javax.swing.*;

public class JColorChooserTest extends JFrame
                    implements ActionListener {
 public static void main(String[] args) {
  new JColorChooserTest();
 }

 public JColorChooserTest() {
  super("Using JColorChooser");
  Container content = getContentPane();
  content.setBackground(Color.white);
  content.setLayout(new FlowLayout());
  JButton colorButton
   = new JButton("Choose Background Color");
  colorButton.addActionListener(this);
  content.add(colorButton);
  setSize(300, 100);
  setVisible(true);
 }

 public void actionPerformed(ActionEvent e) {
  // Args are parent component, title, initial color
  Color bgColor
  = JColorChooser.showDialog(this, "Choose Background Color",
getBackground());
  if (bgColor != null)

   getContentPane().setBackground(bgColor);
 }
}
```

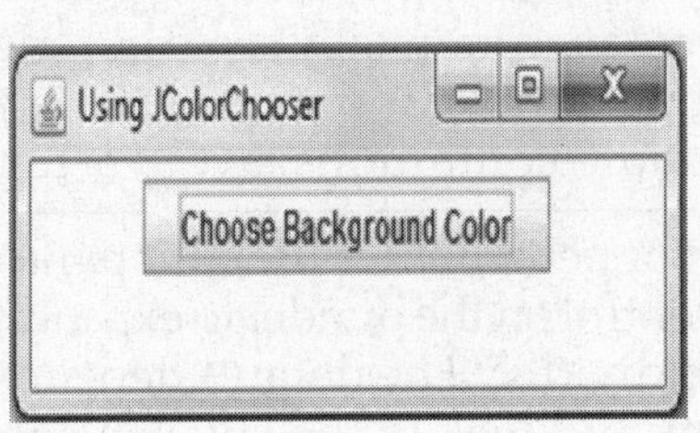

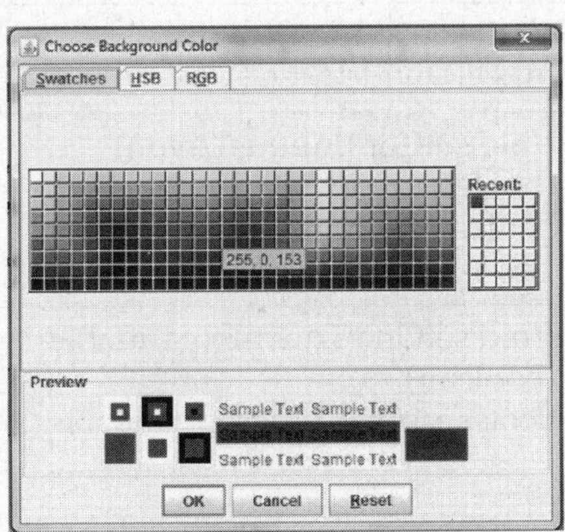

JSplitPane

It is a light weight Swing Container which separates components graphically. Split pane allows two components to be split inside a container, to split more than one component we need to nest split panes and components. The components which we split can be adjusted interactively by user.

To create split panes in java we can use the JSplitPane class which is inherited by JComponent class. Split panes are of two types: Horizontal and vertical, meaning horizontal split pane divides components horizontally and the divider can be moved horizontally. Similarly vertical split pane divides components vertically and divider can be moved vertically.

Setting the Components in a Split Pane

A program can set a split pane's two components dynamically with these four methods:

- Set Left Component
- Set Right Component
- Set Top Component
- Set Bottom Component

S. No.	Constructors	Purpose
1.	JSplitPane () J Split Pane (int) JSplitPane (int, boolean) JSplitPane (int, Component, Component) J Split Pane (int, boolean, Component, Component)	Create a split pane. When present, the int parameter indicates the split pane's orientation, either HORIZONTAL_SPLIT (the default) or VERTICAL_SPLIT. The boolean parameter, when present, sets whether the components continually repaint as the user drags the split pane. If left unspecified, this option (called *continuous layout*) is turned off. The Component parameters set the initial left and right, or top and bottom components, respectively.
2.	Void setOrientation (int) int getOrientation ()	Set or get the split pane's orientation. Use either HORIZONTAL_SPLIT or VERTICAL_SPLIT defined in J Split Pane. If left unspecified, the split pane will be horizontally split.
3.	Void setDividerSize (int) int getDividerSize ()	Set or get the size of the divider in pixels.
4.	Void setContinuousLayout (boolean) boolean isContinuousLayout ()	Set or get whether the split pane's components are continually layed out and painted while the user is dragging the divider. By default, continuous layout is turned off.
5.	Void setOne TouchExpandable (boolean) boolean isOne TouchExpandable ()	Set or get whether the split pane displays a control on the divider to expand/collapse the divider. The default depends on the look and feel. In the Java look and feel, it is off by default.

6.	Void SetTopComponent (Component) void setBottom Component (Component) void setLeft Component (Component) void setRight Component (Component) Component get Bottom Component () Component get Left Component () Component get Right Component ()	Set or get the indicated component. Each method works regardless of the split pane's orientation. Top and left are equivalent, and bottom and right are equivalent.
7.	Void remove (Component) Void remove All ()	Remove the indicated component(s) from the split pane.
8.	Void add (Component)	Add the component to the split pane. You can add only two components to a split pane. The first component added is the top/left component. The second component added is the bottom/right component. Any attempt to add more components results in an exception.
9.	Void set Divider Location (double) void set Divider Location (int) int get Divider Location ()	Set or get the current divider location. When setting the divider location, you can specify the new location as a percentage (double) or a pixel location (int).
10.	Void reset To Preferred Sizes ()	Move the divider such that both components are at their preferred sizes. This is how a split pane divides itself at startup, unless specified otherwise.
11.	Void set Last Divider Location (int) int get Last Divider Location ()	Set or get the previous position of the divider.
12.	Int get Maximum Divider Location () Int get Minimum Divider Location ()	Get the minimum and maximum Locations for the divider. These are set implicitly by setting the minimum sizes of the split pane's two components.
13.	Void set Resize Weight (float) float get Resize Weight ()	Set or get the resize weight for the split pane, a value between 0.0 (the default) and 1.0.

Sample Code

```
import java.awt.Component;
import javax.swing.ImageIcon;
import javax.swing.JFrame;
import javax.swing.JLabel;

import javax.swing.JScrollPane;
import javax.swing.JSplitPane;

public class Main {
  public static void main(String[] args) {
    JFrame frame = new JFrame("SplitPaneFrame");

    JLabel leftImage = new JLabel(new ImageIcon("a.gif"));
    Component left = new JScrollPane(leftImage);
    JLabel rightImage = new JLabel(new ImageIcon("b.gif"));
    Component right = new JScrollPane(rightImage);

    JSplitPane split = new JSplitPane(JSplitPane.HORIZONTAL_SPLIT, left,
right);
    split.setDividerLocation(100);
    frame.getContentPane().add(split);

    frame.setDefaultCloseOperation(JFrame.EXIT_ON_CLOSE);
    frame.setSize(300, 200);
    frame.setVisible(true);
  }
}
```

JEditorPane and JTextPane

Two Swing classes support styled text: JEditorPane and its subclass JTextPane. JEditorPane is sort of a fancy text area that can display text derived from different file formats. The built-in version supports HTML and RTF (Rich Text Format) only, but you can build "editor kits" to handle special purpose applications. In principle, you choose the type of document you want to display by calling set Content Type and specify a custom editor kit via set Editor Kit. Note that unless you extend it, legal choices are "text/html" (the default), "text/plain" (which is also what you get if you supply an unknow type), and "text/rtf".

In practice, however, J Editor Pane is almost always used for displaying HTML. If you have plain text, you might as well use J Text Field. RTF support is pretty primitive. You put content into the J Editor Pane one of four ways.

- The most common way to build a J Editor Pane is via the constructor, where you supply either a URL object or a String corresponding to a URL (which, in applications, could be a file: URL to read off the local disk). Note that this throws I O Exception, so needs to be in a try/catch block.
- Secondly, you can use set Page on a J Editor Pane instance that was created via the empty constructor. The set Page method also takes either a URL object or a String, and also throws I O Exception. This is generally used when the content is determined at run-time by some user action.
- Thirdly, you can use set Text on a J Editor Pane instance, supplying a String that is the actual content.
- Fourthly, but only occasionally, you use read, supplying an InputStream and an HTML Document object.

Editor Panes vs. Text Panes

In order to use editor panes and text panes, you need to understand the text system, which is described in Text Component Features. Several facts about editor panes and text panes are scattered throughout that section. Here we list the facts again and provide a bit more detail. The information here should help you understand the differences between editor panes and text panes, and when to use which.

- An editor pane or a text pane can easily be loaded with text from a URL using the set Page method. The J Editor Pane class also provides constructors that let you initialize an editor pane from a URL. The J Text Pane class has no such constructors. See Using an Editor Pane to Display Text From a URL, for example, that uses this feature to load an uneditable editor pane with HTML-formatted text.

  Be aware that the document and editor kit might change when using the set Page method. For example, if an editor pane contains plain text (the default), and you load it with HTML, the document will change to an HTML Document instance and the editor kit will change to an HTML Editor Kit instance. If your program uses the set Page method, make sure you adjust your code for possible changes to the pane's document and editor kit instances (re-register document listeners on the new document, and so on).
- Editor panes, by default, know how to read, write, and edit plain, HTML, and RTF text. Text panes inherit this capability but impose certain limitations. A text pane insists that its document implement the Styled Document interface. HTML Document and RTF Document are both Styled Documents, so HTML and RTF work as expected within a text pane. If you load a text pane with plain text though, the text pane's document is not a Plain Document as you might expect, but a Default Styled Document.
- To support a custom text format, implement an editor kit that can read, write, and edit text of that format. Then call the register Editor Kit For Content Type method to register your kit with the J Editor Pane class. By registering an editor

kit in this way, all editor panes and text panes in your program will be able to read, write, and edit the new format. However, if the new editor kit is not a Styled Editor Kit, text panes will not support the new format.

- As mentioned previously, a text pane requires its document to implement the Styled Document interface. The Swing text package provides a default implementation of this interface, Default Styled Document, which is the document that text panes use by default. A text pane also requires that its editor kit be an instance of a Styled Editor Kit (or a subclass). Be aware that the read and write methods for Style Editor Kit work with plain text.
- Through their styled document and styled editor kit, text panes provide support for named styles and logical styles. The J Text Pane class itself contains many methods for working with styles that simply call methods in its document or editor kit.
- Through the API provided in the J Text Pane class, you can embed images and components in a text pane. You can embed images in an editor pane too, but only by including the images in an HTML or RTF file.

S. No.	Constructors	Purpose
1.	J Editor Pane (URL) J Editor Pane (String)	Creates an editor pane loaded with the text at the specified URL.
2.	Set Page (URL) set Page (String)	Loads an editor pane (or text pane) with the text at the specified URL.
3.	URL get Page ()	Gets the URL for the editor pane's (or text pane's) current page.
4.	JTextPane () JTextPane (Styled Document)	Creates a text pane. The optional argument specifies the text pane's model.
5.	Styled Document get Styled Document set Styled Document (Styled Document)	Gets or sets the text pane's model.

Sample Code

JTextPane

```
import java.awt.BorderLayout;
import javax.swing.Icon;
import javax.swing.ImageIcon;
import javax.swing.JFrame;
import javax.swing.JLabel;
import javax.swing.JScrollPane;
import javax.swing.JTextPane;
import javax.swing.text.BadLocationException;
import javax.swing.text.DefaultStyledDocument;
import javax.swing.text.Style;
```

```
import javax.swing.text.StyleConstants;
import javax.swing.text.StyleContext;
import javax.swing.text.StyledDocument;

public class JTextPaneWithIcon {
 public static void main(String args[]) {
  JFrame frame = new JFrame("TextPane Example");
  frame.setDefaultCloseOperation(JFrame.EXIT_ON_CLOSE);

  StyleContext context = new StyleContext();
  StyledDocument document = new DefaultStyledDocument(context);

  Style labelStyle = context.getStyle(StyleContext.DEFAULT_STYLE);

  Icon icon = new ImageIcon("b.gif");
  JLabel label = new JLabel(icon);
  StyleConstants.setComponent(labelStyle, label);

  try {
   document.insertString(document.getLength(), "Ignored", labelStyle);
  } catch (BadLocationException badLocationException) {
   System.err.println("Oops");
  }

  JTextPane textPane = new JTextPane(document);
  textPane.setEditable(false);
  JScrollPane scrollPane = new JScrollPane(textPane);
  frame.add(scrollPane, BorderLayout.CENTER);

  frame.setSize(300, 150);
  frame.setVisible(true);
 }
}
```

JEditorPane

Source code

```
import java.net.URL;
import javax.swing.JEditorPane;
import javax.swing.JFrame;
import javax.swing.JScrollPane;
public class LoadingWebPageToJEditorPane {
 public static void main(String[] a)throws Exception {
  JFrame frame = new JFrame();
  frame.setDefaultCloseOperation(JFrame.EXIT_ON_CLOSE);
  JEditorPane editorPane = new JEditorPane();
  editorPane.setPage(new URL("http://www.google.co.in"));
  frame.add(new JScrollPane(editorPane));
  frame.setSize(300, 200);
  frame.setVisible(true);
 }

}
```

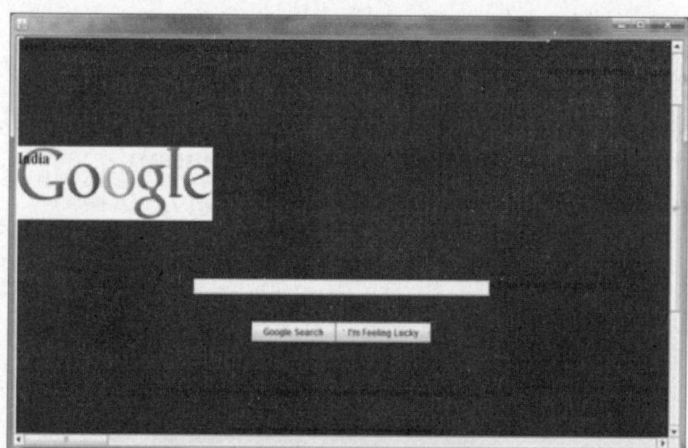

JWindow

A JWindow is a container that can be displayed anywhere on the user's desktop. It does not have the title bar, window-management buttons, or other trimmings associated with a **JFrame**, but it is still a "first-class citizen" of the user's desktop, and can exist anywhere on it.

The JWindow component contains a **JRootPane** as its only child. The content Pane should be the parent of any children of the JWindow. From the older java.awt. Window object you would normally do something like this:

```
window.add(child);
```

However, using JWindow you would code:

```
window. get Content Pane (). add (child);
```

S. No.	Constructors	Purpose
1.	JWindow ()	Creates a window with no specified owner.
2.	JWindow (Frame owner)	Creates a window with the specified owner frame.
3.	JWindow (Graphics Configuration gc)	Creates a window with the specified Graphics Configuration of a screen device.
4.	JWindow (Window owner)	Creates a window with the specified owner window.
5.	JWindow (Window owner, Graphics Configuration gc)	Creates a window with the specified owner window and Graphics Configuration of a screen device.

Sample Code

```
import java.awt.BorderLayout;
import java.awt.event.ActionEvent;
import java.awt.event.ActionListener;
import javax.swing.JButton;
import javax.swing.JFrame;
import javax.swing.JLabel;
import javax.swing.JWindow;
import javax.swing.SwingConstants;

public class JWindowNoTitleBar extends JFrame {
 JWindow window = new JWindow(this);

 public JWindowNoTitleBar() {
  setDefaultCloseOperation(JFrame.EXIT_ON_CLOSE);
  window.getContentPane().add(new JLabel("About"),
BorderLayout.NORTH);
  window.getContentPane().add(new JLabel("Label",
SwingConstants.CENTER),
     BorderLayout.CENTER);
  JButton b = new JButton("Close");
  window.getContentPane().add(b, BorderLayout.SOUTH);
  b.addActionListener(new ActionListener() {
   public void actionPerformed(ActionEvent e) {
    window.setVisible(false);
   }
  });
  window.pack();
  window.setBounds(50, 50, 200, 200);

  b = new JButton("About...");
  b.addActionListener(new ActionListener() {
```

```
        public void actionPerformed(ActionEvent e) {
          window.setVisible(true);
        }
      });
      getContentPane().add(b);
      pack();
    }

    public static void main(String[] args) {
      new JWindowNoTitleBar().setVisible(true);
    }
  }
```

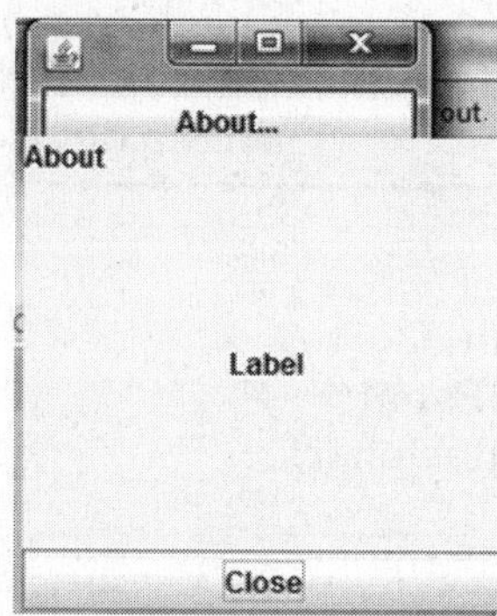

JRootPane

A JRootpane is made up of a glass Pane, an optional menu Bar, and a content Pane. (The JLayeredPane manages the menu Bar and the content Pane.) The glass Pane sits over the top of everything, where it is in a position to intercept mouse movements. Since the glass Pane (like the content Pane) can be an arbitrary component, it is also possible to set up the glass Pane for drawing. Lines and images on the glass Pane can then range over the frames underneath without being limited by their boundaries. Although the menu Bar component is optional, the layered Pane, content Pane, and glass Pane always exist. Attempting to set them to null generates an exception.

To add components to the JRootPane (other than the optional menu bar), you add the object to the contentPane of the JRootPane, like this:

```
rootPane.getContentPane().add(child);
```

The same principle holds true for setting layout managers, removing components, listing children, etc. All these methods are invoked on the content Pane instead of on the JRootPane.

Note: The default layout manager for the content Pane is a Border Layout manager. However, the JRootPane uses a custom Layout Manager. So, when you want to change the layout manager for the components you added to a JRootPane, be sure to use code like this:

```
rootPane.getContentPane().setLayout(new BoxLayout());
```

The custom LayoutManager used by JRootPane ensures that:

1. The glass Pane fills the entire viewable area of the JRootPane (bounds - insets).
2. The layered Pane fills the entire viewable area of the JRootPane (bounds - insets).

3. The menu Bar is positioned at the upper edge of the layered Pane.
4. The content Pane fills the entire viewable area, minus the menu Bar, if present.

S. No.	Constructors	Purpose
1.	JRootPane getRootPane ()	Get the root pane of the applet, dialog, frame, internal frame, or window.
2.	Static JRootPane getRootPane (Component)	If the component contains a root pane, return that root pane. Otherwise, return the root pane (if any) that contains the component.
3.	JRootPane getRootPane ()	Invoke the Swing Utilities getRootPane method for the J Component.
4.	Void setDefaultButton (JButton) J Button getDefaultButton ()	Set or get which button (if any) is the default button in the root pane. A look-and-feel-specific action, such as pressing Enter, causes the button's action to be performed.
5.	JRootPane ()	Creates a JRootPane, setting up its glass Pane, layered Pane, and content Pane.

Sample Code

```
import java.awt.*;
import javax.swing.*;
public class RootExample {
 public static void main(String[] args) {
  JFrame f = new JFrame();
  f.setDefaultCloseOperation(JFrame.EXIT_ON_CLOSE);
  JRootPane root = f.getRootPane();      // XXX Pay attention to these
  Container content = root.getContentPane(); // XXX lines. They get more
  content.add(new JButton("Hello"));      // XXX explanation in the book.
  f.pack();
  f.setVisible(true);
 }
}
```

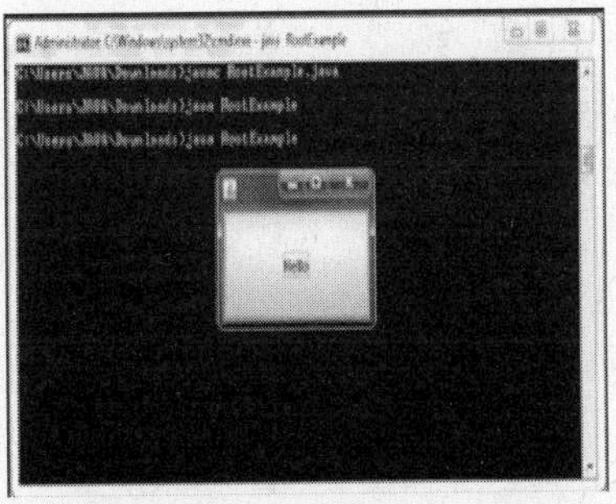

JInternalFrame

A JInternalFrame is confined to a visible area of a container it is placed in. JInternalFrame—a top level swing component that has a contentpane.

- It can be iconified — in this case the icon remains in the main application container.
- It can be maximized — Frame consumes the main application
- It can be closed using standard popup window controls
- It can be layered

S. No.	Constructors	Purpose
1.	JInternalFrame ()	Creates a non-resizable, non-closable, non-maximizable, non-iconifiable JInternal-Frame with no title.
2.	JInternalFrame (String title)	Creates a non-resizable, non-closable, non-maximizable, non-iconifiable JInternalFrame with the specified title.
3.	JInternalFrame (String title, boolean resizable)	Creates a non-closable, non-maximizable, non-iconifiable JInternalFrame with the specified title and resizability.
4.	JInternalFrame (String title, boolean resizable, boolean closable)	Creates a non-maximizable, non-iconifiable JInternalFrame with the specified title, resizability, and closability
5.	JInternalFrame (String title, boolean resizable, boolean closable, boolean maximizable)	Creates a non-iconifiable JInternalFrame with the specified title, resizability, closability, and maximizability.
6.	JInternalFrame (String title, boolean resizable, boolean closable, boolean maximizable, boolean iconifiable)	Creates a JInternalFrame with the specified title, resizability, closability, maximizability

Sample Code

```
import javax.swing.JInternalFrame;
import javax.swing.JDesktopPane;
import javax.swing.JMenu;
import javax.swing.JMenuItem;
import javax.swing.JMenuBar;
import javax.swing.JFrame;
import java.awt.event.*;
import java.awt.*;

public class JInternalFrameDemo extends JFrame {

  JDesktopPane jdpDesktop;
```

```
static int openFrameCount = 0;
public JInternalFrameDemo() {
 super("JInternalFrame Usage Demo");
 // Make the main window positioned as 50 pixels from each edge of the
 // screen.
 int inset = 50;
 Dimension screenSize = Toolkit.getDefaultToolkit().getScreenSize();
 setBounds(inset, inset, screenSize.width - inset * 2,
    screenSize.height - inset * 2);
 // Add a Window Exit Listener
 addWindowListener(new WindowAdapter() {

  public void windowClosing(WindowEvent e) {
   System.exit(0);
  }
 });
 // Create and Set up the GUI.
 jdpDesktop = new JDesktopPane();
 // A specialized layered pane to be used with JInternalFrames
 createFrame(); // Create first window
 setContentPane(jdpDesktop);
 setJMenuBar(createMenuBar());
 // Make dragging faster by setting drag mode to Outline
 jdpDesktop.putClientProperty("JDesktopPane.dragMode", "outline");
}
protected JMenuBar createMenuBar() {
 JMenuBar menuBar = new JMenuBar();
 JMenu menu = new JMenu("Frame");
 menu.setMnemonic(KeyEvent.VK_N);
 JMenuItem menuItem = new JMenuItem("New IFrame");
 menuItem.setMnemonic(KeyEvent.VK_N);
 menuItem.addActionListener(new ActionListener() {

  public void actionPerformed(ActionEvent e) {
   createFrame();
  }
 });
 menu.add(menuItem);
 menuBar.add(menu);
 return menuBar;
}
protected void createFrame() {
 MyInternalFrame frame = new MyInternalFrame();
 frame.setVisible(true);
 // Every JInternalFrame must be added to content pane using JDesktopPane
```

```
    jdpDesktop.add(frame);
    try {
      frame.setSelected(true);
    } catch (java.beans.PropertyVetoException e) {
    }
  }
  public static void main(String[] args) {
    JInternalFrameDemo frame = new JInternalFrameDemo();
    frame.setVisible(true);
  }
  class MyInternalFrame extends JInternalFrame {

    static final int xPosition = 30, yPosition = 30;
    public MyInternalFrame() {
      super("IFrame #" + (++openFrameCount), true, // resizable
          true, // closable
          true, // maximizable
          true);// iconifiable
      setSize(300, 300);
      // Set the window's location.
      setLocation(xPosition * openFrameCount, yPosition
          * openFrameCount);
    }
  }
}
```

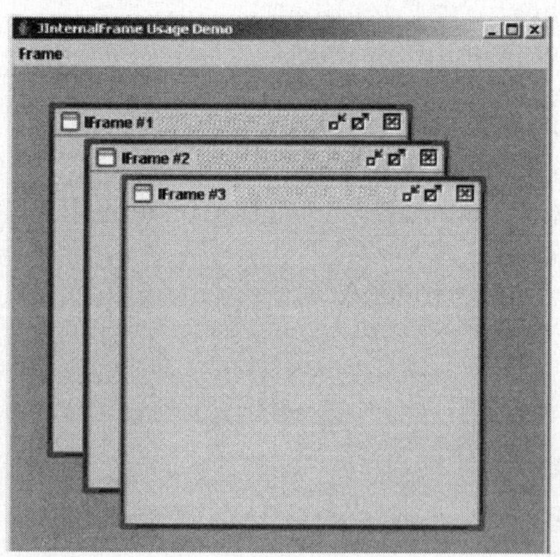

JLayeredPane

JLayeredPane adds depth to a JFC/Swing container, allowing components to overlap each other when needed. An integer object specifies each component's depth in the container, where higher-numbered components sit "on top" of other components.

For convenience, J Layered Pane divides the depth-range into several different layers. Putting a component into one of those layers makes it easy to ensure that components overlap properly, without having to worry about specifying numbers for specific depths:

DEFAULT_LAYER

The standard layer, where most components go. This is the bottommost layer.

PALETTE_LAYER

The palette layer sits over the default layer. Useful for floating toolbars and palettes, so they can be positioned above other components.

MODAL_LAYER

The layer used for modal dialogs. They will appear on top of any toolbars, palettes, or standard components in the container.

POPUP_LAYER

The popup layer displays above dialogs. That way, the popup windows associated with combo boxes, tooltips, and other help text will appear above the component, palette, or dialog that generated them.

DRAG_LAYER

When dragging a component, reassigning it to the drag layer ensures that it is positioned over every other component in the container. When finished dragging, it can be reassigned to its normal layer.

The J Layered Pane methods move To Front (Component), move To Back (Component) and set Position can be used to reposition a component within its layer. The set Layer method can also be used to change the component's current layer.

layeredPane.add(child, JLayeredPane.DEFAULT_LAYER);

or

layeredPane.add(child, new Integer(10));

The layer attribute can also be set on a Component by calling

layeredPaneParent.setLayer(child, 10)

on the J Layered Pane that is the parent of component. The layer should be set *before* adding the child to the parent.

S. No.	Constructors and Methods	Purpose
1.	JLayeredPane()	Create a new J Layered Pane

Sample Code

```
import java.awt.BorderLayout;
import java.awt.Container;

import javax.swing.JDesktopPane;
```

```
import javax.swing.JFrame;
import javax.swing.JInternalFrame;
import javax.swing.JLabel;
import javax.swing.JLayeredPane;

public class JLayeredPaneSample {
 public static void main(String args[]) {
  JFrame f = new JFrame("JDesktopPane Sample");
  f.setDefaultCloseOperation(JFrame.EXIT_ON_CLOSE);
  Container content = f.getContentPane();
  JLayeredPane desktop = new JDesktopPane();
  desktop.setOpaque(false);
  desktop.add(createLayer("Open 1"), JLayeredPane.POPUP_LAYER);
  desktop.add(createLayer("Iconified"), JLayeredPane.DEFAULT_LAYER);
  desktop.add(createLayer("Open 2"), JLayeredPane.PALETTE_LAYER);
  content.add(desktop, BorderLayout.CENTER);
  f.setSize(300, 200);
  f.setVisible(true);
 }

 public static JInternalFrame createLayer(String label) {
  return new SelfInternalFrame(label);
 }

 static class SelfInternalFrame extends JInternalFrame {
  public SelfInternalFrame(String s) {
   getContentPane().add(new JLabel(s), BorderLayout.CENTER);
   setBounds(50, 50, 100, 100);
   setResizable(true);
   setClosable(true);
   setMaximizable(true);
   setIconifiable(true);
   setTitle(s);
   setVisible(true);
  }
 }
}
```

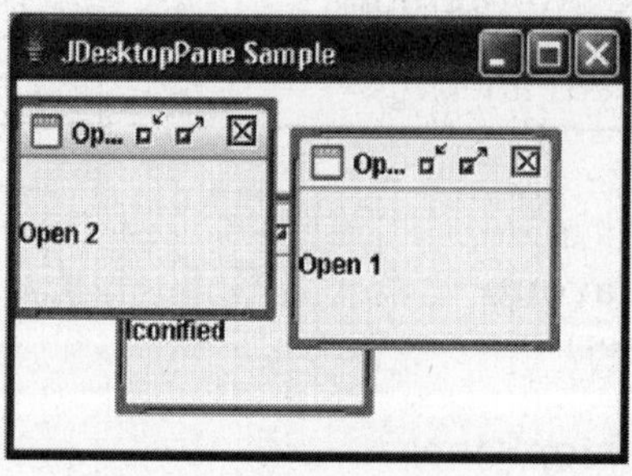

JDesktoppane

A container used to create a multiple-document interface or a virtual desktop. You create JInternalFrame objects and add them to the JDesktopPane. JDesktopPane extends JLayeredPane to manage the potentially overlapping internal frames. It also maintains a reference to an instance of DesktopManager that is set by the UI class for the current look and feel (L&F). Note that JDesktopPane does not support borders.

This class is normally used as the parent of JInternalFrames to provide a pluggable DesktopManager object to the JInternalFrames. The installUI of the L&F specific implementation is responsible for setting the desktopManager variable appropriately. When the parent of a JInternalFrame is a JDesktopPane, it should delegate most of its behavior to the desktopManager (closing, resizing, etc).

S. No.	Constructor and Method	Purpose
1.	JDesktopPane()	Create a new JDesktopPane

Sample Source Code

```
import javax.swing.*;
import javax.swing.event.*;
import java.awt.*;
import java.awt.event.*;

public class AllFrameDesktopContainer{
 JDesktopPane desk;
 JInternalFrame iframe;
 JFrame frame;
 public static void main(String[] args) {
 AllFrameDesktopContainer d = new AllFrameDesktopContainer();
 }

 public AllFrameDesktopContainer(){
 frame = new JFrame("All Frames in a JDesktopPane Container");
 frame.setDefaultCloseOperation(JFrame.EXIT_ON_CLOSE);
 desk = new JDesktopPane();
 try{
 String str = JOptionPane.showInputDialog

 (null, "Enter number of frames :",
 "Roseindia.net", 1);
 int i = Integer.parseInt(str);
 for (int j = 1; j <= i; j++){
 iframe = new JInternalFrame("Internal Frame: " + j, true,

 true, true, true);
 iframe.setBounds(j*20, j*20, 150, 100);
 iframe.setVisible(true);
 desk.add(iframe);
 iframe.setToolTipText("Internal Frame :" + j);
```

```
}
}
catch(NumberFormatException ne){
JOptionPane.showMessageDialog(null, "Please enter number value.",
"demo", 1);
System.exit(0);
}
JMenuBar menubar = new JMenuBar();
JMenu count = new JMenu("Count Total Frames");
count.addMenuListener(new MyAction());
menubar.add(count);
frame.setJMenuBar(menubar);
frame.add(desk);
frame.setSize(400,400);
frame.setVisible(true);
}

public class MyAction implements MenuListener{
public void menuSelected(MenuEvent me){
int i = desk.getAllFrames().length;
JOptionPane.showMessageDialog(null, "Total visible internal frames
are : " + i,
"DEMO", 1);
}

public void menuCanceled(MenuEvent me){}

public void menuDeselected(MenuEvent me){}
}
}
```

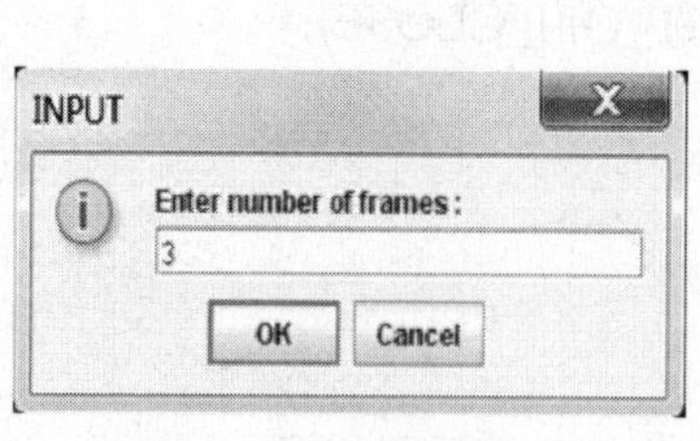

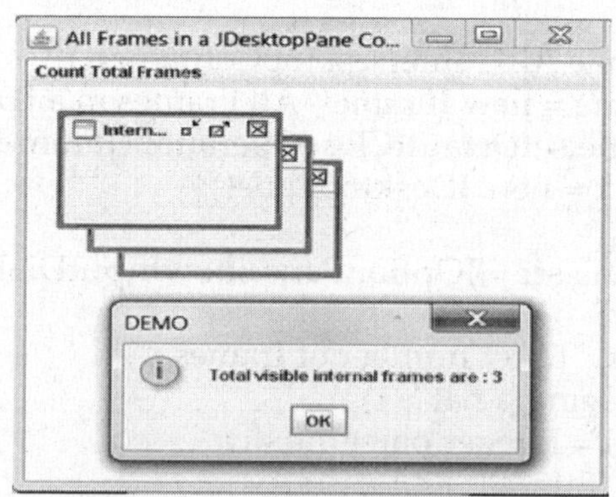

JTable

The **JTable** component is more flexible Java Swing component that allows the user to store, show and edit the data in tabular format. It is a user-interface component that represents the data of two-dimensional tabular format. The Java swing implements tables by using the **JTable** class and a subclass of **JComponent**.

For learning complete JTable component you have to learn the following parts:

- JTable
- Rendering
- Table model
- Observable event model
- Tool tips

Rendering: In computer program, the rendering is a process that generates an image from a model. Model is a specification of three-dimensional objects.

Table model: All tables contains of its data from an object that implements the **TableModel** interface of java swing package. The Table Model interface uses the methods of **JTable** class that integrates a tabular data model.

Observable event model: The event model is a building block of graphical user interfaces (**GUI**). This is a user friendly and most common that is provided by the graphical frameworks. The observable is a system state property that determines by some sequence of physical operations.

Tool tips: The tool tips are most common graphical user interface that is used for displaying the message in small box given by the user appropriate message, when the mouse pointer or cursor goes on it, a message (description) appears without clicking it.

Description of Program

In this Java programming tutorial we are going to implement the simple JTable on the java swing frame with the help of some java methods. To create a JTable component, you need a java swing frame. The JTable based on the frame's panel contains multiple data in a tabular format such as rows and columns.

Description of Code

J Table(Object data [][], Object col [])

This is the constructor of **JTable** class that implements a JTable for displaying the values in tabular format like rows and columns. All rows must have the same length as column names. It takes two arguments: Data and col.

data: It defines all data that will have to create a JTable.

col: It specifies the name of each column of the JTable.

S. No.	Constructors	Purpose
1.	JTable ()	Constructs a default JTable that is initialized with a default data model, a default column model, and a default selection model.
2.	JTable **(int num Rows, int num Columns)**	Constructs a JTable with num Rows and num Columns of empty cells using Default Table Model.
3.	JTable (Object [][] **row Data,** Object [] **column Names)**	Constructs a JTable to display the values in the two-dimensional array, row Data, with column names, column Names.

4.	JTable (Table Model **dm)**	Constructs a JTable that is initialized with dm as the data model, a default column model, and a default selection model.
5.	J Table (Table Model **dm,** Table Column Model **cm)**	Constructs a JTable that is initialized with dm as the data model, cm as the column model, and a default selection model.
6.	J Table (Table Model **dm,** Table Column Model **cm,** List Selection Model **sm)**	Constructs a JTable that is initialized with dm as the data model, cm as the column model, and sm as the selection model.
7.	J Table (Vector **row Data,** Vector **column Names)**	Constructs a JTable to display the values in the Vector of Vectors, row Data, with column names, column Names.

Sample Code

```
import javax.swing.*;
import java.awt.*;

public class JTableComponent{
 public static void main(String[] args)
{
 new JTableComponent();
 }

 public JTableComponent(){
 JFrame frame = new JFrame("Creating
JTable Component Example!");
 JPanel panel = new JPanel();
 String data[][] = {{"vinod","BCA","A"},{"Raju","MCA","b"},
 {"Ranjan","MBA","c"},{"Rinku","BCA","d"}};

  String col[] = {"Name","Course","Grade"};
 JTable table = new JTable(data,col);
 panel.add(table,BorderLayout.CENTER);

frame.add(panel);
 frame.setSize(300,200);
 frame.setVisible(true);
 frame.setDefaultCloseOperation(JFrame.EXIT_ON_CLOSE);
 }
}
```

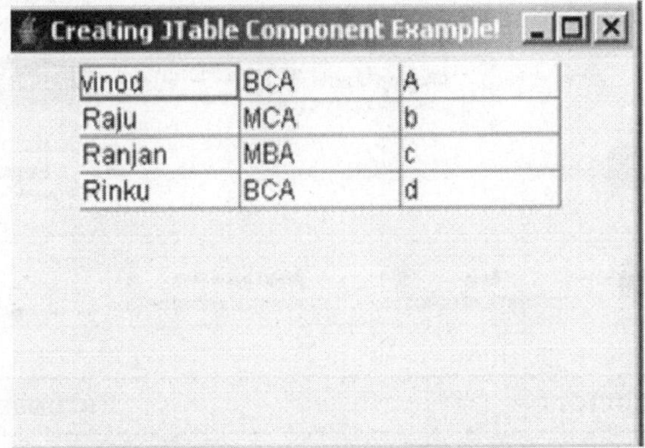

JComponent

The base class for all Swing components except top-level containers. To use a component that inherits from JComponent, you must place the component in a containment hierarchy whose root is a top-level Swing container. Top-level Swing containers — such as JFrame, JDialog, and JApplet — are specialized components that provide a place for other Swing components to paint themselves.

The JComponent class provides:

- The base class for both standard and custom components that use the Swing architecture.
- A "pluggable look and feel" (L&F) that can be specified by the programmer or (optionally) selected by the user at runtime. The look and feel for each component is provided by a *UI delegate* — an object that descends from Component UI.
- Comprehensive keystroke handling.
- Support for tool tips — short descriptions that pop up when the cursor lingers over a component.
- Support for accessibility. JComponent contains all of the methods in the Accessible interface, but it does not actually implement the interface. That is the responsibility of the individual classes that extend JComponent.
- Support for component-specific properties. With the put Client Property (java.lang.Object) and get Client Property (java.lang.Object) methods, you can associate name-object pairs with any object that descends from JComponent.
- An infrastructure for painting that includes double buffering and support for borders.

S. No.	Constructors	Purpose
1.	JLayeredPane()	Default JComponent constructor.

JPopupMenu

Popup menu is the list of menu which is displayed at that point on the frame where you press the right mouse button.

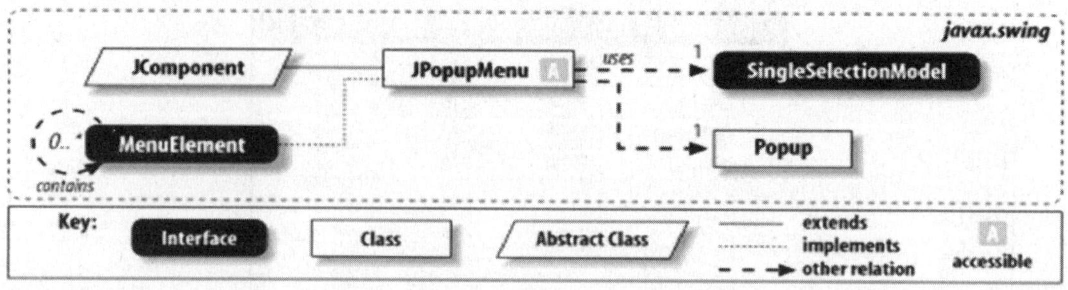

S. No.	Constructors	Purpose
1.	**JPopupMenu**	This is the class which constructs the popup menu using its constructor. This class is helpful to add the object of the **JMenuItem** class which creates a particular menu.
2.	**IsPopupTrigger ()**	This is the method of the **Mouse Event** class of the *java.awt.event.*;* package. This method returns a boolean type value either *true* or *false*. This method returns *true* if the event is generated when the popup is triggered.
3.	**MouseReleased**	Set or get the vertical scroll policy. Scroll Pane Constants defines three values for specifying this policy: VERTICAL_SCROLLBAR_AS_NEEDED (the default), VERTICAL_SCROLLBAR_ALWAYS, and VERTICAL_SCROLLBAR_NEVER.
4.	**Get X ()**	This is the method of the **Mouse Event** class which is imported from the *java.awt.event.*;* package. This method returns the integer type value which is the position on the x-axis for the source component where you click the mouse.
5.	**Get Y ()**	This is also the method of the the **Mouse Event** class. This method returns the vertical positions of the y-coordinate for the source component where you click the mouse.
6.	**Show (me. get Component (), me. get X (), me. get Y())**	This is the method of **JPopupMenu** class which displays the popup menu where you press the right mouse button on the specified location or positions. This position is calculated by the **get X ()** and **get Y ()**.
7.	**Get Component ()**	This is the method of the **Component Event** class of the *java.awt.event.*;* package. This method returns the source component of the generated event.

Sample Code

```
import javax.swing.*;
import java.awt.event.*;

public class PopUpMenu{
 JPopupMenu Pmenu;
 JMenuItem menuItem;
 public static void main(String[] args) {
 PopUpMenu p = new PopUpMenu();
 }

 public PopUpMenu(){
 JFrame frame = new JFrame("Creating a Popup Menu");
 frame.setDefaultCloseOperation(JFrame.EXIT_ON_CLOSE);
 Pmenu = new JPopupMenu();
 menuItem = new JMenuItem("Cut");
 Pmenu.add(menuItem);
 menuItem = new JMenuItem("Copy");
 Pmenu.add(menuItem);
 menuItem = new JMenuItem("Paste");
 Pmenu.add(menuItem);
 menuItem = new JMenuItem("Delete");
 Pmenu.add(menuItem);
 menuItem = new JMenuItem("Undo");
 Pmenu.add(menuItem);
 menuItem.addActionListener(new ActionListener(){
 public void actionPerformed(ActionEvent e){}
 });
 frame.addMouseListener(new MouseAdapter(){
 public void mouseReleased(MouseEvent Me){
 if(Me.isPopupTrigger()){
 Pmenu.show(Me.getComponent(), Me.getX(), Me.getY());
 }
 }
 });
 frame.setSize(400,400);
 frame.setVisible(true);
 }
}
```

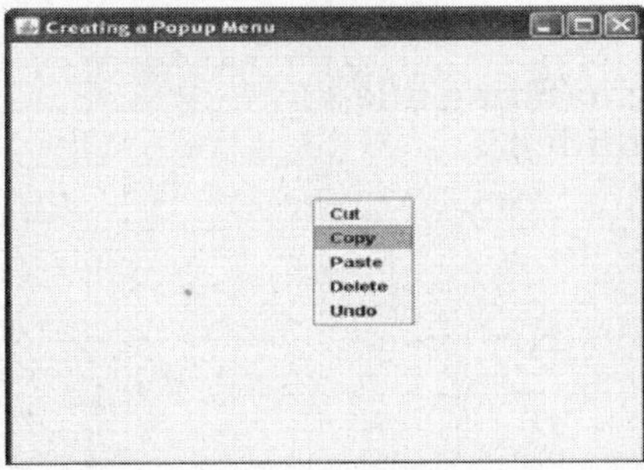

Nested Popup Menus

When you click the right mouse button on the frame; then you get a popup menu. Here, you will show multiple menu items like: line up icon, refresh, properties and new in the list of the menu which represents the popup menu. The new menu item have another popup menu which contains items like: folder, text document and shortcut

createMenu(String)

Sample code

```
import javax.swing.*;
import java.awt.event.*;

public class NestedPopupMenu{
 JPopupMenu Pmenu;
 JMenuItem Mitem;
 public static void main(String[] args) {
 NestedPopupMenu n = new NestedPopupMenu();
 }

 public  NestedPopupMenu(){
 JFrame frame = new JFrame("Creating a Popup Menu with Nested Menus");
 frame.setDefaultCloseOperation(JFrame.EXIT_ON_CLOSE);
 Pmenu = new JPopupMenu();
 Mitem = new JMenuItem("Line Up Icon");
 Pmenu.add(Mitem);
 Mitem = new JMenuItem("Refresh");
 Pmenu.add(Mitem);
 Mitem = new JMenuItem("Properties");
 Pmenu.add(Mitem);
 Pmenu.add(createMenu("New"));
 frame.addMouseListener(new MouseAdapter(){
 public void mouseReleased(MouseEvent Me){
 if(Me.isPopupTrigger()){
 Pmenu.show(Me.getComponent(), Me.getX(), Me.getY());
 }
 }
 });
 frame.setSize(400,400);
 frame.setVisible(true);
 }

 public JMenu createMenu(String title){
 JMenu m = new JMenu(title);
 m.add("Folder");
 m.add("Text Document");
 m.add("Shortcut");
 return m;
 }
}
```

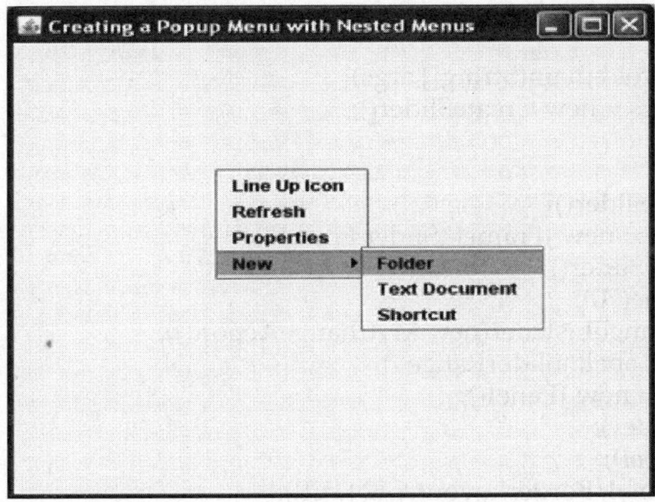

JSlider

A Slider is a Swing tool which you can use for selecting a range. There is a minimum chance of being mistake to illegal input values. If you increase or decrease the slider by selecting, then the actual position of the slider will be displayed on a label.

S. No.	Constructors and Methods	Purpose
1.	JSlider	This is the class which creates the slider for the swing application. For creating the slider this class creates a instance using its constructor JSlider ().
2.	ChangeListener	This is the interface which is used to call state Changed () method which receives the event generated by the slider using add Change Listener () method of the **JSlider** class
3.	ChangeEvent	This is the class which handle the event generated by the JSlider component on change the state
4.	AddChangeListener (object)	This is the method of the **JSlider** class which is used to handle event on change the selected state of the JSlider component.

Sample Code

```
import javax.swing.*;
import javax.swing.event.*;
import java.awt.*;
import java.awt.event.*;

public class CreateSlider{
```

```
JSlider slider;
JLabel label;
public static void main(String[] args){
CreateSlider cs = new CreateSlider();
}

public CreateSlider(){
JFrame frame = new JFrame("Slider Frame");
slider = new JSlider();
slider.setValue(70);
slider.addChangeListener(new MyChangeAction());
label = new JLabel("SliderRange");
JPanel panel = new JPanel();
panel.add(slider);
panel.add(label);
frame.add(panel, BorderLayout.CENTER);
frame.setSize(400, 400);
frame.setVisible(true);
frame.setDefaultCloseOperation(JFrame.EXIT_ON_CLOSE);
}

public class MyChangeAction implements ChangeListener{
public void stateChanged(ChangeEvent ce){
int value = slider.getValue();
String str = Integer.toString(value);
label.setText(str);
}
}
}
```

JApplet

JApplet is the Swing equivalent of the AWT Applet class.

Much like JFrame, JApplet has extensions to allow for

- interposing input,
- special painting behavior,
- supports child components that are managed by a root pane.

Unlike the AWT applet class, JApplet permits the addition of menu bars and toolbars — *A Major Improvement*

S. No.	Constructors and Methods	Purpose
1.	JApplet ()	Create a swing applet instance.

Source Code

doll.java

```
import java.awt.*;
import java.applet.Applet;
import javax.swing.*;
```

```
public class doll extends JApplet
{

public void paint(Graphics g)
{
g.drawOval(50,10,80,40);
g.drawLine(90,50,90,70);
g.drawRect(30,70,110,90);
g.drawLine(60,160,30,190);
g.drawLine(110,160,140,190);
g.drawLine(30,190,140,190);
g.drawLine(70,190,70,230);
g.drawLine(100,190,100,230);
g.drawLine(30,90,10,120);
g.drawLine(140,90,160,120);
}
}
```

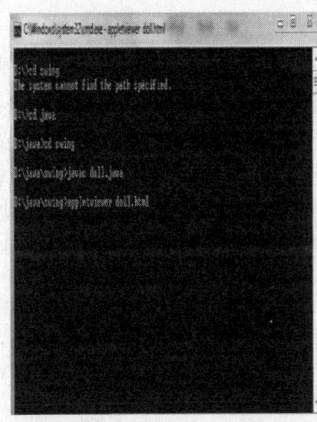

APPLETCODING

doll.html

```
<applet code="doll" width=200 height=300></applet>
```

allout.java

```
import javax.swing.*;
import java.awt.*;
import java.applet.Applet;
public class allout extends JApplet
{
public void paint(Graphics g)
{
g.setColor(Color.BLUE);
g.fillArc(80,40,50,50,0,180);
g.setColor(Color.GRAY);
g.fillRect(80,60,50,20);
g.setColor(Color.PINK);
g.fillRect(80,80,50,70);
g.setColor(Color.BLACK);
g.fillRect(105,80,5,70);
g.setColor(Color.PINK);
g.drawRect(50,30,100,170);
g.drawLine(170,130,250,130);
g.drawLine(250,130,240,150);
g.drawLine(250,130,240,110);
g.setColor(Color.RED);
g.drawString("ALLOUT",260,129);
}
}
```

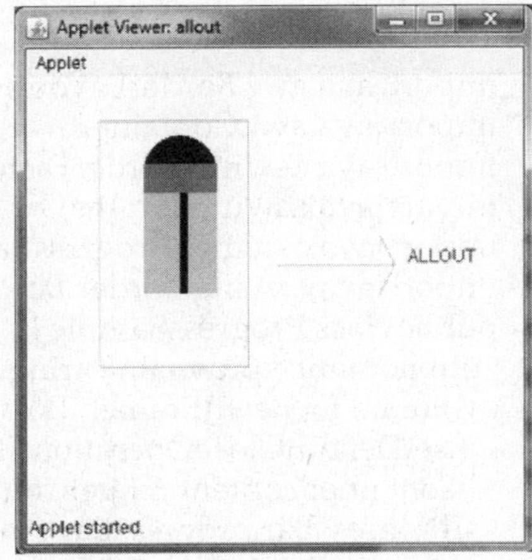

Applet Code

<applet code="allout" width=200 height=300></applet>

JProgressbar

A component that visually displays the progress of some task. As the task progresses towards completion, the progress bar displays the task's percentage of completion. This percentage is typically represented visually by a rectangle which starts out empty and gradually becomes filled in as the task progresses. In addition, the progress bar can display a textual representation of this percentage.

JProgressBar uses a Bounded Range Model as its data model, with the value property representing the "current" state of the task, and the minimum and maximum properties representing the beginning and end points, respectively.

S. No.	Constructors and Methods	Purpose
1.	JProgressBar ()	Creates a horizontal progress bar that displays a border but no progress string.
2.	JProgressBar (Bounded Range Model **newModel)**	Creates a horizontal progress bar that uses the specified model to hold the progress bar's data.
3.	JProgressBar **(int orient)**	Creates a progress bar with the specified orientation, which can be either Swing Constants.VERTICAL or Swing Constants. HORIZONTAL.
4.	JProgressBar **(int min, int max)**	Creates a horizontal progress bar with the specified minimum and maximum.
5.	JProgressBar **(int orient, int min, int max)**	Creates a progress bar using the specified orientation, minimum, and maximum.

Source Code

```
import java.awt.BorderLayout;
import java.awt.Container;
import javax.swing.BorderFactory;
import javax.swing.JFrame;
import javax.swing.JProgressBar;
import javax.swing.border.Border;
public class ProgressSample {
 public static void main(String args[]) {
  JFrame f = new JFrame("JProgressBar Sample");
  f.setDefaultCloseOperation(JFrame.EXIT_ON_CLOSE);
  Container content = f.getContentPane();
  JProgressBar progressBar = new JProgressBar();
  progressBar.setValue(25);
  progressBar.setStringPainted(true);
  Border border = BorderFactory.createTitledBorder("Reading...");
```

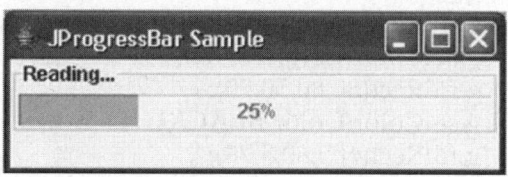

```
    progressBar.setBorder(border);
    content.add(progressBar, BorderLayout.NORTH);
    f.setSize(300, 100);
    f.setVisible(true);
  }
}
```

JSeparator

JSeparator provides a general purpose component for implementing divider lines—most commonly used as a divider between menu items that breaks them up into logical groupings. Instead of using J Separator directly, you can use the J Menu or J Popup Menu add Separator method to create and add a separator. J Separators may also be used elsewhere in a GUI wherever a visual divider is useful.

S. No.	Constructors and Methods	Purpose
1.	JSeparator ()	Creates a new horizontal separator.
2.	JSeparator (int orientation)	Creates a new separator with the specified horizontal or vertical orientation.

Source Code

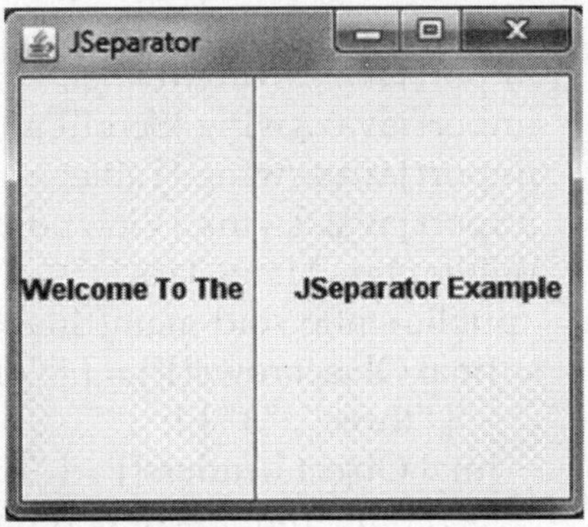

```
import javax.swing.JFrame;
import javax.swing.JLabel;
import javax.swing.JSeparator;
import javax.swing.SwingConstants;
import java.awt.BorderLayout;
public class JSeparatorExample extends JFrame {
public JSeparatorExample() {
JLabel label = new JLabel();
JLabel label1 = new JLabel();
label.setText("Welcome To The ");
label1.setText("JSeparator Example ");
JSeparator separator = new JSeparator();
separator.setOrientation(SwingConstants.VERTICAL);
this.getContentPane().add(label,BorderLayout.WEST);
this.getContentPane().add(separator,BorderLayout.CENTER);
this.getContentPane().add(label1,BorderLayout.EAST);
}
public static void main(String a[]) {
JSeparatorExample separatorExample = new JSeparatorExample();
separatorExample.setSize(250, 200);
separatorExample.setTitle("JSeparator");
separatorExample.setVisible(true);
}
}
```

JSeparator

Welcome To The JSeparator Example

JViewport

The JViewport class is the container that is really responsible for displaying a specific visible region of the component in JScroll Pane. We can set/get a viewport's *view* (the component it contains) using its set View() and get View() methods. We can control how much of this component JViewport displays by setting its *extent* size to a specified Dimension using its set Extent Size() method. We can also specify where the origin (upper left corner) of a JViewport should begin displaying its contained component by providing specific coordinates (as a Point) of the contained component to the set View Position() method. In fact, when we scroll a component in a JScrollPane; this view position is constantly being changed by the scrollbars.

We can retreive JScrollPane's main JViewport by calling its get Viewport() method, or assign it a new one using set Viewport(). We can replace the component in this viewport through JScrollPane's set Viewport View() method, but there is no get Viewport View() counterpart. Instead we must first access its J Scroll Pane's JViewport by calling get Viewport(), and then callget View() on that (as discussed above). Typically, to access a JScrollPane's main child component we would do the following:

Component my Component = jsp get Viewport () get View();

S. No.	Constructors and Methods	Purpose
1.	JViewport ()	Create a JViewport.

Source Code

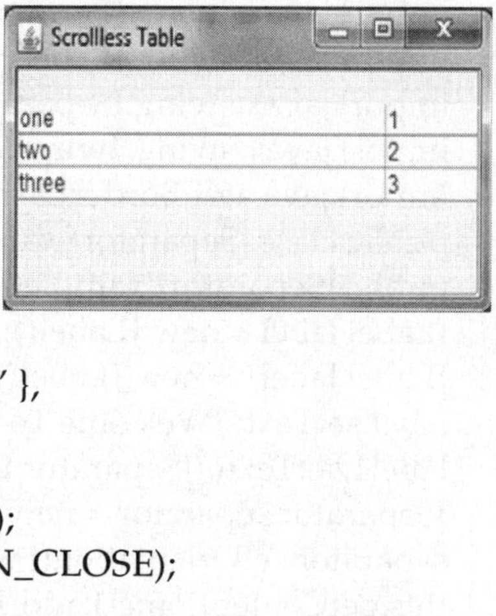

```
import java.awt.BorderLayout;
import javax.swing.JFrame;
import javax.swing.JLabel;
import javax.swing.JScrollPane;
import javax.swing.JTable;
import javax.swing.JViewport;
public class MainClass {
 public static void main(final String args[]) {
  final Object rows[][] = { { "one", "1" }, { "two", "2" },
     { "three", "3" } };
  final Object headers[] = { "English", "Digit" };
     JFrame frame = new JFrame("Scrollless Table");
  frame.setDefaultCloseOperation(JFrame.EXIT_ON_CLOSE);
     JTable table = new JTable(rows,headers);
     JScrollPane scrollPane = new JScrollPane(table);
  JViewport viewport = new JViewport();
  viewport.setView(table);
  scrollPane.setColumnHeaderView( new JLabel("table header here"));
  scrollPane.setRowHeaderView(viewport);
     frame.add(scrollPane, BorderLayout.CENTER);
  frame.setSize(300, 150);
  frame.setVisible(true);
 }
}
```

Scrollless Table

one	1
two	2
three	3

4 Generic Programming

1. MOTIVATION FOR GENERIC PROGRAMMING

Generic programming is a style of computer programming in which algorithms are written in terms of *to-be-specified-later* types that are then *instantiated* when needed for specific types provided as parameters.

As per *Java Language Specification*:

- A **type variable** is an unqualified identifier. Type variables are introduced by generic class declarations, generic interface declarations, generic method declarations, and by generic constructor declarations.
- A **class** is generic if it declares one or more type variables. These type variables are known as the type parameters of the class. It defines one or more type variables that act as parameters. A generic class declaration defines a set of parameterized types, one for each possible invocation of the type parameter section. All of these parameterized types share the same class at runtime.
- An **interface** is generic if it declares one or more type variables. These type variables are known as the type parameters of the interface. It defines one or more type variables that act as parameters. A generic interface declaration defines a set of types, one for each possible invocation of the type parameter section. All parameterized types share the same interface at runtime.
- A **method** is generic if it declares one or more type variables. These type variables are known as the formal type parameters of the method. The form of the formal type parameter list is identical to a type parameter list of a class or interface.
- A **constructor** can be declared as generic, independent of whether the class the constructor is declared is in itself generic. A constructor is generic if it declares one or more type variables. These type of variables are known as the formal type parameters of the constructor. The form of the formal type parameter list is identical to a type parameter list of a generic class or interface.

Overloaded methods are often used to perform similar operations on different types of data. To motivate generic methods, let us begin with an example that contains three overloaded printArray methods. These methods print the string representations of the elements of an Integer array, a Double array and a Character array, respectively. Note that we could have used arrays of primitive types int, double and char in this example. We chose to use arrays of type Integer, Double and Character to set up our generic method example, because only reference types can be used with generic methods and classes.

```
public class OverloadedMethods
{
// method printArray to print Integer array
public static void printArray( Integer[] inputArray)
{
// display array elements
for ( Integer element : inputArray )
        System.out.printf( "%s ", element );
        System.out.println();
}

// end method printArray
// method printArray to print Double array

public static void printArray( Double[] inputArray )
{
// display array elements
for ( Double element : inputArray )

        System.out.printf( "%s ", element );
        System.out.println();
} // end method printArray
// method printArray to print Character array
public static void printArray( Character[] inputArray )
{
// display array elements
for ( Character element : inputArray )
        System.out.printf( "%s ", element );
        System.out.println();
} // end method printArray
public static void main( String args[] )
{
// create arrays of Integer, Double and Character
Integer[] integerArray = { 1, 2, 3, 4, 5, 6 };
Double[] doubleArray = { 1.1, 2.2, 3.3, 4.4, 5.5, 6.6, 7.7 };
Character[] characterArray = { 'H', 'E', 'L', 'L', 'O' };
System.out.println( "Array integerArray contains:" );
printArray( integerArray); // pass an Integer array
System.out.println( "\nArray doubleArray contains:" );
printArray( doubleArray ); // pass a Double array
System.out.println( "\nArray characterArray contains:" );
printArray(characterArray ); // pass a Character array
} // end main
} // end class OverloadedMethods
Array integerArray contains:
1 2 3 4 5 6
```

Array doubleArray contains:

1.1 2.2 3.3 4.4 5.5 6.6 7.7

Array characterArray contains: H E L L O

Generic Class Definitions

Here is an example of a generic class:

```
public class Pair<T, S>
{
public Pair(T f, S s)
{
first = f;
second = s;
}

public T getFirst()
{
return first;
}

public S getSecond()
{
return second;
}

public String toString()
{
return "(" + first.toString() + ", " + second.toString() + ")";
}

private T first;
private S second;
}
```

This generic class can be used in the following ways:

```
Pair<String, String> grade440 = new Pair<String, String>("mike", "A");
Pair<String,  Integer> marks440 = new Pair <String, Integer>("mike", 100);
System.out.println("grade:" + grade440.toString());
System.out.println("marks:" + marks440.toString());
```

Type Erasure

Generics are checked at compile-time for type correctness. The generic type information is then removed via a process called type erasure. For example, List<Integer> will be converted to the raw type (non-generic type) List, which can contain arbitrary objects. However, due to the compile-time check, the resulting code is guaranteed to be type correct, as long as the code generated no unchecked compiler warnings.

As a result, there is no way to tell at runtime which type parameter is used on an object. For example, when you examine an ArrayList at runtime, there is no general way to tell whether it was an ArrayList<Integer> or an ArrayList<Float>. The exception to this is by using **Reflection** on existing list elements. However, if the list is empty or if its elements are subtypes of the parameterized type, even Reflection will not divulge the parameterized type.

The following code demonstrates that the Class objects appear the same.

ArrayList<Integer> li = new ArrayList<Integer>(); ArrayList<Float> lf = new ArrayList<Float>();

if (li.getClass() == lf.getClass()) // evaluates to true

System.out.println("Equal");

Java generics differ from **C++** templates. Java generics generate only one compiled version of a generic class or function regardless of the number of types used. Furthermore, the Java compiler does not need to know which parameterized type is used because the type information is validated at compile-time and erased from the compiled code. Consequently, one cannot instantiate a Java class of a parameterized type because instantiation requires a call to a constructor, which is not possible when the type is unknown at both compile-time and runtime.

```
T instantiateElementType(List<T> arg)
{
return new T(); //causes a compile error
}
```

Because there is only one copy of a generic class, static variables are shared among all the instances of the class, regardless of their type parameter. As a result, the type parameter cannot be used in the declaration of static variables or in static methods. Static variables and static methods are "outside" of the scope of the class's parameterized types.

2. GENERIC CLASSES

A generic class declaration looks like a non-generic class declaration, except that the class name is followed by a type parameter section. As with generic methods, the type parameter section of a generic class can have one or more type parameters separated by commas. These classes are known as parameterized classes or parameterized types because they accept one or more parameters.

Following example illustrates how we can define a generic class:

```
public class Box<T> {

 private T t;

 public void add(T t) {
  this.t = t;
 }
 public T get() {
  return t;
 }
 public static void main(String[] args) {
```

```
      Box<Integer> integerBox = new Box<Integer>();
      Box<String> stringBox = new Box<String>();

      integerBox.add(new Integer(10));
      stringBox.add(new String("Hello World"));

      System.out.printf("Integer Value :%d\n\n", integerBox.get());
      System.out.printf("String Value :%s\n", stringBox.get());
   }
}
```

This would produce the following results:
Integer Value : 10
String Value : Hello World

3. GENERIC METHODS

You can write a single generic method declaration that can be called with arguments of different types. Based on the types of the arguments passed to the generic method, the compiler handles each method call appropriately. Following are the rules to define Generic Methods:

- All generic method declarations have a type parameter section delimited by angle brackets (< and >) that precedes the method's return type (< E > in the next example).
- Each type parameter section contains one or more type parameters separated by commas. A type parameter, also known as a type variable, is an identifier that specifies a generic type name.
- The type parameters can be used to declare the return type and act as placeholders for the types of the arguments passed to the generic method, which are known as actual type arguments.
- A generic method's body is declared like that of any other method. Note that type parameters can represent only reference types, not primitive types (like int, double and char).

Following example illustrates how we can print array of different types using a single Generic method:

```
public class GenericMethodTest
{
  // generic method printArray
  public static < E > void printArray( E[] inputArray )
  {
    // Display array elements
      for ( E element : inputArray ){
        System.out.printf( "%s ", element );
      }
      System.out.println();
  }
```

```
public static void main( String args[] )
{
   // Create arrays of Integer, Double and Character
   Integer[] intArray = { 1, 2, 3, 4, 5 };
   Double[] doubleArray = { 1.1, 2.2, 3.3, 4.4 };
   Character[] charArray = { 'H', 'E', 'L', 'L', 'O' };

   System.out.println( "Array integerArray contains:" );
   printArray( intArray  ); // pass an Integer array

   System.out.println( "\nArray doubleArray contains:" );
   printArray( doubleArray ); // pass a Double array

   System.out.println( "\nArray characterArray contains:" );
   printArray( charArray ); // pass a Character array
}
}
```

This would produce the following results:

```
Array integerArray contains:
1 2 3 4 5 6
Array doubleArray contains:
1.1 2.2 3.3 4.4
Array characterArray contains:
H E L L O
```

Bounded Type Parameters

There may be times when you will want to restrict the kinds of types that are allowed to be passed to a type parameter. For example, a method that operates on numbers might only want to accept instances of Number or its subclasses. This is what bounded type parameters are for.

To declare a bounded type parameter, list the type parameter's name, followed by the extends keyword, followed by its upper bound.

Following example illustrates how extends is used in a general sense to mean either "extends" (as in classes) or "implements" (as in interfaces). This example is Generic method to return the largest of three Comparable objects:

```
public class Maximum

Test
{
  // determines the largest of three Comparable objects
  public static <T extends Comparable<T>> T maximum(T x, T y, T z)
  {
   T max = x; // assume x is initially the largest
   if ( y.compareTo( max ) > 0 ){
     max = y; // y is the largest so far
   }
```

```
  if ( z.compareTo( max ) > 0 ){
    max = z; // z is the largest now
  }
  return max; // returns the largest object
}
public static void main( String args[] )
{
  System.out.printf( "Max of %d, %d and %d is %d\n\n",
          3, 4, 5, maximum( 3, 4, 5 ) );

   System.out.printf( "Maxm of %.1f,%.1f and %.1f is %.1f\n\n",
          6.6, 8.8, 7.7, maximum( 6.6, 8.8, 7.7 ) );

   System.out.printf( "Max of %s, %s and %s is %s\n","pear",
    "apple", "orange", maximum( "pear", "apple", "orange" ) );
  }
}
```

This would produce the following results:
Maximum of 3, 4 and 5 is 5
Maximum of 6.6, 8.8 and 7.7 is 8.8
Maximum of pear, apple and orange is pear

4. GENERIC CODE AND VIRTUAL MACHINE

All objects including generic objects belong to ordinary classes in the virtual machine. The virtual machine does not have objects of generic types. Whenever a generic type is defined, a corresponding raw type automatically provided. The name of the raw type is simply the name of the generic type, with the type parameters removed. The type variable are erased and replaced by their bounding types (or Object for variables without bounds). The raw type replaces type variables with the first bound, or Object if no bounds are given. The code that uses generics compiled using a java compiler will not execute on the pre-5.0 virtual machines.

Example

```
public class Pair
{
    public Pair(Object first,Object second)
    {
            this.first=first;
            this.second=second;
    }
    public Object getFirst()
    {
            return first;
    }
   public  Object getSecond()
                        return second;
            public void setFirst(Object newValue)
```

```
            {
                    first = newValue;
            public void setSecond(Object newValue)
    {
            second=newValue;
    }
        private Object first;
        private Object second;
    }
```

In the example, the raw type is Pair<T>. Here T is an unbound variable. So it is simply replaced by Object. Result is an ordinary class.

If class Interval<T Serializable & Comparable> is used, the raw type replaces T with Serializable, and the compiler inserts casts to Comparable when necessary. For efficiency, tagging interfaces (that is, interfaces without methods) are put at the end of the bounds list.

5. INHERITANCE AND GENERICS

While dealing with generic classes, some rules about inheritance and subtypes are to be known to the programmers. Consider a class and subclass, such as Employee and Manager. But when these are used to create objects of generic types, there will be no relationship between the created objects. That is, there is no relationship between Pair<S> and Pair<T>, no matter how S and T are related.

Example

```
Pair<Manager> man=new Pair(ceo,cfo);
Pair<Employee>emp=man;     //Error
```

In the above example, there is no relationship between man and emp objects, even Employee is the super class of Manager. So assigning man to emp throws compile time error. This restriction is for type safety.

Relation between Employee and Manager Class.

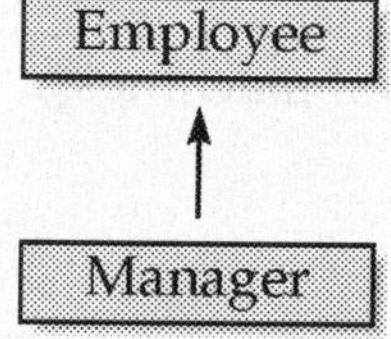

Relation between Pair<Employee> and Pair<Manager>

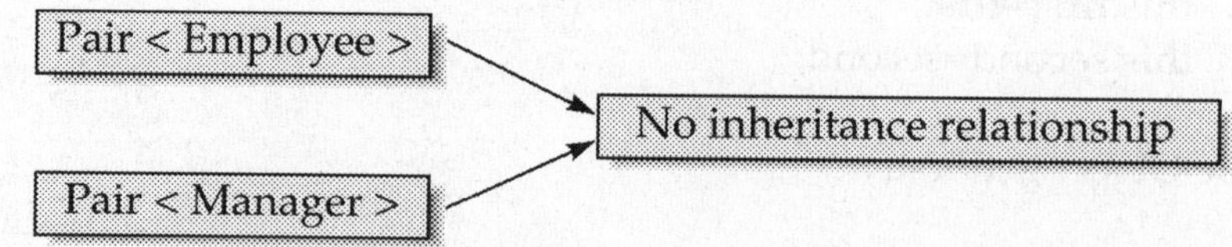

But, parameterized type can be converted into a raw type in java.

Example

```
Pair<Manager> man=new Pair<Manager>(ceo,cfo);
Pair raw = man; //OK
```

Difference between arrays and generics

The main difference between generic types and Java arrays is that we can assign the derived class array to base class array. But this is not possible for generics.

```
Manager[] man ={ceo,cfo };
Employee [] emp = man;// OK
```

Subtype relationships among generic list types

Like ordinary classes, generic classes can extend or implement other generic classes. For example, the class ArrayList<T> implements the interface List<T>. That means, an ArrayList<Manager> can be converted to a List<Manager>. However, as we just saw, an ArrayList<Manager> is not an ArrayList<Manager> is not an ArrayList<Employee> or List<Employee>.

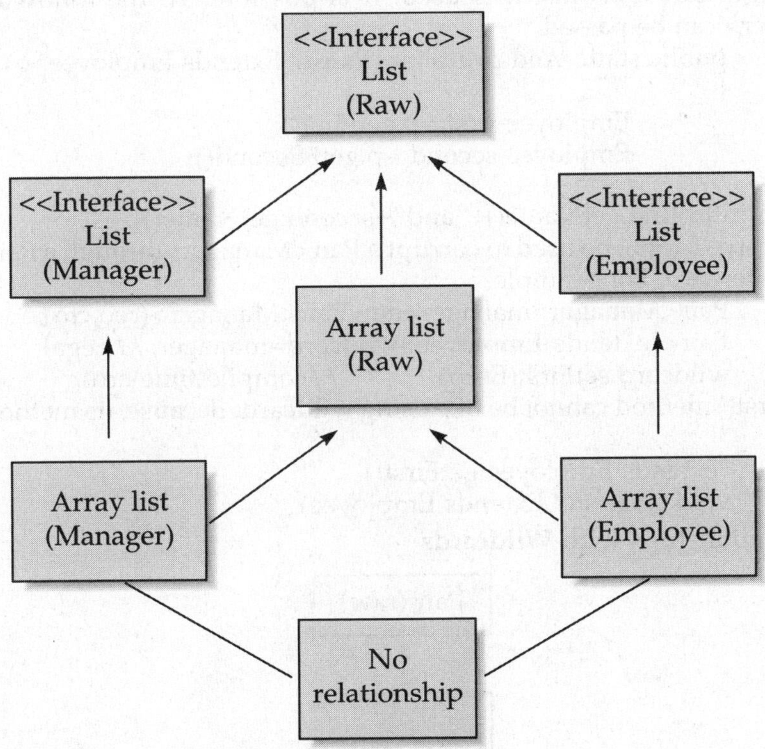

Wildcard Types

For rigid type of systems, generic types are quite unpleasant to use. To overcome this problem, wildcards were introduced by Java designers. Instead of supplying a specific type as the type argument for a generic type, ? can be specified as argument, in which case we have specified the type argument as a wildcard. A wildcard type represents any class or interface type.

Example

```
BinaryTree<?> tree =new BinaryTree<Doubles>();
```

The above example declares variable of a generic type using a wildcard type argument.

```
Pair<? Extends Employee>
```

Here the wildcard type denotes any generic Pair type whose type parameter is a subclass of Employee class, such as Pair<Manager>, but not Pair<String>.

Look at the following code.

```
Public static void printPair(Pair<Employee>p)
{
        Employee first =p.getFirst();
        Employee second =p.getFSecond();
System.out.println(first.getName()+"and"+second.getName());
}
```

This method prints out pairs of employees. Pair<Manager> cannot be passed to this method. But if wildcard is used, it is possible. To the following method, Pair<Manager> can be passed.

```
public static void printPair (Pair<? Extends Employee>p)
{
        Employee first =p.getFirst();
        Employee second =p.getFSecond();

System.out.println(first.getName()+"and"+second.getName());
```

And wildcards cannot be used to corrupt a Pair<Manager>through a Pair<? extends Employee>reference. For example,

```
Pair<Manager>manager=newPair<Manager>(ceo,cfo);
Pair<?extends Employee>wildcard=manager; //Legal
wildcard.setFirst(Emp);          //compile-time error
```

setFirst() method cannot be like using wildcard. Because, its method looks like this.

```
?extends Employee getFirst()
void setFirst(? Extends Employee)
```

Subtype Relationships with Wildcards

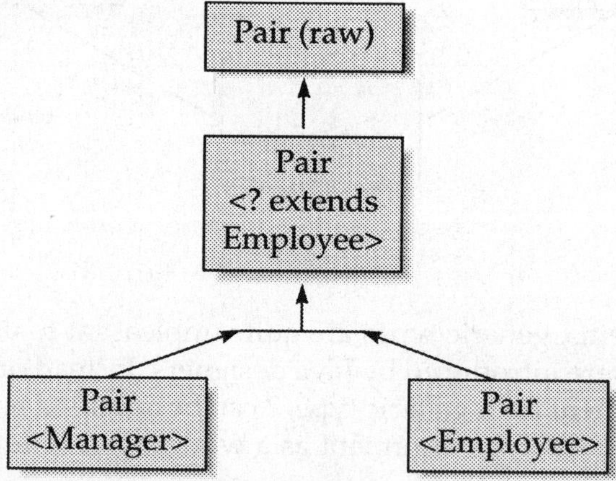

This relationship makes that call impossible.

Supertype Bounds for Wildcards

Wildcards can have supertype bounds, which is not possible to type variables. Super keyword describes the relationship so accurately. A wildcard with a supertype bound gives the opposite behavior of the wildcards. Wildcards with supertype bounds let us write to a generic object, wildcards with subtype bounds let us read from a generic objects.

Example

```
public static minmaxBouns(Manager[] a, Pair<?super Manager> result)
{
        If (a==null | | a.length==0)return;
        Manager min = a[0];
        Manager max =a[0];
        for (int i =1;i<a.length;i==)
        {
                if(min.getBouns()>a[i].getBouns()) min=a[i];
                if(max.getBouns()>a[i].getBouns()) max=a[i];
        }
        result.setFirst(min);
        result.setSecond(max);
}
```

This method will accept any appropriate pair. In this example, <? Super Manager> makes the wildcard restricted to all supertypes of Manager.

Supertype Relationships with Wildcards

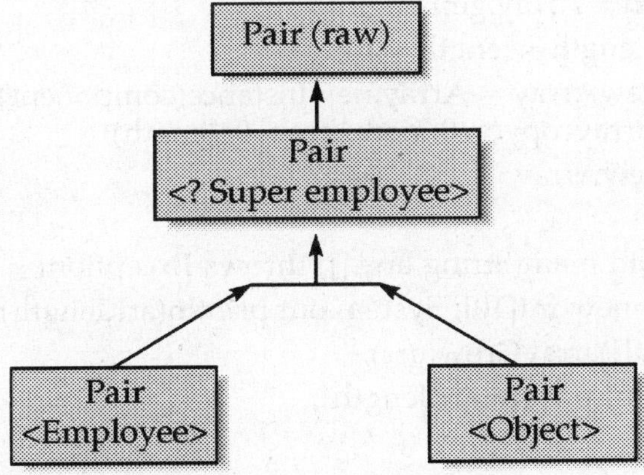

Unbounded Wildcards

Wildcards can be used without bounds also. It is useful for every simple operations.

Example

```
Pair<?>
Pair<?>methods look like this.
? getFirst()
void setFirst(?)
```

The return value of get.First() method can only be assigned to an Object. The setFirst() method can never be called, not even with an Object. Difference between Pair<?> and Pair is that setObject() method of the raw Pair class can be called with any Object.

6. REFLECTION AND GENERICS

The Array class in the java.lang.reflect package allows you to create arrays dynamically. First the given array can be converted to an Object[] array. newInstance() method of Array class, constructs a new array.

Object newarray= Array.newInstance(ComponentType, newlength)

newInstance() method needs two parameters

- **Component Type** of new array

To get component type

1. Get the class object using getClass() method.
2. Confirm that it is really an array using isArray().
3. Use getComponentType method of class Class, to find the right type for the array.

- **Length** of new array

Length is obtained by getLength() method. It returns the length of any array(method is static method, Array.getLengh(array name)).

```
import java.lang.reflect.*;
public class TestArrayRef {
    static Object arrayGrow(Object a){
    Class cl = a.getClass();
            if (!cl.isArray()) return null;
            Class componentType = cl.getComponentType();
            int length = Array.getLength(a);
            int newLength = length + 10;
            Object newArray = Array.newInstance(componentType,newLength);
            System.arraycopy(a, 0, newArray, 0, length);
            return newArray;
    }
    public static void main(String args[]) throws Exception{
            int arr[]=new int[10]; System.out.println(arr.length);
            arr = (int[])arrayGrow(arr);
            System.out.println(arr.length);
    }
}
```

Output:

```
10
20
```

7. EXCEPTIONS HANDLING

Exceptions are such anomalous conditions (or typically an event) which changes the normal flow of execution of a program. Exceptions are used for signaling erroneous (exceptional) conditions which occur during the run time processing. Exceptions may occur in any programming language.

Occurrence of any kind of exception in java applications may result in an abrupt termination of the JVM or simply the JVM crashes which leaves the user unaware of the causes of such anomalous conditions. However, Java provides mechanisms to handle such situations through its superb exception handling mechanism. The Java programming language uses Exception classes to handle such erroneous conditions and exceptional events.

Exception Hierarchy

There are three types of Exceptions:

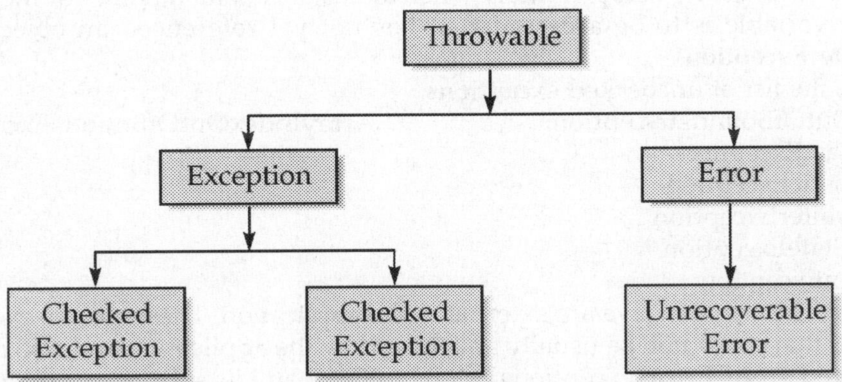

1. **Checked Exceptions:** These are the exceptions which occur during the compile time of the program. The compiler checks at the compile time that whether the program contains handlers for checked exceptions or not. These exceptions do not extend Runtime Exception class and must be handled to avoid a compile-time error by the programmer. These exceptions extend the java.lang.Exception class. These exceptional conditions should be anticipated and recovered by an application. Furthermore, checked exceptions are required to be caught. Remember that all the exceptions are checked exceptions unless and until those indicated by Error, Runtime Exception or their subclasses.

For example, if you call the read Line() method on a Buffered Reader object, then the IO Exception may occur or if you want to build a program that could read a file with a specific name, then you would be prompted to input a file name by the application. Then it passes the name to the constructor for java.io.FileReader and opens the file. However, if you do not provide the name of any existing file, then the constructor throws java.io.FileNotFoundException which abrupt the application to succeed. Hence this exception will be caught by a well-written application and will also prompt to correct the file name.

Here is the list of checked exceptions.

NoSuchFieldException InstantiationException IllegalAccessException
ClassNotFoundException NoSuchMethodException
CloneNotSupportedException InterruptedException

2. **Unchecked Exceptions:** Unchecked exceptions are the exceptions which occur during the runtime of the program. Unchecked exceptions are internal to the application and extend the java.lang.RuntimeException that is inherited from java.lang.Exceptionclass. These exceptions cannot be anticipated and recovered like programming bugs, such as logic errors or improper use of an API. These type of exceptions are also called Runtime exceptions that are usually caused by data errors, like arithmetic overflow, divide by zero, etc.

Let us take the same file name example as described earlier. In that example the file name is passed to the constructor for FileReader. However, the constructor will throwNullPointerException if a logic error causes a null to be passed to the constructor. Well in this case the exception could be caught by the application but it would rather try to eliminate the bug due to which the exception has occurred. You must have encountered the most common exception in your program, i.e. the ArithmeticException. I am sure you must be familiar with the reason of its occurrence, that is, when something tries to divide by zero. Similarly when an instance data member or method of a reference variable is to be accessed that has not yet referenced an object throws NullPointerException.

Here is the list of unchecked exceptions.

IndexOutOfBoundsException ArrayIndexOutOfBoundsException
ClassCastException
ArithmeticException
NullPointerException
IllegalStateException
SecurityException

3. **Error:** The errors in java are external to the application. These are the exceptional conditions that could not be usually anticipated by the application and also could not be recovered from. Error exceptions belong to Error and its subclasses are not subject to the catch or specify requirement. Suppose a file is successfully opened by an application for input but due to some system malfunction could not be able to read that file, then the java.io.IOError would be thrown. This error will cause the program to terminate but if an application wants, then the error might be caught. An error indicates serious problems that a reasonable application should not try to catch. Most such errors are abnormal conditions.

Hence we conclude that Errors and runtime exceptions are together called unchecked exceptions.

Throwing Exceptions

If a method needs to be able to throw an exception, it has to declare the exception(s) thrown in the method signature, and then include a throw-statement in the method. Here is an example:

```
        public void divide(int numberToDivide, int numberToDivideBy)
  throws BadNumberException{
    if(numberToDivideBy == 0){
      throw new BadNumberException("Cannot divide by 0");
    }
    return numberToDivide / numberToDivideBy;
  }
```

When an exception is thrown, the method stops execution right after the "throw" statement. Any statement following the "throw" statement are not executed. In the example above the "return numberToDivide/numberToDivideBy;" statement is not executed if a BadNumberException is thrown. The program resumes execution when the exception is caught somewhere by a "catch" block. Catching exceptions are explained later.

You can throw any type of exception from your code, as long as your method signature declares it. You can also make up your own exceptions. Exceptions are regular Java classes that extends java.lang.Exception, or any of the other built-in exception classes. If a method declares that it throws an exception A, then it is also legal to throw subclasses of A.

Catching Exceptions

If a method calls another method that throws checked exceptions, the calling method is forced to either pass the exception on, or catch it. Catching the exception is done using a try-catch block. Here is an example:

```
public void callDivide(){
  try {
     int result = divide(2,1); System.out.println(result);
  } catch (BadNumberException e) {
     //do something clever with the exception
     System.out.println(e.getMessage());
  }
     System.out.println("Division attempt done");
  }
```

The BadNumberException parameter inside the catch-clause points to the exception thrown from the divide method, if an exception is thrown. If no exeception is thrown by any of the methods called or statements executed inside the try-block, the catch-block is simply ignored. It will not be executed.

If an exception is thrown inside the try-block, for instance from the divide method, the program flow of the calling method, callDivide, is interrupted just like the program flow inside divide. The program flow resumes at a catch-block in the call stack that can catch the thrown exception. In the example above the "System.out.println(result);" statement will not get executed if an exception is thrown from the divide method. Instead program execution will resume inside the "catch (BadNumberException e) { }" block.

If an exception is thrown inside the catch-block and that exception is not caught, the catch-block is interrupted just like the try-block would have been. When the catch block is finished, the program continues with any statement following the catch block. In the example above the "System.out.println("Division attempt done");" statement will always get executed.

Propagating Exceptions

If you cannot do anything about exception where the exception throwing method is called, it may be just allowed to propagate in the exception stack. If done so, the method calling this existing method with exception stack will have to undergo the exception and then complete the process. Here is how the callDivide() method would look in that case.

```
public void callDivide() throws BadNumberException{
int result = divide(2,1);
System.out.println(result);
}
```

Notice how the try-catch block is gone, and the callDivide method now declares that it can throw a BadNumberException. The program execution is still interrupted if an exception is thrown from the divide method. Thus the "System.out.println(result);" method will not get executed if an exception is thrown from the divide method. But now the program execution is not resumed inside the callDivide method. The exception is propagated to the method that calls callDivide. Program execution does not resume until a catch-block somewhere in the call stack catches the exception. All methods in the call stack between the method throwing the exception and the method catching it have their execution stopped at the point in the code where the exception is thrown or propagated.

Example: Catching IOException's

If an exception is thrown during a sequence of statements inside a try-catch block, the sequence of statements is interrupted and the flow of control will skip directly to the catch-block. This code can be interrupted by exceptions in several places:

```
public void openFile(){
  try {
    // constructor may throw FileNotFoundException
    FileReader reader = new FileReader("someFile");

    int i=0;
    while(i != -1){
      //reader.read() may throw IOException i = reader.read();
      System.out.println((char) i );
    }
    reader.close();
    System.out.println("— File End —");
  } catch (FileNotFoundException e) {
    //do something clever with the exception
  } catch (IOException e) {
    //do something clever with the exception
  }
}
```

If the reader.read() method call throws an IOException, the following System.out.println((char) i); is not executed. Neither is the last reader.close() nor the System.out.println("--- File End --- "); statements. Instead the program skips directly to the catch(IOException e){ ... } catch clause. If the new FileReader ("someFile"); constructor call throws an exception, none of the code inside the try-block is executed.

Example: Propagating IOException's

This code is a version of the previous method that throws the exceptions instead of catching them:

```
public void openFile() throws IOException { FileReader reader = new
  FileReader("someFile"); int i=0;
  while(i != -1){
    i = reader.read();
    System.out.println((char) i );
  }
  reader.close();
  System.out.println("— File End —");
}
```

If an exception is thrown from the reader.read() method, then program execution is halted, and the exception is passed on the call stack to the method that called openFile(). If the calling method has a try-catch block, the exception will be caught there. If the calling method also just throws the method on, the calling method is also interrupted at the openFile() method call, and the exception passed on the call stack. The exception is propagated up the call stack like this until some method catches the exception, or the Java Virtual Machine does.

Finally, you can attach a finally-clause to a try-catch block. The code inside the finally clause will always be executed, even if an exception is thrown from within the try or catch block. If your code has a return statement inside the try or catch block, the code inside the finally-block will get executed before returning from the method. Here is how a finally clause looks:

```
public void openFile(){ FileReader reader = null; try {
      reader = new FileReader("someFile");
      int i=0;
      while(i != -1){
         i = reader.read();
         System.out.println((char) i );
      }
} catch (IOException e) {
   //do something clever with the exception
} finally {
   if(reader != null){
      try {
         reader.close();
      } catch (IOException e) {
           //do something clever with the exception
      }
   }
   System.out.println("— File End —");
  }
}
```

No matter whether an exception is thrown or not inside the try or catch block, the code inside the finally-block is executed. The above example shows how the file reader is always closed, regardless of the program flow inside the try or catch block.

Note: If an exception is thrown inside a finally block, and it is not caught, then that finally block is interrupted just like the try-block and catch-block. That is why the previous example had the reader.close() method call in the finally block wrapped in a try-catch block:

```
} finally {
  if(reader != null){
     try {
        reader.close();
     } catch (IOException e) {
         //do something clever with the exception
```

```
      }
    }
    System.out.println("— File End —");
  }
```

That way the System.out.println("— File End ---"); method call will always be executed.

You do not need both a catch and a finally block. You can have one of them or both of them with a try-block, but not none of them. This code does not catch the exception but lets it propagate up the call stack. Due to the finally block, the code still closes the filer reader even if an exception is thrown.

```
public void openFile() throws IOException { FileReader reader = null;
  try {
    reader = new FileReader("someFile");
    int i=0;
    while(i != -1){
      i = reader.read();
      System.out.println((char) i );
    }
  } finally {
      if(reader != null){
        try {
          reader.close();
        } catch (IOException e) {
          //do something clever with the exception
        }
      }
      System.out.println("— File End —");
    }
}
```

Notice how the catch block is gone.

8. STACK TRACE ELEMENTS AND BYTECODE INTERPRETATION

Bytecodes are the machine language of the Java virtual machine. When a JVM loads a class file, it gets one stream of bytecodes for each method in the class. The bytecodes streams are stored in the method area of the JVM. The bytecodes for a method are executed when that method is invoked during the course of running the program. They can be executed by intepretation, just-in-time compiling, or any other technique that was chosen by the designer of a particular JVM.

A method's bytecode stream is a sequence of instructions for the Java virtual machine. Each instruction consists of a one-byte *opcode* followed by zero or more *operands*. The opcode indicates the action to take. If more information is required before the JVM can take the action, that information is encoded into one or more operands that immediately follow the opcode.

Each type of opcode has a mnemonic. In the typical assembly language style, streams of Java bytecodes can be represented by their mnemonics followed by any operand values. For example, the following stream of bytecodes can be disassembled into mnemonics:

```
// Bytecode stream: 03 3b 84 00 01 1a 05 68 3b a7 ff f9
// Disassembly:
iconst_0      // 03
istore_0      // 3b
iinc 0, 1     // 84 00 01
iload_0       // 1a
iconst_2      // 05
imul          // 68
istore_0      // 3b
goto -7       // a7 ff f9
```

The bytecode instruction set was designed to be compact. All instructions, except two that deal with table jumping, are aligned on byte boundaries. The total number of opcodes is small enough so that opcodes occupy only one byte. This helps minimize the size of class files that may be traveling across networks before being loaded by a JVM. It also helps keep the size of the JVM implementation small.

All computation in the JVM centers on the stack. Because the JVM has no registers for storing abitrary values, everything must be pushed onto the stack before it can be used in a calculation. Bytecode instructions therefore operate primarily on the stack. For example, in the above bytecode sequence a local variable is multiplied by two by first pushing the local variable onto the stack with the iload_0 instruction, then pushing two onto the stack withiconst_2. After both integers have been pushed onto the stack, the imul instruction effectively pops the two integers off the stack, multiplies them, and pushes the result back onto the stack. The result is popped off the top of the stack and stored back to the local variable by the istore_0 instruction. The JVM was designed as a stack-based machine rather than a register-based machine to facilitate efficient implementation on register-poor architectures such as the Intel 486.

Primitive Types

The JVM supports seven primitive data types. Java programmers can declare and use variables of these data types, and Java bytecodes operate upon these data types. The seven primitive types are listed in the following table:

Type	Definition
byte	one-byte signed two's complement integer
short	two-byte signed two's complement integer
int	4-byte signed two's complement integer
long	8-byte signed two's complement integer
float	4-byte IEEE 754 single-precision float
double	8-byte IEEE 754 double-precision float
char	2-byte unsigned Unicode character

The primitive types appear as operands in bytecode streams. All primitive types that occupy more than 1 byte are stored in big-endian order in the bytecode stream, which means higher-order bytes precede lower-order bytes. For example, to push the constant value 256 (hex 0100) onto the stack, you would use the sipushopcode followed by a short operand. The short appears in the bytecode stream, shown below, as "01 00" because the JVM is big-endian. If the JVM were little-endian, the short would appear as "00 01".

```
// Bytecode stream: 17 01 00
// Dissassembly:
sipush 256;      // 17 01 00
```

Java opcodes generally indicate the type of their operands. This allows operands to just be themselves, with no need to identify their type to the JVM. For example, instead of having one opcode that pushes a local variable onto the stack, the JVM has several opcodes. Opcodes iload, lload, fload, and dload push local variables of type int, long, float, and double, respectively, onto the stack.

Pushing Constants Onto the Stack

Many opcodes push constants onto the stack. Opcodes indicate the constant value to push in three different ways. The constant value is either implicit in the opcode itself, follows the opcode in the bytecode stream as an operand, or is taken from the constant pool.

Some opcodes by themselves indicate a type and constant value to push. For example, the iconst_1 opcode tells the JVM to push integer value one. Such bytecodes are defined for some commonly pushed numbers of various types. These instructions occupy only 1 byte in the bytecode stream. They increase the efficiency of bytecode execution and reduce the size of bytecode streams. The opcodes that push ints and floats are shown in the following table:

Opcode	Operand(s)	Description
iconst_m1	(none)	pushes int –1 onto the stack
iconst_0	(none)	pushes int 0 onto the stack
iconst_1	(none)	pushes int 1 onto the stack
iconst_2	(none)	pushes int 2 onto the stack
iconst_3	(none)	pushes int 3 onto the stack
iconst_4	(none)	pushes int 4 onto the stack
iconst_5	(none)	pushes int 5 onto the stack
fconst_0	(none)	pushes float 0 onto the stack
fconst_1	(none)	pushes float 1 onto the stack
fconst_2	(none)	pushes float 2 onto the stack

The opcodes shown in the previous table push ints and floats, which are 32-bit values. Each slot on the Java stack is 32 bits wide. Therefore, each time an int or float is pushed onto the stack, it occupies one slot.

The opcodes shown in the next table push longs and doubles. Long and double values occupy 64 bits. Each time a long or double is pushed onto the stack, its value occupies two slots on the stack. Opcodes that indicate a specific long or double value to push are shown in the following table:

Opcode	Operand(s)	Description
lconst_0	(none)	pushes long 0 onto the stack
lconst_1	(none)	pushes long 1 onto the stack
dconst_0	(none)	pushes double 0 onto the stack
dconst_1	(none)	pushes double 1 onto the stack

One other opcode pushes an implicit constant value onto the stack. The aconst_null opcode, shown in the following table, pushes a null object reference onto the stack. The format of an object reference depends upon the JVM implementation. An object reference will somehow refer to a Java object on the garbage-collected heap. A null object reference indicates an object reference variable does not currently refer to any valid object. The aconst_null opcode is used in the process of assigning null to an object reference variable.

Opcode	Operand(s)	Description
aconst_null	(none)	pushes a null object reference onto the stack

Two opcodes indicate the constant to push with an operand that immediately follows the opcode. These opcodes, shown in the following table, are used to push integer constants that are within the valid range for byte or short types. The byte or short that follows the opcode is expanded to an int before it is pushed onto the stack, because every slot on the Java stack is 32 bits wide. Operations on bytes and shorts that have been pushed onto the stack are actually done on their int equivalents.

Opcode	Operand(s)	Description
bipush	byte1	expands byte1 (a byte type) to an int and pushes it onto the stack
sipush	byte1, byte2	expands byte1, byte2 (a short type) to an int and pushes it onto the stack

Three opcodes push constants from the constant pool. All constants associated with a class, such as final variables values, are stored in the class's constant pool. Opcodes that push constants from the constant pool have operands that indicate which constant to push by specifying a constant pool index. The Java virtual machine will look up the constant given the index, determine the constant's type, and push it onto the stack.

The constant pool index is an unsigned value that immediately follows the opcode in the bytecode stream. Opcodes lcd1 and lcd2 push a 32-bit item onto the stack, such as an int or float. The difference between lcd1 and lcd2 is that lcd1 can only refer to constant pool locations one through 255 because its index is just 1 byte. (Constant pool location zero is unused.) lcd2 has a 2-byte index, so it can refer to any constant pool location. lcd2w also has a 2-byte index, and it is used to refer to any constant pool location containing a long or double, which occupy 64 bits. The opcodes that push constants from the constant pool are shown in the following table:

Opcode	Operand(s)	Description
ldc1	indexbyte1	pushes 32-bit constant_pool entry specified by indexbyte1 onto the stack
ldc2	indexbyte1, indexbyte2	pushes 32-bit constant_pool entry specified by indexbyte1, indexbyte2 onto the stack
ldc2w	indexbyte1, indexbyte2	pushes 64-bit constant_pool entry specified by indexbyte1, indexbyte2 onto the stack

Pushing Local Variables Onto the Stack

Local variables are stored in a special section of the stack frame. The stack frame is the portion of the stack being used by the currently executing method. Each stack frame consists of three sections—the local variables, the execution environment, and the operand stack. Pushing a local variable onto the stack actually involves moving a value from the local variables section of the stack frame to the operand section. The operand section of the currently executing method is always the top of the stack, so pushing a value onto the operand section of the current stack frame is the same as pushing a value onto the top of the stack.

The Java stack is a last-in, first-out stack of 32-bit slots. Because each slot in the stack occupies 32 bits, all local variables occupy at least 32 bits. Local variables of type long and double, which are 64-bit quantities, occupy two slots on the stack. Local variables of type byte or short are stored as local variables of type int, but with a value that is valid for the smaller type. For example, an int local variable which represents a byte type will always contain a value valid for a byte (–128 <= value <= 127).

Each local variable of a method has a unique index. The local variable section of a method's stack frame can be thought of as an array of 32-bit slots, each one addressable by the array index. Local variables of type long or double, which occupy two slots, are referred to by the lower of the two slot indexes. For example, a double that occupies slots two and three would be referred to by an index of two.

Several opcodes exist that push int and float local variables onto the operand stack. Some opcodes are defined that implicitly refer to a commonly used local variable position. For example, iload_0 loads the int local variable at position zero. Other local variables are pushed onto the stack by an opcode that takes the local variable index from the first byte following the opcode. The iload instruction is an example of this type of opcode. The first byte following iload is interpreted as an unsigned 8-bit index that refers to a local variable.

Unsigned 8-bit local variable indexes, such as the one that follows the iload instruction, limit the number of local variables in a method to 256. A separate instruction, called wide, can extend an 8-bit index by another 8 bits. This raises the local variable limit to 64 kilobytes. The wide opcode is followed by an 8-bit operand. The wide opcode and its operand can precede an instruction, such as iload, that takes an 8-bit unsigned local variable index. The JVM combines the 8-bit operand of the wide instruction with the 8-bit operand of the iload instruction to yield a 16-bit unsigned local variable index.

The opcodes that push int and float local variables onto the stack are shown in the following table:

Opcode	Operand(s)	Description
iload	vindex	pushes int from local variable position vindex
iload_0	(none)	pushes int from local variable position zero
iload_1	(none)	pushes int from local variable position one
iload_2	(none)	pushes int from local variable position two
iload_3	(none)	pushes int from local variable position three
fload	vindex	pushes float from local variable position vindex
fload_0	(none)	pushes float from local variable position zero
fload_1	(none)	pushes float from local variable position one
fload_2	(none)	pushes float from local variable position two
fload_3	(none)	pushes float from local variable position three

The next table shows the instructions that push local variables of type long and double onto the stack. These instructions move 64 bits from the local variable section of the stack frame to the operand section.

Opcode	Operand(s)	Description
lload	vindex	pushes long from local variable positions vindex and (vindex + 1)
lload_0	(none)	pushes long from local variable positions zero and one
lload_1	(none)	pushes long from local variable positions one and two
lload_2	(none)	pushes long from local variable positions two and three
lload_3	(none)	pushes long from local variable positions three and four
dload	vindex	pushes double from local variable positions vindex and (vindex + 1)

dload_0	(none)	pushes double from local variable positions zero and one
dload_1	(none)	pushes double from local variable positions one and two
dload_2	(none)	pushes double from local variable positions two and three
dload_3	(none)	pushes double from local variable positions three and four

The final group of opcodes that push local variables move 32-bit object references from the local variables section of the stack frame to the operand section. These opcodes are shown in the following table:

Opcode	Operand(s)	Description
aload	vindex	pushes object reference from local variable position vindex
aload_0	(none)	pushes object reference from local variable position zero
aload_1	(none)	pushes object reference from local variable position one
aload_2	(none)	pushes object reference from local variable position two
aload_3	(none)	pushes object reference from local variable position three

Popping to Local Variables

For each opcode that pushes a local variable onto the stack; there exists a corresponding opcode that pops the top of the stack back into the local variable. The names of these opcodes can be formed by replacing "load" in the names of the push opcodes with "store". The opcodes that pop ints and floats from the top of the operand stack to a local variable are listed in the following table. Each of these opcodes moves one 32-bit value from the top of the stack to a local variable.

Opcode	Operand(s)	Description
istore	vindex	pops int to local variable position vindex
istore_0	(none)	pops int to local variable position zero
istore_1	(none)	pops int to local variable position one
istore_2	(none)	pops int to local variable position two
istore_3	(none)	pops int to local variable position three

fstore	vindex	pops float to local variable position vindex
fstore_0	(none)	pops float to local variable position zero
fstore_1	(none)	pops float to local variable position one
fstore_2	(none)	pops float to local variable position two
fstore_3	(none)	pops float to local variable position three

The next table shows the instructions that pop values of type long and double into a local variable. These instructions move a 64-bit value from the top of the operand stack to a local variable.

Opcode	Operand(s)	Description
lstore	vindex	pops long to local variable positions vindex and (vindex + 1)
lstore_0	(none)	pops long to local variable positions zero and one
lstore_1	(none)	pops long to local variable positions one and two
lstore_2	(none)	pops long to local variable positions two and three
lstore_3	(none)	pops long to local variable positions three and four
dstore	vindex	pops double to local variable positions vindex and (vindex + 1)
dstore_0	(none)	pops double to local variable positions zero and one
dstore_1	(none)	pops double to local variable positions one and two
dstore_2	(none)	pops double to local variable positions two and three
dstore_3	(none)	pops double to local variable positions three and four

The final group of opcodes that pops to local variables are shown in the following table. These opcodes pop a 32-bit object reference from the top of the operand stack to a local variable.

Opcode	Operand(s)	Description
astore	vindex	pops object reference to local variable position vindex
astore_0	(none)	pops object reference to local variable position zero
astore_1	(none)	pops object reference to local variable position one
astore_2	(none)	pops object reference to local variable position two
astore_3	(none)	pops object reference to local variable position three

Type Conversions

The Java virtual machine has many opcodes that convert from one primitive type to another. No operands follow the conversion opcodes in the bytecode stream. The value to convert is taken from the top of the stack. The JVM pops the value at the top of the stack, converts it, and pushes the result back onto the stack. Opcodes that convert between int, long, float, and double are shown in the following table. There is an opcode for each possible from-to combination of these four types:

Opcode	Operand(s)	Description
i2l	(none)	converts int to long
i2f	(none)	converts int to float
i2d	(none)	converts int to double
l2i	(none)	converts long to int
l2f	(none)	converts long to float
l2d	(none)	converts long to double
f2i	(none)	converts float to int
f2l	(none)	converts float to long
f2d	(none)	converts float to double
d2i	(none)	converts double to int
d2l	(none)	converts double to long
d2f	(none)	converts double to float

Opcodes that convert from an int to a type smaller than int are shown in the following table. No opcodes exist that convert directly from a long, float, or double to the types smaller than int. Therefore converting from a float to a byte, for example, would require two steps. First the float must be converted to an int with f2i, then the resulting int can be converted to a byte with int2byte.

Opcode	Operand(s)	Description
int2byte	(none)	converts int to byte
int2char	(none)	converts int to char
int2short	(none)	converts int to short

Although opcodes exist that convert an int to primitive types smaller than int (byte, short, and char), no opcodes exist that convert in the opposite direction. This is because any bytes, shorts, or chars are effectively converted to int before being pushed onto the stack. Arithmetic operations upon bytes, shorts, and chars are done by first converting the values to int, performing the arithmetic operations on the ints, and being happy with an int result. This means that if you add 2 bytes you get an int, and if you want a byte result you must explicitly convert the int result back to a byte. For example, the following code would not compile:

```
class BadArithmetic {
    byte addOneAndOne() {
        byte a = 1;
        byte b = 1;
        byte c = a + b;
        return c;
    }
}
```

When presented with the above code, javac objects with the following remark:

BadArithmetic.java(7): Incompatible type for declaration. Explicit cast needed to convert int to byte.

```
        byte c = a + b;
            ^
```

To remedy the situation, the Java programmer must explicitly convert the int result of the addition of a + b back to a byte, as in the following code:

```
class GoodArithmetic {
    byte addOneAndOne() {
        byte a = 1;
        byte b = 1;
        byte c = (byte) (a + b);
        return c;
    }
}
```

This makes javac so happy it drops a GoodArithmetic.class file, which contains the following bytecode sequence for the addOneAndOne() method:

```
iconst_1    // Push int constant 1.
istore_1    // Pop into local variable 1, which is a: byte a = 1;
iconst_1    // Push int constant 1 again.
istore_2    // Pop into local variable 2, which is b: byte b = 1;
iload_1     // Push a (a is already stored as an int in local variable 1).
iload_2     // Push b (b is already stored as an int in local variable 2).
iadd        // Perform addition. Top of stack is now (a + b), an int.
int2byte    // Convert int result to byte (result still occupies 32 bits).
istore_3    // Pop into local variable 3, which is byte c: byte c = (byte) (a + b);
iload_3     // Push the value of c so it can be returned.
ireturn     // Proudly return the result of the addition: return c;
```

Conversion Diversion: A JVM Simulation

The applet below demonstrates a JVM executing a sequence of bytecodes. The bytecode sequence in the simulation was generated by javac for the Convert() method of the class shown below:

```
class Diversion {
  static void Convert() {
    byte imByte = 0;
    int imInt = 125;
    while (true) {
      ++imInt;
```

```
            imByte = (byte) imInt;
            imInt *= -1;
            imByte = (byte) imInt;
            imInt *= -1;
        }
    }
}
```

The actual bytecodes generated by javac for Convert() are shown below:

```
iconst_0      // Push int constant 0.
istore_0      // Pop to local variable 0, which is imByte: byte imByte = 0;
bipush 125    // Expand byte constant 125 to int and push.
istore_1      // Pop to local variable 1, which is imInt: int imInt = 125;
iinc 1 1      // Increment local variable 1 (imInt) by 1: ++imInt;
iload_1       // Push local variable 1 (imInt).
int2byte      // Truncate and sign extend top of stack so it has valid byte value.
istore_0      // Pop to local variable 0 (imByte): imByte = (byte) imInt;
iload_1       // Push local variable 1 (imInt) again.
iconst_m1     // Push integer –1.
imul          // Pop top two ints, multiply, push result.
istore_1      // Pop result of multiply to local variable 1 (imInt): imInt *= –1;
iload_1       // Push local variable 1 (imInt).
int2byte      // Truncate and sign extend top of stack so it has valid byte value.
istore_0      // Pop to local variable 0 (imByte): imByte = (byte) imInt;
iload_1       // Push local variable 1 (imInt) again.
iconst_m1     // Push integer –1.
imul          // Pop top two ints, multiply, push result.
istore_1      // Pop result of multiply to local variable 1 (imInt): imInt *= –1;
goto 5        // Jump back to the iinc instruction: while (true) {}
```

The Convert() method demonstrates the manner in which the JVM converts from int to byte. imInt starts out as 125. Each pass through the while loop, it is incremented and converted to a byte. Then it is multiplied by –1 and again converted to a byte. The simulation quickly shows what happens at the edges of the valid range for the byte type.

The maximum value for a byte is 127. The minimum value is –128. Values of type int that are within this range convert directly to byte. However, as soon as the int gets beyond the valid range for byte, things get interesting.

The JVM converts an int to a byte by truncating and sign extending. The highest order bit, the "sign bit," of longs, ints, shorts, and bytes indicate whether or not the integer value is positive or negative. If the sign bit is zero, the value is positive. If the sign bit is one, the value is negative. Bit 7 of a byte value is its sign bit. To convert an int to a byte, bit 7 of the int is copied to bits 8 through 31. This produces an int that has the same numerical value that the int's lowest order byte would have if it were interpreted as a byte type. After the truncation and sign extension, the int will contain a valid byte value.

The simulation applet shows what happens when an int that is just beyond the valid range for byte types gets converted to a byte. For example, when the imInt variable has a value of 128 (0x00000080) and is converted to byte, the resulting byte value is –128 (0xffffff80). Later, when the imInt variable has a value of –129 (0xffffff7f) and is converted to byte, the resulting byte value is 127 (0x0000007f).

9. ASSERTIONS

An assertion has a Boolean expression that, if evaluated as false, indicates a bug in the code. This mechanism provides a way to detect when a program starts falling into an inconsistent state. Assertions are excellent for documenting assumptions and invariants about a class. Here is a simple example of assertion:

```
BankAccount acct = null;

// ...
// Get a BankAccount object
// ...

// Check to ensure we have one assert acct != null;
```

This asserts that acct is not null. If acct is null, an AssertionError is thrown. Any line that executes after the assert statement can safely assume that acct is not null.

Using assertions helps developers write code that is more correct, more readable, and easier to maintain. Thus, assertions improve the odds that the behavior of a class matches the expectations of its clients.

Note that assertions can be compiled out. In languages such as C/C++, this means using the preprocessor. In C/C++, you can use assertions through the assert macro, which has the following definition in ANSI C:

```
void assert(int expression)
```

The program will be aborted if the expression evaluates to false, and it has no effect if the expression evaluates to true. When testing and debugging is completed, assertions do not have to be removed from the program. However, note that the program will be larger in size and therefore slower to load. When assertions are no longer needed, the line #define NDEBUG is inserted at the beginning of the program. This causes the C/C++ preprocessor to ignore all assertions, instead of deleting them manually.

In other words, this is a requirement for performance reasons. You should write assertions into software in a form that can be optionally compiled. Thus, assertions should be executed with the code only when you are debugging your program—that is, when assertions will really help flush out errors. You can think of assertions as a uniform mechanism that replaces the use of ad hoc conditional tests.

Implementing Assertions in Java Technology

J2SE 1.3 and earlier versions have no built-in support for assertions. They can, however, be provided as an ad hoc solution. Here is an example of how you would roll your own assertion class.

Here we have an assert method that checks whether a Boolean expression is true or false. If the expression evaluates to true, then there is no effect. But if it evaluates to false, the assert method prints the stack trace and the program aborts. In this sample implementation, a second argument for a string is used so that the cause of error can be printed.

Note that in the assert method, I am checking whether the value of NDEBUG is on (true) or off (false). If NDEBUG sets to true, then the assertion is to be executed. Otherwise, it would have no effect. The user of this class is able to set assertions on or off by toggling the value of NDEBUG. Code Sample 1 shows my implementation.

```
public class Assertion {

public static boolean NDEBUG = true;

private static void printStack(String why) { Throwable t = new Throwable(why);
t.printStackTrace();
System.exit(1);
}

  public static void assert(boolean expression, String why) {
  if (NDEBUG && !expression) {
    printStack(why);
  }

  }
}
```

Note: In order for Code Sample 1 to compile, use -source 1.3 because assert is a keyword as of J2SE 1.4. Otherwise, you will get the following error message:

```
C:\CLASSES>javac Assertion.java
Assertion.java:11: as of release 1.4, 'assert' is a keyword, and may not be used as an
identifier
(try -source 1.3 or lower to use 'assert' as an identifier)
public static void assert(boolean expression, String why) {
^
1 error
```

Code Sample 2 demonstrates how to use the Assertion class. In this example, an integer representing the user's age is read. If the age is greater than or equal to 18, the assertion evaluates to true, and it will have no effect on the program execution. But if the age is less than 18, the assertion evaluates to false. The program then aborts, displays the message 'You are too young to vote', and shows the stack trace.

It is important to note that in this example assertions are used to validate user input and that no invariant is being tested or verified. This is merely to demonstrate the use of assertions.

```
import java.util.Scanner;
import java.io.IOException;

public class AssertionTest1 {
  public static void main(String argv[]) throws IOException { Scanner reader =
  new Scanner(System.in); System.out.print("Enter your age: ");
  int age = reader.nextInt();
 //Assertion.NDEBUG=false;
  Assertion.assert(age>=18, "You are too young to vote");
  // use age
  System.out.println("You are eligible to vote");
```

```
  }
}

import java.io.IOException;

public class AssertionTest2 {

public static void main(String argv[]) throws IOException { System.out.print("Enter
your marital status: ");
    int c = System.in.read();

    //Assertion.NDEBUG=false;
    switch ((char) c) {
        case 's':
        case 'S': System.out.println("Single"); break;
        case 'm':
        case 'M': System.out.println("Married"); break;
        case 'd':
        case 'D': System.out.println("Divorced"); break;
        default: Assertion.assert(!true, "Invalid Option"); break;
    }

    }
}
```

In the three examples, if you do not want assertions to be executed as part of the code, uncomment the line

Assert.NDEBUG = false;

Using Assertions

Use the assert statement to insert assertions at particular points in the code. The assert statement can have one of two forms:

```
assert booleanExpression;
assert booleanExpression : errorMessage;
```

The errorMessage is an optional string message that would be shown when an assertion fails.

```
import java.io.*;
public class AssertionTest3 {
public static void main(String argv[]) throws IOException {
System.out.print("Enter your marital status: ");
int c = System.in.read();
switch ((char) c) {
case 's':
case 'S': System.out.println("Single"); break;
    case 'm':
    case 'M': System.out.println("Married"); break;
    case 'd':
```

```
        case 'D': System.out.println("Divorced"); break;
        default: assert !true : "Invalid Option"; break;
      }
    }
  }
```

prompt> javac -source1.4 AssertionTest3.java

If you try to compile your assertion-enabled classes without using the -source 1.4 option, you will get a compiler error saying that assert is a new keyword as of release 1.4.

If you now run the program using the command

prompt> java AssertionTest3

and you enter a valid character, it will work fine. However, if you enter an invalid character, nothing will happen. This is because, by default, assertions are disabled at runtime. To enable assertions, use the switch -enableassertion (or -ea) as follows:

prompt> java -ea AssertionTest3

prompt> java -enableassertion AssertionTest3

Following is a sample run:

C:\CLASSES>java -ea AssertionTest3

Enter your marital status: w

Exception in thread "main" java.lang.AssertionError: Invalid Option at AssertionTest3.main(AssertionTest3.java:15)

When an assertion fails, it means that the application has entered an incorrect state. Possible behaviors may include suspending the program or allowing it to continue to run. A good behavior, however, might be to terminate the application, because it may start functioning inappropriately after a failure. In this case, when an assertion fails, an AssertionError is thrown.

Note: By default, assertions are disabled, so you must not assume that the Boolean expression contained in an assertion will be evaluated. Therefore, your expressions must be free of side effects. The switch -disableassertion (or -da)can be used to disable assertions. This, however, is most useful when you wish to disable assertions on classes from specific packages. For example, to run the program MyClass with assertions disabled in class Hello, you can use the following command:

prompt> java -da:com.javacourses.tests.Hello MyClass

And to disable assertions in a specific package and any subpackages it may have, you can use the following command:

prompt> java -da:com.javacourses.tests... MyClass

Note that the three-dot ellipsis (...) is part of the syntax.

Switches can be combined. For example, to run a program with assertions enabled in the com.javacourses.tests package (and any subpackages) and disabled in the class com.javacourses.ui.phone, you can use the following command:

prompt> java -ea:com.javacourses.tests... -da:com.javacourses.ui.phone MyClass

Note that when switches are combined and applied to packages, they are applied to all classes, including system classes (which do not have class loaders). But if you use them with no arguments (-ea or -da), they do not apply to system classes. In other words, if you use the command

```
prompt> java -ea MyClass
```

then assertions are enabled in all classes except system classes. If you wish to turn assertions on or off in system classes, use the switches -enablesystemassertions (or -esa) and - disablesystemassertions (or -dsa).

Using Assertions for Design by Contract

The assertion facility can help you in supporting an informal design-by-contract style of programming. We will now see examples of using assertions for preconditions, postconditions, and class invariants. The examples are snippets of code from an integer stack, which provides operations such as push to add an item on the stack and pop to retrieve an item from the stack.

Preconditions

In order to retrieve an item from the stack, the stack must not be empty. The condition that the stack must not be empty is a precondition. This precondition can be programmed using assertions as follows:

```
public int pop() {
   // precondition
   assert !isEmpty() : "Stack is empty";
   return stack[—num];
}
```

> **Note:** Because assertions might be disabled in some cases, precondition checking can still be performed by checks inside methods that result in exceptions such as Illegal Argument Exception or Null Pointer Exception. Postconditions

In order to push an item on the stack, the stack must not be full. This is a precondition. To add an item on the stack, we assign the element to be added to the next index in the stack as follows:

```
stack[num++] = element;
```

However, if you make a mistake and you write this statement as

```
stack[num] = element
```

then you have a bug. In this case, we need to ensure that invoking the push operation is working correctly. So the postcondition here is to ensure that the new index in the stack is the old index plus one. Also, we need to make sure that the element has been added on the stack. The following snippet of code shows the push operation with a precondition and a postcondition.

```
public void push(int element) {
// precondition
assert num<capacity : "stack is full";
int oldNum = num;
stack[num] = element;
// postcondition
assert num == oldNum+1 && stack[num-1] == element : "problem with counter";
}
```

Note that if a method has multiple return statements, then postconditions should be evaluated before each of these return statements.

10. LOGGING

The JDK contains the "Java Logging API". Via a logger you can save text to a central place to report on errors, provide additional information about your program, etc. This logging API allows to configure how messages are written by which class with which priority.

Overview of Control Flow

Applications make logging calls on *Logger* objects. Loggers are organized in a hierarchical namespace and child Loggers may inherit some logging properties from their parents in the namespace.

Applications make logging calls on *Logger* objects. These Logger objects allocate *LogRecord* objects which are passed to *Handler* objects for publication. Both Loggers and Handlers may use logging *Levels* and (optionally) *Filters* to decide if they are interested in a particular *LogRecord*. When it is necessary to publish a LogRecord externally, a Handler can (optionally) use a *Formatter* to localize and format the message before publishing it to an I/O stream.

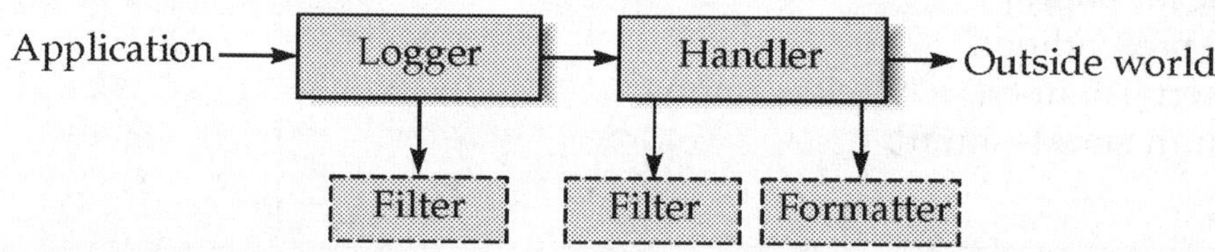

Each Logger keeps track of a set of output Handlers. By default all Loggers also send their output to their parent Logger. But Loggers may also be configured to ignore Handlers higher up the tree.

Some Handlers may direct output to other Handlers. For example, the *MemoryHandler* maintains an internal ring buffer of *LogRecords* and on trigger events it publishes its LogRecords through a target Handler. In such cases, any formatting is done by the last Handler in the chain.

The APIs are structured so that calls on the Logger APIs can be cheap when logging is disabled. If logging is disabled for a given log level, then the Logger can make a cheap comparison test and return. If logging is enabled for a given log level, the Logger is still careful to minimize costs before passing the LogRecord into the Handlers. In particular, localization and formatting (which are relatively expensive) are deferred until the Handler requests them. For example, a MemoryHandler can maintain a circular buffer of LogRecords without having to pay formatting costs.

Log Levels

Each log message has an associated log *Level*. The Level gives a rough guide to the importance and urgency of a log message. Log level objects encapsulate an integer value, with higher values indicating higher priorities. The Level class defines seven standard log levels, ranging from FINEST (the lowest priority, with the lowest value) to SEVERE (the highest priority, with the highest value).

Loggers

As stated earlier, client code sends log requests to Logger objects. Each logger keeps track of a log level that it is interested in, and discards log requests that are below this level.

Loggers are normally named entities, using dot-separated names such as "java.awt". The namespace is hierarchical and is managed by the LogManager. The namespace should typically be aligned with the Java packaging namespace, but is not required to follow it slavishly. For example, a Logger called "java.awt" might handle logging requests for classes in the java.awt package, but it might also handle logging for classes in sun.awt that support the client-visible abstractions defined in the java.awt package.

In addition to named Loggers, it is also possible to create anonymous Loggers that do not appear in the shared namespace. Loggers keep track of their parent loggers in the logging namespace. A logger's parent is its nearest extant ancestor in the logging namespace. The root Logger (named "") has no parent. Anonymous loggers are all given the root logger as their parent. Loggers may inherit various attributes from their parents in the logger namespace. In particular, a logger may inherit:

- Logging level. If a Logger's level is set to be null, then the Logger will use an effective Level that will be obtained by walking up the parent tree and using the first non-null Level.
- Handlers. By default a Logger will log any output messages to its parent's handlers, and so on recursively up the tree.
- Resource bundle names. If a logger has a null resource bundle name, then it will inherit any resource bundle name defined for its parent, and so on recursively up the tree.

Logging Methods

The Logger class provides a large set of convenience methods for generating log messages. For convenience, there are methods for each logging level, named after the logging level name. Thus rather than calling "logger.log(Constants.WARNING,..." a developer can simply call the convenience method "logger.warning(..."

There are two different styles of logging methods, to meet the needs of different communities of users.

First, there are methods that take an explicit source class name and source method name. These methods are intended for developers who want to be able to quickly locate the source of any given logging message. An example of this style is:

```
void warning(String sourceClass, String sourceMethod, String msg);
```

Second, there are a set of methods that do not take explicit source class or source method names. These are intended for developers who want easy-to-use logging and do not require detailed source information.

```
void warning(String msg);
```

For this second set of methods, the Logging framework will make a "best effort" to determine which class and method called into the logging framework and will add this information into the LogRecord. However, it is important to realize that this automatically inferred information may only be approximate. The latest generation of virtual machines perform extensive optimizations when JITing and may entirely remove stack frames, making it impossible to reliably locate the calling class and method.

5 Concurrent Programming

1. THREAD

A *thread* is defined as the path of execution of a program. A thread is a sequence of instructions that is executed to define a unique flow of control. It is the smallest unit of code. For example, a Central Processing Unit (CPU) performs various tasks simultaneously, such as writing and printing a document, installing a software, and displaying the date and time on the status bar. All these processes are handled by separate threads.

A process that is made of one thread is known as *single-threaded* process. A process that creates two or more threads is called a *multithreaded* process. For example, any web browser, such as Internet Explorer is a multithreaded application. Within the browser, you can print a page in the background while you are scrolling through the page. You can play audio files and watch animated images at the same time. Each thread in a multithreaded program runs at the same time and has a different execution path.

A single threaded application can perform only one task at a time. You have to wait for one task to complete before another can start. Threads are used when you have to run various applications that perform large and complex computations. Multithreading helps to perform these operations simultaneously, saving the time of the user. Every program has at least one thread and you can create more threads when necessary.

The microprocessor allocates memory to the processes that you can execute. Each process occupies its own address space or memory. However, all threads in a process occupy the same address space.Java provides built-in support for *multithreaded programming*. A multithreaded program contains two or more parts that can run concurrently. Each part of such a program is called a thread, and each thread defines a separate path of execution.

2. MULTITHREADING AND MULTITASKING

Multitasking is the ability to execute more than one task at the same time. Multitasking can be divided into the following categories:

- Process-based multitasking
- Thread-based multitasking

Process-based Multitasking

A process is a program that is being executed by the processor. The processor-based multitasking feature of Java enables you to switch from one program to another so quickly that it appears as if the programs are executing at the same time. For example, process-based multitasking enables you to run the Java compiler and use the text editor at the same time.

The process-based multitasking feature enables a computer to execute two or more processes concurrently. Processes are the tasks that require separate address space in the computer memory.

Thread-based Multitasking

A single program can contain two or more threads and therefore, perform two or more tasks simultaneously. For example, a text editor can perform writing to a file and print a document simultaneously with separate threads performing that writing and printing actions. Also, in a text-editor, you can format text in a document and print the document at the same time. Threads are called lightweight process because there are fewer overloads when the processor switches from one thread to another. On the other hand, when the processor switches from one process to another process, the overload increases.

A multithreading is a specialized form of multitasking. Multitasking threads require less overhead than multitasking processes. Another term to be defined related to thread is process. A process consists of the memory space allocated by the operating system that can contain one or more threads. A thread cannot exist on its own; it must be a part of a process. A process remains running until all of the non-daemon threads are done executing.

Multithreading enables you to write very efficient programs that make maximum use of the CPU, because idle time can be kept to a minimum.

Benefits of Multithreading

A single-threaded application can perform only one task at a time. In such a situation, you wait for one task to complete so that another can start. A process having more than one thread is said to be a multi-threaded process. The various advantages of multithreading are:

- Improved performance: Provides improvement in the performance of the processor by simultaneous execution of computation and Input/Output (I/O) operations.
- Minimized system resource usage: Minimizes the use of system resources by using threads, which share the same address space and belong to the same process.
- Simultaneous access to multiple applications: Provides access to multiple applications at the same time because of quick content switching among threads.
- Program structure simplification: Simplifies the structure of complex applications, such as multimedia applications. Sub programs can be written for each activity that makes complex program easy to design and code.

Pitfalls of Multithreading

The various disadvantages of multithreading are:

- Race Condition: When two or more threads simultaneously access the same variable, at least one thread tries to write a value in this variable. This is called the *race condition.* This condition is caused by the lack of synchronization between two threads. For example, in a word processor program, there are two threads, one to read a file and the other to write a file. The thread to read a file waits for the thread to write before performing its operation. The race condition arises when the thread to read a file, reads the file, before the thread to write to a file performs its operation.

- Deadlock condition: The *deadlock condition* arises in a computer system when two threads wait for each other to complete their operations before performing their individual action. As a result, the threads become locked and the program fails. For example, consider two threads, Thread A and Thread B. Thread A is waiting for a lock to be released by Thread B, and Thread B is waiting for the lock to be released by Thread A to complete its transaction.
- Lock Starvation: *Lock starvation* occurs when the execution of a thread is postponed because of its low priority. The Java Run-time environment executes threads based on their priority because the CPU can execute only one thread at a time. The thread with a higher priority is executed before the thread with a lower priority.

3. THE THREAD MODEL IN JAVA

Java uses threads to increase the efficiency of CPU by preventing wastage of CPU cycles. In single-threaded systems, an approach called event loop with polling is used. Polling is the process in which a single event is executed at a time.

In the event loop model, a single thread runs in an infinite loop till its operation is completed. When this operation is completed, the event loop dispatches control to the appropriate event-handler. No more processing can happen in the system until the event-handler returns. This results in the wastage of the CPU time.

In a single threaded application when a thread is suspended from execution because it is waiting for a system resource, the entire program stops running. Multithreading eliminates event loop/polling mechanism in Java. The time for which a thread waits for the CPU time can be utilized elsewhere.

The Thread Class

The *java.lang.Thread* class is used to construct and access individual threads in a multithreaded application. You can create a multithreaded application using the Thread class and the Runnable interface.

The Thread class contains various methods that can obtain information about the activities of a thread, such as setting and checking the properties of a thread, causing a thread to wait, and being interrupted or destroyed. You can make applications and classes run in separate threads by extending the Thread class. A few methods defined in the Thread class are:

getPriority ( ): Returns the priority of a thread.
isAlive ( ): Determines whether a thread is running.
sleep ( ): Makes the thread to pause for a period of time.
getName ( ): Returns the name of the thread.
start ( ): Starts a thread by calling the *run ( )* method.

The Main Thread

The first thread to be executed in a multithreaded process is called the main thread. The main thread is created automatically on the start up of a Java program execution. You can access a thread using the *currentThread ( )* method of the Thread class. The following syntax shows how to declare the *currentThread ( )* method:

public static Thread currentThread ( )

In the preceding syntax, the *currentThread( )* method returns a reference to the executing thread object. You can also control the main thread in the same manner as any other thread by obtaining a reference to the main thread. You can use the following code to show the execution of a thread using the methods in the Thread class:

```
class mainThreadDemo
{
    public static void main (String args [ ])
    {
    Thread t = Thread.currentThread ( );
    System.out.println ("The current thread: " +t);
    t.setName ("MainThread");
    System.out.println("The current thread after name change :" +t);
    System.out.println("The current Thread is going to sleep for 10 seconds");
    try
    {
        t.sleep(10000);
    }
Catch(InterruptedException e)
{
    System.out.println("Main thread interrupted");
}
System.out.println("After 10 seconds ....................the current thread is exiting now.");
    }
}
```

In the preceding code, reference to the current thread is obtained by calling the *currentThread( )* method and its reference is stored in the t variable. The *setName( )* method is called to set the name of the thread, and the information about the thread is displayed. The main thread is then made to sleep for 10 seconds by using the *sleep ( )* method, and after 10 seconds the thread terminates. The *sleep ( )* method may throw InterruptedException and it is always written in the try-catch block. The InterruptedException is thrown when another thread interrupts the sleeping thread.

4. LIFE CYCLE OF A THREAD

A thread goes through various stages in its life cycle. For example, a thread is born, started, runs, and then dies. Following diagram shows complete life cycle of a thread.

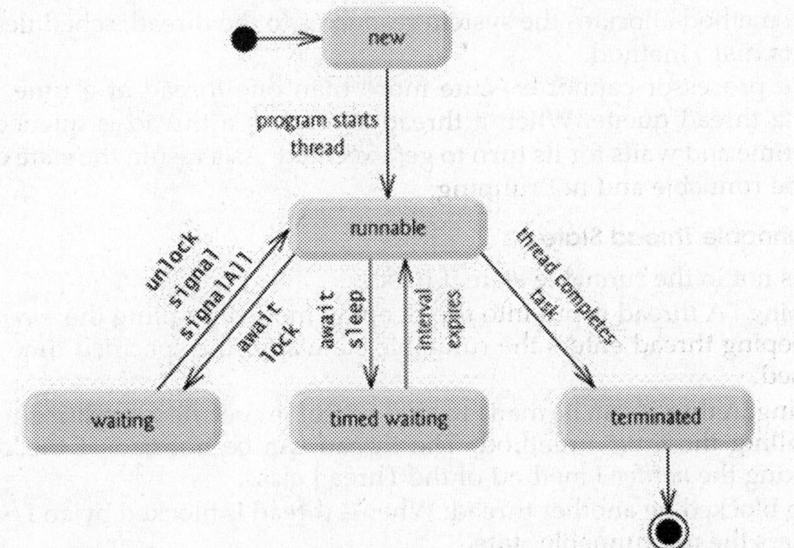

Abovementioned stages are explained here:

- **New:** A new thread begins its life cycle in the new state. It remains in this state until the program starts the thread. It is also referred to as a born thread.
- **Runnable:** After a newly born thread is started, the thread becomes runnable. A thread in this state is considered to be executing its task.
- **Waiting:** Sometimes a thread transitions to the waiting state while the thread waits for another thread to perform a task. A thread transitions back to the runnable state only when another thread signals the waiting thread to continue executing.
- **Timed waiting:** A runnable thread can enter the timed waiting state for a specified interval of time. A thread in this state transitions back to the runnable state when that time interval expires or when the event it is waiting for occurs.
- **Terminated:** A runnable thread enters the terminated state when it completes its task or otherwise terminates.

The New Thread State

When an instance of the Thread class is created, the thread enters the new thread state. The following syntax shows how to instantiate the Thread class:

Thread newThread = new Thread(this, "threadName");

In the preceding syntax, a new thread is created. A *new thread* is an empty object of the Thread class and no system resources, such as memory are allocated to it. You have to invoke the *start( )* method to start the thread. The following syntax shows how to declare the *start( )* method:

newThread.start( );

when the thread is in the new state, no other method except the *start( )* method can be called, otherwise it throws IllegalThreadStateException.

The Runnable Thread State

When the *start( )* method of a thread is invoked, the thread enters the runnable state. The *start( )* method allocates the system resources to the thread, schedules the thread, and calls its *run( )* method.

A single processor cannot execute more than one thread at a time, therefore it maintains a thread queue. When a thread is started, a thread is queued up for the processor time and waits for its turn to get executed. As a result, the state of the thread is said to be runnable and not running.

The Not Runnable Thread State

A thread is not in the runnable state if it is:

- Sleeping : A thread is put into the sleeping mode by calling the *sleep ( )* method. A sleeping thread enters the runnable state after the specified time of sleep has elapsed.
- Waiting: A thread can be made to wait for some specified condition to be satisfied by calling the *wait( )* method. The thread can be notified of the condition by invoking the *notify( )* method of the Thread class.
- Being blocked by another thread: When a thread is blocked by an I/O operation, it enters the not runnable state.

The Dead Thread State

A thread can be dead or alive. A thread enters the dead state when the loop in the *run( )* method is complete. Assigning a null value to a thread object changes the state of the thread to dead. The *isAlive( )* method of the Thread class is used to determine whether a thread has been started or not. You cannot restart a dead thread.

5. CREATING OWN THREADS

You create a thread by instantiating an object of Thread type. You can create a thread in the following ways:

- Implementing Runnable interface
- Extending the Thread class

5.1 Creating Threads By Implementing the Runnable Interface

Applets extend from the Applet class. Since Java does not support multiple inheritance, you cannot inherit a class from the Applet class as well as from the Thread class.

Java provides the Runnable interface to solve this problem. The Runnable interface only consists of the *run ( )* method, which is executed when the thread is activated. In Java applications, you can extend from the Thread class. In other words, when a program needs to inherit from a class other than from the Thread class, you need to implement the Runnable interface. The following syntax shows how to declare the *run ( )* method:

public void run( )

The *run ( )* method contains the code that defines the new thread. When the *run ( )* method is called, another thread starts executing concurrently with the main thread in the program.

The class that implements the Runnable interface creates an instance of the Thread class. The new thread object starts executing by calling the *start ( )* method. The *start ( )* method id declared in the Thread class and when the *start ( )* method executes, the *run ( )* method is called.

You can use the following code to create a thread by implementing the Runnable interface:

```
class NewThread implements Runnable
{
    Thread t;
    NewThread ( )
    {
        t = new Thread (this, "ChildThread");
        System.out.println ("Child Thread:" +t);
        t.start ( );
    }
    public void run ( )
    { // Implementing the run ( ) method of the Runnable interface
        System.out.println ("Child Thread Started");
        System.out.println ("Exiting the Child Thread");
    }
}
class ThreadClass
{
```

```
public static void main (String args [ ])
{
new NewThread ( );
System.out.println ("Main Thread Started");
try
{
        Thread.sleep (5000);
}
catch (InterruptedException e)
{
                System.out.println ("The main thread interrupted");
                System.out.println ("Exiting the main thread");
        }
}
```

In the preceding code, the *NewThread* class implements the *Runnable* interface. A new instance of the *NewThread* class is created in the *ThreadClass.* The constructor of the *NewThread* class creates a new thread by passing two arguments, *ChildThread* and *this.*

The *ChildThread* is the name of the new thread. The *start ( )* method begins the execution of a thread by calling the *run ( )* method. The new thread constructor after calling the *start ( )* method returns to the *main ( )* method. The main thread prints a statement and sleeps for five seconds that is equal to 5000 milliseconds.

The child thread of the main thread prints the statement in the *run ( )* method and transfers the control to the *main ( )* method.

5.2 Creating Threads by Extending the Thread Class

You can also create threads by extending the Thread class. The Thread class is defined in the java.lang package. The Thread class defines several methods that can be overridden by a derived class. You can use the run ( ) method to create threads only if a class does not extend to any other class. The extending class calls the *start ( )* method to begin the child thread execution. You can use the following code to create a thread by extending the Thread class:

```
class ThreadDemo extends Thread
{
        ThreadDemo ( )
        {
                super ("ChildThread");  // Calls the superclass constructor
                System.out.println ("ChildThread:" +this);
                start ( );
        }
        public void run ( )
        {
                System.out.println ("The child thread started");
                System.out.println ("Exiting the child thread");
        }
}
class ThreadDemoClass
{
        public static void main (String args [ ] )
```

```
{
          new ThreadDemo ( );
          System.out.println ("The main thread started");
          System.out.println ("The main thread sleeping");
          try
          {
                    Thread.sleep (1000);
          }
    catch (InterruptedException e)
    {
                    System.out.println ("The main thread interrupted");
          }
          System.out.println ("Exiting the main thread");
    }
}
```

In the preceding code, the *ThreadDemo ( )* class extends the Thread class. The *ThreadDemoClass* creates an object of the *ThreadDemo* class. The constructor of *ThreadDemo* class calls the *super ( )* method that sets the name of the child thread as *child thread.*

The *start ( )* method calls the *run ( )* method for starting the execution of the child thread. The constructor then returns the control to the main thread. The main thread executes two print statements and sleeps for one second.

5.3 Creating Multiple Threads

You can create multiple threads in a program by implementing the Runnable interface or extending the Thread class. You can use the following code to create two child threads other than the main thread:

```
class newThreadClass implements Runnable
{
    String ThreadName;
    newThreadClass (String name)
    {
          ThreadName = name;
          Thread t = new Thread (this, ThreadName);
          System.out.println ("Thread created:" +t);
          t.start ( );
    }
    public void run ( )
    {
          try
          {
                    for (int i=1;i<=5;i++)
                    {
                              System.out.println (ThreadName +"loop:" +i);
                              Thread.sleep (100);
                    }
          }
          catch (InterruptedException e)
```

```
            {
                  System.out.println ("Thread:" +ThreadName+ "interrupted");
            }
            System.out.println (ThreadName +"is exiting");
      }
}
class MultipleThread
{
      public static void main (String args [ ])
      {
      new newThreadClass ("FirstChildThread");
      new newThreadClass ("SecondChildThread");
      try
      {
                  for (int i=1;i<=5;i++)
                  {
                        System.out.println ("Main Thread loop:" +i);
                        Thread.sleep (300);
                  }
            }
            catch (InterruptedException e)
            {
            System.out.println ("Main thread is interrupted");
      }
      System.out.println ("Main thread is terminating now");
}
}
```

In the preceding code, the two child threads and the main thread share the memory of the CPU. The main thread calls the *sleep ( )* method for 300 milliseconds.

6. IDENTIFYING THE THREAD PRIORITIES

The Java Run-time Environment executes threads based on their priority. A CPU can execute only one thread at a time. Therefore, the threads, which are ready for execution, queue up for their turn to get executed by the processor. The threads are scheduled using fixed-priority scheduling. Each thread has a priority that affects its position in the thread queue of the processor. A thread with higher priority runs before threads with low priority.

6.1 Defining Thread Priority

Thread priorities are the integers in the range of 1 to 10 that specify the priority of one thread with respect to the priority of another thread. Execution of multiple threads on a single CPU in a specified order is called scheduling. The Java run-time system executes threads based on their priority.

The threads are scheduled using fixed priority scheduling. In fixed priority scheduling, each thread has a priority that affects its position in the thread queue of the processor. A thread with higher priority runs before threads with low priority.

If Java encounters another thread with higher priority, the current thread is pushed back; and the thread with the higher priority is executed. The Java run-time system selects the runnable thread with the highest priority of execution when a number of threads get ready to execute. The next thread of lower priority starts executing if the higher priority thread stops or becomes not runnable. A thread is pushed back in the queue by another thread if it is waiting for an I/O operation. A thread can also be pushed back in the queue when the time for which the *sleep ( )* method was called on another higher priority thread is over.

6.2 Setting the Thread Priority

You can set the thread priority after it is created using the *setPriority ( )* method declared in the Thread class. The following syntax shows how to declare the *setPriority ( )* method:

public final void setPriority (int newPriority)

In the preceding syntax, the *newPriority* parameter specifies the new priority setting for a thread. The priority levels should be within the range of two constants, MIN_PRIORITY and MAX_PRIORITY. The *setPriority ( )* method throws the IllegalArgumentException if *priorityLevel* is less than MIN_PRIORITY and greater than MAX_PRIORITY.

The thread can be set to a default priority, by specifying the NORM_PRIORITY constant in the *setPriority ( )* method. MAX_PRIORITY is the highest priority that a thread can have, MIN_PRIORITY is the lowest priority a thread can have; and NORM_PRIORITY is the default priority set for a thread. You can use the following code to set priorities to various threads:

```
class ChildThread implements Runnable
{
	Thread t;
	ChildThread (int p)
	{
		t = new Thread (this,"ChildThread");
		t.setPriority (p);
		System.out.println ("Thread created:" +t);
	}
	public void run ( )
	{
		try
		{
			for (int i=1;i<=5;i++)
			{
				System.out.println (t +"loop:" +i);
				Thread.sleep (500);
			}
		}
		catch (InterruptedException obj)
		{
			System.out.println ("Thread:" +t + "interrupted");
		}
}
class PriorityDemo
{
	public static void main (String args [ ])
	{
```

```
        ChildThread obj1 = new ChildThread(Thread.NORM_PRIORITY – 2);
        ChildThread obj2 = new ChildThread(Thread.NORM_PRIORITY + 2);
        ChildThread ob31 = new ChildThread(Thread.NORM_PRIORITY +3);

        // Starting the threads with different priority
        obj1.t.start ( );
        obj2.t.start ( );
        obj3.t.start ( );
        try
        {
                System.out.println ("Main thread waiting for child thread to
finish");
                obj1.t.join ( );
                obj2.t.join ( );
                obj3.t.join ( );
        }
        catch (InterruptedException e)
        {
        System.out.println ("Main thread is interrupted");
    }
    System.out.println (obj1.t  + "is alive ?: "  +obj.t.isAlive ( ));
    System.out.println (ob21.t  + "is alive ?: "  +obj.t.isAlive ( ));
    System.out.println (ob31.t  + "is alive ?: "  +obj.t.isAlive ( ));
    System.out.println ("Main thread is exiting");
 }
}
```

7. THREAD METHODS

Following is the list of important methods available in the thread class.

S. No.	Methods with Description
1.	**public void start()** Starts the thread in a separate path of execution, then invokes the run() method on this Thread object.
2.	**public void run()** If this Thread object was instantiated using a separate Runnable target, the run() method is invoked on that Runnable object.
3.	**public final void setName(String name)** Changes the name of the Thread object. There is also a getName() method for retrieving the name.
4.	**public final void setPriority(int priority)** Sets the priority of this Thread object. The possible values are between 1 and 10.
5.	**public final void setDaemon(boolean on)** A parameter of true denotes this thread as a daemon thread.

6.	**public final void join(long millisec)** The current thread invokes this method on a second thread, causing the current thread to block until the second thread terminates or the specified number of milliseconds passes.
7.	**public void interrupt()** Interrupts this thread, causing it to continue execution if it was blocked for any reason.
8.	**public final boolean isAlive()** Returns true if the thread is alive, which is any time after the thread has been started but before it runs to completion.

The previous methods are invoked on a particular Thread object. The following methods in the Thread class are static. Invoking one of the static methods performs the operation on the currently running thread

S. No.	Methods with Description
1.	**public static void yield()** Causes the currently running thread to yield to any other thread of the same priority that are waiting to be scheduled
2.	**public static void sleep(long millisec)** Causes the currently running thread to block for at least the specified number of milliseconds
3.	**public static boolean holdsLock(Object x)** Returns true if the current thread holds the lock on the given Object.
4.	**public static Thread currentThread()** Returns a reference to the currently running thread, which is the thread that invokes this method.
5.	**public static void dumpStack()** Prints the stack trace for the currently running thread, which is useful when debugging a multithreaded application.

Example

The following ThreadClassDemo program demonstrates some of these methods of the Thread class:

```
// File Name : DisplayMessage.java
// Create a thread to implement Runnable
public class DisplayMessage implements Runnable
{
 private String message;
 public DisplayMessage(String message)
{
 this.message = message;
}
```

```
public void run()
{
 while(true)
 {
  System.out.println(message);
 }
 }
}

// File Name : GuessANumber.java
// Create a thread to extend Thread
public class GuessANumber extends Thread
{
  private int number;
  public GuessANumber(int number)
  {
   this.number = number;
  }
  public void run()
  {
   int counter = 0;
   int guess = 0;
   do
   {
      guess = (int) (Math.random() * 100 + 1);
      System.out.println(this.getName()
               + " guesses " + guess);
      counter++;
   }while(guess != number);
   System.out.println("** Correct! " + this.getName()
               + " in " + counter + " guesses.**");
  }
}
// File Name : ThreadClassDemo.java
public class ThreadClassDemo
{
  public static void main(String [] args)
  {
   Runnable hello = new DisplayMessage("Hello");
   Thread thread1 = new Thread(hello);
   thread1.setDaemon(true);
   thread1.setName("hello");
   System.out.println("Starting hello thread...");
   thread1.start();
    Runnable bye = new DisplayMessage("Goodbye");
   Thread thread2 = new Thread(hello);
```

```
      thread2.setPriority(Thread.MIN_PRIORITY);
      thread2.setDaemon(true);
      System.out.println("Starting goodbye thread...");
      thread2.start();
      System.out.println("Starting thread3...");
      Thread thread3 = new GuessANumber(27);
      thread3.start();
      try
      {
        thread3.join();
      }catch(InterruptedException e)
      {
        System.out.println("Thread interrupted.");
      }
      System.out.println("Starting thread4...");
      Thread thread4 = new GuessANumber(75);   thread4.start();
      System.out.println("main() is ending...");
    }
  }
```

This would produce the following result. You can try this example again and again and you would get different result every time.

```
Starting hello thread...
Starting goodbye thread...
Hello
Hello
Hello
Hello
Hello
Hello
Hello
Hello
Hello
Thread-2 guesses 27
Hello
** Correct! Thread-2 in 102 guesses.**
Hello
Starting thread4...
Hello
Hello
..........remaining result produced.
```

Using the *isAlive ( )* Method

The *isAlive ( )* method is used to check the existence of a thread. Because the main thread is the last thread to be executed, calling the *sleep ( )* method within the *main ( )* method with a long sleep time ensures that all the child threads terminate prior to the main thread.

You can use the *isAlive ( )* method to find the status of a thread. The following syntax shows how to declare the *isAlive ( )* method:

public final boolean isAlive ( );

In the preceding syntax, the *isAlive ( )* method returns true if the thread is running and returns false if the thread is new or is terminated. The thread can be in runnable or not runnable state, if the *isAlive ( )* method returns true. You can use the following code to use the *isAlive ( )* method:

```
class newThreadClass implements Runnable
{
	Thread t;
	newThreadClass ( )
	{
		t = new Thread(this, "ChildThread");
		System.out.println ("Thread created:" +t);
		t.start ( );
	}
	public void run ( )
	{
		try
		{
			for (int i=1;i<=5;i++)
			{
				System.out.println (t +"loop:" +i);
				Thread.sleep (100);
			}
		}
		catch (InterruptedException obj)
		{
			System.out.println ("Thread:" +t + "interrupted");
		}
	}
}
class isAliveDemo
{
	public static void main (String args [ ])
	{
		newThreadClass obj = new newThreadClass ( );
		System.out.println (obj.t + "is alive ?: "  +obj.t.isAlive ( ));
		try
		{
			for (int i=1;i<=5;i++)
			{
				System.out.println ("Main Thread loop:" +i);
				Thread.sleep (200);
			}
		}
```

```
            catch (InterruptedException e)
            {
            System.out.println ("Main thread is interrupted");
        }
        System.out.println (obj.t  + "is alive ?: "  +obj.t.isAlive ( ));
        System.out.println ("Main thread is exiting");
    }
    }
```

In the preceding code, the *isAlive ( )* method is called on the thread to check whether the thread is in running or dead state.

Using the *join ( )* Method

The *join ( )* method waits until the thread on which it is called terminates. It is called the *join ( )* method because the thread calling the *join ( )* method waits until the specified thread joins the calling method. In addition, the *join ( )* method enables you to specify a maximum amount of time that you need to wait for the specified thread to terminate. The following syntax shows how to declare the *join ( )* method:

public final void join ( ) throws InterruptedException

In the preceding syntax, the *join ( )* method throws the InterruptedException if another thread interrupts it. The method returns the control to the calling method when the specified thread dies. You can use the following code to use the *join ( )* method:

```
class ChildThread implements Runnable
{
    Thread t;
    ChildThread ( )
    {
            t = new Thread (this, "ChildThread");
            System.out.println ("Thread created:" +  t);
            t.start ( );
    }
    public void run ( )
    {
            try
            {
                    for (int i=1;i<=5;i++)
                    {
                            System.out.println (t +"loop:" +i);
                            Thread.sleep (500);
                    }
            }
            catch (InterruptedException obj)
            {
                    System.out.println ("Thread:" +t + "interrupted");
            }
    }
}
```

```
class joinDemo
{
	public static void main (String args [ ])
	{
		ChildThread obj = new ChildThread ( );
		System.out.println (obj.t + "is alive ?: " +obj.t.isAlive ( ));
		try
		{
			System.out.println ("Main thread waiting for child thread to
finish");
			Obj.t.join ( );
		}
		catch (InterruptedException e)
		{
		System.out.println ("Main thread is interrupted");
		}
	System.out.println (obj.t + "is alive ?: " +obj.t.isAlive ( ));
	System.out.println ("Main thread is exiting");
	}
}
```

In the preceding code, the *join ( )* method is called on the child thread and it waits till the main thread joins the calling method.

8. MAJOR THREAD CONCEPTS

When two threads need to share data, you must ensure that one thread does not change the data used by the other thread. For example, if you have two threads, one that reads your salary from a file and another that tries to update the salary, data corruption might occur. Java enables you to coordinate the actions of multiple threads by using synchronized statements.

8.1 Synchronizing Threads

Synchronization of threads ensures that if two or more threads need to access a shared resource, then that resource is used by only one thread at a time. You can synchronize your code using the synchronized keyword. You can invoke only one synchronized method for an object at any given time.

Synchronization is based on the concept of monitor. A *monitor,* also known as a *semaphore,* is an object that is used as a mutually exclusive lock. All objects and classes are associated with a monitor and only one thread can own a monitor at a given time. To enter an object's monitor, you need to call a method that has been modified with the synchronized keyword.

The monitor controls the way in which synchronized methods access an object or class. When a thread acquires a lock, it is said to have entered the monitor. The monitor ensures that only one thread has access to the resources at any given time. To enter an object's monitor, you need to call a synchronized method.

When a thread is within a synchronized method, all the other threads that try to call it on the same instance have to wait. During the execution of a synchronized method, the object is locked so that no other synchronized method can be invoked. The monitor is automatically released when the method completes its execution. The monitor can also be released when the synchronized method executes the wait ( ) method. When a thread calls a wait ( ) method, it temporarily releases the locks that it holds. In addition, the thread stops running and is added to the list of waiting threads for that object.

The following code shows the implementation of synchronization among threads.

```
class Thread1
{
    void call ( )
    {
        System.out.println ("first statement");
        try
        {
            Thread.sleep (1000);
        }
        catch (Exception e)
        {
            System.out.println ("Error" +e);
        }
        System.out.println ("second statement");
    }
}
class Thread2 extends Thread
{
    Thread1 t;
    public Thread2 (Thread1 t)
    {
        this.t = t;
    }
    public void run ( )
    {
        t.call ( );
    }
}
public class NotSynchronized
{
    public static void main (String args [ ])
    {
        Thread1 obj1 = new Thread1 ( );
        Thread2 Obja = new Thread2 ( obj1);
        Thread2 Objb = new Thread2 ( obj1);
        Obja.start ( );
        Objb.start( );
    }
}
```

In the preceding code, three classes are used. The first class, *Thread1* has the *call ( )* method that includes two print statements. Within the print statements, the *sleep ( )* method is called on the current object with a lapse time of 1 second.

The second class, *Thread2* extends the Thread class. The *Thread2* class has the *run ( )* method that calls the *call ( )* method with an instance of the *Thread1* class.

The third class, *NotSynchronized* creates two objects of the *Thread2* class and starts the thread corresponding to objects, *Obja* and *Objb*. The *Thread1* and *Thread2* call a method on the same object at the same time.

You can serialize the access to the *call ( )* method to avoid the race condition. The *call ( )* method is used by one thread at a time. This can be done by just prefixing the synchronized keyword to the *call ( )* method. You can use the following code to use the synchronized keyword:

```
class Thread1
{
      synchronized void call ( )
      {
            System.out.println ("first statement");
            try
            {
                  Thread.sleep (1000);
            }
            catch (Exception e)
            {
                  System.out.println ("Error"+e);
            }
            System.out.println("second statement");
      }
}
class Thread2 extends Thread
{
      Thread1 t;
      public Thread2 (Thread1 t)
      {
            this.t = t;
      }
      public void run ( )
      {
            t.call ( );
      }
}
public class SynchronizedThreads
{
      public static void main (String args [ ])
      {
            Thread1 obj1 = new Thread1 ( );
            Thread2 obja = new Thread2 (obj1);
            Thread2 objb = new Thread2 (obj1);
            Obja.start( );
            Objb.start( );
      }
}
```

The Synchronized Statement

Synchronization among threads is achieved by using synchronized statements. The *synchronized statement* is used where the synchronization methods are not used in a class and you do not have access to the source code. You can synchronize the access to an object of this class by placing the calls to the methods defined by it inside a synchronized block. The following syntax shows how to use the synchronized statement:

```
synchronized (obj)
{
        /* statements to be synchronized */
}
```

In the preceding syntax, *obj* is a reference of the object to be synchronized. You can use the following code to use the synchronized block to synchronize various threads:

```
class Thread1
{
        void call ( )
        {
                System.out.println ("first statement");
                try
                {
                        Thread.sleep (1000);
                }
                catch (Exception e)
                {
                        System.out.println ("Error" +e);
                }
                System.out.println ("second statement");
        }
}
class Thread2 extends Thread
{
        Thread1 t;
        public Thread2 (Thread1 t)
        {
                this.t = t;
        }
        public void run ( )
        {
                Synchronized (t)
                {
                t.call ( );
                }
        }
}
public class SynchronizedBlock
{
        public static void main (String args [ ])
        {
```

```
            Thread1 obj1 = new Thread1 ( );
            Thread2 obja = new Thread2 ( obj1);
            Thread2 objb = new Thread2 ( obj1);
            obja.start ( );
            objb.start( );
        }
}
```

Here is another example, using a synchronized block within the run( ) method:

```
// File Name : Callme.java
// This program uses a synchronized block.
class Callme {
  void call(String msg) {
    System.out.print("[" + msg);
    try {
      Thread.sleep(1000);
    } catch (InterruptedException e) {
      System.out.println("Interrupted");
    }
    System.out.println("]");
  }
}

// File Name : Caller.java
class Caller implements Runnable {
  String msg;
  Callme target;
  Thread t;
  public Caller(Callme targ, String s) {
    target = targ;
    msg = s;
    t = new Thread(this);
    t.start();
  }

  // synchronize calls to call()
  public void run() {
    synchronized(target) { // synchronized block
      target.call(msg);
    }
  }
}
// File Name : Synch.java
class Synch {
  public static void main(String args[]) {
    Callme target = new Callme();
    Caller ob1 = new Caller(target, "Hello");
    Caller ob2 = new Caller(target, "Synchronized");
    Caller ob3 = new Caller(target, "World");
```

```
        // wait for threads to end
      try {
         ob1.t.join();
         ob2.t.join();
         ob3.t.join();
      } catch(InterruptedException e) {
         System.out.println("Interrupted");
      }
   }
}
```

This would produce the following result:

[Hello]
[World]
[Synchronized]

8.2 Interthread Communication

Consider the classic queuing problem, where one thread is producing some data and another is consuming it. To make the problem more interesting, suppose that the producer has to wait until the consumer is finished before it generates more data.

In a polling system, the consumer would waste many CPU cycles while it waited for the producer to produce. Once the producer was finished, it would start polling, wasting more CPU cycles waiting for the consumer to finish, and so on. Clearly, this situation is undesirable.

To avoid polling, Java includes an elegant interprocess communication mechanism via the following methods:

- **wait( ):** This method tells the calling thread to give up the monitor and go to sleep until some other thread enters the same monitor and calls notify( ).
- **notify( ):** This method wakes up the first thread that called wait( ) on the same object.
- **notifyAll( ):** This method wakes up all the threads that called wait( ) on the same object. The highest priority thread will run first.

These methods are implemented as **final** methods in Object, so all classes have them. All three methods can be called only from within a **synchronized** context.

These methods are declared within Object. Various forms of wait( ) exist that allow you to specify a period of time to wait.

Threads can communicate with each other. Multithreading eliminates the concept of polling by a mechanism known as *interprocess communication.* For example, there are two threads, thread1 and thread2. The data produced by thread1 is consumed by thread2. Thread1 is the producer, and thread2 is the consumer. The consumer has to repeatedly check if the producer has produced data. This is a waste of CPU time as the consumer occupies CPU time to check whether or not the producer is ready with data. One thread cannot proceed until the other thread has completed its task.

A thread may notify another thread that the task has been completed. This communication between threads is known as interthreaded communication. The various methods used in interthread communication are:

wait ( ): Informs the current thread to leave its control over the monitor and sleep for a specified time until another thread calls the *notify ( )* method. The following syntax shows how to declare the *wait ( )* method:

public final void wait ( ) throws InterruptedException;

notify ( ): Wakes up a single thread that is waiting for the monitor of the object being executed. If multiple threads are waiting, one of them is chosen randomly. The following syntax shows how to declare the *notify ( )* method:

public final void notify ( )

notifyAll ( ): Wakes up all the threads that are waiting for the monitor of the object.

The following code shows the producer–consumer problem, but without interthread communication:

```
class Thread1
{
        int d;
        synchronized void getData ( )
        {
                System.out.println ("Got Data:" +d);
        }
        synchronized void putData (int d)
        {
                this.d = d;
                System.out.println ("Put data:"+d);
        }
}
class producer extends Thread
{
        Thread1 t;
        public producer (Thread1 t)
        {
                this.t = t;
        }
        public void run ( )
        {
        int data = 700;
        while (true)
        {
                System.out.println ("Put Called by producer");
                t.putData(data++);
        }
}
}
class consumer extends Thread
{
        Thread1 t;
        public consumer (Thread1 t)
        {
        this.t = t;
}
public void run ( )
{
        while (true)
```

```
        {
                System.out.println ("Get called by consumer");
        }
    }
    }
    public class ProducerConsumer
    {
        public static void main (String args [ ])
        {
                Thread1 obj1 = new Thread1 ( );
                Producer p = new producer (obj1);
                Consumer c = new consumer (obj1);
                p.start ( );
                c.start ( );
        }
    }
```

The preceding code shows an inaccurate result because the producer repeatedly overrides the consumer.

To avoid such error, *wait ( ), notify ( ),* and *notifyAll ( )* methods are used. All three methods can be called from within synchronized methods. The following code shows the interthread communication using the *wait( )* and *notify ( )* methods:

```
class Thread1
{
int d;
boolean flag = false;
synchronized int getData( )
{
        if (flag = =false)
        {
                try
                {
                        wait ( );
                }
                catch (InterruptedException e)
                {
                                System.out.println ("Exception caught");
                        }
                }
                System.out.println ("Got Data:" +d);
                flag = false;
                notify ( );
                return d;
        }
        synchronized void putData (int d)
        {
                if(flag = = true)
                {
                try
                {
```

```
                        wait ( );
                }
                catch (InterruptedException e)
                {
                                System.out.println ("Exception caught");
                        }
                        this.d = d;
                        System.out.println ("Put data with value:" +d);
                        flag = true;
                        notify ( );
                }
        }
}
class producer implements Runnable
{
        Thread1 t;
        public producer (Thread1 t)
        {
                this.t = t;
                new Thread (this,"producer").start ( );
                System.out.println ("Put called by producer");
        }
        public void run ( )
        {
                int data =0;
                while (true)
                {
                        data = data+1;
                        t.putData(data);
                }
        }
}
class consumer implements Runnable
{
        Thread1 t;
        public consumer (Thread1 t)
        {
                this.t = t;
                new Thread (this,"consumer").start ( );
                System.out.println ("Get called by consumer");
        }
        public void run ( )
        {
                while (true)
                {
                        t.getData( );
                }
        }
}
```

```
public class InterThreadComm
{
    public static void main (String args[ ] )
    {
        Thread1 obj1 = new Thread1 ( );
        producer p = new producer (obj1);
        consumer c = new consumer (obj1);
        System.out.println ("Press ctrl+c to stop");
    }
}
```

The following sample program consists of four classes: Q, the queue that you are trying to synchronize; Producer, the threaded object that is producing queue entries; Consumer, the threaded object that is consuming queue entries; and PC, the tiny class that creates the single Q, Producer, and Consumer.

The proper way to write this program in Java is to use wait( ) and notify( ) to signal in both directions, as shown here:

```
class Q {
  int n;
  boolean valueSet = false;
  synchronized int get() {
    if(!valueSet)
    try {
      wait();
    } catch(InterruptedException e) {
      System.out.println("InterruptedException caught");
    }
    System.out.println("Got: " + n);
    valueSet = false;
    notify();
    return n;
  }

  synchronized void put(int n) {
    if(valueSet)
    try {
      wait();
    } catch(InterruptedException e) {
      System.out.println("InterruptedException caught");
    }
    this.n = n;
    valueSet = true;
    System.out.println("Put: " + n);
    notify();
  }
}

class Producer implements Runnable {
```

```
  Q q;
  Producer(Q q) {
    this.q = q;
    new Thread(this, "Producer").start();
  }

  public void run() {
    int i = 0;
    while(true) {
      q.put(i++);
    }
  }
}

class Consumer implements Runnable {
  Q q;
  Consumer(Q q) {
    this.q = q;
    new Thread(this, "Consumer").start();
  }
  public void run() {
    while(true) {
    q.get();
  }
 }
}
class PCFixed {
  public static void main(String args[]) {
    Q q = new Q();
    new Producer(q);
    new Consumer(q);
    System.out.println("Press Control-C to stop.");
  }
}
```

Inside get( ), wait( ) is called. This causes its execution to suspend until the Producer notifies you that some data is ready.

When this happens, execution inside get( ) resumes. After the data has been obtained, get( ) calls notify( ). This tells Producer that it is okay to put more data in the queue.

Inside put( ), wait( ) suspends execution until the Consumer has removed the item from the queue. When execution resumes, the next item of data is put in the queue, and notify( ) is called. This tells the Consumer that it should now remove it.

Here is some output from this program, which shows the clean synchronous behavior:

```
Put: 1
Got: 1
Put: 2
Got: 2
```

Put: 3
Got: 3
Put: 4
Got: 4
Put: 5
Got: 5

8.3 Thread Deadlock

A special type of error that you need to avoid, that relates specifically to multitasking, is deadlock, which occurs when two threads have a circular dependency on a pair of synchronized objects.

For example, suppose one thread enters the monitor on object X and another thread enters the monitor on object Y. If the thread in X tries to call any synchronized method on Y, it will block as expected. However, if the thread in Y, in turn, tries to call any synchronized method on X, the thread waits forever, because to access X, it would have to release its own lock on Y so that the first thread could complete.

Example

To understand deadlock fully, it is useful to see it in action. The next example creates two classes, A and B, with methods foo( ) and bar( ) respectively, which pause briefly before trying to call a method in the other class.

The main class, named Deadlock, creates an A and a B instance, and then starts a second thread to set up the deadlock condition. The foo( ) and bar( ) methods use sleep( ) as a way to force the deadlock condition to occur.

```
class A {
  synchronized void foo(B b) {
    String name = Thread.currentThread().getName();
    System.out.println(name + " entered A.foo");
    try {
      Thread.sleep(1000);
    } catch(Exception e) {
      System.out.println("A Interrupted");
    }
    System.out.println(name + " trying to call B.last()");
    b.last();
  }
  synchronized void last() {
    System.out.println("Inside A.last");
  }
}
class B {
  synchronized void bar(A a) {
    String name = Thread.currentThread().getName();
    System.out.println(name + " entered B.bar");
    try {
      Thread.sleep(1000);
    } catch(Exception e) {
      System.out.println("B Interrupted");
    }
```

```
    System.out.println(name + " trying to call A.last()");
    a.last();
  }
  synchronized void last() {
    System.out.println("Inside A.last");
  }
}
class Deadlock implements Runnable {
  A a = new A();
  B b = new B();
  Deadlock() {
    Thread.currentThread().setName("MainThread");
    Thread t = new Thread(this, "RacingThread");
    t.start();
    a.foo(b); // get lock on a in this thread.
    System.out.println("Back in main thread");
  }
  public void run() {
    b.bar(a); // get lock on b in other thread.
    System.out.println("Back in other thread");
  }
  public static void main(String args[]) {
    new Deadlock();
  }
}
```

Here is some output from this program:

```
MainThread entered A.foo
RacingThread entered B.bar
MainThread trying to call B.last()
RacingThread trying to call A.last()
```

Because the program has deadlocked, you need to press CTRL-C to end the program. You can see a full thread and monitor cache dump by pressing CTRL-BREAK on a PC.

You will see that RacingThread owns the monitor on **b**, while it is waiting for the monitor on **a**. At the same time, MainThread owns **a** and is waiting to get **b**. This program will never complete.

As this example illustrates, if your multithreaded program locks up occasionally, deadlock is one of the first conditions that you should check for.

Ordering Locks

A common threading trick to avoid the deadlock is to order the locks. By ordering the locks, it gives threads a specific order to obtain multiple locks.

Deadlock Example

Following is the depiction of a deadlock:

```
// File Name ThreadSafeBankAccount.java
public class ThreadSafeBankAccount
{
  private double balance;
```

```
    private int number;
    public ThreadSafeBankAccount(int num, double initialBalance)
    {
      balance = initialBalance;
      number = num;
    }
    public int getNumber()
    {
      return number;
    }
    public double getBalance()
    {
      return balance;
    }
    public void deposit(double amount)
    {
      synchronized(this)
      {
       double prevBalance = balance;
       try
       {
         Thread.sleep(4000);
       }catch(InterruptedException e)
       {}
       balance = prevBalance + amount;
      }
    }
    public void withdraw(double amount)
    {
      synchronized(this)
      {
           double prevBalance = balance;
        try
        {
          Thread.sleep(4000);
        }catch(InterruptedException e)
        {}
        balance = prevBalance - amount;
      }
    }
}

// File Name LazyTeller.java
public class LazyTeller extends Thread
{
    private ThreadSafeBankAccount source, dest;
    public LazyTeller(ThreadSafeBankAccount a,
                ThreadSafeBankAccount b)
    {
```

```
        source = a;
        dest = b;
    }
    public void run()
    {
        transfer(250.00);
    }
    public void transfer(double amount)
    {
        System.out.println("Transferring from "
            + source.getNumber() + " to " + dest.getNumber());
        synchronized(source)
        {
            Thread.yield();
            synchronized(dest)
            {
                System.out.println("Withdrawing from "
                        + source.getNumber());
                source.withdraw(amount);
                System.out.println("Depositing into "
                        + dest.getNumber());
                dest.deposit(amount);
            }
        }
    }
}
public class DeadlockDemo
{
    public static void main(String [] args)
    {
        System.out.println("Creating two bank accounts...");
        ThreadSafeBankAccount checking =
                new ThreadSafeBankAccount(101, 1000.00);
        ThreadSafeBankAccount savings =
                new ThreadSafeBankAccount(102, 5000.00);

        System.out.println("Creating two teller threads...");
        Thread teller1 = new LazyTeller(checking, savings);
        Thread teller2 = new LazyTeller(savings, checking);
        System.out.println("Starting both threads...");
        teller1.start();
        teller2.start();
    }
}
```

This would produce the following results:

Creating two bank accounts...
Creating two teller threads...
Starting both threads...

Transferring from 101 to 102
Transferring from 102 to 101

The problem with the LazyTeller class is that it does not consider the possibility of a race condition, a common occurrence in multithreaded programming.

After the two threads are started, teller1 grabs the checking lock and teller2 grabs the savings lock. When teller1 tries to obtain the savings lock, it is not available. Therefore, teller1 blocks until the savings lock becomes available. When the teller1 thread blocks, teller1 still has the checking lock and does not let it go.

Similarly, teller2 is waiting for the checking lock, so teller2 blocks but does not let go of the savings lock. This leads to one result: Deadlock!

Deadlock Solution Example

Here transfer() method, in a class named OrderedTeller, instead of arbitrarily synchronizing on locks, this transfer() method obtains locks in a specified order based on the number of the bank account.

```
// File Name ThreadSafeBankAccount.java
public class ThreadSafeBankAccount
{
  private double balance;
  private int number;
  public ThreadSafeBankAccount(int num, double initialBalance)
  {
    balance = initialBalance;
    number = num;
  }
  public int getNumber()
  {
    return number;
  }
  public double getBalance()
  {
    return balance;
  }
  public void deposit(double amount)
  {
    synchronized(this)
    {
      double prevBalance = balance;
      try
      {
        Thread.sleep(4000);
      }catch(InterruptedException e)
      {}
      balance = prevBalance + amount;
    }
  }
  public void withdraw(double amount)
  {
    synchronized(this)
```

```
    {
        double prevBalance = balance;
      try
      {
        Thread.sleep(4000);
      }catch(InterruptedException e)
      {}
      balance = prevBalance - amount;
    }
  }
}
// File Name OrderedTeller.java
public class OrderedTeller extends Thread
{
  private ThreadSafeBankAccount source, dest;
  public OrderedTeller(ThreadSafeBankAccount a,
              ThreadSafeBankAccount b)
  {
    source = a;
    dest = b;
  }
  public void run()
  {
    transfer(250.00);
  }
  public void transfer(double amount)
  {
    System.out.println("Transferring from " + source.getNumber()
       + " to " + dest.getNumber());
    ThreadSafeBankAccount first, second;
    if(source.getNumber() < dest.getNumber())
    {
      first = source;
      second = dest;
    }
    else
    {
      first = dest;
      second = source;
    }
    synchronized(first)
    {
      Thread.yield();
      synchronized(second)
      {
        System.out.println("Withdrawing from "
              + source.getNumber());
        source.withdraw(amount);
        System.out.println("Depositing into "
```

```
            + dest.getNumber());
        dest.deposit(amount);
      }
    }
  }
}

// File Name DeadlockDemo.java
public class DeadlockDemo
{
  public static void main(String [] args)
  {
    System.out.println("Creating two bank accounts...");
    ThreadSafeBankAccount checking =
            new ThreadSafeBankAccount(101, 1000.00);
    ThreadSafeBankAccount savings =
            new ThreadSafeBankAccount(102, 5000.00);

    System.out.println("Creating two teller threads...");
    Thread teller1 = new OrderedTeller(checking, savings);
    Thread teller2 = new OrderedTeller(savings, checking);
    System.out.println("Starting both threads...");
    teller1.start();
    teller2.start();
  }
}
```

This would remove deadlock problem and would produce the following results:

```
Creating two bank accounts...
Creating two teller threads...
Starting both threads...
Transferring from 101 to 102
Transferring from 102 to 101
Withdrawing from 101
Depositing into 102
Withdrawing from 102
Depositing into 101
```

8.4 Thread Control: Suspend, Stop and Resume

While the suspend( ), resume( ), and stop( ) methods defined by **Thread** class seem to be a perfectly reasonable and convenient approach to managing the execution of threads, they must not be used for new Java programs and obsolete in newer versions of Java.

The following example illustrates how the wait( ) and notify( ) methods that are inherited from Object can be used to control the execution of a thread.

This example is similar to the program in the previous section. However, the deprecated method calls have been removed. Let us consider the operation of this program.

The NewThread class contains a boolean instance variable named suspendFlag, which is used to control the execution of the thread. It is initialized to false by the constructor.

The run( ) method contains a synchronized statement block that checks suspendFlag. If that variable is true, the wait( ) method is invoked to suspend the execution of the thread. The mysuspend( ) method sets suspendFlag to true. The myresume( ) method sets suspendFlag to false and invokes notify( ) to wake up the thread. Finally, the main( ) method has been modified to invoke the mysuspend( ) and myresume( ) methods.

Example

```
// Suspending and resuming a thread for Java 2
class NewThread implements Runnable {
  String name; // name of thread
  Thread t;
  boolean suspendFlag;
  NewThread(String threadname) {
    name = threadname;
    t = new Thread(this, name);
    System.out.println("New thread: " + t);
    suspendFlag = false;
    t.start(); // Start the thread
  }
  // This is the entry point for thread.
  public void run() {
    try {
    for(int i = 15; i > 0; i—) {
      System.out.println(name + ": " + i);
      Thread.sleep(200);
      synchronized(this) {
        while(suspendFlag) {
          wait();
        }
      }
     }
    } catch (InterruptedException e) {
      System.out.println(name + " interrupted.");
    }
    System.out.println(name + " exiting.");
  }
  void mysuspend() {
    suspendFlag = true;
  }
  synchronized void myresume() {
    suspendFlag = false;
    notify();
  }
}
class SuspendResume {
```

```
  public static void main(String args[]) {
    NewThread ob1 = new NewThread("One");
    NewThread ob2 = new NewThread("Two");
    try {
      Thread.sleep(1000);
      ob1.mysuspend();
      System.out.println("Suspending thread One");
      Thread.sleep(1000);
      ob1.myresume();
      System.out.println("Resuming thread One");
      ob2.mysuspend();
      System.out.println("Suspending thread Two");
      Thread.sleep(1000);
      ob2.myresume();
      System.out.println("Resuming thread Two");
    } catch (InterruptedException e) {
      System.out.println("Main thread Interrupted");
    }
    // wait for threads to finish
    try {
      System.out.println("Waiting for threads to finish.");
      ob1.t.join();
      ob2.t.join();
    } catch (InterruptedException e) {
      System.out.println("Main thread Interrupted");
    }
    System.out.println("Main thread exiting.");
  }
}
```

Here is the output produced by the above program:

```
New thread: Thread[One, 5, main]
One: 15
New thread: Thread[Two, 5, main]
Two: 15
One: 14
Two: 14
One: 13
Two: 13
One: 12
Two: 12
One: 11
Two: 11
Suspending thread One
Two: 10
Two: 9
Two: 8
Two: 7
Two: 6
```

```
Resuming thread One
Suspending thread Two
One: 10
One: 9
One: 8
One: 7
One: 6
Resuming thread Two
Waiting for threads to finish.
Two: 5
One: 5
Two: 4
One: 4
Two: 3
One: 3
Two: 2
One: 2
Two: 1
One: 1
Two exiting.
One exiting.
Main thread exiting.
```

9. USING MULTITHREADING

The key to utilizing multithreading support effectively is to think concurrently rather than serially. For example, when you have two subsystems within a program that can execute concurrently, make them individual threads.

With the careful use of multithreading, you can create very efficient programs. A word of caution is in order, however: If you create too many threads, you can actually degrade the performance of your program rather than enhance it.

6 Java Library

1. JAVA I/O STREAMING

The interfaces involved are:

- DataInput
- DataOutput
- FilenameFilter

DataInput

DataInput is an interface describing streams that can read input in a machine-independent format.

The interface java.io.DataInput is generally represented as:

*public interface **DataInput** extends Object*

Methods	Usage	Use
readFully	Public abstract void readFully (byte b[]) throws IO Exception	Reads bytes, blocking until all bytes are read. b—the buffer into which the data is read
readFully	Public abstract void readFully (byte b[], int off, int len) throws Exception	Reads bytes, blocking until all bytes are read. b—the buffer IO into which the data is read, off—the start offset of the data, len—the maximum number of bytes to read.
skipBytes	Public abstract int skipBytes (int n) throws IO Exception	Skips bytes, block until all bytes are skipped. n—the number of bytes to be skipped. Returns the actual number of bytes skipped.
readBoolean	Public abstract boolean readBoolean() throws IO Exception	Reads in a boolean. Returns the boolean read.
readByte	Public abstract byte readByte() throws IO Exception	Reads an 8 bit byte. Returns the 8 bit byte read.
readUnsigned Byte	Public abstract int readUnsignedByte() throws IO Exception	Reads an unsigned 8 bit byte. Returns the 8 bit byte read.

readShort	Public abstract short readShort() throws IO Exception	Reads 16 bit short. Returns the read 16 bit short.
readUnsigned short	Public abstract int readUnsigned short() throws IO Exception	Reads an unsigned 16 bit short. Returns the read 16 bit short.
readChar	Public abstract char readChar() throws IO Exception	Reads a 16 bit char. Returns the read 16 bit char.
readInt	Public abstract int readInt() throws IO Exception	Reads a 32 bit int. Returns the read 32 bit integer.
readLong	Public abstract long readLong() throws IO Exception	Reads a 64 bit long. Returns the read 64 bit long.
readFloat	Public abstract float readFloat() throws IO Exception	Reads a 32 bit float. Returns the read 32 bit float.
readDouble	Public abstract double readDouble() throws IO Exception	Reads a 64 bit double. Returns the read 64 bit double.
readLine	Public abstract String read Line() throws IO Exception	Reads the line
readUTF	Public abstract String read UTF() throws IO Exception	Reads the UTF

Data Output

DataOutput is an interface describing streams that can write output in a machine-independent format.

public interface ***DataOutput*** *extends Object*

Throws IO exception if I/O error has occurred.

Methods	Usage	Use
write	Public abstract void write (int b) throws IO Exception	Writes a byte. Will block until the byte is actually written. b—the byte to be written.
write	Public abstract void write (byte b[]) throws IO Exception	Writes an array of bytes. b—the data to be written.
write	Public abstract void write (byte b[], int off, int len) throws IO Exception	Writes a sub array of bytes. b—the data to be written, off—the start offset in the data, len–the number of bytes that are written.
writeBoolean	Public abstract void writeBoolean (boolean v) throws IO Exception	Writes a boolean. v—the boolean to be written.
writeByte	Public abstract void writeByte (int v) throws IO Exception	Writes an 8 bit byte. v—the byte value to be written.

writeShort	Public abstract void writeShort (int v) throws IO Exception	Writes a 16 bit short. v—the short value to be written.
writeChar	Public abstract void writeChar (int v) throws IO Exception	Writes a 16 bit char. v—the char value to be written.
writeInt	Public abstract void writeInt (int v) throws IO Exception	Writes a 32 bit int. v—the integer value to be written.
writeLong	Public abstract void writeLong (long v) throws IO Exception	Writes a 64 bit long. v—the long value to be written.
writeFloat	Public abstract void writeFloat (float v) throws IO Exception	Writes a 32 bit float. v—the float value to be written
writeDouble	Public abstract void writeDouble(double v) throws IO Exception	Writes a 64 bit double. v—the double value to be written.
writeBytes	Public abstract void writeBytes (String s) throws IO Exception	Writes a String as a sequence of bytes. s-the String of bytes to be written.
writeChars	Public abstract void writeChars (String s) throws IO Exception	Writes a String as a sequence of chars. s-the String of chars to be written.
writeUTF	Public abstract void writeUTF (String str) throws IOException	Writes a String in UTF format. str-the String in UTF format

Filename Filter

The interface java.io.FilenameFilter is a filter interface for file names. It is generally represented as

*public interface **FilenameFilter** extends Object*

Methods	Usage	Use
accept	Public abstract boolean accept (File dir, String name)	Determines whether a name should be included in a file list. dir—the directory in which the file was found, name—the name of the file. **Returns** true if name should be included in file list; false otherwise.

The classes in java.io include:

- Buffered Input Stream
- Buffered Output Stream
- Byte Array Input Stream
- Byte Array Output Stream
- Data Input Stream
- Data Output Stream

- File
- File Input Stream
- File Output Stream
- Filter Input Stream
- Filter Output Stream
- Input Stream
- Line Number Input Stream
- Output Stream
- Piped Input Stream
- Piped Output Stream
- Print Stream
- Pushback Input Stream
- Random Access File
- Sequence Input Stream
- Stream Tokenizer
- String Buffer Input Stream

1.1 Buffered Input Stream

```
java.lang.Object
   |
   +——java.io.InputStream
         |
         +——java.io.FilterInputStream
               |
               +——java.io.BufferedInputStream
```

A buffered input stream. This stream lets you read in characters from a stream without causing a read every time. The data is read into a buffer, subsequent reads result in a fast buffer access.

*public class **BufferedInputStream** extends FilterInputStream*

The purpose of I/O buffering is to improve system performance. Rather than reading a byte at a time, a large number of bytes are read together the first time the read() method is invoked. When an attempt is made to read subsequent bytes, they are taken from the buffer, not the underlying input stream. This improves data access time and can reduce the number of times an application blocks for input.

Constructors

1. public Buffered Input Stream (Input Stream in)

Creates a new buffered stream with a default buffer size. in—the input stream.

2. public Buffered Input Stream (Input Stream in, int size)

Creates a new buffered stream with the specified buffer size. in—the input stream, size—the buffer size

Methods	Usage	Use
Read	Public synchronized int read () throws IO Exception	Reads a byte of data. This method will block if no input is available. Returns the byte read, or –1 if the end of the stream is reached.

Read	Public synchronized int read (byte b[], int off, int len) throws IO Exception	Reads into an array of bytes. Blocks until some input is available. b—the buffer into which the data is read, off-the start offset of the data, len—the maximum number of bytes read. Returns the actual number of bytes read, –1 is returned when the end of the stream is reached.
Skip	Public synchronized long skip (long n) throws IO Exception	Skips n bytes of input. n—the number of bytes to be skipped. Returns the actual number of bytes skipped.
Available	Public synchronized int available () throws IO Exception	Returns the number of bytes that can be read without blocking. This total is the number of bytes in the buffer and the number of bytes available from the input stream.Returns the number of available bytes.
Mark	Public synchronized void mark (int readlimit)	Marks the current position in the input stream. A subsequent call to the reset () method will reposition the stream at the last marked position so that subsequent reads will re-read the same bytes. The stream promises to allow readlimit bytes to be read before the mark position gets invalidated readlimit—the maximum limit of bytes allowed to be read before the mark position becomes invalid.
Reset	Public synchronized void reset () throws IO Exception	Repositions the stream to the last marked position. If the stream has not been marked, or if the mark has been invalidated, an IO Exception is thrown. Stream marks are intended to be used in situations where you need to read ahead a little to see what's in the stream. Often this is most easily done by invoking some general parser. If the stream is of the type handled by the parser, it just chugs along happily. If the stream is *not* of that type, the

		parser should toss an exception when it fails. If an exception gets tossed within readlimit bytes, the parser will allow the outer code to reset the stream and to try another parser.
Mark Supported	Public boolean mark Supported ()	Returns a boolean indicating if this stream type supports mark/reset.

1.2 Buffered Output Stream

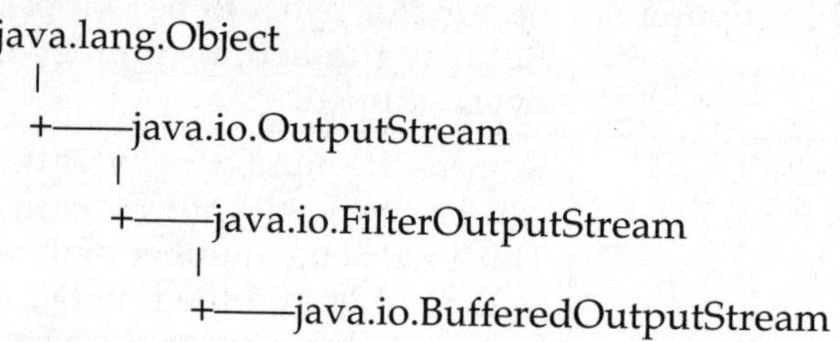

A buffered output stream. This stream lets you write characters to a stream without causing a write every time. The data is first written into a buffer. Data is written to the actual stream only when the buffer is full, or when the stream is flushed.

*public class **BufferedOutputStream** extends FilterOutputStream*

Constructors

1. Public Buffered Output Stream (Output Stream out)

Creates a new buffered stream with a default buffer size. out—the output stream.

2. public BufferedOutputStream(OutputStream out, int size)

Creates a new buffered stream with the specified buffer size. out— the output stream, size—the buffer size

Methods	Usage	Use
Write	Public synchronized void write (int b) throws IO Exception	Writes a byte. This method will block until the byte is actually written. b—the byte to be written.
Write	Public synchronized void write (byte b[], int off, int len) throws IO Exception	Writes a sub array of bytes. b—the data to be written, off—the start offset in the data, len—the number of bytes that are written.
Flush	Public synchronized void flush() throws IO Exception	Flushes the stream. This will write any buffered output bytes.

1.3 Byte Array Input Stream

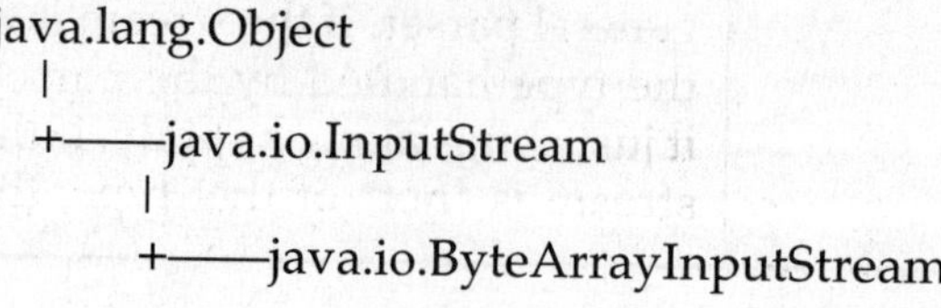

This class implements a buffer that can be used as an InputStream.

*public class **ByteArrayInputStream** extends InputStream*

Constructors

1. public ByteArrayInputStream(byte buf[])

Creates an ByteArrayInputStream from the specified array of bytes. buf—the input buffer (not copied)

2. public ByteArrayInputStream(byte buf[],int offset, int length)

Creates an ByteArrayInputStream from the specified array of bytes. buf—the input buffer (not copied), offset—the offset of the first byte to read, length—the number of bytes to read.

Methods	Usage	Use
Read	Public synchronized int read ()	Reads a byte of data. Returns the byte read, or –1 if the end of the stream is reached.
Read	Public synchronized int read (byte b[], int off, int len)	Reads into an array of bytes. b—the buffer into which the data is read, off—the start offset of the data, len—the maximum number of bytes read. Returns the actual number of bytes read; –1 is returned when the end of the stream is reached.
Skip	Public synchronized long skip (long n)	Skips n bytes of input. n—the number of bytes to be skipped. Returns the actual number of bytes skipped.
Available	Public synchronized int available ()	Returns the number of available bytes in the buffer.
Reset	Public synchronized void reset ()	Resets the buffer to the beginning.

1.4 Byte Array Output Stream

```
java.lang.Object
   |
   +——java.io.OutputStream
        |
        +——java.io.ByteArrayOutputStream
```

- Class java.io.ByteArrayOutputStream implements a buffer that can be used as an OutputStream. The buffer automatically grows when data is written to the stream. The data can be retrieved using toByteArray() and toString().

*public class **ByteArrayOutputStream** extends OutputStream*

Constructors

1. public ByteArrayOutputStream()
Creates a new ByteArrayOutputStream.
2. public ByteArrayOutputStream(int size)
Creates a new ByteArrayOutputStream with the specified initial size.size—the initial size.

Methods	Usage	Use
write	Public synchronized void write (int b)	Writes a byte to the buffer. b-the byte
write	Public synchronized void write (byte b[], int off, int len)	Writes bytes to the buffer. b-the data to be written, off-the start offset in the data, len-the number of bytes that are written.
writeto	Public synchronized void writeTo (Output Stream out) throws IO Exception	Writes the contents of the buffer to another stream. out-the output stream to write to.
reset	Public synchronized void reset()	Resets the buffer so that you can use it again without throwing away the already allocated buffer.
toByteArray	Public synchronized byte[] toByteArray ()	Returns a copy of the input data.
size	Public int size ()	Returns the current size of the buffer.
toString	Public String toString ()	Converts input data to a string. Returns the string.
toString	Public String toString (int hibyte)	Converts input data to a string. The top 8 bits of each 16 bit Unicode character are set to hibyte. hibyte—the bits set.

1.5 Data Input Stream

Class java.io.DataInputStream lets the user to read primitive Java data types from a stream in a portable way. Primitive data types are well understood types with associated operations. For example, integers are considered primitive data types.

```
java.lang.Object
   |
   +——java.io.InputStream
         |
         +——java.io.FilterInputStream
               |
               +——java.io.DataInputStream
```

public class **DataInputStream** extends FilterInputStream implements DataInput

Constructors

- *public DataInputStream(InputStream in)*—this default constructor creates a new DataInputStream.in—the input stream.

Methods

Methods	Usage	Use
read	Public final int read (byte b[]) throws IO Exception	Reads data into an array of bytes. This method blocks until some input is available. b—the buffer into which the data is read. Returns the actual number of bytes read, –1 is returned when the end of the stream is reached.
read	Public final int read (byte b[], int off, int len) throws IO Exception	Reads data into an array of bytes. This method blocks until some input is available. b—the buffer into which the data is read, off—the start offset of the data, len—the maximum number of bytes read. Returns the actual number of bytes read, –1 is returned when the end of the stream is reached.
readFully	Public final void readFully (byte b[]) throws IO Exception	Reads bytes, blocking until all bytes are read. b—the buffer into which the data is read.
readFully	Public final void readFully (byte b[], int off, int len) throws IO Exception	Reads bytes, blocking until all bytes are read. b—the buffer into which the data is read, off—the start offset of the data, len—the maximum number of bytes read.
skipBytes	Public final int skipBytes (int n) throws IO Exception	Skips bytes, block until all bytes are skipped. n—the number of bytes to be skipped. Returns the actual number of bytes skipped.
readBoolean	Public final boolean readBoolean() throws IO Exception	Reads in a boolean. Returns the boolean read.
readByte	Public final byte readByte() throws IO Exception	Reads an 8 bit byte. Returns the 8 bit byte read.
readUnsigned Byte	public final int readUnsigned Byte() throws IO Exception	Reads an unsigned 8 bit byte. Returns the 8 bit byte read.

readShort	Public final short readShort () throws IO Exception	Reads 16 bit short. Returns the read 16 bit short.
readUnsigned Short	Public final int readUnsigned Short() throws IO Exception	Reads 16 bit short. Returns the read 16 bit short.
readChar	Public final char readChar() throws IO Exception	Reads a 16 bit char. Returns the read 16 bit char.
readInt	Public final int readInt() throws IO Exception	Reads a 32 bit int. Returns the read 32 bit integer.
readLong	Public final long readLong() throws IO Exception	Reads a 64 bit long. Returns the read 64 bit long.
readFloat	Public final float readFloat() throws IO Exception	Reads a 32 bit float. Returns the read 32 bit float.
readDouble	Public final double readDouble() throws IO Exception	Reads a 64 bit double. Returns the read 64 bit double.
readLine	Public final String read Line() throws IO Exception	Reads in a line that has been terminated by a \n, \r, \r\n or EOF. Returns a String copy of the line.
readUTF	Public final String readUTF() throws IO Exception	Reads a UTF format String. Returns the String.
readUTF	Public final static String readUTF (DataInput in) throws IO Exception	Reads a UTF format String from the given input stream. Returns the String.

1.6 Data Output Stream

Class java.io.DataOutputStream lets you write primitive Java data types to a stream in a portable way. Primitive data types are well understood types with associated operations. For example, an integer is considered to be a good primitive data type. The data can be converted back using a DataInputStream.

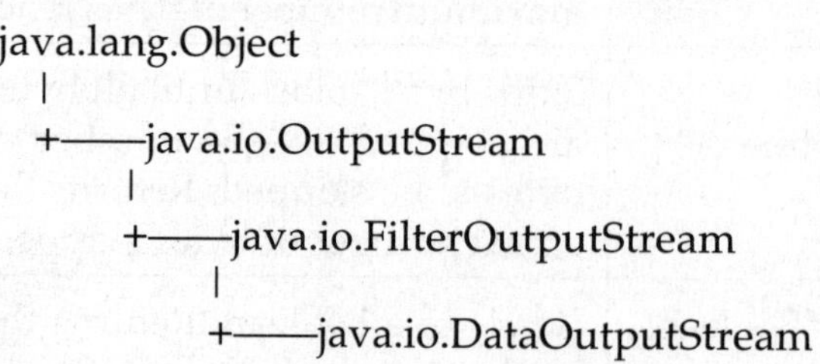

public class **DataOutputStream** extends FilterOutputStream implements DataOutput

Constructors

- *public DataOutputStream(OutputStream out)*

Creates a new DataOutputStream. out—the output stream.

Methods

Methods	Usage	Use
write	public synchronized void write(int b) throws IO Exception	Writes a byte. Will block until the byte is actually written. b—the byte to be written.
write	public synchronized void write (byte b[], int off, int len) throws IO Exception	Writes a sub array of bytes. b—the data to be written, off—the start offset in the data, len—the number of bytes that are written
flush	public void flush() throws IO Exception	Flushes the stream. This will write any buffered output bytes.
write Boolean	public final void writeBoolean (boolean v) throws IO Exception	Writes a boolean. v—the boolean to be written.
writebyte	public final void writeByte(int v) throws IO Exception	Writes an 8 bit byte. v—the byte value to be written.
writeShort	public final void writeShort (int v) throws IO Exception	Writes a 16 bit short. v—the short value to be written.
writeChar	public final void writeChar (int v) throws IO Exception	Writes a 16 bit char. v—the char value to be written.
writeInt	public final void writeInt (int v) throws IO Exception	Writes a 32 bit int. v—the integer value to be written.
writeLong	public final void writeLong (long v) throws IO Exception	Writes a 64 bit long. v—the long value to be written.
writeFloat	public final void writeFloat (float v) throws IO Exception	Writes a 32 bit float. v—the float value to be written.
writeDouble	public final void writeDouble (double v) throws IO Exception	Writes a 64 bit double. v—the double value to be written.
writeBytes	public final void writeBytes (String s) throws IO Exception	Writes a String as a sequence of bytes. s—the String of bytes to be written.
writeChars	public final void writeChars (String s) throws IO Exception	Writes a String as a sequence of chars. s—the String of chars to be written.
writeUTF	public final void writeUTF (String str) throws IO Exception	Writes a String in UTF format. str—the String in UTF format.
size	public final intsize()	Returns the number of bytes written. The number of bytes written thus far.

1.7 File

- Class java.io. File represents a file name of the host file system. The file name can be relative or absolute. It must use the file name conventions of the host platform.

The intention is to provide an abstraction that deals with most of the system-dependent file name features such as the separator character, root, device name, etc. Not all features are currently fully implemented.

Note that whenever a file name or path is used, it is assumed that the host's file name conventions are used.

```
java.lang.Object
   |
   +——java.io.File
```

public class **File** extends Object

Constructors

1. public File(String path)

Creates a File object.path—the file path. **Throws** Null Pointer Exception, if the file path is equal to null.

2. public File (String path, String name)

Creates a File object from the specified directory path—the directory path, name—the file name

3. public File (File dir, String name)

Creates a File object (given a directory File object). dir—the directory, name—the file name

Methods

Methods	Usage	Use
getName	public String getName()	Gets the name of the file. This method does not include the directory. Returns the file name.
getPath	Public String get Path()	Gets the path of the file. Returns the file path.
getAbsolutePath	Public String getAbsolutePath()	Gets the absolute path of the file. Returns the absolute file path.
getParent	Public String getParent()	Gets the name of the parent directory. Returns the parent directory, or null if one is not found.
exists	Public boolean exists()	Returns a boolean indicating whether or not a file exists.
canWrite	Public boolean canWrite()	Returns a boolean indicating whether or not a writable file exists.

canRead	Public boolean canRead()	Returns a boolean indicating whether or not a readable file exists.
isFile	Public boolean isFile()	Returns a boolean indicating whether or not a normal file exists.
isDirectory	Public boolean isDirectory()	Returns a boolean indicating whether or not a directory file exists.
isAbsolute	Public boolean isAbsolute()	Returns a boolean indicating if the file name is absolute.
lastModified	Public long lastModified() throws IO Exception	Returns the last modification time. The return value should only be used to compare modification dates. It is meaningless as an absolute time.
length	Public long length () throws IO Exception	Returns the length of the file.
mkdir	Public boolean mkdir() throws IO Exception	Creates a directory and returns a boolean indicating the success of the creation.
rename to	Public boolean rename to (File dest) throws IO Exception	Renames a file and returns a boolean indicating whether or not this method was successful. dest—the new file name.
mkdirs	Public boolean mkdirs() throws IO Exception	Creates all directories in this path. This method returns true if all directories in this path are created.
List	Public String[] list() throws IO Exception	Lists the files in a directory. Works only on directories. Returns an array of file names. This list will include all files in the directory except the equivalent of "." and "..".
list	Public String[] list (FilenameFilter filter) throws IO Exception	Uses the specified filter to list files in a directory filter—the filter used to select file names. Returns the filter selected files in this directory.
hashCode	Public int hashCode()	Computes a hashcode for the file.
equals	Public boolean equals (Object obj)	Compares this object against the specified object. obj—the object to compare with. Returns true if the objects are the same; false otherwise.
toString	Public String toString()	Returns a String object representing this file's path.

1.8 File Input Stream

- Class java io File Input Stream

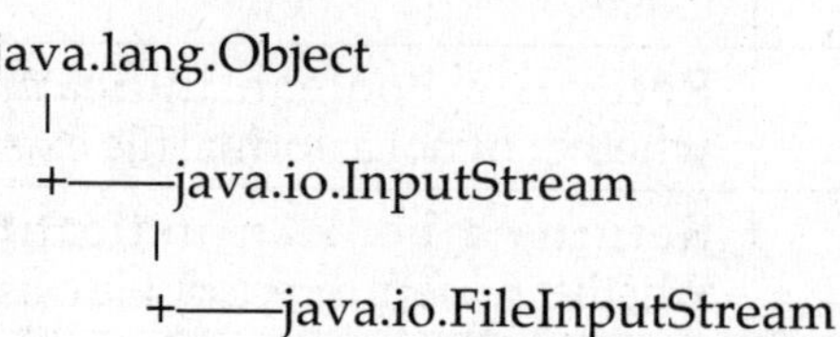

public class **FileInputStream** extends InputStream
File input stream, can be constructed from a file descriptor or a file name.

Constructors

1. Public File Input Stream (String name) throws File Not Found Exception
Creates an input file with the specified system dependent file name.
2. Public File Input Stream (File file) throws File Not Found Exception
Creates an input file from the specified File object.
3. Public File Input Stream(int fd) throws IO Exception
Creates an input file with the specified system dependent file descriptor. fd—the system dependent file descriptor.

Methods

Methods	Usage	Use
read	Public int read() throws IO Exception	Reads a byte of data. This method will block if no input is available. Returns the byte read, or –1 if the end of the stream is reached.
read	Public int read (byte b[]) throws IO Exception	Reads data into an array of bytes. This method blocks until some input is available. b—the buffer into which the data is read. Returns the actual number of bytes read, –1 is returned when the end of the stream is reached.
read	Public int read (byte b[], int off, int len) throws IO Exception	Reads data into an array of bytes. This method blocks until some input is available. b—the buffer into which the data is read, off—the start offset of the data, len—the maximum number of bytes read. Returns the actual number of bytes read, –1 is returned when the end of the stream is reached.

skip	Public long skip (long n) throws IO Exception	Skips n bytes of input. n—the number of bytes to be skipped. Returns the actual number of bytes skipped.
available	Public int available() throws IO Exception	Returns the number of bytes that can be read without blocking. Returns the number of available bytes, which is initially equal to the file size.
close	Public void close() throws IO Exception	Closes the input stream. This method must be called to release any resource associated
get FD	Public final int get FD()	Returns the file descriptor associated with this stream. Returns the file descriptor.
finalize	Protected void finalize() throws IO Exception	Closes the stream when garbage is collected.

1.9 File Output Stream

File output stream can be constructed from a file descriptor or a file name.

```
java.lang.Object
  |
  +——java.io.OutputStream
       |
       +——java.io.FileOutputStream
```

public class **FileOutputStream** extends OutputStream

Constructors

1. Public File Output Stream (String name) throws IO Exception

Creates an output file with the specified system dependent file name. name—the system dependent file name.

2. Public File Output Stream (int fd) throws IO Exception

Creates an output file with the specified system dependent file descriptor.fd—the system dependent file descriptor.

3. Public File Output Stream (File file) throws IO Exception

Creates an output file with the specified File object.file—the file to be opened for reading.

Methods

Methods	Usage	Use
write	Public void write(int b) throws IO Exception	Writes a byte of data. This method will block until the byte is actually written. b—the byte to be written.
write	Public void write(byte b[]) throws IO Exception	Writes an array of bytes. Will block until the bytes are actually written. b—the data to be written
write	Public void write(byte b[], int off, int len) throws IO Exception	Writes a sub array of bytes. b—the data to be written, off—the start offset in the data, len—the number of bytes that are written
close	Public void close() throws IO Exception	Closes the stream. This method must be called to release any resources associated with the stream.
getFD	Public final int getFD()	Returns the file descriptor associated with this stream. Returns the file descriptor.
initialize	Protected void finalize() throws IO Exception	Closes the stream when garbage is collected.

1.10 Filter Input Stream

```
java.lang.Object
   |
   +——java.io.InputStream
         |
         +——java.io.FilterInputStream
```

Abstract class representing a filtered input stream of bytes. This class is the basis for enhancing input stream functionality. It allows multiple input stream filters to be chained together, each providing additional functionality.

public class **FilterInputStream** extends InputStream

Constructors

1. protected FilterInputStream(InputStream in)
Creates an input stream filter. in—the input stream.

Methods

Methods	Usage	Use
read	Public int read() throws IO Exception	Reads a byte. Will block if no input is available. Returns the byte read, or –1 if the end of the stream is reached.
read	Public int read(byte b[]) throws IO Exception	Reads into an array of bytes. Blocks until some input is available. b—the buffer into which the data is read. Returns the actual number of bytes read, –1 is returned when the end of the stream is reached.
read	Public int read(byte b[], int off, int len) throws IO Exception	Reads into an array of bytes. Blocks until some input is available. This method should be overridden in a subclass for efficiency (the default implementation reads 1 byte at a time). b—the buffer into which the data is read, off—the start offset of the data, len—the maximum number of bytes read. Returns the actual number of bytes read, –1 is returned when the end of the stream is reached.
skip	Public long skip(long n) throws IO Exception	Skips bytes of input. n—bytes to be skipped. Returns actual number of bytes skipped.
available	Public int available() throws IO Exception	Returns the number of bytes that can be read without blocking. Returns the number of available bytes.
close	Public void close() throws IO Exception	Closes the input stream. Must be called to release any resources associated with the stream.
mark	Public synchronized void mark (int readlimit)	Marks the current position in the input stream. A subsequent call to reset() will reposition the stream at the last marked position so that subsequent reads will re-read the same bytes. The stream promises to allow

		readlimit bytes to be read before the mark position gets invalidated. readlimit—the maximum limit of bytes allowed to be read before the mark position becomes invalid.
reset	Public synchronized void reset() throws IO Exception	Repositions the stream to the last marked position. If the stream has not been marked, or if the mark has been invalidated, an IO Exception is thrown. Stream marks are intended to be used in situations where you need to read ahead a little to see what is in the stream. Often this is most easily done by invoking some general parser. If the stream is of the type handled by the parse, it just chugs along happily. If the stream is NOT of that type, the parser should toss an exception when it fails, which, if it happens within readlimit bytes, allows the outer code to reset the stream and try another parser.
markSupported	public boolean markSupported()	Returns true if this stream type supports mark/reset

1.11 Filter Output Stream

```
java.lang.Object
 |
 +——java.io.OutputStream
     |
     +——java.io.FilterOutputStream
```

Abstract class representing a filtered output stream of bytes. This class is the basis for enhancing output stream functionality. It allows multiple output stream filters to be chained together, each providing additional functionality.

public class **FilterOutputStream** extends OutputStream

Constructors

- Public Filter Output Stream (Output Stream out)

Creates an output stream filter.out—the output stream.

Methods

Methods	Usage	Use
write	Public void write(int b) throws IO Exception	Writes a byte. Will block until the byte is actually written. b—the byte.
write	Public void write (byte b[]) throws IO Exception	Writes an array of bytes. Will block until the bytes are actually written. b—the data to be written.
write	Public void write (byte b[], int off, int len) throws IO Exception	Writes a sub array of bytes. To be efficient it should be over-ridden in a subclass. b—the data to be written, off—the start offset in the data, len—the number of bytes that are written.
flush	Public void flush() throws IO Exception	Flushes the stream. This will write any buffered output bytes.
close	Public void close() throws IO Exception	Closes the stream. This method must be called to release any resources associated with the stream.

1.12 Input Stream

```
java.lang.Object
  |
  +——java.io.InputStream
```

An abstract class representing an input stream of bytes. All InputStreams are based on this class.

public class **InputStream** extends Object

Constructor

- public Input Stream ()

Methods

Methods	Usage	Use
read	Public abstract int read() throws IO Exception	Reads a byte of data. This method will block if no input is available. Returns the byte read, or –1 if the end of the stream is reached.

read	Public int read(byte b[]) throws IO Exception	Reads into an array of bytes. This method will block until some input is available. b-the buffer into which the data is read. Returns the actual number of bytes read, –1 is returned when the end of the stream is reached.
read	Public int read(byte b[], int off, int len) throws IO Exception	Reads into an array of bytes. This method will block until some input is available. b—the buffer into which the data is read, off—the start offset of the data, len—the maximum number of bytes read. Returns the actual number of bytes read, –1 is returned when the end of the stream is reached.
skip	Public long skip(long n) throws IO Exception	Skips n bytes of input. n—the number of bytes to be skipped. Returns the actual number of bytes skipped.
available	Public int available() throws IO Exception	Returns the number of bytes that can be read without blocking. Returns the number of available bytes.
close	Public void close() throws IO Exception	Closes the input stream. Must be called to release any resources associated with the stream.
mark	Public synchronized void mark (int readlimit)	Marks the current position in the input stream. A subsequent call to reset() will reposition the stream at the last marked position so that subsequent reads will re-read the same bytes. The stream promises to allow readlimit bytes to be read before the mark position gets invalidated. readlimit—the maximum limit of bytes allowed to be read before the mark position becomes invalid.

1.13 Line Number Input Stream

java.lang.Object
|
+——java.io.InputStream
|
+——java.io.FilterInputStream
|
+——java.io.LineNumberInputStream

An input stream that keeps track of line numbers.

public class **LineNumberInputStream** extends FilterInputStream

Constructor

- public Line Number Input Stream (Input Stream in)

Constructs a new Line Number Input Stream initialized with the specified input stream. in—the input stream

Methods

Methods	Usage	Use
read	Public int read () throws IO Exception	Reads a byte of data. The method will block if no input is available. Returns the byte read, or –1 if the end of the stream is reached.
read	Public int read (byte b[], int off, int len) throws IO Exception	Reads into an array of bytes. This method will blocks until some input is available. b—the buffer into which the data is read, off—the start offset of the data, len—the maximum number of bytes read. Returns the actual number of bytes read, –1 is returned when the end of the stream is reached.
setLine Number	Public void setLineNumber (int line number)	Sets the current line number. Line Number—the line number to be set.
getLine Number	Public int getLineNumber ()	Returns the current line number.
skip	Public long skip (long n) throws IO Exception	Skips n bytes of input. n—the number of bytes to be skipped. Returns the actual number of bytes skipped.
available	Public int available () throws IO Exception	Returns the number of bytes that can be read without blocking. Returns the number of available bytes.

mark	Public void mark (int readlimit)	Marks the current position in the input stream. A subsequent call to reset () will reposition the stream at the last marked position so that subsequent reads will re-read the same bytes. The stream promises to allow readlimit bytes to be read before the mark number position gets invalidated.
reset	Public void reset () throws IO Exception	Repositions the stream to the last marked position. If the stream has not been marked, or if the mark has been invalidated, an IO Exception is thrown. Stream marks are intended to be used in situations where you need to read ahead a little to see what is in the stream. Often this is most easily done by invoking some general parser. If the stream is of the type handled by the parser, it just chugs along happily. If the stream is NOT of that type, the parser should toss an exception when it fails, which, if it happens within readlimit bytes, allows the outer code to reset the stream and try another parser.

1.14 Output Stream

java.lang.Object
|
+—java.io.OutputStream

Abstract class representing an output stream of bytes. All Output Streams are based on this class.

public class **OutputStream** extends Object

Constructor

- public OutputStream()—default constructor

Methods

Methods	Usage	Use
write	Public abstract void write (int b) throws IO Exception	Writes a byte. This method will block until the byte is actually written. b—the byte

write	Public void write (byte b[]) throws IO Exception	Writes an array of bytes. This method will block until the bytes are actually written.b—the data to be written.
write	Public void write (byte b[], int off, int len) throws IO Exception	Writes a sub array of bytes. b—the data to be written, off—the start offset in the data, len—the number of bytes that are written.
flush	Public void flush () throws IO Exception	Flushes the stream. This will write any buffered output bytes.
close	Public void close () throws IO Exception	Closes the stream. This method must be called to release any resources associated with the stream.

1.15 Piped Input Stream

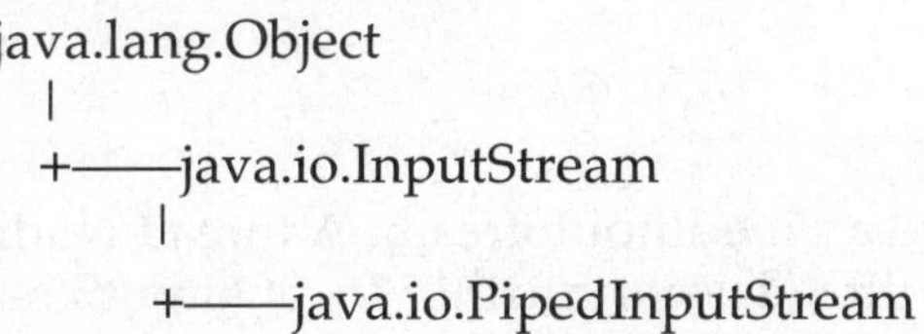

PipedInputStream must be connected to a PipedOutputStream to be useful. A thread reading from a PipedInputStream receives data from a thread writing to the PipedOutputStream it is connected to.

public class **PipedInputStream** extends InputStream

Constructor

- public PipedInputStream (PipedOutputStream src) throws IO Exception

Creates an input file from the specified PiledOutputStream.src—the stream to connect to.

- public PipedInputStream()

Creates an input file that is not connected to anything (yet). It must be connected to a PipedOutputStream before being used.

Methods

Methods	Usage	Use
connect	Public void connect (PipedOutputStream src) throws IO Exception	Connects this input stream to a sender.src—the OutputStream to connect to.
read	Public synchronized int read () throws IO Exception	Reads a byte of data. This method will block if no input is available. Returns the byte read, or –1 if the end of the stream is reached.

read	Public synchronized int read (byte b[], int off, int len) throws IOException	Reads into an array of bytes. Blocks until some input is available. b—the buffer into which the data is read, off—the start offset of the data, len—the maximum number of bytes read. Returns the actual number of bytes read, –1 is returned when the end of the stream is reached.
close	Public void close () throws IOException	Closes the input stream. Must be called to release any resource associated with the stream.

1.16 Piped Output Stream

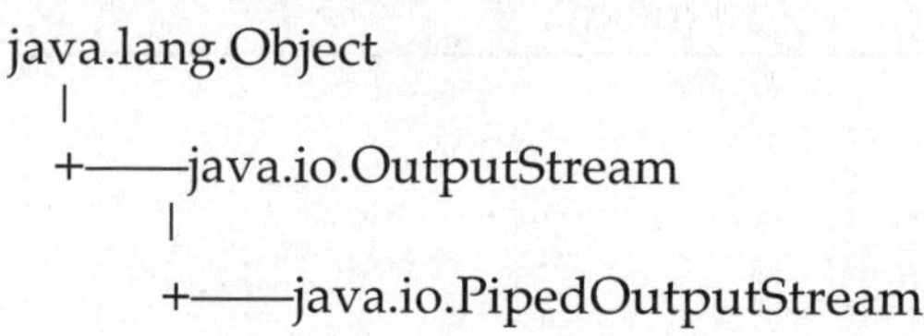

Piped output stream, must be connected to a PipedInputStream. A thread reading from a PipedInputStream receives data from a thread writing to the PipedOutputStream it is connected to.

public class **PipedOutputStream** extends OutputStream

Constructors

- public PipedOutputStream (PipedInputStream snk) throws IOException

Creates an output file connected to the specified PipedInputStream.snk—The InputStream to connect to.

- public PipedOutputStream()

Creates an output file that is not connected to anything (yet). It must be connected before being used.

Methods

Method	Usage	Use
connect	Public void connect (PipedInputStream snk) throws IOException	Connect this output stream to a receiver.snk—The InputStream to connect to.
write	Public void write (int b) throws IOException	Write a byte. This method will block until the byte is actually written. b—the byte to be written.
write	Public void write (byte b[], int off, int len) throws IOException	Writes a sub array of bytes. b—the data to be written, off—the start offset in the data, len—the number of bytes that are written.

close	Public void close () throws IOException	Closes the stream. This method must be called to release any resources associated with the stream.

1.17 Print Stream

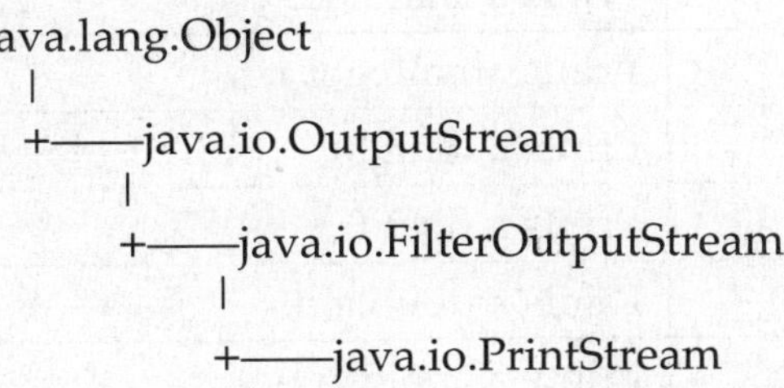

This class implements an output stream that has additional methods for printing. You can specify that the stream should be flushed every time a newline character is written.

The top byte of 16 bit characters is discarded.

public class **PrintStream** extends FilterOutputStream

Example

```
System.out.println("Hello world!");
System.out.print("x = ");
System.out.println(x);
System.out.println("y = " + y);
```

Constructor

public PrintStream(OutputStream out)

Creates a new PrintStream.out—the output stream.

public PrintStream(OutputStream out, boolean autoflush)

Creates a new PrintStream, with auto flushing.out—the output stream, autoflush—if true the stream automatically flushes its output when a newline character is printed.

Methods

Methods	Usage	Use
write	Public void write (int b)	Writes a byte. This method will block until the byte is actually written. b—the byte.
write	Public void write (byte b[], int off, int len)	Writes a sub array of bytes. b-the data to be written, off—the start offset in the data, len—the number of bytes that are written.
flush	Public void flush ()	Flushes the stream. This will write any buffered output bytes.
close	Public void close ()	Closes the stream.

print	Public void print (Object obj)	Prints an object obj.
print	Public synchronized void print (String s)	Prints a String s.
print	Public synchronized void print (char s[])	Prints an array of characters s.
print	Public void print (char c)	Prints a character c.
print	Public void print (int i)	Prints an integer i.
print	Public void print (long l)	Prints a long l.
print	Public void print (float f)	Prints a float f.
print	Public void print (double d)	Prints a double d.
print	Public void print (boolean b)	Prints a boolean b.
println	Public void println ()	Prints a newline.
println	Public synchronized void println (Object obj)	Prints an object followed by a newline obj.
println	Public synchronized void println (String s)	Prints a string s followed by a newline.
println	Public synchronized void println (char s[])	Prints an array of characters s followed by a newline.
println	Public synchronized void println (char c)	Prints a character c followed by a newline.
println	Public synchronized void println (int i)	Prints an integer followed by a newline.
println	Public synchronized void println (long l)	Prints a long l followed by a newline.
println	Public synchronized void println (float f)	Prints a float f followed by a newline.
println	Public synchronized void println (double d)	Prints a double d followed by a newline.
println	Public synchronized void println (boolean b)	Prints a boolean b followed by a newline.

1.18 Push Back Input Stream

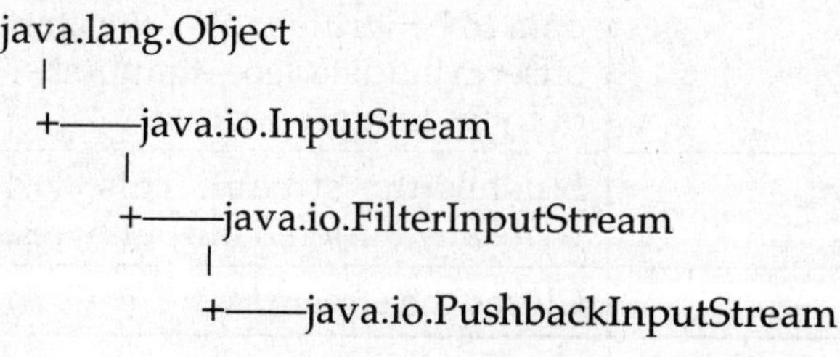

Class java.io.pushbackinputstream is an input stream that has a 1 byte push back buffer.

public class **PushbackInputStream** extends FilterInputStream

Constructor

- public PushbackInputStream(InputStream in)

Creates a PushbackInputStream.in—the input stream.

Methods

Methods	Usage	Use
read	Public int read() throws IOException	Reads a byte of data. This method will block if no input is available. Returns the byte read, or –1 if the end of the stream is reached.
read	Public int read (byte bytes[], int offset, int length) throws IOException	Reads into an array of bytes. This method blocks until some input is available. b—the buffer into which the data is read, off—the start offset of the data, len—the maximum number of bytes read. Returns the actual number of bytes read, –1 is returned when the end of the stream is reached.
unread	Public void unread (int ch) throws IOException	Pushes back a character. ch—the character to push back.
available	Public int available () throws IOException	Returns the number of bytes that can be read without blocking.
mark Supported	Public boolean mark Supported ()	Returns true if this stream type supports mark/reset.

1.19 Random Access File

```
java.lang.Object
   |
   +——java.io.RandomAccessFile
```

Random access files can be constructed from file descriptors, file names, or file objects. This class provides a sense of security by offering methods that allow specified mode accesses of read-only or read-write to files.

public class **RandomAccessFile** extends Object implements DataOutput, DataInput

Constructors

- public RandomAccessFile(String name, String mode) throws IO Exception

Creates a RandomAccessFile with the specified system dependent file name and the specified mode. Mode “r” is for read-only and mode “rw” is for read + write. name—the system dependent file name, mode—the access mode.

• Public RandomAccessFile(int fd) throws IOException

Creates a RandomAccessFile with the specified system dependent file descriptor. fd—the system dependent file descriptor.

• Public RandomAccessFile(File file, String mode) throws IOException

Creates a RandomAccessFile from a specified File object and mode ("r" or "rw"). file—the file object, mode—the access mode.

Methods

Methods	Usage	Use
read	Public int read() throws IOException	Reads a byte of data. This method will block if no input is available. Returns the byte read, or –1 if the end of the stream is reached.
read	Public int read (byte b[], int off, int len) throws IOException	Reads a sub array as a sequence of bytes. b—the data to be written, off—the start offset in the data, len—the number of bytes that are written.
read	Public int read (byte b[]) throws IOException	Reads data into an array of bytes. This method blocks until some input is available. Returns the actual number of bytes read, –1 is returned when the end of the stream is reached.
readFully	Public final void readFully (byte b[]) throws IOException	Reads bytes, blocking until all bytes are read. b—the buffer into which the data is read. Returns the actual number of bytes read, –1 is returned when the end of the stream is reached.
readFully	Public final void readFully (byte b[], int off, int len) throws IOException	Reads bytes, blocking until all bytes are read. b—the buffer into which the data is read, off—the start offset of the data, len—the maximum number of bytes read. Returns the actual number of bytes read, –1 is returned when the end of the stream is reached.
skipBytes	Public int skipBytes (int n) throws IOException	
write	Public void write(int b) throws IOException	Writes a byte of data. This method will block until the byte is actually written. b—the byte to be written.

write	Public void write(byte b[]) throws IOException	Writes an array of bytes. Will block until the bytes are actually written. b—the data to be written.
write	Public void write(byte b[], int off, int len) throws IOException	Writes a sub array of bytes. b—the data to be written, off—the start offset in the data, len—the number of bytes that are written
getFile Pointer	Public long getFilePointer() throws IOException	Returns the current location of the file pointer.
seek	Public void seek(long pos) throws IOException	Sets the file pointer to the specified absolute position.pos—the absolute position.
length	Public long length() throws IOException	Returns the length of the file.
close	Public void close() throws IOException	Closes the file.
read Boolean	Public final boolean readBoolean() throws IOException	Reads a boolean.
readByte	Public final byte readByte() throws IOException	Reads a byte.
read Unsigned Byte()	Public final int readUnsigned Byte() throws IOException	Reads an unsigned 8 bit byte. Returns the 8 bit byte read.
readShort	Public final short readShort() throws IOException	Reads 16 bit short. Returns the read 16 bit short.
read Unsigned Short	Public final int read UnsignedShort() throws IOException	Reads 16 bit short. Returns the read 16 bit short.
readChar	Public final char readChar() throws IOException	Reads a 16 bit char. Returns the read 16 bit char.
readInt	Public final int readInt() throws IOException	Reads a 32 bit int. Returns the read 32 bit integer.
readLong	Public final long readLong() throws IOException	Reads a 64 bit long. Returns the read 64 bit long.
readFloat	Public final float readFloat() throws IOException	Reads a 32 bit float. Returns the read 32 bit float.
readDouble	Public final double read Double() throws IOException	Reads a 64 bit double. Returns the read 64 bit double.
readLine	Public final String readLine() throws IOException	Reads a line terminated by a '\n' or EOF.

readUTF	Public final String readUTF() throws IOException	Reads a UTF formatted String.
writeBoolean	Public final void writeBoolean (boolean v) throws IOException	Writes a boolean. v—the boolean value.
writeByte	Public final void writeByte (int v) throws IOException	Writes a byte. v—the byte.
writeShort	Public final void writeShort (int v) throws IOException	Writes a short. v—the short.
writeChar	Public final void writeChar (int v) throws IOException	Writes a character. v—the char.
writeInt	Public final void writeInt (int v) throws IOException	Writes an integer. v—the integer.
writeLong	Public final void writeLong (long v) throws IOException	Writes a long. v—the long.
writeFloat	Public final void writeFloat (float v) throws IOException	
writeDouble	Public final void writeDouble (double v) throws IOException	
writeBytes	Public final void writeBytes (String s) throws IOException	Writes a String as a sequence of bytes. s—the String.
writeChars	Public final void writeChars (String s) throws IOException	Writes a String as a sequence of chars. s—the String.
writeUTF	Public final void writeUTF (String str) throws IOException	Writes a String in UTF format. str—the String.

1.20 Sequence Input Stream

```
java.lang.Object
  |
  +——java.io.InputStream
        |
        +——java.io.SequenceInputStream
```

Class java.io.SequenceInputStream converts a sequence of input streams into an InputStream.

Its usage in general is:

public class **SequenceInputStream** extends InputStream

Constructors

- public SequenceInputStream(Enumeration e)

Constructs a new SequenceInputStream initialized to the specified list.

- public SequenceInputStream(InputStream s1, InputStream s2)

Constructs a new SequenceInputStream initialized to the two specified input streams. s1—the first input stream, s2—the second input stream.

Methods

Methods	Usage	Use
read	Public int read() throws IOException	Reads a stream, and upon reaching an EOF, flips to the next stream.
read	Public int read(byte buf[], int pos, int len) throws IOException	Reads data into an array of bytes, and upon reaching an EOF, flips to the next stream. buf—the buffer into which the data is read, pos—the start position of the data, len—the maximum number of bytes read.
close	Public void close() throws IOException	Closes the input stream; flipping to the next stream, if an EOF is reached. This method must be called to release any resource associated with the stream.

1.21 Stream Tokenizer

java.lang.Object
 |
 +——java.io.StreamTokenizer

A class to turn an input stream into a stream of tokens. There are a number of methods that define the lexical syntax of tokens.

public class **StreamTokenizer** extends Object

Constructors

- public StreamTokenizer(InputStream I)

Creates a stream tokenizer that parses the specified input stream. By default, it recognizes numbers, Strings quoted with single and double quotes, and all the alphabetics. I-the input stream.

Methods

Methods	Usage	Use
resetSyntax	Public void resetSyntax()	Resets the syntax table so that all characters are special.
wordChars	Public void wordChars (int low, int hi)	Specifies that characters in this range are word characters. low—the low end of the range, hi—the high end of the range.
whitespace Chars	Public void whitespace Chars(int low, int hi)	Specifies that characters in this range are whitespace characters. low—the low end of the range, hi—the high end of the range.
ordinary Chars	Public void ordinaryChar s (int low, int hi)	Specifies that characters in this range are 'ordinary'. Ordinary characters mean that any significance as words, comments, strings, whitespaces or number characters are removed. When these characters are encountered by the parser, they return a ttype equal to the character. low—the low end of the range, hi—the high end of the range.
ordinaryChar	Public void ordinary Char(int ch)	Specifies that this character is 'ordinary': It removes any significance as a word, comment, string, whitespace or number character. When encountered by the parser, it returns a ttype equal to the character. ch—the character.
commentChar	Public void comment Char(int ch)	Specifies that this character starts a single line comment. ch—the character.
quoteChar	Public void quoteChar (int ch)	Specifies that matching pairs of this character delimit String constants. When a String constant is recognized, ttype will be the character that delimits the String, and sval will have the body of the String. ch—the character.
parseNumbers	Public void parseNumbers()	Specifies that numbers should be parsed. This method accepts double precision floating point numbers and returns a ttype of TT_NUMBER with the value in nval.

eolIsSignificant	Public void eolIsSignificant (boolean flag)	If the flag is true, end-of-lines are significant (TT_EOL will be returned by next token). If false, they will be treated as white-space.
slashStar Comments	Public void slashStar Comments(boolean flag)	If the flag is true, recognize C style(/*) comments.
slashSlash Comments	Public void slashSlash Comments(boolean flag)	If the flag is true, recognize C++ style(//) comments.
lowerCase Mode	Public void lowerCaseMode (boolean fl)	Examines a boolean to decide whether TT_WORD tokens are forced to be lower case.fl—the boolean flag
nextToken	Public int nextToken() throws IOException	Parses a token from the input stream. The return value is the same as the value of ttype. Typical clients of this class first set up the syntax tables and then sit in a loop calling nextToken to parse successive tokens until TT_EOF is returned.
pushBack	public void pushBack()	Pushes back a stream token.
lineno	public int lineno()	Return the current line number.
toString	public String toString()	Returns the String representation of the stream token.

1.22 String Buffer Input Stream

```
java.lang.Object
   |
   +——java.io.InputStream
         |
         +——java.io.StringBufferInputStream
```

- Class java.io.StringBufferInputStream implements a String buffer that can be used as an InputStream.

public class StringBufferInputStream extends InputStream

Constructors

- public StringBufferInputStream(String s)

Creates an StringBufferInputStream from the specified array of bytes. s—the input buffer (not copied).

Methods

Methods	Usage	Use
read	Public synchronized int read()	Reads a byte of data. Returns the byte read, or –1 if the end of the stream is reached.
read	Public synchronized int read (byte b[], int off, int len)	Reads into an array of bytes. b—the buffer into which the data is read, off—the start offset of the data, len—the maximum number of bytes read. Returns the actual number of bytes read; –1 is returned when the end of the stream is reached.
skip	Public synchronized long skip (long n)	Skips n bytes of input. n—the number bytes to be skipped. Returns the actual number of bytes skipped.
available	Public synchronized int available()	Returns the number of available bytes in the buffer.
reset	Public synchronized void reset()	Resets the buffer to the beginning.

2. STRING HANDLING

Strings, which are widely used in Java programming, are a sequence of characters. In the Java programming language, strings are objects. The Java platform provides the String class to create and manipulate strings.

Creating Strings

The most direct way to create a string is to write:

String greeting = "Hello world!";

Whenever it encounters a string literal in your code, the compiler creates a String object with its value in this case, "Hello world!'.

As with any other object, you can create String objects by using the new keyword and a constructor. The String class has eleven constructors that allow you to provide the initial value of the string using different sources, such as an array of characters:

```
public class StringDemo{
  public static void main(String args[]){
    char[] helloArray = { 'h', 'e', 'l', 'l', 'o', '.'};
    String helloString = new String(helloArray);
    System.out.println( helloString );
  }
}
```

This would produce the following result:
hello

Note: The String class is immutable, so that once it is created a String object cannot be changed. If there is a necessity to make a lot of modifications to Strings of characters, then you should use String Buffer & String Builder Classes.

String Length

Methods used to obtain information about an object are known as accessor methods. One accessor method that you can use with strings is the length() method, which returns the number of characters contained in the string object.

After the following two lines of code have been executed, len equals 17:

```
import java.io.*;
public class StringDemo{
  public static void main(String args[]){
    String palindrome = "Advanced Java Programming";
    int len = palindrome.length();
    System.out.println( "String Length is : " + len );
  }
}
```

This would produce the following result:

```
D:\ajp_prg>javac StringDemo.java
D:\ajp_prg>java StringDemo
String Length is : 25
```

Concatenating Strings

The String class includes a method for concatenating two strings:

string1.concat(string2);

This returns a new string that is string1 with string2 added to it at the end. You can also use the concat() method with string literals, as in:

"My name is ".concat("Jamal");

Strings are more commonly concatenated with the + operator, as in:

"Hello," + " world" + "!"

which results in:

"Hello, world!"

Let us look at the following example:

```
import java.io.*;
public class StringConcat{
  public static void main(String args[]){
    String string1 = "Java ";
    System.out.println("Advanced " + string1 + "Programming");
  }
}
```

This would produce the following results:

```
D:\ajp_prg>javac StringConcat.java
D:\ajp_prg>java StringConcat
Advanced Java Programming
```

Creating Format Strings

You have printf() and format() methods to print output with formatted numbers. The String class has an equivalent class method, format(), that returns a String object rather than a PrintStream object.

Using String's static format() method allows you to create a formatted string that you can reuse, as opposed to a one-time print statement. For example, instead of:

```
System.out.printf("The value of the float variable is " +
              "%f, while the value of the integer " +
              "variable is %d, and the string " +
              "is %s", floatVar, intVar, stringVar);
```

you can write:

```
String fs;
fs = String.format("The value of the float variable is " +
              "%f, while the value of the integer " +
              "variable is %d, and the string " +
              "is %s", floatVar, intVar, stringVar);
System.out.println(fs);
```

String Methods

Here is the list of methods supported by String class:

Char at() Method

This method returns the character located at the String's specified index. The string indexes start from zero.

a. Syntax

Here is the syntax of this method:

public char charAt(int index)

where, index of the character to be returned, and returns a char at the specified index.

b. Example

```
import java.io.*;
public class CharAtMethod{
  public static void main(String args[]){
    String s = "Strings Methods Demonstration";
    char result = s.charAt(8);
    System.out.println(result);
  }
}
```

This produces the following results:

```
D:\ajp_prg>javac CharAtMethod.java
D:\ajp_prg>java CharAtMethod
M
```

Compare to() Method

There are two variant of this method. The first method compares this String to another Object and the second method compares two strings lexicographically.

a. Syntax

Here is the syntax of this method:

```
int compareTo(Object o)
```

or

```
int compareTo(String anotherString)
```

where, Object o is the Object to be compared, another string is the string to be compared.

Returns the value 0 if the argument is a string lexicographically equal to this string; a value less than 0 if the argument is a string lexicographically greater than this string; and a value greater than 0 if the argument is a string lexicographically less than this string.

b. Example

```
import java.io.*;
public class CompareTo{
  public static void main(String args[]){
    String str1 = "STRING METHOD DEMONSTRATION";
    String str2 = "String Method Demonstration";
    String str3 = "Integers Method Demonstration";

    int result = str1.compareTo( str2 );
    System.out.println(result);

    result = str2.compareTo( str3 );
    System.out.println(result);

    result = str3.compareTo( str1 );
    System.out.println(result);
  }
}
```

This produces the following result:

```
D:\ajp_prg>javac CompareTo.java
D:\ajp_prg>java CompareTo
-32
10
-10
```

Compare to Ignore Case() Method

This method compares two strings lexicographically, ignoring case differences.

a. Syntax

Here is the syntax of this method:

int compareToIgnoreCase(String str)

where, str is the String to be compared.

Returns a negative integer, zero, or a positive integer as the the specified String is greater than, equal to, or less than this String, ignoring case considerations.

Example

```
import java.io.*;
public class CompareToIgnoreCase{
  public static void main(String args[]){
    String str1 = "STRING METHOD DEMONSTRATION";
    String str2 = "String Method Demonstration";
    String str3 = "Integers Method Demonstration";

    int result = str1.compareToIgnoreCase( str2 );
    System.out.println(result);

    result = str2.compareToIgnoreCase( str3 );
    System.out.println(result);

    result = str3.compareToIgnoreCase( str1 );
    System.out.println(result);
  }
}
```

This produces the following result:

```
    D:\ajp_prg>javac CompareToIgnoreCase.java
    D:\ajp_prg>java CompareToIgnoreCase
0
10
–10
```

Concat() Method

This method appends one String to the end of another. The method returns a String with the value of the String passed into the method appended to the end of the String used to invoke this method.

a. Syntax

public String concat(String s)

where s is the String that is concatenated to the end of this String.

It returns a string that represents the concatenation of this object's characters followed by the string argument's characters.

b. Example

```
public class Test{
  public static void main(String args[]){
    String s = "Strings are immutable";
```

```
        s = s.concat(" all the time");
        System.out.println(s);
    }
}
```

This produces the following result:
Strings are immutable all the time

Content Equals() Method

This method returns true if and only if this String represents the same sequence of characters as the specified StringBuffer.

a. Syntax

public boolean contentEquals(StringBuffer sb)

where, sb is the StringBuffer to compare to.

This method returns true if and only if this String represents the same sequence of characters as the specified StringBuffer, otherwise false.

b. Example

```
import java.io.*;
public class ContentEquals{
  public static void main(String args[]){
    String str1 = "Java Programming";
    String str2 = "Advanced Java Programming";
    StringBuffer str3 = new StringBuffer( "Java Programming");

    boolean  result = str1.contentEquals( str3 );
    System.out.println(result);

    result = str2.contentEquals( str3 );
    System.out.println(result);
  }
}
```

This produces the following result:

```
D:\ajp_prg>javac ContentEquals.java
D:\ajp_prg>java ContentEquals
true
false
```

Copy Value of() Method

This method has two different forms:

- **Public static String copyValueOf(char[] data):** Returns a String that represents the character sequence in the array specified.
- **Public static String copyValueOf(char[] data, int offset, int count):** Returns a String that represents the character sequence in the array specified.

a. Syntax

public static String copyValueOf(char[] data)

or

public static String copyValueOf(char[] data, int offset, int count)

where, data: The character array, offset: Initial offset of the subarray, count : length of the subarray. This method returns a String that contains the characters of the character array.

Example

```
import java.io.*;
public class CopyValueOf{
  public static void main(String args[]){
    char[] Str1 = "Java Programming!!";
    String Str2;
    Str2 = copyValueOf(Str1);
    System.out.println("Returned String " + Str2);
    Str2 = copyValueOf(Str1, 5, 10);
    System.out.println("Returned String " + Str2);
  }
}
```

This produces the following results:

```
D:\ajp_prg>javac CopyValueOf.java
D:\ajp_prg>java CopyValueOf
Java Programming!!
Progra
```

Ends with() Method

This method tests if this string ends with the specified suffix.

a. Syntax

public boolean endsWith(String suffix)

This method returns true if the character sequence represented by the argument is a suffix of the character sequence represented by this object; false otherwise. Note that the result will be true if the argument is the empty string or is equal to this String object as determined by the equals(Object) method.

b. Example

```
import java.io.*;
public class EndsWith{
  public static void main(String args[]){
    String Str = new String("Advanced Java Programming!!");
    boolean retVal;
    retVal = Str.endsWith( "Programming!!" );
    System.out.println("Returned  Value " + retVal );
    retVal = Str.endsWith( "ava" );
    System.out.println("Returned  Value " + retVal );
```

```
    }
}
```

This produces the following results:

```
D:\ajp_prg>javac EndsWith.java
D:\ajp_prg>java EndsWith

Returned  Value true
Returned  Value false
```

Equals() Method

This method compares this string to the specified object. The result is true if and only if the argument is not null and is a String object that represents the same sequence of characters as this object. Returns true if the String are equal; false otherwise.

a. Syntax

Here is the syntax of this method:
public boolean equals(Object anObject)

b. Example

```
import java.io.*;
public class EqualCheck{
   public static void main(String args[]){
      String Str1 = new String("Java programming!!");
      String Str2 = new String("JAVA PROGRAMMING!!");
      String Str3 = new String("Advanced Java programming!!");
      String Str4=Str2;
      boolean retVal;
System.out.println("Str1:\t"+Str1);
System.out.println("Str2:\t"+Str2);
System.out.println("Str3:\t"+Str3);
System.out.println("Str4:\t"+Str4);
      retVal = Str1.equals( Str2 );
      System.out.println("\nStr1 compared with Str2 " + retVal );
      retVal = Str1.equals( Str3 );
      System.out.println("\nStr1 compared with Str3" + retVal );
retVal = Str2.equals( Str4 );
      System.out.println("\nStr2 compared with Str4" + retVal );
   }
}
```

This produces the following results:

```
D:\ajp_prg>javac EqualCheck.java
D:\ajp_prg>java EqualCheck

Str1:  Java programming!!
Str2:  JAVA PROGRAMMING!!
Str3:  Advanced Java programming!!
```

Str4: JAVA PROGRAMMING!!

Str1 compared with Str2 false
Str1 compared with Str3false
Str2 compared with Str4true

Equals Ignore Case() Method

This method compares this String to another String, ignoring case considerations. Two strings are considered equal ignoring case if they are of the same length, and corresponding characters in the two strings are equal ignoring case. Returns true if the argument is not null and the Strings are equal ignoring case; false otherwise.

a. Syntax

Here is the syntax of this method:

```
public boolean equalsIgnoreCase(String anotherString)
```

b. Example

```
import java.io.*;
public class EqualIgnoreCase{
  public static void main(String args[]){
     String Str1 = new String("Advanced Java Programming!!");
    String Str2 = Str1;
    String Str3 = new String("Advanced Java Programming!!");
    String Str4 = new String("ADVANCED JAVA PROGRAMMING!!");
    boolean retVal;
System.out.println("Str1:\t"+Str1);
System.out.println("Str2:\t"+Str2);
System.out.println("Str3:\t"+Str3);
System.out.println("Str4:\t"+Str4);
    retVal = Str1.equalsIgnoreCase( Str2 );
    System.out.println("\nStr1 compared with Str2 " + retVal );
    retVal = Str1.equalsIgnoreCase( Str3 );
    System.out.println("\nStr1 compared with Str3" + retVal );
retVal = Str2.equalsIgnoreCase( Str4 );
    System.out.println("\nStr2 compared with Str4" + retVal );
  }
}
```

This produces the following results:

```
D:\ajp_prg>java EqualIgnoreCase
Str1: Advanced Java Programming!!
Str2: Advanced Java Programming!!
Str3: Advanced Java Programming!!
Str4: ADVANCED JAVA PROGRAMMING!!

Str1 compared with Str2 true
Str1 compared with Str3 true
Str2 compared with Str4 true
```

Get Bytes() Method

This method has the following two forms:

- **Get Bytes (String charset name):** Encodes this String into a sequence of bytes using the named charset, storing the result into a new byte array.
- **Get Bytes():** Encodes this String into a sequence of bytes using the platform's default charset, storing the result into a new byte array.

a. Syntax

Here is the syntax of this method:

```
public byte[] getBytes(String charsetName) throws UnsupportedEncodingException
```

or

```
public byte[] getBytes()
```

b. Example

```
import java.io.*;

public class GetBytes{
  public static void main(String args[]){
    String Str1 = new String("Advanced Java Programming");
System.out.println("Given string:" +Str1);
    try{
      byte[] Str2 = Str1.getBytes();
      System.out.println("Byte Representation" + Str2 );

      Str2 = Str1.getBytes( "UTF-8" );
      System.out.println("UTF-8 Character set " + Str2 );

      Str2 = Str1.getBytes( "ISO-8859-1" );
      System.out.println("ISO-8859-1 Character set " + Str2 );
    }catch( UnsupportedEncodingException e){
      System.out.println("Unsupported character set");
    }
  }
}
```

This produces the following results:

```
      D:\ajp_prg>javac GetBytes.java
      D:\ajp_prg>java GetBytes
Given string:Advanced Java Programming
      Byte Representation[B@10b62c9
      UTF-8 Character set [B@82ba41
      ISO-8859-1 Character set [B@923e30
```

Get Chars() Method

This method copies characters from this string into the destination character array. It does not return any value but throws IndexOutOfBoundsException.

a. Syntax

```
public void getChars(int srcBegin, int srcEnd, char[] dst, int dstBegin)
```

where, srcBegin : index of the first character in the string to copy, srcEnd : index after the last character in the string to copy, dst : the destination array, dstBegin : the start offset in the destination array.

b. Example

```
import java.io.*;

public class GetChar{
  public static void main(String args[]){
    String Str1 = new String("Welcome to AJP");
    char[] Str2 = new char[7];

    try{
      Str1.getChars(2, 9, Str2, 0);
      System.out.print("Copied Value " );
      System.out.println(Str2 );

    }catch( Exception ex){
      System.out.println("Raised exception...");
    }
  }
}
```

This produces the following results:

```
D:\ajp_prg>javac GetChar.java
D:\ajp_prg>java GetChar

Copied Value lcome t
```

Hash Code() Method

This method returns a hash code for this string. The hash code for a String object is computed as:

s[0]*31^(n-1) + s[1]*31^(n-2) + ... + s[n-1]

Using int arithmetic, where s[i] is the ith character of the string, n is the length of the string, and ^ indicates exponentiation. (The hash value of the empty string is zero). It returns a hash code value for this object.

a. Syntax

Here is the syntax of this method:

public int hashCode()

b. Example

```
import java.io.*;
public class HashCode{
  public static void main(String args[]){
    String Str = new String("Welcome to AJP");
    System.out.println("Hashcode for Str :" + Str.hashCode() );
  }
}
```

This produces the following results:
D:\ajp_prg>javac HashCode.java
D:\ajp_prg>java HashCode
Hashcode for Str :556484800

Index of() Method

This method has the following different variants:

- **Public int index of (int ch):** Returns the index within this string of the first occurrence of the specified character or –1 if the character does not occur.
- **Public int index of (int ch, int fromIndex):** Returns the index within this string of the first occurrence of the specified character, starting the search at the specified index or –1 if the character does not occur.
- **Int index of (String str):** Returns the index within this string of the first occurrence of the specified substring. If it does not occur as a substring, –1 is returned.
- **int indexOf(String str, int fromIndex):** Returns the index within this string of the first occurrence of the specified substring, starting at the specified index. If it does not occur, –1 is returned.

a. Syntax

Public int index of (int ch)
or
Public int index of (int ch, int fromIndex)
or
Int index of (String str)
or
Int index of (String str, int fromIndex)

where, ch is a character, fromIndex is the index to start the search from, str is a string.

b. Example

```
import java.io.*;
public class IndexOf{
  public static void main(String args[]){
    String Str = new String("Welcome to java programming");
    String SubStr1 = new String("java" );
    String SubStr2 = new String("programming" );
    System.out.print("\nStr\t"+Str);
    System.out.print("\nSubStr1\t"+SubStr1);
    System.out.print("\nSubStr2\t"+SubStr2);
    System.out.print("\nStr.indexOf( 'o' ):"+Str.indexOf( 'o' ));
    System.out.print("\nStr.indexOf( 'o', 5 ) :" +Str.indexOf( 'o', 5 ));
    System.out.print("\nStr.indexOf( SubStr1 ) :" + Str.indexOf( SubStr1 ));
    System.out.print("\nStr.indexOf( SubStr1, 3) :" + Str.indexOf( SubStr1, 3));
    System.out.print("\nStr.indexOf( SubStr2 ):" +Str.indexOf( SubStr2 ));
  }
}
```

This produces the following results:

```
      D:\ajp_prg>javac IndexOf.java
      D:\ajp_prg>java IndexOf
Str   Welcome to java programming
```

```
SubStr1 java
SubStr2 programming
Str.indexOf( 'o' ):4
Str.indexOf( 'o', 5 ) :9
Str.indexOf( SubStr1 ) :11
Str.indexOf( SubStr1, 3) :11
Str.indexOf( SubStr2 ):16
```

Intern() Method

This method returns a canonical representation for the string object. It follows that for any two strings s and t, s.intern() == t.intern() is true if and only if s.equals(t) is true. Returns a canonical representation for the string object.

a. Syntax

```
public String intern()
```

b. Example

```
import java.io.*;
public class internCanonical{
  public static void main(String args[]){
    String Str1 = new String("Welcome to Java ");
    String Str2 = new String("WELCOME TO AJP");
    System.out.print("\nStr1"+Str1);
    System.out.print("\nStr2"+Str2);
    System.out.print("\nCanonical representation of Str1:" +Str1.intern());
    System.out.print("\nCanonical representation of Str2:" +Str2.intern());
  }
}
```

This produces the following result:

```
    D:\ajp_prg>javac internCanonical.java
    D:\ajp_prg>java internCanonical
Str1Welcome to Java
Str2WELCOME TO AJP
Canonical representation of Str1:Welcome to Java
Canonical representation of Str2:WELCOME TO AJP
```

Last Index of() Method

This method has the following variants:

- **Int last index of (int ch):** Returns the index within this string of the last occurrence of the specified character or –1 if the character does not occur.
- **Public int last index of (int ch, int from index):** Returns the index of the last occurrence of the character in the character sequence represented by this object that is less than or equal to from Index, or –1 if the character does not occur before that point.
- **Public int last index of (String str):** If the string argument occurs one or more times as a substring within this object, then it returns the index of the first character of the last such substring is returned. If it does not occur as a substring, –1 is returned.

- **Public int last index of (String str, int from index):** Returns the index within this string of the last occurrence of the specified substring, searching backward starting at the specified index.

a. Syntax:

int lastIndexOf(int ch)
or
public int lastIndexOf(int ch, int fromIndex)
or
public int lastIndexOf(String str)
or
public int lastIndexOf(String str, int fromIndex)

b. Example

```
import java.io.*;
public class LastIndexOf{
  public static void main(String args[]){
    String Str = new String("Welcome Java Programming");
    String SubStr1 = new String("Java" );
    String SubStr2 = new String("Programming" );
    System.out.print("\nString 1:"+Str);
    System.out.print("\nSubstring 1:"+SubStr1);
    System.out.print("\nSubstring 1:"+SubStr1);
    System.out.print("\nStr.lastIndexOf( 'o' ):"+Str.lastIndexOf( 'o' ));
    System.out.print("\nStr.lastIndexOf( 'o', 5 ) :" +Str.lastIndexOf( 'o', 5 ));
    System.out.print("\nStr.lastIndexOf( SubStr1 ):" +Str.lastIndexOf( SubStr1 ));
    System.out.print("\nStr.lastIndexOf( SubStr1, 15 ) :" +Str.lastIndexOf( SubStr1, 15 ));
    System.out.print("\nStr.lastIndexOf( SubStr2 ):" +Str.lastIndexOf( SubStr2 ));
  }
}
```

This produces the following results:

```
D:\ajp_prg>javac LastindexOf.java
D:\ajp_prg>java LastIndexOf
String 1:Welcome Java Programming
Substring 1:Java
Substring 1:Java
Str.lastIndexOf( 'o' ):15
Str.lastIndexOf( 'o', 5 ) :4
Str.lastIndexOf( SubStr1 ):8
Str.lastIndexOf( SubStr1, 15 ) :8
Str.lastIndexOf( SubStr2 ):13
```

Length() Method

This method returns the length of this string. The length is equal to the number of 16-bit Unicode characters in the string.

a. Syntax

Here is the syntax of this method:

public int length()

Returns the length of the sequence of characters represented by this object.

b. Example

```
import java.io.*;
public class StringLength{
  public static void main(String args[]){
    String Str1 = new String("Welcome to Advanced java Programming");
    String Str2 = new String("BPL" );
    System.out.print("\nString 1:"+Str1);
    System.out.print("\nString 2;"+Str2);
    System.out.print("\nString Length Str1:" +Str1.length());
    System.out.print("\nString Length Str2:" +Str2.length());
    }
}
```

This produces the following results:

```
D:\ajp_prg>javac StringLength.java
D:\ajp_prg>java StringLength
String 1:Welcome to Advanced java Programming
String 2;BPL
String Length Str1:36
String Length Str2:3
```

Matches() Method

This method tells whether or not this string matches the given regular expression. An invocation of this method of the form str.matches(regex) yields exactly the same result as the expression Pattern.matches(regex, str).

a. Syntax

```
public boolean matches(String regex)
```

where, regex is the regular expression to which this string is to be matched.

It returns true if, and only if, this string matches the given regular expression.

b. Example

```
import java.io.*;
public class StringMatches{
  public static void main(String args[]){
    String Str = new String("Welcome to Advanced Java Programming");
    System.out.print("Return Value :" );
    System.out.println(Str.matches("(.*)Programming(.*)"));
    System.out.print("Return Value :" );
    System.out.println(Str.matches("Programming"));
    System.out.print("Return Value :" );
    System.out.println(Str.matches("Welcome(.*)"));
  }
}
```

This produces the following results:

```
D:\ajp_prg>javac StringMatches.java
D:\ajp_prg>java StringMatches
```

Return Value :true
Return Value : false
Return Value : true

Region Matches() Method

This method has two variants which can be used to test if two string regions are equal.

a. Syntax

```
public boolean regionMatches(int toffset,  String other, int ooffset,int len)
```

or

```
public boolean regionMatches(boolean ignoreCase, int toffset,
                             String other, int ooffset,int len)
```

where, toffset is the starting offset of the subregion in this string, other is the string argument, ooffset is the starting offset of the subregion in the string argument, len is the number of characters to compare, ignoreCase is if true, ignore case when comparing characters.

This method returns true if the specified subregion of this string matches the specified subregion of the string argument; false otherwise. Whether the matching is exact or case insensitive depends on the ignoreCase argument.

b. Example

```
import java.io.*;
public class RegionMatches{
  public static void main(String args[]){
    String Str1 = new String("Welcome to Advanced Java Programming");
    String Str2 = new String("Programming");
    String Str3 = new String("PROGRAMMING");
System.out.print("\nString 1:"+Str1);
System.out.print("\nString 2:"+Str2);
System.out.print("\nString 3:"+Str3);
 System.out.print("\nStr1.regionMatches(11, Str2, 0, 9) :"+Str1.regionMatches(11,
Str2, 0, 9));
 System.out.print("\nStr1.regionMatches(11, Str3, 0, 9):"+Str1.regionMatches(11,
Str3, 0, 9));
System.out.print("\nStr1.regionMatches(true, 11, Str3, 0, 9):"+Str1.regionMatches
(true, 11, Str3, 0, 9));
  }
}
```

This produces the following results:

```
    D:\ajp_prg>javac RegionMatches.java
    D:\ajp_prg>java RegionMatches
String 1:Welcome to Advanced Java Programming
String 2:Programming
String 3:PROGRAMMING
Str1.regionMatches(11, Str2, 0, 9) :false
Str1.regionMatches(11, Str3, 0, 9):false
Str1.regionMatches(true, 11, Str3, 0, 9):false
```

Replace() Method

This method returns a new string resulting from replacing all occurrences of oldChar in this string with newChar.

a. Syntax

public String replace(char oldChar, char newChar)

It returns a string derived from this string by replacing every occurrence of oldChar with newChar.

b. Example

```
import java.io.*;
public class StringReplace{
  public static void main(String args[]){
    String Str = new String("Welcome to Advanced Java Programming");
    System.out.println("Given String\n:"+Str);
    System.out.println("Str.replace('o', 'B'):"+Str.replace('o', 'B'));
    System.out.println("Str.replace('a', 'P'):"+Str.replace('a', 'P'));
    System.out.println("Str.replace('g', 'L'):"+Str.replace('g', 'L'));
  }
}
```

This produces the following results:

```
    D:\ajp_prg>javac StringReplace.java
    D:\ajp_prg>java StringReplace
Given String
:Welcome to Advanced Java Programming
Str.replace('o', 'B'):WelcBme tB Advanced Java PrBgramming
Str.replace('a', 'P'):Welcome to AdvPnced JPvP ProgrPmming
Str.replace('g', 'L'):Welcome to Advanced Java ProLramminL
```

Replace All()

This method replaces each substring of this string that matches the given regular expression with the given replacement.

a. Syntax

public String replaceAll(String regex, String replacement)

where, regex is the regular expression to which this string is to be matched, replacement is the string which would replace found expression.

b. Example

```
import java.io.*;
public class ReplaceAll{
  public static void main(String args[]){
    String Str = new String("Welcome to Advanced Java Programming");
    System.out.println("Given String:"+Str);
      System.out.println("Str.replaceAll:"+Str.replaceAll("(.*)Programming(.*)",
"AMROOD" ));
  }
}
```

This produces the following results:

```
D:\ajp_prg>javac ReplaceAll.java
D:\ajp_prg>java ReplaceAll
Given String:Welcome to Advanced Java Programming
Str.replaceAll:AMROOD
```

Replace First() Method

This method replaces the first substring of this string that matches the given regular expression with the given replacement.

a. Syntax

```
public String replaceFirst(String regex, String replacement)
```

where, regex is the regular expression to which this string is to be matched, replacement is the string which would replace found expression.

b. Example

```
import java.io.*;
public class ReplaceFirst{
  public static void main(String args[]){
    String Str = new String("Welcome to Advanced Java Programming");
    System.out.println("Given String:"+Str);
  System.out.println("replaceFirst with Regular expression :"+Str.replaceFirst
("(.*)Java(.*)", "BPL" ));
    System.out.print("Simple replaceFirst :" +Str.replaceFirst("Java", "BPL" ));
  }
}
```

This produces the following results:

```
D:\ajp_prg>javac ReplaceFirst.java
D:\ajp_prg>java ReplaceFirst
Given String:Welcome to Advanced Java Programming
replaceFirst with Regular expression :BPL
Simple replaceFirst :Welcome to Advanced BPL Programming
```

Split() Method

This method has two variants and splits this string around matches of the given regular expression.

a. Syntax

Here is the syntax of this method:

```
public String[] split(String regex, int limit)
```

or

```
public String[] split(String regex)
```

where, regex is the delimiting regular expression, limit is the result threshold which means how many strings to be returned.

The method returns the array of strings computed by splitting this string around matches of the given regular expression.

b. Example

```
import java.io.*;
public class StringSplit{
  public static void main(String args[]){
   String Str = new String("Welcome-Advanced-Java-Programming");
   System.out.println("Str.split(-, 2) :" );
    for (String retval: Str.split("-", 2)){
      System.out.println(retval);
    }
  System.out.println("***");
    System.out.println("");
    System.out.println("Str.split(-, 3) :" );
    for (String retval: Str.split("-", 3)){
      System.out.println(retval);
    }
  System.out.println("***");
    System.out.println("");
    System.out.println("Str.split(-, 0) :" );
    for (String retval: Str.split("-", 0)){
      System.out.println(retval);
    }
  System.out.println("***");
    System.out.println("");
    System.out.println("Str.split(-,1) :" );
    for (String retval: Str.split("-",1)){
      System.out.println(retval);
  }
  System.out.println("***");
  }
}
```

This produces the following results:

```
      D:\ajp_prg>javac StringSplit.java
      D:\ajp_prg>java StringSplit
Str.split(–, 2) :
Welcome
Advanced-Java-Programming
***
Str.split(–, 3) :
Welcome
Advanced
Java-Programming
***
Str.split(–, 0) :
Welcome
Advanced
Java
Programming
***
Str.split(–,1) :
Welcome-Advanced-Java-Programming
***
```

Starts with() Method

This method has two variants and tests if a string starts with the specified prefix beginning a specified index or by default at the beginning.

a. Syntax

Here is the syntax of this method:

public boolean startsWith(String prefix, int toffset)

or

public boolean startsWith(String prefix)

where, prefix is the prefix to be matched and toffset is the position to begin looking in the string.

It returns true if the character sequence represented by the argument is a prefix of the character sequence represented by this string; false otherwise.

b. Example

```
import java.io.*;
public class StartsWith{
  public static void main(String args[]){
    String Str = new String("Welcome to Advanced Java Programming");
    System.out.println("Given String:"+Str);
    System.out.println("Str.startsWith(Welcome):"+Str.startsWith("Welcome") );
    System.out.println("Str.startsWith(Java):" +Str.startsWith("Java") );
    System.out.print("Str.startsWith(Advanced, 11):" +Str.startsWith("Advanced", 11) );
  }
}
```

This produces the following results:

```
      D:\ajp_prg>javac StartsWith.java
      D:\ajp_prg>java StartsWith
Given String : Welcome to Advanced Java Programming
Str.startsWith(Welcome):true
Str.startsWith(Java):false
Str.startsWith(Advanced, 11):true
```

Subsequence() Method

This method returns a new character sequence that is a subsequence of this sequence.

a. Syntax

public CharSequence subSequence(int beginIndex, int endIndex)

where, beginIndex is the begin index, inclusive and endIndex is the end index, exclusive.

b. Example

```
import java.io.*;
public class SubSequence{
  public static void main(String args[]){
    String Str = new String("Advanced Java Programming");
```

```
      System.out.println("Given String:"+Str);
      System.out.println("Str.subSequence(0, 10)"+Str.subSequence(0, 10) );
      System.out.println("Str.subSequence(10, 15) "+Str.subSequence(10, 15) );
   }
}
```

This produces the following results:

```
      D:\ajp_prg>javac SubSequence.java
      D:\ajp_prg>java SubSequence
Given String:Advanced Java Programming
Str.subSequence(0, 10)Advanced J
Str.subSequence(10, 15) ava P
```

Substring() Method

This method has two variants and returns a new string that is a substring of this string. The substring begins with the character at the specified index and extends to the end of this string or up to endIndex - 1, if second argument is given.

a. Syntax:

Here is the syntax of this method:

public String substring(int beginIndex)

or

public String substring(int beginIndex, int endIndex)

b. Example

```
import java.io.*;
public class SubString{
   public static void main(String args[]){
      String Str = new String("Advanced Java Programming");
      System.out.println("Given String:"+Str);
      System.out.println("Str.substring(10) :" +Str.substring(10) );
       System.out.println("Str.substring(10, 15) "+Str.substring(10, 15) );
   }
}
```

This produces the following results:

```
      D:\ajp_prg>javac SubString.java
      D:\ajp_prg>java SubString
Given String:Advanced Java Programming
Str.substring(10) :ava Programming
Str.substring(10, 15) ava P
```

To Char Array() Method

This method converts this string to a new character array.

a. Syntax

public char[] toCharArray()

This method returns a newly allocated character array whose length is the length of this string and whose contents are initialized to contain the character sequence represented by this string.

b. Example

```
import java.io.*;
public class ToCharArray{
  public static void main(String args[]){
    String Str = new String("Advanced Java Programming");
    System.out.print("StrtoCharArray :" );
 System.out.println(Str.toCharArray());
  }
}
```

This produces the following results:

```
      D:\ajp_prg>javac ToCharArray.java
      D:\ajp_prg>java ToCharArray
StrtoCharArray :Advanced Java Programming
```

To Lower Case() Method

This method has two variants. The first variant converts all of the characters in this String to lower case using the rules of the given Locale. This is equivalent to calling toLowerCase(Locale.getDefault()). The second variant takes locale as an argument to be used while converting into lower case.

Syntax

Here is the syntax of this method:

```
public String toLowerCase()
```

or

```
public String toLowerCase(Locale locale)
```

The method returns the String, converted to lowercase.

To Upper Case() Method

This method has two variants. The first variant converts all of the characters in this String to upper case using the rules of the given Locale. This is equivalent to calling toUpperCase(Locale.getDefault()). The second variant takes locale as an argument to be used while converting into upper case.

Syntax

Here is the syntax of this method:

```
public String toUpperCase()
```

or

```
public String toUpperCase(Locale locale)
```

The method returns the String, converted to uppercase.

Trim() Method

This method returns a copy of the string, with leading and trailing whitespace omitted.

a. Syntax

Here is the syntax of this method:

```
public String trim()
```

It returns a copy of this string with leading and trailing white space removed, or if it has no leading or trailing white space.

b. Example (toUpper,toLower,trim):

```
import java.io.*;
public class upperlowertrim{
  public static void main(String args[]){
    String Str = new String("advanced java programming");
    String Str1=new String("BPL");
    System.out.println("Str:"+Str);
    System.out.println("Str1:"+Str1);
//str to uppercase
    System.out.println("Str.toUpperCase() :"+Str.toUpperCase() );
//str1 lowercase
    System.out.println("Str1.toLowerCase() :"+Str1.toLowerCase() );
//str trimmed
    System.out.println("Str.trim():"+Str.trim());

  }
}
```

The output appears as below:

```
D:\ajp_prg>javac upperlowertrim.java
D:\ajp_prg>java upperlowertrim
Str:advanced java programming
Str1:BPL
Str.toUpperCase() :ADVANCED JAVA PROGRAMMING
Str1.toLowerCase() :bpl
Str.trim():advanced java programming
```

Value of() Method

This method has following variants which depends on the passed parameters. This method returns the string representation of the passed argument.

- **value of (boolean b):** Returns the string representation of the boolean argument.
- **value of (char c) :** Returns the string representation of the char argument.
- **value of (char[] data) :** Returns the string representation of the char array argument.
- **value of (char[] data, int offset, int count) :** Returns the string representation of a specific subarray of the char array argument.
- **value of (double d) :** Returns the string representation of the double argument.
- **value of (float f) :** Returns the string representation of the float argument.
- **value of (int i) :** Returns the string representation of the int argument.
- **value of (long l) :** Returns the string representation of the long argument.
- **value of (Object obj) :** Returns the string representation of the Object argument.

a. Syntax

Here is the syntax of this method:

- *static String valueOf(boolean b)*
- *static String valueOf(char c)*
- *static String valueOf(char[] data)*
- *static String valueOf(char[] data, int offset, int count)*

- *static String valueOf(double d)*
- *static String valueOf(float f)*
- *static String valueOf(int i)*
- *static String valueOf(long l)*
- *static String valueOf(Object obj)*

b. Example

```
import java.io.*;
public class Test{
  public static void main(String args[]){
    double d = 102939939.939;
    boolean b = true;
    long l = 1232874;
    char[] arr = {'a', 'b', 'c', 'd', 'e', 'f','g' };
    System.out.println("Return Value : " + String.valueOf(d) );
    System.out.println("Return Value : " + String.valueOf(b) );
    System.out.println("Return Value : " + String.valueOf(l) );
    System.out.println("Return Value : " + String.valueOf(arr) );
  }
}
```

This produces the following results:

```
Return Value : 1.02939939939E8
Return Value : true
Return Value : 1232874
Return Value : abcdefg
```

Methods	Usage	Use
charAt	char charAt(int index)	Returns the character at the specified index.
compareTo	int compareTo(Object o)	Compares this String to another Object.
compareTo	int compareTo (String anotherString)	Compares two strings lexicographically.
compareTo IgnoreCase	int compareToIgnore Case(String str)	Compares two strings lexicographically, ignoring case differences.
concat	String concat(String str)	Concatenates the specified string to the end of this string.
content Equals	boolean contentEquals (StringBuffer sb)	Returns true if and only if this String represents the same sequence of characters as the specified StringBuffer.
copyValueOf	static String copyValueOf (char[] data)	Returns a String that represents the character sequence in the array specified.

copyValueOf	static String copyValue Of(char[] data, int offset, int count)	Returns a String that represents the character sequence in the array specified.
endsWith	boolean endsWith (String suffix)	Tests if this string ends with the specified suffix.
equalsIgnore Case	boolean equals (Object anObject)	Compares this string to the specified object.
getBytes	boolean equalsIgnore Case(String anotherString)	Compares this String to another String, ignoring case considera- tions.
getBytes	byte getBytes()	Encodes this String into a sequence of bytes using the platform's default charset, storing the result into a new byte array.
getBytes	byte[] getBytes(String charsetName)	Encodes this String into a sequence of bytes using the named charset, storing the result into a new byte array.
getChars	void getChars(int srcBegin, int srcEnd, char[] dst, int dstBegin)	Copies characters from this string into the destination charac- ter array.
hashCode	int hashCode()	Returns a hash code for this string.
indexOf	int indexOf(int ch)	Returns the index within this string of the first occurrence of the specified character.
indexOf	int indexOf(int ch, int from Index)	Returns the index within this string of the first occurrence of the specified character, starting the search at the specified index.
indexOf	int indexOf(String str)	Returns the index within this string of the first occurrence of the specified substring.
indexOf	int indexOf(String str, int fromIndex)	Returns the index within this string of the first occurrence of the specified substring, starting at the specified index.
intern	String intern()	Returns a canonical represen- tation for the string object.
lastIndexOf	int lastIndexOf(int ch)	Returns the index within this string of the last occurrence of the specified character.

lastIndexOf	int lastIndexOf(int ch, int fromIndex)	Returns the index within this string of the last occurrence of the specified character, searching backward starting at the specified index.
lastIndexOf	int lastIndexOf(String str)	Returns the index within this string of the rightmost occurrence of the specified substring.
lastIndexOf	int lastIndexOf(String str, int fromIndex)	Returns the index within this string of the last occurrence of the specified substring, searching backward starting at the specified index.
length	int length()	Returns the length of this string.
matches	boolean matches(String regex)	Tells whether or not this string matches the given regular expression.
region Matches	boolean regionMatches (boolean ignoreCase, int toffset, String other, int ooffset, int len)	Tests if two string regions are equal.
region Matches	boolean regionMatches(int toffset, String other, int ooffset, int len)	Tests if two string regions are equal.
replace	String replace(char oldChar, char newChar)	Returns a new string resulting from replacing all occurrences of oldChar in this string with newChar.
replaceAll	String replaceAll(String regex, String replacement)	Replaces each substring of this string that matches the given regular expression with the given replacement.
replaceFirst	String replaceFirst(String regex, String replacement)	Replaces the first substring of this string that matches the given regular expression with the given replacement.
split	String[] split(String regex)	Splits this string around matches of the given regular expression.
split	String[] split(String regex, int limit)	Splits this string around matches of the given regular expression.
startsWith	boolean startsWith(String prefix)	Tests if this string starts with the specified prefix.

startsWith	boolean startsWith(String prefix, int toffset)	Tests if this string starts with the specified prefix beginning a specified index.
subSequence	CharSequence subSequence (int beginIndex, int endIndex)	Returns a new character sequence that is a subsequence of this sequence.
substring	String substring(int begin Index)	Returns a new string that is a substring of this string.
substring	String substring(int begin Index, int endIndex)	Returns a new string that is a substring of this string.
toCharArray	char[] toCharArray()	Converts this string to a new character array.
toLowerCase	String toLowerCase()	Converts all of the characters in this String to lower case using the rules of the default locale.
toLowerCase	String toLowerCase (Locale locale)	Converts all of the characters in this String to lower case using the rules of the given Locale.
toString	String toString()	This object (which is already a string!) is itself returned.
toUppercase	String toUpperCase()	Converts all of the characters in this String to upper case using the rules of the default locale.
toUppercase	String toUpperCase (Locale locale)	Converts all of the characters in this String to upper case using the rules of the given Locale.
trim	String trim()	Returns a copy of the string, with leading and trailing whitespace omitted.
valueOf	static String valueOf (primitive data type x)	Returns the string representation of the passed data type argument.

3. JAVA LANGUAGE CLASS

- package java.lang
- INTERFACE java.lang
- Interface java.lang.Runnable

public interface **Runnable** extends Object

This interface is designed to provide a common protocol for Objects that wish to execute code while they are active. For example, Runnable is implemented by class Thread. Being active simply means that a thread has been started and has not yet been stopped.

In addition, Runnable provides the means for a class to be active while not subclassing Thread. A class that implements Runnable can run without subclassing Thread by instantiating a Thread instance and passing itself in as the target. In most cases, the Runnable interface should be used if you are only planning to override the run() method and no other Thread methods. This is important because classes should not be subclassed unless the programmer intends on modifying or enhancing the fundamental behavior of the class.

Method

- run

public abstract void run()

The method that is executed when a Runnable object is activated. The run() method is the "soul" of a Thread. It is in this method that all of the action of a Thread takes place.

- CLASSES- java.lang

Class java.lang.Boolean

```
java.lang.Object
   |
   +——java.lang.Boolean
```

public final class **Boolean** extends Object

The Boolean class provides an object wrapper for Boolean data values, and serves as a place for boolean-oriented operations. A wrapper is useful because most of Java's utility classes require the use of objects. Since booleans are not objects in Java, they need to be "wrapped" in a Boolean instance.

Constructor

- public Boolean(boolean value)

The default constructor, construct a Boolean object initialized to the specified boolean value.

Methods

Methods	Usage	Use
booleanValue	public boolean booleanValue()	Returns the value of this Boolean object as a boolean.
toString	public String toString()	Returns a new String object representing this Boolean's value.
equals	public boolean equals (Object obj)	Compares this object against the specified object.Returns, true if the objects are the same; false otherwise.
getBoolean	public static boolean get Boolean(String name)	Gets a Boolean from the properties.

Class java.lang.Character

java.lang.Object
 |
 +——java.lang.Character

public final class **Character** extends Object

The Character class provides an object wrapper for Character data values and serves as a place for character-oriented operations. A wrapper is useful because most of Java's utility classes require the use of objects. Since characters are not objects in Java, they need to be "wrapped" in a Character instance.

Constructor

- public Character(char value)

The default constructor constructs a Character object with the specified value.

Methods

Methods	Usage	Use
isLowerCase	public static boolean is LowerCase(char ch)	Determines if the specified character is ISO-LATIN-1 lower case. Returns, true if the character is lower case; false otherwise.
isUpperCase	public static boolean is UpperCase(char ch)	Determines if the specified character is ISO-LATIN-1 upper case. Returns, true if the character is upper case; false otherwise.
isDigit	public static boolean is Digit(char ch)	Determines if the specified character is a ISO-LATIN-1 digit. Returns, true if this character is a digit; false otherwise.
isSpace	public static boolean is Space(char ch)	Determines if the specified character is ISO-LATIN-1 white space according to Java. Returns,true if the character is white space; false otherwise.
toLowerCase	public static char toLower Case(char ch)	Returns the lower case character value of the specified ISO LATIN-1 character. Characters that are not upper case letters are returned unmodified.
toUpperCase	public static char toUpper Case(char ch)	Returns the upper case character value of the specified ISO-LATIN-1 character. Characters that are not lower case letters are returned unmodified.

digit	public static int digit (char ch, int radix)	Returns the numeric value of the character digit using the specified radix. If the character is not a valid digit, it returns –1.
forDigit	public static char forDigit(int digit, int radix)	Returns the character value for the specified digit in the specified radix. If the digit is not valid in the radix, the 0 character is returned.
charValue	public char charValue()	Returns the value of this Character object.
toString	public String toString()	Returns a String object representing this character's value.

Class java.lang.Class

```
java.lang.Object
   |
   +——java.lang.Class
```

public final class **Class** extends Object

Class objects contain runtime representations of classes. Every object in the system is an instance of some Class, and for each Class there is one of these descriptor objects. A Class descriptor is not modifiable at runtime.

The following example uses a Class object to print the Class name of an object:

```
void printClassName(Object obj) {
        System.out.println("The class of " + obj + " is " + obj.getClass().getName());
    }
```

Methods

Methods	Usage	Use
forName	public static Class for Name(String className) throws ClassNotFound Exception	Returns the runtime Class descriptor for the specified Class.
newInstance	public Object newInstance() throws InstantiationException, IllegalAccessException	Creates a new instance of this Class.Returns, the new instance of this Class.
getName	public String getName()	Returns the name of this Class.
getSuperclass	public Class getSuperclass()	Returns the superclass of this Class.

getInterfaces	public Class[] getInterfaces()	Returns the interfaces of this Class. An array of length 0 is returned if this Class implements no interfaces.
getClass Loader	public ClassLoader get ClassLoader()	Returns the Class loader of this Class. Returns null if this Class does not have a Class loader.
isInterface	public boolean isInterface()	Returns a boolean indicating whether or not this Class is an interface.
toString	public String toString()	Returns the name of this Class or this interface. The word "class" is prepended if it is a Class; the word "interface" is prepended if it is an interface.

Class java.lang.Number

```
java.lang.Object
|
  +——java.lang.Number
```

public class **Number** extends Object

Number is an abstract superclass for numeric scalar types. Integer, Long, Float and Double are subclasses of Number that bind to a particular numeric representation.

Constructor

Public Number() – default constructor

Methods

Methods	Usage	Use
intValue	Public abstract int intValue()	Returns the value of the number as an int. This may involve rounding if the number is not already an integer.
longValue	Public abstract long longValue()	Returns the value of the number as a long. This may involve rounding if the number is not already a long.
floatValue	Public abstract float floatValue()	Returns the value of the number as a float. This may involve rounding if the number is not already a float.

doubleValue	public abstract double doubleValue()	Returns the value of the number as a double. This may involve rounding if the number is not already a double.

Class java.lang.Double

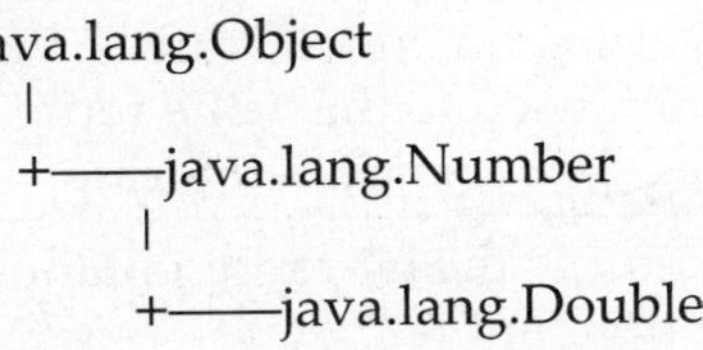

public final class **Double** extends Number

The Double class provides an object wrapper for Double data values and serves as a place for double-oriented operations. A wrapper is useful because most of Java's utility classes require the use of objects. Since doubles are not objects in Java, they need to be "wrapped" in a Double instance.

Constructor

- public Double(double value)

Constructs a Double wrapper for the specified double value.

Methods

Methods	Usage	Use
isNaN	public static boolean is NaN(double v)	Returns true if the specified number is the special Not-a-Number (NaN) value.
isInfinite	public static boolean is Infinite(double v)	Returns true if the specified number is infinitely large in magnitude.
isNaN	public boolean isNaN()	Returns true if this Double value is the special Not-a-Number (NaN) value.
isInfinite	public boolean isInfinite()	Returns true if this Double value is infinitely large in magnitude.
toString	public String toString()	Returns a String representation of this Double object.
intValue	public int intValue()	Returns the integer value of this Double (by casting to an int).
longValue	public long longValue()	Returns the long value of this Double (by casting to a long).
floatValue	public float floatValue()	Returns the float value of this Double.

doubleValue	public double doubleValue()	Returns the double value of this Double.
hashCode	public int hashCode()	Returns a hashcode for this Double.
equals	public boolean equals (Object obj)	Compares this object against the specified object.
valueOf	public static Double valueOf(String s) throws NumberFormatException	Returns a new Double value initialized to the value represented by the specified String.
doubleToLongBits (double value)	public static long doubleTo LongBits	Returns the bit representation of a double-float value
longBitsToDouble	public static double long BitsToDouble(long bits)	Returns the double-float corresponding to a given bit representation.

Throws: NumberFormatException
If the String cannot be parsed.

Class java.lang.Float

```
java.lang.Object
   |
   +——java.lang.Number
        |
        +——java.lang.Float
```

public final class **Float** extends Number

The Float class provides an object wrapper for Float data values, and serves as a place for float-oriented operations. A wrapper is useful because most of Java's utility classes require the use of objects. Since floats are not objects in Java, they need to be "wrapped" in a Float instance.

Constructor

- public Float(float value)

Constructs a Float wrapper for the specified float value.

- public Float(double value)

Constructs a Float wrapper for the specified double value.

Methods

Methods	Usage	Use
toString	public static String to String(float f)	Returns a String representation for the specified float value.
valueOf	public static Float valueOf (String s) throws Number FormatException	Returns the floating point value represented by the specified String.

isNaN	public static boolean is NaN(float v)	Returns true if the specified number is the special Not-a-Number (NaN) value.
isInfinite	public static boolean is Infinite(float v)	Returns true if the specified number is infinitely large in magnitude.
isNaN	public boolean isNaN()	Returns true if this Float value is Not-a-Number (NaN).
isInfinite	public boolean isInfinite()	Returns true if this Float value is infinitely large in magnitude.
toString	public String toString()	Returns a String representation of this Float object.
intValue	public int intValue()	Returns the integer value of this Float (by casting to an int).
longValue	public long longValue()	Returns the long value of this Float (by casting to a long).
floatValue	public float floatValue()	Returns the float value of this Float object.
doubleValue	public double doubleValue()	Returns the double value of this Float.
hashCode	public int hashCode()	Returns a hashcode for this Float.
equals	public boolean equals (Object obj)	Compares this object against some other object.
floatToIntBits	public static int floatToInt Bits(float value)	Returns the bit representation of a single-float value
intBitsToFloat	public static float intBitsTo Float(int bits)	Returns the single-float corres-ponding to a given bit represen-tation.

Throws: Number Format Exception
If the String does not contain a parsable Float.

Class java.lang.Integer

```
java.lang.Object
   |
   +——java.lang.Number
         |
         +——java.lang.Integer
```

public final class **Integer** extends Number

The Integer class is a wrapper for integer values. In Java, integers are not objects and most of the Java utility classes require the use of objects. Thus, if you needed to store an integer in a hashtable, you would have to "wrap" an Integer instance around it.

Constructors

- public Integer(int value)

Constructs an Integer object initialized to the specified int value.

- public Integer(String s) throws NumberFormatException

Constructs an Integer object initialized to the value specified by the String parameter. The radix is assumed to be 10.

Methods

Methods	Usage	Use
toString	public static String toString (int i, int radix)	Returns a new String object representing the specified integer in the specified radix.
toString	public static String toString (int i)	Returns a new String object representing the specified integer. The radix is assumed to be 10.
parseInt	public static int parseInt (String s, int radix) throws NumberFormatException	Assuming the specified String represents an integer, returns that integer's value. Throws an exception if the String cannot be parsed as an int.
parseInt	public static int parseInt (String s) throws Number FormatException	Assuming the specified String represents an integer, returns that integer's value. Throws an exception if the String cannot be parsed as an int. The radix is assumed to be 10.
valueOf	public static Integer value Of(String s, int radix) throws NumberFormatException	Assuming the specified String represents an integer, returns a new Integer object initialized to that value. Throws an exception if the String cannot be parsed as an int.
valueOf	public static Integer valueOf(String s) throws NumberFormatException	Assuming the specified String represents an integer, returns a new Integer object initialized to that value. Throws an exception if the String cannot be parsed as an int. The radix is assumed to be 10.
intValue	public int intValue()	Returns the value of this Integer as an int.
longValue	public long longValue()	Returns the value of this Integer as a long.

floatValue	public float floatValue()	Returns the value of this Integer as a float.
doubleValue	public double doubleValue()	Returns the value of this Integer as a double.
toString	public String toString()	Returns a String object representing this Integer's value.
hashCode	public int hashCode()	Returns a hashcode for this Integer.
equals	public boolean equals (Object obj)	Compares this object to the specified object. **Returns,** true if the objects are the same; false otherwise.
getInteger	public static Integer get Integer(String nm)	Gets an Integer property. If the property does not exist, it will return 0.
getInteger	public static Integer getInteger (String nm, int val)	Gets an Integer property. If the property does not exist, it will return val. Deals with hexadecimal and octal numbers.
getInteger	public static Integer getInteger (String nm, Integer val)	Gets an Integer property. If the property does not exist, it will return val. Deals with hexadecimal and octal numbers.

Throws: NumberFormatException, if the String does not contain a parsable integer.

Class java.lang.Long

```
java.lang.Object
   |
   +——java.lang.Number
          |
          +——java.lang.Long
```

public final class **Long** extends Number

The Long class provides an object wrapper for Long data values and serves as a place for long-oriented operations. A wrapper is useful because most of Java's utility classes require the use of objects. Since longs are not objects in Java, they need to be "wrapped" in a Long instance.

Constructors

- public Long(long value)

Constructs a Long object initialized to the specified value.

- public Long(String s) throws NumberFormatException

Constructs a Long object initialized to the value specified by the String parameter. The radix is assumed to be 10.

Methods

Methods	Usage	Use
toString	public static String toString (long i, int radix)	Returns a new String object representing the specified long in the specified radix.
toString	public static String toString (long i)	Returns a new String object representing the specified integer. The radix is assumed to be 10.
parseLong	public static long parse Long(String s, int radix) throws NumberFormat Exception	Assuming the specified String represents a long, returns that long's value. Throws an exception if the String cannot be parsed as a long.
parseLong	public static long parse Long(String s) throws NumberFormatException	Assuming the specified String represents a long, return that long's value. Throws an exception if the String cannot be parsed as a long. The radix is assumed to be 10.
valueOf	public static Long value Of(String s, int radix) throws NumberFormatException	Assuming the specified String represents a long, returns a new Long object initialized to that value. Throws an exception if the String cannot be parsed as a long.
valueOf	public static Long value Of(String s) throws Number FormatException	Assuming the specified String represents a long, returns a new Long object initialized to that value. Throws an exception if the String cannot be parsed as a long. The radix is assumed to be 10.
intValue	public int intValue()	Returns the value of this Long as an int.
longValue	public long longValue()	Returns the value of this Long as a long.
floatValue	public float floatValue()	Returns the value of this Long as a float.
doubleValue	public double doubleValue()	Returns the value of this Long as a double.
toString	public String toString()	Returns a String object representing this Long's value.
hashCode	public int hashCode()	Computes a hashcode for this Long.

equals	public boolean equals (Object obj)	Compares this object against the specified object.
getLong	public static Long getLong (String nm)	Get a Long property. If the property does not exist, it will return 0.
getLong	public static Long getLong (String nm, long val)	Get a Long property. If the property does not exist, it will return val. Deals with hexadecimal and octal numbers.
getLong	public static Long getLong (String nm, Long val)	Get a Long property. If the property does not exist, it will return val. Deals with hexadecimal and octal numbers.

Throws: Number Format Exception
If the String does not contain a parsable long.

Class java.lang.Math

```
java.lang.Object
   |
   +——java.lang.Math
```

public final class **Math** extends Object

The standard Math library. For the methods in this Class, error handling for out-of-range or immeasurable results are platform dependent. This class cannot be subclassed or instantiated because all methods and variables are static.

Pi is equivalent to 3.14159265358979323846f in Java.

Methods

Methods	Usage	Use
sin	public static double sin (double a)	Returns the trigonometric sine of an angle.
cos	public static double cos (double a)	Returns the trigonometric cosine of an angle.
tan	public static double tan (double a)	Returns the trigonometric tangent of an angle.
asin	public static double asin (double a)	Returns the arc sine of a, in the range of –Pi/2 through Pi/2.
acos	public static double acos (double a)	Returns the arc cosine of a, in the range of 0.0 through Pi.
atan	public static double atan (double a)	Returns the arc tangent of a, in the range of –Pi/2 through Pi/2.

exp	public static double exp (double a)	Returns the exponential number e(2.718...) raised to the power of a.
log	public static double log (double a) throws Arithmetic Exception	Returns the natural logarithm (base e) of a.Where, a is a number greater than 0.0
sqrt	public static double sqrt (double a) throws Arithmetic Exception	Returns the square root of a.
IEEEremainder	public static double IEE Eremainder(double f1, double f2)	Returns the remainder of f1 divided by f2 as defined by IEEE 754.
ceil	public static double ceil (double a)	Returns the "ceiling" or smallest whole number greater than or equal to a.
floor	public static double floor (double a)	Returns the "floor" or largest whole number less than or equal to a.
rint	public static double rint (double a)	Converts a double value into an integral value in double format.
atan2	public static double atan2 (double a, double b)	Converts rectangular coordinates (a, b) to polar (r, theta). This method computes the phase theta by computing an arc tangent of b/a in the range of –Pi to Pi. Returns, the polar coordinates (r, theta).
pow	public static double pow (double a, double b) throws ArithmeticException	Returns the number a raised to the power of b.
round	public static int round(float a)	Rounds off a float value by first adding 0.5 to it and then returning the largest integer that is less than or equal to this new value.
round	public static long round (double a)	Rounds off a double value by first adding 0.5 to it and then returning the largest integer that is less than or equal to this new value.
random	public static synchronized double random()	Generates a random number between 0.0 and 1.0.

abs	public static int abs(int a)	Returns the absolute integer value of a.
abs	public static long abs(long a)	Returns the absolute long value of a.
abs	public static float abs(float a)	Returns the absolute float value of a.
abs	public static double abs (double a)	Returns the absolute double value of a.
max	public static int max(int a, int b)	Takes two int values, a and b, and returns the greater number of the two.
max	public static long max (long a, long b)	Takes two long values, a and b, and returns the greater number of the two.
max	public static float max(float a, float b)	Takes two float values, a and b, and returns the greater number of the two.
max	public static double max (double a, double b)	Takes two double values, a and b, and returns the greater number of the two.
min	public static int min(int a, int b)	Takes two integer values, a and b, and returns the smallest number of the two.
min	public static long min(long a, long b)	Takes two long values, a and b, and returns the smallest number of the two.
min	public static float min(float a, float b)	Takes two float values, a and b, and returns the smallest number of the two.
min	public static double min (double a, double b)	Takes two double values, a and b, and returns the smallest number of the two.

Throws: Arithmetic Exception, If a is less than or equal to 0.0.

Class java.lang.Object

The root of the Class hierarchy. Every Class in the system has Object as its ultimate parent. Every variable and method defined here is available in every Object.

public class **Object**

Constructor

- public Object() – Default Constructor

Methods

Methods	Usage	Use
getClass	public final Class getClass()	Returns the Class of this Object.
hashCode	public int hashCode()	Returns a hashcode for this Object.
equals	public boolean equals (Object obj)	Compares two Objects for equality. Returns a boolean that indicates whether this Object is equivalent to the specified Object.
copy	protected void copy(Object src)	Copies the contents of the specified Object into this Object.
clone	protected Object clone()	Creates a clone of this Object. A new instance is allocated and the copy() method is called to copy the contents of this Object into the clone.
toString	public String toString()	Returns a String that represents the value of this Object.
notify	public final void notify()	Notifies a single waiting thread on a change in condition of another thread. The thread effecting the change notifies the waiting thread using notify().
notifyAll	public final void notifyAll()	Notifies all of the threads waiting for a condition to change.
wait	public final void wait(long timeout)	Causes a thread to wait until it is notified or the specified timeout expires.
wait	public final void wait(long timeout, int nanos)	More accurate wait. The method wait() can only be called from within a synchronized method.
wait	public final void wait()	Causes a thread to wait forever until it is notified.

The method wait() can only be called from within a synchronized method

- Throws: ClassCastException, if obj is not of the same type as this Object.
- Throws: OutOfMemoryError if there is not enough memory.
- Throws: InternalError, InterruptedException, if the current thread is not the owner of the Object's monitor.

Class java.lang.Process

java.lang.Object
 |
 +——java.lang.Process

An instance of class Process is returned by variants of the exec method in class System. From the Process instance, it is possible to get the standin and/or standout of the subprocess, kill the subprocess, wait for it to terminate, and to retrieve the final exit value of the process.

Dropping the last reference to a Process instance does NOT kill the subprocess. There is no requirement that the subprocess execute asynchronously with the existing Java process.

public class **Process** extends Object

Constructor

- public Process()- Default Constructor

Methods

Methods	Usage	Use
getOutputStream	public abstract OutputStream getOutputStream()	Returns a Stream connected to the input of the child process. This stream is traditionally buffered.
getInputStream	public abstract InputStream getInputStream()	Returns a Stream connected to the output of the child process. This stream is traditionally buffered.
getErrorStream	public abstract InputStream getErrorStream()	Returns the an InputStream connected to the error stream of the child process. This stream is traditionally unbuffered.
waitFor	public abstract int waitFor()	Waits for the subprocess to complete. If the subprocess has has not yet terminated, the calling thread will be blocked until the subprocess exits.
exitValue	public abstract int exitValue()	Returns the exit value for the subprocess.
destroy	public abstract void destroy()	Kills the subprocess.

Throws: IllegalThreadStateException, if the subprocess has not yet terminated.

- Class java.lang.Runtime

java.lang.Object
 |
 +——java.lang.Runtime

public class **Runtime** extends Object

Methods

Methods	Usage	Use
getRuntime	public static Runtime get Runtime()	Return the runtime.
exit	public void exit(int status)	Exits the virtual machine with an exit code. This method does not return, use with caution.
exec	public Process exec(String command) throws IOException	Executes the process
exec	public Process exec(String cmdarray[]) throws IOException	Executes the processes specified in amdarray
freeMemory	public long freeMemory()	Returns the number of free bytes in system memory.
totalMemory	public long totalMemory()	Returns the total number of bytes in system memory.
gc	public void gc()	Runs the garbage collector.
runFinalization	public void runFinalization()	Runs the finalization methods of any object pending finalization.
trace Instructions	public void traceInstructions (boolean on)	Enables/Disables tracing of instructions.
traceMethod Calls	public void traceMethod Calls(boolean on)	Enables/Disables tracing of method calls.
load	public synchronized void load(String filename) throws UnsatisfiedLinkError	Loads a dynamic library, given a complete path name.
loadLibrary	public synchronized void loadLibrary(String libname) throws UnsatisfiedLinkError	Loads a dynamic library with the specified library name.
getLocalized InputStream	public InputStream get LocalizedInputStream (InputStream in)	Localize an input stream. A localized input stream will automatically translate the input from the local format to UNICODE.
getLocalized OutputStream	public OutputStream get LocalizedOutputStream (OutputStream out)	Localize an output stream. A localized output stream will automatically translate the output from UNICODE to the local format.

Throws: UnsatisfiedLinkError, if the file does not exist.

Class java.lang.SecurityManager

java.lang.Object
|
+——java.lang.SecurityManager

Security Manager. An abstract class that can be subclassed to implement a security policy. It allows the inspection of the classloaders on the execution stack.
public class **SecurityManager** extends Object

Constructors

- protected SecurityManager()

Constructs a new SecurityManager.

Methods

Methods	Usage	Use
getClass Context	protected Class[] getClass Context()	Gets the context of this Class.
currentClass Loader	protected ClassLoader currentClassLoader()	The current ClassLoader on the execution stack.
classDepth	protected int classDepth (String name)	Return the position of the stack frame containing the first occurrence of the named class.
classLoader Depth	protected int classLoader Depth()	Returns the stack depth of the most recently executing method from a class defined using a non system class loader.
inClass	protected boolean inClass (String name)	Returns true if the specified String is in this Class.
inClass Loader	protected boolean in ClassLoader()	Returns a boolean indicating whether or not the current ClassLoader is equal to null.
checkCreate ClassLoader	public void checkCreate ClassLoader()	Checks to see if the ClassLoader has been created.
checkAccess	public void checkAccess (Thread g)	Checks to see if the specified Thread is allowed to modify the Thread group.
checkAccess	public void checkAccess (ThreadGroup g)	Checks to see if the specified Thread group is allowed to modify this group.
checkExit	public void checkExit (int status)	Checks to see if the system has exited the virtual machine with an exit code. Where, status - exit status, 0 if successful, other values indicate various error types.

checkExec	public void checkExec (String cmd)	Checks to see if the system command is executed by trusted code.
checkLink	public void checkLink (String lib)	Checks to see if the specified linked library exists.
checkRead	public void checkRead(int fd)	Checks to see if an input file with the specified system dependent file descriptor gets created.
checkRead	public void checkRead (String file)	Checks to see if an input file with the specified system dependent file name gets created.
checkWrite	public void checkWrite(int fd)	Checks to see if an output file with the specified system dependent file descriptor gets created.
checkWrite	public void checkWrite (String file)	Checks to see if an output file with the specified system dependent file name gets created.
checkConnect	public void checkConnect (String host, int port)	Checks to see if a socket has connected to the specified port on the the specified host.
checkListen	public void checkListen (int port)	Checks to see if a server socket is listening to the specified local port that it is bounded to.
checkAccept	public void checkAccept (String host, int port)	Checks to see if a socket connection to the specified port on the specified host has been accepted.
check Properties Access	public void checkProperties Access()	Checks to see who has access to the System properties.
checkTop Level Window	public boolean checkTop LevelWindow()	Checks to see if top-level windows can be created by the caller.
checkPackage Access	public void checkPackage Access(String pkg)	Check if an applet can access a package.
checkPackage Definition	public void checkPackage Definition(String pkg)	Check if an a pplet can define classes in a package.
checkSet Factory	public void checkSetFactory()	Check if an applet can set a networking-related object factory.

Throws: SecurityException, if the security manager cannot be created, if security error occurs, if the current Thread is not allowed to access this Thread group, if the library does not exist.

Class java.lang.String

java.lang.Object
  |
  +——java.lang.String

A general class of objects to represent character Strings. Strings are constant, their values cannot be changed after creation. The compiler makes sure that each String constant actually results in a String object. Because String objects are immutable, they can be shared. For example:

```
String str = "abc";
```

is equivalent to:

```
char data[] = {'a', 'b', 'c'};
String str = new String(data);
```

Here are some more examples of how strings can be used:

```
System.out.println("abc");
String cde = "cde";
System.out.println("abc" + cde);
String c = "abc".substring(2,3);

String d = cde.substring(1, 2);
```

public final class **String** extends Object

Constructors

- public String()

Constructs a new empty String.

- public String(String value)

Constructs a new String that is a copy of the specified String.

- public String(char value[])

Constructs a new String whose initial value is the specified array of characters. The character array is **NOT** copied, so **DO NOT** modify the array after the String is created!

- public String(char value[], int offset, int count)

Constructs a new String whose initial value is the specified sub array of characters. The length of the new string will be count characters starting at offset within the specified character array. The character array is**NOT** copied, so **DO NOT** modify the array after the String is created!

Throws StringIndexOutOfBoundsException, if the offset and count arguments are invalid.

- public String(byte ascii[], int hibyte, int offset, int count) **throws** StringIndexOutOfBoundsException

Constructs a new String whose initial value is the specified sub array of bytes. The high-byte of each character can be specified, it should usually be 0. The length of the new String will be count characters starting at offset within the specified character array.

- public String(byte ascii[],int hibyte)

Constructs a new String whose initial value is the specified array of bytes. The byte array transformed into Unicode chars using hibyte as the upper byte of each character.

Methods

Methods	Usage	Use
length	public int length()	Returns the length of the String. The length of the String is equal to the number of 16 bit Unicode characters in the String.
charAt	public char charAt(int index)	Returns the character at the specified index. An index ranges from 0 to length() –1. Throws tringIndexOutOfBounds Exception, if the index is not in the range of 0 to length()–1.
getChars	public void getChars(int src Begin, int srcEnd, char dst[], int dstBegin)	Copies characters from this String into the specified character array. The characters of the specified substring (determined by srcBegin and srcEnd) are copied into the character array, starting at the array's dstBegin location.
getBytes	public void getBytes(int src Begin, int srcEnd, char dst[], int dstBegin)	Copies characters from this String into the specified byte array. Copies the characters of the specified substring (determined by srcBegin and srcEnd) into the byte array, starting at the array's dstBegin location.
equals	public boolean equals(Object anObject)	Compares this String to the specified object. Returns true if the object is equal to this String; that is, has the same length and the same characters in the same sequence. Returns true if the Strings are equal; false otherwise.
equalsIgnore Case	public boolean equalsIgnore Case(String anotherString)	Compares this String to another object. Returns true if the object is equal to this String; that is, has the same length and the same characters in the same sequence. Upper case characters are folded to lower case before they are compared. Returns true if the Strings are equal, ignoring case; false otherwise.

compareTo	public int compareTo(String anotherString)	Compares this String to another specified String. Returns an integer that is less than, equal to, or greater than zero. The integer's value depends on whether this String is less than, equal to, or greater than anotherString.
region Matches	public boolean regionMatches (int toffset, String other, int ooffset, int len)	Determines whether a region of this String matches the specified region of the specified String. Returns true if the region matches with the other; false otherwise.
region Matches	public boolean regionMatches (boolean ignoreCase, toffset, String other, int ooffset, int len)	Determines whether a region of this String matches the specified region of the specified String. If the boolean ignoreCase is true, upper case characters are considered equivalent to lower case letters.
startsWith	public boolean startsWith (String prefix, int toffset)	Determines whether this String starts with some prefix. Returns true if the String starts with the specified prefix; false otherwise.
startsWith	public boolean startsWith (String prefix)	Determines whether this String starts with some prefix.
endsWith	public boolean endsWith (String suffix)	Determines whether the String ends with some suffix.
hashCode	public int hashCode()	Returns a hashcode for this String. This is a large number composed of the character values in the String.
indexOf	public int indexOf(int ch)	Returns the index within this String of the first occurrence of the specified character. This method returns –1 if the index is not found.
indexOf	public int indexOf(int ch, int fromIndex)	Returns the index within this String of the first occurrence of the specified character, starting the search at fromIndex. This method returns –1 if the index is not found.

lastIndexOf	public int lastIndexOf(int ch)	Returns the index within this String of the last occurrence of the specified character. The String is searched backwards starting at the last character. This method returns –1 if the index is not found.
lastIndexOf	public int lastIndexOf(int ch, int fromIndex)	Returns the index within this String of the last occurrence of the specified character. The String is searched backwards starting at fromIndex. This method returns –1 if the index is not found.
indexOf	public int indexOf(String str)	Returns the index within this String of the first occurrence of the specified substring. This method returns –1 if the index is not found.
indexOf	public int indexOf(String str, int fromIndex)	Returns the index within this String of the first occurrence of the specified substring. The search is started at fromIndex. This method returns –1 if the index is not found.
lastIndexOf	public int lastIndexOf (String str)	Returns the index within this String of the last occurrence of the specified substring. The String is searched backwards. This method returns –1 if the index is not found.
lastIndexOf	public int lastIndexOf(String str, int fromIndex)	Returns the index within this String of the last occurrence of the specified substring. The String is searched backwards starting at fromIndex. This method returns –1 if the index is not found.
substring	public String substring(int beginIndex)	Returns the substring of this String. The substring is specified by a beginIndex (inclusive) and the end of the string.
substring	public String substring(int beginIndex, int endIndex)	Returns the substring of a String. The substring is specified by a beginIndex (inclusive) and an endIndex (exclusive).Throws StringIndexOutOfBounds Exception, if the beginIndex or the endIndex is out of range.

concat	public String concat(String str)	Concatenates the specified string to the end of this String.
replace	public String replace(char oldChar, char newChar)	Converts this String by replacing all occurrences of oldChar with newChar.
toLowerCase	public String toLowerCase()	Converts all of the characters in this String to lower case.
toUpperCase	public String toUpperCase()	Converts all of the characters in this String to upper case.
trim	public String trim()	Trims leading and trailing whitespace from this String.
toString	public String toString()	Converts this String to a String.
toCharArray	public char[] toCharArray()	Converts this String to a character array. This creates a new array.
valueOf	public static String valueOf (Object obj)	Returns a String that represents the String value of the object. The object may choose how to represent itself by implementing the toString() method.
valueOf	public static String valueOf (char data[])	Returns a String that is equivalent to the specified character array. Uses the original array as the body of the String (i.e. it does not copy it to a new array).
valueOf	public static String valueOf (char data[], int offset, int count)	Returns a String that is equivalent to the specified character array. Uses the original array as the body of the String (i.e. it does not copy it to a new array).
copyValueOf	public static String copyValue Of(char data[],int offset, int count)	Returns a String that is equivalent to the specified character array. It creates a new array and copies the characters into it.
copyValueOf	public static String copyValue Of(char data[])	Returns a String that is equivalent to the specified character array. It creates a new array and copies the characters into it.

valueOf	public static String value Of(boolean b)	Returns a String object that represents the state of the specified boolean.
valueOf	public static String value Of(char c)	Returns a String object that contains a single character
valueOf	public static String value Of(int i)	Returns a String object that represents the value of the specified integer.
valueOf	public static String value Of(long l)	Returns a String object that represents the value of the specified long.
valueOf	public static String value Of(float f)	Returns a String object that represents the value of the specified float.
valueOf	public static String value Of(double d)	Returns a String object that represents the value of the specified double.
intern	public String intern()	Returns a String that is equal to this String but which is guaranteed to be from the unique String pool.

Class java.lang.System

```
java.lang.Object
  |
  +——java.lang.System
```

This Class provides a system-independent interface to system functionality. One of the more useful things provided by this Class are the standard input and output streams. The standard input streams are used for reading character data. The standard output streams are used for printing. For example:

System.out.println("Hello World!");

This Class cannot be instantiated or subclassed because all of the methods and variables are static.

public final class **System** extends Object

Methods

Methods	Usage	Use
setSecurityManager	public static void setSecurity Manager(Security Manager s)	Sets the System security. This value can only be set once.
getSecurityManager	public static SecurityManager getSecurityManager()	Gets the system security interface.

currentTimeMillis	public static long current TimeMillis()	Returns the current time in milliseconds GMT since the epoch (00:00:00 UTC, January 1, 1970).
arraycopy	public static void arraycopy (Object src, int src_position, Object dst, int dst_position, int length)	Copies an array from the source array, beginning at the specified position, to the specified position of the destination array.
getProperties	public static Properties get Properties()	Gets the System properties.
setProperties	public static void set Properties(Properties props)	Sets the System properties to the specified properties.
getProperty	public static String get Property(String key)	Gets the System property indicated by the specified key.
getProperty	public static String get Property(String key, String def)	Gets the System property indicated by the specified key and def.
getenv	public static String getenv (String name)	Obsolete. Gets an environment variable. An environment vari able is a system dependent external variable that has a string value.
exit	public static void exit (int status)	Exits the virtual machine with an exit code. This method does not return, use with caution.
gc	public static void gc()	Runs the garbage collector.
runFinalization	public static void run Finalization()	Runs the finalization methods of any objects pending finalization.
load	public static void load (String filename)	Loads a dynamic library, given a complete path name.
loadLibrary	public static void load Library(String libname)	Loads a dynamic library with the specified library name.

Throws: ArrayIndexOutOfBoundsException, if copy would cause access of data outside array bounds and ArrayStoreException, if an element in the src array could not be stored into the destination array due to a type mismatch.

SecurityException if the SecurityManager has already been set.

Throws: UnsatisfiedLinkError

If the file or library does not exist.

Class java.lang.StringBuffer

java.lang.Object
 |
 +——java.lang.StringBuffer

This Class is a growable buffer for characters. It is mainly used to create Strings. The compiler uses it to implement the "+" operator. For example:

"a" + 4 + "c"

is compiled to:

new StringBuffer().append("a").append(4).append("c").toString()

Note that the method toString() does not create a copy of the internal buffer. Instead the buffer is marked as shared. Any further changes to the buffer will cause a copy to be made.

public class **StringBuffer** extends Object

Constructors

- public StringBuffer()

Constructs an empty String buffer.

- public StringBuffer(int length)

Constructs an empty String buffer with the specified initial length.

- public StringBuffer(String str)

Constructs a String buffer with the specified initial value.

Methods

Methods	Usage	Use
length	public int length()	Returns the length (character count) of the buffer.
capacity	public int capacity()	Returns the current capacity of the String buffer. The capacity is the amount of storage available for newly inserted characters; beyond which an allocation will occur.
copyWhen Shared	public void copyWhenShared()	Copies the buffer value if it is shared.
ensure Capacity	public synchronized void ensure Capacity(int minimum Capacity)	Ensures that the capacity of the buffer is at least equal to the specified minimum.
setLength	public synchronized void set Length(int newLength)	Sets the length of the String. If the length is reduced, characters are lost. If the length is extended, the values of the new characters are set to 0.

charAt	public synchronized char char At(int index)	Returns the character at the specified index. An index ranges from 0 to length() –1.
getChars	public synchronized void get Chars(int srcBegin, int srcEnd, char dst[], int dstBegin)	Copies the characters of the specified substring (determined by srcBegin and srcEnd) into the character array, starting at the array's dstBegin location. Both srcBegin and srcEnd must be legal indexes into the buffer.
setCharAt	public synchronized void setCharAt(int index, char ch)	Changes the character at the specified index to be ch.
append	public synchronized String Buffer append(Object obj)	Appends an object to the end of this buffer. Returns the StringBuffer itself, NOT a new one.
append	public synchronized String Buffer append(String str)	Appends a String to the end of this buffer.
append	public synchronized String Buffer append(char str[])	Appends an array of characters to the end of this buffer.
append	public synchronized String Buffer append(char str[], int offset, int len)	Appends a part of an array of characters to the end of this buffer.
append	public StringBuffer append (boolean b)	Appends a boolean to the end of this buffer.
append	public synchronized String Buffer append(char c)	Appends a character to the end of this buffer.
append	public StringBuffer append (int i)	Appends an integer to the end of this buffer.
append	public StringBuffer append (long l)	Appends a long to the end of this buffer.
append	public StringBuffer append (float f)	Appends a float to the end of this buffer.
append	public StringBuffer append (double d)	Appends a double to the end of this buffer.
insert	public synchronized String Buffer insert(int offset, Object obj)	Inserts an object into the String buffer.
insert	public synchronized String Buffer insert(int offset, String str)	Inserts a String into the String buffer.

insert	public synchronized String Buffer insert(int offset, char str[])	Inserts an array of characters into the String buffer.
insert	public StringBuffer insert (int offset, boolean b)	Inserts a boolean into the String buffer.
insert	public synchronized String Buffer insert(int offset, char c)	Inserts a character into the String buffer.
insert	public StringBuffer insert(int offset, int i)	Inserts an integer into the String buffer.
insert	public StringBuffer insert(int offset, long l)	Inserts a long into the String buffer.
insert	public StringBuffer insert(int offset, float f)	Inserts a float into the String buffer.
insert	public StringBuffer insert(int offset, double d)	Inserts a double into the String buffer.
toString	public synchronized String toString()	Converts to a String representing the data in the buffer.

Throws: StringIndexOutOfBoundsException, if the length is invalid, if there is an invalid index into the buffer.

Throws: ArrayIndexOutOfBoundsException, if the index or offset is invalid.

Class java.lang.Thread

```
java.lang.Object
   |
   +——java.lang.Thread
```

public class **Thread** extends Object implements Runnable

A Thread is a single sequential flow of control within a process. This simply means that while executing within a program, each thread has a beginning, a sequence, a point of execution occurring at any time during runtime of the thread and of course, an ending. Thread objects are the basis for multi-threaded programming. Multi-threaded programming allows a single program to conduct concurrently running threads that perform different tasks.

To create a new thread of execution, declare a new class which is a subclass of Thread and then override the run() method with code that you want executed in this Thread. An instance of the Thread subclass should be created next with a call to the start() method following the instance. The start() method will create the thread and execute the run() method. For example:

```
class PrimeThread extends Thread {
        public void run() {
                // compute primes...
        }
    }
```

To start this thread you need to do the following:

```
PrimeThread p = new PrimeThread();
p.start();
...
```

Another way to create a thread is by using the Runnable interface. This way any object that implements the Runnable interface can be run in a thread. For example:

```
class Primes implements Runnable {
    public void run() {
        // compute primes...
    }
}
```

To start this thread you need to do the following:

```
Primes p = new Primes();
new Thread(p).start();
...
```

The virtual machine runs until all Threads that are not daemon Threads have died. A Thread dies when its run() method returns, or when the stop() method is called.

When a new Thread is created, it inherits the priority and the daemon flag from its parent (i.e. the Thread that created it).

Constructors

- public Thread()

Constructs a new Thread. Threads created this way must have overridden their run() method to actually do anything. An example illustrating this method being used is shown.

```
import java.lang.*;
class plain01 implements Runnable {
  String name;
  plain01() {
    name = null;
  }
  plain01(String s) {
    name = s;
  }
  public void run() {
    if (name == null)
      System.out.println("A new thread created");
    else
      System.out.println("A new thread with name " + name + " created");
  }
}

class threadtest01 {
  public static void main(String args[] ) {
    int failed = 0 ;

    Thread t1 = new Thread();
    if(t1 != null) {
      System.out.println("new Thread() succeed");
```

```
    } else {
      System.out.println("new Thread() failed");
      failed++;
    }
}
```

- public Thread(Runnable target)

Constructs a new Thread which applies the run() method of the specified target.

- public Thread(ThreadGroup group, Runnable target)

Constructs a new Thread in the specified Thread group that applies the run() method of the specified target.

- public Thread(String name)

Constructs a new Thread with the specified name.

- public Thread(ThreadGroup group, String name)

Constructs a new Thread in the specified Thread group with the specified name.

- public Thread(Runnable target, String name)

Constructs a new Thread with the specified name and applies the run() method of the specified target.

- public Thread(ThreadGroup group, Runnable target, String name)

Constructs a new Thread in the specified Thread group with the specified name and applies the run() method of the specified target.

b. Methods

Methods	Usage	Use
currentThread	public static Thread current Thread()	Returns a reference to the currently executing Thread object.
yield	public static void yield()	Causes the currently executing Thread object to yield. If there are other runnable Threads, they will be scheduled next.
sleep	public static void sleep (long millis) throws InterruptedException	Causes the currently executing Thread to sleep for the specified number of milliseconds.
sleep	public static void sleep (long millis, int nanos) throws InterruptedException	Sleep, in milliseconds and additional nanosecond.
start	public synchronized void start()	Starts this Thread. This will cause the run() method to be called. This method will return immediately.
run	public void run()	The actual body of this Thread. This method is called after the Thread is started. You must either override this method by subclassing class Thread, or you must create the Thread with a Runnable target.

stop	public final void stop()	Stops a Thread by tossing an object. By default this routine tosses a new instance of ThreadDeath to the target Thread.
stop	public final synchronized void stop(Object o)	Stops a Thread by tossing an object. Normally, users should just call the stop() method without any argument.
interrupt	public void interrupt()	Send an interrupt to a thread.
interrupted	public static boolean interrupted()	Ask if you have been interrupted.
isInterrupted	public boolean isInterrupted()	Ask if another thread has been interrupted.
destroy	public void destroy()	Destroy a thread, without any cleanup, i.e. just toss its state; any monitors it has locked remain locked. A last resort.
isAlive	public final boolean isAlive()	Returns a boolean indicating if the Thread is active. Having an active Thread means that the Thread has been started and has not been stopped.
suspend	public final void suspend()	Suspends this Thread's execution.
resume	public final void resume()	Resumes this Thread execution. This method is only valid after suspend() has been invoked.
setPriority	public final void setPriority (int newPriority)	Sets the Thread's priority.
getPriority	public final int getPriority()	Gets and returns the Thread's priority.
setName	public final void setName (String name)	Sets the Thread's name.
getName	public final String getName()	Gets and returns this Thread's name.
getThread Group	public final ThreadGroup get ThreadGroup()	Gets and returns this Thread group.
activeCount	public static int activeCount()	Returns the current number of active Threads in this Thread group.

enumerate	public static int enumerate (Thread tarray[])	Copies, into the specified array, references to every active Thread in this Thread's group.
countStack Frames	public int countStackFrames()	Returns the number of stack frames in this Thread. The Thread must be suspended when this method is called.
join	public final synchronized void join(long millis) throws InterruptedException	Waits for this Thread to die. A timeout in milliseconds can be specified. A timeout of 0 milli-seconds means to wait forever.
join	public final synchronized void join(long millis, int nanos) throws InterruptedException	Waits for the Thread to die, with more precise time.
join	public final void join() throws InterruptedException	Waits forever for this Thread to die.
dumpStack	public static void dumpStack()	A debugging procedure to print a stack trace for the current Thread.
setDaemon	public final void setDaemon (boolean on)	Marks this Thread as a daemon Thread or a user Thread. When there are only daemon Threads left running in the system, Java exits.
isDaemon	public final boolean is Daemon()	Returns the daemon flag of the Thread.
checkAccess	public void checkAccess()	Checks whether the current Thread is allowed to modify this Thread.
toString	public String toString()	Returns a String representation of the Thread.

Throws: IllegalThreadStateException, if the thread was already started, or not suspended.

Throws: IllegalArgumentException, if the priority is not within the range MIN_PRIORITY, MAX_PRIORITY.

Throws: SecurityException, if the current Thread is not allowed to access this Thread group.

Class java.lang.ThreadGroup

```
java.lang.Object
   |
   +——java.lang.ThreadGroup
```

A group of Threads. A Thread group can contain a set of Threads as well as a set of other Thread groups. A Thread can access its Thread group, but it cannot access the parent of its Thread group. This makes it possible to encapsulate a Thread in a Thread group and stop it from manipulating Threads in the parent group.

public class **ThreadGroup** extends Object

- public ThreadGroup(String name)

Creates a new ThreadGroup. Its parent will be the Thread group of the current Thread.

- public ThreadGroup(ThreadGroup parent, String name)

Creates a new ThreadGroup with a specified name in the specified Thread group.

Methods

Methods	Usage	Use
getName	public final String getName()	Gets the name of this Thread group.
getParent	public final ThreadGroup get Parent()	Gets the parent of this Thread group.
getMaxPriority	public final int getMaxPriority()	Gets the maximum priority of the group. Threads that are part of this group cannot have a higher priority than the maximum priority.
isDaemon	public final boolean isDaemon()	Returns the daemon flag of the Thread group. A daemon Thread group is automatically destroyed when it is found empty after a Thread group or Thread is removed from it.
setDaemon	public final void setDaemon (boolean daemon)	Changes the daemon status of this group.
setMaxPriority	public final synchronized void setMaxPriority(int pri)	Sets the maximum priority of the group. Threads that are already in the group can have a higher priority than the set maximum.
parentOf	public final boolean parentOf (ThreadGroup g)	Checks to see if this Thread group is a parent of or is equal to another Thread group.
checkAccess	public final void checkAccess()	Checks to see if the current Thread is allowed to modify this group.
activeCount	public synchronized int active Count()	Returns an estimate of the number of active Threads in the Thread group.

enumerate	public int enumerate (Thread list[])	Copies, into the specified array, references to every active Thread in this Thread group. You can use the activeCount() method to get an estimate of how big the array should be.
enumerate	public int enumerate (Thread list[], boolean recurse)	Copies, into the specified array, references to every active Thread in this Thread group. You can use the activeCount() method to get an estimate of how big the array should be.
activeGroup Count	public synchronized int active GroupCount()	Returns an estimate of the number of active groups in the Thread group.
enumerate	public int enumerate(Thread Group list[])	Copies, into the specified array, references to every active Thread group in this Thread group. You can use the activeGroupCount() method to get an estimate of how big the array should be.
enumerate	public int enumerate(Thread Group list[], boolean recurse)	Copies, into the specified array, references to every active Thread group in this Thread group. You can use the activeGroupCount() method to get an estimate of how big the array should be.
stop	public final synchronized void stop()	Stops all the Threads in this Thread group and all of its sub groups.
suspend	public final synchronized void suspend()	Suspends all the Threads in this Thread group and all of its sub groups.
resume	public final synchronized void resume()	Resumes all the Threads in this Thread group and all of its sub groups.
destroy	public final synchronized void destroy()	Destroys a Thread group. This does NOT stop the Threads in the Thread group.
list	public synchronized void list()	Lists this Thread group. Useful for debugging only.
toString	public String toString()	Returns a String representation of the Thread group.

Throws: NullPointerException, if the given thread group is equal to null.

Throws: SecurityException, if the current Thread is not allowed to access this Thread group.

Throws: IllegalThreadStateException, if the Thread group is not empty or if the Thread group was already destroyed.

Class java.lang.Throwable

```
java.lang.Object
   |
   +——java.lang.Throwable
```

public class **Throwable** extends Object

An object signalling that an exceptional condition has occurred. All exceptions are a subclass of Exception. An exception contains a snapshot of the execution stack, this snapshot is used to print a stack backtrace. An exception also contains a message string. Here is an example of how to catch an exception:

```
try {
        int a[] = new int[2];
        a[4];
    } catch (ArrayIndexOutOfBoundsException e) {
        System.out.println("an exception occurred: " + e.getMessage());
        e.printStackTrace();
}
```

Constructors

- public Throwable()

Constructs a new Throwable with no detail message. The stack trace is automatically filled in.

- public Throwable(String message)

Constructs a new Throwable with the specified detail message. The stack trace is automatically filled in.

Methods

Methods	Usage	Use
getMessage	public String getMessage()	Gets the detail message of the Throwable. A detail message is a String that describes the Throwable that has taken place.
toString	public String toString()	Returns a short description of the Throwable.
printStack Trace	public void printStackTrace()	Prints the Throwable and the Throwable's stack trace.
fillInStack Trace	public Throwable fillIn StackTrace()	Fills in the excecution stack trace. This is useful only when rethrowing a Throwable.

For example:

```
try {
        a = b / c;
    } catch(ArithmeticThrowable e) {
            a = Number.MAX_VALUE;
        throw e.fillInStackTrace();
}
```

4. JAVA UTILITY CLASSES

INTERFACE JAVA.UTIL.

Interface java.util.Enumeration

public interface **Enumeration** extends Object

The Enumeration interface specifies a set of methods that may be used to enumerate, or count through, a set of values. The enumeration is consumed by use; its values may only be counted once.

For example, to print all elements of a Vector v:

```
for (Enumeration e = v.elements() ; e.hasMoreElements() ;) {
   System.out.println(e.nextElement());
}
```

The Enumeration interface defines the methods by which you can enumerate (obtain one at a time) the elements in a collection of objects. Although not deprecated, Enumeration is considered obsolete for new code. However, it is used by several methods defined by the legacy classes such as Vector and Properties, is used by several other API classes, and is currently in widespread use in application code.

The methods declared by Enumeration are summarized in the following table:

Methods	Usage	Use
hasMore Elements	boolean hasMoreElements( )	When implemented, it must return true while there are still more elements to extract, and false when all the elements have been enumerated.
nextElement	Object nextElement( )	This returns the next object in the enumeration as a generic Object reference.

Example

Following is the example showing usage of Enumeration.

```
import java.util.Vector;
import java.util.Enumeration;
public class EnumerationExample {
  public static void main(String args[]) {
    Enumeration days;
    Vector dayNames = new Vector();
    dayNames.add("Sunday");
    dayNames.add("Monday");
    dayNames.add("Tuesday");
```

```
        dayNames.add("Wednesday");
        dayNames.add("Thursday");
        dayNames.add("Friday");
        dayNames.add("Saturday"};
        days = dayNames.elements();
        while (days.hasMoreElements()){
          System.out.println(days.nextElement());
        }
      }
    }
```

This would produce the following results:

Sunday
Monday
Tuesday
Wednesday
Thursday
Friday
Saturday

Interface java.util.Observer

public interface **Observer** extends Object

When implemented, this interface allows all classes to be observable by instances of class Observer.

Method

- **update**

  public abstract void update(Observable o,Object arg)

This is called if observers in the observable list need to be updated, where o—the list of observers, arg—the argument being notified.

CLASSES IN JAVA.UTIL

BitSet

The BitSet class implements a group of bits, or flags, that can be set and cleared individually.

This class is very useful in cases where you need to keep up with a set of boolean values; you just assign a bit to each value and set or clear it as appropriate. A BitSet class creates a special type of array that holds bit values. The BitSet array can increase in size as needed. This makes it similar to a vector of bits.

The BitSet defines two constructors. The first version creates a default object:

- *Bitset( )*
- *Bitset(int Size)*

The second version allows you to specify its initial size, i.e. the number of bits that it can hold. All bits are initialized to zero.

BitSet implements the Cloneable interface and defines the methods listed in table below:

Methods	Usage	Use
and	void and(BitSet bitSet)	ANDs the contents of the invoking BitSet object with those specified by bitSet. The result is placed into the invoking object.

andNot	void andNot(BitSet bitSet)	For each 1 bit in bitSet, the corresponding bit in the invoking BitSet is cleared.
cardinality	int cardinality( )	Returns the number of set bits in the invoking object.
clear	void clear( )	Zeros all bits.
clear	void clear(int index)	Zeros the bit specified by index.
clear	void clear(int startIndex, int endIndex)	Zeros the bits from startIndex to endIndex.1.
clone	Object clone( )	Duplicates the invoking BitSet object.
equals	boolean equals(Object bitSet)	Returns true if the invoking bit set is equivalent to the one passed in bitSet. Otherwise, the method returns false.
flip	void flip(int index)	Reverses the bit specified by index.
flip	void flip(int startIndex, int endIndex)	Reverses the bits from startIndex to endIndex.1.
get	boolean get(int index)	Returns the current state of the bit at the specified index.
get	BitSet get(int startIndex, int endIndex)	Returns a BitSet that consists of the bits from startIndex to endIndex.1. The invoking object is not changed.
hashCode	int hashCode( )	Returns the hash code for the invoking object.
intersects	boolean intersects(BitSet bitSet)	Returns true if at least one pair of corresponding bits within the invoking object and bitSet are 1.
isEmpty	boolean isEmpty( )	Returns true if all bits in the invoking object are zero.
length	int length( )	Returns the number of bits required to hold the contents of the invoking BitSet. This value is determined by the location of the last 1 bit.
nextClearBit	int nextClearBit(int startIndex)	Returns the index of the next cleared bit, (that is, the next zero bit), starting from the index specified by startIndex

nextClearBit	int nextSetBit(int startIndex)	Returns the index of the next set bit (that is, the next 1 bit), starting from the index specified by startIndex. If no bit is set, .1 is returned.
or	void or(BitSet bitSet)	ORs the contents of the invoking BitSet object with that specified by bitSet. The result is placed into the invoking object.
set	void set(int index)	Sets the bit specified by index.
set	void set(int index, boolean v)	Sets the bit specified by index to the value passed in v. true sets the bit, false clears the bit.
set	void set(int startIndex, int endIndex)	Sets the bits from startIndex to endIndex.1.
set	void set(int startIndex, int endIndex, boolean v)	Sets the bits from startIndex to endIndex.1, to the value passed in v. true sets the bits, false clears the bits.
size	int size( )	Returns the number of bits in the invoking BitSet object.
toString	String toString( )	Returns the string equivalent of the invoking BitSet object.
xor	void xor(BitSet bitSet)	XORs the contents of the invoking BitSet object with that specified by bitSet. The result is placed into the invoking object

Example

The following program illustrates several of the methods supported by this data structure:

```
import java.util.BitSet;

class BitSetExample {
  public static void main(String args[]) {
    BitSet bits1 = new BitSet(16);
    BitSet bits2 = new BitSet(16);

       // set some bits
    for(int i=0; i<16; i++) {
      if((i%2) == 0) bits1.set(i);
      if((i%5) != 0) bits2.set(i);
    }
    System.out.println("Initial pattern in bits1: ");
```

```
        System.out.println(bits1);
        System.out.println("\nInitial pattern in bits2: ");
        System.out.println(bits2);

        // AND bits
        bits2.and(bits1);
        System.out.println("\nbits2 AND bits1: ");
        System.out.println(bits2);

        // OR bits
        bits2.or(bits1);
        System.out.println("\nbits2 OR bits1: ");
        System.out.println(bits2);

        // XOR bits
        bits2.xor(bits1);
        System.out.println("\nbits2 XOR bits1: ");
        System.out.println(bits2);
    }
}
```

This would produce the following results:

```
Initial pattern in bits1:
{0, 2, 4, 6, 8, 10, 12, 14}
Initial pattern in bits2:
{1, 2, 3, 4, 6, 7, 8, 9, 11, 12, 13, 14}
bits2 AND bits1:
{2, 4, 6, 8, 12, 14}
bits2 OR bits1:
{0, 2, 4, 6, 8, 10, 12, 14}
bits2 XOR bits1:
{}
```

Vector

The Vector class is similar to a traditional Java array, except that it can grow as necessary to accommodate new elements.

Like an array, elements of a Vector object can be accessed via an index into the vector.

The nice thing about using the Vector class is that you do not have to worry about setting it to a specific size upon creation; it shrinks and grows automatically when necessary.

Vector implements a dynamic array. It is similar to ArrayList, but with two differences:

- Vector is synchronized.
- Vector contains many legacy methods that are not part of the collections framework.

Vector proves to be very useful if you do not know the size of the array in advance, or you just need one that can change sizes over the lifetime of a program.

The Vector class supports four constructors. The first form creates a default vector, which has an initial size of 10:

Vector( )

The second form creates a vector whose initial capacity is specified by size:

Vector(int size)

The third form creates a vector whose initial capacity is specified by size and whose increment is specified by incr. The increment specifies the number of elements to allocate each time that a vector is resized upward:

Vector(int size, int incr)

The fourth form creates a vector that contains the elements of collection c:

Vector(Collection c)

Apart from the methods inherited from its parent classes, Vector defines the following methods:

Methods	Usage	Use
add	void add(int index, Object element)	Inserts the specified element at the specified position in this Vector.
add	boolean add(Object o)	Appends the specified element to the end of this Vector.
addAll	boolean addAll(Collection c)	Appends all of the elements in the specified Collection to the end of this Vector, in the order that they are returned by the specified Collection's Iterator.
addAll	boolean addAll(int index, Collection c)	Inserts all of the elements in the specified Collection into this Vector at the specified position.
addElement	void addElement(Object obj)	Adds the specified component to the end of this vector, increasing its size by one.
capacity	int capacity()	Returns the current capacity of this vector.
clear	void clear()	Removes all of the elements from this Vector.
clone	Object clone()	Returns a clone of this vector.
conatins	boolean contains(Object elem)	Tests if the specified object is a component in this vector.
containsAll	boolean containsAll (Collection c)	Returns true if this Vector contains all of the elements in the specified Collection.
copyInto	void copyInto(Object[] anArray)	Copies the components of this vector into the specified array.

elementAt	Object elementAt(int index)	Returns the component at the specified index.
elements	Enumeration elements()	Returns an enumeration of the components of this vector.
ensure Capacity	void ensureCapacity(int minCapacity)	Increases the capacity of this vector, if necessary, to ensure that it can hold at least the number of components specified by the minimum capacity argument.
equals	boolean equals(Object o)	Compares the specified Object with this Vector for equality.
firstElement	Object firstElement()	Returns the first component (the item at index 0) of this vector.
get	Object get(int index)	Returns the element at the specified position in this Vector.
hashCode	int hashCode()	Returns the hash code value for this Vector.
indexof	int indexOf(Object elem)	Searches for the first occurrence of the given argument, testing for equality using the equals method.
indexOf	int indexOf(Object elem, int index)	Searches for the first occurrence of the given argument, beginning the search at index, and testing for equality using the equals method.
insertElement is Empty	void insertElementAt(Object obj, int index)	Inserts the specified object as a component in this vector at the specified index.
isEmpty	boolean isEmpty()	Tests if this vector has no components.
lastElement	Object lastElement()	Returns the last component of the vector.
lastIndexOf	int lastIndexOf(Object elem)	Returns the index of the last occurrence of the specified object in this vector.
lastIndexOf	int lastIndexOf(Object elem, int index)	Searches backwards for the specified object, starting from the specified index, and returns an index to it.

remove	Object remove(int index)	Removes the element at the specified position in this Vector.
remove	boolean remove(Object o)	Removes the first occurrence of the specified element in this Vector, if the Vector does not contain the element, it is unchanged.
removeAll	boolean removeAll(Collection c)	Removes from this Vector all of its elements that are contained in the specified Collection.
removeAll	void removeAllElements()	Removes all components from this vector and sets its size to zero.
remove Element	boolean removeElement (Object obj)	Removes the first (lowest-indexed) occurrence of the argument from this vector.
remove ElementAt	void removeElementAt(int index)	removeElementAt(int index)
removeRange	protected void removeRange (int fromIndex, int toIndex)	Removes from this List all of the elements whose index is between fromIndex, inclusive and toIndex, exclusive.
retainAll	boolean retainAll(Collection c)	Retains only the elements in this Vector that are contained in the specified Collection.
set	Object set(int index, Object element)	Replaces the element at the specified position in this Vector with the specified element.
setElementAt	void setElementAt(Object obj, int index)	Sets the component at the specified index of this vector to be the specified object.
setSize	void setSize(int newSize)	Sets the size of this vector.
size	int size()	Returns the number of components in this vector.
subList	List subList(int fromIndex, int toIndex)	Returns a view of the portion of this List between fromIndex, inclusive, and toIndex, exclusive.
toArray	Object[] toArray()	Returns an array containing all of the elements in this Vector in the correct order.
toArray	Object[] toArray(Object[] a)	Returns an array containing all of the elements in this Vector in

		the correct order; the runtime type of the returned array is that of the specified array.
toString	String toString()	Returns a string representation of this Vector, containing the String representation of each element.
trimToSize	void trimToSize()	Trims the capacity of this vector to be the vector's current size.

Example

The following program illustrates several of the methods supported by this collection:

```
import java.util.*;
class VectorDemo {
  public static void main(String args[]) {
    // initial size is 3, increment is 2
    Vector v = new Vector(3, 2);
    System.out.println("Initial size: " + v.size());
    System.out.println("Initial capacity: " +
    v.capacity());
    v.addElement(new Integer(1));
    v.addElement(new Integer(2));
    v.addElement(new Integer(3));
    v.addElement(new Integer(4));
    System.out.println("Capacity after four additions: " + v.capacity());
    v.addElement(new Double(5.45));
    System.out.println("Current capacity: " + v.capacity());
    v.addElement(new Double(6.08));
    v.addElement(new Integer(7));
    System.out.println("Current capacity: " + v.capacity());
    v.addElement(new Float(9.4));
    v.addElement(new Integer(10));
    System.out.println("Current capacity: " + v.capacity());
    v.addElement(new Integer(11));
    v.addElement(new Integer(12));
    System.out.println("First element: " + (Integer)v.firstElement());
    System.out.println("Last element: " + (Integer)v.lastElement());
    if(v.contains(new Integer(3)))
      System.out.println("Vector contains 3.");
    // enumerate the elements in the vector.
    Enumeration vEnum = v.elements();
    System.out.println("\nElements in vector:");
    while(vEnum.hasMoreElements())
      System.out.print(vEnum.nextElement() + " ");
    System.out.println();
  }
}
```

This would produce the following results:
Initial size: 0
Initial capacity: 3
Capacity after four additions: 5
Current capacity: 5
Current capacity: 7
Current capacity: 9
First element: 1
Last element: 12
Vector contains 3.
Elements in vector:
1 2 3 4 5.45 6.08 7 9.4 10 11 12

Stack

The Stack class implements a last-in-first-out (LIFO) stack of elements.

You can think of a stack literally as a vertical stack of objects; when you add a new element, it gets stacked on top of the others.

When you pull an element off the stack, it comes off the top. In other words, the last element you added to the stack is the first one to come back off.

Stack is a subclass of Vector that implements a standard last-in, first-out stack.

Stack only defines the default constructor, which creates an empty stack. Stack includes all the methods defined by Vector, and adds several of its own.

Stack( )

Apart from the methods inherited from its parent class Vector, Stack defines the following methods:

Methods	Usage	Use
empty	boolean empty()	Tests if this stack is empty. Returns true if the stack is empty, and returns false if the stack contains elements.
peek	Object peek( )	Returns the element on the top of the stack, but does not remove it.
pop	Object pop( )	Returns the element on the top of the stack, removing it in the process.
push	Object push(Object element)	Pushes element onto the stack, element is also returned.
search	int search(Object element)	Searches for element in the stack. If found, its offset from the top of the stack is returned. Otherwise, .1 is returned.

Example

The following program illustrates several of the methods supported by this collection:

```
import java.util.*;
class StackDemo {
  static void showpush(Stack st, int a) {
    st.push(new Integer(a));
    System.out.println("push(" + a + ")");
    System.out.println("stack: " + st);
  }
  static void showpop(Stack st) {
    System.out.print("pop -> ");
    Integer a = (Integer) st.pop();
    System.out.println(a);
    System.out.println("stack: " + st);
  }
  public static void main(String args[]) {
    Stack st = new Stack();
    System.out.println("stack: " + st);
    showpush(st, 42);
    showpush(st, 66);
    showpush(st, 99);
    showpop(st);
    showpop(st);
    showpop(st);
    try {
      showpop(st);
    } catch (EmptyStackException e) {
      System.out.println("empty stack");
    }
  }
}
```

This would produce the following results:

```
stack: [ ]
push(42)
stack: [42]
push(66)
stack: [42, 66]
push(99)
stack: [42, 66, 99]
pop -> 99
stack: [42, 66]
pop -> 66
stack: [42]
pop -> 42
stack: [ ]
pop -> empty stack
```

Dictionary

The Dictionary class is an abstract class that defines a data structure for mapping keys to values. This is useful in cases where you want to be able to access data via a particular key rather than an integer index. Since the Dictionary class is abstract, it provides only the framework for a key-mapped data structure rather than a specific implementation. Dictionary is an abstract class that represents a key/value storage repository and operates much like Map.

Given a key and value, you can store the value in a Dictionary object. Once the value is stored, you can retrieve it by using its key. Thus, like a map, a dictionary can be thought of as a list of key/value pairs.

The abstract methods defined by Dictionary are listed below:

Methods	Usage	Use
elements	Enumeration elements( )	Returns an enumeration of the values contained in the dictionary.
Get	Object get(Object key)	Returns the object that contains the value associated with key. If key is not in the dictionary, a null object is returned.
isEmpty	boolean isEmpty( )	Returns true if the dictionary is empty, and returns false if it contains at least one key.
keys	Enumeration keys( )	Returns an enumeration of the keys contained in the dictionary.
Put	Object put(Object key, Object value)	Inserts a key and its value into the dictionary. Returns null if key is not already in the dictionary; returns the previous value associated with key if key is already in the dictionary.
remove	Object remove(Object key)	Removes key and its value. Returns the value associated with key. If key is not in the dictionary, a null is returned.
size	int size( )	Returns the number of entries in the dictionary.

The Dictionary class is obsolete. You should implement the Map interface to obtain key/value storage functionality.

Hashtable

The Hashtable class provides a means of organizing data based on some user-defined key structure. For example, in an address list hashtable you could store and sort data based on a key such as ZIP code rather than on a person's name. The specific meaning of keys in regard to hashtables is totally dependent on the usage of the hashtable and the data it contains.

Hashtable was part of the original java.util and is a concrete implementation of a Dictionary.

However, Java 2 reengineered hashtable so that it also implements the map interface. Thus, hashtable is now integrated into the collections framework. It is similar to hashmap, but is synchronized. Like hashmap, hashtable stores key/value pairs in a hashtable. When using a hashtable, you specify an object that is used as a key, and the value that you want linked to that key. The key is then hashed, and the resulting hash code is used as the index at which the value is stored within the table.

The hashtable defines four constructors. The first version is the default constructor:

Hashtable( )

The second version creates a hashtable that has an initial size specified by size:

Hashtable(int size)

The third version creates a hashtable that has an initial size specified by size and a fill ratio specified by fillRatio.

This ratio must be between 0.0 and 1.0, and it determines how full the hashtable can be before it is resized upward.

Hashtable(int size, float fillRatio)

The fourth version creates a hashtable that is initialized with the elements in m.

The capacity of the hashtable is set to twice the number of elements in m. The default load factor of 0.75 is used.

Hashtable(Map m)

Apart from the methods defined by map interface, hashtable defines following methods:

Methods	Usage	Use
clear	void clear( )	Resets and empties the hashtable.
clone	Object clone( )	Returns a duplicate of the invoking object.
contains	boolean contains(Object value)	Returns true if some value equal to value exists within the hash table. Returns false if the value is not found.
containsKey	boolean containsKey(Object key)	Returns true if some key equal to key exists within the hashtable. Returns false if the key is not found.
containsValue	boolean containsValue (Object value)	Returns true if some value equal to value exists within the hashtable. Returns false if the value is not found.
elements	Enumeration elements( )	Returns an enumeration of the values contained in the hashtable.

get	Object get(Object key)	Returns the object that contains the value associated with key. If key is not in the hashtable, a null object is returned.
isEmpty	boolean isEmpty( )	Returns true if the hash table is empty; returns false if it contains at least one key.
keys	Enumeration keys( )	Returns an enumeration of the keys contained in the hashtable.
put	Object put(Object key, Object value)	Inserts a key and a value into the hashtable. Returns null if key is not already in the hashtable; returns the previous value associated with key if key is already in the hashtable.
rehash	void rehash( )	Increases the size of the hashtable and rehashes all of its keys.
remove	Object remove(Object key)	Removes key and its value. Returns the value associated with key. If key is not in the hashtable, a null object is returned.
size	int size( )	Returns the number of entries in the hashtable.
toString	String toString( )	Returns the string equivalent of a hashtable.

Example

The following program illustrates several of the methods supported by this data structure:

```
import java.util.*;

class HashTableDemo {
  public static void main(String args[]) {
    // Create a hash map
    Hashtable balance = new Hashtable();
    Enumeration names;
    String str;
    double bal;

    balance.put("Jamal", new Double(3434.34));
    balance.put("Mahnaz", new Double(123.22));
    balance.put("Ayan", new Double(1378.00));
    balance.put("Daisy", new Double(99.22));
    balance.put("Qadir", new Double(-19.08));
```

```
    // Show all balances in hashtable.
    names = balance.keys();
    while(names.hasMoreElements()) {
      str = (String) names.nextElement();
      System.out.println(str + ": " +
      balance.get(str));
    }
    System.out.println();
    // Deposit 1,000 into Jamal's account
    bal = ((Double)balance.get("Jamal")).doubleValue();
    balance.put("Jamal", new Double(bal+1000));
    System.out.println("Jamal's new balance: " +
    balance.get("Jamal"));
  }
}
```

This would produce the following results:

```
Qadir: -19.08
Jamal: 3434.34
Mahnaz: 123.22
Daisy: 99.22
Ayan: 1378.0

Jamal's new balance: 4434.34
```

Properties

Properties is a subclass of hashtable. It is used to maintain lists of values in which the key is a String and the value is also a String. The Properties class is used by many other Java classes. For example, it is the type of object returned by System.getProperties( ) when obtaining environmental values. Properties defines the following instance variable. This variable holds a default property list associated with a Properties object.

Properties defaults;

The Properties defines two constructors. The first version creates a Properties object that has no default values:

Properties( )

The second creates an object that uses propDefault for its default values. In both cases, the property list is empty:

Properties(Properties propDefault)

Apart from the methods defined by Hashtable, Properties defines the following methods:

Methods	Usage	Use
getproperty	String getProperty(String key)	Returns the value associated with key. A null object is returned if key is neither in the list nor in the default property list.
getProperty	String getProperty(String key, String defaultProperty)	Returns the value associated with key. defaultProperty is returned if key is neither in the list nor in the default property list.

list	void list(PrintStream streamOut)	Sends the property list to the output stream linked to streamOut.
list	void list(PrintWriter streamOut)	Sends the property list to the output stream linked to streamOut.
load	void load(InputStream streamIn) throws IOException	Inputs a property list from the input stream linked to streamIn.
property Names	Enumeration propertyNames( )	Returns an enumeration of the keys. This includes those keys found in the default property list, too.
setProperty	Object setProperty(String key, String value)	Associates value with key. Returns the previous value associated with key, or returns null if no such association exists.
store	void store(OutputStream streamOut, String description)	After writing the string specified by description, the property list is written to the output stream linked to streamOut.

Example

The following program illustrates several of the methods supported by this data structure:

```
import java.util.*;
class PropDemo {
  public static void main(String args[]) {
    Properties capitals = new Properties();
    Set states;
    String str;
    capitals.put("Illinois", "Springfield");
    capitals.put("Missouri", "Jefferson City");
    capitals.put("Washington", "Olympia");
    capitals.put("California", "Sacramento");
    capitals.put("Indiana", "Indianapolis");

    // Show all states and capitals in hashtable.
    states = capitals.keySet(); // get set-view of keys
    Iterator itr = states.iterator();
    while(itr.hasNext()) {
      str = (String) itr.next();
      System.out.println("The capital of " +
        str + " is " + capitals.getProperty(str) + ".");
    }
    System.out.println();

    // look for state not in list — specify default
```

```
    str = capitals.getProperty("Florida", "Not Found");
    System.out.println("The capital of Florida is "
      + str + ".");
  }
}
```

This would produce the following results:

The capital of Missouri is Jefferson City.
The capital of Illinois is Springfield.
The capital of Indiana is Indianapolis.
The capital of California is Sacramento.
The capital of Washington is Olympia.
The capital of Florida is Not Found.

Date

A wrapper for a date. This class lets you manipulate dates in a system independent way. To print today's date use:

```
Date d = new Date();
System.out.println("today = " + d);
```

To find out what day corresponds to a particular date:

```
Date d = new Date(63, 0, 16);          // January 16, 1963
System.out.println("Day of the week: " + d.getDay());
```

The date can be set and examined according to the local time zone into the year, month, day, hour, minute and second.

While the API is intended to reflect UTC, Coordinated Universal Time, it does not do so exactly. This inexact behavior is inherited from the time system of the underlying OS. All modern OS's that I (jag) am aware of assume that 1 day = 24*60*60 seconds. In UTC, about once a year there is an extra second, called a "leap second" added to a day to account for the wobble of the earth. Most computer clocks are not accurate enough to be able to reflect this distinction. Some computer standards are defined in GMT, which is equivalent to UT, Universal Time. GMT is the "civil" name for the standard, UT is the "scientific" name for the same standard. The distinction between UTC and UT is that the first is based on an atomic clock and the second is based on astronomical observations, which for all practical purposes is an invisibly fine hair to split. An interesting source of further information is the US Naval Observatory, particularly the Directorate of Time and their definitions of Systems of Time.

Constructors

- public Date()
  - o Creates today's date/time.
- public Date(long date)
  - o Creates a date. The fields are normalized before the Date object is created. The argument does not have to be in the correct range. For example, the 32nd of January is correctly interpreted as the 1st of February. You can use this to figure out what day a particular date falls on.
- public Date(int year,int month,int date)
  - o Creates a date. The fields are normalized before the Date object is created. The arguments do not have to be in the correct range. For example, the 32nd of January is correctly interpreted as the 1st of February. You can use this to figure out what day a particular date falls on. Where, year—a year after 1900, month—a month between 0 and 11, date—day of the month between 1 and 31.

- public Date(int year,int month,int date,int hrs,int min)
  Creates a date. The fields are normalized before the Date object is created. The arguments do not have to be in the correct range. For example, the 32nd of January is correctly interpreted as the 1st of February. You can use this to figure out what day a particular date falls on. Where, year—a year after 1900, month—a month between 0 and 11, date—day of the month between 1 and 31, hrs— hours between 0 and 23, min—minutes between 0 and 59.
- public Date(int year,int month,int date,int hrs,int min,int sec)
  Creates a date. The fields are normalized before the Date object is created. The arguments do not have to be in the correct range. For example, the 32nd of January is correctly interpreted as the 1st of February. You can use this to figure out what day a particular date falls on.
- public Date(String s)
  Creates a date from a string according to the syntax accepted by parse().

Methods	Usage	Use
UTC	public static long UTC(int year, int month,int date, int hrs, int min, int sec)	Calculates a UTC value from YMDHMS. Interprets the parameters in UTC, not in the local time zone.
parse	public static long parse(String s)	Given a string representing a time, parse it and return the time value. It accepts many syntaxes, but most importantly, in accepts the IETF standard date syntax: "Sat, 12 Aug 1995 13:30:00 GMT". If no time zone is specified, the local time zone is assumed.
getYear	public int getYear()	Returns the year after 1900.
setYear	public void setYear(int year)	Sets the year.
getMonth	public int getMonth()	Returns the month. This method assigns months with the values 0–11, with January beginning at value 0.
setMonth	public void setMonth(int month)	Sets the month.
getDate	public int getDate()	Returns the day of the month. This method assigns days with the values of 1 to 31.
setDate	public void setDate(int date)	Sets the date.
getDay	public int getDay()	Returns the day of the week. This method assigns days of the week with the values 0–6, with 0 being Sunday.

setDay	public void setDay(int day)	Sets the day of the week.
getHours	public int getHours()	Returns the hour. This method assigns the value of the hours of the day to range from 0 to 23, with midnight equal to 0.
setHours	public void setHours(int hours)	Sets the hours.
getMinutes	public int getMinutes()	Returns the minute. This method assigns the minutes of an hour to be any value from 0 to 59.
setMinutes	public void setMinutes (int minutes)	Sets the minutes.
getSeconds	public int getSeconds()	Returns the second. This method assigns the seconds of a minute to values of 0–59.
setSeconds	public void setSeconds (int seconds)	Sets the seconds.
getTime	public long getTime()	Returns the time in milliseconds since the epoch.
setTime	public void setTime(long time)	Sets the time.
before	public boolean before (Date when)	Checks whether this date comes before the specified date.
after	public boolean after (Date when)	Checks whether this date comes after the specified date.
equals	public boolean equals (Object obj)	Compares this object against the specified object.
hashCode	public int hashCode()	Computes a hashCode.
toString	public String toString()	Converts a date to a String, using the UNIX ctime conventions.
toLocale String	public String toLocaleString()	Converts a date to a String, using the locale conventions.
toGMTString	public String toGMTString()	Converts a date to a String, using the Internet GMT conventions.
getTimezone Offset	public int getTimezoneOffset()	Return the time zone offset in minutes for the current locale that is appropriate for this time. This value would be a constant except for daylight savings time.

Class java.util.Observable

This class should be subclassed by observable object, or "data" in the Model-View paradigm. An Observable object may have any number of Observers. Whenever the Observable instance changes, it notifies all of its observers. Notification is done by calling the update() method on all observers.

Constructor

- **Observable**

public Observable()- Default constructor

Methods

Methods	Usage	Use
addObserver	public synchronized void addObserver(Observer o)	Adds an observer to the observer list.
delete Observer	public synchronized void deleteObserver(Observer o)	Deletes an observer from the observer list.
notify Observers	public void notify Observers()	Notifies all observers if an observable change occurs.
notify Observers	public synchronized void notifyObservers(Object arg)	Notifies all observers of the specified observable change which occurred.
delete Observers	public synchronized void deleteObservers()	Deletes observers from the observer list.
setChanged	protected synchronized void setChanged()	Sets a flag to note an observable change.
clearChanged	protected synchronized void clearChanged()	Clears an observable change.
hasChanged	public synchronized boolean hasChanged()	Returns a true boolean if an observable change has occurred.
count Observers	public synchronized int count Observers()	Counts the number of observers.

Class java.util.Random

A Random class generates a stream of pseudo-random numbers.

Constructors

To create a new random number generator, use one of the following methods:

- public Random()

Creates a new random number generator. Its seed will be initialized to a value based on the current time.

- public Random(long seed)

Creates a new random number generator using a single long seed.

Methods

Methods	Usage	Use
setSeed	public synchronized void setSeed(long seed)	Sets the seed of the random number generator using a single long seed.
nextInt	public int nextInt()	Generates a pseudorandom uniformally distributed int value.
nextLong	public long nextLong()	Generate a pseudorandom uniformally distributed long value.
nextFloat	public float nextFloat()	Generates a pseudorandom uniformally distributed float value between 0.0 and 1.0.
nextDouble	public double nextDouble()	Generates a pseudorandom uniformally distributed double value between 0.0 and 1.0.
nextGaussian	public synchronized double nextGaussian()	Generates a pseudorandom Gaussian distributed double value with mean 0.0 and standard deviation 1.0.

Class java.util.StringTokenizer

StringTokenizer is a class that controls simple linear tokenization of a String. The set of delimiters, which defaults to common whitespace characters, may be specified at creation time or on a per-token basis.

Example usage:

```
String s = "this is a test";
StringTokenizer st = new StringTokenizer(s);
while (st.hasMoreTokens()) {
        println(st.nextToken());
}
```

Prints the following on the console:

```
this
is
a
test
```

Constructors

- public StringTokenizer(String str,String delim,boolean returnTokens)

Constructs a StringTokenizer on the specified String, using the specified delimiter set. str—the input String, delim—the delimiter String, returnTokens—returns delimiters as tokens or skip them.

- public StringTokenizer(String str, String delim)

Constructs a StringTokenizer on the specified String, using the specified delimiter set.

- public StringTokenizer(String str)

Constructs a StringTokenizer on the specified String, using the default delimiter set (which is " \t\n\r").

Methods

Methods	Usage	Use
hasMore Tokens	public boolean hasMoreTokens()	Returns true if more tokens exist.
nextToken	public String nextToken()	Returns the next token of the String.
nextToken	public String nextToken (String delim)	Returns the next token, after switching to the new delimiter set. The new delimiter set remains the default after this call.
hasMore Elements	public boolean hasMore Elements()	Returns true if the Enumeration has more elements.
nextElement	public Object nextElement()	Returns the next element in the Enumeration.
countTokens	public int countTokens()	Returns the next number of tokens in the String using the current deliminter set.

5. JAVA COLLECTIONS FRAMEWORK

The collections framework defines several algorithms that can be applied to collections and maps. These algorithms are defined as static methods within the Collections class. Several of the methods can throw a **ClassCastException**, which occurs when an attempt is made to compare incompatible types, or an **UnsupportedOperationException**, which occurs when an attempt is made to modify an unmodifiable collection.

Collections define three static variables: EMPTY_SET, EMPTY_LIST, and EMPTY_MAP. All are immutable.

The methods defined in collection framework's algorithm are summarized in the following table:

Methods	Usage	Use
binarySearch	static int binarySearch(List list, Object value, Comparator c)	Searches for value in list ordered according to c. Returns the position of value in list, or –1 if value is not found.
binarySearch	static int binarySearch(List list, Object value)	Searches for value in list. The list must be sorted. Returns the position of value in list, or –1 if value is not found.

copy	static void copy(List list1, List list2)	Copies the elements of list2 to list1.
enumeration	static Enumeration enumeration (Collection c)	Returns an enumeration over c.
fill	static void fill(List list, Object obj)	Assigns obj to each element of list.
indexOfSub List	static int indexOfSubList(List list, List subList)	Searches list for the first occurrence of subList. Returns the index of the first match, or .1 if no match is found.
indexOfSub List	static int lastIndexOfSubList (List list, List subList)	Searches list for the last occurrence of subList. Returns the index of the last match, or .1 if no match is found.
list	static ArrayList list (Enumeration enum)	Returns an ArrayList that contains the elements of enum.
max	static Object max(Collection c, Comparator comp)	Returns the maximum element in c as determined by comp.
max	static Object max(Collection c)	Returns the maximum element in c as determined by natural ordering. The collection need not be sorted.
min	static Object min(Collection c, Comparator comp)	Returns the minimum element in c as determined by comp. The collection need not be sorted.
min	static Object min(Collection c)	Returns the minimum element in c as determined by natural ordering.
nCopies	static List nCopies(int num, Object obj)	Returns num copies of obj contained in an immutable list. num must be greater than or equal to zero.
replaceAll	static boolean replaceAll(List list, Object old, Object new)	Replaces all occurrences of old with new in list. Returns true if at least one replacement occurred. Returns false, otherwise.
reverse	static void reverse(List list)	Reverses the sequence in list.
reverseOrder	static Comparator reverse Order( )	Returns a reverse comparator
rotate	static void rotate(List list, int n)	Rotates list by n places to the right. To rotate left, use a negative value for n.

shuffle	static void shuffle(List list, Random r)	Shuffles (i.e., randomizes) the elements in list by using r as a source of random numbers.
shuffle	static void shuffle(List list)	Shuffles (i.e., randomizes) the elements in list.
singleton	static Set singleton(Object obj)	Returns obj as an immutable set. This is an easy way to convert a single object into a set.
singletonList	static List singletonList (Object obj)	Returns obj as an immutable list. This is an easy way to convert a single object into a list.
singletonMap	static Map singletonMap (Object k, Object v)	Returns the key/value pair k/v as an immutable map. This is an easy way to convert a single key/value pair into a map.
sort	static void sort(List list, Comparator comp)	Sorts the elements of list as determined by comp.
sort	static void sort(List list)	Sorts the elements of list as determined by their natural ordering.
swap	static void swap(List list, int idx1, int idx2)	Exchanges the elements in list at the indices specified by idx1 and idx2.
synchronized Collection	static Collection synchronized Collection(Collection c)	Returns a thread-safe collection backed by c.
synchronized List	static List synchronized List(List list)	Returns a thread-safe list backed by list.
synchronized Map	static Map synchronized Map(Map m)	Returns a thread-safe map backed by m.
synchronized Set	static Set synchronizedSet(Set s)	Returns a thread-safe set backed by s.
synchronized SortedMap	static SortedMap synchronized SortedMap(SortedMap sm)	Returns a thread-safe sorted set backed by sm.
synchronized SortedSet	static SortedSet synchronized SortedSet(SortedSet ss)	Returns a thread-safe set backed by ss.
unmodifiable Collection	static Collection unmodifiable Collection(Collection c)	Returns an unmodifiable collection backed by c.
unmodifiable List	static List unmodifiableList (List list)	Returns an unmodifiable list backed by list.
unmodifiable Map	static Map unmodifiable Map(Map m)	Returns an unmodifiable map backed by m.

unmodifiable Set	static Set unmodifiableSet (Set s)	Returns an unmodifiable set backed by s.
unmodified SortedMap	static SortedMap unmodifiable SortedMap(SortedMap sm)	Returns an unmodifiable sorted map backed by sm.
unmodified SortedSet	static SortedSet unmodifiable SortedSet(SortedSet ss)	Returns an unmodifiable sorted set backed by ss.

Example

Following is the example which demonstrate various algorithms.

```
import java.util.*;
class AlgorithmsDemo {
 public static void main(String args[]) {
  // Create and initialize linked list
  LinkedList ll = new LinkedList();
  ll.add(new Integer(-8));
   ll.add(new Integer(20));
   ll.add(new Integer(-20));
   ll.add(new Integer(8));
        // Create a reverse order comparator
   Comparator r = Collections.reverseOrder();
   // Sort list by using the comparator
   Collections.sort(ll, r);
   // Get iterator
   Iterator li = ll.iterator();
   System.out.print("List sorted in reverse: ");
   while(li.hasNext()){
     System.out.print(li.next() + " ");
   }   System.out.println();
   Collections.shuffle(ll);
   // display randomized list
   li = ll.iterator();
   System.out.print("List shuffled: ");
   while(li.hasNext()){
     System.out.print(li.next() + " ");
   }   System.out.println();
   System.out.println("Minimum: " + Collections.min(ll));
   System.out.println("Maximum: " + Collections.max(ll));
 }
}
```

This would produce the following results:

List sorted in reverse: 20 8 -8 -20

List shuffled: 20 -20 8 -8

Minimum: -20

Maximum: 20

5.1 Interfaces

The Set Interface

A Set is a Collection that cannot contain duplicate elements. It models the mathematical set abstraction. The Set interface contains only methods inherited from Collection and adds the restriction that duplicate elements are prohibited. Set also adds a stronger contract on the behavior of the equals and hashCode operations, allowing Set instances to be compared meaningfully even if their implementation types differ.

The methods declared by Set are summarized in the following table:

Methods	Use
add( )	Adds an object to the collection
clear( )	Removes all objects from the collection
contains( )	Returns true if a specified object is an element within the collection
isEmpty( )	Returns true if the collection has no elements
iterator( )	Returns an Iterator object for the collection which may be used to retrieve an object
remove( )	Removes a specified object from the collection
size( )	Returns the number of elements in the collection

Example

Set have its implementation in various classes like HashSet, TreeSet, HashSet, LinkedHashSet, following is the example to explain Set functionality:

```
import java.util.*;
public class SetDemo {
  public static void main(String args[]) {
  int count[]={34, 22,10,60,30,22};
  Set<Integer> set = new HashSet<Integer>();
  try{
  for(int i=0; i<5; i++){
    set.add(count[i]);
  }
  System.out.println(set);

  TreeSet sortedSet=new TreeSet<Integer>(set);
  System.out.println("The sorted list is:");
    System.out.println(sortedSet);

    System.out.println("The First element of the set is: "+
                    (Integer)sortedSet.first());
    System.out.println("The last element of the set is: "+
                    (Integer)sortedSet.last());
  }
    catch(Exception e){}
```

```
    }
}
```

This would produce the following results:

```
java SetDemo
[34, 30, 60, 10, 22]
The sorted list is:
[10, 22, 30, 34, 60]
The First element of the set is: 10
The last element of the set is: 60
```

The SortedSet Interface

The SortedSet interface extends Set and declares the behavior of a set sorted in ascending order. In addition to those methods defined by Set, the SortedSet interface declares the methods summarized in the table below:

Several methods throw a NoSuchElementException when no items are contained in the invoking set. A ClassCastException is thrown when an object is incompatible with the elements in a set.

A NullPointerException is thrown if an attempt is made to use a null object and null is not allowed in the set.

Methods	Usage	Use
comparator	Comparator comparator( )	Returns the invoking sorted set's comparator. If the natural ordering is used for this set, null is returned.
first	Object first( )	Returns the first element in the invoking sorted set.
headSet	SortedSet headSet(Object end)	Returns a SortedSet containing those elements less than end that are contained in the invoking sorted set. Elements in the returned sorted set are also referenced by the invoking sorted set.
last	Object last( )	Returns the last element in the invoking sorted set.
subSet	SortedSet subSet(Object start, Object end)	Returns a SortedSet that includes those elements between start and end.1. Elements in the returned collection are also referenced by the invoking object.
tailSet	SortedSet tailSet(Object start)	Returns a SortedSet that contains those elements greater than or equal to start that are contained in the sorted set. Elements in the returned set are also referenced by the invoking object.

Example

SortedSet have its implementation in various classes like TreeSet, following is the example for a TreeSet class:

```
public class SortedSetTest {
    public static void main(String[] args) {
        // Create the sorted set
        SortedSet set = new TreeSet();
        // Add elements to the set
        set.add("b");
        set.add("c");
        set.add("a");
        // Iterating over the elements in the set
        Iterator it = set.iterator();
        while (it.hasNext()) {
            // Get element
            Object element = it.next();
            System.out.println(element.toString());
        }
    }
}
```

This would produce the following results:

```
a
b
c
```

The Map Interface

The Map interface maps unique keys to values. A key is an object that you use to retrieve a value at a later date.

- Given a key and a value, you can store the value in a Map object. After the value is stored, you can retrieve it by using its key.
- Several methods throw a NoSuchElementException when no items exist in the invoking map.
- A ClassCastException is thrown when an object is incompatible with the elements in a map.
- A ClassCastException is thrown when an object is incompatible with the elements in a map.
- A NullPointerException is thrown if an attempt is made to use a null object and null is not allowed in the map.
- An UnsupportedOperationException is thrown when an attempt is made to change an unmodifiable map.

Methods	Usage	Use
Clear	void clear( )	Removes all key/value pairs from the invoking map.
key	boolean containsKey(Object k)	Returns true if the invoking map contains k as a key. Otherwise, returns false.

Value	boolean containsValue(Object v)	Returns true if the map contains v as a value. Otherwise, returns false.
entrySet	Set entrySet( )	Returns a Set that contains the entries in the map. The set contains objects of type Map.Entry. This method provides a set-view of the invoking map.
equals	boolean equals(Object obj)	Returns true if obj is a Map and contains the same entries. Otherwise, returns false.
get	Object get(Object k)	Returns the value associated with the key k.
hashCode	int hashCode( )	Returns the hash code for the invoking map.
Empty	boolean isEmpty( )	Returns true if the invoking map is empty. Otherwise, returns false.
keySet	Set keySet( )	Returns a Set that contains the keys in the invoking map. This method provides a set-view of the keys in the invoking map.
put	Object put(Object k, Object v)	Puts an entry in the invoking map, overwriting any previous value associated with the key. The key and value are k and v, respectively. Returns null if the key did not already exist. Otherwise, the previous value linked to the key is returned.
putAll	void putAll(Map m)	Puts all the entries from m into this map.
remove	Object remove(Object k)	Removes the entry whose key equals k.
size	int size( )	Returns the number of key/value pairs in the map.
values	Collection values( )	Returns a collection containing the values in the map. This method provides a collection-view of the values in the map.

Example

Map has its implementation in various classes like HashMap, following is the example to explain map functionality:

```
import java.util.*;
public class CollectionsDemo {
  public static void main(String[] args) {
    Map m1 = new HashMap();
    m1.put("Jamal", "8");
    m1.put("Mahnaz", "31");
    m1.put("Ayan", "12");
    m1.put("Daisy", "14");
    System.out.println();
    System.out.println(" Map Elements");
    System.out.print("\t" + m1);
  }
}
```

This would produce the following results:

```
Map Elements
     {Mahnaz=31, Ayan=12, Daisy=14, Jamal=8}
```

The Map.Entry Interface

The Map.Entry interface enables you to work with a map entry. The **entrySet( )** method declared by the Map interface returns a Set containing the map entries. Each of these set elements is a Map.Entry object. Following table summarizes the methods declared by this interface.

Methods	Usage	Use
equals	boolean equals(Object obj)	Returns true if obj is a Map.Entry whose key and value are equal to that of the invoking object.
getKey	Object getKey( )	Returns the key for this map entry.
getValue	Object getValue( )	Returns the value for this map entry.
hashCode	int hashCode( )	Returns the hash code for this map entry.
setValue	Object setValue(Object v)	Sets the value for this map entry to v. A ClassCastException is thrown if v is not the correct type for the map. A NullPointerException is thrown if v is null and the map does not permit null keys. An UnsupportedOperation Exception is thrown if the map cannot be changed.

Example

Following is the example showing how **Map.Entry** can be used:

```
import java.util.*;
class HashMapDemo {
  public static void main(String args[]) {
    // Create a hash map
    HashMap hm = new HashMap();
    // Put elements to the map
    hm.put("Jamal", new Double(3434.34));
    hm.put("Mahnaz", new Double(123.22));
    hm.put("Ayan", new Double(1378.00));
    hm.put("Daisy", new Double(99.22));
    hm.put("Qadir", new Double(-19.08));

    // Get a set of the entries
    Set set = hm.entrySet();
    // Get an iterator
    Iterator i = set.iterator();
    // Display elements
    while(i.hasNext()) {
      Map.Entry me = (Map.Entry)i.next();
      System.out.print(me.getKey() + ": ");
      System.out.println(me.getValue());
    }
    System.out.println();
    // Deposit 1000 into Jamal's account
    double balance = ((Double)hm.get("Jamal")).doubleValue();
    hm.put("Jamal", new Double(balance + 1000));
    System.out.println("Jamal's new balance: " +
    hm.get("Jamal"));
  }
}
```

This would produce the following results:

```
Daisy 99.22
Qadir: -19.08
Jamal: 3434.34
Ayan: 1378.0
Mahnaz: 123.22
Jamal.s current balance: 4434.34
```

The SortedMap Interface

The SortedMap interface extends Map. It ensures that the entries are maintained in ascending key order. Several methods throw a NoSuchElementException when no items are in the invoking map. A ClassCastException is thrown when an object is incompatible with the elements in a map. A NullPointerException is thrown if an attempt is made to use a null object when null is not allowed in the map.

The methods declared by SortedMap are summarized in the following table:

Methods	Usage	Use
comparator	Comparator comparator( )	Returns the invoking sorted map's comparator. If the natural ordering is used for the invoking map, null is returned.
firstKey	Object firstKey( )	Returns the first key in the invoking map.
headMap	SortedMap headMap (Object end)	Returns a sorted map for those map entries with keys that are less than end.
lastKey	Object lastKey( )	Returns the last key in the invoking map.
subMap	SortedMap subMap(Object start, Object end)	Returns a map containing those entries with keys that are greater than or equal to start and less than end.
tailMap	SortedMap tailMap(Object start)	Returns a map containing those entries with keys that are greater than or equal to start.

Example

SortedMap have its implementation in various classes like TreeMap, following is the example to explain SortedMap functionality:

```
import java.util.*;
class TreeMapDemo {
  public static void main(String args[]) {
    // Create a hash map
    TreeMap tm = new TreeMap();
    // Put elements to the map
    tm.put("Jamal", new Double(3434.34));
    tm.put("Mahnaz", new Double(123.22));
    tm.put("Ayan", new Double(1378.00));
    tm.put("Daisy", new Double(99.22));
    tm.put("Qadir", new Double(-19.08));

       // Get a set of the entries
    Set set = tm.entrySet();
    // Get an iterator
    Iterator i = set.iterator();
    // Display elements
    while(i.hasNext()) {
      Map.Entry me = (Map.Entry)i.next();
      System.out.print(me.getKey() + ": ");
```

```
        System.out.println(me.getValue());
      }
      System.out.println();
      // Deposit 1000 into Jamal's account
      double balance = ((Double)tm.get("Jamal")).doubleValue();
      tm.put("Jamal", new Double(balance + 1000));
      System.out.println("Jamal's new balance: " +
      tm.get("Jamal"));
    }
  }
```

This would produce the following results:

```
Ayan: 1378.0
Daisy 99.22
Mahnaz: 123.22
Qadir: -19.08
Jamal: 3434.34
Jamal.s current balance: 4434.34
```

Iterator

Often, you will want to cycle through the elements in a collection. For example, you might want to display each element.

The easiest way to do this is to employ an iterator, which is an object that implements either the Iterator or the ListIterator interface.

Iterator enables you to cycle through a collection, obtaining or removing elements. ListIterator extends Iterator to allow bidirectional traversal of a list, and the modification of elements.

Before you can access a collection through an iterator, you must obtain one. Each of the collection classes provides an iterator( ) method that returns an iterator to the start of the collection. By using this iterator object, you can access each element in the collection, one element at a time.

In general, to use an iterator to cycle through the contents of a collection, follow these steps:

1. Obtain an iterator to the start of the collection by calling the collection's iterator( ) method.
2. Set up a loop that makes a call to hasNext( ). Have the loop iterate as long as hasNext( ) returns true.
3. Within the loop, obtain each element by calling next( ).

For collections that implement List, you can also obtain an iterator by calling ListIterator.

The Methods Declared by Iterator:

Methods	Usage	Use
hasNext	boolean hasNext( )	Returns true if there are more elements. Otherwise, returns false.
next	Object next( )	Returns the next element. Throws NoSuchElement Exception if there is not a next element.
remove	void remove( )	Removes the current element. Throws IllegalStateException if an attempt is made to call remove( ) that is not preceded by a call to next( ).

The Methods Declared by List Iterator

Methods	Usage	Use
add	void add(Object obj)	Inserts obj into the list in front of the element that will be returned by the next call to next( ).
hasNext	boolean hasNext( )	Returns true if there is a next element. Otherwise, returns false.
hasPrevious	boolean hasPrevious( )	Returns true if there is a previous element. Otherwise, returns false.
next	Object next( )	Returns the next element. A NoSuchElementException is thrown if there is not a next element.
nextIndex	int nextIndex( )	Returns the index of the next element. If there is not a next element, returns the size of the list.
previous	Object previous( )	Returns the previous element. A NoSuchElementException is thrown if there is not a previous element.
previousIndex	int previousIndex( )	Returns the index of the previous element. If there is not a previous element, returns -1.
remove	void remove( )	Removes the current element from the list. An IllegalState Exception is thrown if remove( ) is called before next( ) or previous( ) is invoked.
set	void set(Object obj)	Assigns obj to the current element. This is the element last returned by a call to either next( ) or previous( ).

Example

Here is an example demonstrating both Iterator and ListIterator. It uses an ArrayList object, but the general principles apply to any type of collection. Of course, ListIterator is available only to those collections that implement the List interface.

```
import java.util.*;
class IteratorDemo {
  public static void main(String args[]) {
    // Create an array list
    ArrayList al = new ArrayList();
    // add elements to the array list
    al.add("C");
```

```
    al.add("A");
    al.add("E");
    al.add("B");
    al.add("D");
    al.add("F");
    // Use iterator to display contents of al
    System.out.print("Original contents of al: ");
    Iterator itr = al.iterator();
    while(itr.hasNext()) {
      Object element = itr.next();
      System.out.print(element + " ");
    }    System.out.println();
   // Modify objects being iterated
    ListIterator litr = al.listIterator();
    while(litr.hasNext()) {
      Object element = litr.next();
      litr.set(element + "+");
    }
    System.out.print("Modified contents of al: ");
    itr = al.iterator();
    while(itr.hasNext()) {
      Object element = itr.next();
      System.out.print(element + " ");
    }    System.out.println();
    // Now, display the list backwards
    System.out.print("Modified list backwards: ");
    while(litr.hasPrevious()) {
      Object element = litr.previous();
      System.out.print(element + " ");
    }
    System.out.println();
  }
}
```

This would produce the following results:
Original contents of al: C A E B D F
Modified contents of al: C+ A+ E+ B+ D+ F+
Modified list backwards: F+ D+ B+ E+ A+ C+

Comparator Interface

Both TreeSet and TreeMap store elements in sorted order. However, it is the comparator that defines precisely what *sorted order* means.

The Comparator interface defines two methods: compare( ) and equals( ). The compare( ) method, shown here, compares two elements for order:

The Compare Method

```
int compare(Object obj1, Object obj2)
```

obj1 and obj2 are the objects to be compared. This method returns zero if the objects are equal. It returns a positive value if obj1 is greater than obj2. Otherwise, a negative value is returned.

By overriding compare( ), you can alter the way that objects are ordered. For example, to sort in reverse order, you can create a comparator that reverses the outcome of a comparison.

The Equals Method

The equals( ) method, shown here, tests whether an object equals the invoking comparator:

boolean equals(Object obj)

obj is the object to be tested for equality. The method returns true if obj and the invoking object are both Comparator objects and use the same ordering. Otherwise, it returns false.

Overriding equals( ) is unnecessary, and most simple comparators will not do so.

Example

```
class Dog implements Comparator<Dog>, Comparable<Dog>{
  private String name;
  private int age;
  Dog(){
  } Dog(String n, int a){
    name = n;
    age = a;
  }
  public String getDogName(){
    return name;
  }
  public int getDogAge(){
    return age;
  }
  // Overriding the compareTo method
  public int compareTo(Dog d){
    return (this.name).compareTo(d.name);
  }
  // Overriding the compare method to sort the age
   public int compare(Dog d, Dog d1){
    return d.age - d1.age;
  }
}
public class Example{
  public static void main(String args[]){
    // Takes a list o Dog objects
    List<Dog> list = new ArrayList<Dog>();
    list.add(new Dog("Shaggy",3));
    list.add(new Dog("Lacy",2));
    list.add(new Dog("Roger",10));
    list.add(new Dog("Tommy",4));
```

```
      list.add(new Dog("Tammy",1));
      Collections.sort(list);// Sorts the array list
      for(Dog a: list)//printing the sorted list of names
        System.out.print(a.getDogName() + ", ");
      // Sorts the array list using comparator
      Collections.sort(list, new Dog());
      System.out.println(" ");
      for(Dog a: list)//printing the sorted list of ages
        System.out.print(a.getDogName() +"  : "+ a.getDogAge() + ", ");
    }
  }
```

This would produce the following results:

Lacy, Roger, Shaggy, Tammy, Tommy,Tammy : 1, Lacy : 2, Shaggy : 3, Tommy : 4, Roger : 10,

Note: Sorting of the Arrays class is same as the Collections.

5.2 The Collection Classes

Java provides a set of standard collection classes that implement Collection interfaces. Some of the classes provide full implementations that can be used as-is and others are abstract class, providing skeletal implementations that are used as starting points for creating concrete collections.

The standard collection classes are summarized in the following table:

Methods	Use
AbstractCollection	Implements most of the Collection interface.
AbstractList	Extends AbstractCollection and implements most of the ListInterface
AbstractSequentialList	Extends AbstractList for use by a collection that uses sequential rather than random access of its elements.
LinkedList	Implements a linked list by extending AbstractSequentialList.
ArrayList	Implements a dynamic array by extending AbstractList.
AbstractSet	Extends AbstractCollection and implements most of the Set interface.
HashSet	Extends AbstractSet for use with a hashtable.
LinkedHashSet	Extends HashSet to allow insertion-order iterations.
TreeSet	Implements a set stored in a tree. Extends AbstractSet.
AbstractMap	Implements most of the Map interface.
HashMap	Extends AbstractMap to use a hashtable.

TreeMap	Extends AbstractMap to use a tree.
WeakHashMap	Extends AbstractMap to use a hash table with weak keys.
LinkedHashMap	Extends HashMap to allow insertion-order iterations.
IdentityHashMap	Extends AbstractMap and uses reference equality when comparing documents.

The *AbstractCollection, AbstractSet, AbstractList, AbstractSequentialList* and *AbstractMap* classes provide skeletal implementations of the core collection interfaces, to minimize the effort required to implement them.

The Linkedlist Class

The LinkedList class extends AbstractSequentialList and implements the List interface. It provides a linked list data structure.

The LinkedList class supports two constructors. The first constructor builds an empty linked list:

LinkedList( )

The following constructor builds a linked list that is initialized with the elements of the collection c.

LinkedList(Collection c)

Apart from the methods inherited from its parent classes, LinkedList defines the following methods:

Methods	Usage	Use
add	void add(int index, Object element)	Inserts the specified element at the specified position index in this list. Throws IndexOutOfBoundsException if the specified index is out of range (index < 0 || index > size()).
add	boolean add(Object o)	Appends the specified element to the end of this list.
addAll	boolean addAll(Collection c)	Appends all of the elements in the specified collection to the end of this list, in the order that they are returned by the specified collection's iterator. Throws NullPointerException if the specified collection is null.
addAll	boolean addAll(int index, Collection c)	Inserts all of the elements in the specified collection into this list, starting at the specified position. Throws NullPointerException if the specified collection is null.

addFirst	void addFirst(Object o)	Inserts the given element at the beginning of this list.
clear	void clear()	Removes all of the elements from this list.
clone	Object clone()	Returns a shallow copy of this LinkedList.
contains	boolean contains(Object o)	Returns true if this list contains the specified element. More formally, returns true if and only if this list contains at least one element e such that (o==null ? e==null : o.equals(e)).
Get	Object get(int index)	Returns the element at the specified position in this list. Throws IndexOutOfBounds Exception if the specified index is out of range (index < 0 || index >= size()).
getFirst	Object getFirst()	Returns the first element in this list. Throws NoSuchElement Exception if this list is empty.
getLast	Object getLast()	Returns the last element in this list. Throws NoSuchElement Exception if this list is empty.
indexOf	int indexOf(Object o)	Returns the index in this list of the first occurrence of the specified element, or –1 if the List does not contain this element.
lastIndexOf	int lastIndexOf(Object o)	Returns the index in this list of the last occurrence of the specified element, or –1 if the list does not contain this element.
listIterator	ListIterator listIterator (int index)	Returns a list-iterator of the elements in this list (in proper sequence), starting at the specified position in the list. Throws IndexOutOfBounds Exception if the specified index is out of range (index < 0 || index >= size()).
remove	Object remove(int index)	Removes the element at the specified position in this list. Throws NoSuchElement Exception if this list is empty.

remove	boolean remove(Object o)	Removes the first occurrence of thespecified element in this list. ThrowsNoSuchElement Exception if this list is empty. Throws IndexOutOfBounds Exception if the specified index is out of range (index < 0 || index >= size()).
removeFirst	Object removeFirst()	Removes and returns the first element from this list. Throws NoSuchElement Exception if this list is empty.
removeLast	Object removeLast()	Removes and returns the last element from this list. Throws NoSuchElementException if this list is empty.
set	Object set(int index, Object element)	Replaces the element at the specified position in this list with the specified element. Throws IndexOutOfBoundsException if the specified index is out of range (index < 0 || index >= size()).
size	int size()	Returns the number of elements in this list.
toArray	Object[] toArray()	Returns an array containing all of the elements in this list in the correct order. Throws Null PointerException if the specified array is null.
toArray	Object[] toArray(Object[] a)	Returns an array containing all of the elements in this list in the correct order; the runtime type of the returned array is that of the specified array.

Example

The following program illustrates several of the methods supported by LinkedList:

```
import java.util.*;
class LinkedListDemo {
  public static void main(String args[]) {
    // create a linked list
    LinkedList ll = new LinkedList();
    // add elements to the linked list
    ll.add("F");
```

```
        ll.add("B");
        ll.add("D");
        ll.add("E");
        ll.add("C");
        ll.addLast("Z");
        ll.addFirst("A");
        ll.add(1, "A2");
        System.out.println("Original contents of ll: " + ll);
        // remove elements from the linked list
        ll.remove("F");
        ll.remove(2);
        System.out.println("Contents of ll after deletion: " + ll);
            // remove first and last elements
        ll.removeFirst();
        ll.removeLast();
        System.out.println("ll after deleting first and last: " + ll);
        // get and set a value
        Object val = ll.get(2);
        ll.set(2, (String) val + " Changed");
        System.out.println("ll after change: " + ll);
    }
}
```

This would produce the following results:

```
Original contents of ll: [A, A2, F, B, D, E, C, Z]
Contents of ll after deletion: [A, A2, D, E, C, Z]
ll after deleting first and last: [A2, D, E, C]
ll after change: [A2, D, E Changed, C]
```

The ArrayList Class

The ArrayList class extends AbstractList and implements the List interface. ArrayList supports dynamic arrays that can grow as needed.

Standard Java arrays are of a fixed length. After arrays are created, they cannot grow or shrink, which means that you must know in advance how many elements an array will hold.

Array lists are created with an initial size. When this size is exceeded, the collection is automatically enlarged. When objects are removed, the array may be shrunk.

The ArrayList class supports three constructors. The first constructor builds an empty array list.:

ArrayList( )

The following constructor builds an array list that is initialized with the elements of the collection c.

ArrayList(Collection c)

The following constructor builds an array list that has the specified initial capacity. The capacity is the size of the underlying array that is used to store the elements.

The capacity grows automatically as elements are added to an array list.

ArrayList(int capacity)

Apart from the methods inherited from its parent classes, ArrayList defines the following methods:

Methods	Usage	Use
add	Void add(int index, Object element)	Inserts the specified element at the specified position index in this list. Throws IndexOutOfBounds Exception if the specified index is out of range (index < 0 || index > size()).
add	Boolean add(Object o)	Appends the specified element to the end of this list.
addAll	Boolean addAll(Collection c)	Appends all of the elements in the specified collection to the end of this list, in the order that they are returned by the specified collection's iterator. Throws NullPointerException if the specified collection is null.
addAll	Boolean addAll(int index, Collection c)	Inserts all of the elements in the specified collection into this list, starting at the specified position. Throws NullPointerException if the specified collection is null.
clear	Void clear()	Removes all of the elements from this list.
clone	Object clone()	Returns a shallow copy of this ArrayList.
contains	Boolean contains(Object o)	Returns true if this list contains the specified element. More formally, returns true if and only if this list contains at least one element e such that (o==null ? e==null : o.equals(e)).
ensure Capacity	Void ensureCapacity(int minCapacity)	Increases the capacity of this ArrayList instance, if necessary, to ensure that it can hold at least the number of elements specified by the minimum capacity argument.
get	Object get(int index)	Returns the element at the specified position in this list. Throws IndexOutOfBounds Exception if the specified index is out of range (index < 0 || index >= size()).

indexof	Int indexOf(Object o)	Returns the index in this list of the first occurrence of the specified element, or –1 if the List does not contain this element.
lastindexOf	Int lastIndexOf(Object o)	Returns the index in this list of the last occurrence of the specified element, or –1 if the list does not contain this element.
remove	Object remove(int index)	Removes the element at the specified position in this list. Throws IndexOutOfBounds Exception if index out of range (index < 0 || index >= size()).
removeRange	Protected void removeRange (int fromIndex, int toIndex)	Removes from this List all of the elements whose index is between fromIndex, inclusive and toIndex, exclusive.
set	Object set(int index, Object element)	Replaces the element at the specified position in this list with the specified element. Throws IndexOutOfBoundsException if the specified index is out of range (index < 0 || index >= size()).
size	Int size()	Returns the number of elements in this list.
toArray	Object[] toArray()	Returns an array containing all of the elements in this list in the correct order. Throws Null PointerException if the specified array is null.
toArray	Object[] toArray(Object[] a)	Returns an array containing all of the elements in this list in the correct order; the runtime type of the returned array is that of the specified array.
trimToSize	Void trimToSize()	Trims the capacity of this ArrayList instance to be the list's current size.

Example

The following program illustrates several of the methods supported by ArrayList:

```
import java.util.*;
class ArrayListDemo {
```

```
public static void main(String args[]) {
  // create an array list
  ArrayList al = new ArrayList();
  System.out.println("Initial size of al: " +
  al.size());

  // add elements to the array list
  al.add("C");
  al.add("A");
  al.add("E");
  al.add("B");
  al.add("D");
  al.add("F");
  al.add(1, "A2");
  System.out.println("Size of al after additions: " +
    al.size());

  // display the array list
  System.out.println("Contents of al: " + al);
  // Remove elements from the array list
  al.remove("F");
  al.remove(2);
  System.out.println("Size of al after deletions: " +
    al.size());
  System.out.println("Contents of al: " + al);
 }
}
```

This would produce the following results:
Initial size of al: 0
Size of al after additions: 7
Contents of al: [C, A2, A, E, B, D, F]
Size of al after deletions: 5
Contents of al: [C, A2, E, B, D]

The HashSet Class

HashSet extends AbstractSet and implements the Set interface. It creates a collection that uses a hashtable for storage.

A hashtable stores information by using a mechanism called hashing. In hashing, the informational content of a key is used to determine a unique value, called its hash code.

The hash code is then used as the index at which the data associated with the key is stored. The transformation of the key into its hash code is performed automatically.

The HashSet class supports four constructors. The first form constructs a default hash set:

HashSet( )

The following constructor form initializes the hash set by using the elements of c.

HashSet(Collection c)

The following constructor form initializes the capacity of the hash set to capacity. The capacity grows automatically as elements are added to the Hash.

HashSet(int capacity)

The fourth form initializes both the capacity and the fill ratio (also called load capacity) of the hash set from its arguments:

HashSet(int capacity, float fillRatio)

Here the fill ratio must be between 0.0 and 1.0, and it determines how full the hash set can be before it is resized upward. Specifically, when the number of elements is greater than the capacity of the hash set multiplied by its fill ratio, the hash set is expanded.

Apart from the methods inherited from its parent classes, HashSet defines the following methods:

Methods	Usage	Use
add	Boolean add(Object o)	Adds the specified element to this set if it is not already present.
clear	Void clear()	Removes all of the elements from this set.
clone	Object clone()	Returns a shallow copy of this HashSet instance: The elements themselves are not cloned.
contains	Boolean contains(Object o)	Returns true if this set contains the specified element
isEmpty	Boolean isEmpty()	Returns true if this set contains no elements.
iterator	Iterator iterator()	Returns an iterator over the elements in this set.
remove	Boolean remove(Object o)	Removes the specified element from this set if it is present.
size	Int size()	Returns the number of elements in this set (its cardinality).

Example

The following program illustrates several of the methods supported by HashSet:

```
import java.util.*;
class HashSetDemo {
  public static void main(String args[]) {
    // create a hash set
    HashSet hs = new HashSet();
    // add elements to the hash set
    hs.add("B");
    hs.add("A");
    hs.add("D");
```

```
      hs.add("E");
      hs.add("C");
      hs.add("F");
      System.out.println(hs);
   }
}
```

This would produce the following results:
[A, F, E, D, C, B]

The LinkedHashSet Class

This class extends HashSet, but adds no members of its own. LinkedHashSet maintains a linked list of the entries in the set, in the order in which they were inserted. This allows insertion-order iteration over the set. That is, when cycling through a LinkedHashSet using an iterator, the elements will be returned in the order in which they were inserted. The hash code is then used as the index at which the data associated with the key is stored. The transformation of the key into its hash code is performed automatically.

The LinkedHashSet class supports four constructors. The first form constructs a default hash set:

LinkedHashSet( )

The following constructor form initializes the hash set by using the elements of c.

LinkedHashSet(Collection c)

The following constructor form initializes the capacity of the hash set to capacity. The capacity grows automatically as elements are added to the Hash.

LinkedHashSet(int capacity)

The fourth form initializes both the capacity and the fill ratio (also called load capacity) of the hash set from its arguments:

LinkedHashSet(int capacity, float fillRatio)

Example

The following program illustrates several of the methods supported by LinkedHashSet:

```
import java.util.*;
class HashSetDemo {
  public static void main(String args[]) {
    // create a hash set
    LinkedHashSet hs = new LinkedHashSet();
    // add elements to the hash set
    hs.add("B");
    hs.add("A");
    hs.add("D");
    hs.add("E");
    hs.add("C");
    hs.add("F");
    System.out.println(hs);  }}
```

This would produce the following results:
[B, A, D, E, C, F]

The TreeSet Class

TreeSet provides an implementation of the Set interface that uses a tree for storage. Objects are stored in sorted, ascending order. Access and retrieval times are quite fast, which makes TreeSet an excellent choice when storing large amounts of sorted information that must be found quickly. The TreeSet class supports four constructors. The first form constructs an empty tree set that will be sorted in ascending order according to the natural order of its elements:

TreeSet( )

The second form builds a tree set that contains the elements of c.

TreeSet(Collection c)

The third form constructs an empty tree set that will be sorted according to the comparator specified by comp.

TreeSet(Comparator comp)

The fourth form builds a tree set that contains the elements of ss:

TreeSet(SortedSet ss)

Apart from the methods inherited from its parent classes, TreeSet defines the following methods:

Methods	Usage	Use
add	Void add(Object o)	Adds the specified element to this set if it is not already present.
addAll	Boolean addAll(Collection c)	Adds all of the elements in the specified collection to this set.
clear	Void clear()	Removes all of the elements from this set.
clone	Object clone()	Returns a shallow copy of this TreeSet instance.
comparator	Comparator comparator()	Returns the comparator used to order this sorted set, or null if this tree set uses its elements natural ordering.
contains	Boolean contains(Object o)	Returns true if this set contains the specified element.
first	Object first()	Returns the first (lowest) element currently in this sorted set.

Example

The following program illustrates several of the methods supported by this collection:

```
import java.util.*;
class TreeSetDemo {
  public static void main(String args[]) {
    // Create a tree set
```

```
    TreeSet ts = new TreeSet();
  // Add elements to the tree set
  ts.add("C");
  ts.add("A");
  ts.add("B");
  ts.add("E");
  ts.add("F");
  ts.add("D");
  System.out.println(ts);
   }
  }
```

This would produce the following results:

[A, B, C, D, E, F]

The HashMap Class

The HashMap class uses a hashtable to implement the Map interface. This allows the execution time of basic operations, such as get( ) and put( ), to remain constant even for large sets.

The HashMap class supports four constructors. The first form constructs a default hash map:

HashMap( )

The second form initializes the hash map by using the elements of m:

HashMap(Map m)

The third form initializes the capacity of the hash map to capacity:

HashMap(int capacity)

The fourth form initializes both the capacity and fill ratio of the hashmap by using its arguments:

HashMap(int capacity, float fillRatio)

Apart from the methods inherited from its parent classes, HashMap defines the following methods:

Methods	Usage	Use
clear	Void clear()	Removes all mappings from this map.
clone	Object clone()	Returns a shallow copy of this HashMap instance: the keys and values themselves are not cloned.
containsKey	Boolean containsKey (Object key)	Returns true if this map contains a mapping for the specified key.
containsValue	Boolean containsValue (Object value)	Returns true if this map maps one or more keys to the specified value.
entrySet	Set entrySet()	Returns a collection view of the mappings contained in this map.
get	Object get(Object key)	Returns the value to which the specified key is mapped in this identity hash map, or null if the map contains no mapping for this key.

isEmpty	Boolean isEmpty()	Returns true if this map contains no key-value mappings.
keySet	Set keySet()	Returns a set view of the keys contained in this map.
put	Object put(Object key, Object value)	Associates the specified value with the specified key in this map.
putAll	PutAll(Map m)	Copies all of the mappings from the specified map to this map. These mappings will replace any mappings that this map had for any of the keys currently in the specified map.
remove	Object remove (Object key)	Removes the mapping for this key from this map if present.
size	Int size()	Returns the number of key-value mappings in this map.
values	Collection values()	Returns a collection view of the values contained in this map.

Example

The following program illustrates several of the methods supported by this collection:

```
import java.util.*;
class HashMapDemo {
  public static void main(String args[]) {
    // Create a hash map
    HashMap hm = new HashMap();
    // Put elements to the map
    hm.put("Jamal", new Double(3434.34));
    hm.put("Mahnaz", new Double(123.22));
    hm.put("Ayan", new Double(1378.00));
    hm.put("Daisy", new Double(99.22));
    hm.put("Qadir", new Double(-19.08));

    // Get a set of the entries
    Set set = hm.entrySet();
    // Get an iterator
    Iterator i = set.iterator();
    // Display elements
    while(i.hasNext()) {
      Map.Entry me = (Map.Entry)i.next();
      System.out.print(me.getKey() + ": ");
      System.out.println(me.getValue());
    }
```

```
System.out.println();
// Deposit 1000 into Jamal's account
double balance = ((Double)hm.get("Jamal")).doubleValue();
hm.put("Jamal", new Double(balance + 1000));
System.out.println("Jamal's new balance: " +
hm.get("Jamal"));
}
}
```

This would produce the following results:

```
Daisy 99.22
Qadir: -19.08
Jamal: 3434.34
Ayan: 1378.0
Mahnaz: 123.22
Jamal's current balance: 4434.34
```

The TreeMap Class

The TreeMap class implements the Map interface by using a tree. A TreeMap provides an efficient means of storing key/value pairs in sorted order, and allows rapid retrieval.

You should note that, unlike a hashmap, a tree map guarantees that its elements will be sorted in ascending key order.

The HashMap class supports four constructors. The first form constructs a default hashmap:

HashMap( )

The second form initializes the hashmap by using the elements of m:

HashMap(Map m)

The third form initializes the capacity of the hashmap to capacity:

HashMap(int capacity)

The fourth form initializes both the capacity and fill ratio of the hashmap by using its arguments:

HashMap(int capacity, float fillRatio)

The TreeMap class supports four constructors. The first form constructs an empty tree map that will be sorted by using the natural order of its keys:

TreeMap( )

The second form constructs an empty tree-based map that will be sorted by using the Comparator comp:

TreeMap(Comparator comp)

The third form initializes a tree map with the entries from m, which will be sorted by using the natural order of the keys:

TreeMap(Map m)

The fourth form initializes a tree map with the entries from sm, which will be sorted in the same order as sm:

*TreeMap(SortedMap sm)*Apart from the methods inherited from its parent classes, TreeMap defines the following methods:

Methods	Usage	Use
clear	Void clear()	Removes all mappings from this TreeMap.
clone	Object clone()	Returns a shallow copy of this TreeMap instance.
comparator	Comparator comparator()	Returns the comparator used to order this map, or null if this map uses its keys' natural order.
containsKey	Boolean containsKey (Object key)	Returns true if this map contains a mapping for the specified key.
containsValue	Boolean containsValue (Object value)	Returns true if this map maps one or more keys to the specified value.
entrySet	Set entrySet()	Returns a set view of the mappings contained in this map.
firstKey	Object firstKey()	Returns the first (lowest) key currently in this sorted map.
get	Object get(Object key)	Returns the value to which this map maps the specified key.
headMap	SortedMap headMap (Object toKey)	Returns a view of the portion of this map whose keys are strictly less than toKey.
keySet	Set keySet()	Returns a Set view of the keys contained in this map.
lastKey	Object lastKey()	Returns the last (highest) key currently in this sorted map.
put	Object put(Object key, Object value)	Associates the specified value with the specified key in this map.
putAll	Void putAll(Map map)	Copies all of the mappings from the specified map to this map.
remove	Object remove(Object key)	Removes the mapping for this key from this TreeMap if present.
size	Int size()	Returns the number of key-value mappings in this map.
subMap	SortedMap subMap(Object from Key, Object toKey)	Returns a view of the portion of this map whose keys range from fromKey, inclusive, to toKey, exclusive.
tailMap	SortedMap tailMap(Object fromKey)	Returns a view of the portion of this map whose keys are greater than or equal to fromKey.
values	Collection values()	Returns a collection view of the values contained in this map.

Example

The following program illustrates several of the methods supported by this collection:

```
import java.util.*;
class TreeMapDemo {
  public static void main(String args[]) {
    // Create a hash map
    TreeMap tm = new TreeMap();
    // Put elements to the map
    tm.put("Jamal", new Double(3434.34));
    tm.put("Mahnaz", new Double(123.22));
    tm.put("Ayan", new Double(1378.00));
    tm.put("Daisy", new Double(99.22));
    tm.put("Qadir", new Double(-19.08));

      // Get a set of the entries
    Set set = tm.entrySet();
    // Get an iterator
    Iterator i = set.iterator();
    // Display elements
    while(i.hasNext()) {
      Map.Entry me = (Map.Entry)i.next();
      System.out.print(me.getKey() + ": ");
      System.out.println(me.getValue());
    }
    System.out.println();
    // Deposit 1000 into Jamal's account
    double balance = ((Double)tm.get("Jamal")).doubleValue();
    tm.put("Jamal", new Double(balance + 1000));
    System.out.println("Jamal's new balance: " +
    tm.get("Jamal"));
  }
}
```

This would produce the following results:

```
Ayan                     :  1378.0
Daisy                    :   99.22
Mahnaz                   :  123.22
Qadir                    :  –19.08
Jamal                    : 3434.34
Jamal's current balance  : 4434.34
```

The WeakHashMap Class

WeakHashMap is an implementation of the Map interface that stores only weak references to its keys. Storing only weak references allows a key-value pair to be garbagecollected when its key is no longer referenced outside of the WeakHashMap.

This class provides the easiest way to harness the power of weak references. It is useful for implementing "registry-like" data structures, where the utility of an entry vanishes when its key is no longer reachable by any thread.

The WeakHashMap functions identically to the HashMap with one very important exception: If the Java memory manager no longer has a strong reference to the object specified as a key, then the entry in the map will be removed.

Weak Reference: If the only references to an object are weak references, the garbage collector can reclaim the object's memory at any time. It does not have to wait until the system runs out of memory. Usually, it will be freed the next time the garbage collector runs.

The WeakHashMap class supports four constructors. The first form constructs a new, empty WeakHashMap with the default initial capacity (16) and the default load factor (0.75):

WeakHashMap()

The second form constructs a new, empty WeakHashMap with the given initial capacity and the default load factor, which is 0.75:

WeakHashMap(int initialCapacity)

The third form constructs a new, empty WeakHashMap with the given initial capacity and the given load factor.

WeakHashMap(int initialCapacity, float loadFactor)

The fourth form constructs a new WeakHashMap with the same mappings as the specified Map:

WeakHashMap(Map t)

Apart from the methods inherited from its parent classes, TreeMap defines the following methods:

Method	Usage	Use
clear	Void clear()	Removes all mappings from this map.
containsKey	Boolean containsKey (Object key)	Returns true if this map contains a mapping for the specified key.
containsValue	Boolean containsValue (Object value)	Returns true if this map maps one or more keys to the specified value.
entrySet	Set entrySet()	Returns a collection view of the mappings contained in this map.
get	Object get(Object key)	Returns the value to which the specified key is mapped in this weak hash map, or null if the map contains no mapping for this key.

isEmpty	Boolean isEmpty()	Returns true if this map contains no key-value mappings.
keySet	Set keySet()	Returns a set view of the keys contained in this map.
put	Object put(Object key, Object value)	Associates the specified value with the specified key in this map.
putAll	Void putAll(Map m)	Copies all of the mappings from the specified map to this map. These mappings will replace any mapping that this map had for any of the keys currently in the specified map.
remove	Object remove (Object key)	Removes the mapping for this key from this map, if present.
size	Int size()	Returns the number of key-value mappings in this map.
values	Collection values()	Returns a collection view of the values contained in this map.

Example

The following program illustrates several of the methods supported by this collection:

```
import java.util.*;
public class WeakHashMap {
  private static Map map;
  public static void main (String args[]) {
    map = new WeakHashMap();
    map.put(new String("Maine"), "Augusta");
        Runnable runner = new Runnable() {
      public void run() {
        while (map.containsKey("Maine")) {
          try {           Thread.sleep(500);
          } catch (InterruptedException ignored) {
          }
          System.out.println("Thread waiting");
          System.gc();
        }
      }
    };
```

```
        Thread t = new Thread(runner);
        t.start();
        System.out.println("Main waiting");
        try {
          t.join();
        } catch (InterruptedException ignored) {
        }
    }
}
```

This would produce the following results:

Main waiting
Thread waiting

If you do not include the call to System.gc(), the system may never run the garbage collector as not much memory is used by the program. For a more active program, the call would be unnecessary.

The LinkedHashMap Class

This class extends HashMap and maintains a linked list of the entries in the map, in the order in which they were inserted. This allows insertion-order iteration over the map. That is, when iterating a LinkedHashMap, the elements will be returned in the order in which they were inserted. You can also create a LinkedHashMap that returns its elements in the order in which they were last accessed.

The LinkedHashMap class supports five constructors. The first form constructs a default LinkedHashMap:

LinkedHashMap( )

The second form initializes the LinkedHashMap with the elements from m:

LinkedHashMap(Map m)

The third form initializes the capacity:

LinkedHashMap(int capacity)

The fourth form initializes both capacity and fill ratio. The meaning of capacity and fill ratio are the same as for HashMap:

LinkedHashMap(int capacity, float fillRatio)

The last form allows you to specify whether the elements will be stored in the linked list by insertion order, or by order of last access. If Order is true, then access order is used. If Order is false, then insertion order is used.

LinkedHashMap(int capacity, float fillRatio, boolean Order)

Apart from the methods inherited from its parent classes, LinkedHashMap defines the following methods:

Methods	Usage	Use
clear	Void clear()	Removes all mappings from this map.
containsKey	Boolean containsKey (Object key)	Returns true if this map maps one or more keys to the specified value.
get	Object get(Object key)	Returns the value to which this map maps the specified key.
removeEldest Entry	Protected boolean remove Eldest Entry(Map.Entry eldest)	Returns true if this map should remove its eldest entry.

Example

The following program illustrates several of the methods supported by this collection:

```
import java.util.*;class LinkedHashMapDemo {
  public static void main(String args[]) {
    // Create a hash map
    LinkedHashMap lhm = new LinkedHashMap();
    // Put elements to the map
    lhm.put("Jamal", new Double(3434.34));
    lhm.put("Mahnaz", new Double(123.22));
    lhm.put("Ayan", new Double(1378.00));
    lhm.put("Daisy", new Double(99.22));
    lhm.put("Qadir", new Double(-19.08));
        // Get a set of the entries
    Set set = lhm.entrySet();
    // Get an iterator
    Iterator i = set.iterator();
    // Display elements
    while(i.hasNext()) {
      Map.Entry me = (Map.Entry)i.next();
      System.out.print(me.getKey() + ": ");
      System.out.println(me.getValue());
    }
    System.out.println();
    // Deposit 1000 into Jamal's account
    double balance = ((Double)lhm.get("Jamal")).doubleValue();
    lhm.put("Jamal", new Double(balance + 1000));
    System.out.println("Jamal's new balance: " +
    lhm.get("Jamal"));
  }
}
```

This would produce the following results:
Jamal: 3434.34
Mahnaz: 123.22
Ayan: 1378.0
Daisy: 99.22
Qadir: –19.08
Jamal's new balance: 4434.34

IdentityHashMap Class

This class implements AbstractMap. It is similar to HashMap except that it uses reference equality when comparing elements. This class is not a general-purpose Map implementation! While this class implements the Map interface, it intentionally violates Map's general contract, which mandates the use of the equals method when comparing objects. This class is designed for use only in the rare cases wherein reference-equality semantics are required. This class provides constant-time performance for the basic operations (get and put), assuming the system identity hash function (System.identityHashCode(Object)) disperses elements properly among the buckets.

This class has one tuning parameter (which affects performance but not semantics): expected maximum size. This parameter is the maximum number of key-value mappings that the map is expected to hold. The IdentityHashMap class supports three constructors. The first form constructs a new, empty identity hash map with a default expected maximum size (21):

IdentityHashMap()

The second form constructs a new, empty map with the specified expected maximum size:

IdentityHashMap(int expectedMaxSize)

The third form constructs a new identity hash map containing the keys-value mappings in the specified map:

IdentityHashMap(Map m)

Apart from the methods inherited from its parent classes, IdentityHashMap defines the following methods:

Methods	Usage	Use
clear	Void clear()	Removes all mappings from this map.
clone	Object clone()	Returns a shallow copy of this identity hash map: the keys and values themselves are not cloned.
containsKey	Boolean containsKey (Object key)	Tests whether the specified object reference is a key in this identity hash map.
containsValue	Boolean containsValue (Object value)	Tests whether the specified object reference is a value in this identity hash map.
entrySet	Set entrySet()	Returns a set view of the mappings contained in this map.

equals	Boolean equals (Object o)	Compares the specified object with this map for equality.
get	Object get(Object key)	Returns the value to which the specified key is mapped in this identity hash map, or null if the map contains no mapping for this key.
hashCode	Int hashCode()	Returns the hash code value for this map.
isEmpty	Boolean isEmpty()	Returns true if this identity hashmap contains no key-value mappings.
keySet	Set keySet()	Returns an identity-based set view of the keys contained in this map.
put	Object put(Object key, Object value)	Associates the specified value with the specified key in this identity hashmap.
putAll	Void putAll(Map t)	Copies all of the mappings from the specified map to this map. These mappings will replace any mappings that this map had for any of the keys currently in the specified map.
remove	Object remove(Object key)	Removes the mapping for this key from this map if present.
size	Int size()	Returns the number of key-value mappings in this identity hashmap.
values	Collection values()	Returns a collection view of the values contained in this map.

Example

The following program illustrates several of the methods supported by this collection:

```
import java.util.*;
class IdentityHashMapDemo {
  public static void main(String args[]) {
    // Create a hash map
    IdentityHashMap ihm = new IdentityHashMap();
    // Put elements to the map
    ihm.put("Jamal", new Double(3434.34));
    ihm.put("Mahnaz", new Double(123.22));
    ihm.put("Ayan", new Double(1378.00));
    ihm.put("Daisy", new Double(99.22));
    ihm.put("Qadir", new Double(-19.08));
       // Get a set of the entries
    Set set = ihm.entrySet();
    // Get an iterator
```

```
    Iterator i = set.iterator();
    // Display elements
    while(i.hasNext()) {
      Map.Entry me = (Map.Entry)i.next();
      System.out.print(me.getKey() + ": ");
      System.out.println(me.getValue());
    }    System.out.println();
    // Deposit 1000 into Jamal's account
    double balance = ((Double)ihm.get("Jamal")).doubleValue();
    ihm.put("Jamal", new Double(balance + 1000));
    System.out.println("Jamal's new balance: " +
    ihm.get("Jamal"));
  }
}
```

This would produce the following results:

```
Ayan                 :  1378.0
Jamal                :  3434.34
Qadir                :  –19.08
Mahnaz               :  123.22
Daisy                :  99.22
Jamal's new balance  :  4434.34
```

7 Network Programming in Java

1. NETWORKING BASICS

A network is a collection of devices that share a common communication protocol and a common communication medium like network cables, dial-up connections, and wireless links. Networks are merely a collection of computers to limit the range of hardware that can use them. For example, printers may be shared across a network, allowing more than one machine to gain access to their services. Other types of devices can also be connected to a network; these devices can provide access to information, or offer services that may be controlled remotely. Many approaches are done towards connecting noncomputing devices to networks. While the technology is still evolving, attention is made in moving toward a network centric as opposed to a computing-centric model. Services and devices can be distributed across a network rather than being bound to individual machines. In the same way, users can move from machine to machine, logging on as if they were sitting at their own familiar terminal. As home networks become easier to use and more affordable, even regular household appliances such as telephones, televisions, and home stereo systems are connected to local networks or even to the Internet. Network and software standards already exist to help devices and hardware talk to each other over networks and to allow instant plug-and-play functionality. Devices and services can be added and removed from the network without the need for complex administration and configuration. It is anticipated that over the course of the next few years, users will become just as comfortable and familiar with network-centric computing as they are with the Internet. In addition to devices that provide services are devices that keep the network going. Depending on the complexity of a network and its physical architecture, elements forming, it may include network cards, routers, hubs, and gateways.

- *Network cards* are hardware devices added to a computer to allow it to communicate to a network. The most common network card in use today is the Ethernet card. Network cards usually connect to a network cable, which is the link to the network and the medium through which data is transmitted. However, other media exist, such as dial-up connections through a phone line, and wireless links.
- *Routers* are machines that act as switches. These machines direct packets of data to the next "hop" in their journey across a network.
- *Hubs* provide connections that allow multiple computers to access a network.
- *Gateways* connect one network to another—for example, a local area network to the Internet. While routers and gateways are similar, a router does not have to bridge multiple networks. In some cases, routers are also gateways.

1.1 Network Communication

The following words express the way how the networks communicate. Networks consist of connections between computers and devices. These connections are most commonly physical connections, such as wires and cables, through which electricity is sent. However, many other media exist. For example, it is possible to use infrared and radio as a communication medium for transmitting data wirelessly, or fiber-optic cables that use light rather than electricity. Such connections carry data between one point in the network and another. This data is represented as bits of information (either "on" or "off," a "zero" or a "one"). Whether through a physical medium such as a cable, through the air, or using light, this raw data is passed across various points in the network called nodes; a node could represent a computer, another type of hardware device such as a printer, or a piece of networking equipment that relays this information onward to other nodes in the network or to an entirely different network. Of course, for data to be successfully delivered to individual nodes, these nodes must be clearly identifiable.

Each node in a network is typically represented by an address, just as a street name and number, town or city, and zip code identifies individual homes and offices. The manufacturer of the network interface card (NIC) installed in such devices is responsible for ensuring that no two card addresses are alike, and chooses a suitable addressing scheme. Each card will have this address stored permanently, so that it remains fixed—it cannot be manually assigned or modified, although some operating systems will allow these addresses to be faked in the event of an accidental conflict with another card's address. Because of the wide variety of NICs(Network Interface Cards) , many addressing schemes are used. For example, Ethernet network cards are assigned a unique 48-bit number to distinguish one card from another. Usually, a number is assigned to each card, and manufacturers are allocated batches of numbers. This system is strictly regulated by industry, of course—two cards with the same address would cause headaches for network administrators. The physical address is referred to by many names, including:

- Hardware address
- Ethernet address
- Media Access Control (MAC) address
- NIC address

These addresses are used to send information to the appropriate node. If two nodes share the same address, they would be competing for the same information and one would inevitably lose out, or both would receive the same data. Often, machines are known by more than one type of address that distinguishes it from other hosts on the Internet, or it may have more than one network card. Within a local area network, machines can use physical addresses to communicate. However, since there are many types of these addresses, they are not appropriate for internetwork communication.

Data transmission using packets are to be discussed in the forecoming words. Sending individual bits of data from node to node is not very cost effective, as a fair bit of overhead is involved in relaying the necessary address information every time a byte of data is transmitted. Most networks, instead, group data into packets. Packets consist of a header and data segment.The header contains addressing information (such as the sender and the recipient), checksums to ensure that a packet has not been corrupted, as well as other useful information that is needed for transmission across the network. The data segment contains sequences of bytes, comprising the actual data being sent from one node to another. Since the header information is needed only for transmission, applications are interested only in the data segment. Maximum data would be combined into a packet, in order to minimize the overhead of the headers. However, if information needs to be sent quickly, packets may be dispatched when nearly empty. Depending on the type of packet and protocol being used, packets may also be padded out to fit a fixed length of bytes.

When a node on the network is ready to transmit a packet, a direct connection to the destination node is usually not available. Instead, intermediary nodes carry packets from one location to another, and this process is repeated indefinitely until the packet reaches its destination. Due to network conditions such as congestion or network failures, packets may take arbitrary routes, and sometimes they may be lost in transit or arrive out of sequence. This may seem like a chaotic way of communicating, but still there are ways to guarantee delivery and sequencing. Indeed, the properties of guaranteed delivery and sequential order are often irrelevant to certain types of applications like streaming video and audio, where it is more important to present current video frames and audio segments than to retransmit lost ones. When these properties are necessary, networking software can keep track of lost packets and out-of sequence data for applications.

Packet transmission and transmission of raw bits of information are low-level processes, while most network programming deals with high-level transmission of data. Rather than simultaneously covering the gamut of transmission from raw bytes to packets and then to actual program data, it is helpful to conceive of these different types of communication as comprising individual layers.

The concept of layers was introduced to acknowledge and address the complexity of networking theory. The most popular approach to network layering is the Open Systems Interconnection (OSI) model, created by the International Standards Organization (ISO). This model groups network operations into seven parts, from the most basic physical layer through to the application layer, where software applications such as Web clients and e-mail servers communicate. Under the OSI model, each of the seven layers into which communication is grouped can be referred to by a number or by a descriptive name. Each of the seven layers is illustrated in the following figure.

Seven Layers of the OSI Reference Model

Each of the layers is responsible for some form of communication task, but each task is narrowly defined and usually relies on the services of one or more layers beneath it. In some systems, one or more layers may be absent, while in other systems all layers are used. Frequently, though, only a subset of the seven layers is employed by an operating system. Generally, programmers limit themselves to working with one layer at a time; details of the layers below the layer concerned are thus hidden from view. When writing software for one layer, programmers need not concern with the issues of other layers. Breaking the network into layers leads to a much simpler system. The division of network protocols and services into layers not only helps simplify networking protocols by breaking them into smaller, more manageable units, but also offers greater flexibility.

By dividing protocols into layers, protocols can be designed for interoperability. Software that uses Layer *n* can communicate with software running on another machine that supports Layer *n*, regardless of the details of Layer *n*–1, Layer *n*–2, and so on. Lower-level layers, for example, can be substituted and replaced without having to modify or redesign higher-level layers, or recompile application software. For example, a network layer protocol can work with an Ethernet network and a token ring network, even though at the physical and data link layers, two different protocols and hardware

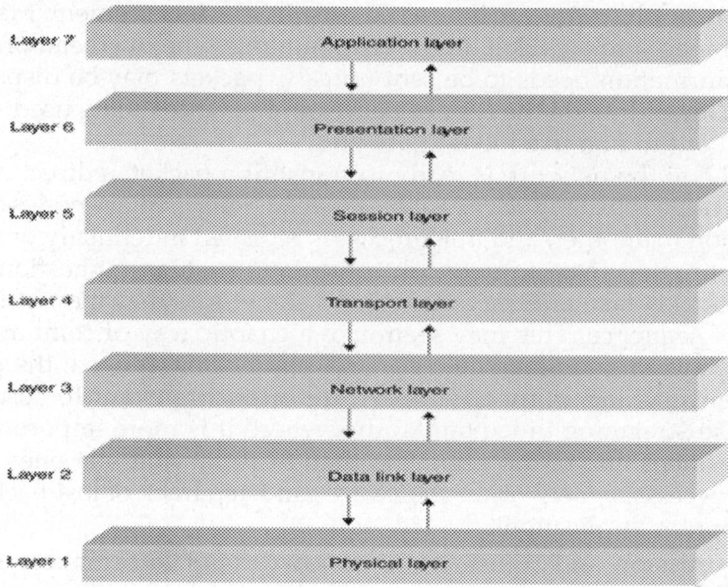

devices are being used. In a world of heterogeneous networks, this is an important quality, as it makes networks interoperable.

a. Layer 1—Physical Layer

The physical layer is networking communication at its most basic level. The physical layer governs the very lowest form of communication between network nodes. At this level, networking hardware, such as cards and cables, transmit a sequence of bits between two nodes. Java programmers do not work at this level—it is the domain of hardware driver developers and electrical engineers. At this layer, no real attempt is made to ensure error-free data transmission. Errors can occur for a variety of reasons, such as a spike in voltage due to interference from an outside source, or line noise in networks that use analog transmission media.

b. Layer 2—Data Link Layer

The data link layer is responsible for providing a more reliable transfer of data, and for grouping data together into frames. Frames are similar to data packets, but are blocks of data specific to a single type of hardware architecture, whereas data packets are used at a higher level and can move from one type of network to another. Frames have checksums to detect errors in transmission, and typically a "start" and "end" marker to alert hardware to the division between one frame and another. Sequences of frames are transmitted between network nodes, and if a frame is corrupted it will be discarded. The data link layer helps to ensure that broken data frames will not be passed to higher layers, confusing applications. However, the data link layer does not normally guarantee retransmission of corrupted frames.

c. Layer 3—Network Layer

Moving up from the data link layer, which sends frames over a network. The network layer deals with data packets, rather than frames, with importance to network address and routing. Packets are sent across the network, and in the case of the Internet, all around the world. Unless traveling to a node in an adjacent network where there is only one choice, these packets will often take alternative routes determined by routers.

Communication at this level is still very low level; network programmers are rarely required to write software services for this layer.

d. Layer 4—Transport Layer

The fourth layer, the transport layer, is concerned with controlling how data is transmitted. This layer deals with issues such as automatic error detection and correction, and flow control limiting the amount of data sent to prevent overload.

e. Layer 5—Session Layer

The purpose of the session layer is to facilitate application-to-application data exchange, and the establishment and termination of communication sessions. Session management involves a variety of tasks, including establishing a session, synchronizing a session, and reestablishing a session that has been abruptly terminated. Not every type of application will require this type of service, as the additional overhead of connection-oriented communication can increase network delays and bandwidth consumption. Some applications will instead choose to use a connectionless form of communication.

f. Layer 6—Presentation Layer

The sixth layer deals with data representation and data conversion. Different machines use different types of data representation like an integer might be represented by 8 bits on one system and 16 bits on another. Some protocols may want to compress data, or encrypt it. Whenever data types are being converted from one format to another, the presentation layer handles these types of tasks.

g. Layer 7—Application Layer

The final OSI layer is the application layer, where the vast majority of programmers write code. Application layer protocols dictate the semantics of how requests for services are made, such as requesting a file or checking for e-mail. In Java, almost all network software written will be for the application layer, although the services of some lower layers may also be called upon.

1.2 Internet Architecture

The most important revolution in networking history has been the evolution of the Internet, a worldwide collection of smaller networks that share a common communication suite (TCP/IP). Over the years, the Internet has been extended to include what we have today; it has evolved from a defense communications project called ARPANET into a worldwide collection of networks that spans both the commercial and noncommercial domains. Contributions to the design of the Internet came from both the original ARPANET developers and from academic and commercial researchers who offered suggestions and improvements that helped to shape what it is today. The Internet is an open system, built on common network, transport, and application layer protocols, while granting the flexibility to connect a variety of computers, devices, and operating systems to it. Whether an individual is running a PC, Unix, Macintosh, or Palm handheld computer, the complexities of communication and translation are handled transparently for users by the TCP/IP suite of protocols.

1.2.1 Design of the Internet

The Internet as we know it today is the result of many decades of innovation and experimentation. The protocols that make up the TCP/IP suite have been carefully designed, tested, and improved upon over the years. Some of the major goals were to achieve:

- *Resource sharing between networks,* by creating network protocols that support internetwork communication or "internetting." The various protocols that make up the Internet must support a variety of networking gateways.
- *Hardware and software independence,* by creating network protocols that would be interoperable with any CPU architecture, operating system, and networking card.
- *Reliability and robustness,* by creating network protocols that would be fault tolerant, so that regardless of the state of intermediary networks, data could be rerouted if necessary in order to reach its destination. Because the Internet started as a defense research project, robustness in the event of catastrophic network failure was extremely important. Damaged networks can be circumvented so that the Internet at large remains accessible.
- *"Good" protocols that are efficient and simple,* by creating network protocols that exhibited quality design principles, such as the concepts of communication sockets, network ports, and so on. Though such a design goal seems intuitive now, designers had to make a conscious effort to develop TCP/IP for long-term and high-volume use, and to make it as simple as possible to use.

The ease of interconnection between computers and networks connected to the Internet has been brought about by common protocols that are independent of specific hardware and software architectures, are robust and fault tolerant, and are efficient and simple to learn. As a result, the TCP/IP protocol suite emerged. Each of the major protocols involved are detailed below.

a. Internet Protocol (IP)

The Internet Protocol (IP) is a Layer 3, i.e. network layer protocol that is used to transmit data packets over the Internet. It is undoubtedly the most widely used networking protocol in the world. Regardless of what type of networking hardware is used, it will almost certainly support IP networking. IP acts as a bridge between networks of different types, forming a worldwide network of computers and smaller subnetworks. Indeed, many organizations use the IP and related protocols within their local area networks, as it can be applied equally well internally and externally.

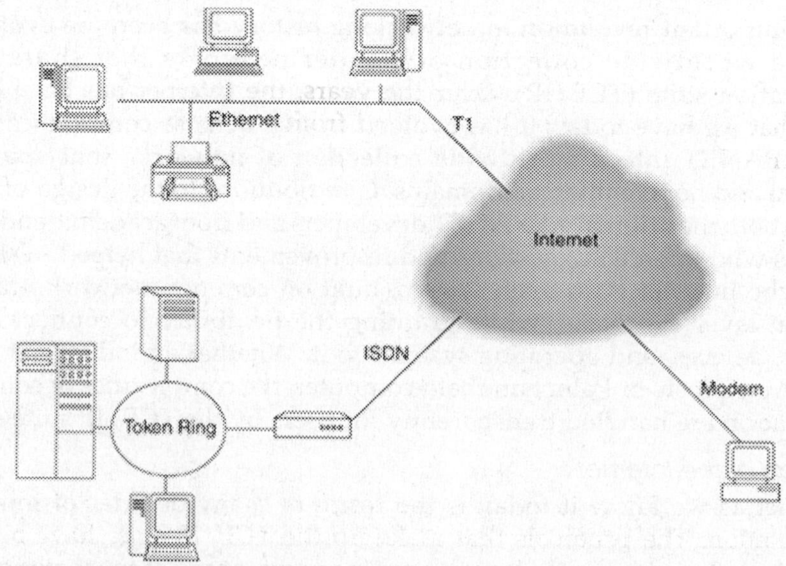

Support for IP networking among various physical networks

The Internet Protocol is a packet-switching network protocol. Information is exchanged between two hosts in the form of IP packets, also known as IP datagrams. Each datagram is treated as a discrete unit, unrelated to any other previously sent packet—there are no "connections" between machines at the network layer. Instead, a series of datagrams are sent and higher-level protocols at the transport layer provide connection services.

IP Datagram Format

The IP datagram carries with it essential information for controlling how it will be delivered. This information is stored inside the datagram header, which is followed by the actual data being sent. The various header fields, and their sizes, are shown in the figure below.

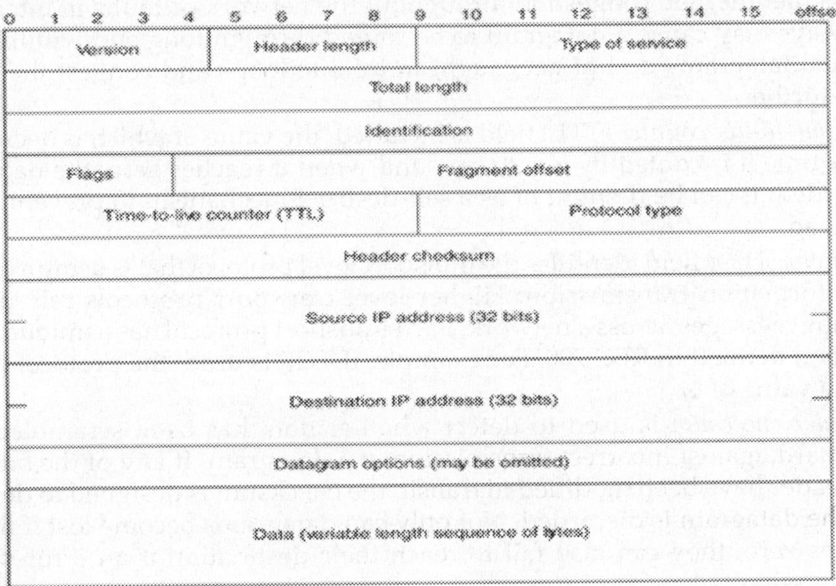

Format of an IPv4 datagram packet

A thorough knowledge of each individual IP datagram header field is not required for everyday programming. A rough understanding of how IP datagrams works will assist readers in understanding how Internet communication takes place; therefore a brief description of these header fields is offered.

- The *version* field describes which version of the Internet Protocol is being used. Currently, Internet Protocol version 4 (referred to as IPv4) is in common use. Future versions of the Internet Protocol will feature additional security, and include an expanded IP address space greater than the current 32-bit address range to allow more devices to have their own addresses.
- The *header length* field specifies the length of the header, in multiples of 32 bits. When no datagram options are specified, the minimum value for this will be 5, leaving a minimum header length of 160 bits. However, when additional options are used, this value can be greater.
- The *type of service* field requests that a specific level of service be offered to the datagram. Some applications may require quick responses to reduce network delays, greater reliability, or higher throughput.

- The *total length* field states the total length of the datagram including both header and data. A maximum value of 65,536 bytes is usually imposed, but many networks may only support smaller sizes. All networks are guaranteed to support a minimum of 576 bytes.
- The *identification* field allows datagrams that are part of a sequence to be uniquely identified. This field can be thought of as a sequence number, allowing ordering of datagrams that arrive out of sequence. Sometimes when packets are sent between network gateways, one gateway will support only smaller packets.
- The *flags* field controls whether these datagrams may be fragmented, sent as smaller pieces and later reassembled or transmitted as such. Fields marked "do not fragment" are discarded and are undeliverable. As datagrams are routed across the Internet, congestion throughout the network or faults in intermediate gateways may cause a datagram to be routed through long and winding paths. So that datagrams do not get caught in infinite loops and congest the network even further.
- The *time-to-live counter* (TTL) field is included, the value of which is decremented every time it is routed by a gateway, and when it reaches zero, the datagram is discarded. It can be thought of as a self-destruct mechanism to prevent network overload.
- The *protocol type* field identifies the transport level protocol that is using a datagram for information transmission. Higher-level transport protocols rely on IP for sending messages across a network. Each transport protocol has a unique protocol number, defined in RFC 790. For example, if TCP is used, the protocol field will have a value of 6.
- A *header checksum* is used to detect whether data has been scrambled and to safeguard against incorrect transmission of a datagram. If any of the bits within the header have been modified in transit, the checksum is designed to detect this, and the datagram is discarded. Not only can datagrams become lost if their TTL reaches zero, they can also fail to reach their destination if an error occurs in transmission.
- The next two fields contain addressing information. The *source IP address* field and *destination IP address* fields are stored as two separate 32-bit values. Note that there is no authentication mechanism to prove that a datagram originated from the specified source address. Though not common, it is possible to use the technique of "IP spoofing" to make it appear that a datagram originated from a specific address, such as a trusted host. The final field within the datagram header is an optional field that is not always present. The *datagram options* field is of variable length, and contains flags to control security settings, routing information, and time stamping of individual datagrams. The length of the options field must be a multiple of 32—if not, extra bits are added as padding.

IP Addressing

The addressing of IP datagrams is an important issue, as applications require a way to deliver packets to specific machines and to identify the sender. Each host machine under the Internet Protocol has a unique address, the IP address. The IP address is a four-byte (32-bit) address, which is usually expressed in dotted decimal format (e.g. 192.168.0.6). Although a physical address will normally be issued to a machine, once outside the local network in which it resides, the physical address is not very useful. Even if somehow every machine could be located by its physical address, if the address changed for any reason, such as installation of a new networking connection, or reassignment of the network interface by the administrator, then the machine would no longer be locatable. Instead, a new type of address is introduced, that is not bound to a particular physical location. IP address is a number that uniquely identifies a machine on the Internet. Typically, one machine has a single IP address, but it can have multiple addresses. A machine could, for example, have more than one network card, or could be assigned multiple IP addresses, known as virtual addresses, so that it can appear to the outside world as many different machines. Machines connected to the Internet can send data to that IP address, and routers and gateways ensure delivery of the message. To map between a physical network address and an IP address, host machines and routers on a local network can use the Address Resolution Protocol (ARP) and Reverse Address Resolution Protocol (RARP). Such details, however, are more the domain of network administrators than of programmers. In normal programming, only the IP address is needed—the physical address is neither useful nor accessible in Java.

Host Name

While numerical address values serve the purposes of computers, they are not designed with people in mind. Users who can remember thousands of 32-bit IP addresses in dotted decimal format and store them in their head are few and far between. A much simpler addressing mechanism is to associate an easy-to-remember textual name with an IP address. This text name is known as the hostname. For example, companies on the Internet usually choose a .com address, such as www.microsoft.com, or java.sun.com.

b. Internet Control Message Protocol (ICMP)

Though the IP might seem to be an ineffective means of transmitting information, it is actually highly efficient leaving the provision of an error-control mechanism to other protocols if they require it. Since the Internet Protocol provides absolutely no guarantee of datagram delivery, there is an obvious need for error-control mechanisms in many situations. One such mechanism is the Internet Control Message Protocol (ICMP), which is used in conjunction with the Internet.

Protocol, to report errors whenever they occur. The relationship between these two protocols is strong. When IP must notify another host of an error, it uses ICMP. ICMP, on the other hand, uses IP to send the error message. When minor errors occur, such as a corrupt header in a datagram, the datagram will be discarded without warning since the sender address in the header cannot be trusted. Therefore, a host cannot rely solely upon ICMP to guarantee delivery—the services of ICMP are more informational, to prevent wasted bandwidth if errors are likely to be repeated. No guarantee is offered that ICMP messages will be sent, or that they will reach their intended destination.

The ICMP defines five error messages:

1. *Destination Unreachable.* As datagrams are passed from gateway to gateway, they will travel closer and closer to their final destination. If a fault in the network occurs, a gateway may be unable to pass the datagram on to its destination. In this case, the "destination unreachable" ICMP message is sent back to the original host.

2. *Parameter Problem.* When a gateway determines that there is a problem with any of the header parameters of an IP datagram and is unable to process them, the datagram is discarded and the sending host may be notified via a "parameter problem" ICMP message.

3. *Redirect.* When a shorter path, or alternate route, is available, a gateway may send a "redirect" ICMP message to the router that passed on a datagram.

4. *Source Quench.* When too many datagram packets hit a router, gateway, or host, it may become overloaded and be unable to accept more packets. This occurs when the buffer allocated for datagram storage becomes full, and datagrams cannot be removed from the buffer as fast as they are coming in. Rather than allowing datagrams to be discarded, an attempt is made to reduce the number of incoming datagrams, by sending a "source quench" ICMP message.

5. *Time Exceeded.* Whenever the TTL value of a datagram reaches zero, it is discarded. When this occurs, a "time exceeded" ICMP message may be sent. In addition to error messages, ICMP supports several informational messages. These are not generated in response to error conditions, and are instead used to pass control information.

Additional ICMP messages include:

- *Echo Request/Echo Reply.* Used to determine whether a host is alive and can be reached. In response to an "echo request" ICMP message, the recipient sends back an "echo reply" ICMP message. Although no guarantee of message delivery is offered, repeated requests can be made if no response is received. If the host is unreachable, then the last gateway dealing with the message should send back a "destination unreachable" ICMP message. The "echo request" and "echo reply" messages are used by the "ping" application to test if a remote host is accessible.
- *Address Mask Request/Address Mask Reply.* Though not part of the original ICMP specification, functionality to determine the address mask (also known as a subnet mask) is added to the protocol in RFC 950. The address mask controls which bits of an IP address correspond to a host, and which bits determine the network/ subnet portion. A host can send an "address mask request" ICMP message, and receive an "address mask reply" ICMP message.

While ICMP is a useful protocol to be aware of, only a few network applications will make use of it, as its functionality is limited to diagnostic and error notification. One of the most well-known applications that use ICMP is the ping network application, used to determine if a host is active and what the delay is between sending a packet and receiving a response.

c. Transmission Control Protocol

The Transmission Control Protocol (TCP) is a Layer 4, i.e. transport layer protocol that provides guaranteed delivery and ordering of bytes. TCP uses the Internet Protocol to send TCP segments, which contain additional information that allows it to order packets and resend them if they are damaged. TCP also adds an extra layer of abstraction, by using a communications port. A communications port is a numerical value, usually in the range of 0–65,535, that can be used to distinguish one application or service from another. An IP address can be thought of as the location of a block of apartments, and the port as the apartment number. One host machine can have many applications connected to one or more ports. An application could connect to a Web server running on a particular host, and also to an e-mail server to check for new mail. Ports make all of this possible. TCP's main advantage is that it guarantees delivery and ordering of data, providing a simpler programming interface. However, this simplicity comes at a cost, reducing network performance. For faster communication, the User Datagram Protocol may be used.

d. User Datagram Protocol

The User Datagram Protocol (UDP) is also a Layer 4 , i.e. transport layer protocol, that applications can be used to send packets of data across the Internet, as opposed to

TCP, which sends a sequence of bytes. Raw access to IP datagrams is not very useful, as there is no easy way to determine which application a packet is for. Like TCP, UDP supports a port number, so it can be used to send datagrams to specific applications and services. Unlike TCP, UDP does not guarantee delivery of packets, or that they will arrive in the correct order. In fact, UDP differs very little from IP datagrams, save for the introduction of a port number. It may seem puzzling why anyone would want to use an unreliable packet delivery system. The additional error checking of TCP adds overhead and delays, so UDP might be seen to offer better performance. Error-free transmission comes at a cost, and UDP can be used as an alternative.

1.2.2 *Internet Application Protocols*

While network and transport layer protocols are certainly interesting, for network programmers the real excitement lies in the application layer. At the application layer are network protocols that do real work, rather than just facilitating communication? This section deals with protocols for accessing and sending e-mail, transferring files, reading Web pages, and much more.

a. Telnet

Telnet is a service that allows users to open a remote-terminal session to a specific machine. This allows Unix users, for example, to access their account from terminal servers or desktop machines. Since Unix servers are intended to support multiple users, a telnet session is often used, as only one person can access the machine from the local terminal (using a keyboard and monitor). Telnet allows many users to connect over the network and to access their accounts as if they were doing so locally. Telnet services use TCP port 23.

b. File Transfer Protocol (FTP)

The ability to transfer files is extremely important. Even before the World Wide Web, people distributed images, documents, and software using the File Transfer Protocol (FTP). FTP allows a user to log in using a special username and password, or to attempt an anonymous log-in by using the username of "anonymous". FTP servers will often grant different access permissions depending on the user. For example, an anonymous account might be unable to write a file to the server, but may be able to read all files. FTP uses two TCP ports for communication—port 21 is used to control sessions and port 20 is used for the actual transfer of file contents.

c. Post Office Protocol Version 3 (POP3)

E-mail has become a vital part of modern life. With the exception of Web-based e-mail or specialized accounts, the majority of people access their e-mail using the Post Office Protocol, version 3 (POP3), which uses TCP port 110. Messages are stored on a server, retrieved by an email client, and then deleted from the server. This allows users to read mail offline, without being connected to the Internet.

d. Internet Message Access Protocol (IMAP)

While many browsers and e-mail clients support only POP3, some also support the Internet Message Access Protocol (IMAP). This protocol is less popular, as it requires a continual connection to the mail server, and thus increases bandwidth consumption and disk usage since messages are not stored on the user's system. IMAP allows users to create folders on the mail server, and also allows online searching of mail. IMAP uses TCP port 143.

e. Simple Mail Transfer Protocol (SMTP)

The Simple Mail Transfer Protocol allows messages to be delivered over the Internet. The separation between retrieving mail and sending mail might be perceived as a bit strange. However, separation actually simplifies the process considerably, allowing different mail-retrieval protocols to be used and enabling custom mail accounts. SMTP uses TCP port 25.

f. HyperText Transfer Protocol (HTTP)

HTTP is one of the most popular protocols in use on the Internet today; it made the World Wide Web possible. HTTP is an extremely important protocol, and Java includes good HTTP support. HTTP uses TCP port 80.

g. Finger

Finger is a handy protocol that allows someone to look up a person's account and find out certain information, such as when they last logged in and checked their mail. Typically, only Unix servers support finger. Unfortunately, many administrators disable finger access for security reasons, and so it is no longer as prevalent as it was. Finger uses TCP port 79.

h. Network News Transport Protocol (NNTP)

The Network News Transport Protocol allows users to access Usenet newsgroups. Usenet is a collection of discussion forums on a colorful and diverse number of topics, ranging from political and social commentary, to fan discussions about television programs, movies, and actors, to computing and business. Online services such as DejaNews (http://www.dejanews.com/usenet/) provide a Web-based interface, but newsgroups can also be accessed via newsreader software that uses NNTP. NNTP uses TCP port 119.

i. WHOIS

The WHOIS protocol allows users to look up information about a domain name (such as awl.com, or microsoft.com). You can find some surprisingly useful information by doing this, such as the address of a company, who registered the domain name, and contact details for the registration. WHOIS uses TCP port 43.

1.3 TCP/IP Protocol Suite Layers

Earlier in the chapter, the seven OSI network layers were discussed. However, not all of these layers are used in Internet programming. The TCP/IP suite of protocols can be mapped to a subset of the OSI layers, as shown in the figure below.

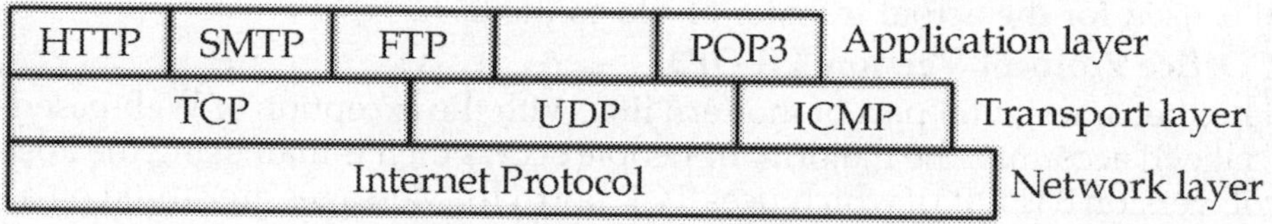

TCP/IP stack divided by layers

Each layer is stacked upon another layer, using encapsulation. Data passes from the top application layer, down to the transport layer, and then flows on to the network layer. At this stage, the data is sent across the Internet, and will reach a local area network or dial-up connection. Below the network layer, the data will flow to the data link layer and finally to the physical layer. Starting from the higher-level layers, protocol requests are encapsulated into the container of the previous layer.

A network is just another possible source of input data, and another place where data can be output. In Java, network communication can be done using input streams and output streams, just as streams can be used to communicate with the user or to work with files. Nevertheless, opening a network connection between two computers is a bit tricky, since there are two computers involved and they have to somehow agree to open a connection. And when each computer can send data to the other, synchronizing communication can be a problem. But the fundamentals are the same as for other forms of I/O. One of the standard Java packages is called java.net. This package includes several classes that can be used for networking. Two different styles of network I/O are supported. One of these, which is fairly high-level, is based on the World Wide Web, and provides the sort of network communication capability that is used by a Web browser when it downloads pages for you to view. The main classes for this style of networking are java.net.URL and java.net.URLConnection. An object of type URL is an abstract representation of a Universal Resource Locator, which is an address for an HTML document or other resource on the Web. A URLConnection represents a network connection to such a resource.

The second style of I/O, which is more general and much more important, views the network at a lower level. It is based on the idea of a socket. A socket is used by a program to establish a connection with another program on a network. Communication over a network involves two sockets, one on each of the computers involved in the communication. Java uses a class called java.net.Socket to represent sockets that are used for network communication. The term "socket" presumably comes from an image of physically plugging a wire into a computer to establish a connection to a network, but it is important to understand that a socket, as the term is used here, is simply an object belonging to the class Socket. In particular, a program can have several sockets at the same time, each connecting it to another program running on some other computer on the network. All these connections use the same physical network connection. This section gives a brief introduction to these basic networking classes, and shows how they relate to input and output streams.

Unlike local area networks (LANs), the Internet is a vast collection of machines and devices spread out across the nation and the world. When there are hundreds of millions (and eventually many billions) of computers and devices attached to a network, the need to identify and locate a specific one is obviously important. Indeed, one of the most fundamental concepts in network programming is that of the network address. Without it, there would be no way of identifying the sender of a data packet or where the packet must be sent.

A new class is introduced from the Java API to cover the representation of an IP address. Even readers comfortable with the theory will benefit from reading the practical sections of this chapter, as they form the foundation for the chapters that follow.

1.4 Address and DNS

a. Local Area Network Addresses

Devices connected to a LAN have their own unique physical or hardware address. This assists other machines on the network in delivering data packets to the correct location. The address is useful only in the context of a LAN, however—a machine cannot be located on the Internet by using its physical address, which does not indicate the location of the machine. Indeed, machines often move from location to location, in the case of laptop or palmtop computers. Java network programmers do not need to be concerned with the details of how data is routed within a LAN. Indeed, Java does not provide access to the lower-level data link protocols used by LANs. Supporting the wide range of protocols available would be a mammoth task. Since each type of protocol uses a different type of address and has different characteristics, different code would need to be written for each and every type of network. Instead, Java provides support for TCP/IP, which can be thought of as the glue that binds networks together. No matter what type of LAN is used—if one is used at all—software can be written for it in Java providing it supports TCP/IP. Individual machines have unique addresses for the transmission and receipt of IP datagrams, in addition to their normal network or physical addresses.

b. Internet Protocol Addresses

Devices having a direct Internet connection are allocated a unique identifier known as an IP address. IP addresses may be static in that they are bound permanently to a certain machine or device, or dynamic as leased to a particular machine or device for a certain period, for example, in the case of an Internet service provider [ISP] that offers a pool of modems for dial-up connections. Dynamically assigned IP addresses are typically used when many devices require Internet access for limited periods of time. Thus addresses can be allocated from a pool of remaining addresses on a case-by-case basis. However, an IP address can be bound to a single machine only; it cannot be shared concurrently. This address is used by the Internet Protocol to route IP datagram to the correct location. Without an address, a machine cannot be contacted; hence, all machines must have a unique IP address.

Structure of the IP Address

Under the Internet Protocol version 4, referred to as IPv4, the IP address is a 32-bit number made up of four octets (a series of 8 bits). While computers read an IP address as a sequence of bits, people see them in dotted decimal format (for example, 127.0.0.1). There are five classes of IP addresses (A through E), and each class is allocated an address range, as shown in the following table. Each class of IP addresses is structured differently, allowing for greater versatility in how private networks and host machines are allocated. Private networks are allocated a network ID that is a unique identifier for a specific network. Control of and responsibility for how machines within a network are allocated falls to the network administrator, so that host IDs can be assigned as needed.

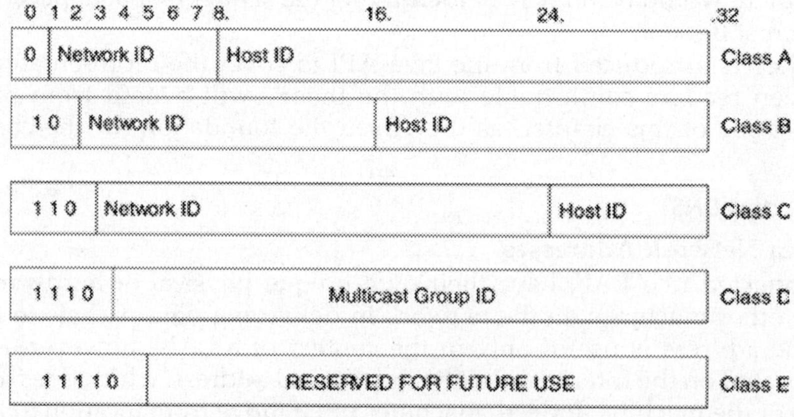

Address structures by class

With a finite range of addresses, this becomes particularly important as the number of allocated addresses increases. Class A addresses, for example, have a very small network ID field but a large host ID.

Range of IP addresses by class

Type	Address range
Class A	0. 0. 0. 0 – 127.255.255.255
Class B	128.0.0.0 – 191.255.255.255
Class C	192.0.0.0 – 223.255.255.255
Class D	224.0.0.0 – 239.255.255.255

This means that a private network with a Class A address can assign IP addresses to a large number of machines, whereas a Class C address can support a smaller number, but there are more network IDs available. Figure on previous page shows the breakup of network and host ID fields by class. Note that two address classes (D and E) are not allocated a network ID field. Class D is used for simultaneously broadcasting data to a large number of machines that are grouped together (a technique known as multicasting). Class E is reserved for future use and allocation.

Obtaining an IP Address

The central body responsible for allocating blocks of IP addresses is the Internet Corporation or Assigned Names and Numbers (ICANN), building on the work of an earlier organization, the Internet Assigned Numbers Authority (IANA). A person setting up a private network would be allocated either a Class A, B, or C address, and could then assign host IP addresses to the machines on that particular network. A discussion of the process of setting up a private network is beyond the scope of this book, of course, but readers interested in further information will find the ICANN and IANA organization Web sites useful, at http://www.icann.org/ and http://www.iana.org/.

The most common way to obtain an IP address is to have one assigned to you by a network administrator, ISP, or other network service. When you establish a dial-up connection, you will usually be assigned an IP address. This address is normally dynamically assigned from a pool of available addresses, and when you reconnect you will usually get a different address. In an intranet, a network administrator may assign a specific address to your machine, or you may have a dynamically assigned address allocated by a Dynamic Host Control Protocol (DHCP) server. DHCP provides addresses on demand; if machines are going online and offline frequently, a smaller pool of addresses can be used.

Programmers should be aware of some special IP addresses as well. The first, and most important from a network programming perspective, is known as the loopback or localhost address. When writing and debugging network software, programmers often want to connect to the local machine for testing purposes. Regardless of whether a connection to the Internet exists via a dial-up service or work is being done offline, the local machine may be accessed using the loopback address; this address is 127.0.0.1.

Another set of useful IP addresses are those reserved for private networking. In an intranet environment, it may be desirable to configure all machines with a unique IP address, without having them exposed to the public Internet. The Internet Assigned Number Authority (IANA) has reserved three sets of addresses for use within a local intranet environment, as described in RFC 1918. If you plan on setting up a LAN of your own, you can pluck addresses from these ranges without worrying about a collision conflict with a host on the Internet. On the Internet, routers will not forward data for these addresses, so they can be safely used locally. It must be noted, however, that this is the only time when IP addresses can be safely picked, unless you are allocated a block of addresses.The following table shows the IP address ranges for Classes A through C for intranet usage.

Class A, B, and C Address Ranges for Intranet Usage

Type	Address Range
Class A	10.0.0.0 – 0.255.255.255
Class B	172.16.0.0 – 172.31.255.255
Class C	192.168.0.0 – 192.168.255.255

c. The Domain Name System

While IP addresses comprise an efficient system for network administrators, most people find memorizing them to be an impossible task. People generally find words much easier to recall than the dotted decimal format of an IP address. It is easier to remember a name such as Amazon or Sun than a set of numbers designating such a dotted decimal IP address.

The domain name system (DNS) makes the Internet user-friendly, by associating a textual name with an IP address. Any entity, be it commercial, government, or private, can apply for a domain name, which can be used by people to locate that entity on the Internet. Simple text names, as opposed to arbitrary numbers, are used for identification.

Given the vast number of machines connected to the Internet, the number of domain-name-to-IPaddress mappings is too great for any one system to handle. Even if there were a large enough system to store all of these mappings, it would be quickly overloaded by requests. Furthermore, in the event of a system breakdown, isolating the problem to a fraction of the Internet would be preferable to having the entire system grind to a shuddering halt. Ironically, however, when the Internet was originally conceived, host-to-IP-address mappings were stored in a single file, called hosts.txt, which was downloaded and mirrored across the Internet. This file still exists in many operating systems and can be used to override DNS mappings or cover up ones that are missing or unresolvable by the local DNS servers. In some cases, they can be used to cover up not having a local DNS server on an intranet.

The DNS (outlined in RFC 1034/1035) is a more sophisticated and robust system. It can be thought of as a distributed database, in which responsibility for accepting new registrations, and returning the addresses of existing registrations, is spread out across many different hosts. Different categories, such as commercial and educational, are handled by different registry servers. Furthermore, international registries handle their own mappings (called country-code top-level domains), and can be subdivided into further categories. This forms a hierarchical structure, a small subset of which is shown in the following figure. Since the range of address categories changes rapidly, only a small number are shown.

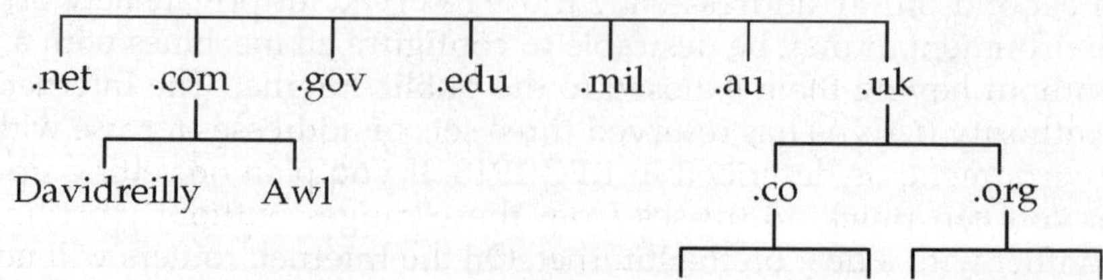

This hierarchical structure is broken up by the type of address (either .net, .com, .gov, .edu, .mil, or one of the newer addresses such as .info or .biz) or by the country (.au, .uk, among many others). For example, to access the site www.davidreilly.com, a request to resolve this name would be sent to the .com DNS server. Some countries have their own way of organizing DNS records, for example, as seen in the above figure, the United Kingdom (.uk) uses .co rather than .com.

Domain Name Resolution

When software applications need to look up a hostname as ftp.earmind.com, they do not contact the .com registry directly. Your network administrator, or ISP, configures your system to access the application's DNS server, which handles the lookup process. Often, however, the DNS server does not even need to send out a query, as the same sites will be requested either by multiple users or a single one. This works much the same as a Web browser caching pages, and prevents excessive overload of the network. When a request for a hostname such as www.earmind.com is made, the operating system of the client computer contacts the local DNS server on the LAN. It is responsible for locating the domain server earmind.com, and interrogating it to find the IP address of the hostname (www.earmind.com). To do this, it must query the "root" level domain server, which refers it to the .com server. Finally, the IP address is returned to the user.

Internet Addressing with Java

By now it should be clear that a host on the Internet can be represented either in dotted decimal format as an IP address, or as a hostname such as admin.earmind.com. Under Java, such addresses are represented by the java.net.InetAddress class. This class can fill a variety of tasks, from resolving an IP address to looking up the hostname. In the next section, this important class is examined in detail.

i. The java.net.InetAddress Class

The InetAddress class is used to represent IP addresses within a Java networking application. Unlike most other classes, there are no public constructors for that of InetAddress. Instead, there are two static methods that return InetAddress instances. Those and the other major methods of this class are covered in the list below; all are public unless otherwise noted.

Methods	Usage	Use
getHost Name	Public String getHostName() throws java.lang.Security Manager	Gets the hostname for this address; also the key in the above hashtable. If the host is equal to null, then this address refers to any of the local machine's available network addresses.
getAddress	Public byte[] getAddress()	Returns the raw IP address in network byte order. The highest order byte position is in addr[0]. An array of bytes is returned so we are prepared for 64-bit IP addresses.

hashCode	Public int hashCode()	Returns a hashcode for this InetAddress.
equals	Public boolean equals (Object obj)	Compares this object against the specified object.
toString	Public String toString()	Converts the InetAddress to a String.
getByName	Public static synchronized InetAddress getByName (String host) throws UnknownHostException	Returns a network address for the indicated host. A host name of null refers to default address for the local machine. A local cache is used to speed access to addresses. If all addresses for host are needed, use the getAllByName() method.
getAllBy Name	Public static synchronized InetAddress[] getAllByName(String host) throws UnknownHost Exception	Given a hostname, return an array of all the corresponding InetAddresses.
getLocalHost	Static InetAddress getLocal Host() throws java.net. UnknownHost Exception	Returns the local host.
getHost Address	String getHostAddress()	Returns the IP address of the InetAddress in dotted decimal format.
isMulticast Address	Boolean isMulticastAddress()	Returns "true" if the InetAddress is a multicast address, also known as a Class D address.

ii. Using InetAddress to Determine Localhost Address

The first and most simple example of InetAddress is to find out the IP address of the current machine. If a direct connection to the Internet exists, a meaningful result will be obtained, but dial-up users and those without any Internet connection (such as in an intranet environment) may get the loopback address of 127.0.0.1. The short example program given below shows how it is possible to determine the address.

Code for LocalHostDemo

```
import java.net.*;
public class LocalHostDemo
{
public static void main(String args[])
{
System.out.println ("Looking up local host");
try
{
```

```
// Get the local host
InetAddress localAddress =InetAddress.getLocalHost();
System.out.println ("IP address : " +localAddress.getHostAddress() );
}
catch (UnknownHostException uhe)
{
System.out.println ("Error - unable to resolve localhost");
}
}
}
```

How LocalHostDemo Works

The LocalHostDemo application starts by prompting the user that an IP address lookup will be performed (this is important if there is any delay in determining the IP address). The networking operation must be enclosed within a try/catch block, because it is possible that no IP address will be found and an exception will be thrown. Using the static method InetAddress.getLocalHost(), we obtain an object representing an IP address. To display the address in dotted decimal notation, the InetAddress.getHostAddress() method is used.

Running LocalHostDemo

This application requires no command-line parameters. Running it will display information about the host machine. To run, type:

```
java LocalHostDemo
```

Using InetAddress to Find Out About Other Addresses

The previous example familiarized you with the InetAddress class. Below is a more complex example, which resolves hostnames to IP addresses and then attempts to perform a reverse lookup of the IP address.

Code for NetworkResolverDemo

```
import java.net.*;
public class NetworkResolverDemo
{
public static void main(String args[])
{
if (args.length != 1)
{
System.err.println ("Syntax - NetworkResolverDemo host");
System.exit(0);
}
System.out.println ("Resolving " + args[0]);
try
{
// Resolve host and get InetAddress
InetAddress addr = InetAddress.getByName( args[0] );
System.out.println ("IP address : " +addr.getHostAddress() );
System.out.println ("Hostname : " +addr.getHostName() );
}
```

```
catch (UnknownHostException uhe)
{
System.out.println ("Error - unable to resolve hostname" );
}
}
}
```

iii. How NetworkResolverDemoWorks

In the previous example, we used a static method of InetAddress to return the local address. Most commonly, however, we will be working with other systems on the network, so we need to learn how to use InetAddress to resolve their hostnames and obtain an IP address. In this example, we use the static method, InetAddress.getByName() to return an InetAddress instance. We then display the IP address and hostname, using the same methods in LocalHostDemo.

Running NetworkResolverDemo

This application requires as a parameter either a hostname or an IP address. To run, type:

```
java NetworkResolverDemo hostname
```

1.5 Data Streams

Communication over networks, with files, and even between applications, is represented in Java by streams. Stream-based communication is central to almost any type of Java application. The concept of streams is especially important when dealing with networking applications. Almost all network communication (except UDP communication) is conducted over streams, so it is essential that programmers be familiar with this concept. Readers with previous input/output (I/O) stream experience may choose to skim through this section, but will likely benefit from a revised look at the topic, as changes to streams were introduced in JDK1.1 to deal with text-based reading and writing. A thorough knowledge of I/O streams is critical for almost any type of Java programmer, and is required for an understanding of the later chapters of this book. In addition, since so much of network communication involves the transmission of text, knowledge of the new text readers and writers is also necessary.

Byte-level communication is represented in Java by data streams, which are conduits through which information—bytes of data—is sent and received. A simple analogy is a pipe through which material such as water may be moved from one location to another. Provided that the pipe is installed correctly, what goes in one end comes out the other (see figure below).

Communication stream sends data from one point to another

When designing a system, the correct stream must be selected; when building a system, the type of stream used is not important, as a consistent interface is provided. The end effect of the stream, not how it was constructed, is of interest. Streams may be chained together, to provide an easier and more manageable interface. If, for example, the data needed to be processed in a particular way, a second stream could connect to an existing stream, to provide for processing of the data. For example, bytes might be converted from one form into another (such as a number), or a stream of bytes may be interpreted as sequences of characters, as shown in the figure below. In Java, streams take a flexible, one-size-fits-all approach—they are fairly interchangeable, and can be applied on top of another stream, or even several other streams.

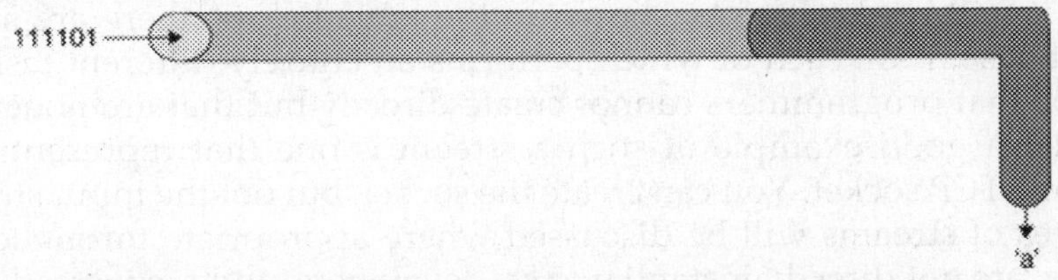

Streams may be fitted together for additional functionality

Interchangeability is an extremely important attribute of streams, as it simplifies programming considerably. Streams are divided into two categories—input streams that may be read from and output streams that may be written to. Provided you do not try to read from an output stream, or write to an input stream, you can safely attach any "filter" stream (a stream that filters data in some fashion, such as processing it or converting it from bytes into a different form) to any low level stream (such as a file or network stream). Readers should be aware that, although streams are usually one-way, multiple streams can be used together (e.g. an input stream and an output stream) for two-way communication.

At the heart of network programming is the transmission of data and the sending of sequences of bytes. A stream-based approach to communication simplifies programming, as a standard interface for data transmission is offered regardless of the type of data being sent and received, or of the mechanism by which data is sent across the network. Of course, streams are not limited to networking—they can read from and be written to data structures, files, and other applications. However, as will be seen in later chapters, there is more than one way to send data over a network, and I/O streams provide a consistent interface for working with network communication.

As mentioned earlier, streams provide communication of data at the byte level, and are used for either reading or writing. Streams for reading inherit from a common superclass, the java.io.InputStream class, shown in the figure below.

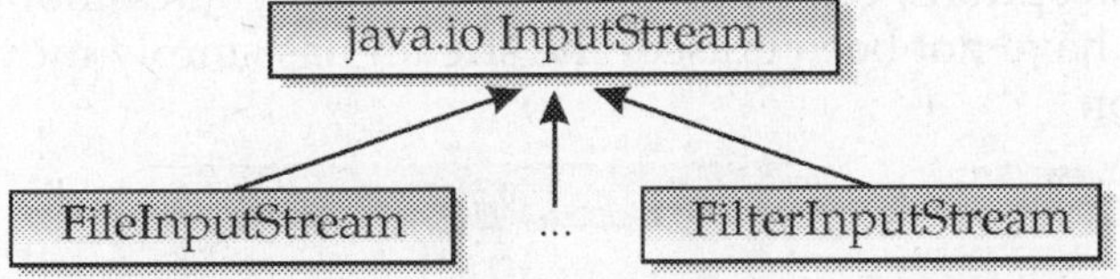

Input streams inherit from InputStream, although not always directly.

Likewise, streams used for writing data inherit from the superclass java.io. OutputStream, shown in the figure below. These are abstract classes; they cannot be instantiated. Instead, an appropriate subclass for the task at hand is created. Several streams inherit directly from either InputStream or OutputStream, but most inherit from a filter stream.

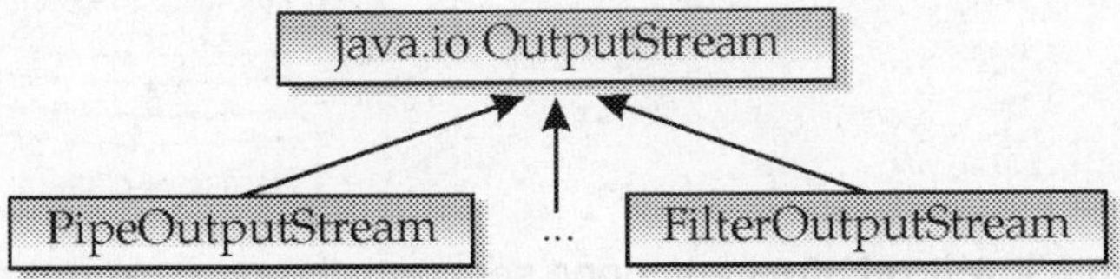

Output streams inherit from OutputStream, although not always directly.

a. Reading from an Input Stream

Many input streams are provided by the java.io package, and choosing the right low-level stream is a fairly straightforward task, since the name of the stream matches the data source it will read from. As shown in the table below, there are six low-level streams to choose from, each of which performs an entirely different task. There are other streams that programmers cannot create directly but that are nonetheless low-level streams. A good example of such a stream is one that represents a network connection to a TCP socket. You can create the socket, but not the input stream to read it. These types of streams will be discussed where appropriate throughout the text. Such streams, are not directly instantiated by developers, and are instead returned by invoking a method of a networking object.

Low-level input streams of the java.io package

Low-level input stream	Purpose of stream
ByteArrayInputStream	Reads bytes of data from an in-memory array
FileInputStream	Reads bytes of data from a file on the local file system
PipedInputStream	Reads bytes of data from a thread pipe
StringBufferInputStream	Reads bytes of data from a string
SequenceInputStream	Reads bytes of data from two or more low-level streams, switching from one stream to the next when the end of the stream is reached
System.in	Reads bytes of data from the user console

When a low-level input stream is created, it will read from a source of information that supplies it with data. Input streams act as consumers of information—they devour bytes of information as they read them. That is, bytes are read from a file sequentially—subject to a few exceptions, once they have been read, you cannot go back and read them again. They have not been erased; the stream has simply moved on to the next byte of information.

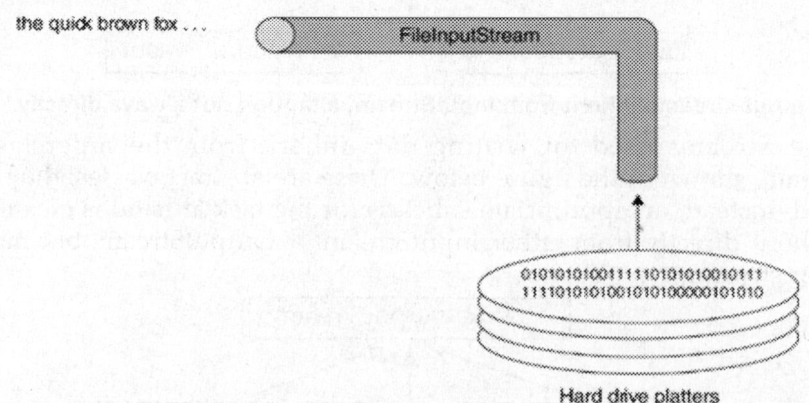

Reading from a low-level stream is done sequentially—you cannot normally go back to the beginning and read again

Of course, the sequential nature of input streams is not necessarily a bad thing. We would not want to store the entire contents of a five-megabyte file in memory just so that a program had the option of going back to an earlier point in time. However, some high-level filter streams support a limited push-back ability, allowing you to jump back to a specific point in the stream, on the condition that the point is marked. This functionality is useful at times for looking ahead to see the contents of a stream, but is not supported by many filter streams.

A further point that readers should be aware of is that reading from an input stream uses blocking I/O. *Blocking I/O* is a term applied to any form of input or output that does not immediately return from an operation. For example, reading from a file will block indefinitely, until the hard drive is accessible and the drive read head moves to the correct location to retrieve a byte of data. This is usually fairly quick, but if you are reading a byte at a time (rather than, say, a chunk of 1,024 bytes, or a kilobyte), then the number of read operations that may block can add up to some significant delays. When reading from a network connection, where delays can range from milliseconds to tens of seconds—or, worse still, not at all—then blocking I/O can cause performance problems. Methods and examples related to streams were already discussed in part 6.

a. Writing to an Output Stream

A number of output streams is available in the java.io package for a variety of tasks, such as writing to data structures including strings and arrays, or to files or communication pipes. As seen in the table, six important low-level output streams may be written to (in addition to filter streams that may be connected to these low-level streams). As mentioned earlier, there are other streams which may be written to that developers cannot create and instantiate directly, but that nonetheless will be encountered. For

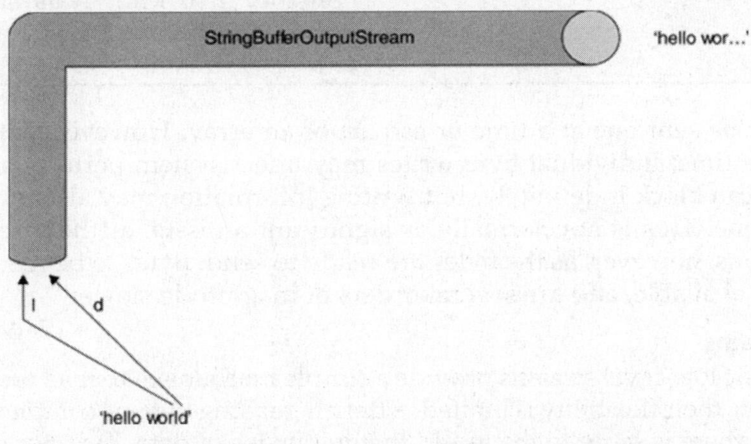

Output streams send data one byte at a time, sequentially.

example, a networking stream such as a connection to a TCP service is not created directly, but is provided by invoking the appropriate method on a socket. These types of streams have been covered in part 6. Output streams work somewhat differently from input streams, as might reasonably be expected due to their different purposes. While an input stream is a data consumer, an output stream is a data producer—it literally creates bytes of information and transmits them to something else (such as a file or data structure or network connection). Like input streams, data is communicated sequentially; that is, the first byte in will be the first byte out . This approach is analogous to a FIFO queue, in which data that is sent may be read back in order, without regard to sequencing. Unlike some specialized filter input streams, which allow you to "go back *n*" bytes within a sequence, once data is sent to an output stream, it cannot be undone. Consider it as like an e-mail message—once sent it cannot be taken back.

Low-level output streams of the java.io package

Low-level output stream	Purpose of stream
ByteArrayOutputStream	Writes bytes of data to an array of bytes.
FileOutputStream	Writes bytes of data to a local file.
PipedOutputStream	Writes bytes of data to a communication pipe, which will be connected to a java.io.PipedInput Stream.
StringBufferOutputStream	Writes bytes to a string buffer (a substitute data structure for the fixed-length string).
System.err	Writes bytes of data to the error stream of the user console, also known as standard error. In addition, this stream is cast to a PrintStream.
System.out	Writes bytes of data to the user console, also known as standard output. In addition, this stream is cast to a PrintStream.

Bytes may be sent one at a time or as part of an array. However, when bytes are read one at a time, individual byte writes may affect system performance. Reading information can block indefinitely, but writing information may also block for small amount of time. This is not normally as significant an issue as the case of blocking read operations, however, as the bytes are ready to send. Bytes to be read may not be immediately available, and are several orders of magnitude slower.

c. Filter Streams

While the basic low-level streams provide a simple mechanism to read and write bytes of information, their flexibility is limited. After all, reading bytes is complex (for people, at least), and there is more to the world that just bytes of data. Text, for example, is a sequence of characters, and other forms of data like numbers take up more than a single byte. Byte-level communication can also be inefficient, and data buffering can improve performance. To overcome these limitations, filter streams are used. Filter streams add additional functionality to an existing stream, by processing data in some form (such as buffering for performance) or offering additional methods that allow data to be accessed in a different manner (for example, reading a line of text rather than a sequence of bytes). Filters make life easier for programmers, as they can work with familiar constructs such as strings, lines of text, and numbers, rather than individual bytes. Instead of the programmer writing a string one character at a time and converting each character to an int value for the OutputStream.write(int) method, the filter stream does this for them.

Filter input streams of the java.io package

Filter input stream	Purpose of stream
BufferedInputStream	Buffers access to data, to improve efficiency.
DataInputStream	Reads primitive data types, such as an int, a float, a double, or even a line of text, from an input stream.
LineNumberInputStream	Maintains a count of which line is being read, based on interpretation of end-of-line characters. Handles both Unix and Windows end-of-line sequences.
PushBackInputStream	Allows a byte of data to be pushed into the head of the stream.

Useful Filter Output Streams of the java.io Package

Filter output stream purpose of stream

Filter output stream	Purpose of stream
BufferedOutputStream	Provides buffering of data writes, to improve efficiency
DataOutputStream	Writes primitive datatypes, such as bytes and numbers
PrintStream	Offers additional methods for writing lines of text, and other datatypes as text

d. Readers and Writers

While input streams and output streams may be used to read and write text as well as bytes of information and primitive data types, a better alternative is to use readers and writers. Readers and writers were introduced in JDK1.1 to better support Unicode character streams.

Unicode is an extended character set. Most people think of characters as being composed of 8 bits of data, offering a range of 256 possible characters. Low ASCII (0–127) characters are followed by high ASCII characters (128–255). The high ASCII characters represent characters and symbols such as those used in foreign languages or punctuation. However, people quickly realized that even 256 characters were not enough to handle the many characters used in languages around the world. This is where Unicode came in. Unicode characters are represented by 16 bits, allowing for a maximum of 65,536 possible characters—an enormous number. Unicode characters are supported by Java, although many developers are unaware of their use. Java also supports a modified form called UTF-8. This is a variable-width encoding format; some characters are a single byte and others multiple bytes.

The Importance of Readers and Writers

Readers and writers are a better alternative than input streams and output streams when used on text data. For those dealing solely with primitive data types, use of input streams and output streams may by all means be continued. However, if applications are processing text information only, use of a reader and / or a writer, to better support Unicode characters, should be considered. Readers and writers support Unicode character sequences. Internationalization may not seem like a significant concern for those of us with an English-speaking background. However, in a growing global economy, internationalization may become more important, and the Y2K bug (which, while overrated, required significant cost and effort to repair) certainly taught us that software should be written with an eye toward the future.

Low-level reader streams of the java.io package

Low-level reader	Purpose of reader
CharArrayReader	Reads from a character array
FileReader	Reads from a file on the local file system.
FileInputStreamPipedReader	Reads a sequence of characters from a thread communications pipe, exactly like a PipedInputStream
StringReader	Reads a sequence of characters from a String, as if it were a StringBuffer InputStream
InputStreamReader	Bridges the divide between an input stream and a reader, by reading from the input stream

Useful filter readers of the java.io package

Filter reader	Purpose of stream
BufferedReader	Buffers access to data, to improve efficiency
FilterReader	Provides a class to extend when creating filters
PushBackReader	Allows text data to be pushed back into the reader's stream
LineNumberReader	Buffered reader subclass, which maintains a count of which line it is On

Low-Level Writers of the java.io package

Low-level writer purpose of writer

Low-level writer	Purpose of writer
CharArrayWriter	Writes to a variable length character array (and resizes as characters are added)
FileWriter	Writes to a file on the local file system.
FileOutputStreamPipedWriter	Writes characters to a thread communications pipe
PipedOutputStream	Writes to a Java pipe as a stream of bytes
StringWriter	Writes characters to a string buffer
OutputStreamWriter	Writes to a legacy output stream

Filter writers of the java.io package

Filter writer	Purpose of writer
BufferedWriter	Buffers write requests, like a Buffered OutputStream
FilterWriter	Abstract class for developing filter writers
PrintWriter	Provides convenient PrintStream-like functionality

The methods explaining their usage has already been discussed in the previous chapters.

2. URLS AND URLCONNECTIONS

The URL class is used to represent resources on the World Wide Web. Every resource has an address, which identifies it uniquely and contains enough information for a Web browser to find the resource on the network and retrieve it. The address is called a "url" or "universal resource locator." An object belonging to the URL class represents such an address. Once you have a URL object, you can use it to open a URLConnection to the resource at that address. A URL is ordinarily specified as a string, such as "http://earmind.com/devotional.html". There are also relative URL's. A relative URL specifies the location of a resource relative to the location of another url, which is called the base or context for the relative url. For example, if the context is given by the URL http://earmind.com/, then the incomplete, relative URL "devotional.html" would really refer to http http://earmind.com/devotional.html. An object of the class URL is not simply a string, but it can be constructed from a string representation of a URL. A URL object can also be constructed from another URL object, representing a context, and a string that specifies a URL relative to that context. These constructors have prototypes

- *public URL(String urlName) throws MalformedURLException*
- *public URL(URL context, String relativeName) throws MalformedURLException*

Note that these constructors will throw an exception of type MalformedURLException if the specified strings do not represent legal URL's. The MalformedURLException class is a subclass of IOException, and it requires mandatory exception handling. That is, you must call the constructor inside a try...catch statement that handles the exception or in a subroutine that is declared to throw the exception. The second constructor is especially convenient when writing applets. In an applet, two methods are available that provide useful URL contexts. The method getDocumentBase(), defined in the Applet and JApplet classes, returns an object of type URL. This URL represents the location from which the HTML page that contains the applet was downloaded. This allows the applet to go back and retrieve other files that are stored in the same location as that document. For example, URL url = new URL(getDocumentBase(), "data.txt"); constructs a URL that refers to a file named data.txt on the same computer and in the same directory as the source file for the web page on which the applet is running. Another method, getCodeBase(), returns a URL that gives the location of the applet class file (which is not necessarily the same as the location of the document). Once you have a valid URL object, you can call its openConnection() method to set up a connection. This method returns a URLConnection. The URLConnection object can, in turn, be used to create an InputStream for reading data from the resource represented by the URL. This is done by calling its getInputStream() method. For example:

```
URL url = new URL(urlAddressString);
URLConnection connection = url.openConnection();
InputStream in = connection.getInputStream();
```

The openConnection() and getInputStream() methods can both throw exceptions of type IOException. Once the InputStream has been created, you can read from it in the usual way, including wrapping it in another input stream type, such as TextReader, or using a Scanner. Reading from the stream can, of course, generate exceptions. One of the other useful instance methods in the URLConnection class is getContentType(), which returns a String that describes the type of information available from the URL. The return value can be null if the type of information is not yet known or if it is not possible to determine the type. The type might not be available until after the input stream has been created, so you should generally call getContentType() after getInputStream(). The string returned by getContentType() is in a format called a mime type. Mime types include "text/plain", "text/html", "image/jpeg", "image/gif", and many others. All mime types contain two parts: A general type, such as "text" or "image", and a more specific type within that general category, such as "html" or "gif". If you are only interested in text data, for example, you can check whether the string returned by getContentType() starts with "text". (Mime types were first introduced to describe the content of email messages. The name stands for "Multipurpose Internet Mail Extensions." They are now used almost universally to specify the type of information in a file or other resource.) Let us look at a short example that uses all this to read the data from a URL. This subroutine opens a connection to a specified URL, checks that the type of data at the URL is text, and then copies the text onto the screen. Many of the operations in this subroutine can throw exceptions. They are handled by declaring that the subroutine "throws IOException" and leaving it up to the main program to decide what to do when an error occurs.

```
static void readTextFromURL( String urlString ) throws IOException

{
/* Open a connection to the URL, and get an input stream for reading data from the URL. */
```

```
URL url = new URL(urlString);
URLConnection connection = url.openConnection();
InputStream urlData = connection.getInputStream();
/* Check that the content is some type of text. */
String contentType = connection.getContentType();
if (contentType == null || contentType.startsWith("text") == false)
throw new IOException("URL does not seem to refer to a text file.");
/* Copy lines of text from the input stream to the screen, until
end-of-file is encountered (or an error occurs). */
BufferedReader in; // For reading from the connection's input stream.
in = new BufferedReader( new InputStreamReader(urlData) );
while (true) {
String line = in.readLine();
if (line == null)
break;
System.out.println(line);
}
} // end readTextFromURL()
```

A complete program that uses this subroutine can be found in the file ReadURL.java. When using the program, note that you have to specify a complete url, including the "http://" at the beginning.

```
import java.net.URL;
import java.net.URLConnection;
import java.io.BufferedReader;
import java.io.IOException;
import java.io.InputStream;
import java.io.InputStreamReader;
/** The url must be specified on the command line. It must be a complete url, including
 * the "protocol" at the beginnning (probably "http://"). */
public class ReadURL {
  public static void main(String[] args) {
    if (args.length == 0) {
      System.out.println("Please specify a URL on the command line.");
      return;
    }
    try {
      readTextFromURL(args[0]);
    }
    catch (IOException e) {
      System.out.println("\n*** Sorry, an error has occurred ***\n");
      System.out.println(e);
      System.out.println();
    }
  }
  /*** This subroutine attempts to copy text from the specified URL onto the screen. */
  static void readTextFromURL( String urlString ) throws IOException {
```

```
/* Open a connection to the URL, and get an input stream for reading data from the URL. */

    URL url = new URL(urlString);
    URLConnection connection = url.openConnection();
    InputStream urlData = connection.getInputStream();
    /* Check that the content is some type of text. Note: If
      getContentType() method were called before getting the input
      stream, it is possible for contentType to be null only because
      no connection can be made. The getInputStream() method will
      throw an error if no connection can be made. */

    String contentType = connection.getContentType();
    if (contentType == null || contentType.startsWith("text") == false)
      throw new IOException("URL does not seem to refer to a text file.");

/* Copy lines of text from the input stream to the screen, until end-of-file is encountered */

    BufferedReader in; // For reading from the connection's input stream.
    in = new BufferedReader( new InputStreamReader(urlData) );

    while (true) {
      String line = in.readLine();
      if (line == null)
        break;
      System.out.println(line);
    }

  } // end readTextFromURL()

} // end class ReadURL
```

3. TCP/IP AND CLIENT/SERVER

Communication over the Internet is based on a pair of protocols called the Transmission Control Protocol and the Internet Protocol, which are collectively referred to as TCP/IP. For two programs to communicate using TCP/IP, each program must create a socket, as discussed earlier in this section, and those sockets must be connected. Once such a connection is made, communication takes place using input streams and output streams. Each program has its own input stream and its own output stream. Data written by one program to its output stream is transmitted to the other computer. There it enters the input stream of the program at the other end of the network connection. When that program reads data from its input stream, it is receiving the data that was transmitted to it over the network. The hard part, then, is making a network connection in the first place. Two sockets are involved. To get things started, one program must create a socket that will wait passively until a connection request comes in from another socket. The waiting socket is said to be listening for a connection. On the other side of the connection-to-be, another program creates a socket that sends out a connection request to the listening socket. When the listening socket receives the connection request, it responds, and the connection is established. Once that is done, each program can obtain an input stream and an output stream for sending data over the connection. Communication takes place through these streams until one program or the other closes the connection.

A program that creates a listening socket is sometimes said to be a server, and the socket is called a serverSocket. A program that connects to a server is called a client, and the socket that it uses to make a connection is called a client socket. The idea is that the server is out there somewhere on the network, waiting for a connection request from some client. The server can be thought of as offering some kind of service, and the client gets access to that service by connecting to the server. This is called the client / server model of network communication. In many actual applications, a server program can provide connections to several clients at the same time. When a client connects to a server's listening socket, that socket does not stop listening. Instead, it continues listening for additional client connections at the same time that the first client is being serviced. To do this, it is necessary to use threads. The URL class that was discussed at the beginning of this section uses a client socket behind the scenes to do any necessary network communication. On the other side of that connection is a server program that accepts a connection request from the URL object, reads a request from that object for some particular file on the server computer, and responds by transmitting the contents of that file over the network back to the URL object. After transmitting the data, the server closes the connection. A client program has to have some way to specify which computer, among all those on the network, it wants to communicate with. Every computer on the Internet has an IP address which identifies it uniquely among all the computers on the net. Many computers can also be referred to by domain names such as earmind.com or www.a2z.co. Traditional (or IPv4) IP addresses are 32-bit integers. They are usually written in the so-called "dotted decimal" form, such as 69.9.161.200, where each of the four numbers in the address represents an 8-bit integer in the range 0 through 255. A new version of the Internet Protocol, IPv6, is currently being introduced. IPv6 addresses are 128-bit integers and are usually written in hexadecimal form (with some colons and maybe some extra information thrown in). In actual use, IPv6 addresses are still fairly rare. A computer can have several IP addresses, and can have both IPv4 and IPv6 addresses. Usually, one of these is the loopback address, which can be used when a program wants to communicate with another program on the same computer. The loopback address has IPv4 address 127.0.0.1 and can also, in general, be referred to using the domain name localhost . In addition, there can be one or more IP addresses associated with physical network connections. Your computer probably has some utility for displaying your computer's IP addresses. The following Java program, ShowMyNetwork.java, that does the same thing.

```
import java.net.*;
import java.util.Enumeration;

/**
 * This short program lists information about available network interfaces
 * on the computer on which it is run.  The name of each interface is
 * output along with a list of one or more IP addresses for that
 * interface.  The names are arbitrary names assigned by the operating
```

```
 * system to the interfaces. The addresses can include both IPv4 and
 * IPv6 addresses. The list should include the local loopback interface
 * (usually referred to as "localhost") as well as the interface
 * corresponding to any network card that has been installed and configured.
 */
public class ShowMyNetwork {
  public static void main(String[] args) {
    Enumeration netInterfaces;
    System.out.println();
    try {
      netInterfaces = NetworkInterface.getNetworkInterfaces();
    }
    catch (Exception e){
      System.out.println();
      System.out.println("Sorry, an error occurred while looking for network");
      System.out.println("interfaces. The error was:");
      System.out.println(e);
      return;
    }
    if (! netInterfaces.hasMoreElements() ) {
      System.out.println("No network interfaces found.");
      return;
    }
    System.out.println("Network interfaces found on this computer:");
    while (netInterfaces.hasMoreElements()) {
      NetworkInterface net = (NetworkInterface)netInterfaces.nextElement();
      String name = net.getName();
      System.out.print("  " + name + " : ");
      Enumeration inetAddresses = net.getInetAddresses();
      while (inetAddresses.hasMoreElements()) {
        InetAddress address = (InetAddress)inetAddresses.nextElement();
        System.out.print(address + "  ");
      }
      System.out.println();
    }
    System.out.println();
  } // end main()
}
```

Output

```
en1 : /192.168.1.47 /fe80:0:0:0:211:24ff:fe9c:5271%5
lo0 : /127.0.0.1 /fe80:0:0:0:0:0:0:1%1 /0:0:0:0:0:0:0:1%0
```

The first thing on each line is a network interface name, which is really meaningful only to the computer's operating system. The output also contains the IP addresses for that interface. In this example, lo0 refers to the loopback address, which has IPv4 address 127.0.0.1 as usual. The most important number here is 192.168.1.47, which is the IPv4 address that can be used for communication over the network. Now, a single computer might have several programs doing network communication at the same time, or one program communicating with several other computers. To allow for this possibility, a network connection is actually identified by a port number in combination with an IP address. A port number is just a 16-bit integer. A server does not simply listen for connections—it listens for connections on a particular port. A potential client must know both the Internet address (or domain name) of the computer on which the server is running and the port number on which the server is listening. A Web server, for example, generally listens for connections on port 80; other standard Internet services also have standard port numbers. (The standard port numbers are all less than 1024, and are reserved for particular services. If you create your own server programs, you should use port numbers greater than 1024.)

3.1 Sockets

To implement TCP/IP connections, the java.net package provides two classes, ServerSocket and Socket. A ServerSocket represents a listening socket that waits for connection requests from clients. A Socket represents one endpoint of an actual network connection. A Socket can be a client socket that sends a connection request to a server. But a Socket can also be created by a server to handle a connection request from a client. This allows the server to create multiple sockets and handle multiple connections. A ServerSocket does not itself participate in connections; it just listens for connection requests and creates Sockets to handle the actual connections.

When you construct a ServerSocket object, you have to specify the port number on which the server will listen. The specification for the constructor is public ServerSocket(int port) throws IOException. The port number must be in the range 0 through 65535, and should generally be greater than 1024. (A value of 0 tells the server socket to listen on any available port). The constructor might throw a SecurityException if a smaller port number is specified. An IOException can occur if, for example, the specified port number is already in use.

As soon as a ServerSocket is created, it starts listening for connection requests. The accept() method in the ServerSocket class accepts such a request, establishes a connection with the client, and returns a Socket that can be used for communication with the client. The accept() method has the form

```
public Socket accept() throws IOException
```

When you call the accept() method, it will not return until a connection request is received or until some error occurs. The method is said to block while waiting for the connection. You can call accept() repeatedly to accept multiple connection requests. The ServerSocket will continue listening for connections until it is closed, using its close() method, or until some error occurs, or until the program is terminated in some way. Suppose that you want a server to listen on port 1728, and suppose that you have written a method provideService(Socket) to handle the communication with one client. Then the basic form of the server program would be:

```
try {
ServerSocket server = new ServerSocket(1728);
while (true) {
Socket connection = server.accept();
provideService(connection);
}
}
catch (IOException e) {
System.out.println("Server shut down with error: " + e);
}
```

On the client side, a client socket is created using a constructor in the Socket class. To connect to a server on a known computer and port, the following constructor may be used.

public Socket(String computer, int port) throws IOException

The first parameter can be either an IP number or a domain name. This constructor will block until the connection is established or until an error occurs. Once you have a connected socket, no matter how it was created, you can use the Socket methods getInputStream() and getOutputStream() to obtain streams that can be used for communication over the connection. These methods return objects of type InputStream and OutputStream, respectively. Keeping all this in mind, here is the outline of a method for working with a client connection:

```
/*** Open a client connection to a specified server computer and port number on
the server, and then do communication through the connection.*/
void doClientConnection(String computerName, int serverPort) {
Socket connection;
InputStream in;
OutputStream out;
try {
connection = new Socket(computerName,serverPort);
in = connection.getInputStream();
out = connection.getOutputStream();
}
catch (IOException e) {
System.out.println("Attempt to create connection failed with error: " + e);
return;
}
.
. // Use the streams, in and out, to communicate with server.
.
try {
connection.close();
// (Alternatively, you might depend on the server to close the connection.)
}
catch (IOException e) {
}
} // end doClientConnection()
```

Some working examples that uses client-server technology are discussed in the foregoing examples.

3.2 A Trivial Client/Server

The first example consists of two programs. The source code files for the programs are DateClient.java and DateServer.java. One is a simple network client and the other is a matching server. The client makes a connection to the server, reads one line of text from the server, and displays that text on the screen. The text sent by the server consists of the current date and time on the computer where the server is running. In order to open a connection, the client must know the computer on which the server is running and the port on which it is listening. The server listens on port number 32007. The port number could be anything between 1025 and 65535, as long the server and the client use the same port. Port numbers between 1 and 1024 are reserved for standard services and should not be used for other servers. The name or IP number of the computer on which the server is running must be specified as a command-line argument. For example, if the server is running on a computer named math.hws.edu, then you would typically run the client with the command "java DateClient math.hws.edu". Here is the complete client program:

```
import java.net.*;
import java.io.*;
/**
* This program opens a connection to a computer specified
* as the first command-line argument. The connection is made to
* the port specified by LISTENING PORT. The program reads one
* line of text from the connection and then closes the
* connection. It displays the text that it read on
* standard output. This program is meant to be used with
* the server program, DateServer, which sends the current
* date and time on the computer where the server is running.
*/
public class DateClient {
public static final int LISTENING PORT = 32007;
public static void main(String[] args) {
String hostName; // Name of the server computer to connect to.
Socket connection; // A socket for communicating with server.
BufferedReader incoming; // For reading data from the connection.
/* Get computer name from command line. */
if (args.length > 0)
hostName = args[0];
else {
// No computer name was given. Print a message and exit.
System.out.println("Usage: java DateClient <server host name>");
return;
}
/* Make the connection, then read and display a line of text. */
try {
connection = new Socket( hostName, LISTENING PORT );
incoming = new BufferedReader(new InputStreamReader(connection.getInputStream()) );
String lineFromServer = incoming.readLine();
if (lineFromServer == null) {
// A null from incoming.readLine() indicates that end-of-stream was encountered.
throw new IOException("Connection was opened, " +"but server did not send any data.");
}
System.out.println();
System.out.println(lineFromServer);
```

```
System.out.println();
incoming.close();
}
catch (Exception e) {
System.out.println("Error: " + e);
}
} // end main()
} //end class DateClient
```

Note that all the communication with the server is done in a try...catch statement. This will catch the IOExceptions that can be generated when the connection is opened or closed and when data is read from the input stream. The connection's input stream is wrapped in a BufferedReader, which has a readLine() method that makes it easy to read one line of text. In order for this program to run without error, the server program must be running on the computer to which the client tries to connect. By the way, it is possible to run the client and the server program on the same computer. For example, you can open two command windows, start the server in one window and then run the client in the other window. To make things like this easier, most computers will recognize the domain name localhost and the IP number 127.0.0.1 as referring to "this computer." This means that the command "java DateClient localhost" will tell the DateClient program to connect to a server running on the same computer. If that command does not work, try "java DateClient 127.0.0.1". The server program that corresponds to the DateClient client program is called DateServer. The DateServer program creates a ServerSocket to listen for connection requests on port 32007. After the listening socket is created, the server will enter an infinite loop in which it accepts and processes connections. This will continue until the program is killed in some way—for example, by typing a CONTROL-C in the command window where the server is running. When a connection is received from a client, the server calls a subroutine to handle the connection. In the subroutine, any Exception that occurs is caught, so that it will not crash the server. Just because a connection to one client has failed for some reason, it does not mean that the server should be shut down; the error might have been the fault of the client. The connection-handling subroutine creates a PrintWriter for sending data over the connection. It writes the current date and time to this stream and then closes the connection. (The standard class java.util.Date is used to obtain the current time. An object of type Date represents a particular date and time. The default constructor, "new Date()", creates an object that represents the time when the object is created.) The complete server program is as follows:

```
import java.net.*;
import java.io.*;
import java.util.Date;
/**
* This program is a server that takes connection requests on
* the port specified by the constant LISTENING PORT. When a
* connection is opened, the program sends the current time to
* the connected socket. The program will continue to receive
* and process connections until it is killed (by a CONTROL-C,
* for example). Note that this server processes each connection
* as it is received, rather than creating a separate thread
```

```
* to process the connection.
*/
public class DateServer {
public static final int LISTENING PORT = 32007;
public static void main(String[] args) {
ServerSocket listener; // Listens for incoming connections.
Socket connection; // For communication with the connecting program.
/* Accept and process connections forever, or until some error occurs.
(Note that errors that occur while communicating with a connected
program are caught and handled in the sendDate() routine, so
they will not crash the server.) */
try {
listener = new ServerSocket(LISTENING PORT);
System.out.println("Listening on port " + LISTENING PORT);
while (true) {
// Accept next connection request and handle it.
connection = listener.accept();
sendDate(connection);
}
}
catch (Exception e) {
System.out.println("Sorry, the server has shut down.");
System.out.println("Error: " + e);
return;
}
} // end main()
/**
* The parameter, client, is a socket that is already connected to another program.
Get an output stream for the connection, send the current time, and close the connection.
*/
private static void sendDate(Socket client) {
try {
System.out.println("Connection from " +client.getInetAddress().toString() );
Date now = new Date(); // The current date and time.
PrintWriter outgoing; // Stream for sending data.
outgoing = new PrintWriter( client.getOutputStream() );
outgoing.println( now.toString() );
outgoing.flush(); // Make sure the data is actually sent!
client.close();
}
catch (Exception e){
System.out.println("Error: " + e);
}
} // end sendDate()
} //end class DateServer
```

When you run DateServer in a command-line interface, it will sit and wait for connection requests and report them as they are received. To make the DateServer service permanently available on a computer, the program really should run as a daemon. A daemon is a program that runs continually on a computer, independently of any user. The computer can be configured to start the daemon automatically as soon as the computer boots up. It then runs in the background, even while the computer is being used for other purposes. For example, a computer that makes pages available on the World Wide Web runs a daemon that listens for requests for pages and responds by transmitting the pages. It is just a souped-up analog of the DateServer program! However, the question of how to set up a program as a daemon is not one. I want to go into here. For testing purposes, it is easy enough to start the program by hand, and, in any case, my examples are not really robust enough or full-featured enough to be run as serious servers. Note that after calling out.println() to send a line of data to the client, the server program calls out.flush(). The flush() method is available in every output stream class. Calling it ensures that data that has been written to the stream is actually sent to its destination. You should generally call this function every time you use an output stream to send data over a network connection. If you do not do so, it is possible that the stream will collect data until it has a large batch of data to send. This is done for efficiency, but it can impose unacceptable delays when the client is waiting for the transmission. It is even possible that some of the data might remain untransmitted when the socket is closed, so it is especially important to call flush() before closing the connection. This is one of those unfortunate cases where different implementations of Java can behave differently. If you fail to flush your output streams, it is possible that your network application will work on some types of computers but not on others.

3.3 Simple Chat

In the DateServer example, the server transmits information and the client reads it. It is also possible to have two-way communication between client and server. As a first example, we will look at a client and server that allow a user on each end of the connection to send messages to the other user. The program works in a command-line interface where the users type in their messages. In this example, the server waits for a connection from a single client and then closes down its listener so that no other clients can connect. After the client and server are connected, both ends of the connection work in much the same way. The user on the client end types a message, and it is transmitted to the server, which displays it to the user on that end. Then the user of the server types a message that is transmitted to the client. Then the client user types another message, and so on. This continues until one user or the other enters "quit" when prompted for a message. When that happens, the connection is closed and both programs terminate. The client program and the server program are very similar. The techniques for opening the connections differ, and the client is programmed to send the first message while the server is programmed to receive the first message. The client and server programs can be found in the files CLChatClient.java and CLChatServer.java. (The name "CLChat" stands for "command-line chat.") Here is the source code for the server:

```
import java.net.*;
import java.io.*;
/**
public class CLChatServer {
/*** Port to listen on, if none is specified on the command line.*/
static final int DEFAULT PORT = 1728;
/**
```

```
* Handshake string. Each end of the connection sends this string to the
* other just after the connection is opened. This is done to confirm that
* the program on the other side of the connection is a CLChat program.
*/
static final String HANDSHAKE = "CLChat";
/*** This character is prepended to every message that is sent.*/
static final char MESSAGE = '0';
/*** This character is sent to the connected program when the user quits.*/
static final char CLOSE = '1';
public static void main(String[] args) {
int port; // The port on which the server listens.
ServerSocket listener; // Listens for a connection request.
Socket connection; // For communication with the client.
BufferedReader incoming; // Stream for receiving data from client.
PrintWriter outgoing; // Stream for sending data to client.
String messageOut; // A message to be sent to the client.
String messageIn; // A message received from the client.
BufferedReader userInput; // A wrapper for System.in, for reading
// lines of input from the user.
/* get the port number from the command line, or use the default port if none is specified.*/
if (args.length == 0)
port = DEFAULT PORT;
else {
try {
port= Integer.parseInt(args[0]);
if (port < 0 || port > 65535)
throw new NumberFormatException();
}
catch (NumberFormatException e) {
System.out.println("Illegal port number, " + args[0]);
return;
}
}
/* Wait for a connection request. When it arrives, close down the listener. Create
streams for communication and exchange the handshake. */
try {
listener = new ServerSocket(port);
System.out.println("Listening on port " + listener.getLocalPort());
connection = listener.accept();
listener.close();
incoming = new BufferedReader(
new InputStreamReader(connection.getInputStream()) );
outgoing = new PrintWriter(connection.getOutputStream());
outgoing.println(HANDSHAKE); // Send handshake to client.
outgoing.flush();
messageIn = incoming.readLine(); // Receive handshake from client.
if (! HANDSHAKE.equals(messageIn) ) {
throw new Exception("Connected program is not a CLChat!");
}
```

```
System.out.println("Connected. Waiting for the first message.");
}
catch (Exception e) {
System.out.println("An error occurred while opening connection.");
System.out.println(e.toString());
return;
}
/* Exchange messages with the other end of the connection until one side
or the other closes the connection. This server program waits for
the first message from the client. After that, messages alternate
strictly back and forth. */
try {
userInput = new BufferedReader(new InputStreamReader(System.in));
System.out.println("NOTE: Enter 'quit' to end the program.\n");
while (true) {
System.out.println("WAITING...");
messageIn = incoming.readLine();
if (messageIn.length() > 0) {
// The first character of the message is a command. If
// the command is CLOSE, then the connection is closed.
// Otherwise, remove the command character from the
// message and proceed.
if (messageIn.charAt(0) == CLOSE) {
System.out.println("Connection closed at other end.");
connection.close();
break;
}
messageIn = messageIn.substring(1);
}
System.out.println("RECEIVED: " + messageIn);
System.out.print("SEND: ");
messageOut = userInput.readLine();
if (messageOut.equalsIgnoreCase("quit")) {
// User wants to quit. Inform the other side
// of the connection, then close the connection.
outgoing.println(CLOSE);
outgoing.flush(); // Make sure the data is sent!
connection.close();
System.out.println("Connection closed.");

break;
}
outgoing.println(MESSAGE + messageOut);
outgoing.flush(); // Make sure the data is sent!
if (outgoing.checkError()) {
throw new IOException("Error occurred while transmitting message.");
}
}
}
```

```
catch (Exception e) {
System.out.println("Sorry, an error has occurred. Connection lost.");
System.out.println("Error: " + e);
System.exit(1);
}
} // end main()
} //end class CLChatServer
```

This program is a little more robust than DateServer. For one thing, it uses a handshake to make sure that a client who is trying to connect is really a CLChatClient program. A handshake is simply information sent between client and server as part of setting up the connection, before any actual data is sent. In this case, each side of the connection sends a string to the other side to identify itself. A protocol is a detailed specification of what data and messages can be exchanged over a connection, how they must be represented, and what order they can be sent in. When you design a client/server application, the design of the protocol is an important consideration. Another aspect of the CLChat protocol is that after the handshake, every line of text that is sent over the connection begins with a character that acts as a command. If the character is 0, the rest of the line is a message from one user to the other. If the character is 1, the line indicates that a user has entered the "quit" command, and the connection is to be shut down. Remember that if you want to try out this program on a single computer, you can use two command-line windows. In one, give the command "java CLChatServer" to start the server. Then, in the other, use the command "java CLChatClient localhost" to connect to the server that is running on the same machine.

4. USER DATAGRAM PROTOCOL

The User Datagram Protocol (UDP) is a commonly used transport protocol employed by many types of applications. UDP is a connectionless transport protocol, meaning that it does not guarantee either packet delivery or that packets arrive in sequential order. Although control over the ultimate destination of a UDP packet rests with the computer that sends it, how it reaches that destination is an arbitrary process.

The packets may travel along different paths, as selected by the various network routers that distribute traffic flow seemingly whimsically—depending on factors such as network congestion, priority of routes, and cost of transmission. This means that a packet can arrive out of sequence, if it encounters a faster route than the previous packet or if the previous packet encounters some other form of delay. No two packets are guaranteed the same route, and if a particular route is heavily congested, the packet may be discarded entirely. Each packet has a time-to-live (TTL) counter, which is updated when the packet is routed along to the next point in the network. When the timer expires, it will be discarded, and the recipient of the packet will not be notified. If a packet does arrive, however, it will always arrive intact. Packets that are corrupt or only partially delivered are discarded. Given the potential for loss of data packets, it may seem odd that anyone would even consider using such an unreliable, seemingly anarchical system. In fact, there are many advantages to using UDP that may not be apparent at the first glance.

- UDP communication can be more efficient than guaranteed-delivery data streams. If the amount of data is small and the data is sent frequently, it may make sense to avoid the overhead of guaranteed delivery.

- Unlike TCP streams, which establish a connection, UDP causes fewer overheads. If the amount of data being sent is small and the data is sent infrequently, the overhead of establishing a connection might not be worth it. UDP may be preferable in this case, particularly if data is being sent from a large number of machines to central one, in which case the sum total of all these connections might cause significant overload.
- Real-time applications that demand up-to-the-second or better performance may be candidates for UDP, as there are fewer delays due to the error checking and flow control of TCP. UDP packets can be used to saturate available network bandwidth to deliver large amounts of data such as streaming video/audio, or data for a multiplayer network game. In addition, if some data is lost, it can be replaced by the next set of packets with updated information, eliminating the need to resend old data that is now out of date.
- UDP sockets can receive data from more than one host machine. If several machines must be communicated with, then UDP may be more convenient than other mechanisms such as TCP.
- Some network protocols specify UDP as the transport mechanism, requiring its use.

Java supports the User Datagram Protocol in the form of two classes:

- java.net.DatagramPacket
- java.net.DatagramSocket

4.1 DatagramPacket Class

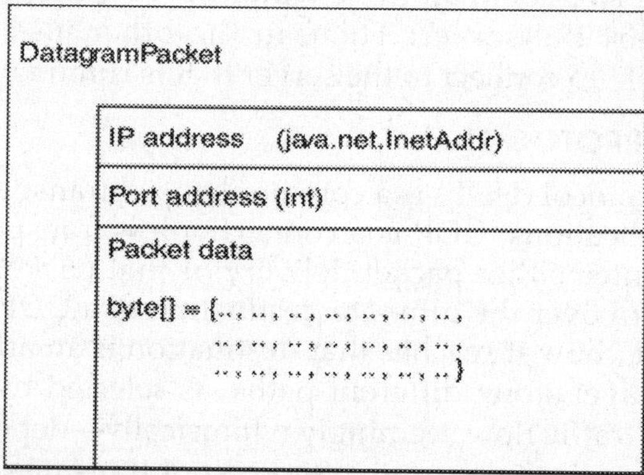

The meaning of the data stored in a DatagramPacket is determined by its context. When a DatagramPacket has been read from a UDP socket, the IP address of the packet represents the address of the sender. However, when a DatagramPacket is used to send a UDP packet, the IP address stored in DatagramPacket represents the address of the recipient.

Creating DatagramPacket

There are two reasons to create a new DatagramPacket:

1. To send data to a remote machine using UDP
2. To receive data sent by a remote machine using UDP

Constructors

The choice of which DatagramPacket constructor to use is determined by its intended purpose. Either constructor requires the specification of a byte array, which will be used to store the UDP packet contents, and the length of the data packet. To create a

DatagramPacket for receiving incoming UDP packets, the following constructor should be used:

DatagramPacket(byte[] buffer, int length)

For example:

DatagramPacket packet = new DatagramPacket(new byte[256], 256);

To send a DatagramPacket to a remote machine, it is preferable to use the following constructor:

DatagramPacket(byte[] buffer, int length, InetAddress dest_addr, int dest_port)

For example:

InetAddress addr = InetAddress.getByName("192.168.0.1");

DatagramPacket packet = new DatagramPacket (new byte[128], 128, addr, 2000);

Using a DatagramPacket

The DatagramPacket class provides some important methods that allow the remote address, remote port, data (as a byte array), and length of the packet to be retrieved. As of JDK1.1, there are also methods to modify these, via a corresponding set method. This means that a received packet can be reused. For example, a packet's contents can be replaced and then sent back to the sender. This saves having to reset addressing information—the address and port of the packet are already set to those of the sender.

Methods

Methods	Usage	Use
getAddress	InetAddress getAddress()	Returns the IP address from which a datagrampacket was sent, or (if the packet is going to be sent to a remote machine), the destination IP address.
getData	Byte[] getData()	Returns the contents of the datagrampacket, represented as an array of bytes.
getLength	Int getLength()	Returns the length of the data stored in a datagrampacket. This can be less than the actual size of the data buffer.
getport	Int getPort()	Returns the port number from which a datagrampacket was sent, or (if the packet is going to be sent to a remote machine), the destination port number.
setAddress	Void setAddress (InetAddress addr)	Assigns a new destination address to a datagrampacket.
setData	Void setData(byte[] buffer)	Assigns a new data buffer to the datagrampacket. Remember to make the buffer long enough, to prevent data loss.

setLength	Void setLength(int length)	Assigns a new length to the datarampacket. Remember that the length must be less than or equal to the maximum size of the data buffer, or an illegalargumentexception will be thrown.
setPort	Void setPort(int port)	Assigns a new destination port to a datagrampacket.

4.2 DatagramSocket Class

The DatagramSocket class provides access to a UDP socket, which allows UDP packets to be sent and received. A DatagramPacket is used to represent a UDP packet, and must be created prior to receiving any packet. The same DatagramSocket can be used to receive packets as well as to send them. However, read operations are blocking, meaning that the application will continue to wait until a packet arrives. Since UDP packets do not guarantee delivery, this can cause an application to stall if the sender does not resubmit packets.

Creating a DatagramSocket

A DatagramSocket can be used to both send and receive packets. Each DatagramSocket binds to a port on the local machine, which is used for addressing packets. The port number need not match the port number of the remote machine, but if the application is a UDP server, it will usually choose a specific port number. If the DatagramSocket is intended to be a client, and does not need to bind to a specific port number, a blank constructor can be specified.

Constructors

To create a client DatagramSocket, the following constructor is used:

DatagramSocket() throws java.net.SocketException

To create a server Datagram Socket, the following constructor is used, which takes as a parameter the port to which the UDP service will be bound:

DatagramSocket(int port) throws java.net.SocketException

Although rarely used, there is a third constructor for DatagramSocket, introduced in JDK1.1. If a machine is known by several IP addresses, you can specify the IP address and port to which a UDP service should be bound. It takes as parameters the port to which the UDP service will be bound, as well as the InetAddress of the service. This constructor is:

DatagramSocket (int port, InetAddress addr) throws java.net.SocketException

Using a DatagramSocket

DatagramSocket is used to receive incoming UDP packets and to send outgoing UDP packets. It provides methods to send and receive packets, as well as to specify a timeout value when nonblocking I/O is being used, to inspect and modify maximum UDP packet sizes, and to close the socket.

Methods

Methods	Usage	Use
close	Void close()	Closes a socket, and unbinds it from the local port.

connect	Void connect(InetAddress remote_addr int remote_port)	Restricts access to the specified remote address and port.
disconnect	Void disconnect()	Disconnects the DatagramSocket and removes any restrictions imposed on it by an earlier connect operation.
getInet Address	InetAddress getInetAddress()	Returns the remote address to which the socket is connected, or null if no such connection exists.
getPort	Int getPort()	Returns the remote port to which the socket is connected, or –1 if no such connection exists.
getLocal Address	InetAddress getLocalAddress()	Returns the local address to which the socket is bound.
getLocalPort	Int getLocalPort()	Returns the local port to which the socket is bound.
getReceived BufferSize	Int getReceiveBufferSize() throws java.net.SocketException	Returns the maximum buffer size used for incoming UDP packets.
getSend BufferSize	Int getSendBufferSize() throws java.net.Socket Exception	Returns the maximum buffer size used for outgoing UDP packets.
getSoTimeout	Int getSoTimeout() throws java.net.SocketException	Returns the value of the timeout socket option. This value is used to determine the number of milliseconds a read operation will block before throwing a java.io.InterruptedIOException.
receive	Void receive(DatagramPacket packet) throws java.io. IOException	Reads a UDP packet and stores the contents in the specified packet.

The address and port fields of the packet will be overwritten with the sender address and port fields, and the length field of the packet will contain the length of the original packet, which can be less than the size of the packet's byte-array. If a timeout value has not been specified by using DatagramSocket.setSoTimeout(int duration), this method will block indefinitely. If a timeout value has been specified, a java.io.InterruptedIOException will be thrown if the time is exceeded.

- void send(DatagramPacket packet) throws java.io.IOException—sends a UDP packet, represented by the specified packet parameter.
- void setReceiveBufferSize(int length) throws java.net. SocketException—sets the maximum buffer size used for incoming UDP packets. Whether the specified length will be adhered to is dependent on the operating system.

- void setSendBufferSize(int length) throws java.net.SocketException—sets the maximum buffer size used for outgoing UDP packets. Whether the specified length will be adhered to is dependent on the operating system.
- void setSoTimeout(int duration) throws java.net.SocketException—sets the value of the timeout socket option. This value is the number of milliseconds a read operation will block before throwing a java.io.InterruptedIOException.

4.3 Listening for UDP Packets

Before an application can read UDP packets sent to it by remote machines, it must bind a socket to a local UDP port using DatagramSocket, and create a DatagramPacket that will act as a container for the UDP packet's data. The figure below shows the relationship between a UDP packet, the various Java classes used to process it, and the actual application.

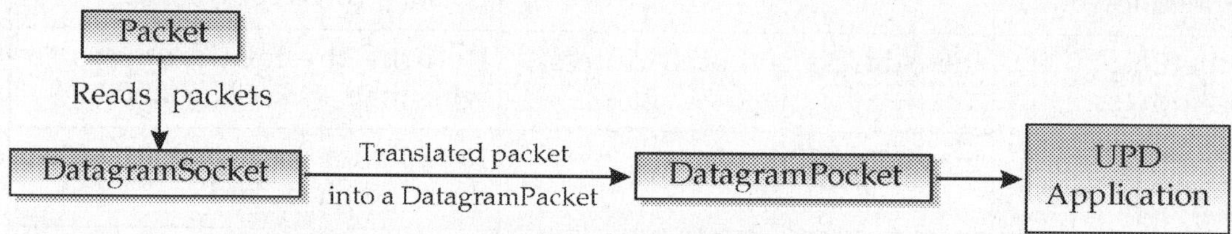

UDP packets are received by a DatagramSocket and translated into DatagramPacket object

When an application wishes to read UDP packets, it calls the DatagramSocket.receive method, which copies a UDP packet into the specified DatagramPacket. The contents of the DatagramPacket are processed, and the process is repeated as needed. The following code snippet illustrates this process:

```
DatagramPacket packet = new DatagramPacket (new byte[256], 256);
DatagramSocket socket = new DatagramSocket(2000);
boolean finished = false;
while (! finished )
{
socket.receive (packet);
// process the packet
}
socket.close();
```

When processing the packet, the application must work directly with an array of bytes. If, however, your application is better suited to reading text, you can use classes from the Java I/O package to convert between a byte array and another type of stream or reader. By hooking a ByteArrayInputStream to the contents of a datagram and then to another type of InputStream or an InputStreamReader, you can access the contents of UDP packets relatively easily. Many developers prefer to use Java I/O streams to process data, using a DataInputStream or a BufferedReader to access the contents of byte arrays.

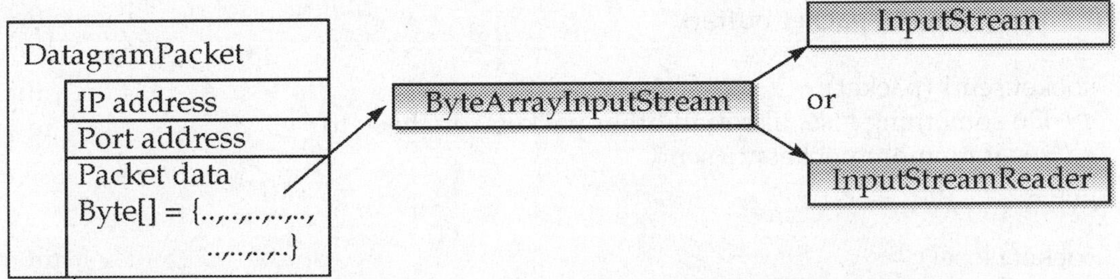

Reading from a UDP packet is simplified by applying input streams

For example, to hook up a DataInputStream to the contents of a DatagramPacket, the following code can be used:

```
ByteArrayInputStream bin = new ByteArrayInputStream( packet.getData() );
DataInputStream din = new DataInputStream (bin);
// Read the contents of the UDP packet
.......
```

4.4 Sending UDP Packets

The same interface (DatagramSocket) employed to receive UDP packets is also used to send them. When sending a packet, the application must create a DatagramPacket, set the address and port information, and write the data intended for transmission to its byte array. If replying to a received packet, the address and port information will already be stored, and only the data need to be overwritten. Once the packet is ready for transmission, the send method of DatagramSocket is invoked, and a UDP packet is sent.

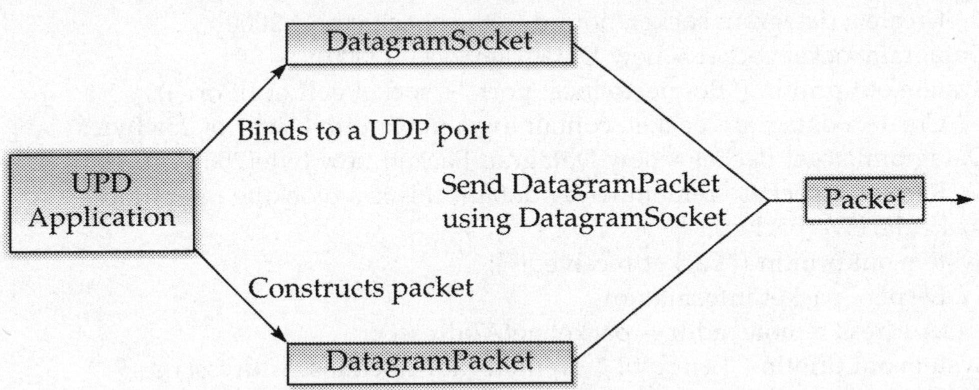

Packets are sent using a DatagramSocket

The following code snippet illustrates this process:

```
DatagramSocket socket = new DatagramSocket(2000);
DatagramPacket packet = new DatagramPacket (new byte[256], 256);
packet.setAddress ( InetAddress.getByName ( somehost ) );
packet.setPort ( 2000 );
boolean finished = false;
while (finished)
```

```
{
// Write data to packet buffer
.........
socket.send (packet);
// Do something else, like read other packets, or check to
// see if no more packets to send
.........
}
socket.close();
```

4.5 User Datagram Protocol Example

To demonstrate how UDP packets are sent and received, we will compile and run two small examples. The first will bind to a local port, read a packet, and display its contents and addressing information. The second example will send the packet read by the first.

Code for PacketReceiveDemo

```
import java.net.*;
import java.io.*;
public class PacketReceiveDemo
{
public static void main (String args[])
{
try
{
System.out.println ("Binding to local port 2000");
// Create a datagram socket, bound to the specific port 2000
DatagramSocket socket = new DatagramSocket(2000);
System.out.println ("Bound to local port "+ socket.getLocalPort());
// Create a datagram packet, containing a maximum buffer of 256 bytes
DatagramPacket packet = new DatagramPacket( new byte[256], 256 );
// Receive a packet - remember by default this is a blocking operation
socket.receive(packet);
System.out.println ("Packet received!");
// Display packet information
InetAddress remote_addr = packet.getAddress();
System.out.println ("Sent by : " +remote_addr.getHostAddress() );
System.out.println ("Sent from: " +packet.getPort());
// Display packet contents, by reading from byte array
ByteArrayInputStream bin = new ByteArrayInputStream (packet.getData());
// Display only up to the length of the original UDP packet
for (int i=0; i < packet.getLength(); i++)
{
int data = bin.read();
if (data == -1)
break;
```

```
else
System.out.print ( (char)data) ;
}
socket.close();
}
catch (IOException ioe)
{
System.err.println ("Error - " + ioe);
}
}
}
```

How PacketReceiveDemo Works

Most of the code is self-explanatory, or is similar to code snippets shown earlier. However, readers may benefit from a closer examination. The application starts by binding to a specific port, 2000. Applications offering a service generally bind to a specific port. When acting as a receiver, your application should choose a specific port number, so that a sender can send UDP packets to this port. Next, the application prepares a DatagramPacket for storing UDP packets, and creates a new buffer for storing packet data.

```
// Create a datagram socket, bound to the specific port 2000
DatagramSocket socket = new DatagramSocket(2000);
System.out.println ("Bound to local port " + socket.getLocalPort());
// Create a datagram packet, containing a maximum buffer of 256 bytes
DatagramPacket packet = new DatagramPacket( new byte[256], 256);
```

Now the application is ready to read a packet. The read operation is blocking, so until a packet arrives, the server will wait. When a packet is successfully delivered to the application, the addressing information for the packet is displayed so that it can be determined where it came from.

```
// Receive a packet - remember by default this is a blocking operation
socket.receive(packet);
// Display packet information
InetAddress remote_addr = packet.getAddress();
System.out.println ("Sent by : " + remote_addr.getHostAddress() );
System.out.println ("Send from: " + packet.getPort());
```

To provide easy access to the contents of the UDP packet, the application uses a ByteArrayInputStream to read from the packet. Reading one character at a time, the program displays the contents of the packet and then finishes. Note that Unicode characters, which are represented by more than just a single byte, cannot be written out in this fashion (readers and writers would be more appropriate if internationalization support is required).

```
// Display packet contents, by reading from byte array
ByteArrayInputStream bin = new ByteArrayInputStream(packet.getData());
// Display only up to the length of the original UDP packet
for (int i=0; i < packet.getLength(); i++)
{
```

```
int data = bin.read();
if (data == -1)
break;
else
System.out.print ( (char) data) ;
}
```

Code for PacketSendDemo

```
import java.net.*;
import java.io.*;
public class PacketSendDemo
{
public static void main (String args[])
{
int argc = args.length;
// Check for valid number of parameters
if (argc != 1)
{
System.out.println ("Syntax :");
System.out.println ("java PacketSendDemo hostname");
return;
}
String hostname = args[0];
try
{
System.out.println ("Binding to a local port");
// Create a datagram socket, bound to any available  local port
DatagramSocket socket = new DatagramSocket();
System.out.println ("Bound to local port "+ socket.getLocalPort());
// Create a message to send using a UDP packet
ByteArrayOutputStream bout = new ByteArrayOutputStream();
PrintStream pout = new PrintStream (bout);
pout.print ("Greetings!");
// Get the contents of our message as an array of bytes
byte[] barray = bout.toByteArray();
// Create a datagram packet, containing our byte array
DatagramPacket packet = new DatagramPacket( barray, barray.length );
System.out.println ("Looking up hostname "+ hostname );
// Lookup the specified hostname, and get an InetAddress
InetAddress remote_addr =InetAddress.getByName(hostname);
System.out.println ("Hostname resolved as " +remote_addr.getHostAddress());
// Address packet to sender
packet.setAddress (remote_addr);
// Set port number to 2000
packet.setPort (2000);
// Send the packet - remember no guarantee of delivery
```

```
socket.send(packet);
System.out.println ("Packet sent!");
}
catch (UnknownHostException uhe)
{
System.err.println ("Can't find host " +hostname);
}
catch (IOException ioe)
{
System.err.println ("Error - " + ioe);
}
}
}
```

How PacketSendDemo Works

The second example uses UDP to talk to the first example. This example acts as the sender, dispatching a UDP packet to the receiver, which contains an ASCII text-greeting message. Though it uses some similar classes (DatagramSocket, DatagramPacket), they are employed in a slightly different way. The application starts by binding a UDP socket to a local port, which will be used to send the data packet. Unlike the receiver demonstration, it does not matter which local port is being used. In fact, any free port is a candidate, and you may find that running the application several times will result in different port numbers. After binding to a port, the port number is displayed to demonstrate this.

```
// Create a datagram socket, bound to any available local port
DatagramSocket socket = new DatagramSocket();
System.out.println ("Bound to local port " +socket.getLocalPort());
```

Before sending any data, we need to create a DatagramPacket. First, a ByteArrayOutputStream is used to create a sequence of bytes. Once this is complete, the array of bytes is passed to the DatagramPacket constructor.

```
// Create a message to send using a UDP packet
ByteArrayOutputStream bout = new ByteArrayOutputStream();
PrintStream pout = new PrintStream (bout);
pout.print ("Greetings!");
// Get the contents of our message as an array of bytes
byte[] barray = bout.toByteArray();
// Create a datagram packet, containing our byte array
DatagramPacket packet = new DatagramPacket( barray, barray.length );
```

Now that the packet has some data, it needs to be correctly addressed. As with a postal message, if it lacks correct address information it cannot be delivered. We start by obtaining an InetAddress for the remote machine, and then display its IP address. This InetAddress is passed to the setAddress method of DatagramPacket, ensuring that it will arrive at the correct machine. However, we must go one step further and specify a port number. In this case, port 2000 is matched, as the receiver will be bound to that port.

```
System.out.println ("Looking up hostname " + hostname );
// Lookup the specified hostname, and get an InetAddress
```

```
InetAddress remote_addr = InetAddress.getByName(hostname);
System.out.println ("Hostname resolved as " +remote_addr.getHostAddress());
// Address packet to sender
packet.setAddress (remote_addr);
// Set port number to 2000
packet.setPort (2000);
```

The final step, after all this work, is to send the packet. This is the easiest step of all—simply invoke the send method of DatagramSocket. Again, remember: There is no guarantee of delivery, so it is possible for a packet to become lost in transit. A more robust application would try to read an acknowledgment and resend the message if it had become lost.

```
// Send the packet - remember no guarantee of delivery
socket.send(packet);
```

Running the UDP Examples

To run these examples, you will need to open two console windows. The first application to be run is the receiver, which will wait for a UDP packet.

There are no parameters for the receiver, so to run it use:

```
java PacketReceiveDemo
```

In a second window, you then need to run the sender. This application could be run from any computer on a local network or the Internet (providing there is not a firewall between the two hosts). If you would like, you can also run it from the same machine. It takes a single parameter, the hostname of the remote machine:

```
java PacketSendDemo myhostname
```

4.6 Building a UDP Client/Server

The previous example illustrates the technical details of how an individual packet may be sent and received. But applications need a series of packets, not just one. The next example shows how to build a UDP server, a long-running system that is capable of serving many requests during its lifetime. The type of service that is provided is an echo service, which echoes back the contents of a packet. The echo service runs on a well-known port, port 7, and if it is known that a system has an echo server installed, the server may be accessed by clients to see if a system is up and running (similar to the ping application). The example below demonstrates how to write an echo client that will send packets to the server as well as read the results back.

Some systems have an echo service already running in the background, and security restrictions may prevent a service from binding to a well-known port. If this is the case, you will need to change the port number to a different number (and in the case of security restrictions, you should select a number above port 1024) in both the client and the server. The same port number must be used for communication.

Building an Echo Service

The following example involves building an echo service, which transmits any packet it receives straight back to the sender. The code uses no new networking classes or methods, but employs a special technique. It loops continuously to serve one client after another. Though only one UDP packet will be processed at a time, the delay between receiving a packet and dispatching it again is negligible, resulting in the illusion of concurrent processing.

```
import java.net.*;
import java.io.*;
public class EchoServer
{
// UDP port to which service is bound
public static final int SERVICE_PORT = 7;
// Max size of packet, large enough for almost any client
public static final int BUFSIZE = 4096;
// Socket used for reading and writing UDP packets
private DatagramSocket socket;
public EchoServer()
{
try
{
// Bind to the specified UDP port, to listen
// for incoming data packets
socket = new DatagramSocket( SERVICE_PORT );
System.out.println ("Server active on port " +socket.getLocalPort() );
}
catch (Exception e)
{
System.err.println ("Unable to bind port");
}
}
public void serviceClients()
{
// Create a buffer large enough for incoming packets
byte[] buffer = new byte[BUFSIZE];
for (;;)
{
try
{
// Create a DatagramPacket for reading UDP packets
DatagramPacket packet = new DatagramPacket ( buffer, BUFSIZE );
// Receive incoming packets
socket.receive(packet);
System.out.println ("Packet received from " +packet.getAddress() + ":"
+packet.getPort() + " of length " + packet.getLength() );
// Echo the packet back - address and port are already set for us !
socket.send(packet);
}
catch (IOException ioe)
{
System.err.println ("Error : " + ioe);
}
```

```
}
}
public static void main(String args[])
{
EchoServer server = new EchoServer();
server.serviceClients();
}
}
```

Building an Echo Client

The following client can be used with the echo service and can easily be adapted to support other services. Repeated packets are sent to the echo service, and a timeout is caught to prevent the service from stalling if a packet becomes lost, and the client then waits to receive it. Remember that packet loss in an intranet environment is unlikely, but with slow network connections on the Internet it is quite possible.

```
import java.net.*;
import java.io.*;
public class EchoClient
{
// UDP port to which service is bound
public static final int SERVICE_PORT = 7;
// Max size of packet
public static final int BUFSIZE = 256;
public static void main(String args[])
{
if (args.length != 1)
{
System.err.println ("Syntax - java EchoClient hostname");
return;
}
String hostname = args[0];
// Get an InetAddress for the specified hostname
InetAddress addr = null;
try
{
// Resolve the hostname to an InetAddr
addr = InetAddress.getByName(hostname);
}
catch (UnknownHostException uhe)
{
System.err.println ("Unable to resolve host");
return;
}
try
```

```
{
// Bind to any free port
DatagramSocket socket = new DatagramSocket();
// Set a timeout value of two seconds
socket.setSoTimeout (2 * 1000);
for (int i = 1 ; i <= 10; i++)
{
// Copy some data to our packet
String message = "Packet number " + i ;
char[] cArray = message.toCharArray();
byte[] sendbuf = new byte[cArray.length];
for (int offset = 0; offset <
cArray.length ; offset++)
{
sendbuf[offset] = (byte)
cArray[offset];
}
// Create a packet to send to the UDP server
DatagramPacket sendPacket = new DatagramPacket(sendbuf, cArray.length,
addr, SERVICE_PORT);
System.out.println ("Sending packet to " + hostname);
// Send the packet
socket.send (sendPacket);
System.out.print ("Waiting for packet.... ");
// Create a small packet for receiving UDP packets
byte[] recbuf = new byte[BUFSIZE];
DatagramPacket receivePacket = new DatagramPacket(recbuf,BUFSIZE);
// Declare a timeout flag
boolean timeout = false;
// Catch any InterruptedIOException that is thrown
// while waiting to receive a UDP packet
try
{
socket.receive (receivePacket);
}
catch (InterruptedIOException ioe)
{
timeout = true;
}
if (!timeout)
{
System.out.println ("packet received!");
System.out.println ("Details : " +
receivePacket.getAddress());
// Obtain a byte input stream to read the UDP packet
```

```
ByteArrayInputStream bin = new ByteArrayInputStream (
receivePacket.getData(), 0, receivePacket.getLength() );
// Connect a reader for easier access
BufferedReader reader = new BufferedReader ( new InputStreamReader ( bin )
);
// Loop indefinitely
for (;;)
{
String line = reader.readLine();
// Check for end of data
if (line == null)
break;
else
System.out.println (line);
}
}
else
{
System.out.println ("packet lost!");
}
// Sleep for a second, to allow user to see packet
try
{
Thread.sleep(1000);
} catch (InterruptedException ie) { }
}
}
catch (IOException ioe)
{
System.err.println ("Socket error " + ioe);
}
}
}
```

Running the Echo Client and Server

Before clients can send requests, the echo server must be active. Otherwise, UDP packets will be sent and then ignored, as there is no program running to read them. However, the client will not stall if a response is not sent back; it will send packets again after two seconds of waiting. This is important, as servers may be inactive or packets may become lost in transmission. To run the echo server, type the following:

```
java EchoServer
```

To run the echo client (either on the same machine as the server or a different one), type the following:

```
java EchoClient hostname
```

where hostname (or localhost, if running locally) is the location of the echo service.

4.7 UDP—Summary

While the UDP is sometimes the best alternative for certain classes of applications, because of its unique properties, it does present some challenges to developers. These challenges can be met, however, by structuring data transmission to overcome the limitations of UDP. Below we examine these limitations and how they may be overcome.

a. Lack of Guaranteed Delivery

Packets sent via UDP may become lost in transit—each additional hop between one router and another introduces more delays and increases the likelihood that a packet may be discarded when its TTL reaches zero. Furthermore, UDP packets can become damaged or lost if the physical network connection they are being routed through goes down. Since Internet packets are being transmitted across a public network, composed of a diverse range of network infrastructures, it is likely that packets will become lost at some point in a connection. Of course, in some applications the loss of individual packets may not have a noticeable effect. For example, a video stream might lose a few frames of picture, but provided that most of the frames arrive, the loss is bearable. However, if a file is being transferred, then the file contents will become garbled, and the loss of packets becomes unacceptable. If guaranteed delivery is required, the best alternative is to avoid packet-based communication altogether and use a more suitable transport mechanism like the Transmission Control Protocol. Nonetheless, if the use of UDP is called for, one solution is for the party receiving packets to send an acknowledgment packet (also referred to as an ACK) back to the sender. The absence of an ACK indicates that a packet was lost and should be retransmitted. Some transport systems send back an ACK for individual packets or for a range of packets. Although it does add additional complexity, acknowledgment of a range of packets makes for a more efficient use of bandwidth. Some systems also use a negative-acknowledgment packet (NAK) to indicate that a specific packet was lost, which triggers immediate retransmission of that packet.

b. Lack of Guaranteed Packet Sequencing

Applications that require sequential access to data should include a sequence number in the contents of a datagram packet. If a packet arrives out of order, it can be buffered until the earlier packets have caught up. Sequencing adds a small amount of complexity, but does make a system more reliable—you always know which packet you are dealing with! Duplicate packets must be discarded, and missing packets requested again.

c. Lack of Flow Control

The face of the Internet is changing rapidly, as network connections move from dial-up and wireless modems to broadband communication through cable modems and ISDN lines. Some systems can handle a large amount of data, whereas others still have extremely limited bandwidth. To avoid flooding a system with more data than it can handle, the technique of flow control is used. Flow control places a limit on how much data is sent, and helps to prevent systems from becoming overloaded (and bandwidth from being wasted). Imagine flow control as a water limiter that restricts the amount of liquid flowing through a showerhead. There are many flow-control techniques, ranging from limiting the number of packets sent per second to limiting the number of packets that have not yet been acknowledged. The settings for the former are hard to determine, as they vary depending on the receiver. The latter is probably the best choice—by limiting the number of unacknowledged packets, control is placed in the hands of the receiver. If the receiver can get and respond to packets quickly, then more packets will be acknowledged and more will be sent. If packets are flooding the network, fewer responses will come back and a throttle is placed on the flow. Some systems may also elect to use a variable flow limit, which can be customized to take into account the length of time for acknowledgments to come back. Since UDP does not offer direct control over flow control, for large-scale applications it may be appropriate to limit the number of packets sent to a host (for example, *n* packets per second, where *n* is a number suitable for the transmission line and speed of the recipient machine).

5. TRANSMISSION CONTROL PROTOCOL

The Transmission Control Protocol (TCP) is a stream-based method of network communication that is far different from any discussed previously. This chapter discusses TCP streams and how they operate under Java.

TCP provides an interface to network communications that is radically different from the User Datagram Protocol (UDP) discussed in Chapter 5. The properties of TCP make it highly attractive to network programmers, as it simplifies network communication by removing many of the obstacles of UDP, such as ordering of packets and packet loss. While UDP is concerned with the transmission of packets of data, TCP focuses instead on establishing a network connection, through which a stream of bytes may be sent and received. In the previous topic on UDP, we saw that packets may be sent through a network using various paths and may arrive at different times. This benefits performance and robustness, as the loss of a single packet does not necessarily disrupt the transmission of other packets. Nonetheless, such a system creates extra work for programmers who need to guarantee delivery of data. TCP eliminates this extra work by guaranteeing delivery and order, providing for a reliable byte communication stream between client and server that supports two-way communication. It establishes a "virtual connection" between two machines, through which streams of data may be sent.

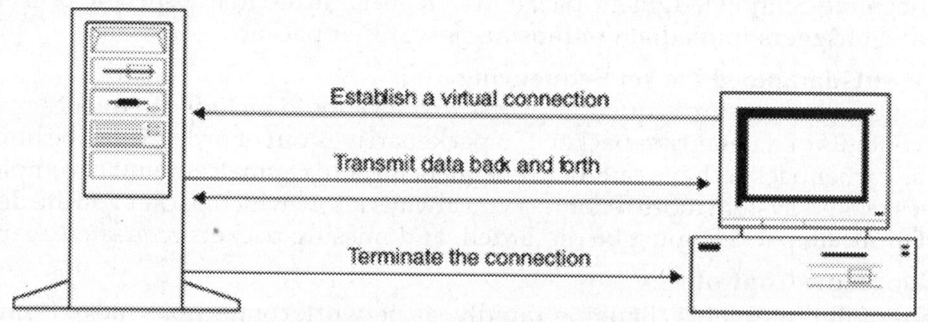

TCP establishes a virtual connection to transmit data

TCP uses a lower-level communications protocol, the Internet Protocol (IP), to establish the connection between machines. This connection provides an interface that allows streams of bytes to be sent and received, and transparently converts the data into IP datagram packets. A common problem with datagrams, is that they do not guarantee that packets arrive at their destination. TCP takes care of this problem. It provides guaranteed delivery of bytes of data. Of course, it is always possible that network errors will prevent delivery, but TCP handles the implementation issues such as resending packets, and alerts the programmer only in serious cases such as if there is no route to a network host or if a connection is lost. The virtual connection between two machines is represented by a socket. Sockets, allow data to be sent and received;

there are substantial differences between a UDP socket and a TCP socket, however. First, TCP sockets are connected to a single machine, whereas UDP sockets may transmit or receive data from multiple machines. Second, UDP sockets only send and receive packets of data, whereas TCP allows transmission of data through byte streams represented as an InputStream and OutputStream. They are converted into datagram packets for transmission over the network, without requiring the programmer to intervene.

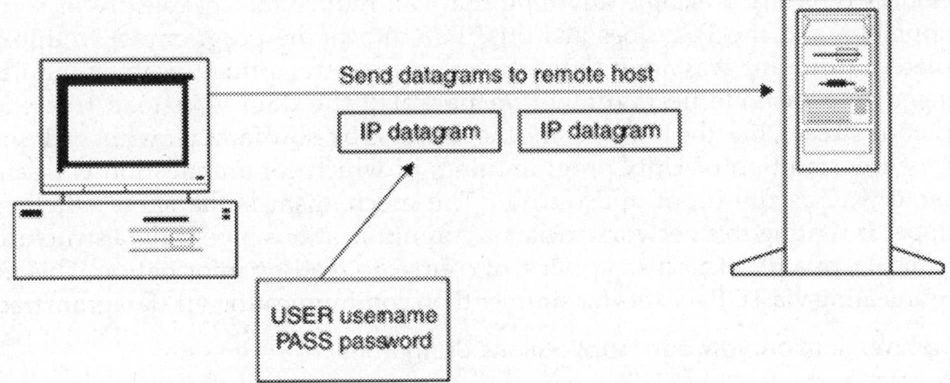

TCP deals with streams of data such as protocol commands, but converts streams into IP datagrams for transport over the network

Advantages of TCP over UDP

Several advantages of using TCP over UDP are briefly summarized below.

a. Automatic Error Control

Data transmission over TCP streams is more dependable than transmission of packets of information via UDP. Under TCP, data packets sent through a virtual connection include a checksum to ensure that they have not been corrupted, just like UDP. However, delivery of data is guaranteed by the TCP— data packets lost in transit are retransmitted. You may be wondering just how this is achieved—after all, IP and UDP do not guarantee delivery; neither do they give any warning when datagram packets are dropped. Whenever a collection of data is sent by TCP using data grams, a timer is started. Recall our UDP examples from Chapter 5, in which the DatagramSocket.setSoTimeout method was used to start a timer for a receive() operation. In TCP, if the recipient sends an acknowledgment, the timer is disabled. But if an acknowledgment is not received before the time runs out, the packet is retransmitted. This means that any data written to a TCP socket will reach the other side without the need for further intervention by programmers. All of the codes for error control is handled by TCP.

b. Reliability

Since the data sent between two machines participating in a TCP connection is transmitted by IP datagrams, the datagram packets will frequently arrive out of order. This would throw for a loop any program reading information from a TCP socket, as the order of the byte stream would be disrupted and frequently unreliable. Fortunately, issues such as ordering are handled by TCP—each datagram packet contains a sequence number that is used to order data. Later packets arriving before earlier packets will be held in a queue until an ordered sequence of data is available. The data will then be passed to the application through the interface of the socket.

c. Ease of Use

While storing information in datagram packets is certainly not beyond the reach of programmers, it does not lead to the most efficient way of communication between computers. There is added complexity, and it can be argued that the task of designing and creating software within a deadline provides complexity enough for programmers. Developers typically welcome anything that can reduce the complexity of software development, and the TCP does just this. TCP allows the programmer to think in a completely different way, one that is much more streamlined. Rather than being packaged into discrete units (datagram packets), the data is instead treated as a continuous stream, like the I/O streams the reader is by now familiar with. TCP sockets continue the tradition of Unix programming, in which communication is treated in the same way as file input and output. The mechanism is the same whether the developer is writing to a network socket, a communications pipe, a data structure, the user console, or a file. This also applies, of course, to reading information. This makes communicating via TCP sockets far simpler than communicating via datagram packets.

5.1 Communication Between Applications Using Ports

It is clear that there are significant differences between TCP and UDP, but there is also an important similarity between these two protocols. Both share the concept of a communications port, which distinguishes one application from another. Many services and clients run on the same port, and it would be impossible to sort out which one was which without distributing them by port number. When a TCP socket establishes a connection to another machine, it requires two very important pieces of information to connect to the remote end—the IP address of the machine and the port number. In addition, a local IP address and port number will be bound to it, so that the remote machine can identify which application established the connection. After all, you would not want your e-mail to be accessible by another user running software on the same system.

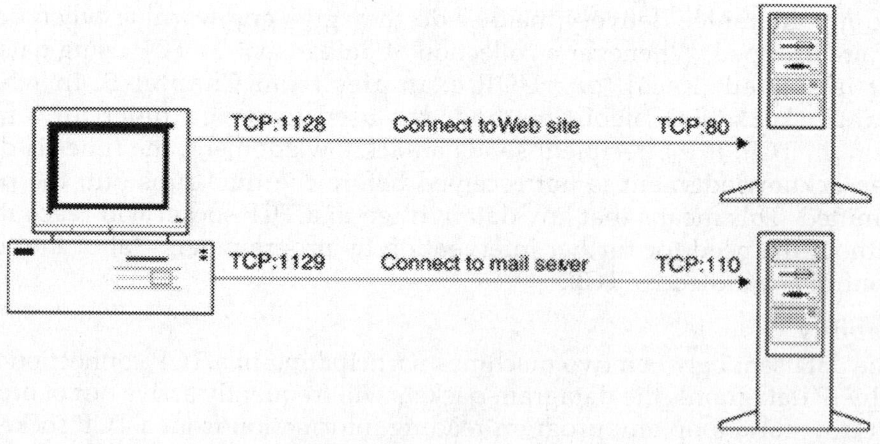

Local ports identify the application establishing a connection from other programs, allowing multiple TCP applications to run on the same machine

Ports in TCP are just like ports in UDP—they are represented by a number in the range 1–65535. Ports below 1024 are restricted to use by well-known services such as HTTP, FTP, SMTP, POP3, and telnet. The table below lists a few of the well-known services and their associated port numbers.

Protocols and their associated ports

Well-known services	**Service ports**
Telnet	23
Simple Mail Transfer Protocol	25
HyperText Transfer Protocol	80
Post Office Protocol	3110

5.2 Socket Operations

TCP sockets can perform a variety of operations. They can:

- Establish a connection to a remote host
- Send data to a remote host
- Receive data from a remote host
- Close a connection

In addition, there is a special type of socket that provides a service that will bind to a specific port number. This type of socket is normally used only in servers, and can perform the following operations:

- Bind to a local port
- Accept incoming connections from remote hosts
- Unbind from a local port

These two sockets are grouped into different categories, and are used by either a client or a server (since some clients may also be acting as servers, and some servers as clients). However, it is normal practice for the role of client and server to be separate.

5.3 TCP and the Client/Server Paradigm

In network programming (and often in other forms of communication, such as database programming), applications that use sockets are divided into two categories, the client and the server. You are probably familiar with the phrase *client/server programming,* although the exact meaning of the phrase may be unclear to you. This paradigm is the subject of the discussion below.

5.3.1 The Client/Server Paradigm

The client/server paradigm divides software into two categories, clients and servers. A client is software that initiates a connection and sends requests, whereas a server is software that listens for connections and processes requests. In the context of UDP programming, no actual connection is established, and UDP applications may both initiate and receive requests on the same socket. In the context of TCP, where connections are established between machines, the client/server paradigm is much more relevant. When software acts as a client, or as a server, it has a rigidly defined role that fits easily into a familiar mental model. Either the software is initiating requests, or it is processing them. Switching between these roles makes for a more complex system. Even if switching is permitted, at any given time one software program must be the client and one software program must be the server. If they both try to be clients at the same time, no server exists to process the requests! The client/server paradigm is an important theoretical concept that is widely used in practical applications. There are other communications models as well, such as peer to peer, in which either party may initiate communication. However, the client/server concept is a popular choice due to its simplicity and is used in most network programming.

5.3.2 Network Clients

Network clients initiate connections and usually take charge of network transactions. The server is there to fulfill the requests of the client—a client does not fulfill the requests of a server. Although the client is in control, some power still resides in the server, of course. A client can tell a server to delete all files on the local file system, but the server is not necessarily compelled to carry out that action. The network client speaks to the server using an agreed-upon standard for communication, the network protocol. For example, an HTTP client uses a set of commands different from a mail client, and has a completely different purpose. Connecting an HTTP client to a mail server, or a mail client to an HTTP server, will result not only in an error message but also in an error message that the client will not understand. For this reason, as part of the protocol specification, a port number is used so that the client can locate the server. A Web server typically runs on port 80, and while some servers can run on nonstandard ports, the convention for a URL is not to list a port, as it is assumed that port 80 is used. For more information on ports, see Section 6.1.2.

Network Servers

The role of the network server is to bind to a specific port (which is used by the client to locate the server), and to listen for new connections. While the client is temporary, and runs only when the user chooses, the server must run continually (even if no clients are actually connected) in the hope that someone, at some time, will want its services. The server is often referred to as a daemon process, to use Unix parlance. It runs indefinitely, and is normally automatically started when the host computer of the server is started. So the server waits, and waits, and waits, until a client establishes a connection to the server port. Some servers can handle only a single connection at a time, while others can handle many connections concurrently, through the use of threads. When a connection is being processed, the server is submissive. It waits for the client to send requests, and dutifully processes them (though the server is free to respond with an error message, particularly if the request violates some important precept of the protocol or presents a security risk). Some protocols, like HTTP/1.0, normally allow only one request per connection, whereas others, such as POP3, support a sequence of requests. Servers will answer the client request by sending either a response or an error message—the format of which varies from protocol to protocol. Learning a network protocol (when writing either a client or a server) is a little like learning a new language, as the syntax changes. Typically, though, the number of commands is much smaller, making things a little easier. The behavior of the server is determined in part by the protocol and in part by the developer.

5.4 TCP Sockets and Java

Java offers good support for TCP sockets, in the form of two socket classes, java.net.Socket and java.net.ServerSocket. When writing client software that connects to an existing service, the Socket class should be used. When writing server software that binds to a local port in order to provide a service, the ServerSocket class should be employed. This is different from the way a DatagramSocket works with UDP—the function of connecting to servers, and the function of accepting data from clients, is split into a separate class under TCP.

5.4.1 Socket Class

The Socket class represents client sockets, and is a communication channel between two TCP communications ports belonging to one or two machines. A socket may connect to a port on the local system, avoiding the need for a second machine, but most network software will usually involve two machines. TCP sockets cannot communicate with more than two machines, however. If this functionality is required, a client application should establish multiple socket connections, one for each machine.

Constructors

There are several constructors for the java.net.Socket class. Two constructors, which allowed a boolean parameter to specify whether UDP or TCP sockets were to be used, have been deprecated. These constructors should not be used and are not listed here—if UDP functionality is required, use a DatagramSocket.

The easiest way to create a socket is to specify the hostname of the machine and the port of the service. For example, to connect to a Web server on port 80, the following code might be used:

```
try
{
// Connect to the specified host and port
Socket mySocket = new Socket ( "www.awl.com", 80);
// ......
}
catch (Exception e)
{
System.err.println ("Err – " + e);
}
```

However, a wide range of constructors is available, for different situations. Unless otherwise specified, all constructors are public.

- protected Socket () creates an unconnected socket using the default implementation provided by the current socket factory.
- Socket (InetAddress address int port) throws java.io.IOException, java.lang.SecurityException creates a socket connected to the specified IP address and port. If a connection cannot be established, or if connecting to that host violates a security restriction, an exception is thrown.
- Socket (InetAddress address, int port, InetAddress localAddress int localPort) throws java.io.IOException, java.lang.SecurityException creates a socket connected to the specified address and port, and is bound to the specified local address and local port. By default, a free port is used, but this method allows you to specify a specific port number, as well as a specific address, in the case of multihomed hosts (i.e. a machine where the localhost is known by two or more IP addresses).
- Protected Socket (SocketImpl implementation) creates an unconnected socket using the specified socket implementation.
- Socket (String host, int port) throws java.net.UnknownHostException, java.io.IOException, java.lang.SecurityException creates a socket connected to the specified host and port. This method allows a string to be specified, rather than an InetAddress. If the hostname could not be resolved, a connection could not be established, or a security restriction is violated, an exception is thrown.

- Socket (String host, int port, InetAddress localAddress, int localPort) throws java.net.UnknownHostException, java.io.IOException, java.lang.Security Exception creates a socket connected to the specified host and port, and bound to the specified local port and address. This allows a hostname to be specified as a string, and not an InetAddress instance, as well as allowing a specific local address and port to be bound to. These local parameters are useful for multihomed hosts (i.e. a machine where the localhost is known by two or more IP addresses). If the hostname cannot be resolved, a connection cannot be established, or a security restriction is violated, an exception is thrown.

Creating a Socket

Under normal circumstances, a socket is connected to a machine and port when it is created. Although there is a blank constructor that does not require a hostname or port, it is protected and cannot be called from normal applications. Furthermore, there is not a connect() method that allows you to specify these details at a later point in time, so under normal circumstances the socket will be connected when created. If the network is fine, the call to a socket constructor will return as soon as a connection is established, but if the remote machine is not responding, the constructor method may block for an indefinite amount of time. This varies from system to system, depending on a variety of factors such as the operating system being used and the default network timeout (some machines on a local intranet, for example, seem to respond faster than some Internet machines, depending on network settings). You cannot ever guarantee how long a socket may block for, but this is abnormal behavior and would not happen frequently. Nonetheless, in mission-critical systems it may be appropriate to place such calls in a second thread, to prevent an application from stalling.

At a lower level, sockets are produced by a socket factory, which is a special class responsible for creating the appropriate socket implementation. Under normal circumstances, a standard java.net.Socket will be produced, but in special situations, such as special networking environments in which custom sockets are used (for example, to break through a firewall by using a special proxy server), socket factories may actually return a socket subclass. The details of socket factories are best left to experienced developers who are familiar with the intricacies of Java networking and have a definite purpose for creating custom sockets and socket factories. For more information on this topic, consult the Java API documentation for the java.net.SocketFactory and java.net.SocketImplFactory class.

Using a Socket

Sockets can perform a variety of tasks, such as reading information, sending data, closing a connection, and setting socket options. In addition, the following methods are provided to obtain information about a socket, such as address and port locations:

Methods

Methods	Usage	Use
close	Void close() throws java.io. IOException	Closes the socket connection. Closing a connect may or may not allow remaining data to be sent, depending on the value of the SO_LINGER socket option.

getInet Address()	InetAddress getInet	Returns the address of the Address remote machine that is connected to the socket.
getInput Stream	InputStream getInput Stream() throws java.io.	Returns an input stream, which IOException reads from the application this socket is connected to.
getOutput Stream	OutputStream getOutput Stream() throws java.io. IOException	Returns an output stream, which writes to the application that this socket is connected to.
getKeepalive	Boolean getKeepAlive() throws java.net.Socket Exception	Returns the state of the SO_KEEPALIVE socket option.
getLocal address	InetAddress getLocal Address()	Returns the local address associated with the socket (useful in the case of multihomed machines).
getLocalport	Int getLocalPort()	Returns the port number that the socket is bound to on the local machine.
getport	Int getPort()	Returns the port number of the remote service to which the socket is connected.
getReceive BufferSize	Int getReceiveBufferSize() throws java.net.Socket Exception	Returns the receive buffer size used by the socket, determined by the value of the SO_RCVBUF socket option.
getSend BufferSize	Int getSendBufferSize() throws java.net.Socket Exception	Returns the end buffer size used by the socket, determined by the value of the SO_SNDBUF socket option.
getSoLinger	Int getSoLinger() throws java.net.Socket Exception	Returns the value of the SO_LINGER socket option, which controls how long unsent data will be queued when a connection is terminated.
getSoTimeout	Int getSoTimeout() throws java.net.Socket Exception	Returns the value of the SO_TIMEOUT socket option, which controls how many milliseconds a read operation will block for.
getTcpno Delay	Boolean getTcpNoDelay() throws java.net.Socket Exception	Returns "true" if the TCP_NODELAY socket option is set, which controls whether Nagle's algorithm is enabled.

setKeepAlive	Void setKeepAlive(boolean onFlag) throws java.net. SocketException	Enables or disables the SO_KEEPALIVE socket option.
setReceive BufferSize	Void setReceiveBufferSize (int size) throws java.net. SocketException	Modifies the value of the SO_RCVBUF socket option, which recommends a buffer size for the operating system's network code to use for receiving incoming data.
setSendBuffer Size	Void setSendBufferSize (int size) throws java.net. SocketException	Modifies the value of the SO_SNDBUF socket option, which recommends a buffer size for the operating system's network code to use for sending incoming data.
setSoLinger	Static void setSocketImpl Factory (SocketImplFactory factory) throws java.net.Socket Exception, java.io.IOException java. lang.SecurityException	Assigns a socket implementation factory for the JVM, which may already exist, or may violate security restrictions, either of which causes an exception to be thrown. Only one factory can be specified, and this factory will be used whenever a socket is created.
setSoTimeout	Void setSoLinger(boolean onFlag, int duration) throws java.net.Socket Exception java.lang. Illegal ArgumentException	Enables or disables the SO_LINGER socket option and specifies a duration in seconds. If a negative value is specified, an exception is thrown.
setSoTimeout	Void setSoTimeout(int duration) throws java.net. SocketException	Modifies the value of the SO_TIMEOUT socket option, which controls how long (in milliseconds) a read operation will block. If a timeout does occur, a java.io.IOInterrupted Exception is thrown whenever a read operation occurs on the socket's input stream.
setTcpNo Delay	Void setTcpNoDelay (boolean onFlag) throws java.net.SocketException	Enables or disables the TCP_NODELAY socket option, which determines whether Nagle's algorithm is used.
shutdown Input	Void shutdownInput() throws java.io.IOException	Closes the input stream associated with this socket and discards any further information that is sent.

shutdown Output	Void shutdownOutput() throws java.io.IOException	Closes the output stream associated with this socket. Any data previously written, but not yet sent, will be flushed, followed by a TCP connection termination sequence, which notifies the application that no more data will be available. Further writes to the socket will cause an IOException to be thrown.

5.4.2 Reading from and Writing to TCP Sockets

Creating client software that uses TCP for communication is extremely easy in Java, no matter what operating system is being used. The Java Networking API provides a consistent, platform-neutral interface that allows client applications to connect to remote services. Once a socket is created, it is connected and ready to read/write by using the socket's input and output streams. These streams do not need to be created; they are provided by the Socket. getInputStream() and Socket.getOutputStream() methods. As was shown in Chapter 4 on I/O streams, filtered streams provide easy I/O access.

A filter can easily be connected to a socket stream, to make for simpler programming. The following code snippet demonstrates a simple TCP client that connects a BufferedReader to the socket input stream, and a PrintStream to the socket output stream.

```
try
{
// Connect a socket to some host machine and port
Socket socket = new Socket ( somehost, someport );
// Connect a buffered reader
BufferedReader reader = new BufferedReader (new InputStreamReader
(socket.getInputStream() ) );
// Connect a print stream
PrintStream pstream = new PrintStream( socket.getOutputStream() );
}
catch (Exception e)
{
System.err.println ("Error – " + e);
}
```

5.4.3 Socket Options

Socket options are settings that modify how sockets work, and they can affect the performance of applications. Support for socket options was introduced in Java 1.1, and some refinements have been made in later versions. Generally, socket options should not be changed unless there is a good reason for doing so, as changes may negatively affect application and network performance. The one exception to this caveat is the SO_TIMEOUT option— virtually every TCP application should handle timeouts gracefully rather than stalling if the application the socket is connected to fails to transmit data when required.

a. SO_Keepalive

By default, no data is sent between two connected sockets unless an application has data to send. This means that an idle socket may not have data submitted for minutes, hours, or even days in the case of long-lived processes. When the keepalive socket option is enabled, the other end of the socket is probed to verify it is still active. To enable keepalive, the Socket.setSoKeepAlive(boolean) method is called with a value of "true".

```
// Enable SO_KEEPALIVE
socketname.setSoKeepAlive(true);
```

Although keepalive does have some advantages, many developers advocate controlling timeouts and dead sockets at a higher level, in application code. It should also be kept in mind that keepalive does not allow you to specify a value for probing socket endpoints. A better solution than keepalive, and one that developers are advised to use, is to instead modify the timeout socket option.

b. SO_Rcvbuf

The receive buffer socket option controls the buffer used for receiving data. Changes can be made to the size by calling the Socket.setReceiveBufferSize(int) method. For example, to increase the receive buffer size to 4,096 bytes, the following code would be used.

```
// modify receive buffer size
socketname.setReceiveBufferSize(4096);
```

The current buffer size can be determined by invoking the Socket. getReceive Buffer Size() method. A better choice for buffering is to use a BufferedInput Stream/Buffered Reader.

c. So_Sndbuf

The send buffer socket option controls the size of the buffer used for sending data. By calling the Socket.setSendBufferSize(int) method, you can attempt to change the buffer size, but requests to change the size may be rejected by the operating system.

```
// Set the send buffer size to 4096 bytes
someSocket.setSendBufferSize(4096);
```

To determine the size of the current send buffer, you can call the Socket. getSendBufferSize() method, which returns an int value.

```
// Get the default size
int size = someSocket.getSendBufferSize();
```

Changing buffer size will be more effective with the DatagramSocket class. When buffering writes, the preferable choice is to use a BufferedOutputStream or a BufferedWriter.

d. So_Linger

When a TCP socket connection is closed, it is possible that data may be queued for delivery and not yet sent (particularly if an IP datagram becomes lost in transit and must be resent). The linger socket option controls the amount of time during which unsent data may be sent, after which it is discarded completely. It is possible to enable/disable the linger option entirely, or to modify the duration of a linger, by using the Socket.setSoLinger(boolean onFlag, int duration)

method:

```
// Enable linger, for fifty seconds
socketname.setSoLinger(true, 50);
```

e. Tcp_Nodelay

This socket option is a flag, the state of which controls whether Nagle's algorithm (RFC 896) is enabled or not. Because TCP data is sent over the network using IP datagrams, a fair bit of overhead exists for each packet, such as IP and TCP header information. If only a few bytes at a time are sent in each packet, the size of the header information will far exceed that of the data. The solution is Nagle's algorithm, which states that TCP may send only one datagram at a time. When an acknowledgment comes back for each IP datagram, a new packet is sent containing any data that has been queued up. This limits the amount of bandwidth being consumed by packet header information, but at a not insignificant cost—network latency. Since data is being queued, it is not dispatched immediately, so systems that require quick response times are slowed. Disabling Nagle's algorithm may improve performance, but if used by too many clients, network performance is reduced.

Nagle's algorithm is enabled or disabled by invoking the Socket.setTcpNoDelay (boolean state) method. For example, to deactivate the algorithm, the following code would be used:

```
// Disable Nagle's algorithm for faster response times
socketname.setTcpNoDelay(false);
```

To determine the state of Nagle's algorithm and the TCP_NODELAY flag, the Socket.getTcpNoDelay() method is used:

```
// Get the state of the TCP_NODELAY flag
boolean state = socketname.getTcpNoDelay();
```

f. So_Timeout

This timeout option is the most useful socket option. By default, I/O operations are blocking. An attempt to read data from an InputStream will wait indefinitely until input arrives. If the input never arrives, the application stalls and in most cases becomes unusable (unless multithreading is used). Users are not fond of unresponsive applications, and find such application behavior annoying, to say the least. A more robust application will anticipate such problems and take corrective action.

In a local intranet environment during testing, network problems are rare, but on the Internet stalled applications are probable. Server applications are not immune—a server connection to a client uses the Socket class as well, and can just as easily stall. For this reason, all applications (be they client or server) should handle network timeouts gracefully.

When the SO_TIMEOUT option is enabled, any read request to the InputStream of a socket starts a timer. When no data arrives in time and the timer expires, a java.io.InterruptedIOException is thrown, which can be caught to check for a timeout. What happens then is up to the application developer—a retry attempt might be made, the user might be notified, or the connection aborted. The duration of the timer is controlled by calling the Socket. setSoTimeout(int) method, which accepts as a parameter the number of milliseconds to wait for data. For example, to set a five-second timeout, the following code would be used:

```
// Set a five second timeout
someSocket.setSoTimeout ( 5 * 1000 );
```

Once enabled, any attempt to read could potentially throw an InterruptedIOException, which is extended from the java.io.IOException class. Since read attempts can already throw an IOException, no further code is required to handle the exception—however, some applications may want to specifically trap timeout-related exceptions, in which case an additional exception handler may be added.

```
try
{
Socket s = new Socket (...);
s.setSoTimeout ( 2000 );
// do some read operation ...
}
catch (InterruptedIOException iioe)
{
timeoutFlag = true; // do something special like set a flag
}
catch (IOException ioe)
{
System.err.println ("IO error " + ioe);
System.exit(0);
}
```

To determine the length of the TCP timer, the Socket.getSoTimeout() method, which returns an int, can be used. A value of zero indicates that timeouts are disabled, and read operations will block indefinitely.

```
// Check to see if timeout is not zero
if ( someSocket.getSoTimeout() == 0)
someSocket.setSoTimeout (500);
```

5.5 Creating a TCP Client

With a little knowledge on Socket Class, the TCP client may be discussed in detail. The client to be discussed is a daytime client, which, as its name suggests, connects to a daytime server to read the current day and time. Establishing a socket connection and reading from it is a fairly simple process, requiring very little code. By default, the daytime service runs on port 13. Not every machine has a daytime server running, but a Unix server would be a good system to run the client against.

Code for DaytimeClient

```
import java.net.*
import java.io.*;
public class DaytimeClient
{
public static final int SERVICE_PORT = 13;
public static void main(String args[])
{
// Check for hostname parameter
if (args.length != 1)
{
System.out.println ("Syntax - DaytimeClient host");
return;
}
// Get the hostname of server
String hostname = args[0];
```

```
try
{
// Get a socket to the daytime service
Socket daytime = new Socket (hostname, SERVICE_PORT);
System.out.println ("Connection established");
// Set the socket option just in case server stalls
daytime.setSoTimeout ( 2000 );
// Read from the server
BufferedReader reader = new BufferedReader (new InputStreamReader
(daytime.getInputStream()));
System.out.println ("Results : " + reader.readLine());
// Close the connection

daytime.close();
}
catch (IOException ioe)
{
System.err.println ("Error " + ioe);
}
}
}
```

How DaytimeClient Works

The daytime application is straightforward, and uses concepts discussed earlier in the chapter. A socket is created, an input stream is obtained, and timeouts are enabled in the rare event that a server as simple as daytime fails during a connection. Rather than connecting a filtered stream, a buffered reader is connected to the socket input stream, and the results are displayed to the user.

Finally, the client terminates after closing the socket connection. This is about as simple a socket client as you can get—complexity comes from implementing network protocols, not from network-specific coding.

Running DaytimeClient

Running the application is easy. Simply specify the hostname of a machine running the daytime service as a command-line parameter and run it. If you use a nonstandard port for the daytime server (discussed later), remember to change the port number in the client and recompile. For example, to run the client against a server running on the local machine, the following command would be used:

```
java DaytimeClient localhost
```

5.5 ServerSocket Class

A special type of socket, the server socket, is used to provide TCP services. Client sockets bind to any free port on the local machine, and connect to a specific server port and host. The difference with server sockets is that they bind to a specific port on the local machine, so that remote clients may locate a service. Client socket connections will connect to only one machine, whereas server sockets are capable of fulfilling the requests of multiple clients.

The way it works is simple—clients are aware of a service running on a particular port. They establish a connection, and within the server, the connection is accepted. Multiple connections can be accepted at the same time, or a server may choose to accept only one connection at any given moment. Once accepted, the connection is represented as a normal socket, in the form of a Socket object—once you have mastered the Socket class, it becomes almost as simple to write servers as it does clients. The only difference between a server and a client is that the server binds to a specific port, using a ServerSocket object. This ServerSocket object acts as a factory for client connections—you do not need to create instances of the Socket class yourself. These connections are modeled as a normal socket, so you can connect input and output filter streams (or even a reader and writer) to the connection.

5.5.1 Creating a ServerSocket

Once a server socket is created, it will be bound to a local port and ready to accept incoming connections. When clients attempt to connect, they are placed into a queue. Once all free space in the queue is exhausted, further clients will be refused.

Constructors

The simplest way to create a server socket is to bind to a local address, which is specified as the only parameter, using a constructor. For example, to provide a service on port 80 (usually used for Web servers), the following snippet of code would be used:

```
try
{
// Bind to port 80, to provide a TCP service (like HTTP)
ServerSocket myServer = new ServerSocket ( 80 );
// ......
}
catch (IOException ioe)
{
System.err.println ("I/O error – " + ioe);
}
```

This is the simplest form of the ServerSocket constructor, but there are several others that allow additional customization. All of these constructors are marked as public.

- *ServerSocket(int port) throws java.io.IOException, java.lang.SecurityException*— binds the server socket to the specified port number, so that remote clients may locate the TCP service. If a value of zero is passed, any free port will be used—however, clients will be unable to access the service unless notified somehow of the port number. By default, the queue size is set to 50, but an alternate constructor is provided that allows modification of this setting. If the port is already bound, or security restrictions (such as security policies or operating system restrictions on well-known ports) prevent access, an exception is thrown.
- *ServerSocket(int port, int numberOfClients) throws java.io.IOException, java.lang.SecurityException*— inds the server socket to the specified port number and allocates sufficient space to the queue to support the specified number of client sockets. This is an overloaded version of the ServerSocket(int port) constructor, and if the port is already bound or security restrictions prevent access, an exception is thrown.

- *ServerSocket(int port, int numberOfClients, InetAddress address) throws java.io.IOException, java.lang.SecurityException*— binds the server socket to the specified port number, and allocates sufficient space to the queue to support the specified number of client sockets. This is an overloaded version of the ServerSocket(int port, int numberOfClients) constructor that allows a server socket to bind to a specific IP address, in the case of a multihomed machine. For example, a machine may have two network cards, or may be configured to represent itself as several machines by using virtual IP addresses. Specifying a null value for the address will cause the server socket to accept requests on all local addresses. If the port is already bound or security restrictions prevent access, an exception is thrown.

5.5.2 Using a ServerSocket

While the Socket class is fairly versatile, and has many methods, the Server Socket class does not really do that much, other than accept connections and act as a factory for Socket objects that model the connection between client and server. The most important method is the accept() method, which accepts client connection requests, but there are several others that developers may find useful.

Methods

All methods are public unless otherwise noted.

- *Socket accept() throws java.io.IOException, java.lang.SecurityException*—waits for a client to request a connection to the server socket, and accepts it. This is a blocking I/O operation, and will not return until a connection is made (unless the timeout socket option is set). When a connection is established, it will be returned as a Socket object. When accepting connections, each client request will be verified by the default security manager, which makes it possible to accept certain IP addresses and block others, causing an exception to be thrown. However, servers do not need to rely on the security manager to block or terminate connections—the identity of a client can be determined by calling the getInetAddress() method of the client socket.
- *void close() throws java.io.IOException*—closes the server socket, which unbinds the TCP port and allows other services to use it.
- *InetAddress getInetAddress()*—returns the address of the server socket, which may be different from the local address in the case of a multihomed machine (i.e. a machine whose localhost is known by two or more IP addresses).
- *int getLocalPort()*— returns the port number to which the server socket is bound.
- *int getSoTimeout() throws java.io.IOException*—returns the value of the timeout socket option, which determines how many milliseconds an accept() operation can block for. If a value of zero is returned, the accept operation blocks indefinitely.
- *void implAccept(Socket socket) throws java.io.IOException*—this method allows ServerSocket subclasses to pass an unconnected socket subclass, and to have that socket object accept an incoming request. Using the implAccept method to accept the connection, an overridden ServerSocket.accept() method can return a connected socket. A few developers will want to subclass the ServerSocket, and using this should be avoided unless required.
- *static void setSocketFactory (SocketImplFactory factory) throws java.io.IOException, java.net.SocketException, java.lang.Security Exception*—assigns a server socket factory for the JVM. This is a static method, and should be called only once during the lifetime of a JVM. If assigning a new socket factory is prohibited, or one has already been assigned, an exception is thrown.

- *void setSoTimeout(int timeout) throws java.net.SocketException*— assigns a timeout value (specified in milliseconds) for the blocking accept() operation.

If a value of zero is specified, timeouts are disabled and the operation will block indefinitely. Providing timeouts are enabled, however, whenever the accept() method is called a timer starts. When the timer expires, a java.io.InterruptedIOException is thrown, which then allows a server to take further actions.

Accepting and Processing Requests from TCP Clients

The most important function of a server socket is to accept client sockets. Once a client socket is obtained, the server can perform all the "real work" of server programming, which involves reading from and writing to the socket to implement a network protocol. The exact data that is sent or received is dependent on the details of the protocol. For example, a mail server that provides access to stored messages would listen to commands and send back message contents. A telnet server would listen for keystrokes and pass these to a log-in shell, and send back output to the network client. Protocol-specific actions are less network- and more programming-oriented. The following snippet shows how client sockets are accepted, and how I/O streams may be connected to the client:

```
// Perform a blocking read operation, to read the next socket connection
Socket nextSocket = someServerSocket.accept();
// Connect a filter reader and writer to the stream
BufferedReader reader = new BufferedReader (new InputStreamReader
(nextSocket.getInputStream() ) );
PrintWriter writer = new PrintWriter(new OutputStreamWriter
(nextSocket.getOutputStream() ) );
```

From then on, the server may conduct the tasks needed to process and respond to client requests, or may choose to leave this task for code executing in another thread. Remember that just like any other form of I/O operation in Java, code will block indefinitely while reading a response from a client—so to service multiple clients concurrently, threads must be used. In simple cases, however, multiple threads of execution may not be necessary, particularly if requests are responded to quickly and take a little time to process.

Creating fully-fledged client/server applications that implement popular Internet protocols involves a fair amount of effort, especially for those new to network programming. It also draws on other skills, such as multi-threaded programming, discussed in the next chapter. For now, we will focus on a simple, bare-bones TCP server that executes as a single-threaded application.

5.6 Creating a TCP Server

One of the most enjoyable parts of networking is writing a network server. Clients send requests and respond to data sent back, but the server performs most of the real work. This next example is of a daytime server (which you can test using the client described in Section 6.5).

Code for DaytimeServer

```
import java.net.*;
import java.io.*;
public class DaytimeServer
{
```

```
public static final int SERVICE_PORT = 13;
public static void main(String args[])
{
try
{
// Bind to the service port, to grant clients access to the TCP daytime service
ServerSocket server = new ServerSocket(SERVICE_PORT);
System.out.println ("Daytime service started");
// Loop indefinitely, accepting clients
for (;;)
{
// Get the next TCP client
Socket nextClient = server.accept();
// Display connection details
System.out.println ("Received request from " + nextClient.getInetAddress() +
":" + nextClient.getPort() );
// Don't read, just write the message
OutputStream out = nextClient.getOutputStream();
PrintStream pout = new PrintStream (out);
// Write the current date out to the user
pout.print( new java.util.Date() );
// Flush unsent bytes
out.flush();
// Close stream
out.close();
// Close the connection
nextClient.close();
}
}
catch (BindException be)
{
System.err.println ("Service already running on port " + SERVICE_PORT );
}
catch (IOException ioe)
{
System.err.println ("I/O error - " + ioe);
}
}
}
```

How DaytimeServer Works

For a server, this is about as simple as it gets. The first step in this server is to create a ServerSocket. If this port is already bound, a BindException will be thrown, as no two servers can share the same port. Otherwise, the server socket is created; the next step is to wait for connections. Since daytime is a very simple protocol and our first example of a TCP server should be a simple one, we use here a single-threaded server. A for loop that loops indefinitely is commonly used in simple TCP servers, or a while loop whose expression always evaluates to true. Inside this loop, the first line you will find is the server.accept() method, which blocks until a client attempts to connect. This method returns a socket that represents the connection to the client. For logging, the IP address and port of the connection is sent to System.out. You will see this every time someone logs in and gets the time of day.

Daytime is a response-only protocol, so we do not need to worry about reading any input. We obtain an OutputStream and then wrap it in a PrintStream to make it easier to work with. Determining the date and time using the java. util. Date class, we send it over the TCP stream to the client. Finally, we flush all data in the print stream and close the connection by calling close() on the socket.

Running DaytimeServer

Running the server is very simple. The server has no command-line parameters. For this server example to run on UNIX, you will need to modify the SERVICE_PORT variable to a number above 1,024, unless you turn off the default daytime process and run this example as root. On Windows or other operating systems, this will not be a problem. To run the server on the local machine, the following command would be used:

```
java DaytimeServer
```

5.7 Exception Handling: Socket-Specific Exceptions

As a medium for communication, networks are fraught with problems. With so many machines connected to the global Internet, the prospect of encountering a host whose hostname cannot be resolved, one that is disconnected from the network, or one that locks up during a connection, is very likely in the lifetime of a software application. It is important, therefore, to be aware of the conditions that might cause such problems to arise in an application and to deal with them gracefully. Of course, not every application will require precise control, and in simple applications you will probably want to handle everything with a generic handler. For those more advanced applications, however, it is important to be aware of the socket-specific exceptions that can be thrown at runtime.

All socket-specific exceptions extend from SocketException, so by simply catching that exception, you catch all of the socket-specific ones and write a single generic handler. In addition, SocketException extends from java.io.IOException if you want to provide a catchall for any I/O exception.

a. SocketException

The java.net.SocketException represents a generic socket error, which can represent a range of specific error conditions.

b. BindException

The java.net.BindException represents an inability to bind a socket to a local port. The most common reason for this will be that the local port is already in use.

c. ConnectException

The java.net.ConnectException occurs when a socket cannot connect to a specific remote host and port.

d. NoRouteToHostException

The java.net.NoRouteToHostException is thrown when, due to a network error, it is impossible to find a route to the remote host. The cause of this may be local, may be a temporary gateway or router problem, or may be the fault of the remote network to which the socket is trying to connect. Another common cause of this is that firewalls and routers are blocking the client software, which is usually a permanent condition.

e. InterruptedIOException

The java.net.InterruptedIOException occurs when a read operation is blocked for sufficient time to cause a network timeout, as discussed earlier in the chapter. Handling timeouts is a good way to make your code more robust and reliable.

6. NETWORK PROGRAMMING AND THREADS

In the previous section, we looked at several examples of network programming. Those examples showed how to create network connections and communicate through them, but they did not deal with one of the fundamental characteristics of network programming, the fact that network communication is fundamentally asynchronous. From the point of view of a program on one end of a network connection, messages can arrive from the other side of the connection at any time; the arrival of a message is an event that is not under the control of the program that is receiving the message. Certainly, it is possible to design a network communication protocol that proceeds in a synchronous, step-by-step process from beginning to end—but whenever the process gets to a point in the protocol where it needs to read a message from the other side of the connection, it has to wait for that message to arrive. Essentially, the process has to wait for a message-arrival event to occur before it can proceed. While it is waiting for the message, we say that the process is blocked. Perhaps an event-oriented networking API would be a good approach to dealing with the asynchronous nature of network communication, but that is not the approach that is taken in Java (or, typically, in other languages). Instead, a serious network program in Java uses threads. A thread is a separate computational process that can run in parallel with other threads. When a program uses threads to do network communication, it is possible that some threads will be blocked, waiting for incoming messages, but other threads will still be able to continue performing useful work.

A Threaded GUI Chat Program

The command-line chat programs, CLChatClient.java and CLChatServer.java, from the previous section use a straight-through, step-by-step protocol for communication. After a user on one side of a connection enters a message, the user must wait for a reply from the other side of the connection. An asynchronous chat program would be much nicer. In such a program, a user could just keep typing lines and sending messages without waiting for any response. Messages that arrive—asynchronously—from the other side would be displayed as soon as they arrive. It is not easy to do this in a command-line interface, but it is a natural application for a graphical user interface. The basic idea for a GUI chat program is to create a thread whose job is to read messages that arrive from the other side of the connection. As soon as the message arrives, it is displayed to the user; then, the message-reading thread blocks until the next incoming message arrives. While it is blocked, however, other threads can continue to run. In particular, the GUI event-handling thread that responds to user actions keeps running; that thread can send outgoing messages as soon as the user generates them.

The sample program GUIChat.java is an example of this. GUIChat is a two-way network chat program that allows two users to send messages to each other over the network. In this chat program, each user can send messages at any time, and incoming messages are displayed as soon as they are received. The GUIChat program can act as both the client end and the server end of a connection. When GUIChat is started, a window appears on the screen. This window has a "Listen" button that the user can click to create a server socket that will listen for an incoming connection request; this makes the program act as a server. It also has a "Connect" button that the user can click to send a connection request; this makes the program act as a client. As usual, the server listens on a specified port number. The client needs to know the computer on which the server is running and the port on which the server is listening. There are input boxes in the GUIChat window where the user can enter this information. Once a connection has been established between two GUIChat windows, each user can send messages to the other. The window has an input box where the user types the message. Pressing return while typing in this box sends the message. This means that the sending of the message is handled by the usual event-handling thread, in response to an event generated by a user action. Messages are received by a separate thread that just sits around waiting for incoming messages. This thread blocks while waiting for a message to arrive; when a message does arrive, it displays that message to the user. The window contains a large transcript area that displays both incoming and outgoing messages, along with other information about the network connection. I urge you to compile the source code, GUIChat.java, and try the program. To make it easy to try it on a single computer, you can make a connection between one window and another window on the same computer, using "localhost" or "127.0.0.1" as the name of the computer.

The program uses a nested class, ConnectionHandler, to handle most network-related tasks. ConnectionHandler is a subclass of Thread. The ConnectionHandler thread is responsible for opening the network connection and then for reading incoming messages once the connection has been opened. (By putting the connection-opening code in a separate thread, we make sure that the GUI is not blocked while the connection is being opened. Like reading incoming messages, opening a connection is a blocking operation that can take some time to complete.)

A ConnectionHandler is created when the user clicks the "Listen" or "Connect" button. The "Listen" button should make the thread act as a server, while "Connect" should make it act as a client. To distinguish these two cases, the ConnectionHandler class has two constructors:

```
/*** Listen for a connection on a specified port. The constructor does not
perform any network operations; it just sets some instance variables and starts
the thread. Note that the thread will only listen for one connection, and then
will close its server socket.*/
ConnectionHandler(int port) {
state = ConnectionState.LISTENING;
this.port = port;
postMessage("\nLISTENING ON PORT " + port + "\n");
start();
}
/*** Open a connection to specified computer and port. The constructor does
not perform any network operations; it just sets some instance variables and
starts the thread.*/
ConnectionHandler(String remoteHost, int port) {
state = ConnectionState.CONNECTING;
this.remoteHost = remoteHost;
this.port = port;
postMessage("\nCONNECTING TO " + remoteHost + " ON PORT " + port + "\n");
start();
}
```

Here, state is an instance variable whose type is defined by an enumerated type enum ConnectionState { LISTENING, CONNECTING, CONNECTED, CLOSED };

The values of this enum represent different possible states of the network connection. It is often useful to treat a network connection as a state machine (see Subsection 6.5.4), since the response to various events can depend on the state of the connection when the event occurs. Setting the state variable to LISTENING or CONNECTING tells the thread whether it should act as a server or as a client. Note that the postMessage() method posts a message to the transcript area of the window, where it will be visible to the user.

Once the thread has been started, it executes the following run() method:

```
/*** The run() method that is executed by the thread. It opens a connection as a
client or as a server (depending on which constructor was used). */
public void run() {
try {
if (state == ConnectionState.LISTENING) {
// Open a connection as a server.
listener = new ServerSocket(port);
socket = listener.accept();
listener.close();
}
else if (state == ConnectionState.CONNECTING) {
// Open a connection as a client.
socket = new Socket(remoteHost,port);
}
connectionOpened(); // Sets up to use the connection (including
// creating a BufferedReader, in, for reading
// incoming messages).
while (state == ConnectionState.CONNECTED) {
// Read one line of text from the other side of the connection, and report it to
the user.
String input = in.readLine();
if (input == null)
connectionClosedFromOtherSide();
else
received(input); // Report message to user.
}
}
catch (Exception e) {
// An error occurred. Report it to the user, but not if the connection has been
closed (since the error might be the expected error that is generated when a
socket is closed).
if (state != ConnectionState.CLOSED)
postMessage("\n\n ERROR: " + e);
}
finally { // Clean up before terminating the thread.
cleanUp();
}
}
```

This method calls several other methods to do some of its work, but you can see the general outline of how it works. After opening the connection as either a server or client, the run() method enters a while loop in which it receives and processes messages from the other side of the connection until the connection is closed. It is important to understand how the connection can be closed. The GUIChat window has a "Disconnect" button that the user can click to close the connection. The program responds to this event by closing the socket that represents the connection. It is likely that when this happens, the connection-handling thread is blocked in the in.readLine()method, waiting for an incoming message. When the socket is closed by another thread, this method will fail and will throw an exception; this exception causes the thread to terminate. (If the connection-handling thread happens to be between calls to in.readLine() when the socket is closed, the while loop will terminate because the connection state changes from CONNECTED to CLOSED.) Note that closing the window will also close the connection in the same way.

It is also possible for the user on the other side of the connection to close the connection. When that happens, the stream of incoming messages ends, and the in.readLine() on this side of the connection returns the value null, which indicates end-of-stream and acts as a signal that the connection has been closed by the remote user. For a final look into the GUIChat code, consider the methods that send and receive messages. These methods are called from different threads. The send() method is called by the event-handling thread in response to a user action. Its purpose is to transmit a message to the remote user. It uses a PrintWriter, out, that writes to the socket's output stream. Synchronization of this method prevents the connection state from changing in the middle of the send operation:

```
/*** Send a message to the other side of the connection, and post the message to
the transcript. This should only be called when the connection state is
ConnectionState.CONNECTED; if it is called at other times, it is ignored.*/
synchronized void send(String message) {
if (state == ConnectionState.CONNECTED) {
postMessage("SEND: " + message);
out.println(message);
out.flush();
if (out.checkError()) {
postMessage("\nERROR OCCURRED WHILE TRYING TO SEND DATA.");
close(); // Closes the connection.
}
}
}
```

The received() method is called by the connection-handling thread after a message has been read from the remote user. Its only job is to display the message to the user, but again it is synchronized to avoid the race condition that could occur if the connection state were changed by another thread while this method is being executed:

```
/*** This is called by the run() method when a message is received from the other
side of the connection. The message is posted to the transcript, but only if the
connection state is CONNECTED. (This is because a message might be received
```

```
after the user has clicked the "Disconnect" button; that message should not be
seen by the user.) */
synchronized private void received(String message) {
if (state == ConnectionState.CONNECTED)
postMessage("RECEIVE: " + message);
}
```

7. IMPLEMENTING APPLICATION PROTOCOLS

The most enjoyable part of learning network programming is putting the network theory that you have learned into practice, by writing real-life applications that interact with other Internet services or clients. Earlier in this book we covered the theory; now it is time to write some code. In the following sections, we will examine and code three protocols—SMTP, POP3, and HTTP.

7.1 SMTP Client Implementation

The Simple Mail Transfer Protocol is used to send messages of various types between users over a TCP/IP network. It should be noted that this protocol assumes that some other method is used to actually read the messages, thus allowing a more stable, flexible, and robust global e-mail system. By separating delivering messages from reading, things are made much simpler. Together, we will write a basic SMTP client that allows the user to send a text message to a specific e-mail address. The client written here offers a good example of networking, while minimizing such supporting code as that for a user interface. For this reason, simple text-based input is used, and the commands sent to the server are displayed to help the reader understand how the protocol works.

Code for SMTPClientDemo

```
import java.io.*;
import java.net.*;
import java.util.*;
public class SMTPClientDemo
{
protected int port = 25;
protected String hostname = "localhost";
protected String from = "";
protected String to = "";
protected String subject = "";
protected String body = "";
protected Socket socket;
protected BufferedReader br;
protected PrintWriter pw;
// Constructs a new instance of the SMTP Client
public SMTPClientDemo() throws Exception
{
```

```
try
{
getInput();
sendEmail();
}
catch (Exception e)
{
System.out.println ("Error sending message - " + e);
}
}
public static void main(String[] args) throws Exception
{
// Start the SMTP client, so it can send messages
SMTPClientDemo client = new SMTPClientDemo();
}
// Check the SMTP response code for an error message
protected int readResponseCode() throws Exception
{
String line = br.readLine();
System.out.println("< "+line);
line = line.substring(0,line.indexOf(" "));
return Integer.parseInt(line);
}
// Write a protocol message both to the network socket and to the screen
protected void writeMsg(String msg) throws Exception
{
pw.println(msg);
pw.flush();
System.out.println("> "+msg);
}
// Close all readers, streams and sockets
protected void closeConnection() throws Exception
{
pw.flush();
pw.close();
br.close();
socket.close();
}
// Send the QUIT protocol message, and terminate connection
protected void sendQuit() throws Exception
{
System.out.println("Sending QUIT");
writeMsg("QUIT");
readResponseCode();
System.out.println("Closing Connection");
```

```
closeConnection();
}
// Send an email message via SMTP, adhering to the protocol known as RFC
2821
protected void sendEmail() throws Exception
{
System.out.println("Sending message now: Debug below");
System.out.println("————————————————————————");
System.out.println("Opening Socket");
socket = new Socket(this.hostname,this.port);
System.out.println("Creating Reader & Writer");
br = new BufferedReader(new InputStreamReader(socket.getInputStream()));
pw = new PrintWriter(new OutputStreamWriter(socket.getOutputStream()));
System.out.println("Reading first line");
int code = readResponseCode();
if(code != 220) {
socket.close();
throw new Exception("Invalid SMTP Server");
}
System.out.println("Sending helo command");
writeMsg("HELO "+InetAddress.getLocalHost().getHostName());
code = readResponseCode();
if(code != 250)
{
sendQuit();
throw new Exception("Invalid SMTP Server");
}
System.out.println("Sending mail from command");
writeMsg("MAIL FROM:<"+this.from+">");
code = readResponseCode();
if(code != 250)
{
sendQuit();
throw new Exception("Invalid from address");
}
System.out.println("Sending rcpt to command");
writeMsg("RCPT TO:<"+this.to+">");
code = readResponseCode();
if(code != 250)
{
sendQuit();
throw new Exception("Invalid to address");
}
System.out.println("Sending data command");
writeMsg("DATA");
```

```
code = readResponseCode();
if(code != 354)
{
sendQuit();
throw new Exception("Data entry not accepted");
}
System.out.println("Sending message");
writeMsg("Subject: "+this.subject);
writeMsg("To: "+this.to);
writeMsg("From: "+this.from);
writeMsg("");
writeMsg(body);
code = readResponseCode();
sendQuit();
if(code != 250)
throw new Exception("Message may not have been sent correctly");
else
System.out.println("Message sent");
}
// Obtain input from the user
protected void getInput() throws Exception
{
// Read input from user console
String data=null;
BufferedReader br = new BufferedReader (new InputStreamReader(System.in));
// Request hostname for SMTP server
System.out.print("Please enter SMTP server hostname: ");
data = br.readLine();
if (data == null || data.equals("")) hostname="localhost";
else
hostname=data;
// Request the sender's email address
System.out.print("Please enter FROM email address: ");
data = br.readLine();
from = data;
// Request the recipient's email address
System.out.print("Please enter TO email address :");
data = br.readLine();
if(!(data == null || data.equals("")))
to=data;
System.out.print("Please enter subject: ");
data = br.readLine();
subject=data;
System.out.println("Please enter plain-text message ('.' character on a blank line signals end of message):");
```

```
StringBuffer buffer = new StringBuffer();
// Read until user enters a . on a blank line
String line = br.readLine();
while(line != null)
{
// Check for a '.', and only a '.', on a line
if(line.equalsIgnoreCase("."))
{
break;
}
buffer.append(line);
buffer.append("\n");
line = br.readLine();
}
buffer.append(".\n");
body = buffer.toString();
}
}
```

How SMTPClientDemo Works

As can be seen here, the Simple Mail Transfer Protocol is quite straightforward, consisting of a single connection to a mail server using a TCP socket, followed by a series of short protocol commands that specify the details of the e-mail to be sent, as described in RFC 2821. While many network applications will require multiple threads of execution, as a general rule simple clients such as this one do not. This example merely asks for input from the user and then sends the message.

To simplify understanding of this code, the networking code has been separated from the non-networking code (which consists mainly of code for obtaining input from the user).

The basic skeleton of the application is as follows:

```
public class SMTPClientDemo
{
public static void main(String[] args) throws Exception
{
SMTPClientDemo client = new SMTPClientDemo ();
}
// Constructs a new instance of the SMTP Client
public SMTPClientDemo () throws Exception;
// Send an email message via SMTP, adhering to the protocol known as RFC
2821
protected void sendEmail() throws Exception;
// Check the SMTP response code for an error message
protected int readResponseCode() throws Exception;
// Write a protocol message both to the network socket and to the screen
protected void writeMsg(String msg) throws Exception;
// Close all readers, streams and sockets
```

```
protected void closeConnection() throws Exception;
// Send the QUIT protocol message, and terminate connection
protected void sendQuit() throws Exception;
// Obtain input from the user
protected void getInput() throws Exception;
}
```

We will cover the e-mail-specific code, as the remainder of the application is fairly straightforward Java coding. Our class stores the message and network details inside protected variables.

These variables are:

- String hostname
- int port (set to 25, the default for SMTP)
- String from
- String to
- String subject
- String body
- java.net.Socket socket
- BufferedReader br
- PrintWriter pw

The hostname, port, from, to, subject, and body variables are fairly self-explanatory—they are required for locating the mail server to send an SMTP message, and for the e-mail addressing and content details. The socket is used to communicate with the remote TCP server that will relay our mail for us, and the reader/writer objects are used for reading and sending SMTP messages. When the application is run, the main() method is executed by the Java Virtual Machine. This in turn creates an instance of our SMTPClientDemo application, which requests input data from the user and attempts to send an e-mail message. The process has been fairly simple, so far. Now let us turn to the application protocol code.

Contained inside the sendEmail() method is the heart of the program. This is where we connect a local TCP socket to a remote SMTP server, and send across protocol requests. While it is certainly possible to group this process into one single method, for convenience and readability, some of the workload has been distributed across several helper methods that decode SMTP responses, send protocol messages, and terminate connections. The first thing the sendEmail() method does is open a socket connection to the mail server located at port *n*, where *n* is represented by the member variable port. The SMTP normally uses port 25, but if your network/ISP uses a nonstandard port number for this service, you could modify the default value of this to the required port number. If an error occurs, an exception is thrown and caught by our error handler; otherwise a successful connection to the server is established.

Once connected, we obtain input streams and output streams for the socket, and connect these to readers and writers for convenience. According to the SMTP specification, upon connecting, the server will send a response code in greeting, so the application checks for a valid message with the aid of the helper method readResponseCode(). This helps identify that we are really talking to an SMTP service, and not some other service using port 25.

The code for determining the response code is wrapped inside a helper method, called readResponseCode(). It takes no parameters, and returns an int representing the code number. It reads a line of text from the buffered reader that communicates with the SMTP server, outputs the result to the screen for illustrative purposes, and then strips off everything after the first space.

Finally, it converts the result to an integer and passes it back, giving us the SMTP response code.

```
// Check the SMTP response code for an error message
protected int readResponseCode() throws Exception
{
String line = br.readLine();
System.out.println("< "+line);
line = line.substring(0,line.indexOf(" "));
return Integer.parseInt(line);
}
```

When we need to send a message back to the SMTP server, we use the helper method writeMsg(String) which displays the message to the screen for illustrative purposes and then sends it (via our PrintWriter) to the server.

```
// Write a protocol message both to the network socket and to the screen
protected void writeMsg(String msg) throws Exception
{
pw.println(msg);
pw.flush();
System.out.println("> "+msg);
}
```

Using these helper methods, the sendEmail() method can send and receive SMTP messages. It checks, upon connecting, that the server has sent an OK message, which is represented by 220.

```
int code = readResponseCode();
if(code != 220) {
socket.close();
throw new Exception("Invalid SMTP Server");
}
```

The next step in SMTP is to send an identification message, telling the SMTP server who we really are. We send the hostname identification command, "HELO." The format is "HELO" followed by a space and the local hostname. The correct response is 250, which signals OK.

```
writeMsg("HELO "+InetAddress.getLocalHost().getHostName());
code = readResponseCode();
if(code != 250)
{
sendQuit();
throw new Exception("Invalid SMTP Server");
}
```

If the server accepts the identification, the client is free to send a message. Each message has certain key aspects—a "From" address, one or more "To" addresses, a subject line, and a message body (the actual text of the message). The SMTP only requires fields that deal directly with delivery of messages, though other fields can be placed in the message body. For example, it is commonly accepted to put the subject line as the first line, with "Subject:" in front of it. You can include as many other optional fields as you like, but remember that not every mail client will support these fields. Furthermore, you should also repeat the "To" and "From" fields in your message body.

Setting the sender and recipient addressing information under SMTP is fairly straightforward. Two commands are used, the "MAIL FROM:" and "RCPT TO:"

```
writeMsg("MAIL FROM:<"+this.from+">");
// Check response from server
// ........
writeMsg("RCPT TO:<"+this.to+">");
// Check response from server
// ........
```

In this example, only one recipient is supported, but SMTP can handle multiple recipients. You should note that well-configured servers normally place a limit on the number of possible recipients, to prevent them from being used in spamming campaigns, and that some SMTP servers will reject messages if the sender is not part of their local or dial-up network. Once the e-mail addresses are set, we can send the data to the server. Our client sends a simple text message, unencumbered by attachments. The first step to send the data is to signal that we are ready to send a message body, by issuing a "DATA" command. The valid response code for this is 354.

```
System.out.println("Sending data command");
writeMsg("DATA");
code = readResponseCode();
if(code != 354)
{
sendQuit();
throw new Exception("Data entry not accepted");
}
```

Now the client sends the message body. This must include relevant header fields, such as "To,"

"From," "Subject," and any other fields you may choose to add. Once the headers are complete, a blank line is sent, indicating that the message text will follow. After outputting the text, the client sends the message, which is then terminated by a period, then a carriage return/line-feed (which the user enters during data input). This tells the server that the message is complete. writeMsg("Subject:"+this.subject);

```
writeMsg("");
writeMsg(body);
```

```
code = readResponseCode();
sendQuit();
if(code != 250)
throw new Exception("Message may not have been sent correctly");
else
System.out.println("Message sent");
```

Finally, the "QUIT" command is sent, and the connection is terminated by invoking the sendQuit() method. Our transaction with the SMTP server is complete, and the message is on its way.

```
// Send the QUIT protocol message, and terminate connection
protected void sendQuit() throws Exception
{
System.out.println("Sending QUIT");
writeMsg("QUIT");
readResponseCode();
System.out.println("Closing Connection");
closeConnection();
}
```

Running SMTPClientDemo

After compiling, the application can be run by typing:

java SMTPClientDemo

The application will request the following information:

- The name of a valid SMTP server (such as that used by your e-mail program)
- The "From" address of the sender (e.g. your e-mail address)
- The "To" address of the recipient (e.g. your e-mail address, so you know that it was delivered)
- The "Subject" of the message
- The message contents

7.2 POP3 Client Implementation

The previous example showed how to write a protocol implementation of SMTP to send a mail message. In order to read mail, you will need the Post Office Protocol version 3, or POP3 as it is commonly referred to. It is also referred to simply as POP, since the earlier Post Office Protocols are not used any more. This term is less confusing for general Internet users, who may wonder what happened to POP1 and POP2, which have been superseded.

When a person reads their e-mail, they are actually retrieving it from a mail server belonging to an Internet service provider or a local network (such as a corporate or private intranet). There are several major types of mail servers, and each type uses a different network protocol to access email. The most common of these is POP, a protocol that provides a contrast with SMTP.

In examining the POP3 client, you will find that the actual code required for a network implementation is minimal. Much of the actual application involves user input or output. The interface is limited to text I/O; the idea behind this client is to provide a good example of networking while minimizing the need for non-networking code that readers may already be familiar with, such as GUI design. The example will retrieve messages from a mailbox and display their contents, one after another, to the text console screen.

Code for Pop3ClientDemo

```
import java.io.*;
import java.net.*;
import java.util.*;
public class Pop3ClientDemo
{
protected int port = 110;
protected String hostname = "localhost";
protected String username = "";
protected String password = "";
protected Socket socket;
protected BufferedReader br;
protected PrintWriter pw;
// Constructs a new instance of the POP3 client
public Pop3ClientDemo() throws Exception
{
try
{
// Get user input
getInput();
// Get mail messages
displayEmails();
}
catch(Exception e)
{
System.err.println ("Error occured - details follow");
e.printStackTrace();
System.out.println(e.getMessage());
}
}
// Returns TRUE if POP response indicates success, FALSE if failure
protected boolean responseIsOk() throws Exception
{
String line = br.readLine();
System.out.println("< "+line);
return line.toUpperCase().startsWith("+OK");
}
// Reads a line from the POP server, and displays it to screen
protected String readLine(boolean debug) throws Exception
{
String line = br.readLine();
// Append a < character to indicate this is a server protocol response
```

```
if (debug)
System.out.println("< "+line);
else
System.out.println(line);
return line;
}
// Writes a line to the POP server, and displays it to the screen
protected void writeMsg(String msg) throws Exception
{
pw.println(msg);
pw.flush();
System.out.println("> "+msg);
}
// Close all writers, streams and sockets
protected void closeConnection() throws Exception
{
pw.flush();
pw.close();
br.close();
socket.close();
}
// Send the QUIT command, and close connection
protected void sendQuit() throws Exception
{
System.out.println("Sending QUIT");
writeMsg("QUIT");
readLine(true);
System.out.println("Closing Connection");
closeConnection();
}
// Display emails in a message
protected void displayEmails() throws Exception
{
BufferedReader userinput = new BufferedReader( new InputStreamReader
(System.in) );
System.out.println("Displaying mailbox with protocol commands and
responses below");
System.out.println("——————————————————————+"——————————
——————");
// Open a connection to POP3 server
System.out.println("Opening Socket");
socket = new Socket(this.hostname, this.port);
br = new BufferedReader(new InputStreamReader(socket.getInputStream()));
pw = new PrintWriter(new OutputStreamWrite7r(socket.getOutputStream()));
// If response from server is not okay
```

```
if(! responseIsOk())
{
socket.close();
throw new Exception("Invalid POP3 Server");
}
// Login by sending USER and PASS commands
System.out.println("Sending username");
writeMsg("USER "+this.username);
if(!responseIsOk())
{
sendQuit();
throw new Exception("Invalid username");
}
System.out.println("Sending password");
writeMsg("PASS "+this.password);
if(!responseIsOk())
{
sendQuit();
throw new Exception("Invalid password");
}
// Get mail count from server ....
System.out.println("Checking mail");
writeMsg("STAT" );
// ... and parse for number of messages
String line = readLine(true);
StringTokenizer tokens = new StringTokenizer(line," ");
tokens.nextToken();
int messages = Integer.parseInt(tokens.nextToken());
int maxsize = Integer.parseInt(tokens.nextToken());
if (messages == 0)
{
System.out.println ("There are no messages.");
sendQuit();
return;
}
System.out.println ("There are " + messages + " messages.");
System.out.println("Press enter to continue.");
userinput.readLine();
for(int i = 1; i <= messages ; i++)
{
System.out.println("Retrieving message number "+i);
writeMsg("RETR "+i);
System.out.println("——————————————————");
line = readLine(false);
while(line != null && !line.equals("."))
```

```
{
line = readLine(false);
}
System.out.println("———————————————");
System.out.println("Press enter to continue." + "To stop, type Q then enter");
String response = userinput.readLine();
if (response.toUpperCase().startsWith("Q"))
break;
}
sendQuit();
}
public static void main(String[] args) throws Exception
{
Pop3ClientDemo client = new Pop3ClientDemo();
}
// Read user input
protected void getInput() throws Exception
{
String data=null;
BufferedReader br = new BufferedReader (new InputStreamReader(System.in));
System.out.print("Please enter POP3 server hostname:");
data = br.readLine();
if(data == null || data.equals("")) hostname= "localhost";
else
hostname=data;
System.out.print("Please enter mailbox username:");
data = br.readLine();
if(!(data == null || data.equals("")))
username=data;
System.out.print("Please enter mailbox password:");
data = br.readLine();
if(!(data == null || data.equals("")))
password=data;
}
}
```

How Pop3ClientDemo Works

The application is structured similarly to the SMTP client, in that the main() method constructs a new instance of the class, which then performs all of the work. In fact, the chief difference is a call to the displayEmails method instead of to the sendEmail method. In addition, there are several helper methods that assist in conducting communication via POP3. Like the SMTP application, several important variables are requested from the user. To retrieve email messages, you need to know the hostname of the mail server, the username of the account, and the password for accessing it. These details are obtained by the getInput() method of the application, which uses simple text I/O to request details from the user. These are then stored in member variables for later access. We also store the port number of the server (which is fixed to the default port of 110).

```
public class Pop3ClientDemo
{
protected int port = 110;
protected String hostname = "localhost";
protected String username = "";
protected String password = "";
// ......
}
```

The most important part of the application is the Post Office Protocol implementation. This is where we get to write real networking code. To make things clearer and more efficient, some of the protocol code is split into helper methods, which perform tasks such as processing a POP response to see that no error code was issued, and reading/writing protocol commands. The main work is done, however, in the displayEmails() method.

We start by creating a network socket to the POP server, using the Socket class. The next step is to obtain readers and writers connected to the socket stream, so that we can communicate with the server.

```
// Open a connection to POP3 server
System.out.println("Opening Socket");
socket = new Socket(this.hostname, this.port);
br = new BufferedReader(new InputStreamReader(socket.getInputStream()));
pw = new PrintWriter(new OutputStreamWriter(socket.getOutputStream()));
```

Upon establishing a connection, the server sends a POP response indicating that the server is ready for commands. If the client does not receive a valid POP response, either the server is malfunctioning or a non-POP server is operating on that port. For this reason, the client must always check the response code after opening a TCP connection. We use the responseOk() helper method to determine this, which returns a boolean value. Note the negation of the "if" statement—we close the connection only if the response is not okay (indicated by the "!" operator).

```
// If response from server is not okay
if(!responseIsOk())
{
socket.close();
throw new Exception("Invalid POP3 Server");
}
```

Now our application is ready to send POP commands. The first command that is sent is "USER," which identifies the user account that the client is trying to access. We send the command, along with the username, by using the helper method writeMsg(String). This also outputs the command to the text console, so you can see how the protocol is working. We do the same then with the "PASS" command, which is used to authenticate the identity of the user. // Login by sending USER and PASS commands

```
System.out.println("Sending username");
writeMsg("USER "+this.username);
if(!responseIsOk())
{
sendQuit();
throw new Exception("Invalid username");
}
writeMsg("PASS "+this.password);
if(!responseIsOk())
{
sendQuit();
throw new Exception("Invalid password");
}
```

At this point, the application will be ready to send commands, unless user authentication failed, in which case the application will have terminated. Now we can request the number of mail messages available from the server. This is achieved by sending the "STAT" command, which returns the number of messages. We parse the response, looking for the message count, and then convert it to an int value.

```
writeMsg("STAT");
// ... and parse for number of messages
String line = readLine(true);
StringTokenizer tokens = new StringTokenizer(line," ");
tokens.nextToken();
int messages = Integer.parseInt(tokens.nextToken());
int maxsize = Integer.parseInt(tokens.nextToken());
```

Armed with this information, we can determine if the mailbox is empty or if there are messages to be retrieved. If so, the application simply loops and displays the message contents. During each loop, the user is presented with the option of quitting in case the message count is too high.

```
System.out.println ("There are " + messages + " messages.");
System.out.println("Press enter to continue.");
userinput.readLine();
for(int i = 1; i <= messages ; i++)
{
System.out.println("Retrieving message number "+i);
writeMsg("RETR "+i);
System.out.println("————————————");
line = readLine(false);
```

```
while(line != null && !line.equals("."))
{
line = readLine(false);
}
System.out.println("————————————————");
System.out.println("Press enter to continue. To stop, type Q then enter");
String response = userinput.readLine();
if (response.toUpperCase().startsWith("Q"))
break;
}
sendQuit();
```

Finally, we call the sendQuit() method, which shuts down the server connection. The application then quits and the task of reading mail using the Post Office Protocol is complete. As you can see, reading e-mail is a fairly simple task, perhaps even easier than sending it.

Running Pop3ClientDemo

Running the application, too, is quite manageable. After compiling, simply type:

```
java Pop3ClientDemo
```

You will be prompted for the hostname of your mail server, the username, and then the password. For example, a user might type the following:

```
java PopClientDemo
Please enter POP3 server hostname: myserver.myisp.com
Please enter mailbox username: BPL
Please enter mailbox password: AJP
```

7.3 HTTP/1.0 Server Implementation

The HyperText Transfer Protocol (HTTP) originated as a means of sharing documents across the Internet. Since many documents are interrelated, the need to provide a link from one to the other was identified, but given the fact that many researchers from around the world were working independently, a single centralized document server was not the ideal method. By placing a hyperlink over a word (such as a scientific term) or a phrase, users would be able to jump instantly from one document to another, even though the documents could reside on servers located in other countries. Not only that, but hyperlinks could also be made within the same document, such as to a glossary of terms. Published as RFC 1945, HTTP became one of the most quickly adopted protocols, and led to what we now know as the World Wide Web. Pioneered by Tim Berners-Lee and his colleagues at the CERN scientific laboratories as a way to share scientific information, it quickly spreads to other parts of the academic world, and then to commercial and consumer markets.

The first, and most widely supported version, of HTTP is known as HTTP/1.0. This protocol supports a simple set of commands for retrieving resources from a Web server, such as HTML pages, images, documents, and other file types, as well as commands for posting information to the Web server so as to allow for the interactivity and customization of Web pages. This capability is particularly important for Web sites that support advanced features, such as user customization or shopping carts.

The latest version of the protocol is known as HTTP/1.1. It offers many improvements over the previous version, with a wider set of commands to support new features of use both to browser developers and to those who write server-side Web applications. However, older browsers and servers will not support all of these features, and indeed some are not yet in common use at all. This next example demonstrates how to write a multi-threaded HTTP server that responds to requests from a Web browser, fetches files or Web pages, and sends them back to the user. For the purposes of this example, we will use version 1.0 of the HTTP protocol and support only the GET method, which is used for file retrieval. Other methods, such as POST, are useful when designing interactive server-side applications. However, for such uses it is advised that the reader consider using a fully-fledged commercial server, and Java servlets (discussed in later chapters).

Code for WebServerDemo

```
import java.io.*;
import java.net.*;
import java.util.*;
public class WebServerDemo
{
// Directory of HTML pages and other files
protected String docroot;
// Port number of web server
protected int port;
// Socket for the web server
protected ServerSocket ss;
// Handler for a HTTP request
class Handler extends Thread
{
protected Socket socket;
protected PrintWriter pw;
protected BufferedOutputStream bos;
protected BufferedReader br;
protected File docroot;
public Handler(Socket _socket, String _docroot) throws Exception
{
socket=_socket;
// Get the absolute directory of the filepath
docroot=new File(_docroot).getCanonicalFile();
}
public void run()
{
try
{
// Prepare our readers and writers
br = new BufferedReader(new
InputStreamReader(socket.getInputStream()));
bos = new BufferedOutputStream(socket.getOutputStream());
pw = new PrintWriter(new OutputStreamWriter(bos));
```

```
// Read HTTP request from user (hopefully GET /file...... )
String line = br.readLine();
// Shutdown any further input
socket.shutdownInput();
if(line == null)
{
socket.close();
return;
}
if(line.toUpperCase().startsWith("GET"))
{
// Eliminate any trailing ? data, such as for a CGI
// GET request
StringTokenizer tokens = new StringTokenizer(line," ?");
tokens.nextToken();
String req = tokens.nextToken();
// If a path character / or \ is not present, add it to the document root and then
add the file //request, to form a full filename
String name;
if(req.startsWith("/") || req.startsWith("\\"))
name = this.docroot+req;
else
name = this.docroot+File.separator+req;
// Get absolute file path
File file = new File(name).getCanonicalFile();
// Check to see if request doesn't start with our
// document root ....
if(!file.getAbsolutePath().startsWith(this.docroot.getAbsolutePath()))
{
pw.println("HTTP/1.0 403 Forbidden");
pw.println();
}
// ... if it's missing .....
else if(!file.exists())
{
pw.println("HTTP/1.0 404 File Not Found");
pw.println();
}
// ... if it can't be read for security reasons ....
else if(!file.canRead())
{
pw.println("HTTP/1.0 403 Forbidden");
pw.println();
}
// ... if its actually a directory, and not a file ....
```

```
else if(file.isDirectory())
{
sendDir(bos,pw,file,req);
}
// ... or if it's really a file
else
{
sendFile(bos, pw, file.getAbsolutePath());
}
}
// If not a GET request, the server will not support it
else
{
pw.println("HTTP/1.0 501 Not Implemented");
pw.println();
}
pw.flush();
bos.flush();
}
catch(Exception e)
{
e.printStackTrace();
}
try
{
socket.close();
}
catch(Exception e)
{
e.printStackTrace();
}
}
protected void sendFile(BufferedOutputStream bos,
PrintWriter pw, String filename) throws Exception
{
try
{
BufferedInputStream bis = new java.io.BufferedInputStream(new
FileInputStream(filename));
byte[] data = new byte[10*1024];
int read = bis.read(data);
pw.println("HTTP/1.0 200 Okay");
pw.println();
pw.flush();
bos.flush();
```

```
while(read != -1)
{
bos.write(data,0,read);
read = bis.read(data);
}
bos.flush();
}
catch(Exception e)
{
pw.flush();
bos.flush();
}
}
protected void sendDir(BufferedOutputStream bos, PrintWriter pw, File dir,
String req) throws Exception
{
try
{
pw.println("HTTP/1.0 200 Okay");
pw.println();
pw.flush();
pw.print("<html><head><title>Directory of ");
pw.print(req);
pw.print("</title></head><body><h1>Directory of ");
pw.print(req);
pw.println("</h1><table border=\"0\">");
File[] contents=dir.listFiles();
for(int i=0;i<contents.length;i++)
{
pw.print("<tr>");
pw.print("<td><a href=\"");
pw.print(req);
pw.print(contents[i].getName());
if(contents[i].isDirectory())
pw.print("/");
pw.print("\">");
if(contents[i].isDirectory())
pw.print("Dir -> ");
pw.print(contents[i].getName());
pw.print("</a></td>");
pw.println("</tr>");
}
pw.println("</table></body></html>");
pw.flush();
}
```

```
catch(Exception e)
{
pw.flush();
bos.flush();
}
}
}
// Check that a filepath has been specified and a port number
protected void parseParams(String[] args) throws Exception
{
switch(args.length)
{
case 1:
case 0:
System.err.println ("Syntax: <jvm> "+this.getClass().getName()+"docroot
port");
System.exit(0);
default:
this.docroot = args[0];
this.port = Integer.parseInt(args[1]);
break;
}
}
public WebServerDemo(String[] args) throws Exception
{
System.out.println ("Checking for paramaters");
// Check for command line parameters
parseParams(args);
System.out.print ("Starting web server...... ");
// Create a new server socket
this.ss = new ServerSocket(this.port);
System.out.println ("OK");
for (;;)
{
// Accept a new socket connection from our server socket
Socket accept = ss.accept();
// Start a new handler instance to process the request
new Handler(accept, docroot).start();
}
}
// Start an instance of the web server running
public static void main(String[] args) throws Exception
{
WebServerDemo webServerDemo = new WebServerDemo(args);
}
}
```

How WebServerDemo Works

The typical Web server must respond to requests from a number of browsers, and will usually handle more than one request for each browser (for example, the parallel downloading of a number of images in an HTML document). This means that such servers need to handle requests for files concurrently, and the simplest method of doing this is to use multiple threads of execution. Those readers who have not yet covered the topic of threads in Java are advised to consult the previous chapter, which examines multi-threading in detail. This example Web server is designed to be extremely compact and rather plain. It supports the bare minimum of features needed to function as a Web server, namely the HTTP/1.0 GET request. No support for dynamic server-side content, such as CGI scripts or Java servlets, is offered—this is a bare-bones Web server example that illustrates how to write a basic server implementation of an RFC, and is intended primarily as a teaching aid. However, it could be used as the scaffolding for a more ambitious project. For ease of implementation, the server combines all of its code into a single Java source file. This involves the use of an inner class (Handler) that deals with the processing of each incoming HTTP request. Most of this server's work is done inside this handler. The reason for this is to give high performance and to handle blocking I/O correctly so that if one client stalls, no others will be affected (since each request handler operates independently). The outline of the server looks like this:

```
import java.io.*;
import java.net.*;
import java.util.*;
public class WebServerDemo
{
class Handler extends Thread
{
}
}
```

Within the WebServerDemo class are stored several variables that are crucial to the server's operation. These are:

- String docroot
- int port
- ServerSocket ss

The docroot variable points to a location on a hard drive or network drive where the HTML pages and associated files (such as images) are stored for the Web server to read. The port number is used to track which port the server is operating on. Remember that while port 80 is standard for HTTP, it is common to run on nonstandard ports as well, particularly when an existing server is already running. A popular secondary standard is port 8080. Finally, the ServerSocket represents a socket that is bound to a specific port and is listening for incoming client requests. It is from this that we will accept browser requests.

When executed, the main(String[]) method of the server creates a new server instance. Inside the constructor, the server checks for the presence of the necessary command-line parameters (a filepath for the HTML pages and a port number). This code is handled by the parseParams(String[]) method, which involves only simple error-checking and string processing. As for the networking code, an attempt is made to bind a server socket to that port, and then the server loops indefinitely while it waits for socket connections.

```
public WebServerDemo(String[] args) throws Exception
{
System.out.println ("Checking for paramters");
// Check for command line parameters
parseParams(args);
System.out.print ("Starting web server...... ");
// Create a new server socket
this.ss = new ServerSocket(this.port);
System.out.println ("OK");
for (;;)
{
// Accept a new socket connection from our server@@@ socket
Socket accept = ss.accept();
// Start a new handler instance to process the request
new Handler(accept, docroot).start();
}
}
```

As you can see, the code for accepting a connection and launching a new handler thread to process it is fairly simple. One thing that is readily apparent is that writing a server for one particular protocol is not very different from writing one for another. The main HTTP-specific work lies in the actual Handler class.

Let us examine the Handler class in more detail. The handler's constructor is very simple—just enough to store the incoming parameters. We also convert the Web server directory to an absolute location. (Since some operating systems allow directory/drive mappings, it is important to know the actual location, as later we must check for browser attempts to access other, forbidden locations).

```
public Handler(Socket _socket, String _docroot) throws Exception
{
socket=_socket;
// Get the absolute directory of the filepath
docroot=new File(_docroot).getCanonicalFile();
}
```

When the Handler thread is started, the run() method will be executed. This is a lengthy method, so we will break it down into manageable chunks.

As is usual with any networking code, we must wrap it in an exception handler to deal with network errors at runtime. Fortunately, we do not need to do anything about the errors—simply terminating the connection is sufficient. The first step in writing our protocol handler will be to obtain I/O streams and wrap them in a reader/writer.

```
// Prepare our readers and writers
```

```
br = new BufferedReader(new InputStreamReader(socket.getInputStream()));
bos = new BufferedOutputStream(socket.getOutputStream());
pw = new PrintWriter(new OutputStreamWriter(bos));
```

The next step is to read a line of text sent by the browser, which (it is hoped) will contain a valid HTTP command. Of course, under HTTP/1.0, it is possible for commands such as "POST" to be composed of multiple lines, but these commands are not supported by this example. Therefore, we will use the socket's shutdownInput() method, which allows for a graceful shutdown of the input reader and instructs the socket to send back acknowledgment of any data coming in and then to silently discard it.

```
// Read HTTP request from user (hopefully GET /file...... )
String line = br.readLine();
// Shutdown any further input
socket.shutdownInput();
```

The next step is to process the HTTP request and work out if the operation is a valid one. We determine first if the line is null and disconnect if it is. As we accept GET requests, we check for a "GET" command, and process the request. If not, the request method sent by the browser is not supported, and we can send back an "HTTP 501 Not Implemented" response. Finally, the output writer and stream are flushed to make sure that the client receives a response.

```
if(line == null)
{
socket.close();
return;
}
if(line.toUpperCase().startsWith("GET"))
{
// Process GET request .......................
}
else
{
pw.println("HTTP/1.0 501 Not Implemented");
pw.println();
}
pw.flush();
bos.flush();
```

The task of processing the GET request is somewhat more complex, and involves some fancy string processing. While some Java programmers will be comfortable with processing text strings, others may not have come across this before, and may be unaware of the java.util.StringTokenizer class. It allows a string to be easily broken into separate pieces, and we use it in this instance to strip off any data (if it exists) after a "?" character. Remember that additional data can be passed to CGI scripts, or to embedded JavaScript, and we must remove it to determine the actual filename that is being requested. Finally, we add the requested filename to our document root, as well as an OS-specific file separator if needed (such as "/" on Wintel systems or "\" on Unix systems).

```
// Eliminate any trailing ? data, such as for a CGI
// GET request
StringTokenizer tokens = new StringTokenizer (line," ?");
tokens.nextToken();
String req = tokens.nextToken();
// If a path character / or \ is not present, add it to the document root and then
add the file //request, to form a full filename
String name;
if(req.startsWith("/") || req.startsWith("\\"))
name = this.docroot+req;
else
name = this.docroot+File.separator+req;
```

We then convert this requested file to an absolute filename, and verify that it is not out of bounds. This is one part of a sequence of conditions that must be tested for, such as the existence of a file, if it can be accessed and if it is a directory request instead of a file request. Two methods, sendDir and sendFile, are used if it is a successful request—otherwise, the appropriate HTTP error status codes are sent to the browser.

```
// Get absolute file path
File file = new File(name).getCanonicalFile();
// Check to see if request doesn't start with our document root ....
if(!file.getAbsolutePath().startsWith(this.docroot.getAbsolutePath()))
{
pw.println("HTTP/1.0 403 Forbidden");
pw.println();
}
else if(!file.exists())
{
pw.println("HTTP/1.0 404 File Not Found");
pw.println();
}
else if(!file.canRead())
{
pw.println("HTTP/1.0 403 Forbidden");
pw.println();
}
else if(file.isDirectory())
{
sendDir(bos,pw,file,req);
}
```

```
else // assume this is a file
{
sendFile(bos, pw, file.getAbsolutePath());
}
```

Breaking the code into two methods (sendDir and sendFile) simplifies things somewhat, as it leads to easier-to-read code. Since sending a file is pretty much the whole point of HTTP, we will cover this important method first. Reading from a file involves the use of a FileInputStream. We could, if we positively knew that only text data would be encountered, use a FileReader—but often, binary data such as data files and images are sent over HTTP, so we must read and write at the byte level. Starting by sending a valid HTTP 200 response code, we proceed to read the contents of the file and forward it on to the browser. We use the byte array method of reading and writing data for improved efficiency.

```
BufferedInputStream bis = new BufferedInputStream(new
FileInputStream(filename));
byte[] data = new byte[10*1024];
int read = bis.read(data);
pw.println("HTTP/1.0 200 Okay");
pw.println();
pw.flush();
bos.flush();
while(read != -1)
{
bos.write(data,0,read);
read = bis.read(data);
}
bos.flush();
```

The final major method of our HTTP handler that must be discussed is sendDir. The task of sending a directory listing, with hyperlinks to files and underlying subdirectories, involves the use of the File class, to obtain a list of file entries within a directory. For this method, we will be using a bit of simple HTML, using tables, to create a simple yet readable output.

Now, we start with the HTML.

```
pw.print("<html><head><title>Directory of ");
pw.print(req);
pw.print("</title></head><body><h1>Directory of ");
pw.print(req);
pw.println("</h1><table border=\"0\">");
```

To obtain a list of files, we use the File.listFiles() method, which returns an array of File objects. This gives us not only whatever files are stored in our directory, but also the names of any subdirectories. Once obtained, it is a simple process of looping through each element of the array and displaying the output. If the File object is actually representing a directory, we also add a / to the URL and Dir -> to the displayed name of the entry.

```
File[] contents=dir.listFiles();
for(int i=0;i<contents.length;i++)
{
pw.print("<tr>");
pw.print("<td><a href=\"");
pw.print(req);
pw.print(contents[i].getName());
if(contents[i].isDirectory())
pw.print("/");
pw.print("\">");
if(contents[i].isDirectory())
pw.print("Dir -> ");
pw.print(contents[i].getName());
pw.print("</a></td>");
pw.println("</tr>");
}
pw.println("</table></body></html>");
pw.flush();
```

Running WebServerDemo

Unlike other examples, running the server involves a little more complexity. Under some operating systems, such as Unix, many ports are restricted by security settings, or your machine may already have a Web server running. For this reason, you may need to run the server on a nonstandard port (e.g., 8080 rather than 80), although if you are running a plain Wintel system, port 80 should work fine. You must also specify the location of your HTML pages, although for testing you can point it to a directory on your hard drive where you have some images or text documents. After compiling the Web server, you can run it by typing:

```
java WebServerDemo port docroot
```

where port is a TCP port such as 80 or 8080, and docroot is a directory.

Next, using your browser, enter the URL http://hostname:port/, where "hostname" is the name or IP address of your machine, and "port" is the port you specified when running the server. If you are not connected to a network, you can use "localhost" as the hostname.

8. HYPERTEXT TRANSFER PROTOCOL

The HyperText Transfer Protocol (HTTP) is perhaps the most prolific network application protocol in the short history of the Internet. With the possible exception of e-mail, HTTP has changed the face of the Internet more profoundly than any other protocol. Without HTTP, there would be no World Wide Web, no electronic commerce as we know it today, nor the rapid, phenomenal growth of Internet usage. In previous chapters, we examined HTTP as an application protocol and constructed a simple Web server that serves up Web pages.

HTTP is an application-level protocol that uses the Transmission Control Protocol (TCP) as a transport mechanism. HTTP provides access to documents and files stored on a Web server. Web browsers use HTTP to request files or dynamically generated content produced by CGI scripts, and other server-side applications. Hypertext documents contain hyperlinks, which are links to other hypertext documents and files.

The World Wide Web is a collection of hypertext documents, stored on a wide variety of Web servers and accessed by an even larger number of Web browsers and HTTP clients.

When an HTTP client, such as a Web browser or a search engine, needs to access a file, it establishes a TCP connection to the Web server (which, by default, uses TCP port 80 for communication). The client sends a request for a particular file, and receives an HTTP response, which will often include the contents of the file, as shown in Figure 9.1. The response includes a status code (indicating the success or failure of a request), some HTTP header information such as the length of the content and its type, and, if appropriate, the file contents. Each request is for a single file; if subsequent files are needed, additional connections must be made.

8.1 HTTP and Java

Java provides extremely good support for the HyperText Transfer Protocol. While developers are free to write their own HTTP implementations using TCP sockets, the java.net package provides several classes that offer HTTP functionality:

- java.net.URL
- java.net.URLConnection
- java.net.HttpURLConnection

8.1.1 URL Class

The URL class represents one of the most frequently used address types of the Internet, the Uniform Resource Locator (URL). URLs can point to files, Web sites, ftp sites, newsgroups, email addresses, and other resources. Some fictitious examples of non-Web URLs are:

- ftp://records.area51.mil/roswell/subjects/autopsy/
- telnet://localhost:8000/
- mailto:president@whitehouse.gov?subject=My%20Opinion

In the context of HTTP and this chapter, we will be dealing with URLs that point to a Web site, but it is important to remember that other network protocols also use URLs. The URL is composed of several components, each of which can be parsed by the URL class and returned separately. Some components (namely the port and the reference fields) are optional, and will not be present in many URLs. As mentioned earlier, CGI parameters can also be included as part of the path field.

Format of the Universal Resource Locator

Creating a URL

The URL class can be used to parse URLs, or as an identifier of a remote resource that can be employed (in conjunction with the other Java HTTP classes) to retrieve that resource.

Constructors

There are six constructors for the URL class; the choice of which to use depends largely on how much control you require over the URL. For most situations, the following constructor will be used:

- *URL (String url_str) throws java.net.MalformedURLException*

Creates a URL object based on the string parameter. If the URL cannot be correctly parsed, a MalformedURLException will be thrown. However, not every application will have a String representation of a URL. For convenience, a wide range of constructors is offered, which some developers may find easier to use.

- URL (String protocol, String host, String file) throws java.net.MalformedURLException creates a URL object with the specified protocol, host, and file path.
- URL (String protocol, String host, int port, String file) throws java.net.MalformedURLException creates a URL object with the specified protocol, host, port, and file path.
- URL (String protocol, String host, int port, String file, URL StreamHandler handler) throws java.net.MalformedURLException creates a URL object with the specified protocol, host, port, file path, and stream handler.
- URL (URL context, String relative) throws java.net.MalformedURLException creates a URL object using the context of an existing URL and a relative URL. For example, if we had an existing URL of http://somewebsite.com/, and a relative URL of /images/icon.gif, the new URL would be http://somewebsite.com/images/icon.gif.
- URL (URL context, String relative, URLStreamHandler handler) throws java.net.MalformedURLException creates a URL object using the context of an existing URL and a relative URL. Any stream handler specified will override the default stream handler of the context URL.

Using a URL

The URL class provides the following methods to parse a URL and extract individual components (such as the protocol or the hostname of the URL), as well as to open an HTTP connection to the resource that it specifies.

Methods	Usage	Use
set	Protected void set(String protocol, String host, int port, String file, String ref)	Sets the fields of the URL. This is not a public method so that only URLStreamHandlers can modify URL fields.
setURLStream Handler Factory	Public static synchronized void setURLStreamHandler Factory(URLStreamHandler Factory fac)	Sets the URLStreamHandler factory
equals	Boolean equals(Object object)	Compares two URLs for equality. If the object is not an instance of the URL class, or if the object does not point to an identical resource, a value of "false" is returned.
getContent	Object getContent() throws java.io.IOException	Retrieves the contents of the resource located at the URL. The type of object returned will vary, depending on the MIME content type of the remote resource and the available content handlers.
getProtocol of a URL	String getProtocol()	Returns the protocol component

getHost	String getHost()	Returns the hostname component of a URL.
getPort	String getPort()	Returns the port component of a URL. This is an optional component, and if not present a value of –1 will be returned.
getFile	String getFile()	Returns the pathname component of a URL.
getRef	String getRef()	Returns the reference component of a URL. This is an optional component, and a reference may not be present. A null value will be returned if no reference was specified.
hashCode	Public int hashCode()	Returns an identifier for a URL object, for the purpose of hash table indexing.
open Connection	URLConnection open Connection()	Returns a URLConnection object, which can be used to establish a connection to the remote resource.
openStream	InputStream openStream() throws java.io.IOException	Establishes a connection to the remote server where the resource is located, and provides an InputStream that can be used to read the resource's contents.
sameFile	Boolean sameFile (URL url)	Compares two URLs for equality, similar to that of the equals(Object) method.
toString	String toString()	Returns a String representation of a URL. There is no difference between this method and the toExternalForm() method.
toExternal Form	String toExternalForm()	Returns a String representation of a URL. There is no difference between this method and the toString() method.

Parsing with the URL Class

To demonstrate how the URL class is used, and how it can extract the individual components out of a URL, we will examine a small application that creates an instance of the URL class from the command-line parameter passed to it.

Code for URLParser

```
import java.net.*;
```

```
public class URLParser
{
public static void main(String args[])
{
int argc = args.length;
// Check for valid number of parameters
if (argc != 1)
{
System.out.println ("Syntax :");
System.out.println ("java URLParser url");
return;
}
// Catch any thrown exceptions
try
{
// Create an instance of java.net.URL
java.net.URL myURL = new URL ( args[0] );
System.out.println ("Protocol : " +myURL.getProtocol() );
System.out.println ("Hostname : " +myURL.getHost() );
System.out.println ("Port : " +myURL.getPort() );
System.out.println ("Filename : " +myURL.getFile() );
System.out.println ("Reference: " +myURL.getRef() );
}
// MalformedURLException indicates parsing error
catch (MalformedURLException mue)
{
System.err.println ("Unable to parse URL!");
return;
}
}
}
```

How URLParser Works

The example starts by checking that a command-line parameter was passed to the application. A null value for the command-line parameter would cause a runtime exception to be thrown. Next, the application begins a try/catch block, which is required since the URL constructor can throw an instance of java.net.MalformedURLException. If an exception is thrown, it will be caught, and a warning sent to the user before terminating.

```
// Catch any thrown exceptions
try
{
// Create an instance of java.net.URL
java.net.URL myURL = new URL ( args[0] );
// Code to process the URL goes here
...........
}
// MalformedURLException indicates parsing error
catch (MalformedURLException mue)
```

```
{
System.err.println ("Unable to parse URL!");
return;
}
```

The code to create a URL instance is relatively easy. Once you have a URL, you can look at the various components of which it is comprised. The application prints out the protocol, host, port, pathname, and reference of the URL. Note that the port and reference fields may be blank, so a –1 and null value, respectively, would be displayed in this case.

```
System.out.println ("Protocol : " + myURL.getProtocol() );
System.out.println ("Hostname : " + myURL.getHost() );
System.out.println ("Port : " + myURL.getPort() );
System.out.println ("Filename : " + myURL.getFile() );
System.out.println ("Reference: " + myURL.getRef() );
```

Running URLParser

The URLParser application takes a valid URL as its only parameter. A valid URL must contain a protocol, hostname, and pathname components, but it does not necessarily need to point to a valid resource. Thus, the following URLs would be accepted without throwing a MalformedURLException:

- http://thisisnotarealmachine/noristhisarealpath
- http://www.earmind.com:80/#top
- ftp://ftp.earmind.com/devotional/

However, these URLs could not be parsed:

- abcdef://abcdef.com/alphabetsoup/
- ttp://www.earmind.com:80/#top

To run the URLParser, specify the URL as the only parameter, as follows:

java URLParser url

Try running some of the sample URLs, and examine the output for URLs with and without port and reference fields.

Retrieving a Resource with the URL Class

The URL class can be used for more than just parsing, however. There are two URL methods that can assist in retrieving the contents of a remote resource:

1. InputStream URL.openStream()
2. URLConnection URL.openConnection();

For greater control over how the request is made, a URLConnection object created by invoking the URL.openConnection() method would be used. In many situations, however, a simpler way to retrieve the contents of a resource is called for. The openStream() method returns an InputStream, which makes reading a resource simple. Examine the following application, which will fetch a URL passed as a command-line parameter.

Code for FetchURL

```
import java.net.*;
import java.io.*;
public class FetchURL
{
public static void main(String args[]) throws Exception
```

```
{
int argc = args.length;
// Check for valid number of parameters
if (argc != 1)
{
System.out.println ("Syntax :");
System.out.println ("java FetchURLConnection url");
return;
}
// Catch any thrown exceptions
try
{
// Create an instance of java.net.URL
java.net.URL myURL = new URL ( args[0] );
// Fetch the content, and read from an InputStream
InputStream in = myURL.openStream();
// Buffer the stream, for better performance
BufferedInputStream bufIn = new BufferedInputStream(in);
// Repeat until end of file
for (;;)
{
int data = bufIn.read();
// Check for EOF
if (data == -1)
break;
else
System.out.print ( (char) data);
}
// Pause for user
System.out.println ();
System.out.println ("Hit enter to continue");
System.in.read();
}
// MalformedURLException indicates parsing error
catch (MalformedURLException mue)
{
System.err.println ("Unable to parse URL!");
return;
}
// IOException indicates network or I/O error
catch (IOException ioe)
{
System.err.println ("I/O Error : " + ioe);
return;
}
}
}
```

How FetchURL Works

You may notice some similarity between this example and the previous one. Since we are working with the same URL class, the code to check for a valid parameter, to construct the URL, and catch a possible MalformedURLException remains the same. Observant readers may notice an extra catch statement, which handles any network or I/O errors. This is necessary in case the host machine could not be contacted or is invalid. Remember that a MalformedURLException will not be thrown if the hostname is invalid, only if the URL syntax was not followed.

```
// IOException indicates network or I/O error
catch (IOException ioe)
{
System.err.println ("I/O Error : " + ioe);
return;
}
```

Inside the try/catch block, the application creates a new URL object and calls the openStream() method. This establishes a connection to the remote machine (or, if a connection could not be opened, throws an IOException). The method returns an InputStream, which can then be used for reading from the remote resource.

```
// Create an instance of java.net.URL
java.net.URL myURL = new URL ( args[0] );
// Fetch the content, and read from an InputStream
InputStream in = myURL.openStream();
```

While we could read directly from the input stream, one byte at a time, a more efficient alternative is to connect another stream, one which is better suited to the task at hand. In this case, we use a BufferedInputStream and read a byte at a time. However, a DataInputStream or a Reader object could just as easily have been used.

```
// Buffer the stream, for better performance
BufferedInputStream bufIn = new BufferedInputStream(in);
// Repeat until end of file
for (;;)
{
int data = bufIn.read();
// Check for EOF
if (data == -1)
break;
else
System.out.print ( (char) data);
}
// Pause for user
System.out.println ();
System.out.println ("Hit enter to continue");
System.in.read();
```

While this example is quite simple, it provides only limited control over the connection. For example, what is the MIME content type of the resource? By reading from an InputStream, we lose some of the details. The URLConnection class, offers greater control over requests and responses.

Running FetchURL

Running the application is straightforward. Simply specify a valid URL as the only command-line parameter, either of a local machine on your network or a Web site, if you are connected to the Internet.

```
java FetchURL url
```

8.1.2 URLConnection Class

The URLConnection class is used to send HTTP requests and read HTTP responses. URLConnection has methods that allow you to connect to a Web server, to set request header fields, to read response header fields, and of course, to read the contents of the resource.

Creating a URLConnection

There are no public constructors for the URLConnection class. The single constructor for URLConnection is marked as a protected method, meaning that only a class in the java.net package can create one. Instead, you should call the URL.openConnection() method, which will return a URLConnection instance.

```
URL url = new URL ( some_url );
URLConnection connection = url.openConnection();
```

Using a URLConnection

The URLConnection provides many methods, the most important of which are listed below. Some are shortcuts to other methods (such as response header fields). For completeness, even advanced methods are included, although for normal programming you would likely use a small subset of these.

Methods

Methods	Usage	Use
connect	Public abstract void connect() throws IOException	Used to create connection
getURL	Public URL getURL()	Gets the URL for this connection.
getContent length	Public int getContentLength()	Gets the content length. Returns –1 if not known.
getContent Type	Public String getContentType()	Gets the content type. Returns null if not known.
getContent Encoding	Public String getContent Encoding()	Gets the content encoding. Returns null if not known.
getExpiration	Public long getExpiration()	Gets the expiration date of the object. Returns 0 if not known.
getDate	Public long getDate()	Gets the sending date of the object. Returns 0 if not known.

getLast Modified	Public long getLastModified()	Gets the last modified date of the object. Returns 0 if not known.
getHeader Field	Public String getHeaderField (String name)	Gets a header field by name. Returns null if not known.
getHeader FieldInt	Public int getHeaderFieldInt (String name, int Default)	Gets a header field by name. Returns null if not known. The field will be parsed as an integer.
getHeader	Public long getHeader	Gets a header field by name.
FieldDate	FieldDate(String name, long Default)	Returns null if not known. The field will be parsed as a date.
getHeader FieldKey	Public String getHeader FieldKey(int n)	Returns the key for the nth header field. Returns null if there are fewer than n fields.
getHeader Field	Public String getHeaderField (int n)	Returns the value for the nth header field. Returns null if there are fewer than n fields.
getContent	Public Object getContent() throws IOException	Gets the object referred to by this URL.
getInput Stream	Public InputStream getInput Stream() throws IOException	Calls this routine to get an InputStream that reads from the object. Protocol implementors should implement this if appropriate.
getOutput Stream	Public OutputStream get OutputStream() throws IOException	Calls this routine to get an OutputStream that writes to the object.
toString	Public String toString()	Returns the String representation of the URL connection.
setDoInput	Public void setDoInput (boolean doinput)	A URL connection can be used for input and/or output. Set the DoInput flag to true if you intend to use the URL connection for input, false if not.
getDoInput	Public boolean getDoInput()	Returns a flag, indicating whether the connection should be used for input. By default, this value will be "true," unless modified by the setDoInput(boolean) method.
setDoOutput	Public void setDoOutput (boolean dooutput)	A URL connection can be used for input and/or output. Set the DoOutput flag to true if you intend to use the URL connection for output, false if not. The default is false.

getDoOutput	Public boolean getDoOutput()	Returns a flag indicating whether the connection should be used for output. By default, this value will be "false," unless modified by the setDoOutput (boolean) method.
setAllowUser Interaction	Public void setAllowUser Interaction(boolean allowuserinteraction)	Some URL connections occasionally need to have interactions with the user. The allowUser-Interaction flag allows these interactions when true. When it is false, they are not allowed an excel occasionally need to have option is tossed. Default is false.
getAllowUser Interaction	Public boolean getAllowUser Interaction()	Returns the default value for the "allowUserInteraction" field. The state of this default value determines the"allowUser Interaction" value of all future URLConnection objects that are created.
setDefault Allow User Interaction	Public static void set DefaultAllowUserInteraction (boolean defaultallowuser interaction)	Modifies the state of the "allowUserInteraction" field, which indicates whether the URL is being used in a way that supports user interaction, such as providing authentication details.
getDefault AllowUser Interaction	Public static boolean get DefaultAllowUserInteraction()	Set/get the default value of the allowUserInteraction flag.
setUseCaches	Public void setUseCaches (boolean usecaches)	If the UseCaches flag on a connection is true, the connection is allowed to use whatever caches it can. If false, caches are to be ignored. Default is true.
getUseCaches	Public boolean getUseCaches()	Returns a flag indicating whether or not resources will be cached. If "true," caching will be used wherever possible, and if "false," a new request will be sent each time.

setIfModified Since	public void setIfModified Since(long ifmodifiedsince)	Sets the "If-Modified-Since" request header, indicating that a resource should only be fetched if modified after a certain date.
getIfModified Since	Public long getIfModifiedSince()	Some protocols support skipping fetching unless the object is newer than some time.
getDefault UseCaches	Public boolean getDefaultUse Caches()	Set/get the default value of the UseCaches flag. This flag applies to the next, and all following, URLConnections that are created.
setDefault UseCaches	Public void setDefaultUse Caches(boolean defaultusecaches)	Modifies the default value of the cache flag. The state of this default value determines the cache flag value of all future URLConnection objects that are created.
setRequest Property	Public void setRequestProperty (String key,String value)	Set/get a general request property.
getRequest Property	Public String getRequest Property(String key)	Returns the value of the specified property, and null if no such property exists.
setDefault Request Property	Public static void setDefault RequestProperty(String key, String value)	Set/get the default value of a general request property. When a URLConnection is created, it gets initialized with these properties.
getDefault Request Property	Public static String getDefault RequestProperty(String key)	Returns the value of the specified default property, applied to all URLConnection objects. Returns null if no such property exists.
setContent Handler Factory	Public static synchronized void setContentHandler Factory(ContentHandler Factory fac)	Sets the ContentHandler factory.
guessContent TypeFrom Stream	Protected static String guessContentTypeFrom Name(String fname)	A useful utility routine that tries to guess the content-type of an object based upon its extension.
guessContent TypeFrom stream	Protected static String guess ContentTypeFromStream (InputStream is) throws IOException	This is used to check for files have some type that can be determined by inspection. The bytes at the beginning of the file are examined loosely.

getFileName Map	Static FileNameMap getFileNameMap()	Returns an object that implements the FileNameMap interface, which is used to map MIME content types to filenames. The content type for a resource may also be determined by calling the getContentType() method.
getLast Modified	Long getLastModified()	Returns the date of the "Last-Modified header field, expressed as the number of seconds since January 1, 1970 GMT. If no such header field was specified, a value of zero will be returned.
getPermission	Permission getPermission() throws java.io.IOException	Returns a Permission object, representing the security permissions required to access a resource.

Retrieving a Resource with the URLConnection Class

While the URL class does allow you to retrieve a resource by using the URL. openStream() method, information about the resource is lost, has the ability to prevent caching of requests and to specify additional header fields, since only a stream object is returned. The example below shows how to use the URL Connection class to retrieve a resource and to determine its MIME content type and the length of the resource. Since we still use the URL class, some of the code is similar to previous examples.

Code for FetchURLConnection

```
import java.net.*;
import java.io.*;
public class FetchURLConnection
{
public static void main(String args[]) throws Exception
{
int argc = args.length;
// Check for valid number of parameters
if (argc != 1)
{
System.out.println ("Syntax :");
System.out.println ("java FetchURLConnection url");
return;
}
// Catch any thrown exceptions
try
{
// Create an instance of java.net.URL
java.net.URL myURL = new URL ( args[0] );
```

```
// Create a URLConnection object, for this URL
// NOTE : no connection has yet been established
URLConnection connection = myURL.openConnection();
// Now open a connection
connection.connect();
// Display the MIME content-type of the resource (e.g. text/html)
String MIME = connection.getContentType();
System.out.println ("Content-type: " + MIME);
// Display, if available, the content length
int contentLength = connection.getContentLength();
if (contentLength != -1)
{
System.out.println ("Content-length: " +contentLength);
}
// Pause for user
System.out.println ("Hit enter to continue");
System.in.read();
// Read the contents of the resource from the connection
InputStream in = connection.getInputStream();
// Buffer the stream, for better performance
BufferedInputStream bufIn = new BufferedInputStream(in);
// Repeat until end of file
for (;;)
{
int data = bufIn.read();
// Check for EOF
if (data == -1)
break;
else
System.out.print ( (char) data);
}
}
// MalformedURLException indicates parsing error
catch (MalformedURLException mue)
{
System.err.println ("Unable to parse URL!");
return;
}
// IOException indicates network or I/O error
catch (IOException ioe)
{
System.err.println ("I/O Error : " + ioe);
return;
}
}
}
```

How FetchURLConnection Works

We start by checking for a valid parameter, and create a URL instance, as in previous examples. Rather than calling URL.openStream() once we have a URL object, however, the openConnection() method is called instead, returning a URLConnection instance.

```
// Create an instance of java.net.URL
java.net.URL myURL = new URL ( args[0] );
// Create a URLConnection object, for this URL
// NOTE : no connection has yet been established
URLConnection connection = myURL.openConnection();
```

The name of the openConnection() method is somewhat misleading. Although it creates an instance of the URLConnection object, it does not establish an HTTP session with the Web server. This can be advantageous, as it allows us to set any request header fields we need. For a simple example like this, however, this functionality is not needed, so we can connect to the Web server immediately, which sends a request for the URL associated with this connection.

```
// Now open a connection
connection.connect();
```

Once the connection is established and the request sent, a response will be issued by the Web server. This response will include a variety of header fields. The most important field of all is the "Content-Type," which tells an application whether the resource is text, an image, a data file, or some other resource. The application could read this data by calling the URLConnection.getHeaderField(String) method and passing a value of "Content-Type." However, a shortcut method exists that makes for more readable source code.

```
// Display the MIME content-type of the resource (e.g. text/html)
String MIME = connection.getContentType();
System.out.println ("Content-type: " + MIME);
```

Another important piece of information is the length of the resource. Large files can take minutes or even hours to download, and the user benefits from knowing the length of the resource at the beginning of the transaction. This information is provided in the "Content-Length" header field, and a shortcut method exists for this data that converts it to an int value. After displaying the length, the application pauses, to allow the user time to read it before displaying the requested resource.

```
// Display, if available, the content length
int contentLength = connection.getContentLength();
if (contentLength != -1)
{
System.out.println ("Content-length: " + contentLength);
}
// Pause for user
System.out.println ("Hit enter to continue");
System.in.read();
```

The next step is to get an InputStream to the contents of the resource. Just like a URL object, URLConnection provides a method to create an InputStream for reading a resource. For this purpose, the URLConnection.getInputStream() method is used.

```
// Read the contents of the resource from the connection
InputStream in = connection.getInputStream();
```

Once an InputStream has been obtained, the resource is read in the same way as the previous example. The chief difference between FetchURL and FetchURLConnection is that more information about the resource was provided. The URLConnection class can be used for more than just content type and length information, however.

Running FetchURLConnection

This application takes the same command-line parameter as the previous example. Specify a valid URL as the only command-line parameter, either of a local machine on your network, or a Web site, if you are connected to the Internet.

```
java FetchURLConnection url
```

The application will issue an HTTP request, the display, the MIME content type and length of the resource and then the resource itself.

Modifying and Examining Header Fields with URLConnection

In the previous example, you learned how to use the URLConnection to fetch HTTP resources, and how to determine the length and MIME content type of a resource. However, there are many more HTTP response header fields to examine, and you can also modify HTTP request header fields to make it possible for server-side applications (such as CGI scripts, Active Server Pages, or Java servlets) to customize their output. In this next example, you will learn how to modify request properties, and how to get response header fields.

Code for HTTPHeaders

```
import java.net.*;
import java.io.*;
public class HTTPHeaders
{
public static void main(String args[]) throws Exception
{
int argc = args.length;
// Check for valid number of parameters
if (argc != 1)
{
System.out.println ("Syntax :");
System.out.println ("java HTTPHeaders url");
return;
}
// Catch any thrown exceptions
try
{
// Create an instance of java.net.URL
java.net.URL myURL = new URL ( args[0] );
```

```
// Create a URLConnection object, for this URL
// NOTE : no connection has yet been established
URLConnection connection = myURL.openConnection();
// Set some basic request fields
// Set user agent, to identify the application as Netscape compatible
connection.setRequestProperty ("User-Agent","Mozilla/4.0 (compatible; JavaApp)");
// Set our referer field - set to any URL you'd like
connection.setRequestProperty ("Referer","http://www.earmind.com/");
// Set use-caches field, to prevent caching
connection.setUseCaches(false);
// Now open a connection
connection.connect();
// Examine request properties, to verify their settings
System.out.println ("Request properties....");
System.out.println();
System.out.println ("User-Agent: " +connection.getRequestProperty("User-Agent"));
System.out.println ("Referer: " +connection.getRequestProperty("Referer"));
System.out.println (); System.out.println ();
// Examine response properties, to see their settings
System.out.println ("Response properties....");
System.out.println();
int i = 1;
// Search through each header field, until no more exist
while ( connection.getHeaderField ( i ) != null )
{
// Get the name of this header field
String headerName =connection.getHeaderFieldKey(i);
// Get the value of this header field
String headerValue =connection.getHeaderField(i);
// Output header field key, and header field value
System.out.println ( headerName + ": " +headerValue);
// Goto the next element in the set of header fields
i++;
}
// Pause for user
System.out.println ("Hit enter to continue");
System.in.read();
}
// MalformedURLException indicates parsing error
catch (MalformedURLException mue)
{
System.err.println ("Unable to parse URL!");
return;
}
// IOException indicates network or I/O error
catch (IOException ioe)
{
```

```
System.err.println ("I/O Error : " + ioe);
return;
}
}
}
```

How HTTPHeaders Works

Like previous examples, this example uses an instance of the URLConnection class to issue HTTP requests. The chief difference here is that, before any request is sent, custom HTTP request fields are added. These header fields provide additional information to server-side applications, which can then be used to customize the HTTP response. When a Web browser sends a request, it identifies itself by sending a "User-Agent" field in the request. Well-behaved HTTP clients do the same, and it is often advantageous to pose as a Web browser by including the *Mozilla* keyword in the identification string and then appending a legitimate-sounding application name, since CGI scripts and servlets sometimes offer different output depending on whether it is an HTTP agent like a search engine or an actual browser. Other request fields can also be set, such as the referring URL and the cache flag, which determines whether or not a unique request will be sent each time to the server.

```
// Create a URLConnection object, for this URL
// NOTE : no connection has yet been established
URLConnection connection = myURL.openConnection();
// Set some basic request fields
// Set user agent, to identify the application as Netscape compatible
connection.setRequestProperty ("User-Agent", "Mozilla/4.0 (compatible; JavaApp)");
// Set our referer field
connection.setRequestProperty ("Referer","http://www.earmind.com/");
// Set use-caches field, to prevent caching
connection.setUseCaches(false);
```

Once the request settings are made, the URLConnection object can send the request. If a call to the connect() method is made before assigning request properties, then the server will not receive them and they will not take effect. After connecting and sending the request, the application displays the request fields to the user.

```
// Now open a connection
connection.connect();
// Examine request properties, to verify their settings
System.out.println ("Request properties....");
System.out.println();
System.out.println ("User-Agent: " + connection.getRequestProperty("User-Agent"));
System.out.println ("Referer: " + connection.getRequestProperty("Referer"));
System.out.println (); System.out.println ();
```

The next set of header fields displayed by the application is from the server response. In the previous section, the names of the request fields were known. It would be impossible, however, to know the name of every field that might be sent back by a server. Not all servers support the same fields, and some server-side applications may send back custom fields that a client has never before encountered. The URLConnection offers several methods that provide access to request fields, two of which support a numerical index value rather than a key name. This allows us to read the *n*th key, and to iterate through every element in the set of header fields. The program prints out both the name of the field and its contents.

```
// Examine response properties, to see their settings
System.out.println ("Response properties....");
System.out.println();
int i = 1;
// Search through each header field, until no more exist
while ( connection.getHeaderField ( i ) != null )
{
// Get the name of this header field
String headerName = connection.getHeaderFieldKey(i);
// Get the name of this header field
String headerValue = connection.getHeaderField(i);
// Output header field key, and header field value
System.out.println ( headerName + ": " + headerValue);
// Goto the next element in the set of header fields
i++;
}
```

Finally, the application pauses to allow the user to read the header fields, and then terminates. The actual contents of the resource are not displayed, although code for this exists in the previous example.

Running HTTPHeaders

This application takes the same command-line parameter as previous examples. You should pass a valid URL as the only command-line parameter, either of a local machine on your network, or a Web site, if you are connected to the Internet. Try URLs pointing to different Web servers, to see a variation in the type of headers returned.

java HTTPHeaders url

The application will issue an HTTP request, display all request and response headers, and then terminate.

8.1.3 HttpURLConnection Class

One of the problems of reading resources using URL.openStream() or the URL Connection class is that access to HTTP-specific functionality is not available. Any protocol for which a registered protocol handler exists can be fetched in this manner, including the File Transfer Protocol (FTP). But there is no notion of a request method, or a response status code in these protocols. How does an application know whether a resource was found, or a 404 "Not Found" error message was sent?

Indeed, many servers output custom pages, designed more for end users than for HTTP client applications. However, there is no uniform error message placed in the message body that is standard across all Web servers. The solution is to read the response status code, as it is the only appropriate way to determine the success or failure of requests.

Though support for HTTP-specific functionality did not exist in earlier versions of Java, as of JDK1.1 there is a solution in the form of the HttpURL Connection class. This class extends the URLConnection class, and provides additional methods and fields that encapsulate HTTP functionality.

Creating a HttpURLConnection

There are no public constructors for the HttpURLConnection class, just as in the case of URLConnection. The single constructor for HttpURLConnection is marked as a protected method, meaning that only a class in the java.net package can create one. Instead, you should call the URL.openConnection() method, which will return a URLConnection instance. The URLConnection class is the superclass of HttpURLConnection, and if the protocol field of the URL is set to HTTP, this method will actually return an HttpURLConnection instance. To gain access to HTTP-specific functionality, you should test to see whether or not the object is an instance of HttpURLConnection; if so, the object must be cast as such.

```
URL url = new URL ( some_url );
URLConnection connection = url.openConnection();
if (connection instanceof java.net.HttpURLConnection)
{
HttpURLConnection httpConnection = (HttpURLConnection) connection;
// do something with httpConnection
}
```

Even if an HttpURLConnection object is expected, it is good programming practice to test the class type using the *instanceof* keyword. If an unexpected class is returned by the URL.openConnection() method, a runtime error will occur and the application may not handle it gracefully.

Using an HttpURLConnection

The HttpURLConnection class inherits all of the functionality (including fields and methods) of its parent class, URLConnection. It also adds additional functionality, in the form of methods that allow greater access to HTTP features, and static fields that represent common HTTP states.

Fields

The HttpURLConnection class defines many static fields, which represent HTTP status codes. While an application can refer to a status code by a numerical value, these fields may make for more readable code. The following fields are all public static final int fields.

Status Code	Field	Indication
200	HTTP_OK	Request was successful.
201	HTTP_CREATED	A resource was created.
202	HTTP_ACCEPTED	The request was accepted but has not yet been acted upon.

203	HTTP_NOT_AUTHORITATIVE	The set of header fields is not from the original source. This may indicate that it has been cached locally or read from a third-party copy, and may represent a subset or superset of fields.
204	HTTP_NO_CONTENT	Indicating that an entity body was not available. For example, if data was sent to a server-side application but no entity body was needed, this code might be returned.
205	HTTP_RESET	The browser should reset the view of the document, but that no new document is available. For example, an HTML form should be cleared, to allow new data to be input by the user.
206	HTTP_PARTIAL	Indicating that the Web server was able to fulfill a partial GET request for a resource. Partial requests occur when an HTTP client has part, but not all, of a resource, and wishes to request only the missing data. If a resource is modified, the new resource will be sent and this status code will not be issued.
300	HTTP_MULT_CHOICE	The resource can be found at multiple locations, from which the client can choose. When a resource is located elsewhere, a "Location" entity field will be sent, along with a 3xx redirection status code, but in the case of multiple choices of location, this status code will be issued.
301	HTTP_MOVED_PERM	The location of a resource has moved permanently and the client should look for a "Location" field in the HTTP response. The new location of the resource should be used in future.

302	HTTP_MOVED_TEMP	A temporary change has been made to the location of the resource, indicated by a "Location" field in the HTTP response.
303	HTTP_SEE_OTHER	A GET request should be used to fetch the resource, at a location specified by the "Location" field. This is often issued in response to a POST request, which processes the information and redirects to a standard page.
304	HTTP_NOT_MODIFIED	Used to inform the client that a resource has not been modified and that no entity body was sent. This is used in conjunction with the "If-Modified-Since" request field, which performs a conditional GET request.
305	HTTP_USE_PROXY	That informs the client that a proxy server (a server that makes requests on behalf of clients, usually ones trapped within a firewall) must be used to access this resource. The "Location" field indicates the location of the proxy server, which should be used to reissue the request.
400	HTTP_BAD_REQUEST	That is issued in response to an invalid HTTP request, which fails to follow the correct syntax.
401	HTTP_UNAUTHORIZED	Access to the resource requires user authentication.
402	HTTP_PAYMENT_REQUIRED	Used to indicate that payment is required for access to this resource. This status code is reserved for the future, and is not in common use.
403	HTTP_FORBIDDEN	Access to a resource is strictly forbidden.
404	HTTP_NOT_FOUND	Used to notify a client that the resource could not be found or has been permanently removed.

405	HTTP_BAD_METHOD	That tells the client that the request method is not supported by the server. A list of allowed methods will be specified in the "Allow" response header field.
406	HTTP_NOT_ACCEPTABLE	That notifies the client that the response contains content with attributes violating those prescribed in the request.
407	HTTP_PROXY_AUTH	The client must authenticate itself to the proxy. This is a slight, but important, difference between that of status code 401.
408	HTTP_CLIENT_TIMEOUT	That occurs when a client fails to send a request within the required timeframe set by the server.
409	HTTP_CONFLICT	That notifies the client that access to a resource is temporarily unavailable but may be available at a later date. For example, an access conflict might occur if a resource is being modified, and a request to read that resource occurs.
410	HTTP_GONE	A resource has been permanently removed. This is similar to the 404 status code, but indicates that the resource is gone forever.
411	HTTP_LENGTH_REQUIRED	Issued when a client fails to specify a required "Content-Length" field.
412	HTTP_PRECON_FAILED	A precondition placed on the request evaluated to false, and the request could not proceed.
413	HTTP_ENTITY_TOO_LARGE	The request entity was too large to process.
414	HTTP_REQ_TOO_LONG	The request path was too long to process.
415	HTTP_UNSUPPORTED_TYPE	Notifying the client that the format of the request entity was unsupported.
500	HTTP_SERVER_ERROR	Indicating that a server error occurred and the request could not be processed.

501	HTTP_INTERNAL_ERROR	A server did not know how to perform the request.
502	HTTP_BAD_GATEWAY	An error occurred while acting as a gateway or proxy server.
503	HTTP_UNAVAILABLE	The server could not process the request due to a temporary condition such as server overload.
504	HTTP_GATEWAY_TIMEOUT	The server, while acting as a gateway or proxy, did not receive a response in time from another server.
505	HTTP_VERSION	The server did not support that HTTP version of the request.

Methods

New methods added by HttpURLConnection are listed below. To conserve space, we have not listed here methods inherited from the URLConnection class.

Methods	Usage	Use
disconnect	Void disconnect()	If a connection to the Web server is still active, the connection is closed.
getError Stream	InputStream getErrorStream()	Returns an InputStream instance that can be used to read error messages sent by the server. If a connection has not yet been established, or no errors have yet occurred, this method returns null.
getFollow Redirects	Static boolean getFollow Redirects()	Indicates whether HTTP redirects will be automatically followed. Returns "true" if re-direction will occur automatically, and "false" if not.
getRequest Method	String getRequestMethod()	Returns the request method (e.g. GET) being used.
getResponse Code	Int getResponseCode()	Returns the response status code. Applications can hardwire the numerical value of codes, or use the HttpURL Connection fields that define state conditions.
getResponse Message	String getResponseMessage()	Returns the message from the response status line, such as "OK," or "Not Found."

setFollow Redirects	Static void set FollowRedirects(boolean flag) throws java.lang. SecurityException	Determines whether the resource specified in a redirection response will be automatically followed. This must be invoked prior to the connect() method for the setting to take effect. If this violates the settings of the security manager, a SecurityException will be thrown.
setRequest Method	Void setRequestMethod (String method) throws java. net.ProtocolException	Sets the request method for this connection. This must be invoked prior to the connect() method for the setting to take effect. If the method is not supported, a ProtocolException will be thrown. The method name must be capitalized, as the protocol names are case sensitive.
usingProxy	Boolean usingProxy()	Shows whether a proxy server is being used for this connection. Returns "true" if using a proxy server, "false" if not.

Accessing HTTP-Specific Functionality Using HttpURLConnection

The next example shows how to read the status code and message of a response. Most URLs will generate a "normal" response, with a 200 status code and a normal entity body. However, in some situations this will not be the case.

Code for UsingHttpURLConnection

```
import java.net.*;
import java.io.*;
public class UsingHttpURLConnection
{
public static void main(String args[]) throws Exception
{
int argc = args.length;
// Check for valid number of parameters
if (argc != 1)
{
System.out.println ("Syntax :");
System.out.println ("java UsingHttpURLConnection url");
return;
}
// Catch any thrown exceptions
try
```

```
{
// Create an instance of java.net.URL
java.net.URL myURL = new URL ( args[0] );
// Create a URLConnection object, for this URL
// NOTE : no connection has yet been established
URLConnection connection = myURL.openConnection();
// Check to see if connection is a HttpURLConnection instance
if (connection instanceof java.net.HttpURLConnection)
{
// Yes... cast to a HttpURLConnection instance
HttpURLConnection hConnection = (HttpURLConnection) connection;
// Disable automatic redirection, to see the status header
hConnection.setFollowRedirects(false);
// Connect to server
hConnection.connect();
// Check to see if a proxy server is being used
if (hConnection.usingProxy())
{
System.out.println
("Proxy server used to access resource");
}
else
{
System.out.println ("No proxy server used to access resource");
}
// Get the status code
int code = hConnection.getResponseCode();
// Get the status message
String msg = hConnection.getResponseMessage();
// If a 'normal' response
if ( code == HttpURLConnection.HTTP_OK )
{
// Notify user
System.out.println ("Normal response returned : " + code + " " + msg );
}
else
{
// Output status code and message
System.out.println ("Abnormal response returned : " + code + " " + msg );
}
// Pause for user
System.out.println ("Hit enter to continue");
System.in.read();
}
else
{
System.err.println ("Invalid transport protocol - not http!");
return;
}
```

```
}
// MalformedURLException indicates parsing error
catch (MalformedURLException mue)
{
System.err.println ("Unable to parse URL!");
return;
}
// IOException indicates network or I/O error
catch (IOException ioe)
{
System.err.println ("I/O Error : " + ioe);
return;
}
}
}
}
```

How UsingHttpURLConnection Works

The program starts by creating a URL object, and from this, a URLConnection object. If the protocol being used to request the resource is HTTP, then the URLConnection will also be an instance of the HttpURLConnection class. It is not advisable to presume that this will always be the case, however. A guard statement checks to see if it is an HttpURLConnection object and performs a casting operation. If not, an error message is displayed and the program terminates.

```
// Create a URLConnection object, for this URL
// NOTE : no connection has yet been established
URLConnection connection = myURL.openConnection();
// Check to see if connection is a HttpURLConnection instance
if (connection instanceof java.net.HttpURLConnection)
{
// Yes... cast to a HttpURLConnection instance
HttpURLConnection hConnection = (HttpURLConnection) connection;
// Do something with hConnection
...
}
else
{
System.err.println ("Invalid transport protocol - not http!");
return;
}
```

If all proceeds according to plan, the application now has an HttpURL Connection, and the extra HTTP-specific functionality it gives. Before a connection is established, it is possible to modify the properties of the request. For example, a different request method could be used, or the "follow redirection" flag could be modified, so that users can see the redirection status code, automatic redirection is disabled by the application, and then the connection is established.

```
// Disable automatic redirection, to see the status header
hConnection.setFollowRedirects(false);
// Connect to server
hConnection.connect();
```

Once a connection has been established, all sorts of useful information becomes available, such as the status code and message and whether or not a proxy server is being used. The application checks for the presence of a proxy server, and displays it to the user, by using the *boolean HttpURLConnection. usingProxy()* method.

```
// Check to see if a proxy server is being used
if (hConnection.usingProxy())
{
System.out.println ("Proxy server used to access resource");
}
else
{
System.out.println ("No proxy server used to access resource");
}
```

Next, the status code and message are retrieved. This is the most useful information of all, as it tells a client whether or not a request was successful and, if it was not successful, gives an indication of why. The human-readable shortcut for the 200 status code (HttpURLConnection.HTTP_OK) is used to check whether the request was successful. If not, the status code and message are displayed to the user, and an application could take further steps, such as resending a request or following a redirection notice.

```
// Get the status code
int code = hConnection.getResponseCode();
// Get the status message
String msg = hConnection.getResponseMessage();
// If a 'normal' response
if ( code == HttpURLConnection.HTTP_OK )
{
// Notify user
System.out.println ("Normal response returned : " + code + " " +msg );
}
else
{
// Output status code and message
System.out.println ("Abnormal response returned : " + code + " "+ msg );
}
```

Running UsingHttpURLConnection

The UsingHttpURLConnection application accepts as a single parameter the URL of the resource you want to investigate. The syntax is as follows:

```
java UsingHttpURLConnection url
```

9. JAVA MESSAGING SERVICES

The Java Message Service (JMS), which is designed by Sun Microsystems and several other companies under the Java Community Process as JSR 914, is the first enterprise messaging API that has received wide industry support. The Java Message Service (JMS) was designed to make it easy to develop business applications that asynchronously send and receive business data and events. It defines a common enterprise messaging API that is designed to be easily and efficiently supported by a wide range of enterprise messaging products. JMS supports both messaging models: Point-to-point (queuing) and publish-subscribe

JMS was defined to allow Java application to use enterprise messaging systems. More importantly, it provides a common way for Java applications to access such enterprise messaging systems. JMS falls under middleware, and specifically Message-Oriented Middleware (MOM), which is a relatively low-level of abstraction that runs underneath complementary layers such as database and application adapters, event processing, and business process automation. MOM is becoming an essential component for integrating intra-company operations as it allows separate business components to be combined into a reliable, yet flexible, system.

JMS defines a set of interfaces and semantics that allow Java applications to communicate with other messaging implementations. A JMS implementation is known as a JMS provider. JMS makes the learning curve easy by minimizing the set of concepts a Java developer must learn to use enterprise messaging products, and at the same time it maximizes the portability of messaging applications.

9.1 JMS Architecture

A JMS application is composed of the following parts:

- ***A JMS provider***: A messaging system that implements the JMS specification.
- ***JMS clients***: Java applications that send and receive messages.
- ***Messages***: Objects that are used to communicate information between JMS clients.
- ***Administered objects***: Preconfigured JMS objects that are created by an administrator for the use of JMS clients.

Message Delivery Models

JMS supports two different message delivery models:

Point-to-Point (Queue destination): In this model, a message is delivered from a producer to one consumer. The messages are delivered to the destination, which is a queue, and then delivered to one of the consumers registered for the queue. While any number of producers can send messages to the queue, each message is guaranteed to be delivered, and consumed by one consumer. If no consumers are registered to consume the messages, the queue holds them until a consumer registers to consume them.

Publish/Subscribe (Topic destination): In this model, a message is delivered from a producer to any number of consumers. Messages are delivered to the topic destination, and then to all active consumers who have subscribed to the topic. In addition, any number of producers can send messages to a topic destination, and each message can be delivered to any number of subscribers. If there are no consumers registered, the topic destination does not hold messages unless it has durable subscription for inactive consumers. A durable subscription represents a consumer registered with the topic destination that can be inactive at the time the messages are sent to the topic.

9.2 The JMS Programming Model

A JMS application consists of a set of application-defined messages and a set of clients that exchange them. JMS clients interact by sending and receiving messages using the JMS API. A message is composed of three parts: Header, properties, and a body.

- The header, which is required for every message, contains information that is used for routing and identifying messages. Some of these fields are set automatically, by the JMS provider, during producing and delivering a message, and others are set by the client on a message by message basis.
- Properties, which are optional, provide values that clients can use to filter messages. They provide additional information about the data, such as which process created it, the time it was created. Properties can be considered as an extension to the header, and consist of property name/value pairs. Using properties, clients can fine-tune their selection of messages by specifying certain values that act as selection criteria.
- The body, which is also optional, contains the actual data to be exchanged. The JMS specification defined six types or classes of messages that a JMS provider must support:

a. Message: This represents a message without a message body.

b. StreamMessage: A message whose body contains a stream of Java primitive types. It is written and read sequentially.

c. MapMessage: A message whose body contains a set of name/value pairs. The order of entries is not defined.

d. TextMessage: A message whose body contains a Java string...such as an XML message.

e. ObjectMessage: A message whose body contains a serialized Java object.

f. BytesMessage: A message whose body contains a stream of uninterpreted bytes.

i. Producing and Consuming Messages

Here are the necessary steps to develop clients to produce and consume messages. Note that there are some common steps that should not be duplicated if the client is both producing and consuming messages. Figure below depicts the high-level view of the steps.

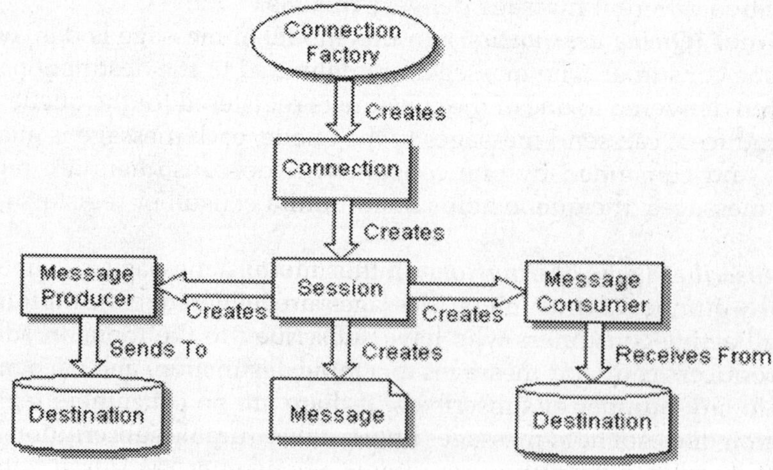

Producing and Consuming Messages

ii. Enterprise Messaging Systems

The Java Message Service was developed by Sun Microsystems to provide a means for Java programs to access enterprise messaging systems. Before we discuss JMS, let us take a look at enterprise messaging systems.

Enterprise messaging systems, often known as message oriented middleware (MOM), provide a mechanism for integrating applications in a loosely coupled, flexible manner. They provide asynchronous delivery of data between applications on a store and forward basis; that is, the applications do not communicate directly with each other, but instead communicate with the MOM, which acts as an intermediary. The MOM provides assured delivery of messages (or at least makes its best effort) and relieves application programmers from knowing the details of remote procedure calls (RPC) and networking/communications protocols. Enterprise messaging systems, often known as message oriented middleware (MOM), provide a mechanism for integrating applications in a loosely coupled, flexible manner. They provide asynchronous delivery of data between applications on a store and forward basis; that is, the applications do not communicate directly with each other, but instead communicate with the MOM, which acts as an intermediary. The MOM provides assured delivery of messages (or at least makes its best effort) and relieves application programmers from knowing the details of remote procedure calls (RPC) and networking/communications protocols.

The Java Message Service was developed by Sun Microsystems to provide a means for Java programs to access enterprise messaging systems. Before we discuss JMS, let us take a look at enterprise messaging systems.

iii. Messaging Flexibility

As shown in the figure below, Application A communicates with Application B by sending a message through the MOM's application programming interface (API).

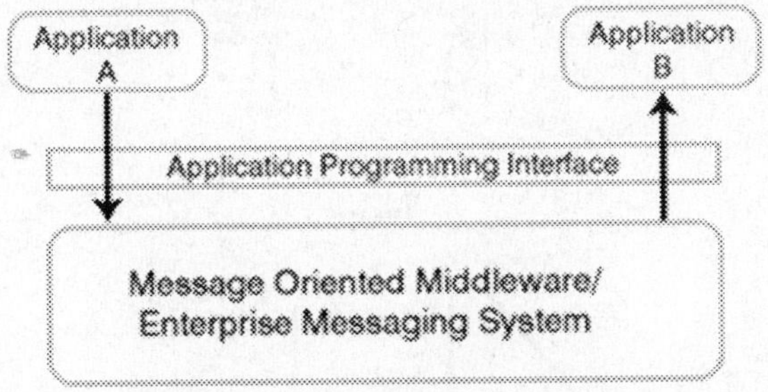

The MOM routes the message to Application B, which may exist on a completely different computer; the MOM handles the network communications. If the network connection is not available, the MOM will store the message until the connection becomes available, and then forward it to Application B. Another aspect of flexibility is that Application B may not even be executing when Application A sends its message. The MOM will hold the message until Application B begins execution and attempts to retrieve its messages. This also prevents Application A from blocking while it waits for Application B to receive the message.

This asynchronous communication requires applications to be designed somewhat differently than most are designed today, but it can be an extremely useful method for time-independent or parallel processing.

iv. Loose Coupling

The real power of enterprise messaging systems lies in the loose coupling of the applications. In the diagram on the previous panel, Application A sends its messages indicating a particular destination, for example, "order processing." Today, Application B provides order processing capabilities.

But, in the future, we can replace Application B with a different order-processing program, and Application A will be none the wiser. It will continue to send its messages to "order processing" and the messages will continue to be processed.

Likewise, we could replace Application A, and as long as the replacement continued to send messages for "order processing," the order-processing program would not need to know there is a new application sending orders.

v. Publish and Subscribe

Originally, enterprise messaging systems were developed to implement a point-to-point model (PTP) in which each message produced by an application is received by one other application. In recent years, a new model has emerged, called publish and subscribe (or pub/sub).

Pub/sub replaces the single destination in the PTP model with a content hierarchy, known as topics. Sending applications publish their messages, indicating that the message represents information about a topic in the hierarchy.

Applications wishing to receive those messages subscribe to that topic. Subscribing to a topic in the hierarchy which contains subtopics allows the subscriber to receive all messages published to the topic and its subtopics.

Multiple applications may both subscribe and publish messages to a topic, and the applications remain anonymous to each other. The MOM acts as a broker, routing the published messages for a topic to all subscribers for that topic.

This figure illustrates the publish and subscribe model.

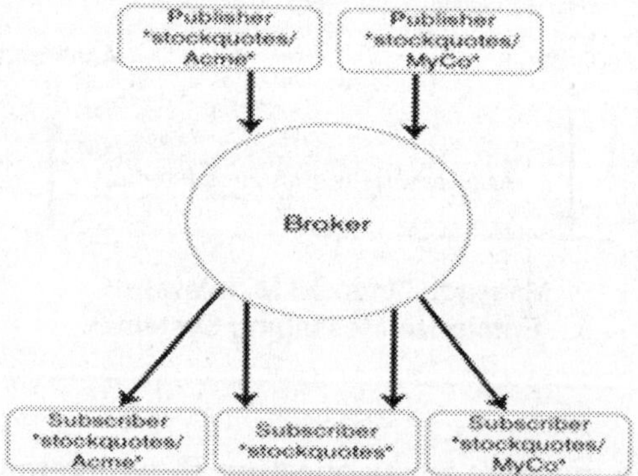

9.3 Objectives of JMS

The Java Message Service specification 1.0.2 states that:

JMS is a set of interfaces and associated semantics that define how a JMS client accesses the facilities of an enterprise messaging product. Prior to JMS, each MOM vendor provided application access to their product through a proprietary API, often available in multiple languages, including the Java language. JMS provides a standard, portable way for Java programs to send and receive messages through a MOM product. Programs written with JMS will be able to run on any MOM that implements the JMS standard.

The key to JMS portability is the fact that the JMS API is provided by Sun as a set of interfaces. Products that want to provide JMS functionality do so by supplying a provider that implements these interfaces. As a developer, you build a JMS application by defining a set of messages and a set of client applications that exchange those messages.

To better understand JMS, it helps to know the objectives set by the authors of the JMS specification.There are many enterprise messaging products on the market today, and several of the companies that produce these products were involved in the development of JMS. These existing systems vary in capability and functionality. The authors knew that JMS would be too complicated and unwieldy if it incorporated all of the features of all existing systems. Likewise, they believed that they could not limit themselves to only the features that all of the systems had in common.

The authors believed that it was important that JMS include all of the functionality required to implement "sophisticated enterprise applications."

The objectives of JMS, as stated in the specification, are to:

- Define a common set of messaging concepts and facilities.
- Minimize the concepts a programmer must learn to use enterprise messaging.
- Maximize the portability of messaging applications.
- Minimize the work needed to implement a provider.
- Provide client interfaces for both point-to-point and pub/sub domains. "Domains" is the JMS term for the messaging models discussed earlier. (Note: A provider need not implement both domains.)

What JMS does not Provide

The following features, common in MOM products, are not addressed by the JMS specification. While acknowledged by the JMS authors as important for the development of robust messaging applications, these features are considered JMS provider-specific.

JMS providers are free to implement these features in any manner they please, if at all:

- Load balancing and fault tolerance
- Error and advisory system messages and notification
- Administration
- Security
- Wire protocol
- Message type repository

JMS overview and architecture

Applications

A JMS application is comprised of the following elements:

- JMS clients. Java programs that send and receive messages using the JMS API.
- Non-JMS clients. It is important to realize that legacy programs will often be part of an overall JMS application and their inclusion must be anticipated in planning.
- Messages. The format and content of messages to be exchanged by JMS and non-JMS clients is integral to the design of a JMS application.
- JMS provider. As was stated previously, JMS defines a set of interfaces for which a provider must supply concrete implementations specific to its MOM product.
- Administered objects. An administrator of a messaging system provider creates objects that are isolated from the proprietary technologies of the provider.

Administered Objects

Providers of MOM products differ significantly in the mechanisms and techniques they use to implement messaging. To keep JMS clients portable, objects that implement the JMS interfaces have to be isolated from the proprietary technologies of a provider.

The mechanism for doing this is administered objects. These objects, which implement JMS interfaces, are created by an administrator of the provider's messaging system and are placed in the JNDI namespace.

The objects are then retrieved by JMS programs and accessed through the JMS interfaces that they implement. The JMS provider must supply a tool that allows creation of administered objects and their placement in the JNDI namespace.

There are two types of administered objects:

- ConnectionFactory: Used to create a connection to the provider's underlying messaging system.
- Destination: Used by the JMS client to specify the destination of messages being sent or the source of messages being received.

While the administered objects themselves are instances of classes specific to a provider's implementation, they are retrieved using a portable mechanism (JNDI) and accessed through portable interfaces (JMS). The JMS program only needs to know the JNDI name and the JMS interface type of the administered object; no provider-specific knowledge is required.

Interfaces

JMS defines a set of high-level interfaces that encapsulate various messaging concepts. In turn, these interfaces are further defined and customized for the two messaging domains — PTP and pub/sub.

The high-level interfaces are:

- ConnectionFactory: An administered object that creates a Connection.
- Connection: An active connection to a provider.
- Destination: An administered object that encapsulates the identity of a message destination, such as where messages are sent to or received from.
- Session: A single-threaded context for sending and receiving messages. For simplicity and because of Sessions control transactions, concurrent access by multiple threads is restricted. Multiple Sessions can be used for multithreaded applications.
- MessageProducer: Used for sending messages.
- MessageConsumer: Used for receiving messages.

The following table identifies the domain-specific interfaces inherited from each high-level interface.

High-level interface	PTP domain	Pub/sub domain
ConnectionFactory	QueueConnection Factory	TopicConnectionFactory
Connection	QueueConnection	TopicConnection
Destination	Queue	Topic
Session	QueueSession	TopicSession
MessageProducer	QueueSender	TopicPublisher
MessageConsumer	QueueReceiver, QueueBrowser	TopicSubscriber

9.4 Developing a JMS Program

A typical JMS program goes through the following steps to begin producing and consuming messages.

1. Look up a ConnectionFactory through JNDI.
2. Look up one or more Destinations through JNDI.
3. Use the ConnectionFactory to create a Connection.
4. Use the Connection to create one or more Sessions.
5. Use a Session and a Destination to create the required MessageProducers and MessageConsumers.
6. Start the Connection.

At this point, messages can begin to flow and the application can receive, process, and send messages, as required.

Messages

At the heart of a messaging system are, of course, messages. JMS provides several message types for different types of content, but all messages derive from the Message interface.

A message is divided into three constituent parts:

- The header is a standard set of fields that are used by both clients and providers to identify and route messages.
- Properties provide a facility for adding optional header fields to a message. If your application needs to categorize or classify a message in a way not provided by the standard header fields, you can add a property to the message to accomplish that categorization or classification. set<Type>Property(...) and Property(...) methods are provided to set and get properties of a variety of Java types, including Object. JMS defines a standard set of properties that are optional for providers to supply.
- The body of the message contains the content to be delivered to a receiving application. Each message interface is specialized for the type of content it supports.

Header fields

The following list gives the name of each header field of Message, its corresponding Java type, and a description of the field.

- JMSMessageID – type string

Uniquely identifies each message that is sent by a provider. This field is set by the provider during the send process; clients cannot determine the JMSMessageID for a message until after it has been sent.

- JMSDestination – type Destination

The Destination to which the message was sent; set by the provider during the send process.

- JMSDeliveryMode – type int

Contains the value DeliveryMode.PERSISTENT or DeliveryMode. NON_PERSISTENT. A persistent message is delivered "once and only once"; a non-persistent message is delivered "at most once." Be aware that "at most once" includes not being delivered at all. A non-persistent message may be lost by a provider during application or system failure. Extra care will be taken to assure that a persistent message is not affected by failures. There is often considerable overhead in sending persistent messages, and the trade-offs between reliability and performance must be carefully considered when deciding the delivery mode of a message.

• JMSTimestamp – type long

The time that the message was delivered to a provider to be sent; set by the provider during the send process.

• JMSExpiration – type long

The time when a message should expire. This value is calculated during the send process as the sum of the time-to-live value of the sending method and the current time. Expired messages should not be delivered by the provider. A value of 0 indicates that the message will not expire.

• JMSPriority – type int

The priority of the message; set by the provider during the send process. A priority of 0 is the lowest priority; a priority of 9 is the highest priority.

• JMSCorrelationID – type string

Typically used to link a response message with a request message; set by the JMS program sending the message. A JMS program responding to a message from another JMS program would copy the JMSMessageID of the message it is responding to into this field, so that the requesting program could correlate the response to the particular request that it made.

• JMSReplyTo – type Destination

Used by a requesting program to indicate where a reply message should be sent; set by the JMS program sending the message.

• JMSType – type string

Can be used by a JMS program to indicate the type of the message. Some providers maintain a repository of message types and will use this field to reference the type definition in the repository; in this case, the JMS program should not use this field.

• JMSRedelivered – type boolean

Indicates that the message was delivered earlier to the JMS program, but that the program did not acknowledge its receipt; set by the provider during receive processing.

Standard Properties

The following list gives the name of each standard property of message, its corresponding Java type, and a description of the property. Support for standard properties by a provider is optional. JMS reserves the "JMSX" property name for these and future JMS-defined properties.

• JMSXUserID – type string

Identity of the user sending the message.

• JMSXApplID – type string

Identity of the application sending the message.

• JMSXDeliveryCount – type int

Number of times delivery of the message has been attempted.

• JMSXGroupID – type string

Identity of the message group to which this message belongs.

• JMSXGroupSeq – type int

Sequence number of this message within the message group.

• JMSXProducerTXID – type string

Identity of the transaction within which this message was produced.

• JMSXConsumerTXID – type string

Identity of the transaction within which this message was consumed.

• JMSXRcvTimestamp – type long

The time JMS delivered the message to the consumer.

• JMSXState – type int

Used by providers that maintain a message warehouse of messages; generally not of interest to JMS producers or consumers.

• JMSX_<vendor_name>

Reserved for provider-specific properties.

Message Body

There are five forms of message body, and each form is defined by an interface that extends message. These interfaces are:

- StreamMessage: Contains a stream of Java primitive values that are filled and read sequentially using standard stream operations.
- MapMessage: Contains a set of name-value pairs; the names are of type string and the values are Java primitives.
- TextMessage: Contains a String.
- ObjectMessage: Contains a Serializable Java object; JDK 1.2 collection classes can be used.
- BytesMessage: Contains a stream of uninterpreted bytes; allows encoding a body to match an existing message format. Each provider supplies classes specific to its product that implement these interfaces. It is important to note that the JMS specification mandates that providers must be prepared to accept and handle a Message object that is not an instance of one of its own Message classes.

While these "alien" objects may not be handled by a provider as efficiently as one of the provider's own implementations, they must be handled to ensure interoperability of all JMS providers.

Transactions

A JMS transaction groups a set of produced messages and a set of consumed messages into an atomic unit of work. If an error occurs during a transaction, the production and consumption of messages that occurred before the error can be "undone."

Session objects control transactions and a Session may be denoted as transacted when it is created. A transacted Session always has a current transaction, that is, there is no begin(); commit() and rollback() end one transaction and automatically begin another.

Distributed transactions may be supported by the Java Transaction API (JTA) XAResource API, though this is optional for providers.

Acknowledgment

Acknowledgment is the mechanism whereby a provider is informed that a message has been successfully received.

If the Session receiving the message is transacted, acknowledgment is handled automatically. If the Session is not transacted, then the type of acknowledgment is determined when the Session is created.

There are three types of acknowledgment:

- Session.DUPS_OK_ACKNOWLEDGE: Lazy acknowledgment of message delivery; reduces overhead by minimizing work done to prevent duplicates; should only be used if duplicate messages are expected and can be handled.
- Session.AUTO_ACKNOWLEDGE: Message delivery is automatically acknowledged upon completion of the method that receives the message.
- Session.CLIENT_ACKNOWLEDGE: Message delivery is explicitly acknowledged by calling the acknowledge() method on the message.

Message selection

JMS provides a mechanism, called a message selector, for a JMS program to filter and categorize the messages it receives.

The message selector is a String that contains an expression whose syntax is based on a subset of SQL92. The message selector is evaluated when an attempt is made to receive a message, and only messages that match the selection criteria of the selector are made available to the program.

Selection is based on matches to header fields and properties; body values cannot be used for selection. The syntax for message selectors is provided in detail in the JMS specification.

JMS and XML

The authors of JMS included the TextMessage message type on the presumption that String messages will be used extensively.

Their reasoning is that XML will be a popular, if not the most popular, means of representing the content of messages. A portable transport mechanism (JMS) coupled with a portable data representation (XML) is proving to be a powerful tool in enterprise application integration (EAI) and other areas of data exchange.

JMS and J2EE

J2EE version 1.2 requires compliant application servers to have the JMS API present, but does not mandate the presence of a JMS provider.

J2EE version 1.3 will require application servers to supply a JMS provider.

Another important development in JMS capabilities is the message-driven bean of the EJB 2.0 specification, which will add asynchronous notification abilities to Enterprise JavaBeans containers. A message-driven bean, which will implement the MessageListener interface, will be invoked by the EJB container on the arrival of a message at a destination designated at deployment time. The message-driven bean will contain the business logic to process the message, including, if needed, the invoking of other enterprise beans.

9.5 JMS—Point-to-Point Interfaces

Introduction

In this section, we will look at each of the important JMS interfaces for point-to-point programming and some of their methods.

In the next section (Point-to-point programming), we will look at some sample code that performs point-to-point message processing.

QueueConnectionFactory

QueueConnectionFactory is an administered object that is retrieved from JNDI to create a connection to a provider. It contains a createQueueConnection() method which returns a QueueConnection object.

QueueConnection

QueueConnection encapsulates an active connection to a provider. Some of its methods are:

- createQueueSession(boolean, int): Returns a QueueSession object. The boolean parameter indicates whether the QueueSession is transacted or not; the int indicates the acknowledgement mode.
- start() (inherited from Connection): Activates the delivery of messages from the provider.

- stop() (inherited from Connection): Temporarily stops delivery of messages; delivery can be restarted with start().
- close() (inherited from Connection): Closes the connection to the provider and releases all resources held in its behalf.

QueueSession

QueueSession is the single-threaded context for sending and receiving PTP messages. Some of its methods are:

- createSender(Queue): Returns a QueueSender object to send messages to the specified Queue.
- createReceiver(Queue): Returns a QueueReceiver object to receive messages from the specified Queue.
- createBrowser(Queue): Returns a QueueBrowser object to browse messages on the specified Queue.
- commit() (inherited from Session): Commits all consumed or produced messages for the current transaction.
- rollback() (inherited from Session): Rolls back all consumed or produced messages for the current transaction.
- create<MessageType>Message(...) (inherited from Session): A variety of methods that return a <MessageType>Message, for example, MapMessage,TextMessage, and so on.

Queue

Queue encapsulates a point-to-point destination. It is an administered object that is retrieved from JNDI.

QueueSender

QueueSender is used to send point-to-point messages. Some of its methods are:

- send(Message): Sends the indicated Message.
- setDeliveryMode(int) (inherited from MessageProducer): Sets the delivery mode for subsequent messages sent; valid values are DeliveryMode. PERSISTENT and DeliveryMode.NON_PERSISTENT.
- setPriority(int) (inherited from MessageProducer): Sets the priority for subsequent messages sent; valid values are 0 through 9.
- setTimeToLive(long) (inherited from MessageProducer): Sets the duration before expiration, in milliseconds, of subsequent messages sent.

QueueReceiver

QueueReceiver is used to receive point-to-point messages. Some of its methods are:

- receive() (inherited from MessageConsumer): Returns the next message that arrives; this method blocks until a message is available.
- receive(long) (inherited from MessageConsumer): Receives the next message that arrives within long milliseconds; this method returns null if no message arrives within the time limit.
- receiveNoWait (inherited from MessageConsumer): Receives the next message if one is immediately available; this method returns null if no message is available.
- setMessageListener(MessageListener) (inherited from MessageConsumer): Sets the MessageListener; the MessageListener object receives messages as they arrive, that is, asynchronously (see MessageListener on page 16).

QueueBrowser

When using QueueReceiver to receive messages, the messages are removed from the queue when they are received. QueueBrowser is used to look at messages on a queue without removing them. The method for doing that is getEnumeration(), which returns a java.util.Enumeration that can be used to scan the messages in the queue; changes to the queue (arriving and expiring of messages) may or may not be visible.

MessageListener

MessageListener is an interface with a single method—onMessage(Message)—that provides asynchronous receipt and processing of messages.

This interface should be implemented by a client class and an instance of that class passed to the QueueReceiver object with the setMessageListener(MessageListener) method. As a message arrives on a queue, it is passed to the object by calling the onMessage(Message) method.

MessageListener objects are used in both the PTP and pub/sub domains.

9.6 Simple JMS Client Applications

The JMS client applications are simple and standalone programs that run outside the server as class files. For the JMS application client, let us create three different basic tasks. They are:

- Creating connection and a session
- Creating message producers and consumers
- Sending and receiving messages

Here we are using J2EE platform for the working of JMS application. The following steps describe the working of JMS application.

Setting your environment to run J2EE clients and applications

A point-to-point example that uses synchronous receives

A publish/subscribe example that uses a message listener

Running JMS client programs on multiple systems

Each step contains two programs. That is, it includes one sends messages and one receives messages.

1. Setting your environment to run J2EE clients and applications, first you must set the environment variable. For windows we are using the following environment variable.

Variable name	Values
(A) %JAVA_HOME%	Directory in which the JavaTM 2 SDK, Standard Edition, version 1.3.1, is installed
(B) %J2EE_HOME%	Directory in which the J2EE SDK 1.3.1 is installed, usually C:\j2sdkee1.3.1
(C) %CLASSPATH%	Include the following .; %J2EE_HOME%\lib\j2ee.jar;%J2EE_HOME%\lib\locale
(D) %PATH%	Include %J2EE_HOME%\bin

2. A point-to-point example that uses synchronous receives

Here we use the receive method to consume messages synchronously. We are using the following steps to perform the PTP example.

Writing the PTP Client Programs

SimpleQueueSender.java sending program performs the following steps.

Performs a Java Naming and Directory InterfaceTM (JNDI) API lookup of the QueueConnectionFactory and queue

- Creates a connection and a session
- Creates a QueueSender
- Creates a TextMessage
- Sends one or more messages to the queue
- Sends a control message to indicate the end of the message stream
- Closes the connection in a finally block, automatically closing the session and QueueSender

SimpleQueueReceiver.java is receiving program and perform the following operations.

- Performs a JNDI API lookup of the QueueConnectionFactory and queue
- Creates a connection and a session
- Creates a QueueReceiver
- Starts the connection, causing message delivery to begin
  Receives the messages sent to the queue until the end-of-message-stream control message is received
- Closes the connection in a finally block, automatically closing the session and QueueReceiver

In the receive method we never define no argument or an argument 0, the method blocks infinitely until a message arrives.

E.g.: Message m = queueReceiver.receive();

Message m = queueReceiver.receive(0);

If we want to use a time synchronous receive, we are using the following codes.

E.g.: Call the receive method with a timeout argument greater than 0:

Message m = queueReceiver.receive(1); // 1 millisecond

Call the receiveNoWait method, which receives a message only if one is available:

Message m = queueReceiver.receiveNoWait();

The SimpleQueueReceiver program uses an indefinite while loop to receive messages, calling receive with a timeout argument. Calling receiveNoWait would have the same effect.

Compiling the PTP clients

For compiling the PTP we are using the following steps.

Make sure that you have set the environment variables

At a command line prompt, compile the two source files:

javac SimpleQueueSender.java

javac SimpleQueueReceiver.java

Starting the JMS Provider

This can be done in two ways.

Use J2EE SDK 1.3.1, the JMS provider is the SDK.

Through command line

J2ee –verbose

Wait until the server displays the message "J2EE server startup complete".

Creating the JMS Administered Objects

For the compilation purpose we are using the j2eeadmin command to create a queue named MyQueue. The last argument specifies the kind of destination to create.

j2eeadmin -addJmsDestination MyQueue queue

"j2eeadmin –listJmsDestination" command helps to check the queue has been created or not.

This example uses the default QueueConnectionFactory object supplied with the J2EE SDK 1.3.1. With a different J2EE product, you might need to create a connection factory yourself.

Running the PTP Clients

Run the clients as follows.

1. Run the SimpleQueueSender program, sending three messages. You need to define a value for jms.properties.

- On a Microsoft Windows system, type the following command:
- java-Djms.properties=%J2EE_HOME%\config\jms_client.properties SimpleQueueSender MyQueue 3
- On a UNIX system, type the following command:
- java-Djms.properties=$J2EE_HOME/config/jms_client.properties SimpleQueueSender MyQueue 3

The output of the program looks like this:

Queue name is MyQueue
Sending message: This is message 1
Sending message: This is message 2
Sending message: This is message 3

2. In the same window, run the SimpleQueueReceiver program, specifying the queue name. The java commands look like this:

- Microsoft Windows systems:
- java-Djms.properties=%J2EE_HOME%\config\jms_client.properties SimpleQueueReceiver MyQueue
- UNIX systems:
- java-Djms.properties=$J2EE_HOME/config/jms_client.properties SimpleQueueReceiver MyQueue

The output of the program looks like this:

Queue name is MyQueue
Reading message: This is message 1
Reading message: This is message 2
Reading message: This is message 3

Deleting the Queue

You can delete the queue you created as follows:

j2eeadmin -removeJmsDestination MyQueue

3. Now try running the programs in the opposite order. Start the SimpleQueueReceiver program. It displays the queue name and then appears to hang, waiting for messages.
4. In a different terminal window, run the SimpleQueueSender program. When the messages have been sent, the SimpleQueueReceiver program receives them and exits.

A publish/subscribe example that uses a message listener

Here we are discussing how to publishing and subscribing programs in a pub/sub that helps to use a message listner to consume message asynchronously. The following section describes how to compile and run the programs using J2EE SDK 1.3.1.

1. Writing the pub/sub client programs

The publishing program, SimpleTopicPublisher.java, performs the following steps: Performs a JNDI API lookup of the TopicConnectionFactory and topic

Creates a connection and a session
Creates a TopicPublisher
Creates a TextMessage
Publishes one or more messages to the topic
Closes the connection, which automatically closes the session and TopicPublisher
The receiving program, SimpleTopicSubscriber.java, performs the following steps:
Performs a JNDI API lookup of the TopicConnectionFactory and topic
Creates a connection and a session
Creates a TopicSubscriber
Creates an instance of the TextListener class and registers it as the message listener for the TopicSubscriber
Starts the connection, causing message delivery to begin
Listens for the messages published to the topic, stopping when the user enters the character q or Q
Closes the connection, which automatically closes the session and TopicSubscriber
The message listener, TextListener.java, follows these steps:
When a message arrives, the onMessage method is called automatically.
The onMessage method converts the incoming message to a TextMessage and displays its content.

2. Compiling the pub/sub clients

To compile the pub / sub example, do the following.

1. Make sure that you have set the environment variables shown in Table 4.1.
2. Compile the programs and the message listener class:
3. javac SimpleTopicPublisher.java
4. javac SimpleTopicSubscriber.java
5. javac TextListener.java

3. Starting the JMS provider

If you did not do so before, start the J2EE server in another terminal window:j2ee -verbose.

Wait until the server displays the message "J2EE server startup complete."

4. Creating the JMS administered objects

For creating the default TopicConnectionFactory object, we compiled the clients, use the j2eeadmin command to create a topic named MyTopic.

E.g.: j2eeadmin -addJmsDestination MyTopic topic

To verify that the queue has been created, use the following command:

j2eeadmin -listJmsDestination

5. Running the pub/sub clients

Run the clients as follows.

i. Run the SimpleTopicSubscriber program, specifying the topic name. You need to define a value for jms.properties.

o On a Microsoft Windows system, type the following command:

o java -Djms.properties=%J2EE_HOME%\config\jms_client.properties SimpleTopicSubscriber MyTopic

o On a UNIX system, type the following command:

o java-Djms.properties=$J2EE_HOME/config/jms_client.properties SimpleTopicSubscriber MyTopic

The program displays the following lines and appears to hang:

Topic name is MyTopic

To end program, enter Q or q, then <return>

ii. In another terminal window, run the SimpleTopicPublisher program, publishing three messages. The java commands look like this:

o Microsoft Windows systems:

o java-Djms.properties=%J2EE_HOME%\config\jms_client.properties SimpleTopicPublisher MyTopic 3

o UNIX systems:

o java-Djms.properties=$J2EE_HOME/config/jms_client.properties SimpleTopicPublisher MyTopic 3

The output of the program looks like this:
Topic name is MyTopic
Publishing message: This is message 1
Publishing message: This is message 2
Publishing message: This is message 3
In the other window, the program displays the following:
Reading message: This is message 1
Reading message: This is message 2
Reading message: This is message 3
Enter Q or q to stop the program.

6. Deleting the topic and stopping the server

You can delete the topic you created as follows:
j2eeadmin -removeJmsDestination MyTopic
If you wish, you can stop the J2EE server as well:
j2ee -stop
Running JMS client programs on multiple systems

JMS client programs can communicate with each other when they are running on different systems in a network.

E.g.: We want to run the SimpleProducer program on one system, earth, and the SimpleSynchConsumer program on another system, jupiter.

Create two new connection factories
Edit the source code
Recompile the source code and update the client JAR files
Before you begin, start the server on both systems:
Start the Application Server on earth and log into the Admin Console.
Start the Application Server on jupiter and log into the Admin Console.
Creating Administered Objects for Multiple Systems
To run these programs, you must do the following:
Create a new connection factory on both earth and jupiter
Create a destination resource and physical destination on both earth and jupiter
Create a new connection factory on jupiter as follows:

In the Admin Console, expand the Resources node, then expand the JMS Resources node.

Select the Connection Factories node.

On the JMS Connection Factories page, click New. The Create JMS Connection Factory page appears.

In the JNDI Name field, type jms/JupiterConnectionFactory.
Choose javax.jms.ConnectionFactory from the Type combo box.
Select the Enabled checkbox.

Click OK.

Create a new connection factory with the same name on earth as follows:

In the Admin Console, expand the Resources node, then expand the JMS Resources node.

Select the Connection Factories node.

On the JMS Connection Factories page, click New. The Create JMS Connection Factory page appears.

In the JNDI Name field, type jms/JupiterConnectionFactory.

Choose javax.jms.ConnectionFactory from the Type combo box.

Select the Enabled checkbox.

In the Additional Properties area, find the AddressList property. In the Value field, replace the name of your current system with the name of the remote system (whatever the real name of jupiter is), as follows: mq://sysname:7676/. If the JMS service on the remote system uses a port number other than the default (7676), change the port number also.

Click OK.

If you have already been working on either earth or jupiter, you have the queue on one system. On the system that does not have the queue, perform the following steps:

1. Use the Admin Console to create a physical destination named PhysicalQueue, just as you did in Creating JMS Administered Objects.
2. Use the Admin Console to create a destination resource named jms/Queue and set its Name property to the value PhysicalQueue.

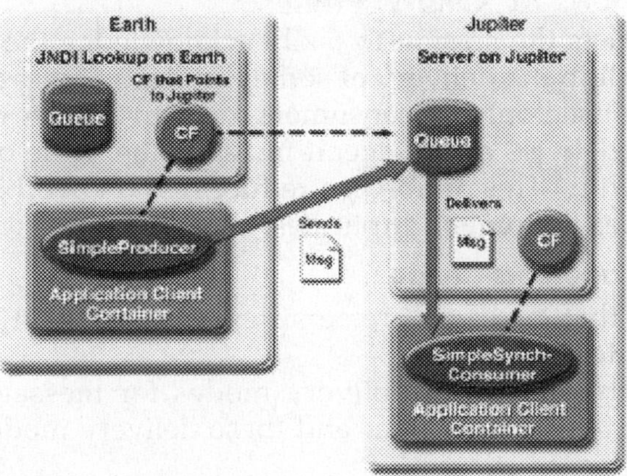

9.7 Developing Robust JMS Applications

The JMS API allows two different ways to achieve various kinds and degrees of reliability. They are:

Using Basic Reliability Mechanism.

Using Advanced Reliability Mechanism.

The below mechanisms are used for extracting the reliable message delivery.

(A) Controlling message acknowledgment

It helps to define different levels of control over message acknowledgment. The successful consumption of JMS message happened in three different stages.

The client receives the message.

The client processes the message.

The message is acknowledged. Acknowledgment is initiated either by the JMS provider or by the client, depending on the session acknowledgment mode.

We are using two different methods for handling acknowledgment.

1. Using transacted session

In this method acknowledgment happens automatically when a transaction is committed and if a transaction id rolled back, all consumed messages are delivered.

2. Using non-transacted session

In this method when and how a message is acknowledged depend on the value specified as the second argument of the createSession method. Here we are using three different types of argument values.

1. Session.AUTO_ACKNOWLEDGE:

The session automatically acknowledges a client's receipt of a message either when the client has successfully returned from a call to receive or when the MessageListener it has called to process the message returns successfully. A synchronous receives in an AUTO_ACKNOWLEDGE session is the one exception to the rule that message consumption is a three-stage process as described earlier.

2. Session.CLIENT_ACKNOWLEDGE:

A client acknowledges a message by calling the message's acknowledge method. In this mode, acknowledgment takes place on the session level: Acknowledging a consumed message automatically acknowledges the receipt of all messages that have been consumed by its session. For example, if a message consumer consumes ten messages and then acknowledges the fifth message delivered, all ten messages are acknowledged.

3. Session.DUPS_OK_ACKNOWLEDGE:

This option instructs the session to lazily acknowledge the delivery of messages. This is likely to result in the delivery of some duplicate messages if the JMS provider fails, so it should be used only by consumers that can tolerate duplicate messages. (If the JMS provider redelivers a message, it must set the value of the JMS Redelivered message header to true.) This option can reduce session overhead by minimizing the work the session does to prevent duplicates.

(B) Specifying message persistence

It helps to define that messages are persistent meaning that they should not lost in the event of a provider failure.

JMS API allows two types of delivery modes for messages to define whether messages are lost if the JMS provider and these delivery modes are the fields of the DeliveryMode interface.

1. PERSISTENT delivery mode

By default it instructs the JMS provider to take extra care to ensure that a message is not in transit in case of a JMS provider failure. A message sent with this delivery mode is logged to stable storage when it is sent.

2. NON_PERSISTENT delivery mode

It does not require the JMS provider to store the message or otherwise guarantee that it is not lost if the provider fails.

The delivery modes define in the two ways.

1. Use setDeliveryMode

Here we are using the MessageProducer interface for implementing this method and it used to set the delivery mode for all messages sent by that producer.

E.g.: producer.setDeliveryMode(DeliveryMode.NON_PERSISTENT);

2. Use the long form of the send or the publish method

It helps to set the delivery mode for a specific message. The second argument set the delivery mode and the third and fourth arguments set the priority level and expiration time.

E.g.: producer.send(message, DeliveryMode.NON_PERSISTENT, 3, 10000);

(C) Setting message priority levels

We can set the priority levels for messages which can affect the order in which the messages are obsolete. It helps to instruct the JMS provider to deliver urgent message first. Here we are accomplishing this operation in two ways.

1. Using set priority method

The messageProducer interface used to set the setPriority method and it helps to set the priority level for all messages sent by that producer.

E.g.: The following call sets a priority level of 7 for a producer.

producer.setPriority(7);

2. Use the long form of the send or the publish method

We can set the priority level for a specific message. The third argument set the priority level. The priority level ranges from 0 to 9 and default priority level is 4.

E.g.: producer.send (message, DeliveryMode.NON_PERSISTENT, 3, 10000);

(D) Allowing messages to expire

It used to specify an expiration time for messages so that they will not be delivered if they are obsolete. By default, a message never expires. If a message will become obsolete after a certain period, however, you may want to set an expiration time. This can be done in two ways.

1. Using setTimeToLive method

The MessageProducer interface uses this method to set a default expiration time for all messages sent by that producer.

E.g.: producer.setTimeToLive(60000);

2. Using the long form of the send or the publish method:

In this method we set an expiration time for a specific message. The fourth argument helps to set the expiration time in milliseconds.

E.g.: producer.send (message, DeliveryMode.NON_PERSISTENT, 3, 10000);

(E) Creating temporary destinations

It helps to make temporary destinations that last only for the duration of the connection if they are created. For creating JMS destinations (queues and topics), we are using three methods:

1. Using administratively rather than programmatically.

2. Using JMS API: The lifetime of this object are only last for the duration of the connection in which they are created.

3. Create dynamically by using the Session.createTemporaryQueue and the Session.createTemporaryTopic methods.

Using Advanced Reliability Mechanism

The below mechanisms are user to extracting the reliable message delivery.

1. Creating durable subscriptions: This type of subscriptions helps to receive messages published while the subscriber is not active. Durable subscriptions offer the reliability of queues to the publish/subscribe message domain. For creating the durable subscriptions we use the following ways.

(A) Ensure that a pub/sub application receives all published messages, use PERSISTENT delivery mode for the publishers. In addition, use durable subscriptions for the subscribers.The Session.createConsumer method create a nondurable subscriber if a topic is specified as the destination. A nondurable subscriber can receive only messages that are published while it is active.

(B) At the cost of higher overhead, we use the Session.createDurableSubscriber method to create a durable subscriber. A durable subscription can have only one active subscriber at a time.

(C) A durable subscriber registers a durable subscription by defining a unique identity that is retained by the JMS provider. For stabling the durable subscriber by setting the following things.

(a) A client ID for the connection

Set the client ID administratively for a client-specific connection factory using the Admin Console. After using this connection factory to create the connection and the session, you call the createDurableSubscriber method with two arguments: The topic and a string that specifies the name of the subscription:

E.g.: String subName = "MySub";
MessageConsumer topicSubscriber =
session.createDurableSubscriber(myTopic, subName);

For closing the subscriber we are using the following syntax.

Syntax: topicSubscriber.close();

(b) A topic and a subscription name for the subscriber

(c) The JMS provider saves the messages sent or published to the topic, like it would store messages sent to a queue. If the program or another application calls createDurableSubscriber using the same connection factory and its client ID, the same topic, and the same subscription name, the subscription is reactivated, and the JMS provider delivers the messages that were published while the subscriber was inactive. To delete a durable subscription, first close the subscriber, and then use the unsubscribe method, with the subscription name as the argument.

topicSubscriber.close();
session.unsubscribe("MySub");

The unsubscribe method deletes the state that the provider maintains for the subscriber.

The figure below shows the differences between durable and non-durable subscriptions.

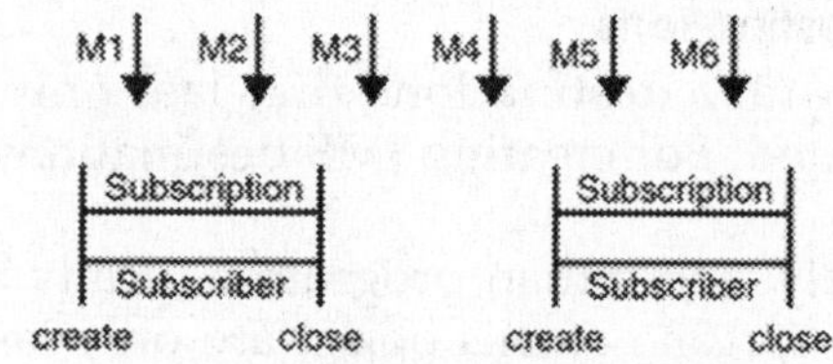

Non-durable subscription

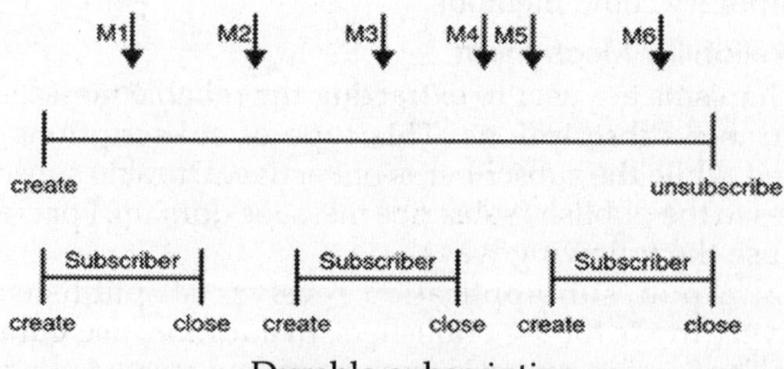

Durable subscription

9.8 JMS APIs in J2EE World

We are using the following steps for developing a J2EE application to use the JMS API directly for asynchronous messaging. We will list all the common APIs that are predominately used in JMS.

1. Import JMS packages: A J2EE application that uses JMS starts with several import statement.

E.g.: import javax.jms.*;   //JMS interfaces
import javax.naming.*;   //Used for JNDI lookup of administered objects.

2. Get an initial context

E.g.: try {
ctx = new InitialContext(env);
}

3. Retrieve administered objects: Retrieve this object from the JNDI namespace. The InitialContext.lookup() method helps to retrieve this object.

E.g.: qcf = (QueueConnectionFactory)ctx.lookup(qcfName);
...
inQueue = (Queue)ctx.lookup(qnameIn);

Or we can use another method for obtaining administratively-defined JMS destination objects by JNDI lookup is to use the Session.createQueue(String) method or Session.createTopic(String) method.

E.g.: Queue q = mySession.createQueue("Q1");

Create a JMS Queue instance that can be used to reference the existing destination Q1.

4. Create a connection to the messaging service provider: The createQueueConnection() method on the factory object helps to make the connection.

E.g.: connection = qcf.createQueueConnection();

The JMS specification defines that connections should be created in the stopped state. Until the connection starts, MessageConsumers that are associated with the connection cannot receive any messages. To start the connection, we use the following command:

connection.start();

5. Create a session for sending or receiving messages: The createQueueSession method is used on the connection to obtain a session and this method takes two parameters.

A boolean that determines whether or not the session is transacted.

A parameter that determines the acknowledge mode.

E.g.: boolean transacted = false;
session = connection.createQueueSession(transacted,
Session.AUTO_ACKNOWLEDGE);

In the above example, the session is not transacted and it should automatically acknowledge received messages. For defining the EJB specification we use the following tags.

i. The transacted flag passed on createQueueSession is ignored inside a global transaction and all work is performed as part of the transaction. Outside of a transaction the transacted flag is used and, if set to true, the application should use session.commit() and session.rollback() to control the completion of the work. In an EJB2.0 module, if the transacted flag is set to true and outside of an XA transaction, then the session is involved in the WebSphere local transaction and the unresolved action attribute of the method applies to the JMS work if it is not committed or rolled back by the application.

ii. Clients cannot use Message.acknowledge() to acknowledge messages. If a value of CLIENT_ACKNOWLEDGE is passed on the createxxxSession call, then messages are automatically acknowledged by the application server and Message.acknowledge() is not used.

6. Send message.

i. Create MessageProducers to create messages. In a point-to-point messaging the MessageProducer is a QueueSender and it is created by passing an output queue object into the createSender method on the session. Also the QueueSender is created for a specific queue.

E.g.: QueueSender queueSender = session.createSender(inQueue);

ii. Use the session to create an empty message and add the data passed:

Methods are provided on the Session object for message creation to avoid referencing the vendor-specific class names for the message types.

E.g.: TextMessage outMessage = session.createTextMessage(outString);

iii. Send the message: To send the message, the message is passed to the send method on the QueueSender.

E.g.: queueSender.send(outMessage);

Receive replies.

Create a correlation ID to link the message sent with any replies.

E.g.: The client receives reply messages that are related to the message that it has sent, by using a provider-specific message ID in a JMSCorrelationID.

messageID = outMessage.getJMSMessageID();

The correlation ID is then used in a message selector, to select only messages that have that ID:

E.g.: String selector = "JMSCorrelationID = '"+messageID+"'";

1. Create a MessageReceiver to receive messages. In a point-to-point the MessageReceiver is a QueueReceiver that is created by passing an input queue object (retrieved earlier) and the message selector into the createReceiver method on the session.

E.g.: QueueReceiver queueReceiver = session.createReceiver(outQueue, selector);

2. Retrieve the reply message: To retrieve a reply message, the receive method on the QueueReceiver is used:

E.g.: Message inMessage = queueReceiver.receive(2000);

The parameter in the receive call is a timeout in milliseconds and this parameter defines how long the method should wait if there is no message available immediately. We should not delay, use the receiveNoWait() method.

3. Act on the message received. When a message is received, you can act on it as needed by the business logic of the client. Some general JMS actions are to check that the message is of the correct type and extract the content of the message. To extract the content from the body of the message, it is necessary to cast from the generic Message class (which is the declared return type of the receive methods) to the more specific subclass, such as TextMessage. It is good practice always to test the message class before casting, so that unexpected errors can be handled gracefully the instanceof operator is used to check that the message received is of the TextMessage type. The message content is then extracted by casting to the TextMessage subclass.

E.g.: if (inMessage instanceof TextMessage)

String replyString = ((TextMessage) inMessage).getText();

If the application needs to create many short-lived JMS objects at the Session level or lower, it is important to close all the JMS resources used. For accomplishing this purpose, we call the close() method on the various classes (QueueConnection, QueueSession, QueueSender, and QueueReceiver) when the resources are no longer required.

E.g.: queueReceiver.close();

```
...
queueSender.close();
...
session.close();
session = null;
...
connection.close();
connection = null;
```

4. Publishing and subscribing to messages: Use JMS Publish/Subscribe support instead of point-to-point messaging.

E.g.: To create a session and connection. The exceptions are that topic resources are used instead of queue resources (such as TopicPublisher instead of QueueSender), as used to publish a message.

E.g.: // Creating a TopicPublisher

```
TopicPublisher pub = session.createPublisher(topic);
...
pub.publish(outMessage);
...
// Closing TopicPublisher
pub.close();
```

9.9 Error Handling Mechanism

JMS runtime errors are reported by exceptions. The majority of methods in JMS throw JMSExceptions to indicate errors. It is good programming practice to catch these exceptions and display them on a suitable output.

Unlike normal Java exceptions, a JMSException can contain another exception embedded in it. The implementation of JMSException does not include the embedded exception in the output of its toString() method. So we need to check explicitly for an embedded exception and print it out.

E.g.: catch (JMSException je)

```
{
System.out.println("JMS failed with "+je);
Exception le = je.getLinkedException();
if (le != null)
{
System.out.println("linked exception "+le);
}
}
```

What Is Messaging?

Messaging is a method of communication between software components or applications. A messaging system is a peer-to-peer facility: A messaging client can send messages to, and receive messages from, any other client. Each client connects to a messaging agent that provides facilities for creating, sending, receiving, and reading messages.

Messaging enables distributed communication that is loosely coupled. A component sends a message to a destination, and the recipient can retrieve the message from the destination. However, the sender and the receiver do not have to be available at the same time in order to communicate. In fact, the sender does not need to know anything about the receiver; nor does the receiver need to know anything about the sender. The sender and the receiver need to know only what message format and what destination to use. In this respect, messaging differs from tightly coupled technologies, such as Remote Method Invocation (RMI), which require an application to know a remote application's methods.

Messaging also differs from electronic mail (e-mail), which is a method of communication between people or between software applications and people. Messaging is used for communication between software applications or software components. Messaging is a mechanism by which data can be passed from one application to another application.

Role of the JMS Provider

The JMS provider handles security of the messages, data conversion and the client triggering. The JMS provider specifies the level of encryption and the security level of the message, the best data type for the non-JMS client.

JMS is the ideal high-performance messaging platform for intrabusiness messaging, with full programmatic control over quality of service and delivery options.

JavaMail provides lowest common denominator, slow, but human-readable messaging using infrastructure already available on virtually every computing platform.

JMS specification defines a transaction mechanisms allowing clients to send and receive groups of logically bounded messages as a single unit of information. A Session may be marked as transacted. It means that all messages sent in a session are considered as parts of a transaction. A set of messages can be committed (commit() method) or rolled back (rollback() method). If a provider supports distributed transactions, it is recommended to use XAResource API.

Synchronous messaging involves a client that waits for the server to respond to a message. So if one end is down, the entire communication will fail.

How does a typical client perform the communication?

1. Use JNDI to locate administrative objects.
2. Locate a single ConnectionFactory object.
3. Locate one or more Destination objects.
4. Use the ConnectionFactory to create a JMS Connection.
5. Use the Connection to create one or more Session(s).
6. Use a Session and the Destinations to create the MessageProducers and MessageConsumers needed.
7. Perform your communication.

What is point-to-point messaging?

With point-to-point message passing the sending application/client establishes a named message queue in the JMS broker/server and sends messages to this queue. The receiving client registers with the broker to receive messages posted to this queue. There is a one-to-one relationship between the sending and receiving clients.

Can two different JMS services talk to each other?

For instance, if A and B are two different JMS providers, can Provider A send messages directly to Provider B? If not, then can a subscriber to Provider A act as a publisher to Provider B?

The answers are no to the first question and yes to the second. The JMS specification does not require that one JMS provider be able to send messages directly to another provider. However, the specification does require that a JMS client must be able to accept a message created by a different JMS provider, so a message received by a subscriber to Provider A can then be published to Provider B. One caveat is that the publisher to Provider B is not required to handle a JMSReplyTo header that refers to a destination that is specific to Provider A.

What is the advantage of persistent message delivery compared to nonpersistent delivery?

If the JMS server experiences a failure, for example, a power outage, any message that it is holding in primary storage potentially could be lost. With persistent storage, the JMS server logs every message to secondary storage. (The logging occurs on the front end, that is, as part of handling the send operation from the message producing client.) The logged message is removed from secondary storage only after it has been successfully delivered to all consuming clients.

Give an example of using the publish / subscribe model.

JMS can be used to broadcast shutdown messages to clients connected to the Weblogic server on a module wise basis. If an application has six modules, each module behaves like a subscriber to a named topic on the server.

Why does not the JMS API provide end-to-end synchronous message delivery and notification of delivery?

Some messaging systems provide synchronous delivery to destinations as a mechanism for implementing reliable applications. Some systems provide clients with various forms of delivery notification so that the clients can detect dropped or ignored messages. This is not the model defined by the JMS API.

JMS API messaging provides guaranteed delivery via the once-and-only-once delivery semantics of PERSISTENT messages. In addition, message consumers can ensure reliable processing of messages by using either CLIENT_ACKNOWLEDGE mode or transacted sessions. This achieves reliable delivery with minimum synchronization and is the enterprise messaging model most vendors and developers prefer.

The JMS API does not define a schema of systems messages (such as delivery notifications). If an application requires acknowledgment of message receipt, it can define an application-level acknowledgment message.

What are the various message types supported by JMS?

Stream Messages?

Group of Java Primitives Map Messages? Name Value Pairs. Name being a string & Value being a java primitive Text Messages? String messages (since being widely used a separate messaging Type has been supported)

Object Messages?

Group of serialize able java object Bytes Message? Stream of uninterrupted bytes

What is the role of JMS in enterprise solution development?

JMS is typically used in the following scenarios:

1. Enterprise Application Integration: Where a legacy application is integrated with a new application via messaging.
2. B2B or Business to Business: Businesses can interact with each other via messaging because JMS allows organizations to cooperate without tightly coupling their business systems.
3. Geographically dispersed units: JMS can ensure safe exchange of data amongst the geographically dispersed units of an organization.

4. One-to-many applications: The applications that need to push data in packet to huge number of clients in a one-to-many fashion are good candidates for the use of JMS. Typical such applications are Auction Sites, Stock Quote Services, etc.

What is the use of Message object?

Message is a light weight message having only header and properties and no payload. Thus if the receivers are to be notified an event, and no data needs to be exchanged, then using Message can be very efficient.

What is the basic difference between Publish Subscribe model and P2P model?

Publish Subscribe model is typically used in one-to-many situation. It is unreliable but very fast. P2P model is used in one-to-one situation. It is highly reliable.

What is the use of BytesMessage?

BytesMessage contains an array of primitive bytes in its payload. Thus it can be used for transfer of data between two applications in their native format which may not be compatible with other Message types. It is also useful where JMS is used purely as a transport between two systems and the message payload is opaque to the JMS client. Whenever you store any primitive type, it is converted into its byte representation and then stored in the payload. There is no boundary line between the different data types stored. Thus you can even read a long as short. This would result in erroneous data and hence it is advisable that the payload be read in the same order and using the same type in which it was created by the sender.

What is the use of StreamMessage?

StreamMessage carries a stream of Java primitive types as its payload. It contains some convenient methods for reading the data stored in the payload. However, StreamMessage prevents reading a long value as short, something that is allowed in case of BytesMessage. This is so because the StreamMessage also writes the type information along with the value of the primitive type and enforces a set of strict conversion rules which actually prevents reading of one primitive type as another.

What is the use of TextMessage?

TextMessage contains instance of java.lang.String as its payload. Thus it is very useful for exchanging textual data. It can also be used for exchanging complex character data such as an XML document.

What is the use of ObjectMessage?

ObjectMessage contains a Serializable java object as its payload. Thus it allows exchange of Java objects between applications. This in itself mandates that both the applications be Java applications. The consumer of the message must typecast the object received to its appropriate type. Thus the consumer should beforehand know the actual type of the object sent by the sender. Wrong type casting would result in ClassCastException. Moreover, the class definition of the object set in the payload should be available on both the machine, the sender as well as the consumer. If the class definition is not available in the consumer machine, an attempt to type cast would result in ClassNotFoundException. Some of the MOMs might support dynamic loading of the desired class over the network, but the JMS specification does not mandate this behavior and would be a value added service if provided by your vendor. And relying on any such vendor specific functionality would hamper the portability of your application. Most of the time the class need to be put in the classpath of both, the sender and the consumer, manually by the developer.

What is the use of MapMessage?

A MapMessage carries name-value pair as its payload. Thus its payload is similar to the java.util.Properties object of Java. The values can be Java primitives or their wrappers.

What is the difference between BytesMessage and StreamMessage?

BytesMessage stores the primitive data types by converting them to their byte representation. Thus the message is one contiguous stream of bytes. While the StreamMessage maintains a boundary between the different data types stored because it also stores the type information along with the value of the primitive being stored. BytesMessage allows data to be read using any type. Thus even if your payload contains a long value, you can invoke a method to read a short and it will return you something. It will not give you a semantically correct data but the call will succeed in reading the first two bytes of data. This is strictly prohibited in the StreamMessage. It maintains the type information of the data being stored and enforces strict conversion rules on the data being read.

How is a java object message delivered to a non-java Client?

It is according to the specification that the message sent should be received in the same format. A non-java client cannot receive a message in the form of java object. The provider in between handles the conversion of the data type and the message is transferred to the other end.

What is MDB and what is the special feature of that?

MDB is Message-driven bean, which very much resembles the Stateless session bean. The incoming and outgoing messages can be handled by the Message-driven bean. The ability to communicate asynchronously is the special feature about the Message-driven bean.

What are the types of messaging?

There are two kinds of messaging.

Synchronous messaging: Synchronous messaging involves a client that waits for the server to respond to a message.

Asynchronous messaging: Asynchronous messaging involves a client that does not wait for a message from the server. An event is used to trigger a message from a server.

What are the core JMS-related objects required for each JMS-enabled application?

Each JMS-enabled client must establish the following: A connection object provided by the JMS server (the message broker). Within a connection, one or more sessions, which provide a context for message sending and receiving. Within a session, either a queue or topic object representing the destination (the message staging area) within the message broker. Within a session, the appropriate sender or publisher or receiver or subscriber object (depending on whether the client is a message producer or consumer and uses a point-to-point or publish/subscribe strategy, respectively). Within a session, a message object (to send or to receive).

What is JMS?

JMS is an acronym used for Java Messaging Service. It is Java's answer to creating software using asynchronous messaging. It is one of the official specifications of the J2EE technologies and is a key technology.

How JMS is different from RPC?

In RPC the method invoker waits for the method to finish execution and return the control back to the invoker. Thus it is completely synchronous in nature. While in JMS, the message sender just sends the message to the destination and continues its own processing. The sender does not wait for the receiver to respond. This is asynchronous behavior.

What are the advantages of JMS?

JMS is asynchronous in nature. Thus not all the pieces need to be up all the time for the application to function as a whole. Even if the receiver is down, the MOM will store the messages on its behalf and will send them once it comes back up. Thus at least a part of application can still function as there is no blocking. Are you aware of any major JMS products available in the market? IBM's MQ Series is one of the most popular product used as Message Oriented Middleware. Some of the other products are SonicMQ, iBus, etc. All the J2EE compliant application servers come built with their own implementation of JMS.

What are the different types of messages available in the JMS API?

Message, TextMessage, BytesMessage, StreamMessage, ObjectMessage, MapMessage are the different messages available in the JMS API.

What are the different messaging paradigms JMS supports?

Publish and Subscribe, i.e. pub/sub and Point to Point, i.e. p2p.

What is the difference between ic and queue?

A ic is typically used for one to many messaging, i.e. it supports publish subscribe model of messaging. While queue is used for one-to-one messaging, i.e. it supports Point to Point Messaging.

8 Multi-Tier Application Development

1. SERVER SIDE PROGRAMMING

J2EE (Java 2 -Enterprise Edition) is a basket of **12** inter-related technologies, which can be grouped as follows for convenience:

Group-1 Web Server & Support Technologies

1. JDBC (Java Database Connectivity)
2. Servlets
3. JSP (Java Server Pages)
4. Java Mail

Group-2 Distributed-Objects Technologies

1. RMI (Remote Method Invocation)
2. Corba-IDL (Corba-using Java with OMG-IDL)
3. RMI-IIOP (Corba in Java without OMG-IDL)
4. EJB (Enterprise Java Beans)

Group-3 Supporting & Advanced Enterprise Technologies

1. JNDI (Java Naming & Directory Interfaces)
2. JMS (Java Messaging Service)
3. JAVA-XML
4. Connectors (for ERP and Legacy systems).

2. JDBC

JDBC stands for Java Database Connectivity, which is a standard Java API for database-independent connectivity between the Java programming language and a wide range of databases.

The JDBC library includes APIs for each of the tasks commonly associated with database usage:

- Making a connection to a database
- Creating SQL or MySQL statements
- Executing that SQL or MySQL queries in the database
- Viewing & Modifying the resulting records

Fundamentally, JDBC is a specification that provides a complete set of interfaces that allows for portable access to an underlying database. Java can be used to write different types of executables, such as:

- Java Applications
- Java Applets
- Java Servlets
- Java ServerPages (JSPs)
- Enterprise JavaBeans (EJBs)

All of these different executables are able to use a JDBC driver to access a database and take advantage of the stored data.

JDBC provides the same capabilities as ODBC, allowing Java programs to contain database-independent code.

2.1 JDBC Architecture

The JDBC API supports both two-tier and three-tier processing models for database access but in general JDBC Architecture consists of two layers:

1. **JDBC API:** This provides the application-to-JDBC Manager connection.
2. **JDBC Driver API:** This supports the JDBC Manager-to-Driver Connection.

The JDBC API uses a driver manager and database-specific drivers to provide transparent connectivity to heterogeneous databases.

The JDBC driver manager ensures that the correct driver is used to access each data source. The driver manager is capable of supporting multiple concurrent drivers connected to multiple heterogeneous databases.

Following is the architectural diagram, which shows the location of the driver manager with respect to the JDBC drivers and the Java application:

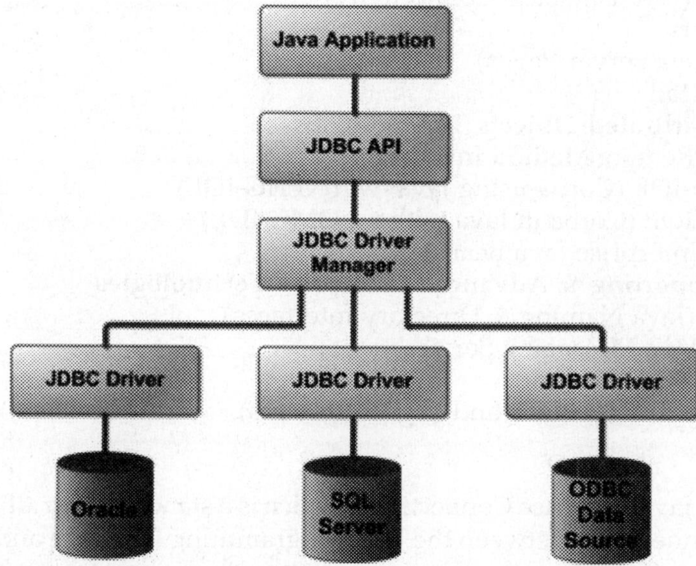

Common JDBC Components

The JDBC API provides the following interfaces and classes:

- **DriverManager:** This interface manages a list of database drivers. Matches connection requests from the java application with the proper database driver using communication subprotocol. The first driver that recognizes a certain subprotocol under JDBC will be used to establish a database Connection.
- **Driver:** This interface handles the communications with the database server. You will interact directly with Driver objects very rarely. Instead, you use DriverManager objects, which manages objects of this type. It also abstracts the details associated with working with Driver objects.
- **Connection:** Interface with all methods for contacting a database. The connection object represents communication context, i.e. all communications with database are through connection object only.

- **Statement:** You use objects created from this interface to submit the SQL statements to the database. Some derived interfaces accept parameters in addition to executing stored procedures.
- **ResultSet:** These objects hold data retrieved from a database after you execute an SQL query using Statement objects. It acts as an iterator to allow you to move through its data.
- **SQLException:** This class handles any errors that occur in a database application.

The JDBC 4.0 Packages

The java.sql and javax.sql are the primary packages for JDBC 4.0. This is the latest JDBC version at the time of writing this tutorial. It offers the main classes for interacting with your data sources.

The new features in these packages include changes in the following areas:

- Automatic database driver loading
- Exception handling improvements
- Enhanced BLOB/CLOB functionality
- Connection and statement interface enhancements
- National character set support
- SQL ROWID access
- SQL 2003 XML data type support
- Annotations

2.2 SQL-Primer

Structured Query Language (SQL) is a standardized language that allows you to perform operations on a database, such as creating entries, reading content, updating content, and deleting entries.

SQL is supported by almost all database you will likely use, and it allows you to write database code independently of the underlying database.

This tutorial gives an overview of SQL, which is a pre-requisite to understand JDBC concepts. This tutorial gives you enough SQL to be able to Create, Read, Update, and Delete (often referred to as **CRUD** operations) data from a database.

Create Database

The CREATE DATABASE statement is used for creating a new database. The syntax is:

```
SQL> CREATE DATABASE DATABASE_NAME;
```

The following SQL statement creates a Database named EMP:

```
SQL> CREATE DATABASE EMP;
```

Drop Database

The DROP DATABASE statement is used for deleting an existing database. The syntax is:

```
SQL> DROP DATABASE DATABASE_NAME;
```

Note: To create or drop a database you should have administrator privilege on your database server. Be careful, deleting a database would loss all the data stored in database.

Create Table

The CREATE TABLE statement is used for creating a new table. The syntax is:

```
SQL> CREATE TABLE table_name
(
   column_name column_data_type,
   column_name column_data_type,
   column_name column_data_type
   ...
);
```

The following SQL statement creates a table named Students with four columns:

```
SQL> CREATE TABLE Students
(
   id INT NOT NULL,
   age INT NOT NULL,
   first VARCHAR(255),
   last VARCHAR(255),
   PRIMARY KEY ( id )
);
```

Drop Table

The DROP TABLE statement is used for deleting an existing table. The syntax is:

```
SQL> DROP TABLE table_name;
```

The following SQL statement deletes a table named Students:

```
SQL> DROP TABLE Students;
```

INSERT Data

The syntax for INSERT looks similar to the following, where column1, column2, and so on represent the new data to appear in the respective columns:

```
SQL> INSERT INTO table_name VALUES (column1, column2, ...);
```

The following SQL INSERT statement inserts a new row in the Students database created earlier:

```
SQL> INSERT INTO Students VALUES (100, 18, 'BPL', 'AJP');
```

SELECT Data

The SELECT statement is used to retrieve data from a database. The syntax for SELECT is:

```
SQL> SELECT column_name, column_name, ...
     FROM table_name
     WHERE conditions;
```

The WHERE clause can use the comparison operators such as =, !=, <, >, <=,and >=, as well as the BETWEEN and LIKE operators.

The following SQL statement selects the age, first and last columns from the Students table where id column is 100:

```
SQL> SELECT first, last, age
     FROM Students
     WHERE id = 100;
```

The following SQL statement selects the age, first and last columns from the Students table where *first* column contains *BPL*:

```
SQL> SELECT first, last, age
     FROM Students
     WHERE first LIKE '%BPL%';
```

UPDATE Data

The UPDATE statement is used to update data. The syntax for UPDATE is:

```
SQL> UPDATE table_name
     SET column_name = value, column_name = value, ...
     WHERE conditions;
```

The WHERE clause can use the comparison operators such as =, !=, <, >, <=,and >=, as well as the BETWEEN and LIKE operators.

The following SQL UPDATE statement changes the age column of the Student whose id is 100:

```
SQL> UPDATE Students SET age=20 WHERE id=100;
```

DELETE Data

The DELETE statement is used to delete data from tables. The syntax for DELETE is:

```
SQL> DELETE FROM table_name WHERE conditions;
```

The WHERE clause can use the comparison operators such as =, !=, <, >, <=,and >=, as well as the BETWEEN and LIKE operators.

The following SQL DELETE statement delete the record of the Student whose id is 100:

```
SQL> DELETE FROM Students WHERE id=100;
```

2.3 Create DB in SQL

Create Database

To create the **EMP** database, use the following steps:

Step 1:

Open a **Command Prompt** and change to the installation directory as follows:

```
C:\>
C:\>cd Program Files\MySQL\bin
C:\Program Files\MySQL\bin>
```

Note: The path to **mysqld.exe** may vary depending on the install location of MySQL on your system. You can also check documentation on how to start and stop your database server.

Step 2
Start the database server by executing the following command, if it is already not running.

```
C:\Program Files\MySQL\bin>mysqld
C:\Program Files\MySQL\bin>
```

Step 3
Create the **EMP** database by executing the following command

```
C:\Program Files\MySQL\bin> mysqladmin create EMP -u root -p
Enter password: ********
C:\Program Files\MySQL\bin>
```

Create Table
To create the **Students** table in EMP database, use the following steps:

Step 1
Open a **Command Prompt** and change to the installation directory as follows:

```
C:\>
C:\>cd Program Files\MySQL\bin
C:\Program Files\MySQL\bin>
```

Step 2
Login to database as follows

```
C:\Program Files\MySQL\bin>mysql -u root -p
Enter password: ********
mysql>
```

Step 3
Create the table **Student** as follows:

```
mysql> use EMP;
mysql> create table Students
   -> (
   -> id int not null,
   -> age int not null,
   -> first varchar (255),
   -> last varchar (255)
   -> );
Query OK, 0 rows affected (0.08 sec)
mysql>
```

Create Data Records

Finally you create a few records in Student table as follows:

```
mysql> INSERT INTO Students VALUES (100, 18, 'BPL', 'AJP');
Query OK, 1 row affected (0.05 sec)
mysql> INSERT INTO Students VALUES (101, 25, 'Karthick', 'Satish');
Query OK, 1 row affected (0.00 sec)
mysql> INSERT INTO Students VALUES (102, 30, 'sathya', 'Sri');
Query OK, 1 row affected (0.00 sec)
mysql> INSERT INTO Students VALUES (103, 28, 'Sumi', 'Sekar');
Query OK, 1 row affected (0.00 sec)
mysql>
```

2.4 Creating JDBC Application

There are following six steps involved in building a JDBC application:

1. **Import the packages:** Requires that you include the packages containing the JDBC classes needed for database programming. Most often, using *import java.sql.** will suffice.
2. **Register the JDBC driver:** Requires that you initialize a driver so you can open a communications channel with the database.
3. **Open a connection:** Requires using the *DriverManager.getConnection()* method to create a Connection object, which represents a physical connection with the database.
4. **Execute a query:** Requires using an object of type Statement for building and submitting an SQL statement to the database.
5. **Extract data from result set:** Requires that you use the appropriate*ResultSet.getXXX()* method to retrieve the data from the result set.
6. **Clean up the environment:** Requires explicitly closing all database resources versus relying on the JVM's garbage collection.

   This sample example can serve as a **template** when you need to create your own JDBC application in the future.

SNIPPET: Creating JDBC application

```
//STEP 1. Import required packages
import java.sql.*;

public class FirstExample {
   // JDBC driver name and database URL
   static final String JDBC_DRIVER = "com.mysql.jdbc.Driver";
   static final String DB_URL = "jdbc:mysql://localhost/EMP";
   // Database credentials
   static final String USER = "username";
```

```
static final String PASS = "password";

public static void main(String[] args) {
Connection conn = null;
Statement stmt = null;  try{
   //STEP 2: Register JDBC driver
    Class.forName("com.mysql.jdbc.Driver");

   //STEP 3: Open a connection
   System.out.println("Connecting to database...");
   conn = DriverManager.getConnection(DB_URL,USER,PASS);

   //STEP 4: Execute a query
   System.out.println("Creating statement...");
   stmt = conn.createStatement();
   String sql;
   sql = "SELECT id, first, last, age FROM Students";
   ResultSet rs = stmt.executeQuery(sql);

   //STEP 5: Extract data from result set
   while(rs.next()){
     //Retrieve by column name
     int id  = rs.getInt("id");
     int age = rs.getInt("age");
     String first = rs.getString("first");
     String last = rs.getString("last");

     //Display values
     System.out.print("ID: " + id);
     System.out.print(", Age: " + age);
     System.out.print(", First: " + first);
     System.out.println(", Last: " + last);
   }
   //STEP 6: Clean-up environment
   rs.close();
   stmt.close();
   conn.close();
}catch(SQLException se){
   //Handle errors for JDBC
   se.printStackTrace();
}catch(Exception e){
   //Handle errors for Class.forName
   e.printStackTrace();
}finally{
   //finally block used to close resources
   try{
```

```
      if(stmt!=null)
         stmt.close();
    }catch(SQLException se2){
    }// nothing we can do
    try{
      if(conn!=null)
         conn.close();
    }catch(SQLException se){
      se.printStackTrace();
    }//end finally try
  }//end try
  System.out.println("Goodbye!");
}//end main
}//end FirstExample
```

Now let us compile above example as follows:

```
C:\>javac FirstExample.java
C:\>
```

When you run **FirstExample**, it produces the following results:

```
C:\>java FirstExample
Connecting to database...
Creating statement...
ID: 100, Age: 18, First: BPL, Last: AJP
ID: 101, Age: 25, First: Karthick, Last: Satish
ID: 102, Age: 30, First: sathya, Last: Sri
ID: 103, Age: 28, First: Sumi, Last: Sekar
C:\>
```

2.5 JDBC Driver

JDBC drivers implement the defined interfaces in the JDBC API for interacting with your database server.

For example, using JDBC drivers enable you to open database connections and to interact with it by sending SQL or database commands, then receiving results with Java.

The *Java.sql* package that ships with JDK contains various classes with their behaviours defined and their actual implementations are done in third-party drivers. Third party vendors implements the *java.sql.Driver* interface in their database driver.

JDBC Drivers Types

JDBC driver implementations vary because of the wide variety of operating systems and hardware platforms in which Java operates. Sun has divided the implementation types into four categories, Types 1, 2, 3, and 4, which is explained below:

Type 1: JDBC-ODBC Bridge Driver

In a Type 1 driver, a JDBC bridge is used to access ODBC drivers installed on each client machine. Using ODBC requires configuring on your system a Data Source Name (DSN) that represents the target database.

When Java first came out, this was a useful driver because most databases only supported ODBC access but now this type of driver is recommended only for experimental use or when no other alternative is available.

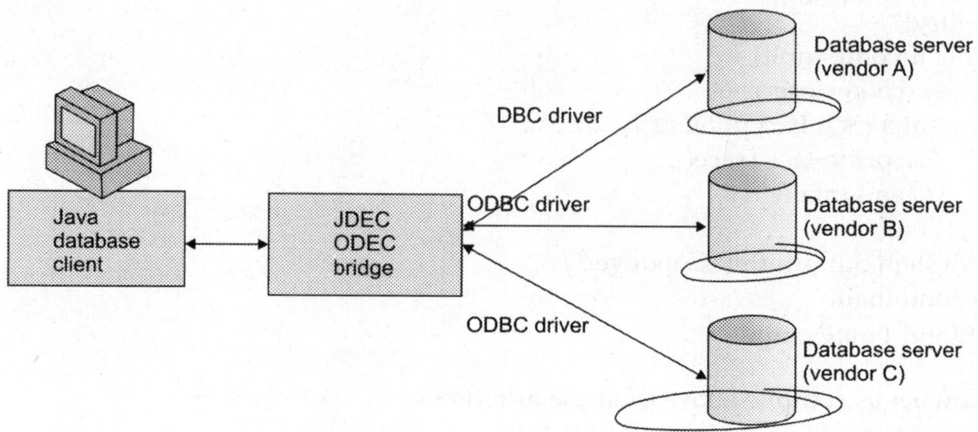

The JDBC-ODBC bridge that comes with JDK 1.2 is a good example of this kind of driver.

Type 2: JDBC-Native API

In a Type 2 driver, JDBC API calls are converted into native C/C++ API calls which are unique to the database. These drivers typically provided by the database vendors and used in the same manner as the JDBC-ODBC Bridge, the vendor-specific driver must be installed on each client machine.

If we change the Database we have to change the native API as it is specific to a database and they are mostly obsolete now but you may realize some speed increase with a Type 2 driver, because it eliminates ODBC's overhead.

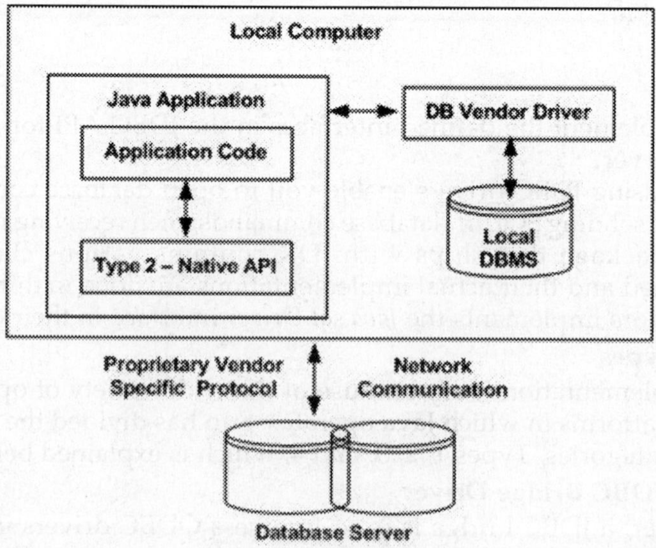

The Oracle Call Interface (OCI) driver is an example of a Type 2 driver.

Type 3: JDBC-Net pure Java

In a Type 3 driver, a three-tier approach is used to accessing databases. The JDBC clients use standard network sockets to communicate with an middleware application server. The socket information is then translated by the middleware application server into the call format required by the DBMS, and forwarded to the database server.

This kind of driver is extremely flexible, since it requires no code installed on the client and a single driver can actually provide access to multiple databases.

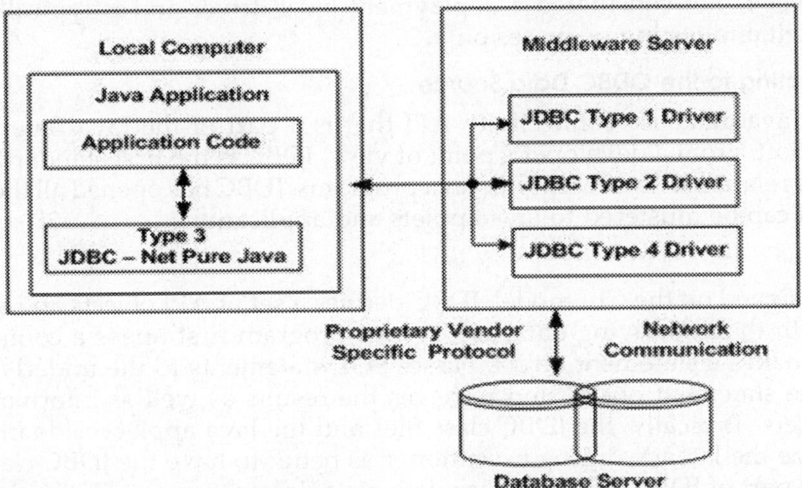

You can think of the application server as a JDBC "proxy," meaning that it makes calls for the client application. As a result, you need some knowledge of the application server's configuration in order to effectively use this driver type.

Your application server might use a Type 1, 2, or 4 driver to communicate with the database, understanding the nuances will prove helpful.

Type 4: 100% pure Java

In a Type 4 driver, a pure Java-based driver that communicates directly with vendor's database through socket connection. This is the highest performance driver available for the database and is usually provided by the vendor itself.

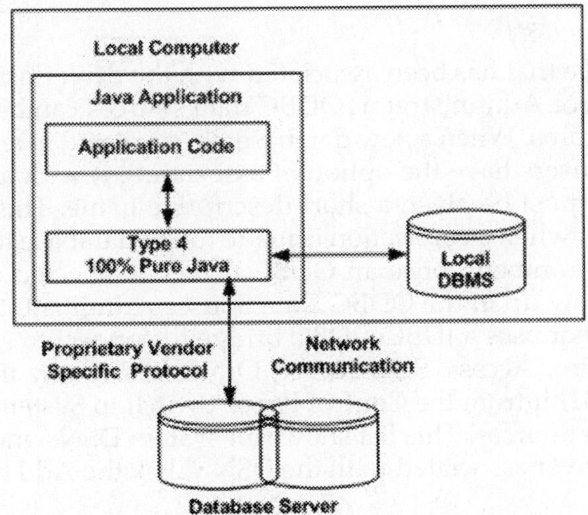

This kind of driver is extremely flexible, you do not need to install special software on the client or server. Further, these drivers can be downloaded dynamically.

MySQL's Connector/J driver is a Type 4 driver. Because of the proprietary nature of their network protocols, database vendors usually supply type 4 drivers.

If you are accessing one type of database, such as Oracle, Sybase, or IBM, the preferred driver type is 4. If your Java application is accessing multiple types of databases at the same time, type 3 is the preferred driver. Type 2 drivers are useful in situations where a type 3 or type 4 driver is not available yet for your database. The type 1 driver is not considered a deployment-level driver and is typically used for development and testing purposes only.

2.6 Connecting to the ODBC Data Source

JDBC is a Java database connectivity API that is a part of the Java Enterprise APIs from Java Soft. From a developer's point of view, JDBC is the first standardized effort to integrate relational databases with Java programs. JDBC has opened all the relational power that can be mustered to Java applets and applications.

2.6.1 Model

JDBC is designed on the CLI model. JDBC defines a set of API objects and methods to interact with the underlying database. A Java program first opens a connection to a database, makes a statement object, passes SQL statements to the underlying DBMS through the statement object, and retrieves the results as well as information about the result sets. Typically, the JDBC class files and the Java applet reside in the client. To minimize the latency during execution, it is better to have the JDBC classes in the client. As a part of JDBC, Java Soft also delivers a driver to access ODBC data sources from JDBC. This driver is jointly developed with Intersolv and is called the JDBC-ODBC bridge. The JDBC-ODBC bridge is implemented as the JdbcOdbc.class and a native library to access the ODBC driver. For the Windows platform, the native library is a DLL (JDBCODBC.DLL).

As JDBC is close to ODBC in design, the ODBC Bridge is a thin layer over JDBC. Internally, this driver maps JDBC methods to ODBC calls and, thus, interacts with any available ODBC driver. The advantage of this bridge is that now JDBC has the capability to access almost all databases, as ODBC drivers are widely available. The JDBC-ODBC Bridge allows JDBC driver to be used as ODBC drivers by converting JDBC method calls into ODBC function calls.

2.6.2 Creating a New DSN

An ODBC data source that has been associated with the driver using software such as the ODBC Data Source Administrator. ODBC data sources can be set up from within some database programs. When a new database file is created in any ODBC supported application system, users have the option of associating it with an ODBC driver. All ODBC data sources must be given a short descriptive name. This name will be used inside Java programs when a connection is made to the database that the source refers to. In Windows environment, once an ODBC driver is selected and the database is created, they will show up in the ODBC Data Source Administrator.

Java application that uses a JDBC-ODBC bridge to connect to a database file either a dbase, Excel, FoxPro, Access, SQL Server, Oracle and many more. First open the ODBC Data Source 32Bit from the Control Panel. Switch to System DSN. System DSN lists the System Data Sources. This list shows all system DSNs, including the name of each DSN and the driver associated with the DSN. Click the Add button to create new Data Source.

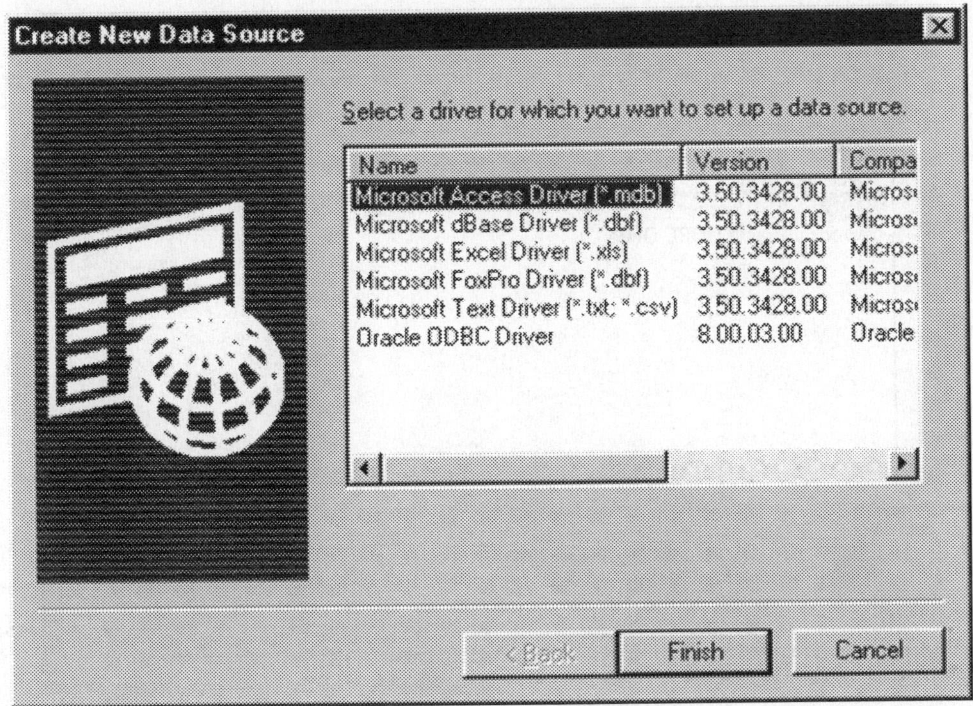

Create New Data Source

Select Oracle ODBC Driver or any other and click the Finish button to finish. This will pop up with a new window Oracle8 ODBC Driver Setup. Give the Data Source name as oraodbc, UserID Scott and click OK to finish.

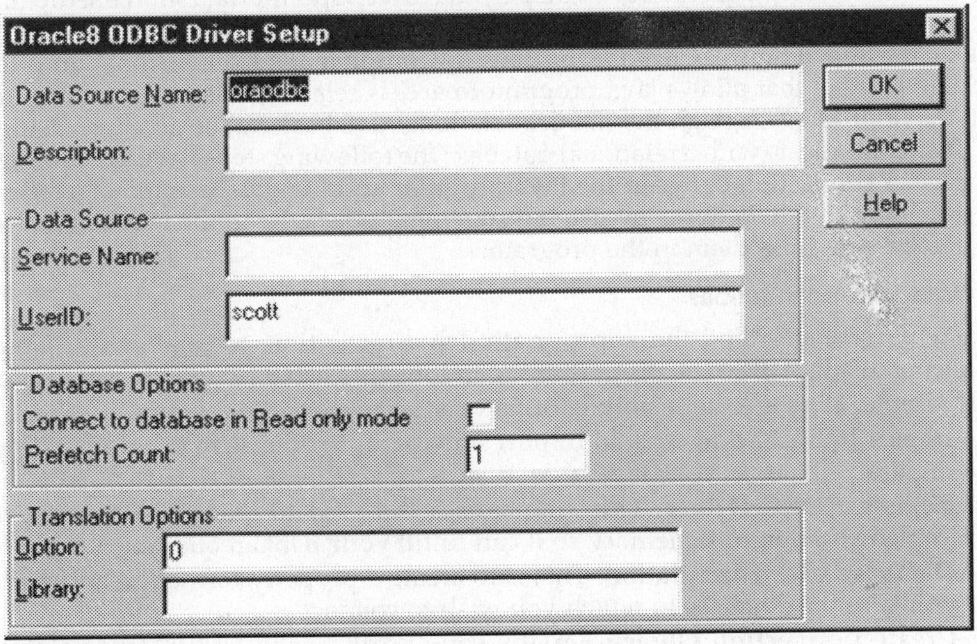

Oracle8 ODBC Driver Setup

Data Source Name: oraodbc

Description:

Data Source

Service Name:

UserID: scott

Database Options

Connect to database in Read only mode

Prefetch Count: 1

Translation Options

Option: 0

Library:

OK

Cancel

Help

Oracle8 ODBC Driver Setup

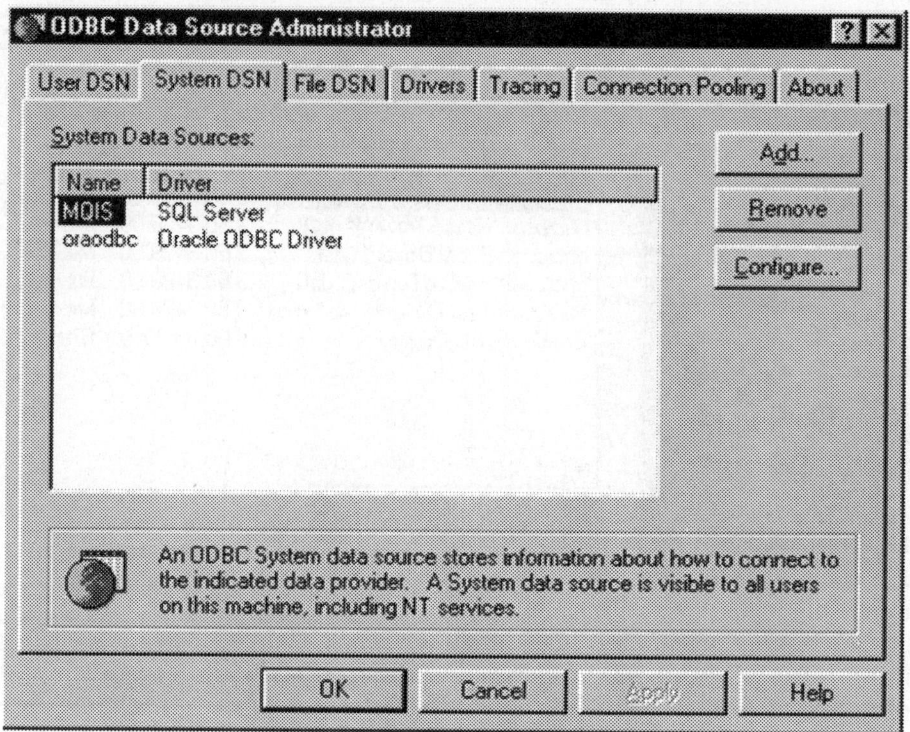

ODBC Data Source Administrator

After finishing Oracle8 ODBC Driver set up, the ODBC Data Source Administrator will be displaying the following. (See the above figure). Now the ODBC Driver Connection has been established. To display the driver-specific data source setup dialog box for a user data source, double-click the system DSN. Now the dialog box will be look as the above figure. Click OK buttons to Complete the Connection. Sun offers a package java.sql that allows java program to access relational database management systems (RDBMS). Through the JDBC a relational database can be accessed using the sql. To communicate with a relational database the following steps have to be followed. Establish a connection between the Java program and the database manager. Send a sql statement to the database by using a statement object. Read the results back from the database and use them in the program.

2.7 Database connections

After you have installed the appropriate driver, it is time to establish a database connection using JDBC. The programming involved to establish a JDBC connection is fairly simple. Here are these simple four steps:

1. **Import JDBC Packages:** Add **import** statements to your Java program to import required classes in your Java code.
2. **Register JDBC Driver:** This step causes the JVM to load the desired driver implementation into memory so it can fulfill your JDBC requests.
3. **Database URL Formulation:** This is to create a properly formatted address that points to the database to which you wish to connect.
4. **Create Connection Object:** Finally, code a call to the *DriverManager* object's *getConnection( )* method to establish actual database connection.

2.7.1 Import JDBC Packages

The **Import** statements tell the Java compiler where to find the classes you reference in your code and are placed at the very beginning of your source code.

To use the standard JDBC package, which allows you to select, insert, update, and delete data in SQL tables, add the following *imports* to your source code:

```
import java.sql.* ; // for standard JDBC program
simport java.math.* ; // for BigDecimal and BigInteger support
```

2.7.2 Register JDBC Driver

You must register your driver in your program before you use it. Registering the driver is the process by which the Oracle driver's class file is loaded into memory so it can be utilized as an implementation of the JDBC interfaces.

You need to do this registration only once in your program. You can register a driver in one of two ways.

Approach (I)—Class.forName()

The most common approach to register a driver is to use Java's **Class.forName()** method to dynamically load the driver's class file into memory, which automatically registers it. This method is preferable because it allows you to make the driver registration configurable and portable.

The following example uses Class.forName( ) to register the Oracle driver:

```
try {
  Class.forName("oracle.jdbc.driver.OracleDriver");
}
catch(ClassNotFoundException ex) {
  System.out.println("Error: unable to load driver class!");
  System.exit(1);
}
```

You can use **getInstance()** method to work around noncompliant JVMs, but then you will have to code for two extra Exceptions as follows:

```
try {
   Class.forName("oracle.jdbc.driver.OracleDriver").newInstance();
}
catch(ClassNotFoundException ex) {
  System.out.println("Error: unable to load driver class!");   System.exit(1);
catch(IllegalAccessException ex) {
  System.out.println("Error: access problem while loading!");
  System.exit(2);catch(InstantiationException ex) {
  System.out.println("Error: unable to instantiate driver!");
  System.exit(3);

}
```

Approach (II)—DriverManager.registerDriver()

The second approach you can use to register a driver is to use the static**DriverManager.registerDriver()** method.

You should use the *registerDriver()* method if you are using a non-JDK compliant JVM, such as the one provided by Microsoft.

The following example uses registerDriver() to register the Oracle driver:

```
try {
   Driver myDriver = new oracle.jdbc.driver.OracleDriver();
   DriverManager.registerDriver( myDriver );
}
catch(ClassNotFoundException ex) {
   System.out.println("Error: unable to load driver class!");
   System.exit(1);
}
```

2.7.3 Database URL Formulation

After you have loaded the driver, you can establish a connection using the**DriverManager.getConnection()** method. For easy reference, let me list the three overloaded DriverManager.getConnection() methods:

1. getConnection(String url)
2. getConnection(String url, Properties prop)
3. getConnection(String url, String user, String password)

Here each form requires a database **URL**. A database URL is an address that points to your database.

Formulating a database URL is where most of the problems associated with establishing a connection occur.

Table 8.1: JDBC driver names and database URL

RDBMS	JDBC driver name	URL format
MySQL	com.mysql.jdbc.Driver	**Jdbc:mysql:**//hostname/ databaseName
ORACLE	oracle.jdbc.driver.OracleDriver	**Jbc:oracle:thin:@**hostname:port Number:databaseName
DB2	COM.ibm.db2.jdbc.net. DB2Driver	**jdbc:db2:**hostname:port Number/databaseName
Sybase	com.sybase.jdbc.SybDriver	**jdbc:sybase:Tds:**hostname: port Number/databaseName

All the highlighted part in URL format is static and you need to change only the remaining part as per your database setup.

2.7.4 Create Connection Object

2.7.4.1 Using a database URL with a username and password

Three forms of **DriverManager.getConnection()** method to create a connection object are explained. The most commonly used form of getConnection() requires you to pass a database URL, a *username,* and a *password*:

Assuming you are using Oracle's **thin** driver, you will specify a host:port:databaseName value for the database portion of the URL.

If you have a host at TCP/IP address 192.0.0.1 with a host name of amrood, and your Oracle listener is configured to listen on port 1521, and your database name is EMP, then complete database URL would be:

```
jdbc:oracle:thin:@amrood:1521:EMP
```

Now you have to call getConnection() method with appropriate username and password to get a**Connection** object as follows:

```
String URL = "jdbc:oracle:thin:@amrood:1521:EMP";
String USER = "username";
String PASS = "password"
Connection conn = DriverManager.getConnection(URL, USER, PASS);
```

2.7.4.2 *Using only a database URL*

A second form of the DriverManager.getConnection( ) method requires only a database URL:

```
DriverManager.getConnection(String url);
```

However, in this case, the database URL includes the username and password and has the following general form:

```
jdbc:oracle:driver:username/password@database
```

So the above connection can be created as follows:

```
String URL = "jdbc:oracle:thin:username/password@amrood:1521:EMP";
Connection conn = DriverManager.getConnection(URL);
```

2.7.4.3 *Using a database URL and a properties object*

A third form of the DriverManager.getConnection( ) method requires a database URL and a properties object:

```
DriverManager.getConnection(String url, Properties info);
```

A properties object holds a set of keyword-value pairs. It is used to pass driver properties to the driver during a call to the getConnection() method.

To make the same connection made by the previous examples, use the following code:

```
import java.util.*;
String URL = "jdbc:oracle:thin:@amrood:1521:EMP";
Properties info = new Properties( );
info.put( "user", "username" );
info.put( "password", "password" );
Connection conn = DriverManager.getConnection(URL, info);
```

2.7.5 Closing JDBC Connections

At the end of your JDBC program, it is required explicitly close all the connections to the database to end each database session. However, if you forget, Java's garbage collector will close the connection when it cleans up stale objects.

Relying on garbage collection, especially in database programming, is very poor programming practice. You should make a habit of always closing the connection with the close() method associated with connection object.

To ensure that a connection is closed, you could provide a finally block in your code. A *finally*block always executes, regardless if an exception occurs or not.

To close above opened connection you should call close() method as follows:

```
conn.close();
```

Explicitly closing a connection conserves DBMS resources, which will make your database administrator happy.

2.8 Statements

As soon as the connection is obtained, we can interact with the database. The JDBC *Statement, CallableStatement,* and *PreparedStatement* interfaces define the methods and properties that enable you to send SQL or PL/SQL commands and receive data from your database. They also define methods that help bridge data type differences between Java and SQL data types used in a database.

Table 8.2 Interfaces and their use

Interfaces	Use
Statement	Use for general-purpose access to your database. Useful when you are using static SQL statements at runtime. The Statement interface cannot accept parameters.
PreparedStatement	Use when you plan to use the SQL statements many times. The PreparedStatement interface accepts input parameters at runtime.
CallableStatement	Use when you want to access database stored procedures. The CallableStatement interface can also accept runtime input parameters.

2.8.1 The Statement Objects—Creating Statement Object

Before you can use a Statement object to execute a SQL statement, you need to create one using the Connection object's createStatement( ) method, as in the following example:

```
Statement stmt = null;
try {
  stmt = conn.createStatement( );
  . . .
}
catch (SQLException e) {
```

```
   . . .
}
finally {
   . . .
}
```

Once you have created a Statement object, you can then use it to execute a SQL statement with one of its three execute methods.

1. **boolean execute(String SQL)**: Returns a boolean value of true if a ResultSet object can be retrieved; otherwise, it returns false. Use this method to execute SQL DDL statements or when you need to use truly dynamic SQL.
2. **int executeUpdate(String SQL)**: Returns the numbers of rows affected by the execution of the SQL statement. Use this method to execute SQL statements for which you expect to get a number of rows affected. For example, an INSERT, UPDATE, or DELETE statement.
3. **ResultSet executeQuery(String SQL)**: Returns a ResultSet object. Use this method when you expect to get a result set, as you would with a SELECT statement.

2.8.2 Closing Statement Object

Just as you close a Connection object to save database resources, for the same reason you should also close the Statement object. A simple call to the close() method will do the job. If you close the Connection object first, it will close the Statement object as well. However, you should always explicitly close the Statement object to ensure proper cleanup.

```
Statement stmt = null;
try {
   stmt = conn.createStatement( );
   . . .}
catch (SQLException e) {
   . . .
}
finally {
   stmt.close();
}
```

2.9 The PreparedStatement Objects

The *PreparedStatement* interface extends the Statement interface which gives you added functionality with a couple of advantages over a generic Statement object.

This statement gives you the flexibility of supplying arguments dynamically.

2.9.1 Creating PreparedStatement Object

```
PreparedStatement pstmt = null;
try {
   String SQL = "Update Students SET age = ? WHERE id = ?";
   pstmt = conn.prepareStatement(SQL);
   . . .
```

```
}
catch (SQLException e) {
  . . .
}
finally {
  . . .
}
```

All parameters in JDBC are represented by the ? symbol, which is known as the parameter marker. You must supply values for every parameter before executing the SQL statement. The **setXXX()** methods bind values to the parameters, where **XXX** represents the Java data type of the value you wish to bind to the input parameter. If you forget to supply the values, you will receive an SQLException.

Each parameter marker is referred to by its ordinal position. The first marker represents position 1, the next position 2, and so forth. This method differs from that of Java array indices, which start at 0. All of the **Statement object's** methods for interacting with the database (a) execute(), (b) executeQuery(), and (c) executeUpdate() also work with the PreparedStatement object. However, the methods are modified to use SQL statements that can take input the parameters.

2.9.2 Closing PreparedStatement Object

Just as you close a Statement object, for the same reason you should also close the PreparedStatement object. A simple call to the close() method will do the job. If you close the Connection object first, it will close the PreparedStatement object as well. However, you should always explicitly close the PreparedStatement object to ensure proper cleanup.

```
PreparedStatement pstmt = null;
try {
  String SQL = "Update Students SET age = ? WHERE id = ?";
  pstmt = conn.prepareStatement(SQL);
  . . .
}
catch (SQLException e) {
  . . .
}
finally {
  pstmt.close();
}
```

2.10 The CallableStatement Objects

Just as a Connection object creates the Statement and PreparedStatement objects, it also creates the CallableStatement object which would be used to execute a call to a database stored procedure.

2.10.1 Creating CallableStatement Object

Suppose, you need to execute the following Oracle stored procedure:

```
CREATE OR REPLACE PROCEDURE getEmpName
```

```
  (EMP_ID IN NUMBER, EMP_FIRST OUT VARCHAR) AS
BEGIN
  SELECT first INTO EMP_FIRST
  FROM Students
  WHERE ID = EMP_ID;
END;
```

Note: Above stored procedure has been written for Oracle, but we are working with MySQL database so let us write the same stored procedure for MySQL as follows to create it in EMP database:

```
DELIMITER $$
DROP PROCEDURE IF EXISTS 'EMP'.'getEmpName' $$
CREATE PROCEDURE 'EMP'.'getEmpName'
  (IN EMP_ID INT, OUT EMP_FIRST VARCHAR(255))
BEGIN
  SELECT first INTO EMP_FIRST
  FROM Students
  WHERE ID = EMP_ID;END $$DELIMITER ;
```

Three types of parameters exist: IN, OUT, and INOUT. The PreparedStatement object only uses the IN parameter. The CallableStatement object can use all three.

Table 8.3 Parameters of CallableStatement

Parameter	Description
IN	A parameter whose value is unknown when the SQL statement is created. You bind values to IN parameters with the setXXX() methods.
OUT	A parameter whose value is supplied by the SQL statement it returns. You retrieve values from the OUT parameters with the getXXX() methods.
INOUT	A parameter that provides both input and output values. You bind variables with the setXXX() methods and retrieve values with the getXXX() methods.

The following code snippet shows how to employ the **Connection.prepareCall()** method to instantiate a **CallableStatement** object based on the preceding stored procedure:

```
CallableStatement cstmt = null;
try {
  String SQL = "{call getEmpName (?, ?)}";
  cstmt = conn.prepareCall (SQL);
```

```
   ...}
catch (SQLException e) {
   ...
}
finally {
   ...
}
```

The String variable SQL represents the stored procedure, with parameter placeholders. Using CallableStatement objects is much like using PreparedStatement objects. You must bind values to all parameters before executing the statement, or you will receive an SQLException. If you have IN parameters, just follow the same rules and techniques that apply to a PreparedStatement object; use the setXXX() method that corresponds to the Java data type you are binding. When you use OUT and INOUT parameters you must employ an additional CallableStatement method, registerOutParameter(). The registerOutParameter() method binds the JDBC data type to the data type the stored procedure is expected to return.

Once you call your stored procedure, you retrieve the value from the OUT parameter with the appropriate getXXX() method. This method casts the retrieved value of SQL type to a Java data type.

2.10.2 Closing CallableStatement Object

Just as you close other Statement object, for the same reason you should also close the CallableStatement object.

A simple call to the close() method will do the job. If you close the Connection object first, it will close the CallableStatement object as well. However, you should always explicitly close the CallableStatement object to ensure proper cleanup.

```
CallableStatement cstmt = null;try {
   String SQL = "{call getEmpName (?, ?)}";
   cstmt = conn.prepareCall (SQL);
   ...
}
catch (SQLException e) {
   ...
}
finally {
   cstmt.close();
}
```

2.11 ResultSet

Statements that read data from a database query return the data in a result set. The SELECT statement is the standard way to select rows from a database and view them in a result set. The *java.sql.ResultSet* interface represents the result set of a database query. A ResultSet object maintains a cursor that points to the current row in the result set. The term "result set" refers to the row and column data contained in a ResultSet object.

The methods of the ResultSet interface can be broken down into three categories:

1. **Navigational methods:** Used to move the cursor around.
2. **Get methods:** Used to view the data in the columns of the current row being pointed to by the cursor.
3. **Update methods:** Used to update the data in the columns of the current row. The updates can then be updated in the underlying database as well.

   The cursor is movable based on the properties of the ResultSet. These properties are designated when the corresponding Statement that generated the ResultSet is created.

   JDBC provides following connection methods to create statements with desired ResultSet:

1. **createStatement(int RSType, int RSConcurrency);**
2. **prepareStatement(String SQL, int RSType, int RSConcurrency);**
3. **prepareCall(String sql, int RSType, int RSConcurrency);**

   The first argument indicate the type of a ResultSet object and the second argument is one of two ResultSet constants for specifying whether a result set is read-only or updatable.

2.11.1 Type of ResultSet

The possible RSType are given below. If you do not specify any ResultSet type, you will automatically get one that is TYPE_FORWARD_ONLY.

Type	Description
ResultSet.TYPE_FORWARD_ONLY	The cursor can only move forward in the result set.
ResultSet.TYPE_SCROLL_INSENSITIVE	The cursor can scroll forwards and backwards, and the result set is not sensitive to changes made by others to the database that occur after the result set was created.
ResultSet.TYPE_SCROLL_SENSITIVE.	The cursor can scroll forwards and backwards, and the result set is sensitive to changes made by others to the database that occur after the result set was created.

2.11.2 Concurrency of ResultSet

The possible RSConcurrency are given below. If you do not specify any Concurrency type, you will automatically get one that is CONCUR_READ_ONLY.

Concurrency	Description
ResultSet.CONCUR_READ_ONLY	Creates a read-only result set. This is the default.
ResultSet.CONCUR_UPDATABLE	Creates an updateable result set.

Our all the examples written so far can be written as follows which initializes a Statement object to create a forward-only, read only ResultSet object:

```
try {
  Statement stmt = conn.createStatement(
                 ResultSet.TYPE_FORWARD_ONLY,
                 ResultSet.CONCUR_READ_ONLY);}catch(Exception ex)
  {
    ....
  }
finally {
    ....
  }
```

2.11.3 Navigating a ResultSet

There are several methods in the ResultSet interface that involve moving the cursor, including:

Methods	Usage	Use
beforeFirst	Public void beforeFirst() throws SQLException	Moves the cursor to just before the first row
afterLast	Public void afterLast() throws SQLException	Moves the cursor to just after the last row
first	Public boolean first() throws SQLException	Moves the cursor to the first row
last	Public void last() throws SQLException	Moves the cursor to the last row
absolute	Public boolean absolute (int row) throws SQLException	Moves the cursor to the specified row
relative	Public boolean relative(int row) throws SQLException	Moves the cursor the given number of rows forward or backwards from where it currently is pointing.
previous	Public boolean previous() throws SQLException	Moves the cursor to the previous row. This method returns false if the previous row is off the result set.
next	Public boolean next() throws SQLException	Moves the cursor to the next row. This method returns false if there are no more rows in the result set.
getRow	Public int getRow() throws SQLException	Returns the row number that the cursor is pointing to.
moveToinsert Row	Public void moveToInsert Row() throws SQLException	Moves the cursor to a special row in the result set that can be used to insert a new row into the database. The current cursor location is remembered.

moveTo CurrentRow	Public void moveToCurrent Row() throws SQLException	Moves the cursor back to the current row if the cursor is currently at the insert row; otherwise, this method does nothing.

2.11.4 Viewing a ResultSet

The ResultSet interface contains dozens of methods for getting the data of the current row.

There is a get method for each of the possible data types, and each get method has two versions:

1. One that takes in a column name.
2. One that takes in a column index.

For example, if the column you are interested in viewing contains an int, you need to use one of the getInt() methods of ResultSet:

Methods	Usage	Use
getInt	public int getInt(String column Name) throws SQLException	Returns the int in the current row in the column named columnName
getInt	public int getInt(int column Index) throws SQLException	Returns the int in the current row in the specified column index. The column index starts at 1, meaning the first column of a row is 1, the second column of a row is 2, and so on.

Similarly there are get methods in the ResultSet interface for each of the eight Java primitive types, as well as common types such as java.lang.String, java.lang.Object, and java.net.URL. There are also methods for getting SQL data types java.sql.Date, java.sql.Time, java.sql.TimeStamp, java.sql.Clob, and java.sql.Blob. Check the documentation for more information about using these SQL data types.

2.11.5 Updating a ResultSet

The ResultSet interface contains a collection of update methods for updating the data of a result set.

As with the get methods, there are two update methods for each data type:

1. One that takes in a column name.
2. One that takes in a column index.

For example, to update a String column of the current row of a result set, you would use one of the following updateString() methods:

Methods	Usage	Use
updateString	public void updateString(int columnIndex, String s) throws SQLException	Changes the String in the specified column to the value of s.

updateString	public void updateString (String columnName, String s) throws SQLException	Similar to the previous method, except that the column is specified by its name instead of its index.

There are update methods for the eight primitive data types, as well as String, Object, URL, and the SQL data types in the java.sql package. Updating a row in the result set changes the columns of the current row in the ResultSet object, but not in the underlying database. To update your changes to the row in the database, you need to invoke one of the following methods.

Methods	**Usage**	**Use**
updateRow	Public void updateRow()	Updates the current row by updating the corresponding row in the database.
deleteRow	Public void deleteRow()	Deletes the current row from the database.
refreshRow	Public void refreshRow()	Refreshes the data in the result set to reflect any recent changes in the database.
cancel Rowupdates	Public void cancelRowUpdates()	Cancels any updates made on the current row.
insertRow	Public void insertRow()	Inserts a row into the database. This method can only be invoked when the cursor is pointing to the insert row.

2.12 Datatypes

The JDBC driver converts the Java data type to the appropriate JDBC type before sending it to the database. It uses a default mapping for most data types. For example, a Java int is converted to an SQL INTEGER. Default mappings were created to provide consistency between drivers.

The following table summarizes the default JDBC data type that the Java data type is converted to when you call the setXXX() method of the PreparedStatement or CallableStatement object or the ResultSet.updateXXX() method.

SQL	**JDBC/Java**	**setXXX**	**updateXXX**
VARCHAR	java.lang.String	setString	updateString
CHAR	java.lang.String	setString	updateString
LONGVARCHAR	java.lang.String	setString	updateString
BIT	boolean	setBoolean	updateBoolean
NUMERIC	java.math.BigDecimal	setBigDecimal	updateBig Decimal

TINYINT	byte	setByte	updateByte
SMALLINT	short	setShort	updateShort
INTEGER	int	setInt	updateInt
BIGINT	long	setLong	updateLong
REAL	float	setFloat	updateFloat
FLOAT	float	setFloat	updateFloat
DOUBLE	double	setDouble	updateDouble
VARBINARY	byte[ ]	setBytes	updateBytes
BINARY	byte[ ]	setBytes	updateBytes
DATE	java.sql.Date	setDate	updateDate
TIME	java.sql.Time	setTime	updateTime
TIMESTAMP	java.sql.Timestamp	setTimestamp	updateTimestamp
CLOB	java.sql.Clob	setClob	updateClob
BLOB	java.sql.Blob	setBlob	updateBlob
ARRAY	java.sql.Array	setARRAY	updateARRAY
REF	java.sql.Ref	SetRef	updateRef
STRUCT	java.sql.Struct	SetStruct	updateStruct

JDBC 3.0 has enhanced support for BLOB, CLOB, ARRAY, and REF data types. The ResultSet object now has updateBLOB(), updateCLOB(), updateArray(), and updateRef() methods that enable you to directly manipulate the respective data on the server.

The setXXX() and updateXXX() methods enable you to convert specific Java types to specific JDBC data types. The methods, setObject() and updateObject(), enable you to map almost any Java type to a JDBC data type.

ResultSet object provides corresponding getXXX() method for each data type to retrieve column value. Each method can be used with column name or by its ordinal position.

Date & Time Data Types

The java.sql.Date class maps to the SQL DATE type, and the java.sql.Time and java.sql.Timestamp classes map to the SQL TIME and SQL TIMESTAMP data types, respectively.

Following examples shows how the Date and Time classes format standard Java date and time values to match the SQL data type requirements.

```
import java.sql.Date;
import java.sql.Time;
import java.sql.Timestamp;
import java.util.*;
public class SqlDateTime {
```

```
public static void main(String[] args) {
  //Get standard date and time
  java.util.Date javaDate = new java.util.Date();
  long javaTime = javaDate.getTime();
  System.out.println("The Java Date is:" +
        javaDate.toString());
  //Get and display SQL DATE
  java.sql.Date sqlDate = new java.sql.Date(javaTime);
  System.out.println("The SQL DATE is: " +
        sqlDate.toString());

  //Get and display SQL TIME
  java.sql.Time sqlTime = new java.sql.Time(javaTime);
  System.out.println("The SQL TIME is: " +
        sqlTime.toString());
  //Get and display SQL TIMESTAMP
  java.sql.Timestamp sqlTimestamp =
  new java.sql.Timestamp(javaTime);
  System.out.println("The SQL TIMESTAMP is: " +
        sqlTimestamp.toString());
 }//end main
}//end SqlDateTime
```

Now let us compile above example as follows:

```
C:\>javac SqlDateTime.java
C:\>
```

When you run **JDBCExample**, it produces the following results:

```
C:\>java SqlDateTime
The Java Date is:Tue Aug 18 13:46:02 GMT+04:00 2009
The SQL DATE is: 2009-08-18
The SQL TIME is: 13:46:02
The SQL TIMESTAMP is: 2009-08-18 13:46:02.828
C:\>
```

2.13 Handling NULL Values

SQL's use of NULL values and Java's use of null are different concepts. There are three ways to handle SQL NULL values.

1. Avoid using getXXX( ) methods that return primitive data types.
2. Use wrapper classes for primitive data types, and use the ResultSet object's wasNull( ) method to test whether the wrapper class variable that received the value returned by the getXXX( ) method should be set to null.
3. Use primitive data types and the ResultSet object's wasNull( ) method to test whether the primitive variable that received the value returned by the getXXX( ) method should be set to an acceptable value that you have chosen to represent a NULL.

Here is one example to handle a NULL value:

```
Statement stmt = conn.createStatement( );
String sql = "SELECT id, first, last, age FROM Students";
ResultSet rs = stmt.executeQuery(sql);

int id = rs.getInt(1);
if( rs.wasNull( ) ) {
   id = 0;
}
```

2.14 Transactions

If your JDBC Connection is in *auto-commit* mode, which it is by default, then every SQL statement is committed to the database upon its completion.

That may be fine for simple applications, but there are three reasons why you may want to turn off auto-commit and manage your own transactions:

1. To increase performance
2. To maintain the integrity of business processes
3. To use distributed transactions

Transactions enable you to control if, and when, changes are applied to the database. It treats a single SQL statement or a group of SQL statements as one logical unit, and if any statement fails, the whole transaction fails.

To enable manual-transaction support instead of the *auto-commit* mode that the JDBC driver uses by default, use the Connection object's **setAutoCommit()** method. If you pass a boolean false to setAutoCommit( ), you turn off auto-commit. You can pass a boolean true to turn it back on again.

For example, if you have a Connection object named conn, code the following to turn off auto-commit:

```
conn.setAutoCommit(false);
```

Commit & Rollback

Once you are done with your changes and you want to commit the changes, then call **commit()**method on connection object is as follows:

```
conn.commit( );
```

Otherwise, to roll back updates to the database made using the Connection named conn, use the following code:

```
conn.rollback( );
```

The following example illustrates the use of a commit and rollback object:

```
try{
   //Assume a valid connection object conn
   conn.setAutoCommit(false);
   Statement stmt = conn.createStatement();
     String SQL = "INSERT INTO Students " +
```

```
        "VALUES (106, 20, 'Rita', 'Tez')";
    stmt.executeUpdate(SQL);
    //Submit a malformed SQL statement that breaks
  String SQL = "INSERTED IN Students " +
        "VALUES (107, 22, 'Sita', 'Singh')";
  stmt.executeUpdate(SQL);
  // If there is no error.
  conn.commit();
}catch(SQLException se){
  // If there is any error.
  conn.rollback();
}
```

In this case none of the above INSERT statement would success and everything would be rolled back.

Using Savepoints

The new JDBC 3.0 Savepoint interface gives you additional transactional control. Most modern DBMS support savepoints within their environments such as Oracle's PL/SQL.

When you set a savepoint, you define a logical rollback point within a transaction. If an error occurs past a savepoint, you can use the rollback method to undo either all the changes or only the changes made after the savepoint.

The Connection object has two new methods that help you manage savepoints:

1. **setSavepoint(String savepointName):** Defines a new savepoint. It also returns a Savepoint object.
2. **releaseSavepoint(Savepoint savepointName):** Deletes a savepoint. Notice that it requires a Savepoint object as a parameter. This object is usually a savepoint generated by the setSavepoint() method.

There is one **rollback (String savepointName)** method which rolls back work to the specified savepoint.

The following example illustrates the use of a Savepoint object:

```
try{
  //Assume a valid connection object conn
  conn.setAutoCommit(false);
  Statement stmt = conn.createStatement();

    //set a Savepoint
  Savepoint savepoint1 = conn.setSavepoint("Savepoint1");
  String SQL = "INSERT INTO Students " +
        "VALUES (106, 20, 'Rita', 'Tez')";
  stmt.executeUpdate(SQL);
   //Submit a malformed SQL statement that breaks
  String SQL = "INSERTED IN Students " +
        "VALUES (107, 22, 'Sita', 'Tez')";
  stmt.executeUpdate(SQL);
  // If there is no error, commit the changes.  conn.commit();
}catch(SQLException se){
```

```
    // If there is any error.
    conn.rollback(savepoint1);
}
```

2.15 Exception Handling

Exception handling allows you to handle exceptional conditions such as program-defined errors in a controlled fashion. When an exception condition occurs, an exception is thrown. The term thrown means that current program execution stops, and control is redirected to the nearest applicable catch clause. If no applicable catch clause exists, then the program's execution ends.

JDBC Exception handling is very similar to Java Excpetion handling but for JDBC, the most common exception you will deal with is **java.sql.SQLException.**

SQLException Methods

A SQLException can occur both in the driver and the database. When such an exception occurs, an object of type SQLException will be passed to the catch clause.

The passed SQLException object has the following methods available for retrieving additional information about the exception:

Methods	Description
getErrorCode( )	Gets the error number associated with the exception.
getMessage( )	Gets the JDBC driver's error message for an error handled by the driver or gets the Oracle error number and message for a database error.
getSQLState( )	Gets the XOPEN SQLstate string. For a JDBC driver error, no useful information is returned from this method. For a database error, the five-digit XOPEN SQLstate code is returned. This method can return null.
getNextException( )	Gets the next Exception object in the exception chain.
printStackTrace( )	Prints the current exception, or throwable, and its backtrace to a standard error stream.
printStackTrace(PrintStream s)	Prints this throwable and its backtrace to the print stream you specify.
printStackTrace(PrintWriter w)	Prints this throwable and its backtrace to the print writer you specify.

By utilizing the information available from the Exception object, you can catch an exception and continue your program appropriately. Here is the general form of a try block:

```
try {
  // Your risky code goes between these curly braces!!!
```

```
}
catch(Exception ex) {
   // Your exception handling code goes between these
    // curly braces, similar to the exception clause
    // in a PL/SQL block.}
finally {
   // Your must-always-be-executed code goes between these
    // curly braces. Like closing database connection.
}
```

Example

Study the following example code to understand the usage of **try....catch...finally** blocks.

```
//STEP 1. Import required packages
import java.sql.*;

public class JDBCExample {
   // JDBC driver name and database URL
   static final String JDBC_DRIVER = "com.mysql.jdbc.Driver";
   static final String DB_URL = "jdbc:mysql://localhost/EMP";

   //  Database credentials
   static final String USER = "username";
   static final String PASS = "password";

    public static void main(String[] args) {
   Connection conn = null;
   try{
      //STEP 2: Register JDBC driver
      Class.forName("com.mysql.jdbc.Driver");

      //STEP 3: Open a connection
      System.out.println("Connecting to database...");
      conn = DriverManager.getConnection(DB_URL,USER,PASS);
      //STEP 4: Execute a query
      System.out.println("Creating statement...");
      Statement stmt = conn.createStatement();
      String sql;
      sql = "SELECT id, first, last, age FROM Students";
      ResultSet rs = stmt.executeQuery(sql);

      //STEP 5: Extract data from result set
      while(rs.next()){
         //Retrieve by column name
         int id  = rs.getInt("id");
         int age = rs.getInt("age");
```

```
      String first = rs.getString("first");
      String last = rs.getString("last");

      //Display values
      System.out.print("ID: " + id);
      System.out.print(", Age: " + age);
      System.out.print(", First: " + first);
      System.out.println(", Last: " + last);
    }
    //STEP 6: Clean-up environment
    rs.close();
    stmt.close();
    conn.close();
  }catch(SQLException se){
    //Handle errors for JDBC
    se.printStackTrace();
  }catch(Exception e){
    //Handle errors for Class.forName
    e.printStackTrace();
  }finally{
    //finally block used to close resources
    try{
      if(conn!=null)
        conn.close();
    }catch(SQLException se){
      se.printStackTrace();
    }//end finally try
  }//end try
  System.out.println("Goodbye!");
}//end main
}//end JDBCExample
```

Now let us compile the above example as follows:

```
C:\>javac JDBCExample.java
C:\>
```

When you run **JDBCExample**, it produces the following result if there is no problem, otherwise corresponding error would be caught and error message would be displayed:

```
C:\>java JDBCExample
Connecting to database...
Creating statement...
ID: 100, Age: 18, First: BPL, Last: AJP
ID: 101, Age: 25, First: Karthick, Last: Satish
ID: 102, Age: 30, First: sathya, Last: Sri
```

```
ID: 103, Age: 28, First: Sumi, Last: Sekar
C:\>
```

Try the above example by passing wrong database name or wrong username or password and check the result.

2.16 Batch Processing

Batch Processing allows you to group related SQL statements into a batch and submit them with one call to the database.

When you send several SQL statements to the database at once, you reduce the amount of communication overhead, thereby improving performance.

- JDBC drivers are not required to support this feature. You should use the*DatabaseMetaData.supportsBatchUpdates()* method to determine if the target database supports batch update processing. The method returns true if your JDBC driver supports this feature.
- The **addBatch()** method of *Statement, PreparedStatement,* and *CallableStatement* is used to add individual statements to the batch. The **executeBatch()** is used to start the execution of all the statements grouped together.
- The **executeBatch()** returns an array of integers, and each element of the array represents the update count for the respective update statement.
- Just as you can add statements to a batch for processing, you can remove them with the**clearBatch()** method. This method removes all the statements you added with the addBatch() method. However, you cannot selectively choose which statement to remove.

Batching with Statement Object

Here is a typical sequence of steps to use Batch Processing with Statement Object:

1. Create a Statement object using either *createStatement()* methods.
2. Set auto-commit to false using *setAutoCommit()*.
3. Add as many as SQL statements you like into batch using *addBatch()* method on created statement object.
4. Execute all the SQL statements using *executeBatch()* method on created statement object.
5. Finally, commit all the changes using *commit()* method.

The following code snippet provides an example of a batch update using Statement object:

```
// Create statement object
Statement stmt = conn.createStatement();

// Set auto-commit to false
conn.setAutoCommit(false);

// Create SQL statement
String SQL = "INSERT INTO Students (id, first, last, age) " +
        "VALUES(200,'Zia', 'AJP', 30)";
// Add above SQL statement in the batch.
stmt.addBatch(SQL);
```

```
// Create one more SQL statement
String SQL = "INSERT INTO Students (id, first, last, age) " +
        "VALUES(201,'Raj', 'Kumar', 35)";
// Add above SQL statement in the batch.
stmt.addBatch(SQL);

// Create one more SQL statement
String SQL = "UPDATE Students SET age = 35 " +
        "WHERE id = 100";
// Add above SQL statement in the batch.
stmt.addBatch(SQL);
// Create an int[] to hold returned values
int[] count = stmt.executeBatch();

//Explicitly commit statements to apply changes
conn.commit();
```

For a better understanding, I would suggest to study Batching - Example Code.

Batching with PrepareStatement Object

Here is a typical sequence of steps to use Batch Processing with PrepareStatement Object:

1. Create SQL statements with placeholders.
2. Create PrepareStatement object using either *prepareStatement()* methods.
3. Set auto-commit to false using *setAutoCommit()*.
4. Add as many as SQL statements you like into batch using *addBatch()* method on created statement object.
5. Execute all the SQL statements using *executeBatch()* method on created statement object.
6. Finally, commit all the changes using *commit()* method.

The following code snippet provides an example of a batch update using PrepareStatement object:

```
// Create SQL statement
String SQL = "INSERT INTO Students (id, first, last, age) " +"VALUES(?, ?, ?, ?)";

// Create PrepareStatement object
PreparedStatement pstmt = conn.prepareStatement(SQL);

//Set auto-commit to false
conn.setAutoCommit(false);

// Set the variables
pstmt.setInt( 1, 400 );
pstmt.setString( 2, "Pappu" );
pstmt.setString( 3, "Singh" );
```

```
pstmt.setInt( 4, 33 );
// Add it to the batch
pstmt.addBatch();

// Set the variables
pstmt.setInt( 1, 401 );
pstmt.setString( 2, "Pawan" );
pstmt.setString( 3, "Singh" );
pstmt.setInt( 4, 31 );
// Add it to the batch
pstmt.addBatch();

//add more batches
.
.
.
.
//Create an int[] to hold returned values
int[] count = stmt.executeBatch();

//Explicitly commit statements to apply changes
conn.commit();
```

2.17 Stored Procedure

Just as a Connection object creates the Statement and PreparedStatement objects, it also creates the CallableStatement object which would be used to execute a call to a database stored procedure.

Creating CallableStatement Object

Suppose, you need to execute the following Oracle stored procedure:

```
CREATE OR REPLACE PROCEDURE getEmpName
   (EMP_ID IN NUMBER, EMP_FIRST OUT VARCHAR) AS
BEGIN
   SELECT first INTO EMP_FIRST
   FROM Students
   WHERE ID = EMP_ID;
END;
```

Note: Above stored procedure has been written for Oracle, but we are working with MySQL database, so let us write the same stored procedure for MySQL as follows to create it in EMP database:

```
DELIMITER $$
DROP PROCEDURE IF EXISTS 'EMP'.'getEmpName' $$
CREATE PROCEDURE 'EMP'.'getEmpName'
```

```
  (IN EMP_ID INT, OUT EMP_FIRST VARCHAR(255))
BEGIN
  SELECT first INTO EMP_FIRST
  FROM Students
  WHERE ID = EMP_ID;
END $$
DELIMITER ;
```

Three types of parameters exist: IN, OUT, and INOUT. The PreparedStatement object only uses the IN parameter. The CallableStatement object can use all three.

Here are the definitions of each:

Parameter	Description
IN	A parameter whose value is unknown when the SQL statement is created. You bind values to IN parameters with the setXXX() methods.
OUT	A parameter whose value is supplied by the SQL statement it returns. You retrieve values from theOUT parameters with the getXXX() methods.
INOUT	A parameter that provides both input and output values. You bind variables with the setXXX() methods and retrieve values with the getXXX() methods.

The following code snippet shows how to employ the **Connection.prepareCall()** method to instantiate a **CallableStatement** object based on the preceding stored procedure:

```
CallableStatement cstmt = null;
try {
  String SQL = "{call getEmpName (?, ?)}";
  cstmt = conn.prepareCall (SQL);
  . . .
}
catch (SQLException e) {
  . . .
}
finally {
  . . .
}
```

The String variable SQL represents the stored procedure, with parameter placeholders.

Using CallableStatement objects is much like using PreparedStatement objects. You must bind values to all parameters before executing the statement, or you will receive an SQLException.

If you have IN parameters, just follow the same rules and techniques that apply to a PreparedStatement object; use the setXXX() method that corresponds to the Java data type you are binding.

When you use OUT and INOUT parameters, you must employ an additional CallableStatement method, registerOutParameter(). The registerOutParameter() method binds the JDBC data type to the data type the stored procedure is expected to return.

Once you call your stored procedure, you retrieve the value from the OUT parameter with the appropriate getXXX() method. This method casts the retrieved value of SQL type to a Java data type.

2.18 JDBC SQL Escape Syntax

The escape syntax gives you the flexibility to use database specific features unavailable to you by using standard JDBC methods and properties.

The general SQL escape syntax format is as follows:

```
{keyword 'parameters'}
```

Here are the following escape sequences which you would find very useful while doing JDBC programming:

2.18.1 d, t, ts Keywords

They help identify date, time, and timestamp literals. As you know, no two DBMSs represent time and date the same way. This escape syntax tells the driver to render the date or time in the target database's format. For example:

```
{d 'yyyy-mm-dd'}
```

Where yyyy = year, mm = month; dd = date. Using this syntax {d '2009-09-03'} is March 9, 2009.

Here is a simple example showing how to INSERT date in a table:

```
//Create a Statement object
stmt = conn.createStatement();
//Insert data ==> ID, First Name, Last Name, DOB
String sql="INSERT INTO STUDENTS VALUES" + "(100,'BPL','AJP', {d '2001-
12-16'})";
stmt.executeUpdate(sql);
```

Similarly, you can use one of the following two syntaxes, either **t** or **ts**:

```
{t 'hh:mm:ss'}
```

Where hh = hour; mm = minute; ss = second. Using this syntax {t '13:30:29'} is 1:30:29 PM.

```
{ts 'yyyy-mm-dd hh:mm:ss'}
```

This is combined syntax of the above two syntax for 'd' and 't' to represent timestamp.

2.18.2 escape Keyword

This keyword identifies the escape character used in LIKE clauses. Useful when using the SQL wildcard %, which matches zero or more characters. For example:

String sql = "SELECT symbol FROM MathSymbols WHERE symbol LIKE '\%' {escape '\'}";stmt.execute(sql);

If you use the backslash character (\) as the escape character, you also have to use two backslash characters in your Java String literal, because the backslash is also a Java escape character.

2.18.3 fn Keyword

This keyword represents scalar functions used in a DBMS. For example, you can use SQL function *length* to the length of a string:

{fn length('Hello World')}

This returns 11, the length of the character string 'Hello World'.

2.18.4 call Keyword

This keyword is used to call stored procedures. For example, for a stored procedure requiring an IN parameter, use the following syntax:

{call my_procedure(?)};

For a stored procedure requiring an IN parameter and returning an OUT parameter, use the following syntax:

{? = call my_procedure(?)};

2.18.5 oj Keyword

This keyword is used to signify outer joins. The syntax is as follows:

{oj outer-join}

Where outer-join = table {LEFT | RIGHT | FULL} OUTERJOIN {table | outer-join} on search-condition. For example:

String sql = "SELECT Students FROM {oj ThisTable RIGHT OUTER JOIN ThatTable on id = '100'}";
stmt.execute(sql);

2.19 Data Streaming

A PreparedStatement object has the ability to use input and output streams to supply parameter data. This enables you to place entire files into database columns that can hold large values, such as CLOB and BLOB data types.

There are the following methods which can be used to stream data:

1. **setAsciiStream():** This method is used to supply large ASCII values.
2. **setCharacterStream():** This method is used to supply large UNICODE values.
3. **setBinaryStream():** This method is used to supply large binary values.

The setXXXStream() method requires an extra parameter, the file size, besides the parameter placeholder. This parameter informs the driver how much data should be sent to the database using the stream.

Consider we want to upload an XML file XML_Data.xml into a database table. Here is the content of this XML file:

<?xml version="1.0"?>

```
<Student>
<id>100</id>
<first>BPL</first>
<last>AJP</last>
<Mark>10000</Mark>
<Dob>18-08-1978</Dob>
<Student>
```

Keep this XML file in the same directory where you are going to run this example.

This example would create a database table XML_Data and then file XML_Data.xml would be uploaded into this table.

Copy and paste following example in JDBCExample.java, compile and run as follows:

```
// Import required packages
import java.sql.*;
import java.io.*;
import java.util.*;

public class JDBCExample {
  // JDBC driver name and database URL
  static final String JDBC_DRIVER = "com.mysql.jdbc.Driver";
   static final String DB_URL = "jdbc:mysql://localhost/EMP";

  //  Database credentials
  static final String USER = "username";
  static final String PASS = "password";

  public static void main(String[] args) {
  Connection conn = null;
  PreparedStatement pstmt = null;
  Statement stmt = null;
  ResultSet rs = null;
  try{
    // Register JDBC driver
    Class.forName("com.mysql.jdbc.Driver");

    // Open a connection
    System.out.println("Connecting to database...");
    conn = DriverManager.getConnection(DB_URL,USER,PASS);

    //Create a Statement object and build table
    stmt = conn.createStatement();
    createXMLTable(stmt);

    //Open a FileInputStream
    File f = new File("XML_Data.xml");
```

```
long fileLength = f.length();
FileInputStream fis = new FileInputStream(f);

//Create PreparedStatement and stream data
String SQL = "INSERT INTO XML_Data VALUES (?,?)";
pstmt = conn.prepareStatement(SQL);
pstmt.setInt(1,100);
pstmt.setAsciiStream(2,fis,(int)fileLength);
pstmt.execute();

//Close input stream
fis.close();
// Do a query to get the row
SQL = "SELECT Data FROM XML_Data WHERE id=100";
rs = stmt.executeQuery (SQL);
// Get the first row
if (rs.next ()){
  //Retrieve data from input stream
  InputStream xmlInputStream = rs.getAsciiStream (1);
  int c;
  ByteArrayOutputStream bos = new ByteArrayOutputStream();
  while (( c = xmlInputStream.read ()) != -1)
    bos.write(c);
  //Print results
  System.out.println(bos.toString());
}
// Clean-up environment
rs.close();
stmt.close();
pstmt.close();
conn.close();
}catch(SQLException se){
  //Handle errors for JDBC
  se.printStackTrace();
}catch(Exception e){
  //Handle errors for Class.forName
  e.printStackTrace();
}finally{
  //finally block used to close resources
  try{
    if(stmt!=null)
      stmt.close();
  }catch(SQLException se2){
  }// nothing we can do
  try{
```

```
      if(pstmt!=null)
         pstmt.close();
    }catch(SQLException se2){
    }// nothing we can do
    try{
      if(conn!=null)
         conn.close();
    }catch(SQLException se){
      se.printStackTrace();
    }//end finally try
  }//end try
  System.out.println("Goodbye!");
}//end main

public static void createXMLTable(Statement stmt)
   throws SQLException{
  System.out.println("Creating XML_Data table..." );
  //Create SQL Statement
  String streamingDataSql = "CREATE TABLE XML_Data " +
                  "(id INTEGER, Data LONG)";
  //Drop table first if it exists.
  try{
    stmt.executeUpdate("DROP TABLE XML_Data");
  }catch(SQLException se){
  }// do nothing
  //Build table.
  stmt.executeUpdate(streamingDataSql);
}//end createXMLTable
}//end JDBCExample
```

Now let us compile the above example as follows:

```
C:\>javac JDBCExample.java
C:\>
```

When you run **JDBCExample**, it produces the following results:

```
C:\>java JDBCExample
Connecting to database...
Creating XML_Data table...
<?xml version="1.0"?>
<Student>
<id>100</id>
<first>BPL</first>
<last>AJP</last>
<Mark>10000</Mark>
```

```
<Dob>18-08-1978</Dob>
<Student>
Goodbye!
C:\>
```

3. SERVLETS

Servlets are pieces of Java source code that add functionality to a web server in a manner similar to the way applets add functionality to a browser. Servlets are designed to support a request/response computing model that is commonly used in web servers. In a request/response model, a client sends a request message to a server and the server responds by sending back a reply message. From the Java Servlet Development Kit (JSDK), the Java Servlet API can be used to create servlets for responding to requests from clients. These servlets can do many tasks, like process HTML forms with a custom servlet or manage middle-tier processing to connect to existing data sources behind a corporate firewall. In addition, servlets can maintain services, like database sessions, between requests to manage resources better than Common Gateway Interface (CGI) technologies. The Java Servlet API is based on several Java interfaces that are provided in standard Java extension (javax) packages. The javax.servlet and javax.servlet.http packages provide interfaces and classes for writing servlets. All servlets must implement the Servlet interface, which defines life-cycle methods. When implementing a generic service, GenericServlet class provided with the Java Servlet API is extended. The HttpServlet class provides methods, such as doGet and doPost, for handling HTTP-specific services.

Detailed explanation of the method adopted by basic servlet to handle GET requests is provided. GET requests are requests made by browsers when the user types in a URL on the address line, follows a link from a Web page, or makes an HTML form that does not specify a METHOD. Servlets can also very easily handle POST requests, which are generated when someone creates an HTML form that specifies METHOD="POST".

```
import java.io.*;
import javax.servlet.*;
import javax.servlet.http.*;
public class SomeServlet extends HttpServlet {
 public void doGet(HttpServletRequest request, HttpServletResponse response)
     throws ServletException, IOException {
   // Use "request" to read incoming HTTP headers (e.g. cookies)
   // and HTML form data (e.g. data the user entered and submitted)

   // Use "response" to specify the HTTP response line and headers
   // (e.g. specifying the content type, setting cookies).

   PrintWriter out = response.getWriter();
   // Use "out" to send content to browser
 }
}
```

To be a servlet, a class should extend HttpServlet and override doGet or doPost (or both), depending on whether the data is being sent by GET or by POST. These methods take two arguments: An HttpServletRequest and an HttpServletResponse.

The HttpServletRequest has methods to find out about incoming information such as FORM data, HTTP request headers, and many. The HttpServletResponse has methods that lets the user specify the HTTP response line (200, 404, etc.), response headers (Content-Type, Set-Cookie, etc.), and most importantly, lets the user obtain a PrintWriter used to send output back to the client. For simple servlets, most of the effort is spent in println statements that generate the desired page. doGet and doPost throws two exceptions, which are required to include them in the declaration. Also the classes in java.io (for PrintWriter, etc.), javax.servlet (for HttpServlet, etc.), and javax.servlet.http (for HttpServletRequest and HttpServletResponse) are to be imported. Finally, the doGet and doPost are called by the service method, and sometimes the service may be overriden directly.

Here is a simple servlet that just generates plain text. The following section will show the more usual case where HTML is generated.

HelloWorld.java

```
package aaa;
import java.io.*;
import javax.servlet.*;
import javax.servlet.http.*;
public class HelloWorld extends HttpServlet {
  public void doGet(HttpServletRequest request, HttpServletResponse
response)throws ServletException, IOException
  {
    PrintWriter out = response.getWriter();
    out.println("Hello World");
  }
}
```

Advantages of Servlets

In any J2EE web applications servlets are an integral part. The server side component of a servlet gives a powerful mechanism for developing server side web applications. It provides an important role in the explosion of Internet, its reusability, performance and scalability. Web developers can create fast and efficient server side programs or applications by using servlets. By using servlets web developers can run these applications in any servlet enabled web servers.

The main advantages of using servlets over CGI are the CGI programs are run outside the web server so a new process should be started before the execution of a CGI program. At a time the CGI program handle only one request. After the execution of a CGI program they return the result in the web server and exit. But in the case of servlets it can handle multiple requests simultaneously. Servlets generate dynamic content or create dynamic web pages that is easy to write and faster to run within web servers. Servlets can access any J2SE and J2EE APIs and it can take the full advantage and capabilities of the java programming language. Servlets are component based, platform independent method to create in the web-based applications, without the performance limitations of CGI programs.

The advantages of servlets are discussed below.

(1) Portability
(2) Powerful
(3) Efficiency
(4) Safety
(5) Integration
(6) Extensibility
(7) Inexpensive
(8) Secure
(9) Performance

Portability

Servlets are highly portable crosswise operating systems and server implementations because the servlets are written in java and follow well-known standardized APIs. Servlets are writing once, run anywhere (WORA) program, because a servlet can be developed on Windows machine running the tomcat server or any other server and later the server can be deployed on any other operating system like Unix. Servlets are extremely portable so it can be run on any platform. Hence servlets are platform independent one. Servlets are written entirely in java.

Powerful

Several things that are difficult or impossible to be implemented with the servlets can be done with CGI. For example, the CGI programs cannot interact directly to the web server but the servlets can directly interact with the web server. Servlets can share data among each other, they make the database connection pools easy to implement. By using the session tracking mechanism, servlets can maintain the session which helps them to maintain information from request to request. Servlets can do many things which are difficult to implement in the CGI programs.

Efficiency

The servlets invocation is highly efficient as compared to CGI programs. The servlet remains in the server's memory as a single object instance, when the servlet gets loaded in the server. The servlets are highly scalable because multiple concurrent requests are handled by separate threads. That is, N number of threads can be handled by using a single servlet class.

Safety

As servlets are written in java, servlets inherit the strong type safety of java language. In java , automatic garbage collection mechanism and a lack of pointers protect the servlets from memory management problems. In servlets errors can be easily handled due to the availability of exception handling mechanism. If any error occurs, then it will throw an exception by using the throw statement.

Integration

Servlets are tightly integrated with the server. Servlet can use the server to translate the file paths, perform logging, check authorization, and MIME type mapping, etc.

Extensibility

The servlet API is designed in such a way that it can be easily extensible. As it stands today, the servlet API supports Http Servlets, but in later date it can be extended for another type of servlets. Java Servlets are developed in java which is robust, well-designed and object-oriented language which can be extended or polymorphed into new objects. So the java servlets take all these advantages and can be extended from existing class to provide the ideal solutions. So servlets are more extensible and reliable.

Inexpensive

There are number of free web servers available for personal use or for commercial purpose. Web servers are relatively expensive. So by using the free available web servers, servlet programming can be done easily.

Secure

Servlets are server side components, so it inherits the security provided by the web server. Servlets are also benefited with Java Security Manager.

Performance

Servlets are faster than CGI programs because each script in CGI produces a new process and these processes takes a lot of time for execution. But in case of servlets it creates only new thread. Due to interpreted nature of java, programs written in java are slow. But the java servlets runs very fast. These are due to the way servlets run on web server. For any program initialization takes significant amount of time. But in case of servlets initialization takes place first time it receives a request and remains in memory till times out or server shut downs. After servlet is loaded, to handle a new request it simply creates a new thread and runs service method of servlet. In comparison to traditional CGI scripts which creates a new process to serve the request.

3.1 Architecture

The following diagram shows the position of Servlets in a Web Application.

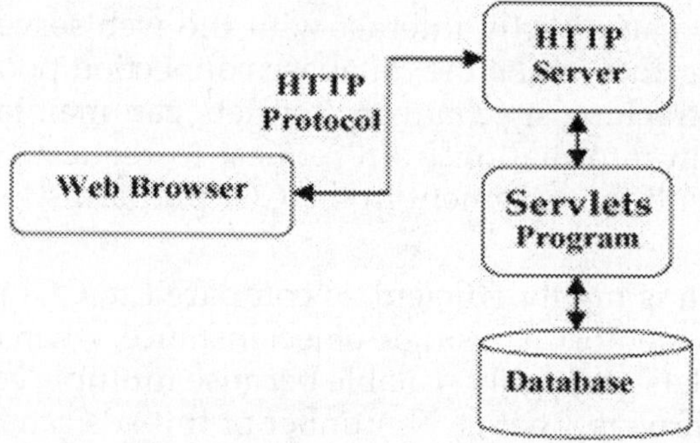

Servlets perform the following major tasks:

1. Read the explicit data sent by the clients (browsers). This includes an HTML form on a Web page or it could also come from an applet or a custom HTTP client program.
2. Read the implicit HTTP request data sent by the clients (browsers). This includes cookies, media types and compression schemes the browser understands, and so forth.
3. Process the data and generate the results. This process may require interacting with a database, executing an RMI or CORBA call, invoking a Web service, or computing the response directly.
4. Send the explicit data (i.e. the document) to the clients (browsers). This document can be sent in a variety of formats, including text (HTML or XML), binary (GIF images), Excel, etc.
5. Send the implicit HTTP response to the clients (browsers). This includes telling the browsers or other clients what type of document is being returned (e.g. HTML), setting cookies and caching parameters, and other such tasks.

3.2 Environment Setup

Java Servlets are Java classes run by a web server that has an interpreter supporting the Java Servlet specification.

Servlets can be created using the **javax.servlet** and **javax.servlet.http** packages, which are a standard part of the Java's enterprise edition, an expanded version of the Java class library that supports large-scale development projects. These classes implement the Java Servlet and JSP specifications. Java servlets have been created and compiled just like any other Java class. After installing the servlet packages and adding them to the Classpath, servlets can be compiled with the JDK's Java compiler or any other current compiler.

A development environment is where the Servlet can be developed, tested and finally run. Like any other Java program, a servlet program is compiled by using the Java compiler **javac** and after compiling the servlet application, it would be deployed in a configured environment to test and run.

This development environment setup involves the following steps:

This step involves downloading an implementation of the Java Software Development Kit (SDK) and setting up PATH environment variable appropriately. It can be downloaded from Oracle's Java site.

Once you download your Java implementation, follow the given instructions to install and configure the setup. Finally set PATH and JAVA_HOME environment variables to refer to the directory that contains java and javac, typically java_install_dir/bin and java_install_dir respectively.

If you are running Windows and installed the SDK in C:\jdk1.5.0_20, you would put the following line in your C:\autoexec.bat file.

```
set PATH=C:\jdk1.5.0_20\bin;%PATH%
set JAVA_HOME=C:\jdk1.5.0_20
```

Alternatively, on Windows NT/2000/XP, you could also right-click on My Computer, select Properties, then Advanced, then Environment Variables. Then, you would update the PATH value and press the OK button.

On Unix (Solaris, Linux, etc.), if the SDK is installed in /usr/local/jdk1.5.0_20 and you use the C shell, you would put the following into your .cshrc file.

```
setenv PATH /usr/local/jdk1.5.0_20/bin:$PATH
setenv JAVA_HOME /usr/local/jdk1.5.0_20
```

Alternatively, if you use an Integrated Development Environment (IDE) like Borland JBuilder, Eclipse, IntelliJ IDEA, or Sun ONE Studio, compile and run a simple program to confirm that the IDE knows where you installed Java.

Setting up Web Server: Tomcat

A number of Web Servers that support servlets are available in the market. Some web servers are freely downloadable and Tomcat is one of them.

Apache Tomcat is an open source software implementation of the Java Servlet and JavaServer Pages technologies and can act as a standalone server for testing servlets and can be integrated with the Apache Web Server. Here are the steps to setup Tomcat on your machine:

- Download latest version of Tomcat from http://tomcat.apache.org/.
- Once you downloaded the installation, unpack the binary distribution into a convenient location. For example, in C:\apache-tomcat-5.5.29 on windows, or /usr/local/apache-tomcat-5.5.29 on Linux/Unix and create CATALINA_HOME environment variable pointing to these locations.

Tomcat can be started by executing the following commands on windows machine:

%CATALINA_HOME%\bin\startup.bat

or

C:\apache-tomcat-5.5.29\bin\startup.bat

Tomcat can be started by executing the following commands on Unix (Solaris, Linux, etc.) machine:

$CATALINA_HOME/bin/startup.sh

or

/usr/local/apache-tomcat-5.5.29/bin/startup.sh

After startup, the default web applications included with Tomcat will be available by visiting **http://localhost:8080/**. If everything is fine, then it should display the following result:

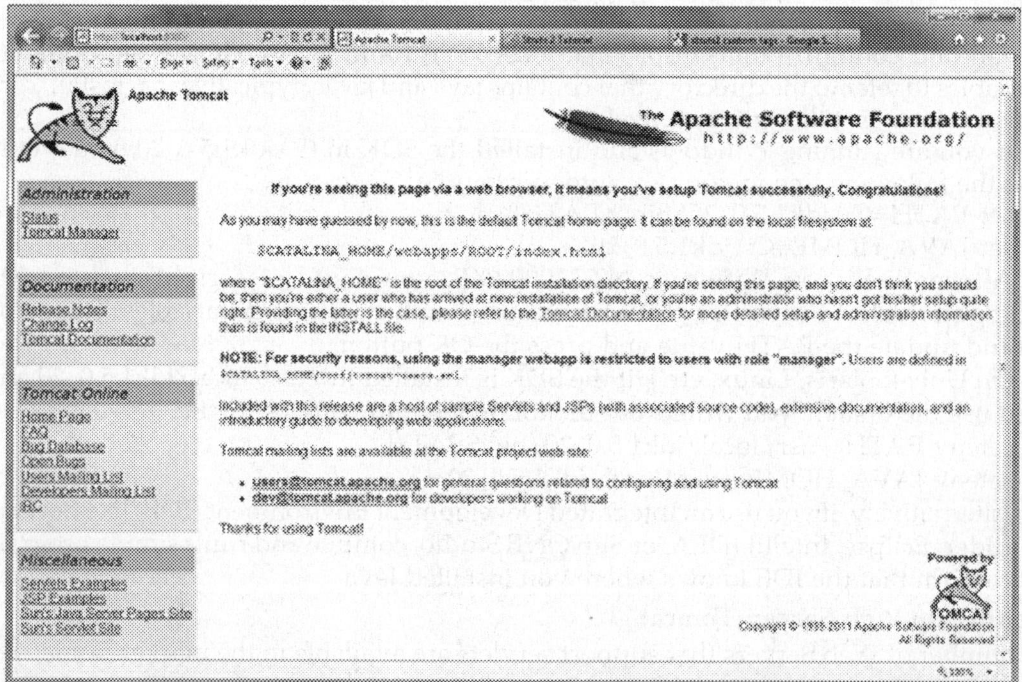

Further information about configuring and running Tomcat can be found in the documentation included here, as well as on the Tomcat web site: http://tomcat.apache.org

Tomcat can be stopped by executing the following commands on windows machine:

C:\apache-tomcat-5.5.29\bin\shutdown

Tomcat can be stopped by executing the following commands on Unix (Solaris, Linux, etc.) machine:

/usr/local/apache-tomcat-5.5.29/bin/shutdown.sh

Setting up CLASSPATH

Since servlets are not part of the Java Platform, Standard Edition, you must identify the servlet classes to the compiler.

If you are running Windows, you need to put the following lines in your C:\autoexec.bat file.

```
set CATALINA=C:\apache-tomcat-5.5.29

set CLASSPATH=%CATALINA%\common\lib\servlet-api.jar;%CLASSPATH%
```

Alternatively, on Windows NT/2000/XP, you could also right-click on My Computer, select Properties, then Advanced, then Environment Variables. Then, you would update the CLASSPATH value and press the OK button.

On Unix (Solaris, Linux, etc.), if you are using the C shell, you would put the following lines into your .cshrc file.

```
setenv CATALINA=/usr/local/apache-tomcat-5.5.29

setenv CLASSPATH $CATALINA/common/lib/servlet-api.jar:$CLASSPATH
```

3.3 Life Cycle

A servlet life cycle can be defined as the entire process from its creation till the destruction. The following are the paths followed by a servlet:

- The servlet is initialized by calling the **init ()** method.
- The servlet calls **service()** method to process a client's request.
- The servlet is terminated by calling the **destroy()** method.
- Finally, servlet is garbage collected by the garbage collector of the JVM.

Now let us discuss the life cycle methods in details.

The init() method

The init method is designed to be called only once. It is called when the servlet is first created, and not called again for each user request. So, it is used for one-time initializations, just as with the init method of applets.

The servlet is normally created when a user first invokes a URL corresponding to the servlet, but you can also specify that the servlet be loaded when the server is first started.

When a user invokes a servlet, a single instance of each servlet gets created, with each user request resulting in a new thread that is handed off to doGet or doPost as appropriate. The init() method simply creates or loads some data that will be used throughout the life of the servlet.

The init method definition looks like this:

```
public void init() throws ServletException {
  // Initialization code...
}
```

The service() method

The service() method is the main method to perform the actual task. The servlet container (i.e. web server) calls the service() method to handle requests coming from the client (browsers) and to write the formatted response back to the client.

Each time the server receives a request for a servlet, the server spawns a new thread and calls service. The service() method checks the HTTP request type (GET, POST, PUT, DELETE, etc.) and calls doGet, doPost, doPut, doDelete, etc. methods as appropriate.

Here is the signature of this method:

```
public void service(ServletRequest request, ServletResponse response)
  throws ServletException, IOException
{ }
```

The service () method is called by the container and service method invokes doGe, doPost, doPut, doDelete, etc. methods as appropriate. So you have nothing to do with service() method but you override either doGet() or doPost() depending on what type of request you receive from the client.

The doGet() and doPost() are most frequently used methods within each service request. Here are the signature of these two methods.

The doGet() Method

A GET request results from a normal request for a URL or from an HTML form that has no METHOD specified and it should be handled by doGet() method.

```
public void doGet(HttpServletRequest request, HttpServletResponse response)
throws ServletException, IOException {
   // Servlet code}
```

The doPost() Method

A POST request results from an HTML form that specifically lists POST as the METHOD and it should be handled by doPost() method.

```
public void doPost(HttpServletRequest request, HttpServletResponse response)
throws ServletException, IOException {
   // Servlet code }
```

The destroy() method

The destroy() method is called only once at the end of the life cycle of a servlet. This method gives your servlet a chance to close database connections, halt background threads, write cookie lists or hit counts to disk, and perform other such cleanup activities.

After the destroy() method is called, the servlet object is marked for garbage collection. The destroy method definition looks like this:

```
public void destroy() {
  // Finalization code...
}
```

The following figure depicts a typical servlet life-cycle scenario.

- First the HTTP requests coming to the server are delegated to the servlet container.
- The servlet container loads the servlet before invoking the service() method.

Then the servlet container handles multiple requests by spawning multiple threads, each thread executing the service() method of a single instance of the servlet.

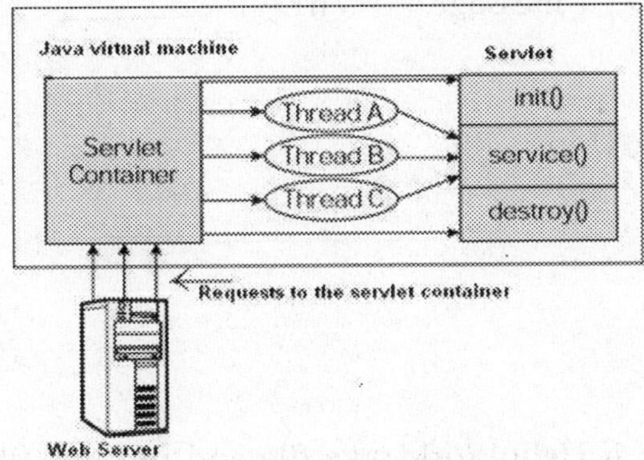

3.4 First Program

Servlets are Java classes which service HTTP requests and implement the **javax.servlet.Servlet** interface. Web application developers typically write servlets that extend javax.servlet.http.HttpServlet, an abstract class that implements the Servlet interface and is specially designed to handle HTTP requests.

Sample Code for Hello World

Following is the sample source code structure of a servlet example to write Hello World:

```
// Import required java libraries
import java.io.*;
import javax.servlet.*;
import javax.servlet.http.*;

// Extend HttpServlet class
public class HelloWorld extends HttpServlet {

  private String message;

  public void init() throws ServletException
  {
    // Do required initialization
    message = "Hello World";
  }

  public void doGet(HttpServletRequest request, HttpServletResponse response)
        throws ServletException, IOException
  {
    // Set response content type
    response.setContentType("text/html");
```

```
        // Actual logic goes here.
        PrintWriter out = response.getWriter();
        out.println("<h1>" + message + "</h1>");
    }

    public void destroy()
    {
        // do nothing.
    }
}
```

Compiling a Servlet

Let us put above code if HelloWorld.java file and put this file in C:\ServletDevel (Windows) or /usr/ServletDevel (Unix), then you would need to add these directories as well in CLASSPATH.

Assuming your environment is setup properly, go in **ServletDevel** directory and compile HelloWorld.java as follows:

```
$ javac HelloWorld.java
```

If the servlet depends on any other libraries, you have to include those JAR files on your CLASSPATH as well. In this example, only servlet-api.jar JAR file since no other library is used in Hello World program.

This command line uses the built-in javac compiler that comes with the Sun Microsystems Java Software Development Kit (JDK). For this command to work properly, you have to include the location of the Java SDK that you are using in the PATH environment variable. It can be done, for example, as

```
$ javac HelloWorld.java  - servlet-api.jar
```

If everything goes fine, above compilation would produce **HelloWorld.class** file in the same directory. Next section would explain how a compiled servlet would be deployed in production.

Servlet Deployment

By default, a servlet application is located at the path <Tomcat-installation-directory>/webapps/ROOT and the class file would reside in <Tomcat-installation-directory>/webapps/ROOT/WEB-INF/classes.

If you have a fully qualified class name of **com.myorg.MyServlet**, then this servlet class must be located in WEB-INF/classes/com/myorg/MyServlet.class.

For now, let us copy HelloWorld.class into <Tomcat-installation-directory>/webapps/ROOT/WEB-INF/classes and create following entries in **web.xml** file located in <Tomcat-installation-directory>/webapps/ROOT/WEB-INF/

```
<servlet>
    <servlet-name>HelloWorld</servlet-name>
    <servlet-class>HelloWorld</servlet-class>
</servlet>
```

```
<servlet-mapping>
    <servlet-name>HelloWorld</servlet-name>
    <url-pattern>/HelloWorld</url-pattern>
</servlet-mapping>
```

Above entries to be created inside <web-app>...</web-app> tags available in web.xml file. There could be various entries in this table already available, but never mind.

You are almost done, now let us start tomcat server using <Tomcat-installation-directory>\bin\startup.bat (on windows) or <Tomcat-installation-directory>/bin/startup.sh (on Linux/Solaris, etc.) and finally type **http://localhost:8080/HelloWorld** in browser's address box. If everything goes fine, you would get the following result:

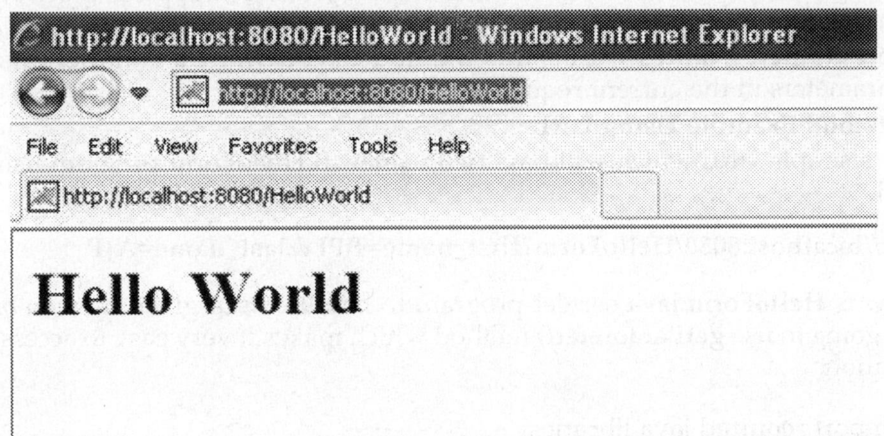

3.5 Form Data

There may be situations where the user needs to pass some information from browser to web server and ultimately to the backend program. The browser uses two methods to pass this information to web server. These methods are GET Method and POST Method.

GET method

The GET method sends the encoded user information appended to the page request. The page and the encoded information are separated by the ? character as follows:

http://localhost:8080/hello?key1=value1&key2=value2

The GET method is the defualt method to pass information from browser to web server and it produces a long string that appears in the browser's Location: box. Never use the GET method if you have password or other sensitive information to pass to the server. The GET method has size limitation: Only 1024 characters can be in a request string.

This information is passed using QUERY_STRING header and will be accessible through QUERY_STRING environment variable and Servlet handles this type of requests using **doGet()** method.

POST method

A generally more reliable method of passing information to a backend program is the POST method. This packages the information in exactly the same way as GET methods, but instead of sending it as a text string after a ? in the URL, it sends it as a separate message. This message comes to the backend program in the form of the standard input which you can parse and use for your processing. Servlet handles this type of requests using **doPost()** method.

Reading Form Data using Servlet

Servlets handles form data parsing automatically using the following methods depending on the situation:

- **getParameter():** You call request.getParameter() method to get the value of a form parameter.
- **getParameterValues():** Call this method if the parameter appears more than once and returns multiple values, for example, checkbox.
- **getParameterNames():** Call this method if you want a complete list of all parameters in the current request.

GET Method Example Using URL

Here is a simple URL which will pass two values to HelloForm program using GET method.

http://localhost:8080/HelloForm?first_name=BPL&last_name=AJP

Below is **HelloForm.java** servlet program to handle input given by web browser. We are going to use **getParameter()** method which makes it very easy to access passed information:

```
// Import required java libraries
import java.io.*;
import javax.servlet.*;
import javax.servlet.http.*;

// Extend HttpServlet class
public class HelloForm extends HttpServlet {

  public void doGet(HttpServletRequest request, HttpServletResponse
response)
        throws ServletException, IOException
  {
    // Set response content type
    response.setContentType("text/html");

    PrintWriter out = response.getWriter();
        String title = "Using GET Method to Read Form Data";
    String docType =
    "<!doctype html public \"-//w3c//dtd html 4.0 " +
    "transitional//en\">\n";
    out.println(docType +
```

```
        "<html>\n" +
        "<head><title>" + title + "</title></head>\n" +
        "<body bgcolor=\"#f0f0f0\">\n" +
        "<h1 align=\"center\">" + title + "</h1>\n" +
        "<ul>\n" +
        "  <li><b>First Name</b>: "
        + request.getParameter("first_name") + "\n" +
        "  <li><b>Last Name</b>: "
        + request.getParameter("last_name") + "\n" +
        "</ul>\n" +
        "</body></html>");
  }
}
```

Assuming your environment is setup properly, compile HelloForm.java as follows:

```
$ javac HelloForm.java
```

If everything goes fine, above compilation would produce **HelloForm.class** file. Next you would have to copy this class file in <Tomcat-installation-directory>/webapps/ROOT/WEB-INF/classes and create following entries in **web.xml** file located in <Tomcat-installation-directory>/webapps/ROOT/WEB-INF/

```
<servlet>
   <servlet-name>HelloForm</servlet-name>
   <servlet-class>HelloForm</servlet-class>
</servlet>

<servlet-mapping>
   <servlet-name>HelloForm</servlet-name>
   <url-pattern>/HelloForm</url-pattern>
</servlet-mapping>
```

Now type *http://localhost:8080/HelloForm?first_name=BPL&last_name=AJP* in your browser's Location:box and make sure you already started tomcat server, before firing above command in the browser. This would generate the following result:

Using GET Method to Read Form Data·
First Name: BPL
Last Name: AJP

GET Method Example Using Form

Here is a simple example which passes two values using HTML FORM and submit button. We are going to use same Servlet HelloForm to handle this input.

```
<html>
<body>
```

```
<form action="HelloForm" method="GET">
First Name: <input type="text" name="first_name">
<br />
Last Name: <input type="text" name="last_name" />
<input type="submit" value="Submit" />
</form>
</body>
</html>
```

Keep this HTML in a file Hello.htm and put it in <Tomcat-installation-directory>/ webapps/ROOT directory. When you would access *http://localhost:8080/Hello.htm*, here is the actual output of the above form.

First Name: [ ]

Last Name: [ ]

Try to enter First Name and Last Name and then click submit button to see the result on your local machine where tomcat is running. Based on the input provided, it will generate similar result as mentioned in the above example.

POST Method Example Using Form

Let us do a little modification in the above servlet, so that it can handle GET as well as POST methods. Below is **HelloForm.java** servlet program to handle input given by web browser using GET or POST methods.

```
// Import required java libraries
import java.io.*;
import javax.servlet.*;
import javax.servlet.http.*;

// Extend HttpServlet class
public class HelloForm extends HttpServlet {

  // Method to handle GET method request.
  public void doGet(HttpServletRequest request,
            HttpServletResponse response)
        throws ServletException, IOException
  {
    // Set response content type
    response.setContentType("text/html");

    PrintWriter out = response.getWriter();
      String title = "Using GET Method to Read Form Data";
    String docType =
    "<!doctype html public \"-//w3c//dtd html 4.0 " +
    "transitional//en\">\n";
    out.println(docType +
```

```
        "<html>\n" +
        "<head><title>" + title + "</title></head>\n" +
        "<body bgcolor=\"#f0f0f0\">\n" +
        "<h1 align=\"center\">" + title + "</h1>\n" +
        "<ul>\n" +
        "  <li><b>First Name</b>: "
        + request.getParameter("first_name") + "\n" +
        "  <li><b>Last Name</b>: "
        + request.getParameter("last_name") + "\n" +
        "</ul>\n" +
        "</body></html>");
  }
  // Method to handle POST method request.
  public void doPost(HttpServletRequest request,
           HttpServletResponse response)
    throws ServletException, IOException {
    doGet(request, response);
  }
}
```

Now compile, deploy the above Servlet and test it using Hello.htm with the POST method as follows:

```
<html>
<body>
<form action="HelloForm" method="POST">
First Name: <input type="text" name="first_name">
<br />
Last Name: <input type="text" name="last_name" />
<input type="submit" value="Submit" />
</form>
</body>
</html>
```

Here is the actual output of the above form, Try to enter First and Last Name and then click submit button to see the result on your local machine where tomcat is running.

First Name: [ ]

Last Name: [ ]

Based on the input provided, it would generate similar result as mentioned in the above examples.

Passing Checkbox Data to Servlet Program

Checkboxes are used when more than one option is required to be selected.

For example, HTML code, CheckBox.htm, for a form with two checkboxes

```
<html>
<body>
<form action="CheckBox" method="POST" target="_blank">
<input type="checkbox" name="maths" checked="checked" /> Maths
<input type="checkbox" name="physics"
  /> Physics<input type="checkbox" name="chemistry" checked="checked" />
                                   Chemistry
<input type="submit" value="Select Subject" />
</form>
</body>
</html>
```

The result of this code is of the following form

Top of Form

Maths Physics Chemistry

Below is CheckBox.java servlet program to handle input given by web browser for checkbox button.

```
// Import required java libraries
import java.io.*;
import javax.servlet.*;
import javax.servlet.http.*;

// Extend HttpServlet class
public class CheckBox extends HttpServlet {

  // Method to handle GET method request.
 public void doGet(HttpServletRequest request,
          HttpServletResponse response)
      throws ServletException, IOException
 {
   // Set response content type
   response.setContentType("text/html");

   PrintWriter out = response.getWriter();
      String title = "Reading Checkbox Data";
   String docType =
   "<!doctype html public \"-//w3c//dtd html 4.0 " +
   "transitional//en\">\n";
   out.println(docType +
        "<html>\n" +
        "<head><title>" + title + "</title></head>\n" +
        "<body bgcolor=\"#f0f0f0\">\n" +
        "<h1 align=\"center\">" + title + "</h1>\n" +
        "<ul>\n" +
        "  <li><b>Maths Flag : </b>: "
```

```
        + request.getParameter("maths") + "\n" +
        "  <li><b>Physics Flag: </b>: "
        + request.getParameter("physics") + "\n" +
        "  <li><b>Chemistry Flag: </b>: "
        + request.getParameter("chemistry") + "\n" +
        "</ul>\n" +
        "</body></html>");
  }
  // Method to handle POST method request.
  public void doPost(HttpServletRequest request,
            HttpServletResponse response)
    throws ServletException, IOException {
    doGet(request, response);
  }
}
```

For the above example, it would display the following results:

Reading Checkbox Data
 Maths Flag : : on
 Physics Flag: : null
 Chemistry Flag: : on

Reading All Form Parameters

Following is the generic example which uses **getParameterNames()** method of HttpServletRequest to read all the available form parameters. This method returns an Enumeration that contains the parameter names in an unspecified order.

Once we have an Enumeration, we can loop down the Enumeration in the standard manner, using *hasMoreElements()* method to determine when to stop and using *nextElement()* method to get each parameter name.

```
// Import required java libraries
import java.io.*;
import javax.servlet.*;
import javax.servlet.http.*;
import java.util.*;

// Extend HttpServlet class
public class ReadParams extends HttpServlet {

  // Method to handle GET method request.
  public void doGet(HttpServletRequest request, HttpServletResponse response)
    throws ServletException, IOException
  {
    // Set response content type
```

```
response.setContentType("text/html");
PrintWriter out = response.getWriter();
      String title = "Reading All Form Parameters";
String docType =
"<!doctype html public \"-//w3c//dtd html 4.0 " +
"transitional//en\">\n";
out.println(docType +
  "<html>\n" +
  "<head><title>" + title + "</title></head>\n" +
  "<body bgcolor=\"#f0f0f0\">\n" +
  "<h1 align=\"center\">" + title + "</h1>\n" +
  "<table width=\"100%\" border=\"1\" align=\"center\">\n" +
  "<tr bgcolor=\"#949494\">\n" +
  "<th>Param Name</th><th>Param Value(s)</th>\n"+
  "</tr>\n");
Enumeration paramNames = request.getParameterNames();

while(paramNames.hasMoreElements()) {
   String paramName = (String)paramNames.nextElement();
   out.print("<tr><td>" + paramName + "</td>\n<td>");
   String[] paramValues =
         request.getParameterValues(paramName);
   // Read single valued data
   if (paramValues.length == 1) {
    String paramValue = paramValues[0];
    if (paramValue.length() == 0)
     out.println("<i>No Value</i>");
    else
     out.println(paramValue);
   } else {
     // Read multiple valued data
     out.println("<ul>");
     for(int i=0; i < paramValues.length; i++) {
       out.println("<li>" + paramValues[i]);
     }
     out.println("</ul>");
   }
 }
 out.println("</tr>\n</table>\n</body></html>");
}
// Method to handle POST method request.
public void doPost(HttpServletRequest request, HttpServletResponse
response)
  throws ServletException, IOException {
  doGet(request, response);
}
}
```

Now, try above servlet with the following form:

```
<html>
<body>
<form action="ReadParams" method="POST" target="_blank">
<input type="checkbox" name="maths" checked="checked" /> Maths
<input type="checkbox" name="physics" /> Physics
<input type="checkbox" name="chemistry" checked="checked" /> Chem
<input type="submit" value="Select Subject" />
</form>
</body>
</html>
```

Now calling servlet using the above form would generate the following result:

Reading All Form Parameters

Param Name	Param Value(s)
Maths	On
Chemistry	On

The above servlet can be used to read any other form's data which is having other objects like text box, radio button or drop down box, etc.

3.6 Client HTTP Request

When a browser requests for a web page, it sends a lot of information to the web server which cannot be read directly because this information travel as a part of header of HTTP request. Following is the important header information which comes from browser side and you would use very frequently in web programming:

Header	Description
Accept	This header specifies the MIME types that the browser or other clients can handle. Values of **image/png** or **image/jpeg** are the two most common possibilities.
Accept-Charset	This header specifies the character sets the browser can use to display the information. For example ISO-8859-1.
Accept-Encoding	This header specifies the types of encodings that the browser knows how to handle. Values of **gzip** or **compress** are the two most common possibilities.
Accept-Language	This header specifies the client's preferred languages in case the servlet can produce results in more than one language. For example, en, en-us, ru, etc.
Authorization	This header is used by clients to identify themselves when accessing password-protected Web pages.
Connection	This header indicates whether the client can handle persistent HTTP connections. Persistent connections permit

	the client or other browser to retrieve multiple files with a single request. A value of **Keep-Alive** means that persistent connections should be used.
Content-Length	This header is applicable only to POST requests and gives the size of the POST data in bytes.
Cookie	This header returns cookies to servers that previously sent them to the browser.
Host	This header specifies the host and port as given in the original URL.
If-Modified-Since	This header indicates that the client wants the page only if it has been changed after the specified date. The server sends a code, 304 which means **Not Modified** header if no newer result is available.
If-Unmodified-Since	This header is the reverse of If-Modified-Since; it specifies that the operation should succeed only if the document is older than the specified date.
Referer	This header indicates the URL of the referring Web page. For example, if you are at Web page 1 and click on a link to Web page 2, the URL of Web page 1 is included in the Referer header when the browser requests Web page 2.
User-Agent	This header identifies the browser or other client making the request and can be used to return different content to different types of browsers.

Methods to read HTTP Header

There are following methods which can be used to read HTTP header in your servlet program. These methods are available with *HttpServletRequest* object.

Methods	Description
Cookie[] getCookies()	Returns an array containing all of the Cookie objects the client sent with this request.
Enumeration getAttributeNames()	Returns an Enumeration containing the names of the attributes available to this request.
Enumeration getHeaderNames()	Returns an enumeration of all the header names this request contains.
Enumeration getParameterNames()	Returns an Enumeration of String objects containing the names of the parameters contained in this request.
HttpSession getSession()	Returns the current session associated with this request, or if the request does not have a session, creates one.
HttpSession getSession(boolean create)	Returns the current HttpSession associated with this request, or if here is no current session and create is true, returns a new session.

Locale getLocale()	Returns the preferred Locale that the client will accept content, based on the Accept-Language header
Object getAttribute(String name)	Returns the value of the named attribute as an Object, or null if no attribute of the given name exists.
ServletInputStream getInputStream()	Retrieves the body of the request as binary data using a ServletInputStream.
String getAuthType()	Returns the name of the authentication scheme used to protect the servlet, for example, "BASIC" or "SSL," or null if the JSP was not protected.
String getCharacterEncoding()	Returns the name of the character encoding used in the body of this request.
String getContentType()	Returns the MIME type of the body of the request, or null if the type is not known.
String getContextPath()	Returns the portion of the request URI that indicates the context of the request.
String getHeader(String name)	Returns the value of the specified request header as a String.
String getMethod()	Returns the name of the HTTP method with which this request was made, for example, GET, POST, or PUT.
String getParameter(String name)	Returns the value of a request parameter as a String, or null if the parameter does not exist.
String getPathInfo()	Returns any extra path information associated with the URL the client sent when it made this request.
String getProtocol()	Returns the name and version of the protocol the request.
String getQueryString()	Returns the query string that is contained in the request URL after the path.
String getRemoteAddr()	Returns the Internet Protocol (IP) address of the client that sent the request.
String getRemoteHost()	Returns the fully qualified name of the client that sent the request.
String getRemoteUser()	Returns the login of the user making this request, if the user has been authenticated, or null if the user has not been authenticated.
String getRequestURL()	Returns the part of this request's URL from the protocol name up to the query string in the first line of the HTTP request.
String getRequestedSessionId()	Returns the session ID specified by the client.

String getServletPath()	Returns the part of this request's URL that calls the JSP.
String[] getParameterValues (String name)	Returns an array of String objects containing all of the values the given request parameter has, or null if the parameter does not exist.
Boolean isSecure()	Returns a boolean indicating whether this request was made using a secure channel, such as HTTPs.
Int getContentLength()	Returns the length, in bytes, of the request body and made available by the input stream, or –1 if the length is not known.
Int getIntHeader(String name)	Returns the value of the specified request header as an int.
Int getServerPort()	Returns the port number on which this request was received.

HTTP Header Request Example

Following is the example which uses **getHeaderNames()** method of HttpServletRequest to read the HTTP header infromation. This method returns an Enumeration that contains the header information associated with the current HTTP request.

Once we have an Enumeration, we can loop down the Enumeration in the standard manner, using *hasMoreElements()* method to determine when to stop and using *nextElement()* method to get each parameter name.

```
// Import required java libraries
import java.io.*;
import javax.servlet.*;
import javax.servlet.http.*;
import java.util.*;

// Extend HttpServlet class
public class DisplayHeader extends HttpServlet {

  // Method to handle GET method request.
  public void doGet(HttpServletRequest request,
            HttpServletResponse response)
        throws ServletException, IOException
  {
     // Set response content type
     response.setContentType("text/html");
     PrintWriter out = response.getWriter();
    String title =
"HTTP Header Request Example";
    String docType =
```

```
    "<!doctype html public \"-//w3c//dtd html 4.0 " +
    "transitional//en\">\n";
    out.println(docType +
      "<html>\n" +
      "<head><title>" + title + "</title></head>\n"+
      "<body bgcolor=\"#f0f0f0\">\n" +
      "<h1 align=\"center\">" + title + "</h1>\n" +
      "<table width=\"100%\" border=\"1\" align=\"center\">\n" +
      "<tr bgcolor=\"#949494\">\n" +
      "<th>Header Name</th><th>Header Value(s)</th>\n"+
      "</tr>\n");

     Enumeration headerNames = request.getHeaderNames();

        while(headerNames.hasMoreElements()) {
      String paramName = (String)headerNames.nextElement();
      out.print("<tr><td>" + paramName + "</td>\n");
      String paramValue = request.getHeader(paramName);
      out.println("<td> " + paramValue + "</td></tr>\n");
    }
    out.println("</table>\n</body></html>");
  }
  // Method to handle POST method request.
  public void doPost(HttpServletRequest request,
            HttpServletResponse response)
    throws ServletException, IOException {
   doGet(request, response);
  }
}
```

Now calling the above servlet would generate the following results:

HTTP Header	Request Example
Header Name	Header Value(s)
Accept	*/*
Accept-language	en-us
User-agent	Mozilla/4.0 (compatible; MSIE 7.0; Windows NT 5.1; Trident/4.0; InfoPath.2; MS-RTC LM 8)
Accept-encoding	gzip, deflate
Host	localhost:8080
Connection	Keep-Alive
Cache-control	no-cache

3.7 Server HTTP Response

When a Web server responds to a HTTP request to the browser, the response typically consists of a status line, some response headers, a blank line, and the document. A typical response looks like this:

```
HTTP/1.1 200 OK
Content-Type: text/html
Header2: ...
...
HeaderN: ...
 (Blank Line)
<!doctype ...>
<html>
<head>...</head>
<body>
...
</body>
</html>
```

The status line consists of the HTTP version (HTTP/1.1 in the example), a status code (200 in the example), and a very short message corresponding to the status code (OK in the example).

Following is a summary of the most useful HTTP 1.1 response headers which go back to the browser from web server side and you would use them very frequently in web programming:

Header	Description
Allow	This header specifies the request methods (GET, POST, etc.) that the server supports.
Cache-Control	This header specifies the circumstances in which the response document can safely be cached. It can have values **public, private** or **no-cache,** etc. Public means document is cacheable. Private means document is for a single user and can only be stored in private (nonshared) caches and no-cache means document should never be cached.
Connection	This header instructs the browser whether to use persistent in HTTP connections or not. A value of **close** instructs the browser not to use persistent HTTP connections and **keep-alive** means using persistent connections.
Content-Disposition	This header lets you request that the browser ask the user to save the response to disk in a file of the given name.
Content-Encoding	This header specifies the way in which the page was encoded during transmission.
Content-Language	This header signifies the language in which the document is written. For example, en, en-us, ru, etc.

Content-Length	This header indicates the number of bytes in the response. This information is needed only if the browser is using a persistent (keep-alive) HTTP connection.
Content-Type	This header gives the MIME (Multipurpose Internet Mail Extension) type of the response document.
Expires	This header specifies the time at which the content should be considered out-of-date and thus no longer be cached.
Last-Modified	This header indicates when the document was last changed. The client can then cache the document and supply a date by an **If-Modified-Since** request header in later requests.
Location	This header should be included with all responses that have a status code in the 300s. This notifies the browser of the document address. The browser automatically reconnects to this location and retrieves the new document.
Refresh	This header specifies how soon the browser should ask for an updated page. You can specify time in number of seconds after which a page would be refreshed.
Retry-After	This header can be used in conjunction with a 503 (Service Unavailable) response to tell the client how soon it can repeat its request.
Set-Cookie	This header specifies a cookie associated with the page.

Methods to Set HTTP Response Header

There are following methods which can be used to set HTTP response header in your servlet program. These method are available with *HttpServletResponse* object.

Methods	Use
String encodeRedirectURL(String url)	Encodes the specified URL for use in the sendRedirect method or, if encoding is not needed, returns the URL unchanged.
String encodeURL(String url)	Encodes the specified URL by including the session ID in it, or, if encoding is not needed, returns the URL unchanged.
Boolean containsHeader(String name)	Returns a boolean indicating whether the named response header has already been set.
Boolean isCommitted()	Returns a boolean indicating if the response has been committed.
Void addCookie(Cookie cookie)	Adds the specified cookie to the response.
Void addDateHeader(String name, long date)	Adds a response header with the given name and date-value.

Void addHeader(String name, String value)	Adds a response header with the given name and value.
Void addIntHeader(String name, int value)	Adds a response header with the given name and integer value.
Void flushBuffer()	Forces any content in the buffer to be written to the client.
Void reset()	Clears any data that exists in the buffer as well as the status code and headers.
Void resetBuffer()	Clears the content of the underlying buffer in the response without clearing headers or status code.
Void sendError(int sc)	Sends an error response to the client using the specified status code and clearing the buffer.
Void sendError(int sc, String msg)	Sends an error response to the client using the specified status.
Void sendRedirect(String location)	Sends a temporary redirect response to the client using the specified redirect location URL.
Void setBufferSize(int size)	Sets the preferred buffer size for the body of the response.
Void setCharacterEncoding (String charset)	Sets the character encoding (MIME charset) of the response being sent to the client, for example, to UTF-8.
Void setContentLength(int len)	Sets the length of the content body in the response. In HTTP servlets, this method sets the HTTP Content-Length header.
Void setContentType(String type)	Sets the content type of the response being sent to the client, if the response has not been committed yet.
Void setDateHeader(String name, long date)	Sets a response header with the given name and date-value.
Void setHeader(String name, String value)	Sets a response header with the given name and value.
Void setIntHeader(String name, value)	Sets a response header with the given int name and integer value.
Void setLocale(Locale loc)	Sets the locale of the response, if the response has not been committed yet.
Void setStatus(int sc)	Sets the status code for this response.

HTTP Header Response Example

You already have seen setContentType() method working in previous examples and following example would also use same method, additionally we would use **setIntHeader()** method to set **Refresh** header.

```
// Import required java libraries
import java.io.*;
import javax.servlet.*;
import javax.servlet.http.*;
import java.util.*;

// Extend HttpServlet class
public class Refresh extends HttpServlet {
  // Method to handle GET method request.
  public void doGet(HttpServletRequest request, HttpServletResponse response)
        throws ServletException, IOException
  {
     // Set refresh, autoload time as 5 seconds
     response.setIntHeader("Refresh", 5);

     // Set response content type
     response.setContentType("text/html");

     // Get current time
     Calendar calendar = new GregorianCalendar();
     String am_pm;
     int hour = calendar.get(Calendar.HOUR);
     int minute = calendar.get(Calendar.MINUTE);
     int second = calendar.get(Calendar.SECOND);
     if(calendar.get(Calendar.AM_PM) == 0)
      am_pm = "AM";
     else
      am_pm = "PM";

     String CT = hour+":"+ minute +":"+ second +" "+ am_pm;

       PrintWriter out = response.getWriter();

     String title = "Auto Refresh Header Setting";
     String docType =
     "<!doctype html public \"-//w3c//dtd html 4.0 " +
     "transitional//en\">\n";
     out.println(docType +
      "<html>\n" +
      "<head><title>" + title + "</title></head>\n"+
      "<body bgcolor=\"#f0f0f0\">\n" +
```

```
        "<h1 align=\"center\">" + title + "</h1>\n" +
        "<p>Current Time is: " + CT + "</p>\n");
    }
    // Method to handle POST method request.
    public void doPost(HttpServletRequest request, HttpServletResponse
  response)
        throws ServletException, IOException {
      doGet(request, response);
    }
  }
```

Now calling the above servlet would display current system time after every 5 seconds as follows. Just run the servlet and wait to see the result:

Auto Refresh Header Setting
Current Time is: 9:44:50 PM

3.8 Http Status Code

The format of the HTTP request and HTTP response messages are similar and will have the following structure:

An initial status line + CRLF (Carriage Return + Line Feed ie. New Line)
Zero or more header lines + CRLF
A blank line i.e. a CRLF
An optioanl message body like file, query data or query output.
For example, a server response header looks as follows:

```
HTTP/1.1 200 OK
Content-Type: text/html
Header2: ...
...
HeaderN: ...
 (Blank Line)
<!doctype ...>
<html>
<head>...</head>
<body>
...
</body>
</html>
```

The status line consists of the HTTP version (HTTP/1.1 in the example), a status code (200 in the example), and a very short message corresponding to the status code (OK in the example).

Methods to Set HTTP Status Code

There are following methods which can be used to set HTTP Status Code in your servlet program. These methods are available with *HttpServletResponse* object.

Methods	Description
Public void setStatus (int statusCode)	This method sets an arbitrary status code. The setStatus method takes an int (the status code) as an argument. If your response includes a special status code and a document, be sure to call setStatus before actually returning any of the content with the *PrintWriter*.
Public void sendRedirect(String url)	This method generates a 302 response along with a *Location* header giving the URL of the new document. public void sendError
getStatus (Int code, String message)	This method sends a status code (usually 404) along with a short message that is automatically formatted inside an HTML document and sent to the client.

HTTP Status Code Example

Following is the example which would send 407 error code to the client browser and browser would show you "Need authentication!!!" message.

```
// Import required java libraries
import java.io.*;
import javax.servlet.*;
import javax.servlet.http.*;
import java.util.*;

// Extend HttpServlet class
public class showError extends HttpServlet {

  // Method to handle GET method request.
  public void doGet(HttpServletRequest request, HttpServletResponse response)
        throws ServletException, IOException
  {
     // Set error code and reason.
     response.sendError(407, "Need authentication!!!" );
  }
  // Method to handle POST method request.
  public void doPost(HttpServletRequest request, HttpServletResponse
response)
     throws ServletException, IOException {
     doGet(request, response);
  }
}
```

Now calling the above servlet would display the following results:

HTTP Status 407—Need Authentication!!!

Type Status report

Message Need authentication!!!

Description The client must first authenticate itself with the proxy (Need authentication!!!).

3.9 Writing Filters

Servlet Filters are Java classes that can be used in Servlet Programming for the following purposes:

- To intercept requests from a client before they access a resource at back end.
- To manipulate responses from server before they are sent back to the client.

There are various types of filters suggested by the specifications:

- Authentication Filters
- Data compression Filters
- Encryption Filters
- Filters that trigger resource access events
- Image Conversion Filters
- Logging and Auditing Filters
- MIME-TYPE Chain Filters
- Tokenizing Filters
- XSL/T filters that transform XML content

Filters are deployed in the deployment descriptor file **web.xml** and then map to either servlet names or URL patterns in your application's deployment descriptor.

When the web container starts up the web application, it creates an instance of each filter that have been declared in the deployment descriptor. The filters execute in the order that they are declared in the deployment descriptor.

Servlet Filter Methods

A filter is simply a Java class that implements the javax.servlet.Filter interface. The javax.servlet.Filter interface defines three methods:

Methods	Description
Public void doFilter (ServletRequest, ServletResponse, FilterChain)	This method is called by the container each time a request/response pair is passed through the chain due to a client request for a resource at the end of the chain.
Public void init(FilterConfig filterConfig)	This method is called by the web container to indicate to a filter that it is being placed into service.
Public void destroy()	This method is called by the web container to indicate to a filter that it is being taken out of service.

Servlet Filter Example

Following is the Servlet Filter Example that would print the clients IP address and current date time. This example would give you basic understanding of Servlet Filter, but you can write more sophisticated filter applications using the same concept:

```
// Import required java libraries
import java.io.*;
```

```
import javax.servlet.*;
import javax.servlet.http.*;
import java.util.*;

// Implements Filter class
public class LogFilter implements Filter {
  public void  init(FilterConfig config)
                 throws ServletException{
    // Get init parameter
     String testParam = config.getInitParameter("test-param");

     //Print the init parameter
     System.out.println("Test Param: " + testParam);
  }
  public void  doFilter(ServletRequest request, ServletResponse response,
FilterChain chain)
             throws java.io.IOException, ServletException {
    // Get the IP address of client machine.
      String ipAddress = request.getRemoteAddr();
    // Log the IP address and current timestamp.
    System.out.println("IP "+ ipAddress + ", Time "+ new Date().toString());

   // Pass request back down the filter chain
     chain.doFilter(request,response);
  }
  public void destroy( ){
     /* Called before the Filter instance is removed
      from service by the web container*/
  }
}
```

Compile **LogFilter.java** in usual way and put your class file in <Tomcat-installation-directory>/webapps/ROOT/WEB-INF/classes.

Servlet Filter Mapping in Web.xml

Filters are defined and then mapped to a URL or Servlet, in much the same way as Servlet is defined and then mapped to a URL pattern. Create the following entry for filter tag in the deployment descriptor file **web.xml**

```
<filter>
  <filter-name>LogFilter</filter-name>
  <filter-class>LogFilter</filter-class>
  <init-param>
   <param-name>test-param</param-name>
   <param-value>Initialization Paramter</param-value>
  </init-param>
</filter>
```

```
<filter-mapping>
  <filter-name>LogFilter</filter-name>
  <url-pattern>/*</url-pattern>
</filter-mapping>
```

The above filter would apply to all the servlets because we specified /* in our configuration. You can specify a particular servlet path if you want to apply filter on a few servlets only.

Now try to call any servlet in usual way and you would see generated log in your web server log. You can use Log4J logger to log above log in a separate file.

Using Multiple Filters

Your web application may define several different filters with a specific purpose. Consider, you define two filters *AuthenFilter* and *LogFilter*. Rest of the process would remain as explained above except you need to create a different mapping as mentioned below:

```
<filter>
  <filter-name>LogFilter</filter-name>
  <filter-class>LogFilter</filter-class>
  <init-param>
    <param-name>test-param</param-name>
    <param-value>Initialization Paramter</param-value>
  </init-param>
</filter>

<filter>
  <filter-name>AuthenFilter</filter-name>
  <filter-class>AuthenFilter</filter-class>
  <init-param>
    <param-name>test-param</param-name>
    <param-value>Initialization Paramter</param-value>
  </init-param>
</filter>

<filter-mapping>
  <filter-name>LogFilter</filter-name>
  <url-pattern>/*</url-pattern>
</filter-mapping>

<filter-mapping>
  <filter-name>AuthenFilter</filter-name>
  <url-pattern>/*</url-pattern>
</filter-mapping>
```

Filters Application Order

The order of filter-mapping elements in web.xml determines the order in which the web container applies the filter to the servlet. To reverse the order of the filter, you just need to reverse the filter-mapping elements in the web.xml file.

For example, the above example would apply LogFilter first and then it would apply AuthenFilter to any servlet but the following example would reverse the order:

```
<filter-mapping>
  <filter-name>AuthenFilter</filter-name>
  <url-pattern>/*</url-pattern>
</filter-mapping>

<filter-mapping>
  <filter-name>LogFilter</filter-name>
  <url-pattern>/*</url-pattern>
</filter-mapping>
```

3.10 Exception Handling

When a servlet throws an exception, the web container searches the configurations in **web.xml** that use the exception-type element for a match with the thrown exception type.

You would have to use the **error-page** element in web.xml to specify the invocation of servlets in response to certain **exceptions** or HTTP **status codes**.

web.xml Configuration

Consider, you have an *ErrorHandler* servlet which would be called whenever there is any defined exception or error. Following would be the entry created in web.xml.

```
<!— servlet definition —>
<servlet>
     <servlet-name>ErrorHandler</servlet-name>
     <servlet-class>ErrorHandler</servlet-class>
</servlet>
<!— servlet mappings —>
<servlet-mapping>
     <servlet-name>ErrorHandler</servlet-name>
     <url-pattern>/ErrorHandler</url-pattern>
</servlet-mapping>

<!— error-code related error pages —>
<error-page>
   <error-code>404</error-code>
   <location>/ErrorHandler</location>
</error-page>
<error-page>
   <error-code>403</error-code>
   <location>/ErrorHandler</location>
</error-page>
```

```
<!— exception-type related error pages —>
<error-page>
    <exception-type>
        javax.servlet.ServletException
    </exception-type >
    <location>/ErrorHandler</location>
</error-page>

<error-page>
    <exception-type>java.io.IOException</exception-type >
    <location>/ErrorHandler</location>
</error-page>
```

If you want to have a generic Error Handler for all the exceptions, then you should define the following error-page instead of defining separate error-page elements for every exception:

```
<error-page>
    <exception-type>java.lang.Throwable</exception-type >
    <location>/ErrorHandler</location>
</error-page>
```

Following are the points to be noted about the above web.xml for Exception Handling:

- The servlet ErrorHandler is defined in usual way as any other servlet and configured in web.xml.
- If there is any error with status code either 404 (Not Found) or 403 (Forbidden), then ErrorHandler servlet would be called.
- If the web application throws either *ServletException* or *IOException*, then the web container invokes the /ErrorHandler servlet.
- You can define different Error Handlers to handle different type of errors or exceptions. The above example is very much generic and hope it serves the purpose to explain you the basic concept.

Request Attributes—Errors/Exceptions

The following is the list of request attributes that an error-handling servlet can access to analyse the nature of error/exception.

Attribute	Description
javax.servlet.error.status_code	This attribute give status code which can be stored and analysed after storing in a java.lang.Integer data type.
javax.servlet.error.exception_type	This attribute gives information about exception type which can be stored and analysed after storing in a java.lang.Class data type.
javax.servlet.error.message	This attribute gives information exact error

	message which can be stored and analysed after storing in a java.lang.String data type.
javax.servlet.error.request_uri	This attribute gives information about URL calling the servlet and it can be stored and analysed after storing in a java.lang.String data type.
javax.servlet.error.exception	This attribute gives information about the exception raised which can be stored and analysed after storing in a java.lang.Throwable data type.
javax.servlet.error.servlet_name	This attribute gives servlet name which can be stored and analysed after storing in a java.lang.String data type.

Error Handler Servlet Example

Following is the Servlet example that would be used as Error Handler in case of any error or exception occurs with your any of the servlet defined.

This example would give you basic understanding of Exception Handling in Servlet, but you can write more sophisticated filter applications using the same concept:

```
// Import required java libraries
import java.io.*;
import javax.servlet.*;
import javax.servlet.http.*;
import java.util.*;

// Extend HttpServlet class
public class ErrorHandler extends HttpServlet {

  // Method to handle GET method request.
  public void doGet(HttpServletRequest request, HttpServletResponse response)
        throws ServletException, IOException
  {
    // Analyze the servlet exception
        Throwable throwable = (Throwable)
    request.getAttribute("javax.servlet.error.exception");
    Integer statusCode = (Integer)
    request.getAttribute("javax.servlet.error.status_code");
    String servletName = (String)
    request.getAttribute("javax.servlet.error.servlet_name");
    if (servletName == null){
      servletName = "Unknown";
    }
    String requestUri = (String)
    request.getAttribute("javax.servlet.error.request_uri");
```

```
if (requestUri == null){
  requestUri = "Unknown";
}

// Set response content type
response.setContentType("text/html");

PrintWriter out = response.getWriter();
     String title = "Error/Exception Information";
String docType =
"<!doctype html public \"-//w3c//dtd html 4.0 " +
"transitional//en\">\n";
out.println(docType +
 "<html>\n" +
 "<head><title>" + title + "</title></head>\n" +
 "<body bgcolor=\"#f0f0f0\">\n");

if (throwable == null && statusCode == null){
  out.println("<h2>Error information is missing</h2>");
  out.println("Please return to the <a href=\"" +
    response.encodeURL("http://localhost:8080/") +
    "\">Home Page</a>.");
}else if (statusCode != null){
  out.println("The status code : " + statusCode);
}else{
  out.println("<h2>Error information</h2>");
  out.println("Servlet Name : " + servletName +
            "</br></br>");
  out.println("Exception Type : " +  throwable.getClass( ).getName( ) +
            "</br></br>");
  out.println("The request URI: " + requestUri +
            "<br><br>");
  out.println("The exception message: " +
                throwable.getMessage( ));
}
out.println("</body>");
out.println("</html>");
}
// Method to handle POST method request.
public void doPost(HttpServletRequest request, HttpServletResponse
response)
   throws ServletException, IOException {
  doGet(request, response);
 }
}
```

Compile **ErrorHandler.java** in usual way and put your class file in <Tomcat-installation-directory>/webapps/ROOT/WEB-INF/classes.

Let us add the following configuration in web.xml to handle exceptions:

```
<servlet>
    <servlet-name>ErrorHandler</servlet-name>
    <servlet-class>ErrorHandler</servlet-class>
</servlet>
<!— servlet mappings —>
<servlet-mapping>
    <servlet-name>ErrorHandler</servlet-name>
    <url-pattern>/ErrorHandler</url-pattern>
</servlet-mapping>
<error-page>
  <error-code>404</error-code>
  <location>/ErrorHandler</location>
</error-page>
<error-page>
  <exception-type>java.lang.Throwable</exception-type >
  <location>/ErrorHandler</location>
</error-page>
```

Now try to use a servlet which raise any exception or type a wrong URL, this would trigger Web Container to call **ErrorHandler** servlet and display an appropriate message as programmed. For example, if you type a wrong URL, then it would display the following result:

The status code : 404

Above code may not work with some web browsers. So try with Mozilla and Safari and it should work.

3.11 Cookies Handling

Cookies are text files stored on the client computer and they are kept for various information tracking purpose. Java Servlets transparently supports HTTP cookies.

There are three steps involved in identifying returning users:

- Server script sends a set of cookies to the browser. For example, name, age, or identification number, etc.
- Browser stores this information on local machine for future use.
- When next time browser sends any request to web server, then it sends those cookies information to the server and server uses that information to identify the user.

3.11.1 The Anatomy of a Cookie

Cookies are usually set in an HTTP header (although JavaScript can also set a cookie directly on a browser). A servlet that sets a cookie might send headers that look something like this:

```
HTTP/1.1 200 OK
Date: Fri, 04 Feb 2000 21:03:38 GMT
Server: Apache/1.3.9 (UNIX) PHP/4.0b3
```

Set-Cookie: name=xyz; expires=Friday, 04-Feb-07 22:03:38 GMT;
path=/; domain=tutorialspoint.com
Connection: close
Content-Type: text/html

As you can see, the Set-Cookie header contains a name value pair, a GMT date, a path and a domain. The name and value will be URL encoded. The expires field is an instruction to the browser to "forget" the cookie after the given time and date.

If the browser is configured to store cookies, it will then keep this information until the expiry date. If the user points the browser at any page that matches the path and domain of the cookie, it will resend the cookie to the server. The browser's headers might look something like this:

GET / HTTP/1.0
Connection: Keep-Alive
User-Agent: Mozilla/4.6 (X11; I; Linux 2.2.6-15apmac ppc)
Host: zink.demon.co.uk:1126
Accept: image/gif, */*
Accept-Encoding: gzip
Accept-Language: en
Accept-Charset: iso-8859-1,*,utf-8
Cookie: name=xyz

A servlet will then have access to the cookie through the request method *request.getCookies()* which returns an array of *Cookie* objects.

3.11.2 Servlet Cookies Methods

Following is the list of useful methods which you can use while manipulating cookies in servlet.

Method	Description
public void setDomain (String pattern)	This method sets the domain to which cookie applies, for example, tutorialspoint.com.
public String getDomain()	This method gets the domain to which cookie applies, for example, tutorialspoint.com.
public void setMaxAge(int expiry)	This method sets how much time (in seconds) should elapse before the cookie expires. If you do not set this, the cookie will last only for the current session.
public int getMaxAge()	This method returns the maximum age of the cookie, specified in seconds. By default, –1 indicating the cookie will persist until browser shutdown.
public String getName()	This method returns the name of the cookie. The name cannot be changed after creation.

public void setValue (String newValue)	This method sets the value associated with the cookie.
public String getValue()	This method gets the value associated with the cookie.
public void setPath(String uri)	This method sets the path to which this cookie applies. If you do not specify a path, the cookie is returned for all URLs in the same directory as the current page as well as all subdirectories.
public String getPath()	This method gets the path to which this cookie applies.
public void setSecure (boolean flag)	This method sets the boolean value indicating whether the cookie should only be sent over encrypted (i.e. SSL) connections.
public void setComment (String purpose)	This method specifies a comment that describes a cookie's purpose. The comment is useful if the browser presents the cookie to the user.
public String getComment()	This method returns the comment describing the purpose of this cookie, or null if the cookie has no comment.

3.11.3 Setting Cookies with Servlet

Setting cookies with servlet involves three steps:

(1) Creating a Cookie object: You call the Cookie constructor with a cookie name and a cookie value, both of which are strings.

```
Cookie cookie = new Cookie("key","value");
```

Keep in mind, neither the name nor the value should contain white space or any of the following characters:

[ ] ( ) = , " / ? @ : ;

(2) Setting the maximum age: You use setMaxAge to specify how long (in seconds) the cookie should be valid. Following would set up a cookie for 24 hours.

```
cookie.setMaxAge(60*60*24);
```

(3) Sending the Cookie into the HTTP response headers: You use **response.addCookie** to add cookies in the HTTP response header as follows:

```
response.addCookie(cookie);
```

Example

Let us modify our Form Example to set the cookies for first and last name.

```
// Import required java libraries
import java.io.*;
import javax.servlet.*;
import javax.servlet.http.*;
// Extend HttpServlet class
```

```
public class HelloForm extends HttpServlet {

   public void doGet(HttpServletRequest request, HttpServletResponse
response)
         throws ServletException, IOException
   {
      // Create cookies for first and last names.
         Cookie firstName = new Cookie("first_name",
               request.getParameter("first_name"));
      Cookie lastName = new Cookie("last_name",
               request.getParameter("last_name"));

      // Set expiry date after 24 Hrs for both the cookies.
      firstName.setMaxAge(60*60*24);
       lastName.setMaxAge(60*60*24);

       // Add both the cookies in the response header.
      response.addCookie( firstName );
      response.addCookie( lastName );

      // Set response content type
      response.setContentType("text/html");

       PrintWriter out = response.getWriter();
      String title = "Setting Cookies Example";
      String docType =
      "<!doctype html public \"-//w3c//dtd html 4.0 " +
      "transitional//en\">\n";
      out.println(docType +
            "<html>\n" +
            "<head><title>" + title + "</title></head>\n" +
            "<body bgcolor=\"#f0f0f0\">\n" +
            "<h1 align=\"center\">" + title + "</h1>\n" +
            "<ul>\n" +
            "  <li><b>First Name</b>: "
            + request.getParameter("first_name") + "\n" +
            "  <li><b>Last Name</b>: "
            + request.getParameter("last_name") + "\n" +
            "</ul>\n" +
            "</body></html>");
   }
}
```

Compile the above servlet **HelloForm** and create appropriate entry in web.xml file and finally try following HTML page to call servlet.

```
<html>
```

```
<body>
<form action="HelloForm" method="GET">
First Name: <input type="text" name="first_name">
<br />
Last Name: <input type="text" name="last_name" />
<input type="submit" value="Submit" />
</form>
</body>
</html>
```

Keep the above HTML content in a file Hello.htm and put it in <Tomcat-installation-directory>/webapps/ROOT directory. When you would access *http://localhost:8080/Hello.htm*, here is the actual output of the above form.

First Name: [ ]

Last Name: [ ]

Try to enter First Name and Last Name and then click submit button. This would display first name and last name on your screen and same time it would set two cookies firstName and lastName which would be passed back to the server when next time you would press Submit button.

Next section would explain you how you would access these cookies back in your web application.

3.11.4 Reading Cookies with Servlet

To read cookies, you need to create an array of *javax.servlet.http.Cookie* objects by calling the **getCookies( )** method of *HttpServletRequest*. Then cycle through the array, and use getName() and getValue() methods to access each cookie and associated value.

1. Example

Let us read cookies which we have set in previous example:

```
// Import required java libraries
import java.io.*;
import javax.servlet.*;
import javax.servlet.http.*;

 // Extend HttpServlet class
public class ReadCookies extends HttpServlet {

  public void doGet(HttpServletRequest request,
            HttpServletResponse response)
        throws ServletException, IOException
  {
   Cookie cookie = null;
   Cookie[] cookies = null;
    // Get an array of Cookies associated with this domain
    cookies = request.getCookies();
   // Set response content type
```

```
response.setContentType("text/html");

 PrintWriter out = response.getWriter();
String title = "Reading Cookies Example";
String docType =
"<!doctype html public \"-//w3c//dtd html 4.0 " +
"transitional//en\">\n";
out.println(docType +
      "<html>\n" +
      "<head><title>" + title + "</title></head>\n" +
      "<body bgcolor=\"#f0f0f0\">\n" );
if( cookies != null ){
   out.println("<h2> Found Cookies Name and Value</h2>");
   for (int i = 0; i < cookies.length; i++){
     cookie = cookies[i];
     out.print("Name : " + cookie.getName( ) + ", ");
     out.print("Value: " + cookie.getValue( )+" <br/>");
   }
}else{
   out.println(
    "<h2>No cookies founds</h2>");
}
out.println("</body>");
out.println("</html>");
  }
}
```

Compile above servlet **ReadCookies** and create appropriate entry in web.xml file. If you would have set first_name cookie as "John" and last_name cookie as "Player", then running *http://localhost:8080/ReadCookies* would display the following result:

2. Found Cookies Name and Value

Name : first_name Value: John
Name : last_name Value: Player

3.11.5 Delete Cookies with Servlet

To delete cookies is very simple. If you want to delete a cookie, then you simply need to follow up the following three steps:

1. Read an already exsiting cookie and store it in Cookie object.
2. Set cookie age as zero using **setMaxAge()** method to delete an existing cookie.
3. Add this cookie back into response header.

Example

Following example would delete and existing cookie named "first_name" and when you would run ReadCookies servlet next time it would return null value for first_name.

```
// Import required java libraries
import java.io.*;
import javax.servlet.*;
```

```
import javax.servlet.http.*;

 // Extend HttpServlet class
public class DeleteCookies extends HttpServlet {
  public void doGet(HttpServletRequest request, HttpServletResponse
response)
        throws ServletException, IOException
  {
   Cookie cookie = null;
   Cookie[] cookies = null;
    // Get an array of Cookies associated with this domain
    cookies = request.getCookies();

   // Set response content type
    response.setContentType("text/html");

    PrintWriter out = response.getWriter();
    String title = "Delete Cookies Example";
    String docType =
    "<!doctype html public \"-//w3c//dtd html 4.0 " +
    "transitional//en\">\n";
    out.println(docType +
          "<html>\n" +
          "<head><title>" + title + "</title></head>\n" +
          "<body bgcolor=\"#f0f0f0\">\n" );
    if( cookies != null ){
      out.println("<h2> Cookies Name and Value</h2>");
      for (int i = 0; i < cookies.length; i++){
        cookie = cookies[i];
        if((cookie.getName( )).compareTo("first_name") == 0 ){
          cookie.setMaxAge(0);
          response.addCookie(cookie);
          out.print("Deleted cookie : " +
                 cookie.getName( ) + "<br/>");
        }
        out.print("Name : " + cookie.getName( ) + ", ");
        out.print("Value: " + cookie.getValue( )+" <br/>");
      }
    }else{
      out.println(
        "<h2>No cookies founds</h2>");
    }
    out.println("</body>");
    out.println("</html>"); }
}
```

Compile above servlet **DeleteCookies** and create appropriate entry in web.xml file. Now running *http://localhost:8080/DeleteCookies* would display the following result:

2. Cookies Name and Value

Deleted cookie: first_name
Name: first_name, Value: John
Name: last_name, Value: Player

Now try to run *http://localhost:8080/ReadCookies* and it would display only one cookie as follows:

3. Found Cookies Name and Value

Name: last_name, Value: Player

You can delete your cookies in Internet Explorer manually. Start at the Tools menu and select Internet Options. To delete all cookies, press Delete Cookies.

3.12 Session Tracking

HTTP is a "stateless" protocol which means each time a client retrieves a Web page, the client opens a separate connection to the Web server and the server automatically does not keep any record of previous client request.

Still there are following three ways to maintain session between web client and web server:

1. Cookies

A webserver can assign a unique session ID as a cookie to each web client and for subsequent requests from the client they can be recognized using the received cookie. This may not be an effective way because many time browser does not support a cookie, hence it is not always recommended to use this procedure to maintain the sessions.

2. Hidden Form Fields

A web server can send a hidden HTML form field along with a unique session ID as follows:

```
<input type="hidden" name="sessionid" value="12345">
```

This entry means that, when the form is submitted, the specified name and value are automatically included in the GET or POST data. Each time when web browser sends request back, then session_id value can be used to keep the track of different web browsers.

This could be an effective way of keeping track of the session but clicking on a regular (<A HREF...>) hypertext link does not result in a form submission, so hidden form fields also cannot support general session tracking.

3. URL Rewriting

Some extra data can be appended at the end of each URL that identifies the session, and the server can associate that session identifier with data it has stored about that session.

For example, with http://google.com/file.htm;sessionid=12345, the session identifier is attached as sessionid=12345 which can be accessed at the web server to identify the client.

URL rewriting is a better way to maintain sessions and works for the browsers when they do not support cookies but here drawback is that every URL has to be generated dynamically to assign a session ID though page is simple static HTML page.

The HttpSession Object

Apart from the abovementioned three ways, servlet provides HttpSession Interface which provides a way to identify a user across more than one page request or visit to a Web site and to store information about that user.

The servlet container uses this interface to create a session between an HTTP client and an HTTP server. The session persists for a specified time period, across more than one connection or page request from the user. HttpSession object can be retrieved by calling the public method **getSession()** of HttpServletRequest, as below:

```
HttpSession session = request.getSession();
```

You need to call *request.getSession()* before you send any document content to the client. Here is a summary of the important methods available through HttpSession object:

Method	Description
public Object getAttribute(String name)	This method returns the object bound with the specified name in this session, or null if no object is bound under the name.
public Enumeration getAttributeNames()	This method returns an Enumeration of String objects containing the names of all the objects bound to this session.
public long getCreationTime()	This method returns the time when this session was created, measured in milliseconds since midnight January 1, 1970 GMT.
public String getId()	This method returns a string containing the unique identifier assigned to this session.
public long getLastAccessedTime()	This method returns the last time the client sent a request associated with this session, as the number of milliseconds since midnight January 1, 1970 GMT.
public int getMaxInactiveInterval()	This method returns the maximum time interval, in seconds, that the servlet container will keep this session open between client accesses.
public void invalidate()	This method invalidates this session and unbinds any objects bound to it.
public boolean isNew()	This method returns true if the client does not yet know about the session or if the client chooses not to join the session.
public void removeAttribute (String name)	This method removes the object bound with the specified name from this session.
public void setAttribute (String name, Object value)	This method binds an object to this session, using the name specified.
public void setMaxInactive Interval(int interval)	This method specifies the time, in seconds, between client requests before the servlet container will invalidate this session.

Session Tracking Example

This example describes how to use the HttpSession object to find out the creation time and the last-accessed time for a session. We would associate a new session with the request if one does not already exist.

```
// Import required java libraries
import java.io.*;
import javax.servlet.*;
import javax.servlet.http.*;
import java.util.*;

// Extend HttpServlet class
public class SessionTrack extends HttpServlet {

  public void doGet(HttpServletRequest request, HttpServletResponse
response)
        throws ServletException, IOException
  {
    // Create a session object if it is already not  created.
    HttpSession session = request.getSession(true);
    // Get session creation time.
    Date createTime = new Date(session.getCreationTime());
    // Get last access time of this web page.
    Date lastAccessTime = new Date(session.getLastAccessedTime());

    String title = "Welcome Back to my website";
    Integer visitCount = new Integer(0);
    String visitCountKey = new String("visitCount");
    String userIDKey = new String("userID");
    String userID = new String("ABCD");
    // Check if this is new comer on your web page.
    if (session.isNew()){
      title = "Welcome to my website";
      session.setAttribute(userIDKey, userID);
    } else {
      visitCount = (Integer)session.getAttribute(visitCountKey);
      visitCount = visitCount + 1;
      userID = (String)session.getAttribute(userIDKey);
    }
    session.setAttribute(visitCountKey,  visitCount);

    // Set response content type
    response.setContentType("text/html");
    PrintWriter out = response.getWriter();
```

```
String docType =
"<!doctype html public \"-//w3c//dtd html 4.0 " +
"transitional//en\">\n";
out.println(docType +
      "<html>\n" +
      "<head><title>" + title + "</title></head>\n" +
      "<body bgcolor=\"#f0f0f0\">\n" +
      "<h1 align=\"center\">" + title + "</h1>\n" +
      "<h2 align=\"center\">Session Infomation</h2>\n" +
      "<table border=\"1\" align=\"center\">\n" +
      "<tr bgcolor=\"#949494\">\n" +
      "  <th>Session info</th><th>value</th></tr>\n" +
      "<tr>\n" +
      "  <td>id</td>\n" +
      "  <td>" + session.getId() + "</td></tr>\n" +
      "<tr>\n" +
      "  <td>Creation Time</td>\n" +
      "  <td>" + createTime +
      "  </td></tr>\n" +
      "<tr>\n" +
      "  <td>Time of Last Access</td>\n" +
      "  <td>" + lastAccessTime +
      "  </td></tr>\n" +
      "<tr>\n" +
      "  <td>User ID</td>\n" +
      "  <td>" + userID +
      "  </td></tr>\n" +
      "<tr>\n" +
      "  <td>Number of visits</td>\n" +
      "  <td>" + visitCount + "</td></tr>\n" +
      "</table>\n" +
      "</body></html>");
}
}
```

Compile above servlet **SessionTrack** and create appropriate entry in web.xml file. Now running *http://localhost:8080/SessionTrack* would display the following result when you would run for the first time:

Welcome to My Website

Session Information

Session info	Value
Id	0AE3EC93FF44E3C525B4351B77ABB2D5
Creation Time	Tue Jun 08 17:26:40 GMT+04:00 2010
Time of Last Access	Tue Jun 08 17:26:40 GMT+04:00 2010
User ID	ABCD
Number of visits	0

Welcome Back to My Website

Session Infomation

Info type	Value
Id	0AE3EC93FF44E3C525B4351B77ABB2D5
Creation Time	Tue Jun 08 17:26:40 GMT+04:00 2010
Time of Last Access	Tue Jun 08 17:26:40 GMT+04:00 2010
User ID	ABCD
Number of visits	1

Deleting Session Data

When the user's session has been set, one of the following possibilities may occur:

1. **Remove a particular attribute:** *removeAttribute()* method can be called to delete the value associated with a particular key.
2. **Delete the whole session:** *invalidate()* method can be called to discard an entire session.
3. **Setting Session timeout:** *setMaxInactiveInterval()* method can be called to set the timeout for a session individually.
4. **Log the user out:** The servers that support servlets 2.4, call **logout** to log the client out of the Web server and invalidate all sessions belonging to all the users.
5. **web.xml Configuration:** While using Tomcat, apart from the abovementioned methods, the session time out is configured in web.xml file as follows.

```
<session-config>
  <session-timeout>15</session-timeout>
</session-config>
```

The timeout is expressed as minutes, and overrides the default timeout which is 30 minutes in Tomcat.

The getMaxInactiveInterval( ) method in a servlet returns the timeout period for that session in seconds. So if your session is configured in web.xml for 15 minutes, getMaxInactiveInterval( ) returns 900.

3.13 Database Access

To start with basic concept, let us create a simple table and create a few records in that table as follows:

Create Table

To create the **Employees** table in TEST database, use the following steps:

Step 1

Open a **Command Prompt** and change to the installation directory as follows:

```
C:\>
C:\>cd Program Files\MySQL\bin
C:\Program Files\MySQL\bin>
```

Step 2

Login to database as follows

```
C:\Program Files\MySQL\bin>mysql -u root -p
Enter password: ********
mysql>
```

Step 3

Create the table **Employee** in **TEST** database as follows:

```
mysql> use TEST;
mysql> create table Employees
  (
    id int not null,
    age int not null,
    first varchar (255),
    last varchar (255)
  );
Query OK, 0 rows affected (0.08 sec)
mysql>
```

Create Data Records

Finally you create a few records in Employee table as follows:

```
mysql> INSERT INTO Employees VALUES (100, 18, 'BPL', 'AJP');
Query OK, 1 row affected (0.05 sec)

mysql> INSERT INTO Employees VALUES (101, 25, 'Karthick', 'Satish');
Query OK, 1 row affected (0.00 sec)

mysql> INSERT INTO Employees VALUES (102, 30, 'sathya', 'Sri');
Query OK, 1 row affected (0.00 sec)

mysql> INSERT INTO Employees VALUES (103, 28, 'Sumi', 'Sekar');
Query OK, 1 row affected (0.00 sec)

mysql>
```

Accessing a Database

Here is an example which shows how to access TEST database using Servlet.

```
// Loading required libraries
import java.io.*;
import java.util.*;
import javax.servlet.*;
import javax.servlet.http.*;
import java.sql.*;
public class DatabaseAccess extends HttpServlet{

    public void doGet(HttpServletRequest request,
```

```
        HttpServletResponse response)
    throws ServletException, IOException
{
  // JDBC driver name and database URL
  static final String JDBC_DRIVER="com.mysql.jdbc.Driver";
   static final String DB_URL="jdbc:mysql://localhost/TEST";

  //  Database credentials
  static final String USER = "root";
  static final String PASS = "password";

  // Set response content type
  response.setContentType("text/html");
  PrintWriter out = response.getWriter();
  String title = "Database Result";
  String docType =
    "<!doctype html public \"-//w3c//dtd html 4.0 " +
    "transitional//en\">\n";
    out.println(docType +
    "<html>\n" +
    "<head><title>" + title + "</title></head>\n" +
    "<body bgcolor=\"#f0f0f0\">\n" +
    "<h1 align=\"center\">" + title + "</h1>\n");
  try{
    // Register JDBC driver
    Class.forName("com.mysql.jdbc.Driver");

    // Open a connection
    conn = DriverManager.getConnection(DB_URL,USER,PASS);

    // Execute SQL query
    stmt = conn.createStatement();
    String sql;
    sql = "SELECT id, first, last, age FROM Employees";
    ResultSet rs = stmt.executeQuery(sql);

    // Extract data from result set
    while(rs.next()){
      //Retrieve by column name
      int id  = rs.getInt("id");
      int age = rs.getInt("age");
      String first = rs.getString("first");
      String last = rs.getString("last");

      //Display values
```

```
        out.println("ID: " + id + "<br>");
        out.println(", Age: " + age + "<br>");
        out.println(", First: " + first + "<br>");
        out.println(", Last: " + last + "<br>");
      }
      out.println("</body></html>");

      // Clean-up environment
      rs.close();
      stmt.close();
      conn.close();
    }catch(SQLException se){
      //Handle errors for JDBC
      se.printStackTrace();
    }catch(Exception e){
      //Handle errors for Class.forName
      e.printStackTrace();
    }finally{
      //finally block used to close resources
      try{
        if(stmt!=null)
          stmt.close();
      }catch(SQLException se2){
      }// nothing we can do
      try{
        if(conn!=null)
        conn.close();
      }catch(SQLException se){
        se.printStackTrace();
      }//end finally try
    } //end try
  }
}
```

Now let us compile above servlet and create the following entries in web.xml

```
....
<servlet>
    <servlet-name>DatabaseAccess</servlet-name>
    <servlet-class>DatabaseAccess</servlet-class>
 </servlet>

 <servlet-mapping>
    <servlet-name>DatabaseAccess</servlet-name>
    <url-pattern>/DatabaseAccess</url-pattern>
 </servlet-mapping>
....
```

Now call this servlet using URL http://localhost:8080/DatabaseAccess which would display the following response:

Database Result

ID: 100, Age: 18, First: BPL, Last: AJP
ID: 101, Age: 25, First: Karthick, Last: Satish
ID: 102, Age: 30, First: sathya, Last: Sri
ID: 103, Age: 28, First: Sumi, Last: Sekar

3.14 File Uploading

A Servlet can be used with an HTML form tag to allow users to upload files to the server. An uploaded file could be a text file or image file or any document.

Creating a File Upload Form

The following HTM code below creates an uploader form. Following are the important points to be noted down:

- The form **method** attribute should be set to **POST** method and GET method can not be used.
- The form **enctype** attribute should be set to **multipart/form-data**.
- The form **action** attribute should be set to a servlet file which would handle file uploading at backend server. Following example is using **UploadServlet** servlet to upload file.
- To upload a single file you should use a single <input .../> tag with attribute type="file". To allow multiple files uploading, include more than one input tags with different values for the name attribute. The browser associates a Browse button with each of them.

```
<html>
<head>
<title>File Uploading Form</title>
</head>
<body>
<h3>File Upload:</h3>
Select a file to upload: <br />
<form action="UploadServlet" method="post"
                  enctype="multipart/form-data">
<input type="file" name="file" size="50" />
<br />
<input type="submit" value="Upload File" />
</form>
</body>
</html>
```

This will display the following result which would allow to select a file from local PC and when user would click at "Upload File", form would be submitted along with the selected file:

File Upload: Select a file to upload:

Note: This is just dummy form and would not work.

Writing Backend Servlet

Following is the servlet **UploadServlet** which would take care of accepting uploaded file and to store it in directory <Tomcat-installation-directory>/webapps/data. This directory name could also be added using an external configuration such as a **context-param** element in web.xml as follows:

```
<web-app>
....
<context-param>
    <description>Location to store uploaded file</description>
    <param-name>file-upload</param-name>
    <param-value>
        c:\apache-tomcat-5.5.29\webapps\data\
    </param-value>
</context-param>
...
</web-app>
```

Following is the source code for UploadServlet which can handle multiple file uploading at a time. Before proceeding you have make sure the followings:

- Following example depends on FileUpload, so make sure you have the latest version of **commons-fileupload.x.x.jar** file in your classpath. You can download it from http://commons.apache.org/fileupload/.
- FileUpload depends on Commons IO, so make sure you have the latest version of **commons-io-x.x.jar** file in your classpath. You can download it from http://commons.apache.org/io/.
- While testing following example, you should upload a file which has less size than *maxFileSize* otherwise file would not be uploaded.
- Make sure you have created directories c:\temp and c:\apache-tomcat-5.5.29\webapps\data well in advance.

```
// Import required java libraries
import java.io.*;
import java.util.*;

import javax.servlet.ServletConfig;
import javax.servlet.ServletException;
import javax.servlet.http.HttpServlet;
import javax.servlet.http.HttpServletRequest;
import javax.servlet.http.HttpServletResponse;

import org.apache.commons.fileupload.FileItem;
import org.apache.commons.fileupload.FileUploadException;
import org.apache.commons.fileupload.disk.DiskFileItemFactory;
```

```
import org.apache.commons.fileupload.servlet.ServletFileUpload;
import org.apache.commons.io.output.*;

public class UploadServlet extends HttpServlet {

private boolean isMultipart;
private String filePath;
private int maxFileSize = 50 * 1024;
private int maxMemSize = 4 * 1024;
private File file ;

public void init( ){
// Get the file location where it would be stored.
filePath =
        getServletContext().getInitParameter("file-upload");
  }
  public void doPost(HttpServletRequest request,
        HttpServletResponse response)
        throws ServletException, java.io.IOException {
    // Check that we have a file upload request
    isMultipart = ServletFileUpload.isMultipartContent(request);
    response.setContentType("text/html");
    java.io.PrintWriter out = response.getWriter( );
    if( !isMultipart ){
      out.println("<html>");
      out.println("<head>");
      out.println("<title>Servlet upload</title>");
      out.println("</head>");
      out.println("<body>");
      out.println("<p>No file uploaded</p>");
      out.println("</body>");
      out.println("</html>");
      return;
    }
    DiskFileItemFactory factory = new DiskFileItemFactory();
    // maximum size that will be stored in memory
    factory.setSizeThreshold(maxMemSize);
    // Location to save data that is larger than maxMemSize.
    factory.setRepository(new File("c:\\temp"));

    // Create a new file upload handler
    ServletFileUpload upload = new ServletFileUpload(factory);
    // maximum file size to be uploaded.
    upload.setSizeMax( maxFileSize );

    try{
```

```
    // Parse the request to get file items.
   List fileItems = upload.parseRequest(request);
    // Process the uploaded file items
   Iterator i = fileItems.iterator();

   out.println("<html>");
   out.println("<head>");
   out.println("<title>Servlet upload</title>");
   out.println("</head>");
   out.println("<body>");
   while ( i.hasNext () )
   {
     FileItem fi = (FileItem)i.next();
     if ( !fi.isFormField () )
     {
       // Get the uploaded file parameters
       String fieldName = fi.getFieldName();
       String fileName = fi.getName();
       String contentType = fi.getContentType();
       boolean isInMemory = fi.isInMemory();
       long sizeInBytes = fi.getSize();
       // Write the file
       if( fileName.lastIndexOf("\\") >= 0 ){
         file = new File( filePath +
         fileName.substring( fileName.lastIndexOf("\\"))) ;
       }else{
         file = new File( filePath +
          fileName.substring(fileName.lastIndexOf("\\")+1)) ;
       }
       fi.write( file ) ;
       out.println("Uploaded Filename: " + fileName + "<br>");
     }
   }
    out.println("</body>");
    out.println("</html>");
  }catch(Exception ex) {
    System.out.println(ex);
  }
  }
  public void doGet(HttpServletRequest request,
               HttpServletResponse response)
     throws ServletException, java.io.IOException {

          throw new ServletException("GET method used with " +
          getClass( ).getName( )+": POST method required.");
  }
}
```

Compile and Running Servlet

Compile above servlet UploadServlet and create required entry in web.xml file as follows.

```
<servlet>
   <servlet-name>UploadServlet</servlet-name>
   <servlet-class>UploadServlet</servlet-class>
</servlet>

<servlet-mapping>
   <servlet-name>UploadServlet</servlet-name>
   <url-pattern>/UploadServlet</url-pattern>
</servlet-mapping>
```

Now try to upload files using the HTML form which you created above. When you would try http://localhost:8080/UploadFile.htm, it would display following result which would help you uploading any file from your local machine.

File Upload: Select a file to upload:

If your servlet script works fine, your file should be uploaded in c:\apache-tomcat-5.5.29\webapps\data\ directory.

3.15 Handling Date

One of the most important advantages of using Servlet is that you can use most of the methods available in core Java. This tutorial would take you through Java provided **Date** class which is available in **java.util** package, this class encapsulates the current date and time.

The Date class supports two constructors. The first constructor initializes the object with the current date and time.

Date( )

The following constructor accepts one argument that equals the number of milliseconds that have elapsed since midnight, January 1, 1970

Date(long millisec)

Once you have a Date object available, you can call any of the following support methods to play with dates:

Methods	Usage	Use
after	boolean after(Date date)	Returns true if the invoking Date object contains a date that is later than the one specified by date, otherwise, it returns false.
before	boolean before(Date date)	Returns true if the invoking Date object contains a date that is earlier than the one specified by date, otherwise, it returns false.
clone	object clone( )	Duplicates the invoking Date object.

compareTo	int compareTo(Date date)	Compares the value of the invoking object with that of date. Returns 0 if the values are equal. Returns a negative value if the invoking object is earlier than date. Returns a positive value if the invoking object is later than date.
compareTo	int compareTo(Object obj)	Operates identically to compareTo(Date) if obj is of class Date. Otherwise, it throws a ClassCastException.
equals	boolean equals(Object date)	Returns true if the invoking Date object contains the same time and date as the one specified by date, otherwise, it returns false.
getTime	long getTime( )	Returns the number of milli-seconds that have elapsed since January 1, 1970.
hashCode	int hashCode( )	Returns a hash code for the invoking object.
setTime	void setTime(long time)	Sets the time and date as specified by time, which represents an elapsed time in milliseconds from midnight, January 1, 1970
toString	string toString( )	Converts the invoking Date object into a string and returns the result.

Getting Current Date and Time

This is very easy to get current date and time in Java Servlet. You can use a simple Date object with *toString()* method to print current date and time as follows:

```
// Import required java libraries
import java.io.*;
import java.util.Date;
import javax.servlet.*;
import javax.servlet.http.*;
// Extend HttpServlet class
public class CurrentDate extends HttpServlet {

  public void doGet(HttpServletRequest request,
            HttpServletResponse response)
      throws ServletException, IOException
```

```
    {
      // Set response content type
      response.setContentType("text/html");

      PrintWriter out = response.getWriter();
      String title = "Display Current Date & Time";
      Date date = new Date();
      String docType =
      "<!doctype html public \"-//w3c//dtd html 4.0 " +
      "transitional//en\">\n";
      out.println(docType +
        "<html>\n" +
        "<head><title>" + title + "</title></head>\n" +
        "<body bgcolor=\"#f0f0f0\">\n" +
        "<h1 align=\"center\">" + title + "</h1>\n" +
        "<h2 align=\"center\">" + date.toString() + "</h2>\n" +
        "</body></html>");
    }
  }
```

Now let us compile above servlet and create appropriate entries in web.xml and then call this servlet using URL http://localhost:8080/CurrentDate. This would produce the following result:

Display Current Date and Time

Mon Jun 21 21:46:49 GMT+04:00 2010

Try to refersh URL http://localhost:8080/CurrentDate and you would find difference in seconds everytime you would refresh.

Date Comparison:

As I mentioned above, you can use all the available Java methods in your Servlet. In case you need to compare two dates, following are the methods:

- You can use getTime( ) to obtain the number of milliseconds that have elapsed since midnight, January 1, 1970, for both objects and then compare these two values.
- You can use the methods before( ), after( ), and equals( ). Because the 12th of the month comes before the 18th, for example, new Date(99, 2, 12).before(new Date (99, 2, 18)) returns true.
- You can use the compareTo( ) method, which is defined by the Comparable interface and implemented by Date.

Date Formatting using SimpleDateFormat

SimpleDateFormat is a concrete class for formatting and parsing dates in a locale-sensitive manner. SimpleDateFormat allows you to start by choosing any user-defined patterns for date-time formatting.

Let us modify the above example as follows:

```
// Import required java libraries
import java.io.*;
```

```
import java.text.*;
import java.util.Date;
import javax.servlet.*;
import javax.servlet.http.*;

// Extend HttpServlet class
public class CurrentDate extends HttpServlet {

public void doGet(HttpServletRequest request,
          HttpServletResponse response)
        throws ServletException, IOException
  {
     // Set response content type
     response.setContentType("text/html");

      PrintWriter out = response.getWriter();
     String title = "Display Current Date & Time";
     Date dNow = new Date( );
     SimpleDateFormat ft =
      new SimpleDateFormat ("E yyyy.MM.dd 'at' hh:mm:ss a zzz");
     String docType =
     "<!doctype html public \"-//w3c//dtd html 4.0 " +
     "transitional//en\">\n";
     out.println(docType +
      "<html>\n" +
      "<head><title>" + title + "</title></head>\n" +
      "<body bgcolor=\"#f0f0f0\">\n" +
      "<h1 align=\"center\">" + title + "</h1>\n" +
      "<h2 align=\"center\">" + ft.format(dNow) + "</h2>\n" +
      "</body></html>");
  }
}
```

Compile the above servlet once again and then call this servlet using URL http://localhost:8080/CurrentDate. This would produce the following result:

Display Current Date and Time

Mon 2010.06.21 at 10:06:44 PM GMT+04:00

Simple DateFormat format codes

To specify the time format use a time pattern string. In this pattern, all ASCII letters are reserved as pattern letters, which are defined as follows:

For a complete list of constant available methods to manipulate date, you can refer to standard Java documentation.

Character	Description	Example
G	Era designator	AD
Y	Year in four digits	2001
M	Month in year	July or 07
D	Day in month	10
H	Hour in A.M./P.M. (1~12)	12
H	Hour in day (0~23)	22
m	Minute in hour	30
s	Second in minute	55
S	Millisecond	234
E	Day in week	Tuesday
D	Day in year	360
F	Day of week in month	2 (second Wed. in July)
w	Week in year	40
W	Week in month	1
a	A.M./P.M. marker	PM
k	Hour in day (1~24)	24
K	Hour in A.M./P.M. (0~11)	10
z	Time zone	Eastern Standard Time
'	Escape for text	Delimiter
"	Single quote	`

3.16 Servlets—Page Redirection

Page redirection is generally used when a document moves to a new location and we need to send the client to this new location or may be because of load balancing, or for simple randomization.

The simplest way of redirecting a request to another page is using method **sendRedirect()** of response object. Following is the signature of this method:

public void HttpServletResponse.sendRedirect(String location)throws IOException

This method sends back the response to the browser along with the status code and new page location. You can also use setStatus() and setHeader() methods together to achieve the same:

```
....
String site = "http://www.newpage.com" ;
response.setStatus(response.SC_MOVED_TEMPORARILY);
response.setHeader("Location", site);
....
```

Example

This example shows how a servlet performs page redirection to an another location:

```
import java.io.*;
import java.sql.Date;
import java.util.*;
import javax.servlet.*;
import javax.servlet.http.*;
public class PageRedirect extends HttpServlet{
    public void doGet(HttpServletRequest request,
            HttpServletResponse response)
        throws ServletException, IOException
  {
    // Set response content type
    response.setContentType("text/html");

    // New location to be redirected
    String site = new String("http://www.photofuntoos.com");

    response.setStatus(response.SC_MOVED_TEMPORARILY);
    response.setHeader("Location", site);
     }
}
```

Now let us compile above servlet and create following entries in web.xml

```
....
<servlet>
    <servlet-name>PageRedirect</servlet-name>
    <servlet-class>PageRedirect</servlet-class>
</servlet>

<servlet-mapping>
    <servlet-name>PageRedirect</servlet-name>
    <url-pattern>/PageRedirect</url-pattern>
</servlet-mapping>....
```

Now call this servlet using URL http://localhost:8080/PageRedirect. This would take you given URL http://www.photofuntoos.com.

3.17 Sample Programs

3.17.1 Hit Counter for a Web Page

Many times you would be interested in knowing total number of hits on a particular page of your website. It is very simple to count these hits using a servlet because the life cycle of a servlet is controlled by the container in which it runs.

Following are the steps to be taken to implement a simple page hit counter which is based on Servlet Life Cycle:

- Initialize a global variable in init() method.
- Increase global variable every time either doGet() or doPost() method is called.

- If required, you can use a database table to store the value of global variable in destroy() method. This value can be read inside init() method when servlet would be initialized next time. This step is optional.
- If you want to count only unique page hits within a session, then you can use isNew() method to check if same page already have been hit within that session. This step is optional.
- You can display value of the global counter to show total number of hits on your website. This step is also optional.

Here I am assuming that the web container will not be restarted. If it is restarted or servlet destroyed, the hit counter will be reset.

Example

This example shows how to implement a simple page hit counter:

```
import java.io.*;
import java.sql.Date;
import java.util.*;
import javax.servlet.*;
import javax.servlet.http.*;

public class PageHitCounter extends HttpServlet{

    private int hitCount;

            public void init()
  {
    // Reset hit counter.
   hitCount = 0;
  }

  public void doGet(HttpServletRequest request,
          HttpServletResponse response)
      throws ServletException, IOException
  {
    // Set response content type
    response.setContentType("text/html");
    // This method executes whenever the servlet is hit
    // increment hitCount
    hitCount++;
    PrintWriter out = response.getWriter();
    String title = "Total Number of Hits";
    String docType =
    "<!doctype html public \"-//w3c//dtd html 4.0 " +
    "transitional//en\">\n";
    out.println(docType +
      "<html>\n" +
      "<head><title>" + title + "</title></head>\n" +
```

```
        "<body bgcolor=\"#f0f0f0\">\n" +
        "<h1 align=\"center\">" + title + "</h1>\n" +
        "<h2 align=\"center\">" + hitCount + "</h2>\n" +
        "</body></html>");

  }
  public void destroy()
  {
      // This is optional step but if you like you
      // can write hitCount value in your database.
  }
}
```

Now let us compile the above servlet and create the following entries in web.xml

```
....
 <servlet>
    <servlet-name>PageHitCounter</servlet-name>
    <servlet-class>PageHitCounter</servlet-class>
</servlet>

<servlet-mapping>
    <servlet-name>PageHitCounter</servlet-name>
    <url-pattern>/PageHitCounter</url-pattern>
</servlet-mapping>
....
```

Now call this servlet using URL http://localhost:8080/PageHitCounter. This would increase counter by one every time this page gets refreshed and it would display the following result:

Total Number of Hits: 210

3.17.2 Hit Counter for a Website

Many times you would be interested in knowing total number of hits on your whole website. This is also very simple in Servlet and we can achieve this using filters.

Following are the steps to be taken to implement a simple website hit counter which is based on Filter Life Cycle:

- Initialize a global variable in init() method of a filter.
- Increase global variable every time doFilter method is called.
- If required, you can use a database table to store the value of global variable in destroy() method of filter. This value can be read inside init() method when filter would be initialized next time. This step is optional.

Here I am assuming that the web container will not be restarted. If it is restarted or servlet destroyed, the hit counter will be reset.

Example

This example shows how to implement a simple website hit counter:

```
// Import required java libraries
import java.io.*;
```

```
import javax.servlet.*;
import javax.servlet.http.*;
import java.util.*;

public class SiteHitCounter implements Filter{
    private int hitCount;
            public void  init(FilterConfig config)
              throws ServletException{
    // Reset hit counter.
    hitCount = 0;
  }

  public void  doFilter(ServletRequest request,
          ServletResponse response,
         FilterChain chain)
          throws java.io.IOException, ServletException {

    // increase counter by one
    hitCount++;

    // Print the counter.
    System.out.println("Site visits count :"+ hitCount );
    // Pass request back down the filter chain
    chain.doFilter(request,response);
  }
  public void destroy()
  {
     // This is optional step but if you like you
     // can write hitCount value in your database.
  }
}
```

Now let us compile the above servlet and create the following entries in web.xml

```
....
<filter>
  <filter-name>SiteHitCounter</filter-name>
  <filter-class>SiteHitCounter</filter-class>
</filter>

<filter-mapping>
  <filter-name>SiteHitCounter</filter-name>
  <url-pattern>/*</url-pattern>
</filter-mapping>
....
```

Now call any URL like URL http://localhost:8080/. This would increase counter by one every time any page gets a hit and it would display following message in the log:

Site visits count : 1
Site visits count : 2
Site visits count : 3
Site visits count : 4
Site visits count : 5
....

3.17.3 Auto Page Refresh

Consider a webpage which is displaying live game score or stock market status or currency exchange ratio. For all such type of pages, you would need to refresh your web page regularly using referesh or reload button with your browser.

Java Servlet makes this job easy by providing you a mechanism where you can make a webpage in such a way that it would refresh automatically after a given interval.

The simplest way of refreshing a web page is using method **setIntHeader()** of response object. Following is the signature of this method:

public void setIntHeader(String header, int headerValue)

This method sends back header "Refresh" to the browser along with an integer value which indicates time interval in seconds.

Auto Page Refresh Example

This example shows how a servlet performs auto page refresh using **setIntHeader()** method to set **Refresh** header.

```
// Import required java libraries
import java.io.*;
import javax.servlet.*;
import javax.servlet.http.*;
import java.util.*;

// Extend HttpServlet class
public class Refresh extends HttpServlet {

  // Method to handle GET method request.
  public void doGet(HttpServletRequest request,
           HttpServletResponse response)
      throws ServletException, IOException
  {
    // Set refresh, autoload time as 5 seconds
    response.setIntHeader("Refresh", 5);

    // Set response content type
    response.setContentType("text/html");

    // Get current time
    Calendar calendar = new GregorianCalendar();
    String am_pm;
    int hour = calendar.get(Calendar.HOUR);
    int minute = calendar.get(Calendar.MINUTE);
```

```
    int second = calendar.get(Calendar.SECOND);
    if(calendar.get(Calendar.AM_PM) == 0)
      am_pm = "AM";
    else
      am_pm = "PM";

    String CT = hour+":"+ minute +":"+ second +" "+ am_pm;
    PrintWriter out = response.getWriter();
    String title = "Auto Page Refresh using Servlet";
    String docType =
    "<!doctype html public \"-//w3c//dtd html 4.0 " +
    "transitional//en\">\n";
    out.println(docType +
      "<html>\n" +
      "<head><title>" + title + "</title></head>\n"+
      "<body bgcolor=\"#f0f0f0\">\n" +
      "<h1 align=\"center\">" + title + "</h1>\n" +
      "<p>Current Time is: " + CT + "</p>\n");
  } // Method to handle POST method request.
  public void doPost(HttpServletRequest request,
              HttpServletResponse response)
    throws ServletException, IOException {
    doGet(request, response);
  }
}
```

Now let us compile the above servlet and create the following entries in web.xml

```
....
 <servlet>
   <servlet-name>Refresh</servlet-name>
     <servlet-class>Refresh</servlet-class>
 </servlet>
 <servlet-mapping>
     <servlet-name>Refresh</servlet-name>
     <url-pattern>/Refresh</url-pattern>
 </servlet-mapping>
....
```

Now call this servlet using URL http://localhost:8080/Refresh which would display current system time after every 5 seconds as follows. Just run the servlet and wait to see the result:

Auto Page Refresh using Servlet
Current Time is: 9:44:50 PM

3.17.4 Sending Email

To send an email using your Servlet is simple enough but to start with you should have **JavaMail API** and **Java Activation Framework (JAF)** installed on your machine.

- You can download latest version of JavaMail (Version 1.2) from Java's standard website.
- You can download latest version of JAF (Version 1.1.1) from Java's standard website.

Download and unzip these files, in the newly created top level directories you will find a number of jar files for both the applications. You need to add **mail.jar** and **activation.jar** files in your CLASSPATH.

Send a Simple Email

Here is an example to send a simple email from your machine. Here it is assumed that your **localhost** is connected to the internet and capable enough to send an email. At the same time make sure all the jar files from Java Email API package and JAF package are available in CLASSPATH.

```
// File Name SendEmail.java
import java.io.*;
import java.util.*;
import javax.servlet.*;
import javax.servlet.http.*;
import javax.mail.*;
import javax.mail.internet.*;
import javax.activation.*;

public class SendEmail extends HttpServlet{

  public void doGet(HttpServletRequest request,
          HttpServletResponse response)
      throws ServletException, IOException
  {
    // Recipient's email ID needs to be mentioned.
    String to = "abcd@gmail.com";

    // Sender's email ID needs to be mentioned
    String from = "web@gmail.com";

    // Assuming you are sending email from localhost
    String host = "localhost";

    // Get system properties
    Properties properties = System.getProperties();
    // Setup mail server
    properties.setProperty("mail.smtp.host", host);

    // Get the default Session object.
    Session session = Session.getDefaultInstance(properties);

   // Set response content type
```

```
      response.setContentType("text/html");
      PrintWriter out = response.getWriter();

      try{
         // Create a default MimeMessage object.
         MimeMessage message = new MimeMessage(session);
         // Set From: header field of the header.
         message.setFrom(new InternetAddress(from));
         // Set To: header field of the header.
         message.addRecipient(Message.RecipientType.TO,
                              new InternetAddress(to));
         // Set Subject: header field
         message.setSubject("This is the Subject Line!");
         // Now set the actual message
         message.setText("This is actual message");
         // Send message
         Transport.send(message);
         String title = "Send Email";
         String res = "Sent message successfully....";
         String docType =
         "<!doctype html public \"-//w3c//dtd html 4.0 " +
         "transitional//en\">\n";
        out.println(docType +
         "<html>\n" +
         "<head><title>" + title + "</title></head>\n" +
         "<body bgcolor=\"#f0f0f0\">\n" +
         "<h1 align=\"center\">" + title + "</h1>\n" +
         "<p align=\"center\">" + res + "</p>\n" +
         "</body></html>");
      }catch (MessagingException mex) {
         mex.printStackTrace();
      }
   }
}
```

Now let us compile the above servlet and create the following entries in web.xml

```
....
 <servlet>
    <servlet-name>SendEmail</servlet-name>
    <servlet-class>SendEmail</servlet-class>
 </servlet>

 <servlet-mapping>
    <servlet-name>SendEmail</servlet-name>
    <url-pattern>/SendEmail</url-pattern>
 </servlet-mapping>
....
```

Now call this servlet using URL http://localhost:8080/SendEmail which would send an email to given email ID *abcd@gmail.com* and would display the following response:

Send Email

Sent message successfully...

If you want to send an email to multiple recipients, then following methods would be used to specify multiple email IDs:

```
void addRecipients(Message.RecipientType type,
            Address[] addresses)
throws MessagingException
```

Here is the description of the parameters:

- **type:** This would be set to TO, CC or BCC. Here CC represents Carbon Copy and BCC represents Black Carbon Copy. For example, *Message.RecipientType.TO*
- **addresses:** This is the array of email ID. You would need to use InternetAddress() method while specifying email IDs

Send an HTML Email

Here is an example to send an HTML email from your machine. Here it is assumed that your **localhost** is connected to the internet and capable enough to send an email. Same time make sure all the jar files from Java Email API package and JAF package ara available in CLASSPATH.

This example is very similar to previous one, except here we are using setContent() method to set content whose second argument is "text/html" to specify that the HTML content is included in the message.

Using this example, you can send as big as HTML content you like.

```
// File Name SendEmail.java
import java.io.*;
import java.util.*;
import javax.servlet.*;
import javax.servlet.http.*;
import javax.mail.*;
import javax.mail.internet.*;
import javax.activation.*;

public class SendEmail extends HttpServlet{

    public void doGet(HttpServletRequest request,
            HttpServletResponse response)
        throws ServletException, IOException
  {
    // Recipient's email ID needs to be mentioned.
    String to = "abcd@gmail.com";

    // Sender's email ID needs to be mentioned
    String from = "web@gmail.com";
```

```
  // Assuming you are sending email from localhost
  String host = "localhost";

  // Get system properties
  Properties properties = System.getProperties();

  // Setup mail server
  properties.setProperty("mail.smtp.host", host);

  // Get the default Session object.
  Session session = Session.getDefaultInstance(properties);

 // Set response content type
  response.setContentType("text/html");
  PrintWriter out = response.getWriter();
  try{
    // Create a default MimeMessage object.
    MimeMessage message = new MimeMessage(session);
    // Set From: header field of the header.
    message.setFrom(new InternetAddress(from));
    // Set To: header field of the header.
    message.addRecipient(Message.RecipientType.TO,
                   new InternetAddress(to));
    // Set Subject: header field
    message.setSubject("This is the Subject Line!");
    // Send the actual HTML message, as big as you like
    message.setContent("<h1>This is actual message</h1>",
             "text/html" );
    // Send message
    Transport.send(message);
    String title = "Send Email";
    String res = "Sent message successfully....";
    String docType =
    "<!doctype html public \"-//w3c//dtd html 4.0 " +
    "transitional//en\">\n";
    out.println(docType +
    "<html>\n" +
    "<head><title>" + title + "</title></head>\n" +
    "<body bgcolor=\"#f0f0f0\">\n" +
    "<h1 align=\"center\">" + title + "</h1>\n" +
    "<p align=\"center\">" + res + "</p>\n" +
    "</body></html>");
  }catch (MessagingException mex) {
    mex.printStackTrace();
  }
 }
}
```

Compile and run above servlet to send HTML message on a given email ID.

3.17.5 Send Attachment in Email

Here is an example to send an email with attachment from your machine. Here it is assumed that your **localhost** is connected to the internet and capable enough to send an email.

```
// File Name SendEmail.java
import java.io.*;
import java.util.*;
import javax.servlet.*;
import javax.servlet.http.*;
import javax.mail.*;
import javax.mail.internet.*;
import javax.activation.*;

public class SendEmail extends HttpServlet{

    public void doGet(HttpServletRequest request,
            HttpServletResponse response)
        throws ServletException, IOException
  {
    // Recipient's email ID needs to be mentioned.
    String to = "abcd@gmail.com";

    // Sender's email ID needs to be mentioned
    String from = "web@gmail.com";

    // Assuming you are sending email from localhost
    String host = "localhost";

    // Get system properties
    Properties properties = System.getProperties();

    // Setup mail server
    properties.setProperty("mail.smtp.host", host);

    // Get the default Session object.
    Session session = Session.getDefaultInstance(properties);
   // Set response content type
    response.setContentType("text/html");
    PrintWriter out = response.getWriter();

    try{
```

```
// Create a default MimeMessage object.
MimeMessage message = new MimeMessage(session);

// Set From: header field of the header.
message.setFrom(new InternetAddress(from));

// Set To: header field of the header.
message.addRecipient(Message.RecipientType.TO,
            new InternetAddress(to));

// Set Subject: header field
message.setSubject("This is the Subject Line!");

// Create the message part
BodyPart messageBodyPart = new MimeBodyPart();

// Fill the message
messageBodyPart.setText("This is message body");

    // Create a multipar message
Multipart multipart = new MimeMultipart();

// Set text message part
multipart.addBodyPart(messageBodyPart);

// Part two is attachment
messageBodyPart = new MimeBodyPart();
String filename = "file.txt";
DataSource source = new FileDataSource(filename);
messageBodyPart.setDataHandler(new DataHandler(source));
messageBodyPart.setFileName(filename);
multipart.addBodyPart(messageBodyPart);

// Send the complete message parts
message.setContent(multipart );

// Send message
Transport.send(message);
String title = "Send Email";
String res = "Sent message successfully....";
String docType =
"<!doctype html public \"-//w3c//dtd html 4.0 " +
"transitional//en\">\n";
out.println(docType +
```

```
        "<html>\n" +
        "<head><title>" + title + "</title></head>\n" +
        "<body bgcolor=\"#f0f0f0\">\n" +
        "<h1 align=\"center\">" + title + "</h1>\n" +
        "<p align=\"center\">" + res + "</p>\n" +
        "</body></html>");
      }catch (MessagingException mex) {
        mex.printStackTrace();
      }
    }
  }
```

Compile and run above servlet to send a file as an attachement along with a message on a given email ID.

User Authentication Part

If it is required to provide user ID and Password to the email server for authentication purpose, then you can set these properties as follows:

```
props.setProperty("mail.user", "myuser");
props.setProperty("mail.password", "mypwd");
```

Rest of the email sending mechanism would remain as explained above.

3.18 Servlets-Packaging

The web application structure involving the WEB-INF subdirectory is standard to all Java web applications and specified by the servlet API specification. Given a top-level directory name of myapp, here is what this directory structure looks like:

```
/myapp
   /images
   /WEB-INF
      /classes
      /lib
```

The WEB-INF subdirectory contains the application's deployment descriptor, named web.xml. All the HTML files live in the top-level directory which is *myapp*. For admin user, you would find ROOT directory as parent directory as myapp.

Creating Servlets in Packages

The WEB-INF/classes directory contains all the servlet classes and other class files, in a structure that matches their package name. For example, if you have a fully qualified class name of **com.myorg.MyServlet**, then this servlet class must be located in the following directory:

```
/myapp/WEB-INF/classes/com/myorg/MyServlet.class
```

Following is the example to create MyServlet class with a package name *com.myorg*

```
// Name your package
package com.myorg;

  // Import required java libraries
import java.io.*;
import javax.servlet.*;
import javax.servlet.http.*;

public class MyServlet extends HttpServlet {

  private String message;

public void init() throws ServletException
{
    // Do required initialization
    message = "Hello World";
  }

  public void doGet(HttpServletRequest request,
            HttpServletResponse response)
        throws ServletException, IOException
  {
    // Set response content type
    response.setContentType("text/html");

     // Actual logic goes here.
    PrintWriter out = response.getWriter();
    out.println("<h1>" + message + "</h1>");
  }

   public void destroy()
  {
    // do nothing.
  }
}
```

Compiling Servlets in Packages

There is nothing much different to compile a class available in package. The simplest way is to keep your java file in fully qualified path, as mentioned above class would be kept in com.myorg. You would also need to add these directory in CLASSPATH.

Assuming your environment is setup properly, go in **<Tomcat-installation-directory>/webapps/ROOT/WEB-INF/classes** directory and compile MyServlet.java as follows:

```
$ javac MyServlet.java
```

If the servlet depends on any other libraries, you have to include those JAR files on your CLASSPATH as well. I have included only servlet-api.jar JAR file because I am not using any other library in Hello World program.

This command line uses the built-in javac compiler that comes with the Sun Microsystems Java Software Development Kit (JDK). For this command to work properly, you have to include the location of the Java SDK that you are using in the PATH environment variable.

If everything goes fine, above compilation would produce **MyServlet.class** file in the same directory. Next section would explain how a compiled servlet would be deployed in production.

Packaged Servlet Deployment

By default, a servlet application is located at the path <Tomcat-installation-directory>/webapps/ROOT and the class file would reside in <Tomcat-installation-directory>/webapps/ROOT/WEB-INF/classes.

If you have a fully qualified class name of **com.myorg.MyServlet**, then this servlet class must be located in WEB-INF/classes/com/myorg/MyServlet.class and you would need to create following entries in **web.xml** file located in <Tomcat-installation-directory>/webapps/ROOT/WEB-INF/

```
<servlet>
    <servlet-name>MyServlet</servlet-name>
    <servlet-class>com.myorg.MyServlet</servlet-class>
  </servlet>

  <servlet-mapping>
    <servlet-name>MyServlet</servlet-name>
    <url-pattern>/MyServlet</url-pattern>
  </servlet-mapping>
```

Above entries to be created inside <web-app>...</web-app> tags available in web.xml file. There could be various entries in this table already available, but never mind.

You are almost done, now let us start tomcat server using <Tomcat-installation-directory>\bin\startup.bat (on windows) or <Tomcat-installation-directory>/bin/startup.sh (on Linux/Solaris, etc.) and finally type **http://localhost:8080/MyServlet** in browser's address box. If everything goes fine, you would get the following result:

Hello World

3.19 Debugging

It is always difficult to testing/debugging a servlets. Servlets tend to involve a large amount of client/server interaction, making errors likely but hard to reproduce.

Here are a few hints and suggestions that may aid you in your debugging.

System.out.println()

System.out.println() is easy to use as a marker to test whether a certain piece of code is being executed or not. We can print out variable values as well. Additionally:

- Since the System object is part of the core Java objects, it can be used everywhere without the need to install any extra classes. This includes Servlets, JSP, RMI, EJB's, ordinary Beans and classes, and standalone applications.

- Compared to stopping at breakpoints, writing to System.out does not interfere much with the normal execution flow of the application, which makes it very valuable when timing is crucial.

Following is the syntax to use System.out.println():
System.out.println("Debugging message");

All the messages generated by above syntax would be logged in web server log file.

Message Logging

It is always great idea to use proper logging method to log all the debug, warning and error messages using a standard logging method. I use log4J to log all the messages.

The Servlet API also provides a simple way of outputting information by using the log() method as follows:

```
// Import required java libraries
import java.io.*;
import javax.servlet.*;
import javax.servlet.http.*;

public class ContextLog extends HttpServlet {
  public void doGet(HttpServletRequest request,
      HttpServletResponse response) throws ServletException,
       java.io.IOException {

     String par = request.getParameter("par1");
     //Call the two ServletContext.log methods
     ServletContext context = getServletContext( );

     if (par == null || par.equals(""))
     //log version with Throwable parameter
     context.log("No message received:",
        new IllegalStateException("Missing parameter"));
     else
        context.log("Here is the visitor's message: " + par);
     response.setContentType("text/html");
     java.io.PrintWriter out = response.getWriter( );
     String title = "Context Log";
     String docType =
     "<!doctype html public \"-//w3c//dtd html 4.0 " +
     "transitional//en\">\n";
     out.println(docType +
      "<html>\n" +
      "<head><title>" + title + "</title></head>\n" +
      "<body bgcolor=\"#f0f0f0\">\n" +
```

```
        "<h1 align=\"center\">" + title + "</h1>\n" +
        "<h2 align=\"center\">Messages sent</h2>\n" +
        "</body></html>");
    }//doGet
  }
```

The ServletContext logs its text messages to the servlet container's log file. With Tomcat these logs are found in <Tomcat-installation-directory>/logs.

The log files do give an indication of new emerging bugs or the frequency of problems. For that reason it is good to use the log() function in the catch clause of exceptions which normally should not occur.

Using JDB Debugger

You can debug servlets with the same jdb commands you use to debug an applet or an application.

To debug a servlet, we can debug sun.servlet.http.HttpServer, then watch as HttpServer executes servlets in response to HTTP requests we make from a browser. This is very similar to how applets are debugged. The difference is that with applets, the actual program being debugged is sun.applet.AppletViewer.

Most debuggers hide this detail by automatically knowing how to debug applets. Until they do the same for servlets, you have to help your debugger by doing the following:

- Set your debugger's classpath so that it can find sun.servlet.http.Http-Server and associated classes.
- Set your debugger's classpath so that it can also find your servlets and support classes, typically server_root/servlets and server_root/classes.

You normally would not want server_root/servlets in your classpath because it disables servlet reloading. This inclusion, however, is useful for debugging. It allows your debugger to set breakpoints in a servlet before the custom servlet loader in HttpServer loads the servlet.

Once you have set the proper classpath, start debugging sun.servlet.http.HttpServer. You can set breakpoints in whatever servlet you are interested in debugging, then use a web browser to make a request to the HttpServer for the given servlet (http://localhost:8080/servlet/ServletToDebug). You should see execution stop at your breakpoints.

Using Comments

Comments in your code can help the debugging process in various ways. Comments can be used in lots of other ways in the debugging process.

The Servlet uses Java comments and single line (// ...) and multiple line (/* ... */) comments can be used to temporarily remove parts of your Java code. If the bug disappears, take a closer look at the code you just commented and find out the problem.

Client and Server Headers

Sometimes when a servlet does not behave as expected, it is useful to look at the raw HTTP request and response. If you are familiar with the structure of HTTP, you can read the request and response and see what exactly is going with those headers.

Important Debugging Tips

Here is a list of some more debugging tips on servlet debugging:

- Be aware that server_root/classes does not reload and that server_root/servlets probably does.

- Ask a browser to show the raw content of the page it is displaying. This can help identify formatting problems. It is usually an option under the View menu.
- Make sure the browser is not caching a previous request output by forcing a full reload of the page. With Netscape Navigator, use Shift-Reload; with Internet Explorer use Shift-Refresh.
- Verify that your servlet's init() method takes a ServletConfig parameter and calls super.init(config) right away.

3.20 Internationalization

Before we proceed, let me explain three important terms:

- **Internationalization (i18n):** This means enabling a website to provide different versions of content translated into the visitor's language or nationality.
- **Localization (l10n):** This means adding resources to a website to adapt it to a particular geographical or cultural region, for example, Hindi translation to a website.
- **Locale:** This is a particular cultural or geographical region. It is usually referred to as a language symbol followed by a country symbol which are separated by an underscore. For example, "en_US" represents english locale for US.

There are number of items which should be taken care while building up a global website. This tutorial would not give you complete detail on this but it would give you a good example on how you can offer your web page in different languages to internet community by differentiating their location, i.e. locale.

A servlet can pickup appropriate version of the site based on the requester's locale and provide appropriate site version according to the local language, culture and requirements. Following is the method of request object which returns Locale object.

java.util.Locale request.getLocale()

Detecting Locale

Following are the important locale methods which you can use to detect requester's location, language and of course locale. All the below methods display country name and language name set in requester's browser.

Method	Usage	Use
getCountry	string getCountry()	This method returns the country/ region code in upper case for this locale in ISO 3166 2-letter format.
getdisplay Country	string getDisplayCountry()	This method returns a name for the locale's country that is appropriate for display to the user.
getLanguage	string getLanguage()	This method returns the language code in lower case for this locale in ISO 639 format.
getDisplay Language	string getDisplayLanguage()	This method returns a name for the locale's language that is appropriate for display to the user.

getISO3 Country	string getISO3Country()	This method returns a three-letter abbreviation for this locale's country.
getISO3 Language	string getISO3Language()	This method returns a three-letter abbreviation for this locale's language.

Example

This example shows how you display a language and associated country for a request:

```
import java.io.*;
import javax.servlet.*;
import javax.servlet.http.*;
import java.util.Locale;

public class GetLocale extends HttpServlet{

    public void doGet(HttpServletRequest request,
            HttpServletResponse response)
        throws ServletException, IOException
  {
    //Get the client's Locale
    Locale locale = request.getLocale();
    String language = locale.getLanguage();
    String country = locale.getCountry();

    // Set response content type
    response.setContentType("text/html");
    PrintWriter out = response.getWriter();

    String title = "Detecting Locale";
    String docType =
    "<!doctype html public \"-//w3c//dtd html 4.0 " +
    "transitional//en\">\n";
    out.println(docType +
      "<html>\n" +
      "<head><title>" + title + "</title></head>\n" +
      "<body bgcolor=\"#f0f0f0\">\n" +
      "<h1 align=\"center\">" + language + "</h1>\n" +
      "<h2 align=\"center\">" + country + "</h2>\n" +
      "</body></html>");
  }
}
```

Languages Setting

A servlet can output a page written in a Western European language such as English, Spanish, German, French, Italian, Dutch, etc. Here it is important to set Content-Language header to display all the characters properly.

Second point is to display all the special characters using HTML entities, for example, "ñ" represents "ñ", and "¡" represents "¡" as follows:

```
import java.io.*;
import javax.servlet.*;
import javax.servlet.http.*;
import java.util.Locale;

public class DisplaySpanish extends HttpServlet{

    public void doGet(HttpServletRequest request,
            HttpServletResponse response)
        throws ServletException, IOException
  {
    // Set response content type
    response.setContentType("text/html");
    PrintWriter out = response.getWriter();
    // Set spanish language code.
    response.setHeader("Content-Language", "es");

    String title = "En Espa&ntilde;ol";
    String docType =
     "<!doctype html public \"-//w3c//dtd html 4.0 " +
     "transitional//en\">\n";
     out.println(docType +
     "<html>\n" +
     "<head><title>" + title + "</title></head>\n" +
     "<body bgcolor=\"#f0f0f0\">\n" +
     "<h1>" + "En Espa&ntilde;ol:" + "</h1>\n" +
     "<h1>" + "&iexcl;Hola Mundo!" + "</h1>\n" +
     "</body></html>");
  }
}
```

Locale Specific Dates

You can use the java.text.DateFormat class and its static getDateTimeInstance( ) method to format date and time specific to locale. Following is the example which shows how to format dates specific to a given locale:

```
import java.io.*;
import javax.servlet.*;
import javax.servlet.http.*;
import java.util.Locale;
```

```
import java.text.DateFormat;
import java.util.Date;

public class DateLocale extends HttpServlet{

    public void doGet(HttpServletRequest request,
            HttpServletResponse response)
        throws ServletException, IOException
  {
    // Set response content type
    response.setContentType("text/html");
    PrintWriter out = response.getWriter();
    //Get the client's Locale
    Locale locale = request.getLocale( );
    String date = DateFormat.getDateTimeInstance(
                            DateFormat.FULL,
                            DateFormat.SHORT,
                            locale).format(new Date( ));
    String title = "Locale Specific Dates";
    String docType =
      "<!doctype html public \"-//w3c//dtd html 4.0 " +
      "transitional//en\">\n";
      out.println(docType +
      "<html>\n" +
      "<head><title>" + title + "</title></head>\n" +
      "<body bgcolor=\"#f0f0f0\">\n" +
      "<h1 align=\"center\">" + date + "</h1>\n" +
      "</body></html>");
  }
}
```

Locale Specific Currency

You can use the java.txt.NumberFormat class and its static getCurrencyInstance( ) method to format a number, such as a long or double type, in a locale specific curreny. Following is the example which shows how to format currency specific to a given locale:

```
import java.io.*;
import javax.servlet.*;
import javax.servlet.http.*;
import java.util.Locale;
import java.text.NumberFormat;
import java.util.Date;
public class CurrencyLocale extends HttpServlet{
    public void doGet(HttpServletRequest request,
            HttpServletResponse response)
        throws ServletException, IOException
```

```
  {
    // Set response content type
    response.setContentType("text/html");
    PrintWriter out = response.getWriter();
    //Get the client's Locale
    Locale locale = request.getLocale( );
    NumberFormat nft = NumberFormat.getCurrencyInstance(locale);
    String formattedCurr = nft.format(1000000);

    String title = "Locale Specific Currency";
    String docType =
      "<!doctype html public \"-//w3c//dtd html 4.0 " +
      "transitional//en\">\n";
      out.println(docType +
      "<html>\n" +
      "<head><title>" + title + "</title></head>\n" +
      "<body bgcolor=\"#f0f0f0\">\n" +
      "<h1 align=\"center\">" + formattedCurr + "</h1>\n" +
      "</body></html>");
  }
}
```

Locale Specific Percentage

You can use the java.txt.NumberFormat class and its static getPercentInstance( ) method to get locale specific percentage. Following is the example which shows how to format percentage specific to a given locale:

```
import java.io.*;
import javax.servlet.*;
import javax.servlet.http.*;
import java.util.Locale;
import java.text.NumberFormat;
import java.util.Date;

public class PercentageLocale extends HttpServlet{

    public void doGet(HttpServletRequest request,
              HttpServletResponse response)
        throws ServletException, IOException
  {
    // Set response content type
    response.setContentType("text/html");
    PrintWriter out = response.getWriter();
    //Get the client's Locale
    Locale locale = request.getLocale( );
```

```
    NumberFormat nft = NumberFormat.getPercentInstance(locale);
    String formattedPerc = nft.format(0.51);

    String title = "Locale Specific Percentage";
    String docType =
      "<!doctype html public \"-//w3c//dtd html 4.0 " +
      "transitional//en\">\n";     out.println(docType +
      "<html>\n" +
      "<head><title>" + title + "</title></head>\n" +
      "<body bgcolor=\"#f0f0f0\">\n" +
      "<h1 align=\"center\">" + formattedPerc + "</h1>\n" +
      "</body></html>");
  }
}
```

4. JAVA SERVER PAGES

4.1 An Overview

JavaServerPages also known as JSPs, are one of the most powerful and simplest way of generating dynamic HTML on the server side by developers insert java code in HTML pages by making use of special JSP tags, most of which start with <% and end with %>.

JSP is a presentation layer technology that sits on top of a Java Servlets model and makes working with HTML easier. Like Server Side Java Script(SSJS), it allows you to mix static HTML content with server-side scripting to produce dynamic output. A JavaServer Pages component is a type of Java servlet that is designed to fulfill the role of a user interface for a Java web application By default, JSP uses Java as it is scripting language; however, the specification allows other languages to be used, just as ASP can use other languages (such as JavaScript and VBScript). While JSP with Java will be more flexible and robust than scripting platforms based on simpler languages like JavaScript and VBScript, Java also has a steeper learning curve than simple scripting languages. To offer the best of both worlds—a robust web application platform and a simple, easy-to-use language and tool set—JSP provides a number of server-side tags that allow developers to perform most dynamic content operations without ever writing a single line of Java code. So developers who are only familiar with scripting, or even those who are simply HTML designers, can use JSP tags for generating simple output without having to learn Java. Advanced scriptures or Java developers can also use the tags, or they can use the full Java language if they want to perform advanced operations in JSP pages.

Using JSP, you can collect input from users through web page forms, present records from a database or another source, and create web pages dynamically. JSP tags can be used for a variety of purposes, such as retrieving information from a database or registering user preferences, accessing JavaBeans components, passing control between pages and sharing information between requests, pages, etc.

JavaServer Pages often serve the same purpose as programs implemented using the Common Gateway Interface (CGI). But JSP offer several advantages in comparison with the CGI.

- Performance is significantly better because JSP allows embedding Dynamic Elements in HTML Pages itself instead of having a separate CGI files.
- JSP are always compiled before it is processed by the server unlike CGI/Perl which requires the server to load an interpreter and the target script each time the page is requested.
- JavaServer Pages are built on top of the Java Servlets API, so like Servlets, JSP also has access to all the powerful Enterprise Java APIs, including JDBC, JNDI, EJB, JAXP, etc.
- JSP pages can be used in combination with servlets that handle the business logic, the model supported by Java servlet template engines.

Finally, JSP is an integral part of J2EE, a complete platform for enterprise class applications. This means that JSP can play a part in the simplest applications to the most complex and demanding.

4.2 Advantages of JSP

Following is the list of other advantages of using JSP over other technologies:

- **vs. Active Server Pages (ASP):** The advantages of JSP are twofold. First, the dynamic part is written in Java, not Visual Basic or other MS specific language, so it is more powerful and easier to use. Second, it is portable to other operating systems and non-Microsoft Web servers.
- **vs. Pure Servlets:** It is more convenient to write (and to modify!) regular HTML than to have plenty of println statements that generate the HTML.
- **vs. Server-Side Includes (SSI):** SSI is really only intended for simple inclusions, not for "real" programs that use form data, make database connections, and the like.
- **vs. JavaScript:** JavaScript can generate HTML dynamically on the client but can hardly interact with the web server to perform complex tasks like database access and image processing, etc.
- **vs. Static HTML:** Regular HTML, of course, cannot contain dynamic information.

4.3 Architecture

The source code of a JSP page is essentially just HTML (or text—or even XML) sprinkled here and there with either special JSP tags and/or Java code enclosed in these tags. The file's extension is .jsp rather than the usual. html or .htm, and it tells the server that this document requires special handling.

The special handling, accomplished with a Web server extension or plug-in, involves four steps:

1. The JSP engine parses the page and creates a Java source file.
2. It then compiles the file produced in Step 1 into a Java class file. The class file created in Step 2 is a servlet, and from this point on, the servlet engine handles the class file in the same manner as all other servlets.
3. The servlet engine loads the servlet class for execution.
4. The servlet executes and streams back the results to the requestor.

Although this process might seem time consuming and expensive, it is much more efficient than it sounds. Steps 1 and 2 occur only once, when you first deploy or update the JSP. The web server needs a JSP engine, i.e. container to process JSP pages. The JSP container is responsible for intercepting requests for JSP pages. This tutorial makes use of Apache which has built-in JSP container to support JSP pages development.

A JSP container works with the Web server to provide the runtime environment and other services a JSP needs. It knows how to understand the special elements that are part of JSPs.

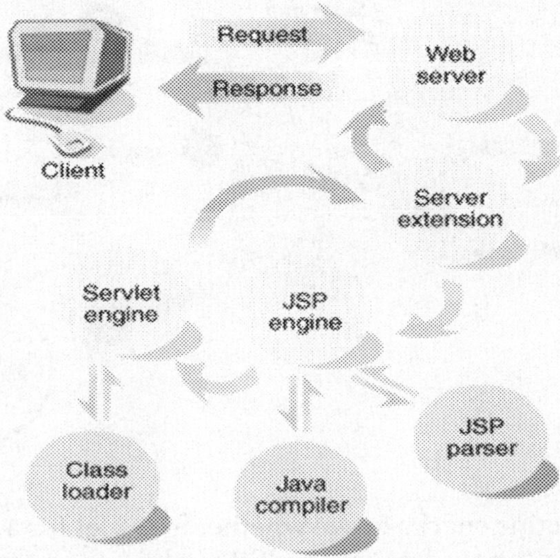

The servlet engine performs Step 3 only upon the first request of that servlet since the last server restart. After that, the class loader loads the class once and is available for the life of that JVM. Finally, some application servers provide page caching, which can further improve the performance and reduce the cost of executing the request. With page caching, even Step 4 may execute only once depending on how dynamic the page data is.

4.4 JSP Processing

The following steps explain how the web server creates the web page using JSP:

- As with a normal page, your browser sends an HTTP request to the web server.
- The web server recognizes that the HTTP request is for a JSP page and forwards it to a JSP engine. This is done by using the URL or JSP page which ends with **.jsp** instead of .html.
- The JSP engine loads the JSP page from disk and converts it into a servlet content. This conversion is very simple in which all template text is converted to println( ) statements and all JSP elements are converted to Java code that implements the corresponding dynamic behavior of the page.
- The JSP engine compiles the servlet into an executable class and forwards the original request to a servlet engine.
- A part of the web server called the servlet engine loads the Servlet class and executes it. During execution, the servlet produces an output in HTML format, which the servlet engine passes to the web server inside an HTTP response.
- The web server forwards the HTTP response to your browser in terms of static HTML content.
- Finally web browser handles the dynamically generated HTML page inside the HTTP response exactly as if it were a static page.

All the abovementioned steps can be shown below in the following diagram:

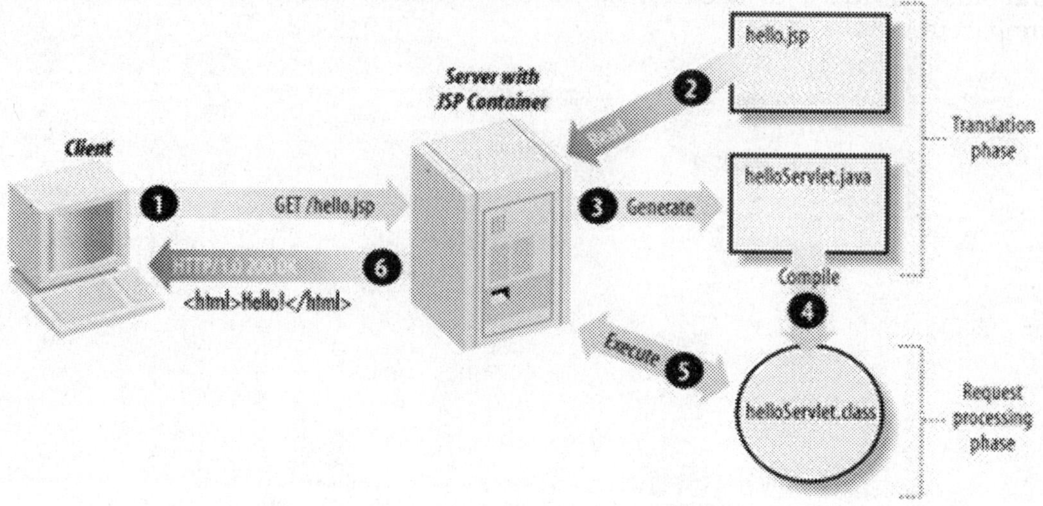

Typically, the JSP engine checks to see whether a servlet for a JSP file already exists and whether the modification date on the JSP is older than the servlet. If the JSP is older than its generated servlet, the JSP container assumes that the JSP has not changed and that the generated servlet still matches the JSP's contents. This makes the process more efficient than with other scripting languages (such as PHP) and therefore faster.

So in a way, a JSP page is really just another way to write a servlet without having to be a Java programming wiz. Except for the translation phase, a JSP page is handled exactly like a regular servlet.

4.5 JSP Life Cycle

The key to understanding the low-level functionality of JSP is to understand the simple life cycle they follow.

A JSP life cycle can be defined as the entire process from its creation till the destruction which is similar to a servlet life cycle with an additional step which is required to compile a JSP into servlet.

The following are the paths followed by a JSP

- Compilation
- Initialization
- Execution
- Cleanup

The three major phases of JSP life cycle are very similar to Servlet Life Cycle and they are as follows:

(1) JSP Compilation

When a browser asks for a JSP, the JSP engine first checks to see whether it needs to compile the page. If the page has never been compiled, or if the JSP has been modified since it was last compiled, the JSP engine compiles the page.

The compilation process involves three steps:

1. Parsing the JSP.
2. Turning the JSP into a servlet.
3. Compiling the servlet.

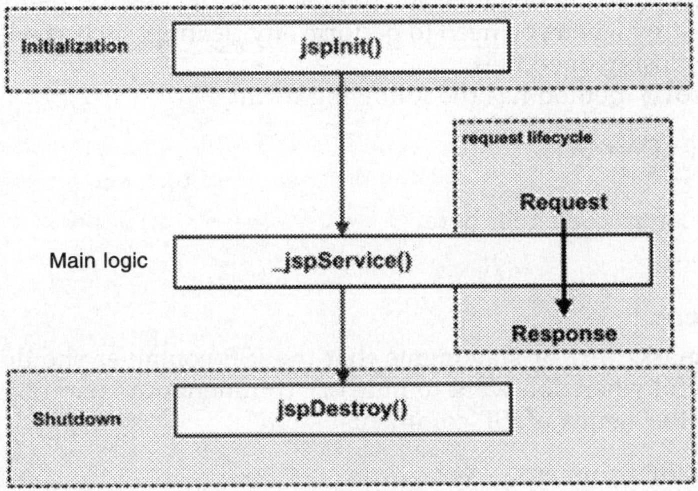

(2) JSP Initialization

When a container loads a JSP, it invokes the jspInit() method before servicing any requests. If you need to perform JSP-specific initialization, override the jspInit() method:

```
public void jspInit(){
// Initialization code...
}
```

Typically initialization is performed only once and as with the servlet init method, you generally initialize database connections, open files, and create lookup tables in the jspInit method.

(3) JSP Execution

This phase of the JSP life cycle represents all interactions with requests until the JSP is destroyed.

Whenever a browser requests a JSP and the page has been loaded and initialized, the JSP engine invokes the **_jspService()** method in the JSP.

The _jspService() method takes an **HttpServletRequest** and an **HttpServletResponse** as its parameters as follows:

```
void _jspService(HttpServletRequest request,
            HttpServletResponse response)
{
  // Service handling code...
}
```

The _jspService() method of a JSP is invoked once per a request and is responsible for generating the response for that request and this method is also responsible for generating responses to all seven of the HTTP method, i.e. GET, POST, DELETE, etc.

(4) JSP Cleanup

The destruction phase of the JSP life cycle represents when a JSP is being removed from use by a container.

The **jspDestroy()** method is the JSP equivalent of the destroy method for servlets. Override jspDestroy when you need to perform any cleanup, such as releasing database connections or closing open files.

The jspDestroy() method has the following form:

```
public void jspDestroy()
{
   // Your cleanup code goes here.
}
```

4.6 JSP Comments

JSP comment marks text or statements that the JSP container should ignore. A JSP comment is useful when you want to hide or "comment out" part of your JSP page.

Following is the syntax of JSP comments:

```
<%— This is JSP comment —%>
```

Following is the simple example for JSP Comments:

```
<html>
<head><title>A Comment Test</title></head>
<body>
<h2>A Test of Comments</h2>
<%— This comment will not be visible in the page source —%>
</body>
</html>
```

This would generate the following result:

A Test of Comments

There are a small number of special constructs you can use in various cases to insert comments or characters that would otherwise be treated specially. Here is a summary:

Syntax	Purpose
<%— comment —%>	A JSP comment. Ignored by the JSP engine.
<!— comment —>	An HTML comment. Ignored by the browser.
<\%	Represents static <% literal.
%\>	Represents static %> literal.
\'	A single quote in an attribute that uses single quotes.
\"	A double quote in an attribute that uses double quotes.

4.7 JSP-Syntax

As said earlier Java Server Pages (JSP) lets you separate the dynamic part of your pages from the static HTML. Let us start examining the same HelloWorld.jsp example again.

```
<HTML>
<BODY>
<% out.println("Hello World"); %>
</BODY>
</HTML>
```

HelloWorld.jsp

Here the regular HTML is writen in the normal manner, using whatever Web-page-building tools you normally use. Then the code for the dynamic parts alone are encoded in special tags starting with "<%" and ending with "%>".

To run the above example,

- Place the HelloWorld.jsp in the examples directory of Web server. (root_directory\JavaWebServer2.0\examples)
- Run the JavaWebServer. Go to root_directory\JavaWebServer2.0\bin Type http.
- Now open the browser and type http://localhost:8080/examples/HelloWorld.jsp in the place where address of the file is required.

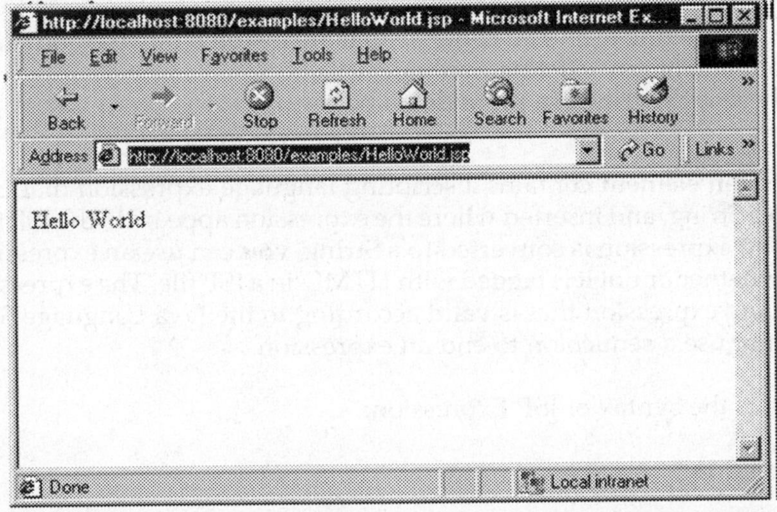

You normally give your file a jsp extension, and typically install it in any place you could place a normal Web page. Although what you write often looks more like a regular HTML file than a Servlet, behind the scenes, the JSP page just gets converted to a normal Servlet, with the static HTML simply being printed to the output stream associated with the servlet's service method. This is normally done the first time the page is requested, and developers can imply request the page themselves when first installing it if they want to be sure that the first real user does not get a momentary delay when the JSP page is translated to a Servlet and the Servlet is compiled and loaded. Note also that many Web servers let you define aliases so that a URL that appears to reference an HTML file really points to a Servlet or JSP page. Aside from the regular HTML, there are three main types of JSP constructs that you embed in a page:

- Scripting elements
- Directives
- Actions.

Scripting elements let you specify Java code that will become part of the resultant Servlet.

Directives let you control the overall structure of the Servlet, and Actions let you specify existing components that should be used, and otherwise control the behavior of the JSP engine.

In many cases, a large percent of your JSP page just consists of static HTML, known as template text. In all respects except one, this HTML looks just like normal HTML, follows all the same syntax rules, and is simply "passed through" to the client by the Servlet created to handle the page. Not only does the HTML look normal, it can be created by whatever tools you are already using for building Web pages. The one minor exception to the "template text is passed straight through" rule is that, if you want to have "<%" in the output, you need to put "<\%" in the template text.

JSP scripting elements let you insert Java code into the Servlet that will be generated from the current JSP page. There are three forms:

1. Expressions of the form <%= expression %> that are evaluated and inserted into the output
2. Script lets of the form <% code %> that are inserted into the Servlets service method, and
3. Declarations of the form <%! code %> that are inserted into the body of the Servlet class, outside of any existing method.

Each of these is described in more detail below.

4.7.1 Expression

A JSP expression element contains a scripting language expression that is evaluated, converted to a String, and inserted where the expression appears in the JSP file. Because the value of an expression is converted to a String, you can use an expression within a line of text, whether or not it is tagged with HTML, in a JSP file. The expression element can contain any expression that is valid according to the Java Language Specification but you cannot use a semicolon to end an expression.

Following is the syntax of JSP Expression:

```
<%= expression %>
```

You can write XML equivalent of the above syntax as follows:

```
<jsp:expression>  expression</jsp:expression>
```

Following is the simple example for JSP Expression:

```
<HTML>
<HEAD>
<TITLE>JSP DATE EXAMPLE</TITLE>
</HEAD>
<BODY>
<BIG>
<H1>DATE</H1>
<H2><%=new java.util.Date() %></H2>
</BIG>
</BODY>
</HTML>
```

This would generate the following result:
Today's date: 11-Sep-2010 21:24:25

4.7.2 The Scriptlet

A scriptlet can contain any number of JAVA language statements, variable or method declarations, or expressions that are valid in the page scripting language.

Following is the syntax of Scriptlet:

```
<% code fragment %>
```

You can write XML equivalent of the above syntax as follows:

```
<jsp:scriptlet>
code fragment
</jsp:scriptlet>
```

Any text, HTML tags, or JSP elements you write must be outside the scriptlet. Following is the simple and first example for JSP:

```
<HTML>
<BODY>
<%String queryData = request.getQueryString();
if (queryData.equals(“hello”))
out.println(queryData+” U are most welcome to JSP”);
else
out.println(“Attached GET data is “+queryData);
%>
</BODY>
</HTML>
```

To run the above example, type http://localhost:8080/examples/Scriptlets.jsp?hello in the browser URL.

Note that code inside a scriptlet gets inserted exactly as written, and any static HTML (template text) before or after a scriptlet gets converted to print statements. This means that scriptlets need not contain complete Java statements, and blocks left open can affect the static HTML outside of the scriptlets. For example, the following JSP fragment, containing mixed template text and scriptlets

```
<% if (Math.random() < 0.5) { %>
Have a <B>nice</B> day!
<% } else { %>
Have a <B>lousy</B> day!
<% } %>
```

will get converted to something like:

```
if (Math.random() < 0.5) {
out.println(“Have a <B>nice</B> day!”);
} else {
out.println(“Have a <B>lousy</B> day!”);
}
```

If you want to use the characters “%>” inside a scriptlet, enter “%\>” instead.

4.7.3 JSP Declarations

A declaration declares one or more variables or methods that you can use in Java code later in the JSP file. You must declare the variable or method before you use it in the JSP file.

Following is the syntax of JSP Declarations:

```
<%! declaration; [ declaration; ]+ ... %>
```

You can write XML equivalent of the above syntax as follows:

```
<jsp:declaration>   code fragment</jsp:declaration>
<HTML>
<BODY>
<%! private int accessCount = 0; %>
ACCESSES TO THE PAGE SINCE SERVER REBOOT:
<%= ++accessCount %>
</BODY>
31.  </HTML>
```

4.7.4 JSP-Directives

JSP directives provide directions and instructions to the container, telling it how to handle certain aspects of JSP processing.

A JSP directive affects the overall structure of the servlet class. It usually has the following form:

```
<%@ directive attribute="value" %>
```

Directives can have a number of attributes which you can list down as key-value pairs and separated by commas.

The blanks between the @ symbol and the directive name, and between the last attribute and the closing %>, are optional.

There are three types of directive tag:

Directive	Description
<%@ page ... %>	Defines page-dependent attributes, such as scripting language, error page, and buffering requirements.
<%@ include ... %>	Includes a file during the translation phase.
<%@ taglib ... %>	Declares a tag library, containing custom actions, used in the page.

1. The page directive

The **page** directive is used to provide instructions to the container that pertain to the current JSP page. You may code page directives anywhere in your JSP page. By convention, page directives are coded at the top of the JSP page.

Following is the basic syntax of page directive:

```
<%@ page attribute="value" %>
```

You can write XML equivalent of the above syntax as follows:

```
<jsp:directive.page attribute="value" />
```

Attributes

Following is the list of attributes associated with page directive:

Attribute	Purpose
buffer	Specifies a buffering model for the output stream.
autoFlush	Controls the behavior of the servlet output buffer.
contentType	Defines the character encoding scheme.
errorPage	Defines the URL of another JSP that reports on Java unchecked runtime exceptions.
isErrorPage	Indicates if this JSP page is a URL specified by another JSP page's errorPage attribute.
extends	Specifies a superclass that the generated servlet must extend
import	Specifies a list of packages or classes for use in the JSP as the Java import statement does for Java classes.
info	Defines a string that can be accessed with the servlet's getServletInfo() method.
isThreadSafe	Defines the threading model for the generated servlet.
language	Defines the programming language used in the JSP page.
session	Specifies whether or not the JSP page participates in HTTP sessions
isELIgnored	Specifies whether or not EL expression within the JSP page will be ignored.
isScriptingEnabled	Determines if scripting elements are allowed for use.

a. The buffer attribute

The **buffer** attribute specifies buffering characteristics for the server output response object.

You may code a value of "none" to specify no buffering so that all servlet output is immediately directed to the response object or you may code a maximum buffer size in kilobytes, which directs the servlet to write to the buffer before writing to the response object.

To direct the servlet to write output directly to the response output object, use the following:

```
<%@ page buffer="none" %>
```

Use the following to direct the servlet to write output to a buffer of size not less than 8 kilobytes:

```
<%@ page buffer="8kb" %>
```

The autoFlush Attribute:

The **autoFlushs** attribute specifies whether buffered output should be flushed automatically when the buffer is filled, or whether an exception should be raised to indicate buffer overflow.

A value of true (default) indicates automatic buffer flushing and a value of false throws an exception.

The following directive causes the servlet to throw an exception when the servlet's output buffer is full:

```
<%@ page autoFlush="false" %>
```

This directive causes the servlet to flush the output buffer when full:

```
<%@ page autoFlush="true" %>
```

Usually, the buffer and autoFlush attributes are coded on a single page directive as follows:

```
<%@ page buffer="16kb" autoflush="true" %>
```

b. The contentType attribute

The contentType attribute sets the character encoding for the JSP page and for the generated response page. The default content type is text/html, which is the standard content type for HTML pages.

If you want to write out XML from your JSP, use the following page directive:

```
<%@ page contentType="text/xml" %>
```

The following statement directs the browser to render the generated page as HTML:

```
<%@ page contentType="text/html" %>
```

The following directive sets the content type as a Microsoft Word document:

```
<%@ page contentType="application/msword" %>
```

You can also specify the character encoding for the response. For example, if you wanted to specify that the resulting page that is returned to the browser uses ISO Latin 1, you would use the following page directive:

```
<%@ page contentType="text/html:charset=ISO-8859-1" %>
```

c. The errorPage attribute

The errorPage attribute tells the JSP engine which page to display if there is an error while the current page runs. The value of the errorPage attribute is a relative URL.

The following directive displays MyErrorPage.jsp when all uncaught exceptions are thrown:

```
<%@ page errorPage="MyErrorPage.jsp" %>
```

d. The isErrorPage attribute

The isErrorPage attribute indicates that the current JSP can be used as the error page for another JSP.

The value of isErrorPage is either true or false. The default value of the isErrorPage attribute is false.

For example, the handleError.jsp sets the isErrorPage option to true because it is supposed to handle errors:

```
<%@ page isErrorPage="true" %>
```

e. The extends attribute

The extends attribute specifies a superclass that the generated servlet must extend.

For example, the following directive directs the JSP translator to generate the servlet such that the servlet extends *somePackage.SomeClass*:

```
<%@ page extends="somePackage.SomeClass" %>
```

The import Attribute:

The import attribute serves the same function as, and behaves like, the Java import statement. The value for the import option is the name of the package you want to import.

To import java.sql.*, use the following page directive:

```
<%@ page import="java.sql.*" %>
```

To import multiple packages you can specify them separated by comma as follows:

```
<%@ page import="java.sql.*,java.util.*"  %>
```

By default, a container automatically imports java.lang.*, javax.servlet.*, javax.servlet.jsp.*, and javax.servlet.http.*.

f. The info attribute

The info attribute lets you provide a description of the JSP. The following is a coding example:

```
<%@ page info="This JSP Page Written By BPL"  %>
```

g. The isThreadSafe attribute

The isThreadSafe option marks a page as being thread-safe. By default, all JSPs are considered thread-safe. If you set the isThreadSafe option to false, the JSP engine makes sure that only one thread at a time is executing your JSP.

The following page directive sets the isThreadSafe option to false:

```
<%@ page isThreadSafe="false" %>
```

h. The language attribute

The language attribute indicates the programming language used in scripting the JSP page.

For example, because you usually use Java as the scripting language, your language option looks like this:

```
<%@ page language="java" %>
```

i. The session attribute

The session attribute indicates whether or not the JSP page uses HTTP sessions. A value of true means that the JSP page has access to a builtin **session** object and a value of false means that the JSP page cannot access the builtin session object.

Following directive allows the JSP page to use any of the builtin object session methods such as session.getCreationTime() or session.getLastAccessTime():

```
<%@ page session="true" %>
```

j. The isELIgnored attribute

The isELIgnored option gives you the ability to disable the evaluation of Expression Language (EL) expressions which has been introduced in JSP 2.0.

The default value of the attribute is true, meaning that expressions, ${...}, are evaluated as dictated by the JSP specification. If the attribute is set to false, then expressions are not evaluated but rather treated as static text.

Following directive set an expressions not to be evaluated:

```
<%@ page isELIgnored="false" %>
```

k. The isScriptingEnabled attribute

The isScriptingEnabled attribute determines if scripting elements are allowed for use.

The default value (true) enables scriptlets, expressions, and declarations. If the attribute's value is set to false, a translation-time error will be raised if the JSP uses any scriptlets, expressions (non-EL), or declarations.

You can set this value to false if you want to restrict usage of scriptlets, expressions (non-EL), or declarations:

```
<%@ page isScriptingEnabled="false" %>
```

2. The include directive

The **include** directive is used to includes a file during the translation phase. This directive tells the container to merge the content of other external files with the current JSP during the translation phase. You may code *include* directives anywhere in your JSP page.

The general usage form of this directive is as follows:

```
<%@ include file="relative url" >
```

The filename in the include directive is actually a relative URL. If you just specify a filename with no associated path, the JSP compiler assumes that the file is in the same directory as your JSP.

You can write XML equivalent of the above syntax as follows:

```
<jsp:directive.include file="relative url" />
```

Example

A good example of **include** directive is including a common header and footer with multiple pages of content.

Let us define the following three files (a) header.jps (b)footer.jsp and (c)main.jsp as follows:

Following is the content of header.jsp:

```
<%!
int pageCount = 0;
```

```
void addCount() {
  pageCount++;
}%>
<% addCount(); %>
<html>
<head>
<title>The include Directive Example</title>
</head>
<body>
<center>
<h2>The include Directive Example</h2>
<p>This site has been visited <%= pageCount %> times.</p>
</center>
<br/><br/>
```

Following is the content of footer.jsp:

```
<br/><br/>
<center>
<p>Copyright © 2010</p>
</center>
</body>
</html>
```

Finally here is the content of main.jsp:

```
<%@ include file="header.jsp" %>
<center>
<p>Thanks for visiting my page.</p>
</center>
<%@ include file="footer.jsp" %>
```

Now let us keep all these files in root directory and try to access main.jsp. This would display the following result:

The include directive example

This site has been visited 1 time.

Thanks for visiting my page.

Try to refresh the main.jsp and you will find page hit count will keep increasing.

Now it is up to your creativity how you design your web pages but my suggestion is to keep dyanamic parts of your website in separate files and then include them in main file so that if tomorrow you need to change a part of your web page you can change it easily.

Now we are going to write one html file and two JSP files. This html file contains some coding which is common to both the JSP files. Instead of rewriting the contents, we are going to include the entire file using 'include' directive in JSP.

File 1: Navigation.html

```
<HTML>
<BODY>
This content will be displayed wherever the file gets inserted.
</BODY>
</HTML>
```

File 2: JSPInclude1.jsp.

```
<HTML>
<HEAD>
<TITLE>JavaServer Pages (JSP) 1.0</TITLE>
</HEAD>
<BODY>
<H1> THIS IS JSP PAGE 1 </H1>
<%@ include file="/Navigation.html" %>
</BODY>
</HTML>
```

File 3: JSPInclude.jsp.

```
<HTML>
<HEAD>
<TITLE>JavaServer Pages (JSP) 1.0</TITLE>
</HEAD>
<BODY>
<H1> THIS IS JSP PAGE 2 </H1>
<%@ include file="/Navigation.html" %>
</BODY>
</HTML>
```

Note that since the "include directive" inserts the files at the time the page is translated, if the navigation bar changes, you need to re-translate all the JSP pages that refer to it. This is a good compromise in a situation like this, since the navigation bar probably changes infrequently, and you want the inclusion process to be as efficient as possible. If, however, the included files changed more often, you could use the jsp:include action(we will see this soon) instead. This includes the file at the time the JSP page is requested. For example, Using Scripting Elements and Directives.

Here is a simple example showing the use of JSP expressions, scriptlets, declarations, and directives.

```
<HTML>
<HEAD>
<TITLE>Using JavaServer Pages</TITLE>
</HEAD>
<BODY>
<H1> USING JSP PAGES </H1>
```

Some dynamic content created using various JSP mechanisms

```
<UL>
<LI><B>Expression.</B><BR>
Your hostname: <%= request.getRemoteHost() %>.
```

```
<LI><B>Scriptlet. </B><BR>
<% out.println("Attached GET data: " +
request.getQueryString()); %>
<LI><B>Declaration(plus expression).</B><BR>
<%! private int accessCount = 0; %>
Accesses to page since server reboot: <%= ++accessCount %>
<LI><B>Directive (plus expression).</B><BR>
<%@ page import = "java.util.*" %>
Current date: <%= new Date() %>
</UL>
</BODY>
</HTML>
```

Listing Dynamic.jsp

Here's the typical result:

JSP—The taglib Directive

The JavaServer Pages API allows you to define custom JSP tags that look like HTML or XML tags and a tag library is a set of user-defined tags that implement custom behavior.

The **taglib** directive declares that your JSP page uses a set of custom tags, identifies the location of the library, and provides a means for identifying the custom tags in your JSP page.

The taglib directive follows the following syntax:

```
<%@ taglib uri="uri" prefix="prefixOfTag" >
```

Where the **uri** attribute value resolves to a location the container understands and the **prefix** attribute informs a container what bits of markup are custom actions.

You can write XML equivalent of the above syntax as follows:

```
<jsp:directive.taglib uri="uri" prefix="prefixOfTag" />
```

When you use a custom tag, it is typically of the form <prefix:tagname>. The prefix is the same as the prefix you specify in the taglib directive, and the tagname is the name of a tag implemented in the tag library.

Example

For example, suppose the **custlib** tag library contains a tag called **hello**. If you wanted to use the hello tag with a prefix of **mytag**, your tag would be **<mytag:hello>** and it will be used in your JSP file as follows:

```
<%@ taglib uri="http://www.example.com/custlib" prefix="mytag" %>
<html>
<body>
<mytag:hello/>
</body>
</html>
```

We would be able to call another piece of code using **<mytag:hello>**.

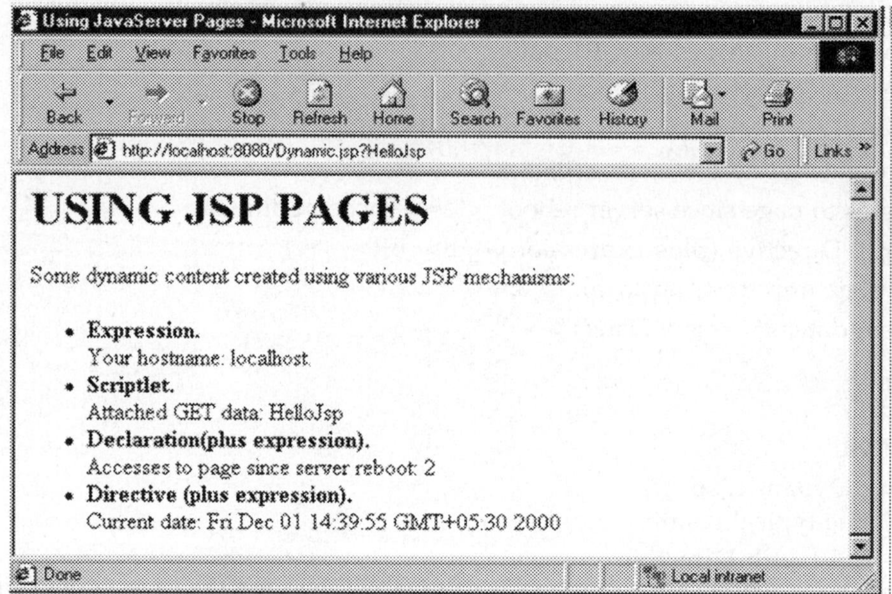

4.7.5 Actions

JSP actions use constructs in XML syntax to control the behavior of the servlet engine. You can dynamically insert a file, reuse JavaBeans components, forward the user to another page, or generate HTML for the Java plugin.

There is only one syntax for the Action element, as it conforms to the XML standard:
<jsp:action_name attribute="value" />

Action elements are basically predefined functions and there are following JSP actions available:

Syntax	Purpose
Jsp:include	Includes a file at the time the page is requested
Jsp:useBean	Finds or instantiates a JavaBean
Jsp:setProperty	Sets the property of a JavaBean
Jsp:getProperty	Inserts the property of a JavaBean into the output
Jsp:forward	Forwards the requester to a new page
Jsp:plugin	Generates browser-specific code that makes an OBJECT or EMBED tag for the Java plugin
Jsp:element	Defines XML elements dynamically.
Jsp:attribute	Defines dynamically defined XML element's attribute.
Jsp:body	Defines dynamically defined XML element's body.
Jsp:text	Use to write template text in JSP pages and documents.

These actions are described in more detail below. Remember that, as with XML in general, the element and attribute names are case sensitive.

Common Attributes

There are two attributes that are common to all Action elements: the **id** attribute and the **scope** attribute.

- **Id attribute:** The id attribute uniquely identifies the Action element, and allows the action to be referenced inside the JSP page. If the Action creates an instance of an object the id value can be used to reference it through the implicit object PageContext.
- **Scope attribute:** This attribute identifies the life cycle of the Action element. The id attribute and the scope attribute are directly related, as the scope attribute determines the lifespan of the object associated with the id. The scope attribute has four possible values: (a) page, (b)request, (c)session, and (d) application.

a. The jsp:include Action

This action lets you insert files into the page being generated. The syntax looks like this:

```
<jsp:include page="relative URL" flush="true" />
```

Unlike the **include** directive, which inserts the file at the time the JSP page is translated into a servlet, this action inserts the file at the time the page is requested.

Following is the list of attributes associated with include action:

Attribute	Description
Page	The relative URL of the page to be included.
Flush	The boolean attribute determines whether the included resource has its buffer flushed before it is included.

This pays a small penalty in efficiency, and precludes the included page from containing general JSP code (it cannot set HTTP headers, for example), but it gains significantly in flexibility. For example, here is a JSP page that inserts four different snippets into a "What's New?" Web page. Each time the headlines change, authors only need to update the four files, but can leave the main JSP page unchanged.

Example

```
<HTML>
<HEAD>
<TITLE>What's New</TITLE>
</HEAD>
<BODY>
<H1>What's New at JspNews.com</H1>
<P>

Here is a summary of our four most recent news stories:

<OL>
<LI><jsp:include page="news/Item1.html" flush="true"/> <LI><jsp:include
page="news/Item2.html" flush="true"/>
<LI><jsp:include page="news/Item3.html" flush="true"/>
<LI><jsp:include page="news/Item4.html" flush="true"/>
</OL>
```

```
</BODY>
</HTML>
```

Listing WhatsNew . jsp

Here's the typical result:

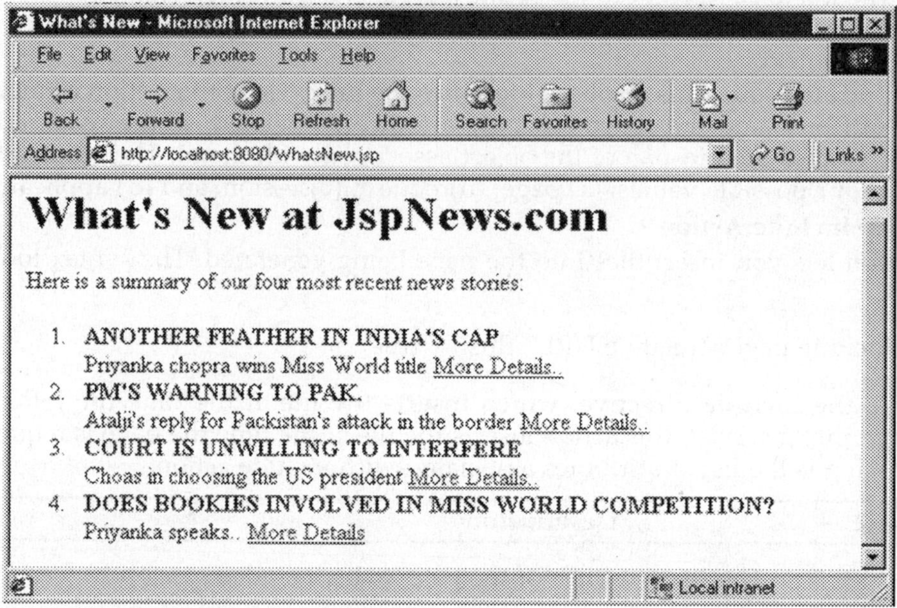

b. The jsp:useBean Action

This action lets you load in a JavaBean to be used in the JSP page. This is a a very useful capability because it lets you exploit the reusability of Java classes without sacrificing the convenience that JSP adds over Servlets alone. The simplest syntax for specifying that a bean should be used is:

```
<jsp:useBean id="name"
class="package.class"
scope="page | request | session | application" />
```

This usually means "instantiate an object of the class specified by class, and bind it to a variable with the name specified by id." The scope attribute represents the life of the object.

Once a bean class is loaded, you can use **jsp:setProperty** and **jsp:getProperty** actions to modify and retrieve bean properties.

Following is the list of attributes associated with useBean action:

Attribute	Description
Class	Designates the full package name of the bean.
Type	Specifies the type of the variable that will refer to the object.
BeanName	Gives the name of the bean as specified by the instantiate () method of the java.beans.Beans class.

c. The jsp:setProperty Action

You use jsp:setProperty to give values to properties of beans that have been referenced earlier. You can do this in two contexts. First, you can use jsp:setProperty after, but outside of, a jsp:useBean element, as below:

```
<jsp:useBean id="myName" ... />
...
<jsp:setProperty name="myName"
property="someProperty" ... />
```

In this case, the jsp:setProperty is executed regardless of whether a new bean was instantiated or an existing bean was found. A second context in which jsp:setProperty can appear is inside the body of a jsp:useBean element, as below:

```
<jsp:useBean id="myName" ... >
...
<jsp:setProperty name="myName"
property="someProperty" ... />
</jsp:useBean>
```

Here, the jsp:setProperty is executed only if a new object was instantiated, not if an existing one was found.

There are four possible attributes of jsp:setProperty:

Attribute Usage

Name

This required attribute designates the bean whose property will be set. The jsp:useBean element must appear before the jsp:setProperty element.

Property

This required attribute indicates the property you want to set. However, there is one special case: a value of "*" means that all request parameters whose names match bean property names will be passed to the appropriate setter methods.

Value

This optional attribute specifies the value for the property. String values are automatically converted to numbers, boolean, Boolean, byte, Byte, char, and Character via the standard valueOf method in the target or wrapper class. For example, a value of "true" for a boolean or Boolean property will be converted via Boolean.valueOf, and a value of "42" for an int or Integer property will be converted via Integer.valueOf. You cannot use both value and param, but it is permissible to use neither. See the discussion of param below.

Param

This optional attribute designates the request parameter from which the property should be derived. If the current request has no such parameter, nothing is done: the system does *not* pass null to the setter method of the property. Thus, you can let the bean itself supply default values, overriding them only when the request parameters say to do so. For example, the following snippet says "set the numberOfItems property to whatever the value of the numItems request parameter is, if there is such a request parameter. Otherwise don't do anything."

```
<jsp:setProperty name="orderBean"
property="numberOfItems"
param="numItems" />
```

If you omit both value and param, it is the same as if you supplied a param name that matches the property name. You can take this idea of automatically using the request property whose name matches the property one step further by supplying a property name of "*" and omitting both value and param. In this case, the server iterates through available properties and request parameters, matching up ones with identical names.

d. The jsp:getProperty Action

The **getProperty** action is used to retrieve the value of a given property and converts it to a string, and finally inserts it into the output.

The getProperty action has only two attributes, both of which are required and simple syntax is as follows:

```
<jsp:useBean id="myName" ... />
...
<jsp:getProperty name="myName" property="someProperty" .../>
```

Following is the list of required attributes associated with setProperty action:

Attribute	Description
Name	The name of the Bean that has a property to be retrieved. The Bean must have been previously defined.
Property	The property attribute is the name of the Bean property to be retrieved.

This element retrieves the value of a bean property, converts it to a string, and inserts it into the output. The two required attributes are name, the name of a bean previously referenced via jsp:useBean, and property, the property whose value should be inserted. Now we are going to see one sample application which use jsp:useBean, jsp:setAttribute and jsp:getAttribute actions. This example does nothing but manipulation of a string via JavaBean.

Example: Using Beans inside a JSP page

File 1: Bean program

```
package hall;
public class SimpleBean {
private String message = "No message specified";
//Bean's getter method.
public String getMessage() {
return(message);
}
// Bean's setter method.
public void setMessage(String message) {
this.message = message;
```

```
}
}
```

Listing SimpleBean.java

File2: BeanJSP.jsp

```
<HTML>
<BODY>
<jsp:useBean id="test" class="hall.SimpleBean" />
<jsp:setProperty name="test"
property="message"
value="Hello WWW" />
<H1> Message:
<I> <jsp:getProperty name="test" property="message" /></I>
</H1>
</BODY>
</HTML>
```

Lising BeanJsp.jsp

To work with the above example,

- The bean .class should reside in the javawebserver2.0\classes\hall directory
- If your .jsp file is inside javawebserver2.0\public_html directory, then type the following in the URL of the browser:

http://localhost:8080/BeanJsp.jsp

Here's the typical output:

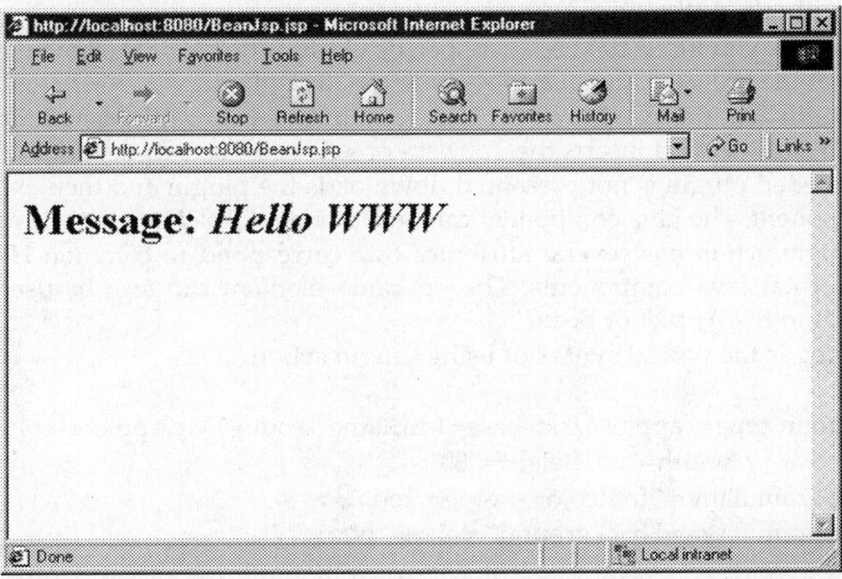

JSP defines a tag, <jsp:request>, that you can use to redirect to an external page in one of two ways, as specified by the FORWARD attribute or the INCLUDE attribute. With the FORWARD attribute, you can redirect to any valid URL. This effectively halts processing of the current page at the point where the redirect occurs, although all processing up to that point will still take place. This is exactly analogous to a typical redirect using CGI, SSJS, ASP, or JavaScript. With the INCLUDE attribute, you cannot only redirect to another page but also come back to the calling page upon completion of processing in the called page. For instance, you could actually call out to another JSP page that generates some HTML dynamically and have that page generate its HTML; upon returning, that HTML would be inserted into the calling page at the point where your <jsp:request> tag occurs. In fact, the called page has no idea that it is being called from another JSP page. It simply sees an HTTP request and responds by returning HTML text.

Keep in mind that you can use the INCLUDE method of redirection to access static HTML pages, JSP pages, servlets, SSJS pages, ASP pages - just about any resource that responds to HTTP requests and generates a response that you want to include in your page. But note that if the resource you access returns a complete HTML page, including <HTML> and <BODY> tags, you may not get the result you intended.

e. The jsp:forward Action

This action lets you forward the request to another page. It has a single attribute page, which should consist of a relative URL.

The simple syntax of this action is as follows:

```
<jsp:forward page="Relative URL"/>
```

This could be a static value, or could be computed at request time, as in the two examples below.

```
<jsp:forward page="/utils/errorReporter.jsp" />
<jsp:forward page="<%= someJavaExpression %>" />
```

f. The jsp:plugin Action

This action lets you insert the browser-specific OBJECT or EMBED element needed to specify that the browser run an applet using the Java plugin.

The **plugin** action is used to insert Java components into a JSP page. It determines the type of browser and inserts the <object> or <embed> tags as needed.

If the needed plugin is not present, it downloads the plugin and then executes the Java component. The Java component can be either an Applet or a JavaBean.

The plugin action has several attributes that correspond to common HTML tags used to format Java components. The <param> element can also be used to send parameters to the Applet or Bean.

Following is the typical syntax of using plugin action:

```
<jsp:plugin type="applet" codebase="dirname" code="MyApplet.class"
                width="60" height="80">
  <jsp:param name="fontcolor" value="red" />
  <jsp:param name="background" value="black" />

  <jsp:fallback>
   Unable to initialize Java Plugin
  </jsp:fallback>
</jsp:plugin>
```

You can try this action using some applet if you are interested. A new element, the <fallback> element, can be used to specify an error string to be sent to the user in case the component fails.

g. The <jsp:body> Action

The <jsp:element>, <jsp:attribute> and <jsp:body> actions are used to define XML elements dynamically. The word dynamically is important, because it means that the XML elements can be generated at request time rather than statically at compile time.

Following is a simple example to define XML elements dynamically:

```
<%@page language="java" contentType="text/html"%>
<html xmlns="http://www.w3c.org/1999/xhtml"
    xmlns:jsp="http://java.sun.com/JSP/Page">

<head><title>Generate XML Element</title></head>
<body>
<jsp:element name="xmlElement">
<jsp:attribute name="xmlElementAttr">
  Value for the attribute
</jsp:attribute>
<jsp:body>
  Body for XML element
</jsp:body>
</jsp:element>
</body>
</html>
```

This would produce the following HTML code at run time:

```
<html xmlns="http://www.w3c.org/1999/xhtml"
    xmlns:jsp="http://java.sun.com/JSP/Page">
 <head><title>Generate XML Element</title></head>
<body>
<xmlElement xmlElementAttr="Value for the attribute">
  Body for XML element
</xmlElement>
</body>
</html>
```

h. The <jsp:text> Action

The <jsp:text> action can be used to write template text in JSP pages and documents. Following is the simple syntax for this action:

```
<jsp:text>Template data</jsp:text>
```

The body of the template cannot contain other elements; it can only contain text and EL expressions (Note: EL expressions are explained in subsequent chapter). Note that in XML files, you cannot use expressions such as ${whatever > 0}, because the greater than signs are illegal. Instead, use the gt form, such as ${whatever gt 0} or an alternative is to embed the value in a CDATA section.

```
<jsp:text><![CDATA[<br>]]></jsp:text>
```

If you need to include a DOCTYPE declaration, for instance for XHTML, you must also use the <jsp:text> element as follows:

```
<jsp:text><![CDATA[<!DOCTYPE html
PUBLIC "-//W3C//DTD XHTML 1.0 Strict//EN"
"DTD/xhtml1-strict.dtd">]]>
</jsp:text>
<head><title>jsp:text action</title></head>
<body>

<books><book><jsp:text>
    Welcome to JSP Programming
</jsp:text></book></books>

</body>
</html>
```

Try the above example with and without <jsp:text> action.

4.8 JSP Implicit Objects

To simplify code in JSP expressions and scriptlets, you are supplied with eight automatically defined variables, sometimes called *implicit objects*. The available variables are request, response, out, session, application, config, pageContext, and page. Details for each are given below.

Objects	Description
request	This is the **HttpServletRequest** object associated with the request.
response	This is the **HttpServletResponse** object associated with the response to the client.
out	This is the **PrintWriter** object used to send output to the client.
session	This is the **HttpSession** object associated with the request.
application	This is the **ServletContext** object associated with application context.
config	This is the **ServletConfig** object associated with the page.
pageContext	This encapsulates use of server-specific features like higher performance **JspWriters**.
page	This is simply a synonym for **this**, and is used to call the methods defined by the translated servlet class.
exception	The **Exception** object allows the exception data to be accessed by designated JSP.

a. The Request Object

The request object is an instance of a javax.servlet.http.HttpServletRequest object. Each time a client requests a page the JSP engine creates a new object to represent that request.

The request object provides methods to get HTTP header information including form data, cookies, HTTP methods etc.

We would see complete set of methods associated with request object in coming chapter: **JSP-Client Request**.

b. The Response Object

The response object is an instance of a javax.servlet.http.HttpServletResponse object. Just as the server creates the request object, it also creates an object to represent the response to the client.

The response object also defines the interfaces that deal with creating new HTTP headers. Through this object the JSP programmer can add new cookies or date stamps, HTTP status codes, etc.

We would see complete set of methods associated with response object in coming chapter: **JSP-Server Response**.

c. Out

This is the PrintWriter used to send output to the client. However, in order to make the response object (see the previous section) useful, this is a buffered version of PrintWriter called JspWriter. The out implicit object is an instance of a javax.servlet.jsp.JspWriter object and is used to send content in a response. Note that you can adjust the buffer size, or even turn buffering off, through use of the buffer attribute of the page directive. Also note that out is used almost exclusively in scriptlets, since JSP expressions automatically get placed in the output stream, and thus rarely need to refer to out explicitly.

Following are the important methods which we would use to write boolean char, int, double, object, String, etc.

Method	Description
out.print(dataType dt)	Print a data type value
out.println(dataType dt)	Print a data type value, then terminate the line with new line character.
out.flush()	Flush the stream.

Example: Using request and out objects. This example just shows how a JSP file processes FORM data using request and out objects.

File 1: A HTML file to collect and send name and password data from the user

```
<HTML>
<BODY >
<FORM TYPE=POST ACTION=result.jsp>
<FONT size=5 COLOR="red">
Type your name and password <br>
Name <input TYPE=text NAME=name > <BR>
Password <input TYPE=password NAME=password> <BR>
<INPUT TYPE=submit NAME=submit VALUE="Submit">
```

```
</FONT>
</FORM>
</BODY>
</HTML>
```

Listing text.html

File 2: A JSP file to process the data sent by text.html.

```
<HTML>
<BODY>
<%
String name=request.getParameter("name");
String password=request.getParameter("password");
if(name.equals(""))
out.println("Name is must..");
if(password.equals(""))
out.println("Password is must..");
%>
</BODY>
</HTML>
```

Listing result.jsp

d. Session

This is the HttpSession object associated with the request. Recall that sessions are created automatically, so this variable is bound even if there was no incoming session reference. TheHttpSession object is used to store objects in between client requests.

Example-Listing session.jsp

This example illustrates the use of session object by tracking the number of visits the user makes. 'Refresh' many times and see how the value changes.

```
<HTML>
<BODY>
<%
//let 'session_count' be an object representing session count
Integer session_count =
(Integer)session.getValue("COUNT");
// if session_count is not found then it is time to create one.
if(session_count==null) {
session_count = new Integer(1);
session.putValue("COUNT",session_count);
}
// increment the session count otherwise.
else {
int new_value = session_count.intValue()+1;
session_count = new Integer(new_value);
session.putValue("COUNT",session_count);
}
out.println("This is "+ session_count +
" time you are visiting this site");
```

```
    %>
    </BODY>
</HTML>
```

e. Application

This is the ServletContext as obtained via

getServletConfig().getContext().

The application object has application scope, which means that it is available to all JSPs until the JSP engine is shut down. The application object is most often used to retrieve environment information. The application object is direct wrapper around the ServletContext object for the generated Servlet and in reality an instance of a javax.servlet.ServletContext object.

This object is a representation of the JSP page through its entire lifecycle. This object is created when the JSP page is initialized and will be removed when the JSP page is removed by the jspDestroy() method.

By adding an attribute to application, you can ensure that all JSP files that make up your web application have access to it.

f. Config

This is the ServletConfig object for this page. The config object is an instantiation of javax.servlet.ServletConfig and is a direct wrapper around the ServletConfig object for the generated servlet.

This object allows the JSP programmer access to the Servlet or JSP engine initialization parameters such as the paths or file locations, etc. The following config method is the only one you might ever use, and its usage is trivial:

Config.getServletName();

This returns the servlet name, which is the string contained in the <servlet-name> element defined in the WEB-INF\web.xml file

g. PageContext

The pageContext object is an instance of a javax.servlet.jsp.PageContext object. The pageContext object is used to represent the entire JSP page.

This object is intended as a means to access information about the page while avoiding most of the implementation details.

This object stores references to the request and response objects for each request. The application, config, session, and out objects are derived by accessing attributes of this object.

The pageContext object also contains information about the directives issued to the JSP page, including the buffering information, the errorPageURL, and page scope.

The PageContext class defines several fields, including PAGE_SCOPE, REQUEST_SCOPE, SESSION_SCOPE, and APPLICATION_SCOPE, which identify the four scopes. It also supports more than 40 methods, about half of which are inherited from the javax.servlet.jsp. JspContext class.

One of the important methods is **removeAttribute**, which accepts either one or two arguments. For example, pageContext.removeAttribute ("attrName") removes the attribute from all scopes, while the following code only removes it from the page scope:

pageContext.removeAttribute("attrName", PAGE_SCOPE);

h. The Page Object

This object is an actual reference to the instance of the page. It can be thought of as an object that represents the entire JSP page.

The page object is really a direct synonym for the **this** object.

i. The Exception Object

The exception object is a wrapper containing the exception thrown from the previous page. It is typically used to generate an appropriate response to the error condition.

4.9 Control-Flow Statements

JSP provides full power of Java to be embeded in your web application. You can use all the APIs and building blocks of Java in your JSP programming including decision making statements, loops, etc.

4.9.1 Decision-Making Statements

The **if...else** block starts out like an ordinary Scriptlet, but the Scriptlet is closed at each line with HTML text included between Scriptlet tags.

```
<%! int day = 3; %>
<html>
<head><title>IF...ELSE Example</title></head>
<body>
<% if (day == 1 | day == 7) { %>
    <p> Today is weekend</p>
<% } else { %>
    <p> Today is not weekend</p>
<% } %>
</body>
</html>
```

This would produce the following result:

Today is not weekend

Now look at the following **switch...case** block which has been written a bit differently using **out.println()** and inside Scriptlet as:

```
<%! int day = 3; %>
<html>
<head><title>SWITCH...CASE Example</title></head>
<body>
<%
switch(day) {
case 0:
   out.println("It\'s Sunday.");
   break;
case 1:
   out.println("It\'s Monday.");
   break;
case 2:
   out.println("It\'s Tuesday.");
   break;
case 3:
   out.println("It\'s Wednesday.");
```

```
  break;
case 4:
  out.println("It\'s Thursday.");
  break;
case 5:
  out.println("It\'s Friday.");
  break;
default:
  out.println("It's Saturday.");
}
%>
</body>
</html>
```

This would produce the following result:
It's Wednesday.

4.9.2 Loop Statements

You can also use three basic types of looping blocks in Java: **for, while, and do…while** blocks in your JSP programming.

Let us look at the following **for** loop example:

```
<%! int fontSize; %>
<html>
<head><title>FOR LOOP Example</title></head>
<body>
<%for ( fontSize = 1; fontSize <= 3; fontSize++){ %>
  <font color="green" size="<%= fontSize %>">
  JSP Tutorial
  </font><br />
<%}%>
</body>
</html>
```

This would produce the following result:

JSP Tutorial
JSP Tutorial
JSP Tutorial

Above example can be written using **while** loop as follows:

```
<%! int fontSize; %>
<html>
<head><title>WHILE LOOP Example</title></head>
<body>
<%while ( fontSize <= 3){ %>
  <font color="green" size="<%= fontSize %>">
  JSP Tutorial
  </font><br />
<%fontSize++;%>
```

```
<%}%>
</body>
</html>
```

This would also produce the following result:

JSP Tutorial
JSP Tutorial
JSP Tutorial

4.9.3 Operators

JSP supports all the logical and arithmetic operators supported by Java. Following table give a list of all the operators with the highest precedence appear at the top of the table, those with the lowest appear at the bottom.

Within an expression, higher precedenace operators will be evaluated first.

<table>
<tr><th>Category</th><th>Operator</th><th>Associativity</th></tr>
<tr><td>Postfix</td><td>() [] . (dot operator)</td><td>Left to right</td></tr>
<tr><td>Unary</td><td>++ - - ! ~</td><td>Right to left</td></tr>
<tr><td>Multiplicative</td><td>* / %</td><td>Left to right</td></tr>
<tr><td>Additive</td><td>+ -</td><td>Left to right</td></tr>
<tr><td>Shift</td><td>>> >>> <<</td><td>Left to right</td></tr>
<tr><td>Relational</td><td>> >= < <=</td><td>Left to right</td></tr>
<tr><td>Equality</td><td>== !=</td><td>Left to right</td></tr>
<tr><td>Bitwise AND</td><td>&</td><td>Left to right</td></tr>
<tr><td>Bitwise XOR</td><td>^</td><td>Left to right</td></tr>
<tr><td>Bitwise OR</td><td>|</td><td>Left to right</td></tr>
<tr><td>Logical AND</td><td>&&</td><td>Left to right</td></tr>
<tr><td>Logical OR</td><td>||</td><td>Left to right</td></tr>
<tr><td>Conditional</td><td>?:</td><td>Right to left</td></tr>
<tr><td>Assignment</td><td>= += -= *= /= %= >>= <<= &= ^= |=</td><td>Right to left</td></tr>
<tr><td>Comma</td><td>,</td><td>Left to right</td></tr>
</table>

4.9.4 Literals

The JSP expression language defines the following literals:

- **Boolean:** True and false
- **Integer:** As in Java
- **Floating point:** As in Java
- **String:** With single and double quotes; " is escaped as \", ' is escaped as \', and \ is escaped as \\.
- **Null:** Null

4.10 Client-Sever JSP

4.10.1 Client Request

When a browser requests for a web page, it sends a lot of information to the web server which cannot be read directly because this information travel as a part of header of HTTP request. Following is the important header information which comes from browser side and you would use very frequently in web programming:

Header	Description
accept	This header specifies the MIME types that the browser or other clients can handle. Values of **image/png** or **image/jpeg** are the two most common possibilities.
accept-Charset	This header specifies the character sets the browser can use to display the information. For example, ISO-8859-1.
accept-Encoding	This header specifies the types of encodings that the browser knows how to handle. Values of **gzip** or **compress** are the two most common possibilities.
accept-Language	This header specifies the client's preferred languages in case the servlet can produce results in more than one language. For example, en, en-us, ru, etc.
authorization	This header is used by clients to identify themselves when accessing password-protected Web pages.
connection	This header indicates whether the client can handle persistent HTTP connections. Persistent connections permit the client or other browser to retrieve multiple files with a single request. A value of **Keep-Alive** means that persistent connections should be used.
content-Length	This header is applicable only to POST requests and gives the size of the POST data in bytes.
cookie	This header returns cookies to servers that previously sent them to the browser.
host	This header specifies the host and port as given in the original URL.
if-Modified-Since	This header indicates that the client wants the page only if it has been changed after the specified date. The server sends a code, 304 which means **Not Modified** header if no newer result is available.
if-Unmodified-Since	This header is the reverse of If-Modified-Since; it specifies that the operation should succeed only if the document is older than the specified date.
referer	This header indicates the URL of the referring Web page. For example, if you are at Web page 1 and click on a link to Web page 2, the URL of Web page 1 is included in the Referer header when the browser requests Web page 2.
user-Agent	This header identifies the browser or other client making the request and can be used to return different content to different types of browsers.

The HttpServletRequest Object

The request object is an instance of a javax.servlet.http.HttpServletRequest object. Each time a client requests a page the JSP engine creates a new object to represent that request.

The request object provides methods to get HTTP header information including form data, cookies, HTTP methods, etc.

There are following important methods which can be used to read HTTP header in your JSP program. These methods are available with *HttpServletRequest* object which represents client request to web servers in Section 2.6.

HTTP Header Request Example

> **Note:** Examples in all sections of JSP appears and produces the same result as in the relevant section of servlets. Differences in writing programs for servlets and JSP could be easily identified on comparison.

Following is the example which uses **getHeaderNames()** method of HttpServletRequest to read the HTTP header infromation. This method returns an Enumeration that contains the header information associated with the current HTTP request.

Once we have an Enumeration, we can loop down the Enumeration in the standard manner, using *hasMoreElements()* method to determine when to stop and using *nextElement()* method to get each parameter name.

```
<%@ page import="java.io.*,java.util.*" %>
<html>
<head>
<title>HTTP Header Request Example</title>
</head>
<body>
<center>
<h2>HTTP Header Request Example</h2>
<table width="100%" border="1" align="center">
<tr bgcolor="#949494">
<th>Header Name</th><th>Header Value(s)</th>
</tr>
<%
  Enumeration headerNames = request.getHeaderNames();
  while(headerNames.hasMoreElements()) {
    String paramName = (String)headerNames.nextElement();
    out.print("<tr><td>" + paramName + "</td>\n");
    String paramValue = request.getHeader(paramName);
    out.println("<td> " + paramValue + "</td></tr>\n");
  }
```

```
%>
</table>
</center>
</body>
</html>
```

Now put the above code in main.jsp and try to access it. This would produce result something as follows:

HTTP Header Request Example

Header Name	Header Value(s)
accept	*/*
accept-language	en-us
user-agent	Mozilla/4.0 (compatible; MSIE 7.0; Windows NT 5.1; Trident/4.0; InfoPath.2; MS-RTC LM 8)
accept-encoding	gzip, deflate
host	localhost:8080
connection	Keep-Alive
cache-control	no-cache

To become more comfortable with other methods, you can try a few more above listed methods in the same fashion.

4.10.2 Server Response

When a Web server responds to a HTTP request to the browser, the response typically consists of a status line, some response headers, a blank line, and the document. A typical response looks like this:

```
HTTP/1.1 200 OK
Content-Type: text/html
Header2: ...
...
HeaderN: ...
  (Blank Line)
<!doctype ...>
<html>
<head>...</head>
<body>
...
</body>
</html>
```

The status line consists of the HTTP version (HTTP/1.1 in the example), a status code (200 in the example), and a very short message corresponding to the status code (OK in the example).

Following is a summary of the most useful HTTP 1.1 response headers which go back to the browser from web server side and you would use them very frequently in web programming:

Header	Description
allow	This header specifies the request methods (GET, POST, etc.) that the server supports.
cache-Control	This header specifies the circumstances in which the response document can safely be cached. It can have values **public, private** or **no-cache,** etc. Public means document is cacheable, private means document is for a single user and can only be stored in private (nonshared) caches and no-cache means document should never be cached.
connection	This header instructs the browser whether to use persistent in HTTP connections or not. A value of **close** instructs the browser not to use persistent HTTP connections and **keep-alive** means using persistent connections.
content-Disposition	This header lets you request that the browser ask the user to save the response to disk in a file of the given name.
content-Encoding	This header specifies the way in which the page was encoded during transmission.
content-Language	This header signifies the language in which the document is written. For example, en, en-us, ru, etc.
content-Length	This header indicates the number of bytes in the response. This information is needed only if the browser is using a persistent (keep-alive) HTTP connection.
content-Type	This header gives the MIME (Multipurpose Internet Mail Extension) type of the response document.
expires	This header specifies the time at which the content should be considered out-of-date and thus no longer be cached.
last-Modified	This header indicates when the document was last changed. The client can then cache the document and supply a date by an **If-Modified-Since** request header in later requests.
location	This header should be included with all responses that have a status code in the 300s. This notifies the browser of the document address. The browser automatically reconnects to this location and retrieves the new document.
refresh	This header specifies how soon the browser should ask for an updated page. You can specify time in number of seconds after which a page would be refreshed.

retry-After	This header can be used in conjunction with a 503 (Service Unavailable) response to tell the client how soon it can repeat its request.
set-Cookie	This header specifies a cookie associated with the page.

The HttpServletResponse Object

The response object is an instance of a javax.servlet.http.HttpServletRequest object. Just as the server creates the request object, it also creates an object to represent the response to the client.

The response object also defines the interfaces that deal with creating new HTTP headers. Through this object the JSP programmer can add new cookies or date stamps, HTTP status codes, etc.

There are following methods which can be used to set HTTP response header in your servlet program. These method are available with *HttpServletResponse* object which represents server response as listed in Section 2.7.

HTTP Header Response Example

Following example would use **setIntHeader()** method to set **Refresh** header to simulate a digital clock:

```
<%@ page import="java.io.*,java.util.*" %>
<html>
<head>
<title>Auto Refresh Header Example</title>
</head>
<body>
<center>
<h2>Auto Refresh Header Example</h2>
<%
  // Set refresh, autoload time as 5 seconds
  response.setIntHeader("Refresh", 5);
  // Get current time
  Calendar calendar = new GregorianCalendar();
  String am_pm;  int hour = calendar.get(Calendar.HOUR);
  int minute = calendar.get(Calendar.MINUTE);
  int second = calendar.get(Calendar.SECOND);
  if(calendar.get(Calendar.AM_PM) == 0)
    am_pm = "AM";
  else
    am_pm = "PM";
  String CT = hour+":"+ minute +":"+ second +" "+ am_pm;
  out.println("Crrent Time: " + CT + "\n");
%>
</center>
</body>
</html>
```

Now put the above code in main.jsp and try to access it. This would display current system time after every 5 seconds as follows. Just run the JSP and wait to see the result:

Auto Refresh Header Example

Current Time is: 9:44:50 PM

To become more comfortable with other methods, you can try a few more above listed methods in the same fashion.

4.10.3 HTTP Status Codes

The format of the HTTP request and HTTP response messages are similar and will have the following structure:

An initial status line + CRLF (Carriage Return + Line Feed, i.e. New Line)
Zero or more header lines + CRLF
A blank line, i.e. a CRLF
An optional message body like file, query data or query output.

For example, a server response header looks as follows:

```
HTTP/1.1 200 OK
Content-Type: text/html
Header2: ...
...
HeaderN: ...
  (Blank Line)
<!doctype ...>
<html>
<head>...</head>
<body>
...
</body>
</html>
```

The status line consists of the HTTP version (HTTP/1.1 in the example), a status code (200 in the example), and a very short message corresponding to the status code (OK in the example).

Following is a list of HTTP status codes and associated messages that might be returned from the Web Server as in Section 2.8.

Methods to Set HTTP Status Code

There are following methods which can be used to set HTTP Status Code in your servlet program. These method are available with *HttpServletResponse* object as tabulated in Section 2.8.

HTTP Status Code Example

Following is the example which would send 407 error code to the client browser and browser would show you "Need authentication!!!" message.

```
<html>
<head>
```

```
<title>Setting HTTP Status Code</title>
</head>
<body>
<%
  // Set error code and reason.
  response.sendError(407, "Need authentication!!!" );
%>
</body>
</html>
```

Now calling the above JSP would display the following result:

HTTP Status 407-Need authentication!!!

type **Status report**

message **Need authentication!!!**

description **The client must first authenticate itself with the proxy (Need authentication!!!).**

Apache Tomcat/5.5.29

To become more comfortable with HTTP status codes, try to set different status codes and their description.

4.10.4 Form Processing

You must have come across many situations when you need to pass some information from your browser to web server and ultimately to your backend program. The browser uses two methods to pass this information to web server. These methods are GET Method and POST Method.

GET method

The GET method sends the encoded user information appended to the page request. The page and the encoded information are separated by the ? character as follows:

```
http://www.test.com/hello?key1=value1&key2=value2
```

The GET method is the defualt method to pass information from browser to web server and it produces a long string that appears in your browser's Location:box. Never use the GET method if you have password or other sensitive information to pass to the server.

The GET method has size limitation: Only 1024 characters can be in a request string.

This information is passed using QUERY_STRING header and will be accessible through QUERY_STRING environment variable which can be handled using getQueryString() and getParameter() methods of request object.

POST method

A generally more reliable method of passing information to a backend program is the POST method.

This method packages the information in exactly the same way as GET methods, but instead of sending it as a text string after a ? in the URL, it sends it as a separate message. This message comes to the backend program in the form of the standard input which you can parse and use for your processing.

JSP handles this type of requests using getParameter() method to read simple parameters and getInputStream() method to read binary data stream coming from the client.

Reading Form Data using JSP

JSP handles form data parsing automatically using the following methods depending on the situation:

- **getParameter():** You call request.getParameter() method to get the value of a form parameter.
- **getParameterValues():** Call this method if the parameter appears more than once and returns multiple values, for example, checkbox.
- **getParameterNames():** Call this method if you want a complete list of all parameters in the current request.
- **getInputStream():** Call this method to read binary data stream coming from the client.

GET Method Example Using URL

Here is a simple URL which will pass two values to HelloForm program using GET method.

http://localhost:8080/main.jsp?first_name=BPL&last_name=AJP

Below is **main.jsp** JSP program to handle input given by web browser. We are going to use **getParameter()** method which makes it very easy to access passed information:

```
<html>
<head>
<title>Using GET Method to Read Form Data</title>
</head>
<body>
<center>
<h1>Using GET Method to Read Form Data</h1>
<ul>
<li><p><b>First Name:</b>
   <%= request.getParameter("first_name")%>
</p></li>
<li><p><b>Last  Name:</b>
   <%= request.getParameter("last_name")%>
</p></li>
</ul>
</body>
</html>
```

Now type *http://localhost:8080/main.jsp?first_name=BPL&last_name=AJP* in your browser's Location:box. This would generate the following result:

Using GET Method to Read Form Data

First Name: BPL

Last Name: AJP

GET Method Example Using Form

Here is a simple example which passes two values using HTML FORM and submit button. We are going to use same JSP main.jsp to handle this input.

```
<html>
<body>
```

```
<form action="main.jsp" method="GET">
First Name: <input type="text" name="first_name">
<br />
Last Name: <input type="text" name="last_name" />
<input type="submit" value="Submit" />
</form>
</body>
</html>
```

Keep this HTML in a file Hello.htm and put it in <Tomcat-installation-directory>/webapps/ROOT directory. When you would access *http://localhost:8080/Hello.htm*, here is the actual output of the above form.

First Name: [ ]

Last Name: [ ]

Try to enter First Name and Last Name and then click submit button to see the result on your local machine where tomcat is running. Based on the input provided, it will generate similar result as mentioned in the above example.

POST Method Example Using Form

Let us do a little modification in the above JSP to handle GET as well as POST methods. Below is **main.jsp** JSP program to handle input given by web browser using GET or POST methods.

In fact there is no change in above JSP because only way of passing parameters is changed and no binary data is being passed to the JSP program. File handling related concepts would be explained in separate chapter where we need to read binary data stream.

```
<html>
<head> Read Form Data</h1>
<ul>
<li><p><b>First Name:</b>
   <%= request.getParameter("first_name")%>
</p></li>
<li><p><b>Last Name:</b>
  <%= request.getParameter("last_name")%>
</p></li>
</ul>
</body>
</html>
```

Following is the content of Hello.htm file:

```
<html>
<body>
<form action="main.jsp" method="POST">
First Name: <input type="text" name="first_name">
<br />
```

```
Last Name: <input type="text" name="last_name" />
<input type="submit" value="Submit" />
</form>
</body>
</html>
```

Now let us keep main.jsp and hello.htm in <Tomcat-installation-directory>/ webapps/ROOT directory. When you would access *http://localhost:8080/Hello.htm*, below is the actual output of the above form.

First Name: [ ]

Last Name: [ ]

Try to enter First and Last Name and then click submit button to see the result on your local machine where tomcat is running.

Based on the input provided, it would generate similar result as mentioned in the above examples.

Passing Checkbox Data to JSP Program

Checkboxes are used when more than one option is required to be selected.

Here is an example of HTML code, CheckBox.htm, for a form with two checkboxes

```
<html>
<body>
<form action="main.jsp" method="POST" target="_blank">
<input type="checkbox" name="maths" checked="checked" /> Maths
<input type="checkbox" name="physics" /> Physics
<input type="checkbox" name="chemistry" checked="checked" />Chemistry
<input type="submit" value="Select Subject" />
</form>
</body>
</html>
```

The result of this code is of the following form

Maths Physics Chemistry

Below is main.jsp JSP program to handle input given by web browser for checkbox button.

```
<html>
<head>
<title>Reading Checkbox Data</title>
</head>
<body>
<center>
<h1>Reading Checkbox Data</h1>
<ul>
<li><p><b>Maths Flag:</b>
  <%= request.getParameter("maths")%>
</p></li>
```

```
<li><p><b>Physics Flag:</b>
  <%= request.getParameter("physics")%>
</p></li>
<li><p><b>Chemistry Flag:</b>
  <%= request.getParameter("chemistry")%>
</p></li>
</ul>
</body>
</html>
```

For the above example, it would display the following result:

Reading Checkbox Data
Maths Flag : : on
Physics Flag: : null
Chemistry Flag: : on

Reading All Form Parameters

Following is the generic example which uses **getParameterNames()** method of HttpServletRequest to read all the available form parameters. This method returns an Enumeration that contains the parameter names in an unspecified order.

Once we have an Enumeration, we can loop down the Enumeration in the standard manner, using *hasMoreElements()* method to determine when to stop and using *nextElement()* method to get each parameter name.

```
<%@ page import="java.io.*,java.util.*" %>
<html>
<head>
<title>HTTP Header Request Example</title>
</head>
<body>
<center>
<h2>HTTP Header Request Example</h2>
<table width="100%" border="1" align="center">
<tr bgcolor="#949494">
<th>Param Name</th><th>Param Value(s)</th>
</tr>
<%
Enumeration paramNames = request.getParameterNames();
  while(paramNames.hasMoreElements()) {
    String paramName = (String)paramNames.nextElement();
    out.print("<tr><td>" + paramName + "</td>\n");
    String paramValue = request.getHeader(paramName);
    out.println("<td> " + paramValue + "</td></tr>\n");
  }
%>
```

```
</table>
</center>
</body>
</html>
```

Following is the content of Hello.htm:

```
<html>
<body>
<form action="main.jsp" method="POST" target="_blank">
<input type="checkbox" name="maths" checked="checked" /> Maths
<input type="checkbox" name="physics" /> Physics
<input type="checkbox" name="chemistry" checked="checked" /> Chem
<input type="submit" value="Select Subject" />
</form>
</body>
</html>
```

Now try calling JSP using above Hello.htm, this would generate a result something like as below based on the provided input:

Reading All Form Parameters

Param Name	Param Value(s)
maths	on
chemistry	on

You can try above JSP to read any other form's data which is having other objects like text box, radio button or drop down box, etc.

4.10.5 Cookies Handling

Cookies are text files stored on the client computer and they are kept for various information tracking purpose. JSP transparently supports HTTP cookies using underlying servlet technology.

There are three steps involved in identifying returning users:

- Server script sends a set of cookies to the browser. For example, name, age, or identification number, etc.
- Browser stores this information on local machine for future use.
- When next time browser sends any request to web server then it sends those cookies information to the server and server uses that information to identify the user or may be for some other purpose as well.

Having discussed about the anatomy of cookie and the methods involved in cookie handling in section 2.11, we may now discuss the remaining topics of Section 2.11 with respect to JSP.

Setting Cookies with JSP

Setting cookies with JSP involves three steps:

(1) Creating a Cookie object: You call the Cookie constructor with a cookie name and a cookie value, both of which are strings.

```
Cookie cookie = new Cookie("key","value");
```

Keep in mind, neither the name nor the value should contain white space or any of the following characters:

[ ] ( ) = , " / ? @ : ;

(2) Setting the maximum age: You use setMaxAge to specify how long (in seconds) the cookie should be valid. Following would set up a cookie for 24 hours.

```
cookie.setMaxAge(60*60*24);
```

(3) Sending the Cookie into the HTTP response headers: You use **response.addCookie** to add cookies in the HTTP response header as follows:

```
response.addCookie(cookie);
```

Example

Let us modify our Form Example to set the cookies for first and last name.

```
<%
  // Create cookies for first and last names.
      Cookie firstName = new Cookie("first_name",
      request.getParameter("first_name"));
  Cookie lastName = new Cookie("last_name",
      request.getParameter("last_name"));

  // Set expiry date after 24 Hrs for both the cookies.
  firstName.setMaxAge(60*60*24);
  lastName.setMaxAge(60*60*24);

  // Add both the cookies in the response header.
  response.addCookie( firstName );
  response.addCookie( lastName );
%>
<html>
<head>
<title>Setting Cookies</title>
</head>
<body>
<center>
<h1>Setting Cookies</h1>
</center>
<ul>
<li><p><b>First Name:</b>
  <%= request.getParameter("first_name")%>
</p></li>
<li><p><b>Last  Name:</b>
  <%= request.getParameter("last_name")%>
</p></li>
```

```
</ul>
</body>
</html>
```

Let us put above code in main.jsp file and use it in the following HTML page:

```
<html>
<body>
<form action="main.jsp" method="GET">
First Name: <input type="text" name="first_name">
<br />
Last Name: <input type="text" name="last_name" />
<input type="submit" value="Submit" />
</form>
</body>
</html>
```

Keep above HTML content in a file hello.jsp and put hello.jsp and main.jsp in <Tomcat-installation-directory>/webapps/ROOT directory. When you would access *http://localhost:8080/hello.jsp*, here is the actual output of the above form.

First Name: [ ]

Last Name: [ ]

Try to enter First Name and Last Name and then click submit button. This would display first name and last name on your screen and at the same time it would set two cookies, firstName and lastName which would be passed back to the server when next time you would press Submit button.

Next section would explain you how you would access these cookies back in your web application.

Reading Cookies with JSP

To read cookies, you need to create an array of *javax.servlet.http.Cookie* objects by calling the **getCookies( )** method of *HttpServletRequest*. Then cycle through the array, and use getName() and getValue() methods to access each cookie and associated value.

Example

Let us read cookies which we have set in previous example:

```
<html>
<head>
<title>Reading Cookies</title>
</head>
<body>
<center>
<h1>Reading Cookies</h1>
</center>
<%
  Cookie cookie = null;
```

```
Cookie[] cookies = null;
// Get an array of Cookies associated with this domain
cookies = request.getCookies();
if( cookies != null ){
  out.println("<h2> Found Cookies Name and Value</h2>");
  for (int i = 0; i < cookies.length; i++){
    cookie = cookies[i];
    out.print("Name : " + cookie.getName( ) + ", ");
    out.print("Value: " + cookie.getValue( )+" <br/>");
  }
}else{
  out.println("<h2>No cookies founds</h2>");
}
%>
</body>
</html>
```

Now let us put above code in main.jsp file and try to access it. If you would have set first_name cookie as "John" and last_name cookie as "Player", then running *http://localhost:8080/main.jsp* would display the following result:

Found Cookies Name and Value

Name : first_name, Value: John
Name : last_name, Value: Player

Delete Cookies with JSP

To delete cookies is very simple. If you want to delete a cookie, then you simply need to follow up the following three steps:

1. Read an already existing cookie and store it in Cookie object.
2. Set cookie age as zero using **setMaxAge()** method to delete an existing cookie.
3. Add this cookie back into response header.

Example

Following example would delete an existing cookie named "first_name" and when you would run main.jsp JSP next time, it would return null value for first_name.

```
<html>
<head>
<title>Reading Cookies</title>
</head>
<body>
<center>
<h1>Reading Cookies</h1>
</center>
<%
  Cookie cookie = null;
  Cookie[] cookies = null;
  // Get an array of Cookies associated with this domain
  cookies = request.getCookies();
  if( cookies != null ){
```

```
        out.println("<h2> Found Cookies Name and Value</h2>");
        for (int i = 0; i < cookies.length; i++){
          cookie = cookies[i];
          if((cookie.getName( )).compareTo("first_name") == 0 ){
            cookie.setMaxAge(0);
            response.addCookie(cookie);
            out.print("Deleted cookie: " +
             cookie.getName( ) + "<br/>");
          }
          out.print("Name : " + cookie.getName( ) + ", ");
          out.print("Value: " + cookie.getValue( )+" <br/>");
        }
      }else{
        out.println(
        "<h2>No cookies founds</h2>");
      }
    %>
    </body>
    </html>
```

Now let us put above code in main.jsp file and try to access it. It would display the following result:

Cookies Name and Value

Deleted cookie : first_name

Name : first_name, Value: John

Name : last_name, Value: Player

Now try to run *http://localhost:8080/main.jsp* once again and it should display only one cookie as follows:

Found Cookies Name and Value

Name : last_name, Value: Player

You can delete your cookies in Internet Explorer manually. Start at the Tools menu and select Internet Options. To delete all cookies, press Delete Cookies.

HTTP is a "stateless" protocol which means each time a client retrieves a Web page, the client opens a separate connection to the Web server and the server automatically does not keep any record of previous client request.

Still there are following three ways to maintain session between web client and web server:

(1) Cookies

A web server can assign a unique session ID as a cookie to each web client and for subsequent requests from the client they can be recognized using the received cookie.

This may not be an effective way because many time browser does not support a cookie, so I would not recommend to use this procedure to maintain the sessions.

(2) Hidden Form Fields

A web server can send a hidden HTML form field along with a unique session ID as follows:

```
<input type="hidden" name="sessionid" value="12345">
```

This entry means that, when the form is submitted, the specified name and value are automatically included in the GET or POST data. Each time when web browser sends request back, then session_id value can be used to keep the track of different web browsers.

This could be an effective way of keeping track of the session but clicking on a regular (<A HREF...>) hypertext link does not result in a form submission, so hidden form fields also cannot support general session tracking.

(3) URL Rewriting

You can append some extra data on the end of each URL that identifies the session, and the server can associate that session identifier with data it has stored about that session.

For example, with http://tutorialspoint.com/file.htm; sessionid=12345, the session identifier is attached as sessionid=12345 which can be accessed at the web server to identify the client.

URL rewriting is a better way to maintain sessions and works for the browsers when they do not support cookies but here drawback is that you would have generate every URL dynamically to assign a session ID though page is simple static HTML page.

The session Object

Apart from the abovementioned three ways, JSP makes use of servlet provided HttpSession Interface which provides a way to identify a user across more than one page request or visit to a Web site and to store information about that user.

By default, JSPs have session tracking enabled and a new HttpSession object is instantiated for each new client automatically. Disabling session tracking requires explicitly turning it off by setting the page directive session attribute to false as follows:

```
<%@ page session="false" %>
```

The JSP engine exposes the HttpSession object to the JSP author through the implicit **session** object. Since **session** object is already provided to the JSP programmer, the programmer can immediately begin storing and retrieving data from the object without any initialization or getSession().

Summary of important methods available through session object has already been discussed in Section 2.12.

4.10.6 Session Tracking

This example describes how to use the HttpSession object to find out the creation time and the last-accessed time for a session. We would associate a new session with the request if one does not already exist.

```
<%@ page import="java.io.*,java.util.*" %>
<%
  // Get session creation time.
  Date createTime = new Date(session.getCreationTime());
  // Get last access time of this web page.
  Date lastAccessTime = new Date(session.getLastAccessedTime());

  String title = "Welcome Back to my website";
  Integer visitCount = new Integer(0);
```

```
    String visitCountKey = new String("visitCount");
    String userIDKey = new String("userID");
    String userID = new String("ABCD");

    // Check if this is new comer on your web page.
    if (session.isNew()){
      title = "Welcome to my website";
      session.setAttribute(userIDKey, userID);
      session.setAttribute(visitCountKey,  visitCount);
    }
     visitCount = (Integer)session.getAttribute(visitCountKey;
    visitCount = visitCount + 1;
    userID = (String)session.getAttribute(userIDKey);
    session.setAttribute(visitCountKey, visitCount);
  %>
  <html>
  <head>
  <title>Session Tracking</title>
  </head>
  <body>
  <center>
  <h1>Session Tracking</h1>
  </center>
  <table border="1" align="center">
   <tr bgcolor="#949494">
    <th>Session info</th>
    <th>Value</th>
  </tr>
   <tr>
    <td>id</td>
    <td><% out.print( session.getId()); %></td>
  </tr>
  <tr>
    <td>Creation Time</td>
    <td><% out.print(createTime); %></td>
  </tr>
   <tr>
    <td>Time of Last Access</td>
    <td><% out.print(lastAccessTime); %></td>
  </tr>
  <tr>
    <td>User ID</td>
    <td><% out.print(userID); %></td>
  </tr>
  <tr>
```

```
    <td>Number of visits</td>
    <td><% out.print(visitCount); %></td>
  </tr>
  </table>
  </body>
  </html>
```

Now put above code in main.jsp and try to access *http://localhost:8080/main.jsp*. It would display the following result when you would run for the first time:

Welcome to My Website

Session Information

Session info	**Value**
Id	0AE3EC93FF44E3C525B4351B77ABB2D5
Creation Time	Tue Jun 08 17:26:40 GMT+04:00 2010
Time of Last Access	Tue Jun 08 17:26:40 GMT+04:00 2010
User ID	ABCD
Number of visits	0

Now try to run the same JSP for second time, it would display the following result.

Welcome Back to My Website

Session	**Information**
Info Type	Value
Id	0AE3EC93FF44E3C525B4351B77ABB2D5
Creation Time	Tue Jun 08 17:26:40 GMT+04:00 2010
Time of Last Access	Tue Jun 08 17:26:40 GMT+04:00 2010
User ID	ABCD
Number of visits	1

Deleting Session Data

When you are done with a user's session data, you have several options:

1. **Remove a particular attribute:** You can call *public void removeAttribute(String name)* method to delete the value associated with a particular key.
2. **Delete the whole session:** You can call *public void invalidate()* method to discard an entire session.
3. **Setting session timeout:** You can call *public void setMaxInactiveInterval(int interval)* method to set the timeout for a session individually.
4. **Log the user out:** The servers that support servlets 2.4, you can call **logout** to log the client out of the web server and invalidate all sessions belonging to all the users.
5. **web.xml configuration:** If you are using Tomcat, apart from the abovementioned methods, you can configure session time out in web.xml file as follows.

```
<session-config>
  <session-timeout>15</session-timeout>
</session-config>
```

The timeout is expressed as minutes, and overrides the default timeout which is 30 minutes in Tomcat.

The getMaxInactiveInterval( ) method in a servlet returns the timeout period for that session in seconds. So if your session is configured in web.xml for 15 minutes, getMaxInactiveInterval( ) returns 900.

4.10.7 File Uploading

A JSP can be used with an HTML form tag to allow users to upload files to the server. An uploaded file could be a text file or binary or image file or any document.

Creating a File Upload Form

The following HTM code below creates an uploader form. Following are the important points to be noted down:

- The form **method** attribute should be set to **POST** method and GET method cannot be used.
- The form **enctype** attribute should be set to **multipart/form-data**.
- The form **action** attribute should be set to a JSP file which would handle file uploading at backend server. Following example is using **uploadFile.jsp** program file to upload file.
- To upload a single file, you should use a single <input .../> tag with attribute type="file". To allow multiple files uploading, include more than one input tags with different values for the name attribute. The browser associates a Browse button with each of them.

```
<html>
<head>
<title>File Uploading Form</title>
</head>
<body>
<h3>File Upload:</h3>
Select a file to upload: <br />
<form action="UploadServlet" method="post"
            enctype="multipart/form-data">
<input type="file" name="file" size="50" />
<br />
<input type="submit" value="Upload File" />
</form>
</body>
</html>
```

This will display following result which would allow to select a file from local PC and when user would click at "Upload File", form would be submitted along with the selected file:

File Upload

Select a file to upload:

Note: Above form is just dummy form and would not work, you should try above code at your machine to make it work.

Writing Backend JSP Script

First let us define a location where uploaded files would be stored. You can hard code this in your program or this directory name could also be added using an external configuration such as a **context-param** element in web.xml as follows:

```
<web-app>
....
<context-param>
    <description>Location to store uploaded file</description>
    <param-name>file-upload</param-name>
    <param-value>
        c:\apache-tomcat-5.5.29\webapps\data\
    </param-value>
 </context-param>
....
</web-app>
```

Following is the source code for UploadFile.jsp which can handle multiple file uploading at a time. Before procedding you have make sure the followings:

- Following example depends on FileUpload, so make sure you have the latest version of **commons-fileupload.x.x.jar** file in your classpath. You can download it from http://commons.apache.org/fileupload/.
- FileUpload depends on Commons IO, so make sure you have the latest version of **commons-io-x.x.jar** file in your classpath. You can download it from http://commons.apache.org/io/.
- While testing following example, you should upload a file which has less size than *maxFileSize*, otherwise file would not be uploaded.
- Make sure you have created directories c:\temp and c:\apache-tomcat-5.5.29\webapps\data well in advance.

```
<%@ page import="java.io.*,java.util.*, javax.servlet.*" %>
<%@ page import="javax.servlet.http.*" %>
<%@ page import="org.apache.commons.fileupload.*" %>
<%@ page import="org.apache.commons.fileupload.disk.*" %>
<%@ page import="org.apache.commons.fileupload.servlet.*" %>
<%@ page import="org.apache.commons.io.output.*" %>
```

```
<%
 File file ;
 int maxFileSize = 5000 * 1024;
 int maxMemSize = 5000 * 1024;
 ServletContext context = pageContext.getServletContext();
 String filePath = context.getInitParameter("file-upload");

 // Verify the content type
 String contentType = request.getContentType();
 if ((contentType.indexOf("multipart/form-data") >= 0)) {

   DiskFileItemFactory factory = new DiskFileItemFactory();
   // maximum size that will be stored in memory
   factory.setSizeThreshold(maxMemSize);
   // Location to save data that is larger than maxMemSize.
   factory.setRepository(new File("c:\\temp"));

   // Create a new file upload handler
   ServletFileUpload upload = new ServletFileUpload(factory);
   // maximum file size to be uploaded.
   upload.setSizeMax( maxFileSize );
   try{
     // Parse the request to get file items.
     List fileItems = upload.parseRequest(request);

     // Process the uploaded file items
     Iterator i = fileItems.iterator();

     out.println("<html>");
     out.println("<head>");
     out.println("<title>JSP File upload</title>");
     out.println("</head>");
     out.println("<body>");
     while ( i.hasNext () )
     {
 FileItem fi = (FileItem)i.next();
       if ( !fi.isFormField () )
       {
       // Get the uploaded file parameters
       String fieldName = fi.getFieldName();
       String fileName = fi.getName();
       boolean isInMemory = fi.isInMemory();
       long sizeInBytes = fi.getSize();
```

```
        // Write the file
        if( fileName.lastIndexOf("\\") >= 0 ){
        file = new File( filePath +
        fileName.substring( fileName.lastIndexOf("\\"))) ;
        }else{
        file = new File( filePath +
        fileName.substring(fileName.lastIndexOf("\\")+1)) ;
        }
        fi.write( file ) ;
        out.println("Uploaded Filename: " + filePath +
        fileName + "<br>");
        }
      }
      out.println("</body>");
      out.println("</html>");
    }catch(Exception ex) {
      System.out.println(ex);
    }
  }else{
    out.println("<html>");
    out.println("<head>");
    out.println("<title>Servlet upload</title>");
    out.println("</head>");
    out.println("<body>");
    out.println("<p>No file uploaded</p>");
    out.println("</body>");
    out.println("</html>");
  }
%>
```

Now try to upload files using the HTML form which you created above. When you would try http://localhost:8080/UploadFile.htm, it would display the following result which would help you uploading any file from your local machine.

File Upload:

Select a file to upload:

If your JSP script works fine, your file should be uploaded in c:\apache-tomcat-5.5.29\webapps\data\ directory.

4.10.8 Handling Date

One of the most important advantages of using JSP is that you can use all the methods available in core Java. Having discussed the methods in detail in Section 2.15, we may now discuss the method usage in JSP.

Getting Current Date & Time

This is very easy to get current date and time in JSP program. You can use a simple Date object with *toString()* method to print current date and time as follows:

```
<%@ page import="java.io.*,java.util.*, javax.servlet.*" %>
<html>
<head>
<title>Display Current Date & Time</title>
</head>
<body>
<center>
<h1>Display Current Date & Time</h1>
</center>
<%
  Date date = new Date();
  out.print( "<h2 align=\"center\">" +date.toString()+"</h2>");
%>
</body>
</html>
```

Now let us keep about code in CurrentDate.jsp and then call this JSP using URL http://localhost:8080/CurrentDate.jsp. This would produce the following result:

Display Current Date & Time
Mon Jun 21 21:46:49 GMT+04:00 2010

Try to refersh URL http://localhost:8080/CurrentDate.jsp and you would find difference in seconds everytime you would refresh.

Date Comparison

As I mentioned above you can use all the available Java methods in your JSP scripts. In case you need to compare two dates, following are the methods:

- You can use getTime( ) to obtain the number of milliseconds that have elapsed since midnight, January 1, 1970, for both objects and then compare these two values.
- You can use the methods before( ), after( ), and equals( ). Because the 12th of the month comes before the 18th, for example, new Date(99, 2, 12).before(new Date (99, 2, 18)) returns true.
- You can use the compareTo( ) method, which is defined by the Comparable interface and implemented by Date.

Date Formatting using SimpleDateFormat

SimpleDateFormat is a concrete class for formatting and parsing dates in a locale-sensitive manner. SimpleDateFormat allows you to start by choosing any user-defined patterns for date-time formatting.

Let us modify the above example as follows:

```
<%@ page import="java.io.*,java.util.*" %>
<%@ page import="javax.servlet.*,java.text.*" %>
<html>
<head>
<title>Display Current Date & Time</title>
</head>
<body>
<center>
<h1>Display Current Date & Time</h1>
</center>
<%
  Date dNow = new Date( );
  SimpleDateFormat ft =
   new SimpleDateFormat ("E yyyy.MM.dd 'at' hh:mm:ss a zzz");
  out.print( "<h2 align=\"center\">" +
            ft.format(dNow) +
              "</h2>");
%>
</body>
</html>
```

Compile above servlet once again and then call this servlet using URL http://localhost:8080/CurrentDate. This would produce the following result:

Display Current Date & Time
Mon 2010.06.21 at 10:06:44 PM GMT+04:00

Simple DateFormat codes
To specify the time format use a time pattern string. In this pattern, all ASCII letters are reserved as pattern letters, which are defined as the following:

Character	Description	Example
G	Era designator	AD
y	Year in four digits	2001
M	Month in year	July or 07
d	Day in month	10
h	Hour in A.M./P.M. (1~12)	12
H	Hour in day (0~23)	22
m	Minute in hour	30

s	Second in minute	55
S	Millisecond	234
E	Day in week	Tuesday
D	Day in year	360
F	Day of week in month	2 (second Wed. in July)
w	Week in year	40
W	Week in month	1
a	A.M./P.M. marker	PM
k	Hour in day (1~24)	24
K	Hour in A.M./P.M. (0~11)	10
z	Time zone	Eastern Standard Time
'	Escape for text	Delimiter
"	Single quote	`

For a complete list of constant available methods to manipulate date, you can refer to standard Java documentation.

4.10.9 Page Redirecting

Page redirection is generally used when a document moves to a new location and we need to send the client to this new location or may be because of load balancing, or for simple randomization.

The simplest way of redirecting a request to another page is using method **sendRedirect()** of response object. Following is the signature of this method:

```
public void response.sendRedirect(String location)
throws IOException
```

This method sends back the response to the browser along with the status code and new page location. You can also use setStatus() and setHeader() methods together to achieve the same redirection:

```
....
String site = "http://www.newpage.com" ;
response.setStatus(response.SC_MOVED_TEMPORARILY);
response.setHeader("Location", site);
....
```

Example

This example shows how a JSP performs page redirection to an another location:

```
<%@ page import="java.io.*,java.util.*" %>
<html>
```

```
<head>
<title>Page Redirection</title>
</head>
<body>
<center>
<h1>Page Redirection</h1>
</center>
<%
  // New location to be redirected
  String site = new String("http://www.photofuntoos.com");
  response.setStatus(response.SC_MOVED_TEMPORARILY);
  response.setHeader("Location", site);
%>
</body>
</html>
```

Now let us put above code in PageRedirect.jsp and call this JSP using URL http://localhost:8080/PageRedirect.jsp. This would take you given URL http://www.photofuntoos.com.

4.10.10 Sample JSP codes

a. Hits Counter

A hit counter tells you about the number of visits on a particular page of your web site. Usually you attach a hit counter with your index.jsp page assuming people first land on your home page.

To implement a hit counter you can make use of Application Implicit object and associated methods getAttribute() and setAttribute().

This object is a representation of the JSP page through its entire lifecycle. This object is created when the JSP page is initialized and will be removed when the JSP page is removed by the jspDestroy() method.

Following is the syntax to set a variable at application level:

*application.setAttribute(String Key, Object Value);

You can use the above method to set a hit counter variable and to reset the same variable. Following is the method to read the variable set by previous method:

application.getAttribute(String Key);

Every time use access your page, you can read current value of hit counter and increase it by one and again set it for future use.

Example

This example shows how you can use JSP to count total number of hits on a particular page. If you want to count total number of hits of your website, then you would have to include same code in all the JSP pages.

```
<%@ page import="java.io.*,java.util.*" %>

<html>
<head>
```

```
<title>Application object in JSP</title>
</head>
<body>
<%
   Integer hitsCount =
      (Integer)application.getAttribute("hitCounter");
   if( hitsCount ==null || hitsCount == 0 ){
      /* First visit */
      out.println("Welcome to my website!");
      hitsCount = 1;
   }else{
      /* return visit */
      out.println("Welcome back to my website!");
      hitsCount += 1;
   }
   application.setAttribute("hitCounter", hitsCount);
%>
<center>
<p>Total number of visits: <%= hitsCount%></p>
</center>
</body>
</html>
```

Now let us put above code in main.jsp and call this JSP using URL http://localhost:8080/main.jsp. This would display hit counter value which would increase every time when you refresh the page. You can try to access the page using different browsers and you will find that hit counter will keep increasing with every hit and would display result something as follows:

Welcome back to my website!
Total number of visits: 12

b. Hit Counter Resets

What about if you re-start your application, i.e. web server, this will reset your application variable and your counter will reset to zero. To avoid this loss, you can implement your counter in professional way which is as follows:

- Define a database table with a single count, let us say hit count. Assign a zero value to it.
- With every hit, read the table to get the value of hit count.
- Increase the value of hit count by one and update the table with new value.
- Display new value of hit count as total page hit counts.
- If you want to count hits for all the pages, implement above logic for all the pages.

c. Auto Refresh

Consider a webpage which is displaying live game score or stock market status or currency exchange ratio. For all such type of pages, you would need to refresh your web page regularly using referesh or reload button with your browser.

JSP makes this job easy by providing you a mechanism where you can make a webpage in such a way that it would refresh automatically after a given interval.

The simplest way of refreshing a web page is using method **setIntHeader()** of response object. Following is the signature of this method:

public void setIntHeader(String header, int headerValue)

This method sends back header "Refresh" to the browser along with an integer value which indicates time interval in seconds.

Auto Page Refresh Example

Following example would use **setIntHeader()** method to set **Refresh** header to simulate a digital clock:

```
<%@ page import="java.io.*,java.util.*" %>
<html>
<head>
<title>Auto Refresh Header Example</title>
</head>
<body>
<center>
<h2>Auto Refresh Header Example</h2>
<%
  // Set refresh, autoload time as 5 seconds
  response.setIntHeader("Refresh", 5);
  // Get current time
  Calendar calendar = new GregorianCalendar();
  String am_pm;
  int hour = calendar.get(Calendar.HOUR);
  int minute = calendar.get(Calendar.MINUTE);
  int second = calendar.get(Calendar.SECOND);
  if(calendar.get(Calendar.AM_PM) == 0)
    am_pm = "AM";
  else
    am_pm = "PM";
  String CT = hour+":"+ minute +":"+ second +" "+ am_pm;
  out.println("Crrent Time: " + CT + "\n");
%>
</center>
</body>
</html>
```

Now put the above code in main.jsp and try to access it. This would display current system time after every 5 seconds as follows. Just run the JSP and wait to see the result:

Auto Refresh Header Example

Current Time is: 9:44:50 PM

To become more comfortable with other methods, you can try a few more above listed methods in the same fashion.

d. Sending Email

To send an email using a JSP is simple enough but to start with you should have **JavaMail API** and **Java Activation Framework (JAF)** installed on your machine.

- You can download latest version of JavaMail (Version 1.2) from Java's standard website.
- You can download latest version of JavaBeans Activation Framework JAF (Version 1.0.2) from Java's standard website.

Download and unzip these files in the newly created top level directories, you will find a number of jar files for both the applications. You need to add **mail.jar** and **activation.jar** files in your CLASSPATH.

(1) Send a Simple Email

Here is an example to send a simple email from your machine. Here it is assumed that your **localhost** is connected to the internet and capable enough to send an email. At the same time make sure all the jar files from Java Email API package and JAF package are available in CLASSPATH.

```
<%@ page import="java.io.*,java.util.*,javax.mail.*"%>
<%@ page import="javax.mail.internet.*,javax.activation.*"%>
<%@ page import="javax.servlet.http.*,javax.servlet.*" %>
<%
  String result;
  // Recipient's email ID needs to be mentioned.
  String to = "abcd@gmail.com";

  // Sender's email ID needs to be mentioned
  String from = "mcmohd@gmail.com";

  // Assuming you are sending email from localhost
  String host = "localhost";

  // Get system properties object
  Properties properties = System.getProperties();

  // Setup mail server
  properties.setProperty("mail.smtp.host", host);

  // Get the default Session object.
  Session mailSession = Session.getDefaultInstance(properties);

  try{
    // Create a default MimeMessage object.
    MimeMessage message = new MimeMessage(mailSession);
    // Set From: header field of the header.
    message.setFrom(new InternetAddress(from));
    // Set To: header field of the header.
    message.addRecipient(Message.RecipientType.TO,
                         new InternetAddress(to));
```

```
      // Set Subject: header field
      message.setSubject("This is the Subject Line!");
      // Now set the actual message
      message.setText("This is actual message");
      // Send message
      Transport.send(message);
      result = "Sent message successfully....";
   }catch (MessagingException mex) {
      mex.printStackTrace();
      result = "Error: unable to send message....";
   }
%>
<html>
<head>
<title>Send Email using JSP</title>
</head>
<body>
<center>
<h1>Send Email using JSP</h1>
</center>
<p align="center">
<%
   out.println("Result: " + result + "\n");
%>
</p>
</body>
</html>
```

Now let us put above code in SendEmail.jsp file and call this JSP using URL http://localhost:8080/SendEmail.jsp which would send an email to given email ID *abcd@gmail.com* and would display the following response:

Send Email using JSP

Result: Sent message successfully...

If you want to send an email to multiple recipients, then following methods would be used to specify multiple email IDs:

```
void addRecipients(Message.RecipientType type,
            Address[] addresses)
throws MessagingException
```

Here is the description of the parameters:

- **type:** This would be set to TO, CC or BCC. Here CC represents Carbon Copy and BCC represents Black Carbon Copy. For example, *Message.RecipientType.TO*
- **addresses:** This is the array of email ID. You would need to use InternetAddress() method while specifying email IDs

(2) Send an HTML Email

Here is an example to send an HTML email from your machine. Here it is assumed that your **localhost** is connected to the internet and capable enough to send an email.

At the same time make sure all the jar files from Java Email API package and JAF package are available in CLASSPATH.

This example is very similar to previous one, except here we are using setContent() method to set content whose second argument is "text/html" to specify that the HTML content is included in the message.

Using this example, you can send as big as HTML content you like.

```
<%@ page import="java.io.*,java.util.*,javax.mail.*"%>
<%@ page import="javax.mail.internet.*,javax.activation.*"%>
<%@ page import="javax.servlet.http.*,javax.servlet.*" %>
<%
  String result;
  // Recipient's email ID needs to be mentioned.
  String to = "abcd@gmail.com";

  // Sender's email ID needs to be mentioned
  String from = "mcmohd@gmail.com";

  // Assuming you are sending email from localhost
  String host = "localhost";

  // Get system properties object
  Properties properties = System.getProperties();

  // Setup mail server
  properties.setProperty("mail.smtp.host", host);

  // Get the default Session object.
  Session mailSession = Session.getDefaultInstance(properties);

  try{
    // Create a default MimeMessage object.
    MimeMessage message = new MimeMessage(mailSession);
    // Set From: header field of the header.
    message.setFrom(new InternetAddress(from));
    // Set To: header field of the header.
    message.addRecipient(Message.RecipientType.TO,
                  new InternetAddress(to));
    // Set Subject: header field
    message.setSubject("This is the Subject Line!");
      // Send the actual HTML message, as big as you like
    message.setContent("<h1>This is actual message</h1>",
                "text/html" );
    // Send message
    Transport.send(message);
    result = "Sent message successfully...";
```

```
  }catch (MessagingException mex) {
    mex.printStackTrace();
    result = "Error: unable to send message....";
  }
%>
<html>
<head>
<title>Send HTML Email using JSP</title>
</head>
<body>
<center>
<h1>Send Email using JSP</h1>
</center>
<p align="center">
<%
  out.println("Result: " + result + "\n");
%>
</p>
</body>
</html>
```

Now try to use above JSP to send HTML message on a given email ID.

(3) Send Attachment in Email

Here is an example to send an email with attachment from your machine:

```
<%@ page import="java.io.*,java.util.*,javax.mail.*"%>
<%@ page import="javax.mail.internet.*,javax.activation.*"%>
<%@ page import="javax.servlet.http.*,javax.servlet.*" %>
<%  String result;
  // Recipient's email ID needs to be mentioned.
  String to = "abcd@gmail.com";
  // Sender's email ID needs to be mentioned
  String from = "mcmohd@gmail.com";

  // Assuming you are sending email from localhost
  String host = "localhost";

  // Get system properties object
  Properties properties = System.getProperties();

  // Setup mail server
  properties.setProperty("mail.smtp.host", host);

  // Get the default Session object.
  Session mailSession = Session.getDefaultInstance(properties);
```

```
try{
  // Create a default MimeMessage object.
  MimeMessage message = new MimeMessage(mailSession);

  // Set From: header field of the header.
  message.setFrom(new InternetAddress(from));

  // Set To: header field of the header.
  message.addRecipient(Message.RecipientType.TO,
                  new InternetAddress(to));

  // Set Subject: header field
  message.setSubject("This is the Subject Line!");

  // Create the message part
   BodyPart messageBodyPart = new MimeBodyPart();

  // Fill the message
  messageBodyPart.setText("This is message body");

      // Create a multipart message
  Multipart multipart = new MimeMultipart();

  // Set text message part
  multipart.addBodyPart(messageBodyPart);

  // Part two is attachment
  messageBodyPart = new MimeBodyPart();
  String filename = "file.txt";
  DataSource source = new FileDataSource(filename);
  messageBodyPart.setDataHandler(new DataHandler(source));
  messageBodyPart.setFileName(filename);
  multipart.addBodyPart(messageBodyPart);

  // Send the complete message parts
  message.setContent(multipart );

  // Send message
  Transport.send(message);
  String title = "Send Email";
  result = "Sent message successfully....";
}catch (MessagingException mex) {
  mex.printStackTrace();
  result = "Error: unable to send message....";
}
```

```
%>
<html>
<head>
<title>Send Attachment Email using JSP</title>
</head>
<body>
<center>
<h1>Send Attachment Email using JSP</h1>
</center>
<p align="center">
<%
   out.println("Result: " + result + "\n");
%>
</p>
</body>
</html>
```

Now try to run above JSP to send a file as an attachment along with a message on a given email ID.

4. User Authentication Part

If it is required to provide user ID and Password to the email server for authentication purpose, then you can set these properties as follows:

```
props.setProperty("mail.user", "myuser");
props.setProperty("mail.password", "mypwd");
```

Rest of the email sending mechanism would remain as explained above.

5. Using Forms to send email

You can use HTML form to accept email parameters and then you can use **request** object to get all the information as follows:

```
String to = request.getParameter("to");
String from = request.getParameter("from");
String subject = request.getParameter("subject");
String messageText = request.getParameter("body");
```

Once you have all the information, you can use abovementioned programs to send email.

4.11 Database Access

This section assumes you have good understanding on how JDBC application works. Before starting with database access through a JSP, make sure you have proper JDBC environment setup along with a database.

To start with basic concept, let us create a simple table and create a few records in that table as follows:

Create Table

To create the **Employees** table in EMP database, use the following steps:

Step 1
Open a **Command Prompt** and change to the installation directory as follows:

```
C:\>
C:\>cd Program Files\MySQL\bin
C:\Program Files\MySQL\bin>
```

Step 2
Login to database as follows

```
C:\Program Files\MySQL\bin>mysql -u root -p
Enter password: ********
mysql>
```

Step 3
Create the table **Employee** in **TEST** database as follows:

```
mysql> use TEST;
mysql> create table Employees
  (
    id int not null,
    age int not null,
    first varchar (255),
    last varchar (255)
  );
Query OK, 0 rows affected (0.08 sec)
mysql>
```

Create Data Records
Finally you create a few records in Employee table as follows:

```
mysql> INSERT INTO Employees VALUES (100, 18, 'BPL', 'AJP');
Query OK, 1 row affected (0.05 sec)

 mysql> INSERT INTO Employees VALUES (101, 25, 'Karthick', 'Satish');
Query OK, 1 row affected (0.00 sec)

mysql> INSERT INTO Employees VALUES (102, 30, 'sathya', 'Sri');
Query OK, 1 row affected (0.00 sec)

mysql> INSERT INTO Employees VALUES (103, 28, 'Sumi', 'Sekar');
Query OK, 1 row affected (0.00 sec)

mysql>
```

SELECT Operation

Following example shows how we can execute SQL SELECT statement using JTSL in JSP programming:

```
<%@ page import="java.io.*,java.util.*,java.sql.*"%>
<%@ page import="javax.servlet.http.*,javax.servlet.*" %>
<%@ taglib uri="http://java.sun.com/jsp/jstl/core" prefix="c"%>
<%@ taglib uri="http://java.sun.com/jsp/jstl/sql" prefix="sql"%>

<html>
<head>
<title>SELECT Operation</title>
</head>
<body>

<sql:setDataSource var="snapshot" driver="com.mysql.jdbc.Driver"
   url="jdbc:mysql://localhost/TEST"
   user="root" password="pass123"/>
<sql:query dataSource="${snapshot}" var="result">
SELECT * from Employees;
</sql:query>

<table border="1" width="100%">
<tr>
  <th>Emp ID</th>
  <th>First Name</th>
  <th>Last Name</th>
  <th>Age</th>
</tr>
<c:forEach var="row" items="${result.rows}">
<tr>
  <td><c:out value="${row.id}"/></td>
  <td><c:out value="${row.first}"/></td>
  <td><c:out value="${row.last}"/></td>
  <td><c:out value="${row.age}"/></td>
</tr>
</c:forEach>
</table>

</body>
</html>
```

Emp ID	First Name	Last Name	Age
100	BPL	AJP	18
101	Karthick	Satish	25
102	Sathya	Sri	30
103	Sumi	Sekar	28

INSERT Operation

Following example shows how we can execute SQL INSERT statement using JTSL in JSP programming:

```
<%@ page import="java.io.*,java.util.*,java.sql.*"%>
<%@ page import="javax.servlet.http.*,javax.servlet.*" %>
<%@ taglib uri="http://java.sun.com/jsp/jstl/core" prefix="c"%>
<%@ taglib uri="http://java.sun.com/jsp/jstl/sql" prefix="sql"%>

<html>
<head>
<title>JINSERT Operation</title>
</head>
<body>

<sql:setDataSource var="snapshot" driver="com.mysql.jdbc.Driver"
   url="jdbc:mysql://localhost/TEST"
   user="root"  password="pass123"/>

<sql:query dataSource="${snapshot}" var="result">
INSERT INTO Employees VALUES (104, 2, 'Nuha', 'AJP');
</sql:query>

<sql:query dataSource="${snapshot}" var="result">
SELECT * from Employees;
</sql:query>

<table border="1" width="100%">
<tr>
  <th>Emp ID</th>
  <th>First Name</th>
  <th>Last Name</th>
  <th>Age</th>
</tr>
<c:forEach var="row" items="${result.rows}">
<tr>
  <td><c:out value="${row.id}"/></td>
  <td><c:out value="${row.first}"/></td>
  <td><c:out value="${row.last}"/></td>
  <td><c:out value="${row.age}"/></td>
</tr>
</c:forEach>
</table>

</body>
</html>
```

Now try to access above JSP, which should display the following result:

Emp ID	First Name	Last Name	Age
100	BPL	AJP	18
101	Karthick	Satish	25
102	Sathya	Sri	30
103	Sumi	Sekar	28
104	Nuha	AJP	2

DELETE Operation

Following example shows how we can execute SQL DELETE statement using JTSL in JSP programming:

```
<%@ page import="java.io.*,java.util.*,java.sql.*"%>
<%@ page import="javax.servlet.http.*,javax.servlet.*" %>
<%@ taglib uri="http://java.sun.com/jsp/jstl/core" prefix="c"%>
<%@ taglib uri="http://java.sun.com/jsp/jstl/sql" prefix="sql"%>

<html>
<head>
<title>DELETE Operation</title>
</head>
<body>

<sql:setDataSource var="snapshot" driver="com.mysql.jdbc.Driver"
   url="jdbc:mysql://localhost/TEST"
   user="root" password="pass123"/>

<c:set var="empId" value="103"/>

<sql:update dataSource="${snapshot}" var="count">
 DELETE FROM Employees WHERE Id = ?
 <sql:param value="${empId}" />
</sql:update>

<sql:query dataSource="${snapshot}" var="result">
  SELECT * from Employees;
</sql:query>

<table border="1" width="100%">
<tr>
  <th>Emp ID</th>
  <th>First Name</th>
  <th>Last Name</th>
  <th>Age</th>
```

```
</tr>
<c:forEach var="row" items="${result.rows}">
<tr>

  <td><c:out value="${row.id}"/></td>
  <td><c:out value="${row.first}"/></td>
  <td><c:out value="${row.last}"/></td>
  <td><c:out value="${row.age}"/></td>
</tr>
</c:forEach>
</table>

</body>
</html>
```

Now try to access above JSP, which should display the following result:

Emp ID	First Name	Last Name	Age
100	BPL	AJP	18
101	Karthick	Satish	25
102	Sathya	Sri	30

UPDATE Operation

Following example shows how we can execute SQL UPDATE statement using JTSL in JSP programming:

```
<%@ page import="java.io.*,java.util.*,java.sql.*"%>
<%@ page import="javax.servlet.http.*,javax.servlet.*" %>
<%@ taglib uri="http://java.sun.com/jsp/jstl/core" prefix="c"%>
<%@ taglib uri="http://java.sun.com/jsp/jstl/sql" prefix="sql"%>

<html>
<head>
<title>DELETE Operation</title>
</head>
<body>

<sql:setDataSource var="snapshot" driver="com.mysql.jdbc.Driver"
   url="jdbc:mysql://localhost/TEST"
   user="root" password="pass123"/>
<c:set var="empId" value="102"/>

<sql:update dataSource="${snapshot}" var="count">
 UPDATE Employees SET last = 'AJP'
 <sql:param value="${empId}" />
</sql:update>
```

```
<sql:query dataSource="${snapshot}" var="result">
 SELECT * from Employees;
</sql:query>

<table border="1" width="100%">
<tr>
  <th>Emp ID</th>
  <th>First Name</th>
  <th>Last Name</th>
  <th>Age</th>
</tr>
<c:forEach var="row" items="${result.rows}"><tr>
  <td><c:out value="${row.id}"/></td>
  <td><c:out value="${row.first}"/></td>
  <td><c:out value="${row.last}"/></td>
  <td><c:out value="${row.age}"/></td>
</tr>
</c:forEach>
</table>

</body>
</html>
```

Now try to access above JSP, which should display the following result:

Emp ID	First Name	Last Name	Age
100	BPL	AJP	18
101	Karthick	Satish	25
102	Sathya	Sri	30

4.12 XML Data

When you send XML data via HTTP, it makes sense to use JSP to handle incoming and outgoing XML documents, for example, RSS documents. As an XML document is merely a bunch of text, creating one through a JSP is no more difficult than creating an HTML document.

Sending XML from a JSP

You can send XML content using JSPs the same way you send HTML. The only difference is that you must set the content type of your page to text/xml. To set the content type, use the <%@page%> tag, like this:

```
<%@ page contentType="text/xml" %>
```

Following is a simple example to send XML content to the browser:

```
<%@ page contentType="text/xml" %>
```

```
<books>
  <book>
    <name>Padam History</name>
    <author>BPL</author>
    <price>100</price>
  </book>
</books>
```

Try to access above XML using different browsers to see the document tree presentation of the above XML.

Processing XML in JSP

Before you proceed with XML processing using JSP, you would need to copy following two XML and XPath related libraries into your <Tomcat Installation Directory>\lib:

1. **XercesImpl.jar:** Download it from http://www.apache.org/dist/xerces/j/
2. **Xalan.jar:** Download it from http://xml.apache.org/xalan-j/index.html

Let us put the following content in books.xml file:

```
<books>
<book>
  <name>Padam History</name>
  <author>BPL</author>
  <price>100</price>
</book>
<book>
  <name>Great Mistry</name>
  <author>NUHA</author>
  <price>2000</price>
</book>
</books>
```

Now try the following main.jsp, keeping in the same directory:

```
<%@ taglib prefix="c" uri="http://java.sun.com/jsp/jstl/core" %>
<%@ taglib prefix="x" uri="http://java.sun.com/jsp/jstl/xml" %>

<html>
<head>
<title>JSTL x:parse Tags</title>
</head>
<body>
<h3>Books Info:</h3>
<c:import var="bookInfo" url="http://localhost:8080/books.xml"/>

 <x:parse xml="${bookInfo}" var="output"/>
<b>The title of the first book is</b>:
```

```
<x:out select="$output/books/book[1]/name" />
<br>
<b>The price of the second book</b>:
<x:out select="$output/books/book[2]/price" />

</body>
</html>
```

Now try to access above JSP using http://localhost:8080/main.jsp, this would produce the following result:

Books Info:
The title of the first book is: **Padam History**
The price of the second book: **2000**

Formatting XML with JSP:
Consider the following XSLT stylesheet style.xsl:

```
<?xml version="1.0"?>
<xsl:stylesheet xmlns:xsl=
"http://www.w3.org/1999/XSL/Transform" version="1.0">

<xsl:output method="html" indent="yes"/>
<xsl:template match="/">
<html>
<body>
<xsl:apply-templates/>
</body>
</html>
</xsl:template>

<xsl:template match="books">
  <table border="1" width="100%">
    <xsl:for-each select="book">
      <tr>
        <td>
          <i><xsl:value-of select="name"/></i>
        </td>
        <td>
          <xsl:value-of select="author"/>
        </td>
        <td>
          <xsl:value-of select="price"/>
        </td>
      </tr>
    </xsl:for-each>
  </table>
</xsl:template>
</xsl:stylesheet>
```

Now consider the following JSP file:

```
<%@ taglib prefix="c" uri="http://java.sun.com/jsp/jstl/core" %>
<%@ taglib prefix="x" uri="http://java.sun.com/jsp/jstl/xml" %>

<html>
<head>
  <title>JSTL x:transform Tags</title>
</head>
<body>
<h3>Books Info:</h3>
<c:set var="xmltext">
  <books>
    <book>
      <name>Padam History</name>
      <author>BPL</author>
      <price>100</price>
    </book>
    <book>
      <name>Great Mistry</name>
      <author>NUHA</author>
      <price>2000</price>
    </book>
  </books>
</c:set>

<c:import url="http://localhost:8080/style.xsl" var="xslt"/>
<x:transform xml="${xmltext}" xslt="${xslt}"/>

</body>
</html>
```

This would produce the following result:
Books Info:

Padam History	BPL	100
Great Mistry	NUHA	2000

4.13 Custom Tags

A custom tag is a user-defined JSP language element. When a JSP page containing a custom tag is translated into a servlet, the tag is converted to operations on an object called a tag handler. The Web container then invokes those operations when the JSP page servlet is executed.

JSP tag extensions let you create new tags that you can insert directly into a JavaServer Page just as you would the built-in tags you learned about in earlier chapter. The JSP 2.0 specification introduced Simple Tag Handlers for writing these custom tags.

To write a customer tab you can simply extend SimpleTagSupport class and override the **doTag()** method, where you can place your code to generate content for the tag.

Create "Hello" Tag

Consider you want to define a custom tag named <ex:Hello> and you want to use it in the following fashion without a body:

```
<ex:Hello />
```

To create a custom JSP tag, you must first create a Java class that acts as a tag handler. So let us create HelloTag class as follows:

```
package com.tutorialspoint;

import javax.servlet.jsp.tagext.*;
import javax.servlet.jsp.*;
import java.io.*;

public class HelloTag extends SimpleTagSupport {
public void doTag() throws JspException, IOException {
   JspWriter out = getJspContext().getOut();
   out.println("Hello Custom Tag!");
  }
}
```

Above code has simple coding where doTag() method takes the current JspContext object using getJspContext() method and uses it to send "Hello Custom Tag!" to the current JspWriter object.

Let us compile the above class and copy it in a directory available in environment variable CLASSPATH. Finally create following tag library file: <Tomcat-Installation-Directory>webapps\ROOT\WEB-INF\custom.tld.

```
<taglib>
  <tlib-version>1.0</tlib-version>
  <jsp-version>2.0</jsp-version>
  <short-name>Example TLD</short-name>
  <tag>
    <name>Hello</name>
    <tag-class>com.tutorialspoint.HelloTag</tag-class>
    <body-content>empty</body-content>
  </tag>
</taglib>
```

Now it is time to use the above defined custom tag **Hello** in our JSP program as follows:

```
<%@ taglib prefix="ex" uri="WEB-INF/custom.tld"%>
<html>
  <head>
```

```
    <title>A sample custom tag</title>
  </head>
  <body>
    <ex:Hello/>
  </body>
</html>
```

Try to call above JSP and this should produce the following result:
Hello Custom Tag!

Accessing the Tag Body

You can include a message in the body of the tag as you have seen with standard tags. Consider you want to define a custom tag named <ex:Hello> and you want to use it in the following fashion with a body:

```
<ex:Hello>
  This is message body
</ex:Hello>
```

Let us make the following changes in our tag code above to process the body of the tag:

```
package com.tutorialspoint;

import javax.servlet.jsp.tagext.*;
import javax.servlet.jsp.*;
import java.io.*;
public class HelloTag extends SimpleTagSupport {

  StringWriter sw = new StringWriter();
  public void doTag()
    throws JspException, IOException
  {
    getJspBody().invoke(sw);
    getJspContext().getOut().println(sw.toString());
  }
}
```

In this case, the output resulting from the invocation is first captured into a StringWriter before being written to the JspWriter associated with the tag. Now accordingly we need to change TLD file as follows:

```
<taglib>
  <tlib-version>1.0</tlib-version>
  <jsp-version>2.0</jsp-version>
  <short-name>Example TLD with Body</short-name>
  <tag>
    <name>Hello</name>
```

```
    <tag-class>com.tutorialspoint.HelloTag</tag-class>
    <body-content>scriptless</body-content>
  </tag></taglib>
```

Now let us call above tag with proper body as follows:

```
<%@ taglib prefix="ex" uri="WEB-INF/custom.tld"%>
<html>
  <head>
    <title>A sample custom tag</title>
  </head>
  <body>
    <ex:Hello>
       This is message body
    </ex:Hello>
  </body>
</html>
```

This will produce the following result:
This is message body

Custom Tag Attributes

You can use various attributes along with your custom tags. To accept an attribute value, a custom tag class needs to implement setter methods, identical to JavaBean setter methods as shown below:

```
package com.tutorialspoint;

import javax.servlet.jsp.tagext.*;
import javax.servlet.jsp.*;
import java.io.*;

public class HelloTag extends SimpleTagSupport {
  private String message;

  public void setMessage(String msg) {
    this.message = msg;
  }

  StringWriter sw = new StringWriter();

  public void doTag()
    throws JspException, IOException
  {
    if (message != null) {
      /* Use message from attribute */
      JspWriter out = getJspContext().getOut();
```

```
            out.println( message );
        }
        else {
            /* use message from the body */
            getJspBody().invoke(sw);
    getJspContext().getOut().println(sw.toString());
        }
    }
}
```

The attribute's name is "message", so the setter method is setMessage(). Now let us add this attribute in TLD file using <attribute> element as follows:

```
<taglib>
  <tlib-version>1.0</tlib-version>
  <jsp-version>2.0</jsp-version>
  <short-name>Example TLD with Body</short-name>
  <tag>
    <name>Hello</name>
    <tag-class>com.tutorialspoint.HelloTag</tag-class>
    <body-content>scriptless</body-content>
    <attribute>
      <name>message</name>
    </attribute>
  </tag>
</taglib>
```

Now let us try following JSP with message attribute as follows:

```
<%@ taglib prefix="ex" uri="WEB-INF/custom.tld"%><html>
  <head>
    <title>A sample custom tag</title>
  </head>
  <body>
    <ex:Hello message="This is custom tag" />
  </body>
</html>
```

This will produce the following result:
This is custom tag
Hope above example makes sense for you. It would be worth to note that you can include the following properties for an attribute:

Property	Purpose
Name	The name element defines the name of an attribute. Each attribute name must be unique for a particular tag.
Required	This specifies if this attribute is required or optional. It would be false for optional.
Rtexprvalue	Declares if a runtime expression value for a tag attribute is valid
Type	Defines the Java class-type of this attribute. By default it is assumed as **String**
Description	Informational description can be provided.
Fragment	Declares if this attribute value should be treated as a **JspFragment**.

Following is the example to specify properties related to an attribute:

```
.....
    <attribute>
      <name>attribute_name</name>
      <required>false</required>
      <type>java.util.Date</type>
      <fragment>false</fragment>
    </attribute>
.....
```

If you are using two attributes, then you can modify your TLD as follows:

```
.....
    <attribute>
      <name>attribute_name1</name>
      <required>false</required>
      <type>java.util.Boolean</type>
      <fragment>false</fragment>
    </attribute>
    <attribute>
      <name>attribute_name2</name>
      <required>true</required>
      <type>java.util.Date</type>
    </attribute>
.....
```

4.14 Expression Language (EL)

JSP Expression Language (EL) makes it possible to easily access application data stored in JavaBeans components. JSP EL allows you to create expressions both **(a)** arithmetic and **(b)** logical. Within a JSP EL expression, you can use integers, floating point numbers, strings, the built-in constants true and false for boolean values, and null.

Simple Syntax

Typically, when you specify an attribute value in a JSP tag, you simply use a string. For example:

```
<jsp:setProperty name="box" property="perimeter" value="100"/>
```

JSP EL allows you to specify an expression for any of these attribute values. A simple syntax for JSP EL is as follows:

```
${expr}
```

Here **expr** specifies the expression itself. The most common operators in JSP EL are . and []. These two operators allow you to access various attributes of Java Beans and built-in JSP objects.

For example, above syntax <jsp:setProperty> tag can be written with an expression like:

```
<jsp:setProperty name="box" property="perimeter"
        value="${2*box.width+2*box.height}"/>
```

When the JSP compiler sees the ${} form in an attribute, it generates code to evaluate the expression and substitues the value of expression.

You can also use JSP EL expressions within template text for a tag. For example, the <jsp:text> tag simply inserts its content within the body of a JSP. The following <jsp:text> declaration inserts <h1>Hello JSP!</h1> into the JSP output:

```
<jsp:text>
<h1>Hello JSP!</h1>
</jsp:text>
```

You can include a JSP EL expression in the body of a <jsp:text> tag (or any other tag) with the same ${} syntax you use for attributes. For example:

```
<jsp:text>
Box Perimeter is: ${2*box.width + 2*box.height}
</jsp:text>
```

EL expressions can use parentheses to group subexpressions. For example, ${(1 + 2) * 3} equals 9, but ${1 + (2 * 3)} equals 7.

To deactivate the evaluation of EL expressions, we specify the isELIgnored attribute of the page directive as below:

```
<%@ page isELIgnored ="true | false" %>
```

The valid values of this attribute are true and false. If it is true, EL expressions are ignored when they appear in static text or tag attributes. If it is false, EL expressions are evaluated by the container.

Basic Operators in EL

JSP Expression Language (EL) supports most of the arithmetic and logical operators supported by Java. Below is the list of most frequently used operators:

Operator	Description
.	Access a bean property or Map entry
[]	Access an array or List element
( )	Group a subexpression to change the evaluation order
+	Addition
-	Subtraction or negation of a value
*	Multiplication
/ or div	Division
% or mod	Modulo (remainder)
== or eq	Test for equality
!= or ne	Test for inequality
< or lt	Test for less than
> or gt	Test for greater than
<= or le	Test for less than or equal
>= or gt	Test for greater than or equal
&& or and	Test for logical AND
|| or or	Test for logical OR
! or not	Unary Boolean complement
empty	Test for empty variable values

Functions in JSP EL

JSP EL allows you to use functions in expressions as well. These functions must be defined in custom tag libraries. A function usage has the following syntax:

```
${ns:func(param1, param2, ...)}
```

Where ns is the namespace of the function, func is the name of the function and param1 is the first parameter value. For example, the function fn:length, which is part of the JSTL library can be used as follows to get the length of a string.

```
${fn:length("Get my length")}
```

To use a function from any tag library (standard or custom), you must install that library on your server and must include the library in your JSP using <taglib> directive as explained in JSTL chapter.

EL Implicit Objects

The JSP expression language supports the following implicit objects:

Implicit object	Description
PageScope	Scoped variables from page scope
RequestScope	Scoped variables from request scope

SessionScope	Scoped variables from session scope
ApplicationScope	Scoped variables from application scope
Param	Request parameters as strings
ParamValues	Request parameters as collections of strings
Header	HTTP request headers as strings
HeaderValues	HTTP request headers as collections of strings
InitParam	Context-initialization parameters
Cookie	Cookie values
PageContext	The JSP PageContext object for the current page

You can use these objects in an expression as if they were variables. Here are a few examples which would clear the concept:

The pageContext Object

The pageContext object gives you access to the pageContext JSP object. Through the pageContext object, you can access the request object. For example, to access the incoming query string for a request, you can use the expression:

```
${pageContext.request.queryString}
```

The Scope Objects

The pageScope, requestScope, sessionScope, and applicationScope variables provide access to variables stored at each scope level.

For example, if you need to explicitly access the box variable in the application scope, you can access it through the applicationScope variable as applicationScope.box.

The param and ParamValues Objects

The param and paramValues objects give you access to the parameter values normally available through the request.getParameter and request.getParameterValues methods.

For example, to access a parameter named order, use the expression ${param.order} or ${param["order"]}.

Following is the example to access a request parameter named username:

```
<%@ page import="java.io.*,java.util.*" %>
<%
  String title = "Accessing Request Param";
%>
<html>
<head>
<title><% out.print(title); %></title>
</head>
<body>
<center>
<h1><% out.print(title); %></h1>
```

```
</center>
<div align="center">
<p>${param["username"]}</p>
</div>
</body>
</html>
```

The param object returns single string values, whereas the paramValues object returns string arrays.

Header and HeaderValues Objects

The header and headerValues objects give you access to the header values normally available through the request.getHeader and request.getHeaders methods.

For example, to access a header named user-agent, use the expression ${header.user-agent} or ${header["user-agent"]}.

Following is the example to access a header parameter named user-agent:

```
<%@ page import="java.io.*,java.util.*" %>
<%
  String title = "User Agent Example";
%>
<html>
<head>
<title><% out.print(title); %></title>
</head>
<body>
<center>
<h1><% out.print(title); %></h1>
</center>
<div align="center">
<p>${header["user-agent"]}</p>
</div>
</body>
</html>
```

This would display something as follows:

User Agent Example

Mozilla/4.0 (compatible; MSIE 8.0; Windows NT 6.1; WOW64; Trident/4.0; SLCC2; .NET CLR 2.0.50727; .NET CLR 3.5.30729; .NET CLR 3.0.30729; Media Center PC 6.0; HPNTDF; .NET4.0C; InfoPath.2)

The header object returns single string values, whereas the headerValues object returns string arrays.

4.15 Exception Handling

When you are writing JSP code, a programmer may leave a coding errors which can occur at any part of the code. You can have the following type of errors in your JSP code:

- **Checked exceptions:** A checked exception is an exception that is typically a user error or a problem that cannot be foreseen by the programmer. For example, if a file is to be opened, but the file cannot be found, an exception occurs. These exceptions cannot simply be ignored at the time of compilation.
- **Runtime exceptions:** A runtime exception is an exception that occurs that probably could have been avoided by the programmer. As opposed to checked exceptions, runtime exceptions are ignored at the time of compliation.
- **Errors:** These are not exceptions at all, but problems that arise beyond the control of the user or the programmer. Errors are typically ignored in your code because you can rarely do anything about an error. For example, if a stack overflow occurs, an error will arise. They are also ignored at the time of compilation.

This tutorial will give you a few simple and elegant ways to handle runtime exception/error occuring in your JSP code.

Using Exception Object

The exception object is an instance of a subclass of Throwable (e.g. java.lang. NullPointerException) and is only available in error pages. Following is the list of important methods available in the Throwable class.

S. No.	Methods with Description
1	**public String getMessage()** Returns a detailed message about the exception that has occurred. This message is initialized in the Throwable constructor.
2	**public Throwable getCause()** Returns the cause of the exception as represented by a Throwable object.
3	**public String toString()** Returns the name of the class concatenated with the result of getMessage()
4	**public void printStackTrace()** Prints the result of toString() along with the stack trace to System.err, the error output stream.
5	**public StackTraceElement [] getStackTrace()** Returns an array containing each element on the stack trace. The element at index 0 represents the top of the call stack, and the last element in the array represents the method at the bottom of the call stack.
6	**public Throwable fillInStackTrace()** Fills the stack trace of this Throwable object with the current stack trace, adding to any previous information in the stack trace.

JSP gives you an option to specify Error Page for each JSP. Whenever the page throws an exception, the JSP container automatically invokes the error page.

Following is an example to specify an error page for a main.jsp. To set up an error page, use the <%@ page errorPage="xxx" %> directive.

```
<%@ page errorPage="ShowError.jsp" %>

<html>
<head>
```

```
<title>Error Handling Example</title>
</head>
<body><%
  // Throw an exception to invoke the error page
  int x = 1;
  if (x == 1)
  {
    throw new RuntimeException("Error condition!!!");
  }
%>
</body>
</html>
```

Now you would have to write one Error Handling JSP ShowError.jsp, which is given below. Notice that the error-handling page includes the directive <%@ page isErrorPage="true" %>. This directive causes the JSP compiler to generate the exception instance variable.

```
<%@ page isErrorPage="true" %>
<html>
<head>
<title>Show Error Page</title>
</head>
<body>
<h1>Opps...</h1>
<p>Sorry, an error occurred.</p>
<p>Here is the exception stack trace: </p>
<pre>
<% exception.printStackTrace(response.getWriter()); %>
</pre>
</body>
</html>
```

Now try to access main.jsp, it should generate something as follows:

```
java.lang.RuntimeException: Error condition!!!
...
Opps...
Sorry, an error occurred.
Here is the exception stack trace:
```

Using JSTL tags for Error Page

You can make use of JSTL tags to write an error page ShowError.jsp. This page has almost same logic which we have used in the above example, but it has better structure and it provides more information:

```
<%@ taglib prefix="c" uri="http://java.sun.com/jsp/jstl/core" %>
<%@page isErrorPage="true" %>
```

```
<html>
<head>
<title>Show Error Page</title>
</head>
<body>
<h1>Opps...</h1>
<table width="100%" border="1">
<tr valign="top">
<td width="40%"><b>Error:</b></td>
<td>${pageContext.exception}</td>
</tr>
<tr valign="top">
<td><b>URI:</b></td>
<td>${pageContext.errorData.requestURI}</td>
</tr>
<tr valign="top">
<td><b>Status code:</b></td>
<td>${pageContext.errorData.statusCode}</td>
</tr>
<tr valign="top">
<td><b>Stack trace:</b></td>
<td>
<c:forEach var="trace"
    items="${pageContext.exception.stackTrace}">
<p>${trace}</p>
</c:forEach>
</td>
</tr>
</table>
</body>
</html>
```

Now try to access main.jsp, it should generate something as follows:
Opps...

Error:	java.lang.RuntimeException: Error condition!!!
URI:/	main.jsp
Status code:	500
Stack trace:	org.apache.jsp.main_jsp._jspService(main_jsp.java:65) org.apache.jasper.runtime.HttpJspBase.service(HttpJspBase.java:68) javax.servlet.http.HttpServlet.service(HttpServlet.java:722) org.apachejasper.servlet.JspServlet.service(JspServlet.java:265) javax.servlet.http.HttpServlet.service(HttpServlet.java:722) ....

Using Try...Catch Block

If you want to handle errors within the same page and want to take some action instead of firing an error page, you can make use of **try...catch** block.

Following is a simple example which shows how to use try…catch block. Let us put following code in main.jsp:

```
<html>
<head>
  <title>Try...Catch Example</title>
</head>
<body>
<%
  try{
    int i = 1;
    i = i / 0;
    out.println("The answer is " + i);
  }
  catch (Exception e){
    out.println("An exception occurred: " + e.getMessage());
  }
%>
</body>
</html>
```

Now try to access main.jsp, it should generate something as follows:
An exception occurred: / by zero

4.16 Debugging

It is always difficult to testing/debugging a JSP and servlets. JSP and Servlets tend to involve a large amount of client/server interaction, making errors likely but hard to reproduce.

Here are a few hints and suggestions that may aid you in your debugging.

Using System.out.println()

System.out.println() is easy to use as a marker to test whether a certain piece of code is being executed or not. We can print out variable values as well. Additionally:

- Since the System object is part of the core Java objects, it can be used everywhere without the need to install any extra classes. This includes Servlets, JSP, RMI, EJB's, ordinary Beans and classes, and standalone applications.
- Compared to stopping at breakpoints, writing to System.out does not interfere much with the normal execution flow of the application, which makes it very valuable when timing is crucial.

Following is the syntax to use System.out.println():

System.out.println("Debugging message");

Following is a simple example of using System.out.print():

```
<%@taglib prefix="c" uri="http://java.sun.com/jsp/jstl/core" %>
<html>
```

```
<head><title>System.out.println</title></head>
<body>
<c:forEach var="counter" begin="1" end="10" step="1">
  <c:out value="${counter-5}"/></br>
  <% System.out.println( "counter= " +
               pageContext.findAttribute("counter") ); %>
</c:forEach>
</body>
</html>
```

Now if you will try to access above JSP, it will produce the following result at browser:

```
–4
–3
–2
–1
0
1
2
3
4
5
```

If you are using Tomcat, you will also find these lines appended to the end of stdout.log in the logs directory.

```
counter = 1
counter = 2
counter = 3
counter = 4
counter = 5
counter = 6
counter = 7
counter = 8
counter = 9
counter = 10
```

This way you can pring variables and other information into system log which can be analyzed to find out the root cause of the problem or for various other reasons.

Using the JDB Logger

The J2SE logging framework is designed to provide logging services for any class running in the JVM. So we can make use of this framework to log any information.

Let us re-write above example using JDK logger API:

```
<%@taglib prefix="c" uri="http://java.sun.com/jsp/jstl/core" %>
<%@page import="java.util.logging.Logger" %>

<html>
<head><title>Logger.info</title></head>
<body>
<% Logger logger=Logger.getLogger(this.getClass().getName());%>
```

```
<c:forEach var="counter" begin="1" end="10" step="1" >
  <c:set var="myCount" value="${counter-5}" />
  <c:out value="${myCount}" /></br>
  <% String message = "counter="
          + pageContext.findAttribute("counter")
          + " myCount="
          + pageContext.findAttribute("myCount");
          logger.info( message );
  %>
</c:forEach>
</body>
</html>
```

This would generate similar result at the browser and in stdout.log, but you will have additional information in stdout.log. Here we are using **info** method of the logger because we are logging message just for informational purpose. Here is a snapshot of stdout.log file:

```
24-Sep-2010 23:31:31 org.apache.jsp.main_jsp _jspService
INFO: counter=1 myCount=-4
24-Sep-2010 23:31:31 org.apache.jsp.main_jsp _jspService
INFO: counter=2 myCount=-3
24-Sep-2010 23:31:31 org.apache.jsp.main_jsp _jspService
INFO: counter=3 myCount=-2
24-Sep-2010 23:31:31 org.apache.jsp.main_jsp _jspService
INFO: counter=4 myCount=-1
24-Sep-2010 23:31:31 org.apache.jsp.main_jsp _jspService
INFO: counter=5 myCount=0
24-Sep-2010 23:31:31 org.apache.jsp.main_jsp _jspService
INFO: counter=6 myCount=1
24-Sep-2010 23:31:31 org.apache.jsp.main_jsp _jspService
INFO: counter=7 myCount=2
24-Sep-2010 23:31:31 org.apache.jsp.main_jsp _jspService
INFO: counter=8 myCount=3
24-Sep-2010 23:31:31 org.apache.jsp.main_jsp _jspService
INFO: counter=9 myCount=4
24-Sep-2010 23:31:31 org.apache.jsp.main_jsp _jspService
INFO: counter=10 myCount=5
```

Messages can be sent at various levels by using the convenience functions **severe(), warning(), info(), config(), fine(), finer(),** and **finest()**. Here finest() method can be used to log finest information and severe() method can be used to log severe information.

Debugging Tools

NetBeans is a free and open-source Java Integrated Development Environment that supports the development of standalone Java applications and Web applications supporting the JSP and servlet specifications and includes a JSP debugger as well.

NetBeans supports the following basic debugging functionalities:

- Breakpoints
- Stepping through code
- Watchpoints

You can refere to NetBeans documentation to understand above debugging functionalities.

Using JDB Debugger

You can debug JSP and servlets with the same **jdb** commands you use to debug an applet or an application.

To debug a JSP or servlet, you can debug sun.servlet.http.HttpServer, then watch as HttpServer executing JSP/servlets in response to HTTP requests we make from a browser. This is very similar to how applets are debugged. The difference is that with applets, the actual program being debugged is sun.applet.AppletViewer.

Most debuggers hide this detail by automatically knowing how to debug applets. Until they do the same for JSP, you have to help your debugger by doing the following:

- Set your debugger's classpath so that it can find sun.servlet.http.Http-Server and associated classes.
- Set your debugger's classpath so that it can also find your JSP and support classes, typically ROOT\WEB-INF\classes.

Once you have set the proper classpath, start debugging sun.servlet.http.HttpServer. You can set breakpoints in whatever JSP you are interested in debugging, then use a web browser to make a request to the HttpServer for the given JSP (http://localhost:8080/JSPToDebug). You should see execution stop at your breakpoints.

Using Comments

Comments in your code can help the debugging process in various ways. Comments can be used in lots of other ways in the debugging process.

The JSP uses Java comments and single line (// ...) and multiple line (/* ... */) comments can be used to temporarily remove parts of your Java code. If the bug disappears, take a closer look at the code you just commented and find out the problem.

Client and Server Headers

Sometimes when a JSP does not behave as expected, it is useful to look at the raw HTTP request and response. If you are familiar with the structure of HTTP, you can read the request and response and see what exactly is going with those headers.

Important Debugging Tips

Here is a list of some more debugging tips on JSP debugging:

- Ask a browser to show the raw content of the page it is displaying. This can help identify formatting problems. It is usually an option under the View menu.
- Make sure the browser is not caching a previous request's output by forcing a full reload of the page. With Netscape Navigator, use Shift-Reload; with Internet Explorer use Shift-Refresh.

4.17 Security

JavaServer Pages and servlets make several mechanisms available to Web developers to secure applications. Resources are protected declaratively by identifying them in the application deployment descriptor and assigning a role to them.

Several levels of authentication are available, ranging from basic authentication using identifiers and passwords to sophisticated authentication using certificates.

Role Based Authentication

The authentication mechanism in the servlet specification uses a technique called role-based security. The idea is that rather than restricting resources at the user level, you create roles and restrict the resources by role.

You can define different roles in file tomcat-users.xml, which is located off Tomcat's home directory in conf. An example of this file is shown below:

```
<?xml version='1.0' encoding='utf-8'?>
<tomcat-users>
<role rolename="tomcat"/>
<role rolename="role1"/>
<role rolename="manager"/>
<role rolename="admin"/>
<user username="tomcat" password="tomcat" roles="tomcat"/>
<user username="role1" password="tomcat" roles="role1"/>
<user username="both" password="tomcat" roles="tomcat,role1"/>
<user username="admin" password="secret" roles="admin,manager"/>
</tomcat-users>
```

This file defines a simple mapping between user name, password, and role. Notice that a given user may have multiple roles, for example, user name="both" is in the "tomcat" role and the "role1" role.

Once you identified and defined different roles, a role-based security restrictions can be placed on different Web Application resources by using the **<security-constraint>** element in web.xml file available in WEB-INF directory.

Following is a sample entry in web.xml:

```
<web-app>
...
  <security-constraint>
    <web-resource-collection>
      <web-resource-name>
        SecuredBookSite
      </web-resource-name>
      <url-pattern>/secured/*</url-pattern>
      <http-method>GET</http-method>
      <http-method>POST</http-method>
    </web-resource-collection>
    <auth-constraint>
      <description>
      Let only managers use this app
      </description>
      <role-name>manager</role-name>
    </auth-constraint>
  </security-constraint>
  <security-role>
```

```
    <role-name>manager</role-name>
  </security-role>
  <login-config>
    <auth-method>BASIC</auth-method>
  </login-config>
...
</web-app>
```

Above entries would mean:

- Any HTTP GET or POST request to a URL matched by /secured/* would be subject to the security restriction.
- A person with manager role is given access to the secured resources.
- Last, the login-config element is used to describe the BASIC form of authentication.

Now if you try browsing to any URL including the /security directory, it would display a dialogue box asking for user name and password. If you provide a user "admin" and password "secrer", then only you would have access on URL matched by /secured/* because above we have defined user admin with manager role who is allowed to access this resource.

Form Based Authentication

When you use the FORM authentication method, you must supply a login form to prompt the user for a username and password. Following is a simple code of login.jsp to create a form for the same purpose:

```
<html>
<body bgcolor="#ffffff">
  <form method="POST" action="j_security_check">
    <table border="0">
    <tr>
    <td>Login</td>
    <td><input type="text" name="j_username"></td>
    </tr>
    <tr>
    <td>Password</td>
    <td><input type="password" name="j_password"></td>
    </tr>
    </table>
    <input type="submit" value="Login!">
    </center>
  </form>
</body>
</html>
```

Here you have to make sure that the login form must contain form elements named j_username and j_password. The action in the <form> tag must be j_security_check. POST must be used as the form method. At the same time you would have to modify <login-config> tag to specify auth-method as FORM:

```
<web-app>
...
  <security-constraint>
    <web-resource-collection>
      <web-resource-name>
        SecuredBookSite
      </web-resource-name>
      <url-pattern>/secured/*</url-pattern>
      <http-method>GET</http-method>
      <http-method>POST</http-method>
    </web-resource-collection>
    <auth-constraint>
      <description>
      Let only managers use this app
      </description>
      <role-name>manager</role-name>
    </auth-constraint>
  </security-constraint>
  <security-role>
   <role-name>manager</role-name>
  </security-role>
  <login-config>
   <auth-method>FORM</auth-method>
   <form-login-config>
     <form-login-page>/login.jsp</form-login-page>
     <form-error-page>/error.jsp</form-error-page>
   </form-login-config>
  </login-config>
...
</web-app>
```

Now when you try to access any resource with URL /secured/*, it would display above form asking for user id and password. When the container sees the "j_security_check" action, it uses some internal mechanism to authenticate the caller.

If the login succeeds and the caller is authorized to access the secured resource, then the container uses a session-id to identify a login session for the caller from that point on. The container maintains the login session with a cookie containing the session-id. The server sends the cookie back to the client, and as long as the caller presents this cookie with subsequent requests, then the container will know who the caller is.

If the login fails, then the server sends back the page identified by the form-error-page setting.

Here j_security_check is the action that applications using form based login have to specify for the login form. In the same form you should also have a text input control called j_username and a password input control called j_password. When you see this, it means that the information contained in the form will be submitted to the server, which will check name and password. How this is done is server specific.

Check Standard Realm Implementations to understand how j_security_check works for Tomcat container.

Programmatic Security in a Servlet/JSP

The HttpServletRequest object provides the following methods, which can be used to mine security information at runtime:

S. No.	Method and Description
1.	**string getAuthType()**The getAuthType() method returns a String object that represents the name of the authentication scheme used to protect the Servlet.
2.	**boolean isUserInRole(java.lang.String role)**The isUserInRole() method returns a boolean value: True if the user is in the given role or false if they are not.
3.	**string getProtocol()**The getProtocol() method returns a String object representing the protocol that was used to send the request. This value can be checked to determine if a secure protocol was used.
4.	**boolean isSecure()**The isSecure() method returns a boolean value representing if the request was made using HTTPs. A value of true means it was and the connection is secure. A value of false means the request was not.
5.	**principle getUserPrinciple()**The getUserPrinciple() method returns a java.security.Principle object that contains the name of the current authenticated user.

For example, a JavaServer Page that links to pages for managers, you might have the following code:

```
<% if (request.isUserInRole("manager")) { %>
<a href="managers/mgrreport.jsp">Manager Report</a>
<a href="managers/personnel.jsp">Personnel Records</a>
<% } %>
```

By checking the user's role in a JSP or servlet, you can customize the Web page to show the user only the items she can access. If you need the user's name as it was entered in the authentication form, you can call getRemoteUser method in the request object.

4.18 Internationalization

Before we proceed, let me explain three important terms:

- **Internationalization (i18n):** This means enabling a website to provide different versions of content translated into the visitor's language or nationality.
- **Localization (l10n):** This means adding resources to a website to adapt it to a particular geographical or cultural region, for example, Hindi translation to a website.
- **locale:** This is a particular cultural or geographical region. It is usually referred to as a language symbol followed by a country symbol which are separated by an underscore. For example "en_US" represents english locale for US.

There are a number of items which should be taken care while building up a global website. This tutorial would not give you complete detail on this but it would give you a good example on how you can offer your web page in different languages to internet community by differentiating their location, i.e. locale.

A JSP can pickup appropriate version of the site based on the requester's locale and provide appropriate site version according to the local language, culture and requirements. Following is the method of request object which returns Locale object.

java.util.Locale request.getLocale()

Detecting Locale:

Following are the important locale methods which you can use to detect requester's location, language and of course locale. All the below methods display country name and language name set in requester's browser.

S. No.	Method and Description
1.	**string getCountry()**This method returns the country/region code in upper case for this locale in ISO 3166 2-letter format.
2.	**string getDisplayCountry()**This method returns a name for the locale's country that is appropriate for display to the user.
3.	**string getLanguage()**This method returns the language code in lower case for this locale in ISO 639 format.
4.	**string getDisplayLanguage()**This method returns a name for the locale's language that is appropriate for display to the user.
5.	**string getISO3Country()**This method returns a three-letter abbreviation for this locale's country.
6.	**string getISO3Language()**This method returns a three-letter abbreviation for this locale's language.

Example

This example shows how you display a language and associated country for a request in a JSP:

```
<%@ page import="java.io.*,java.util.Locale" %>
<%@ page import="javax.servlet.*,javax.servlet.http.* "%>
<%
  //Get the client's Locale
  Locale locale = request.getLocale();
  String language = locale.getLanguage();
  String country = locale.getCountry();
%>
<html>
<head>
<title>Detecting Locale</title>
</head>
<body>
```

```
<center>
<h1>Detecting Locale</h1>
</center>
<p align="center">
<%
  out.println("Language : " + language + "<br />");
  out.println("Country  : " + country  + "<br />");
%>
</p>
</body>
</html>
```

Languages Setting

A JSP can output a page written in a Western European language such as English, Spanish, German, French, Italian, Dutch, etc. Here it is important to set Content-Language header to display all the characters properly.

Second point is to display all the special characters using HTML entities, for example, "ñ" represents "ñ", and "¡" represents "¡" as follows:

```
<%@ page import="java.io.*,java.util.Locale" %>
<%@ page import="javax.servlet.*,javax.servlet.http.* "%>
<%
  // Set response content type
  response.setContentType("text/html");
  // Set spanish language code.
  response.setHeader("Content-Language", "es");
  String title = "En Español";

%>
<html>
<head>
<title><%  out.print(title); %></title>
</head>
<body>
<center>
<h1><%  out.print(title); %></h1>
</center>
<div align="center">
<p>En Español</p>
<p>¡Hola Mundo!</p>
</div>
</body>
</html>
```

Locale Specific Dates

You can use the java.text.DateFormat class and its static getDateTimeInstance( ) method to format date and time specific to locale. Following is the example which shows how to format dates specific to a given locale:

```
<%@ page import="java.io.*,java.util.Locale" %>
<%@ page import="javax.servlet.*,javax.servlet.http.* "%>
<%@ page import="java.text.DateFormat,java.util.Date" %>

<%
   String title = "Locale Specific Dates";
   //Get the client's Locale
   Locale locale = request.getLocale( );
   String date = DateFormat.getDateTimeInstance(
                           DateFormat.FULL,
                           DateFormat.SHORT,
                           locale).format(new Date( ));
%>
<html>
<head>
<title><% out.print(title); %></title>
</head>
<body>
<center>
<h1><% out.print(title); %></h1>
</center>
<div align="center">
<p>Local Date: <%  out.print(date); %></p>
</div>
</body>
</html>
```

Locale Specific Currency

You can use the java.txt.NumberFormat class and its static getCurrencyInstance( ) method to format a number, such as a long or double type, in a locale specific curreny. Following is the example which shows how to format currency specific to a given locale:

```
<%@ page import="java.io.*,java.util.Locale" %>
<%@ page import="javax.servlet.*,javax.servlet.http.* "%>
<%@ page import="java.text.NumberFormat,java.util.Date" %>
<%
   String title = "Locale Specific Currency";
   //Get the client's Locale
   Locale locale = request.getLocale( );
   NumberFormat nft = NumberFormat.getCurrencyInstance(locale);
```

```
    String formattedCurr = nft.format(1000000);
%>
<html>
<head>
<title><% out.print(title); %></title>
</head>
<body>
<center><h1><% out.print(title); %></h1></center><div
align="center"><p>Formatted Currency: <%  out.print(formattedCurr); %></
p></div></body></html>
```

Locale Specific Percentage

You can use the java.txt.NumberFormat class and its static getPercentInstance( ) method to get locale specific percentage. Following is the example which shows how to format percentage specific to a given locale:

```
<%@ page import="java.io.*,java.util.Locale" %>
<%@ page import="javax.servlet.*,javax.servlet.http.* "%>
<%@ page import="java.text.NumberFormat,java.util.Date" %>

<%
    String title = "Locale Specific Percentage";
    //Get the client's Locale
    Locale locale = request.getLocale( );
    NumberFormat nft = NumberFormat.getPercentInstance(locale);
    String formattedPerc = nft.format(0.51);
%>
<html>
<head>
<title><% out.print(title); %></title>
</head>
<body>
<center>
<h1><% out.print(title); %></h1>
</center>
<div align="center">
<p>Formatted Percentage: <%  out.print(formattedPerc); %></p>
</div>
</body>
</html>
```

4.19 JSP Standard Tag Library (JSTL)

The JavaServer Pages Standard Tag Library (JSTL) is a collection of useful JSP tags which encapsulates core functionality common to many JSP applications.

JSTL has support for common, structural tasks such as iteration and conditionals, tags for manipulating XML documents, internationalization tags, and SQL tags. It also provides a framework for integrating existing custom tags with JSTL tags.

The JSTL tags can be classified, according to their functions, into following JSTL tag library groups that can be used when creating a JSP page:

1. Core Tags
2. Formatting tags
3. SQL tags
4. XML tags
5. JSTL Functions

Install JSTL Library

If you are using Apache Tomcat container, then follow the following two simple steps:

1. Download the binary distribution from Apache Standard Taglib and unpack the compressed file.
2. To use the Standard Taglib from its Jakarta Taglibs distribution, simply copy the JAR files in the distribution's 'lib' directory to your application's webapps\ROOT\WEB-INF\lib directory.

To use any of the libraries, you must include a <taglib> directive at the top of each JSP that uses the library.

4.19.1 Core Tags

The core group of tags are the most frequently used JSTL tags. Following is the syntax to include JSTL Core library in your JSP:

```
<%@ taglib prefix="c"
        uri="http://java.sun.com/jsp/jstl/core" %>
```

There are following Core JSTL Tags:

Tag	Description
<c:out >	Like <%= ... >, but for expressions.
<c:set >	Sets the result of an expression evaluation in a 'scope'
<c:remove >	Removes a scoped variable (from a particular scope, if specified).
<c:catch>	Catches any Throwable that occurs in its body and optionally exposes it.
<c:if>	Simple conditional tag which evalutes its body if the supplied condition is true.
<c:choose>	Simple conditional tag that establishes a context for mutually exclusive conditional operations, marked by <when> and <otherwise>
<c:when>	Subtag of <choose> that includes its body if its condition evalutes to 'true'.
<c:otherwise>	Subtag of <choose> that follows <when> tags and runs only if all of the prior conditions evaluated to 'false'.
<c:import>	Retrieves an absolute or relative URL and exposes its contents to either the page, a String in 'var', or a Reader in 'varReader'.

<c:forEach>	The basic iteration tag, accepting many different collection types and supporting subsetting and other functionality.
<c:forTokens>	Iterates over tokens, separated by the supplied delimiters.
<c:param>	Adds a parameter to a containing 'import' tag's URL.
<c:redirect >	Redirects to a new URL.
<c:url>	Creates a URL with optional query parameters

4.19.2 Formatting Tags

The JSTL formatting tags are used to format and display text, the date, the time, and numbers for internationalized websites. Following is the syntax to include Formatting library in your JSP:

```
<%@ taglib prefix="fmt"
        uri="http://java.sun.com/jsp/jstl/fmt" %>
```

Following is the list of Formatting JSTL Tags:

Tag	Description
<fmt:formatNumber>	To render numerical value with specific precision or format.
<fmt:parseNumber>	Parses the string representation of a number, currency, or percentage.
<fmt:formatDate>	Formats a date and/or time using the supplied styles and pattern.
<fmt:parseDate>	Parses the string representation of a date and/or time
<fmt:bundle>	Loads a resource bundle to be used by its tag body.
<fmt:setLocale>	Stores the given locale in the locale configuration variable.
<fmt:setBundle>	Loads a resource bundle and stores it in the named scoped variable or the bundle configuration variable.
<fmt:timeZone>	Specifies the time zone for any time formatting or parsing actions nested in its body.
<fmt:setTimeZone>	Stores the given time zone in the time zone configuration variable
<fmt:message>	To display an internationalized message.
<fmt:requestEncoding>	Sets the request character encoding

4.19.3 SQL Tags

The JSTL SQL tag library provides tags for interacting with relational databases (RDBMSs) such as Oracle, mySQL, or Microsoft SQL Server.

Following is the syntax to include JSTL SQL library in your JSP:

```
<%@ taglib prefix="sql"
      uri="http://java.sun.com/jsp/jstl/sql" %>
```

Following is the list of SQL JSTL Tags:

Tag	Description
<sql:setDataSource>	Creates a simple DataSource suitable only for prototyping
<sql:query>	Executes the SQL query defined in its body or through the sql attribute.
<sql:update>	Executes the SQL update defined in its body or through the sql attribute.
<sql:param>	Sets a parameter in an SQL statement to the specified value.
<sql:dateParam>	Sets a parameter in an SQL statement to the specified java.util.Date value.
<sql:transaction >	Provides nested database action elements with a shared Connection, set up to execute all statements as one transaction.

4.19.4 XML Tags

The JSTL XML tags provide a JSP-centric way of creating and manipulating XML documents. Following is the syntax to include JSTL XML library in your JSP.

The JSTL XML tag library has custom tags for interacting with XML data. This includes parsing XML, transforming XML data, and flow control based on XPath expressions.

```
<%@ taglib prefix="x"
      uri="http://java.sun.com/jsp/jstl/xml" %>
```

Before you proceed with the examples, you would need to copy the following two XML and XPath related libraries into your <Tomcat Installation Directory>\lib:

1. **XercesImpl.jar:** Download it from http://www.apache.org/dist/xerces/j/
2. **xalan.jar:** Download it from http://xml.apache.org/xalan-j/index.html

Following is the list of XML JSTL Tags:

Tag	Description
<x:out>	Like <%= ... >, but for XPath expressions.
<x:parse>	Use to parse XML data specified either via an attribute or in the tag body.
<x:set >	Sets a variable to the value of an XPath expression.
<x:if >	Evaluates a test XPath expression and if it is true, it processes its body. If the test condition is false, the body is ignored.
<x:forEach>	To loop over nodes in an XML document.

<x:choose>	Simple conditional tag that establishes a context for mutually exclusive conditional operations, marked by <when> and <otherwise>
<x:when >	Subtag of <choose> that includes its body if its expression evalutes to 'true'
<x:otherwise >	Subtag of <choose> that follows <when> tags and runs only if all of the prior conditions evaluated to 'false'
<x:transform >	Applies an XSL transformation on a XML document
<x:param >	Use along with the transform tag to set a parameter in the XSLT stylesheet

4.19.5 JSTL Functions

JSTL includes a number of standard functions, most of which are common string manipulation functions. Following is the syntax to include JSTL Functions library in your JSP:

```
<%@ taglib prefix="fn"
       uri="http://java.sun.com/jsp/jstl/functions" %>
```

Following is the list of JSTL Functions:

Function	Description
fn:contains()	Tests if an input string contains the specified substring.
fn:containsIgnoreCase()	Tests if an input string contains the specified substring in a case insensitive way.
fn:endsWith()	Tests if an input string ends with the specified suffix.
fn:escapeXml()	Escapes characters that could be interpreted as XML markup.
fn:indexOf()	Returns the index within a string of the first occurrence of a specified substring.
fn:join()	Joins all elements of an array into a string.
fn:length()	Returns the number of items in a collection, or the number of characters in a string.
fn:replace()	Returns a string resulting from replacing in an input string all occurrences with a given string.
fn:split()	Splits a string into an array of substrings.
fn:startsWith()	Tests if an input string starts with the specified prefix.
fn:substring()	Returns a subset of a string.
fn:substringAfter()	Returns a subset of a string following a specific substring.
fn:substringBefore()	Returns a subset of a string before a specific substring.

fn:toLowerCase()	Converts all of the characters of a string to lower case.
fn:toUpperCase()	Converts all of the characters of a string to upper case.
fn:trim()	Removes white spaces from both ends of a string.

5. APPLET TO APPLET COMMUNICATION

Getting two or more applets within a single Web page to talk to each other has some benefits. Although this applet capability has been around since the earliest version of Java, it is not often used, because there is more emphasis placed on getting applets to communicate with servers.

While this is understandable given the current fashion of client / server programming, it is still a valuable skill for developers to learn. Another reason the technique is not used much is that complicated Web-borne applets are usually shown in a single window. If there is a lot of information to show, the designers simply make the applet larger.

However, in terms of Web page design, it is better in some cases to place small bits of Java-based functionality in different parts of the page, leaving the rest to be filled with text and images. To do this, you need multiple applet windows that are, in some sense, part of the same program.

In this article, we will look into the basic techniques of inter-applet communication. We will develop a nice API that takes care of the work for us. Then we will show a sample application called DumChat, which uses the API.

Method

The secret of inter-applet communication (which we will abbreviate to IAC) is the method AppletContext.getApplets(). This method provides us with an Enumeration of all the applets running on the same page as the calling applet. From this Enumeration, you can take actual Applet objects, allowing you to freely call methods on it.

Clearly, this is enough to get started. However, it does not provide an elegant metaphor for communication. So how should we describe it? Well, how about some kind of message passing? Sounds good.

What we will do first is give names to the applets on the page and then allow them to send text strings to each other using the names as destinations.

Here is an API for this:

```
public void send( String appletName, String message );
protected String rcv();
```

The send() method sends a string to another applet with a given name; the rcv() method returns the next string that has been sent to you.

This API will be implemented in a class called CommunicatingApplet. To use this, you will derive all your applets from this instead of directly from Applet, and the methods in the API will be available to you.

Example

Our simple chat system is called DumChat because, although it allows chatting between two windows, both windows have to be in the same Web page, on the same screen. So it is not very useful, but it does demonstrate our technology.

The runtime structure of the DumChat applet is simple. It has a TextArea for displaying things that have been said, and above that it has a TextField for typing things in. Each time something is typed into the textfield, that message is sent to the other applet. And each applet has a background thread that is waiting for messages to come in and which prints them out when they do.

Here is an excerpt from the code that handles the user typing:

```
// Extract the typed string
String message = ae.getActionCommand();

// Send it to the other applet
send( otherName, message );

// Display it in our window
ta.append( message+"\n" );
tf.setText( "" );
```

The variable otherName contains the name of the other applet; here you can see that sending a message to the other applet is simple — a single call to send takes care of it.

Here is the code for the background thread that waits for incoming messages and displays them:

```
while (true) {
    // Get the next message
    String message = rcv();

    // Print it
    ta.append( message+"\n" );
}
```

Again, very simple. It is worth the effort to put a nice, minimal API over a raw communication mechanism; a more complicated applet would be making lots of calls to send and rcv, and that is better than having lots of bits of code calling Applet.getApplets() and hunting around in that list for otherApplet objects.

```
<applet code="DumChat.class" width=300 height=300>
<param name="communicationname" value="foo">
<param name="other" value="bar">
</applet>
<applet code="DumChat.class" width=300 height=300>
<param name="communicationname" value="bar">
<param name="other" value="foo">
</applet>
```

We have named the applets foo and bar. Very creative of us. Now, each applet needs to have a parameter that tells us what our name is and another one to tell us the name of our peer on the other side of our communication mechanism.

Implementation

Let us take a look inside our implementation of the send() and rcv() methods:

```
// Send 'message' to applet named 'appletName'
public void send( String appletName, String message )
  throws CommunicatingAppletException {

  // Have we managed to find our peer and send the message?
  boolean messageSent = false;

  // Zip through all the applets on the page
  Enumeration e = getAppletContext().getApplets();
  while (e.hasMoreElements()) {

   Applet applet = (Applet)e.nextElement();

   // Ignore applets that aren't part of our little game
   if (!(applet instanceof CommunicatingApplet))
    continue;

   // Grab the applet, if it might possibly be our peer
   CommunicatingApplet capplet = (CommunicatingApplet)applet;

   // If this is the applet we're looking for,
   // give it the message
   if (capplet.getCommunicationName().equals( appletName )) {
    capplet.takeMessage( message );
    messageSent = true;

    // We're done!
    break;
   }
  }
  // The named applet isn't around; throw an Exception
  if (!messageSent)
   throw
    new CommunicatingAppletException(
       "Can't send message "+message );
}
```

Here's how sending works: We sort through the list of applets on a page until we find the one with the target name. If we find it, we stuff the message directly into its queue by calling its takeMessage()method.

Note that we throw an exception if we do not find the target applet. This may or may not be overkill for some applications. Even though our chat example is bogus, it does show a case where we might ignore messages that do not arrive; after all, chat programs do not always warn you when your message does not get to everyone.

Here's the receiving routine:

```
// Receive a waiting message, or block
// until there is one
protected String rcv() {
  synchronized( messages ) {

    // We have a lock on the 'messages' object;
    // wait() until there are messages to be had
    while (true) {
      if (messages.isEmpty())
        try { messages.wait(); }
        catch( InterruptedException ie ) {}
      else
        break;
    }

    // Good, there is one. Remove it from the queue
    // and return it
    String message = (String)messages.elementAt( 0 );
    messages.removeElementAt( 0 );
    return message;
  }
}
```

The trick here is that if there is no message, we have to block until there are some. Which means going into a class wait/condition loop. Once the loop is exited, we know that there is a message, so we take it from the queue and return it to the caller.

Note that we are using synchronization in this code. We are using a Java.util.Vector object to keep the messages in, and we cannot have multiple threads messing with it at the same time; so we control access to it by synchronizing on the Vector itself. We do not want to synchronize on the applet — it would work fine, but you might have other code that is already synchronizing on the applet, and you do not want that interacting. It is best to have a separate object for each synchronization issue in your program, and in this case it was fine to just use the Vector itself.

Ideas for Further Work

The main thing to do is to use this technique for something useful. After all, DumChat is pretty silly, and it does not really suggest what one might use this for.

As mentioned above, you could have a page where you want to use Java to implement website navigational tools. There's a good chance you might want a collection of smaller applets — a collapsible menu on the left, a "What's Hot" ticker at the top, and so on. If these were entirely separate applets, they would likely be duplicating some of the site-layout information they both use, which increases download time. Instead, you can have a master applet telling other applets what to do and what to display.

Our implementation is also quite bare-bones. We are using String objects as our messages, and we might want to have something more sophisticated. We could allow the applets to send arbitrary Objects to each other (and we then send strings if we wanted, or anything else, depending on the application). We could even define a special CommunicatingAppletMessage class, which could contain the actual string or object, and also contain the name of the applet that sent it — like a return address. The receiving applet might need to know where to send a reply. This latter option is probably most elegant from an object-oriented design standpoint.

Another idea is to allow some kind of "discovery" of other applets. We hard-coded the name of the "other" applet into our HTML, but in some cases, our applets might be put on the page in different configurations. Our applet might want to sift through a list of other CommunicatingApplets on the page and see which ones it wants to send things to.

Finally, there might be a problem of subclassing from CommunicatingApplet— after all, there might be some other useful abstract applet base class we want to derive from, and that would make it impossible to use CommunicatingApplet. If this were the case, we would want to create another class, sayAppletPostOffice, and have the various methods be static methods of that class.

Most of the time we have one single applet embedded in the HTML page, which does all the work that is designed. But there are cases where a web-based product consists of more than one applet, each performing its own task and at the same time communicating with each other. There are two scenarios in which this can happen.

Applets in single HTML page in a browser

This is the simple case where more than one applet are embedded in a single HTML page in a browser. These applets will have the same Applet Context due to which they can freely communicate with each other.

The following figure shows the communication between two applets on the same page.

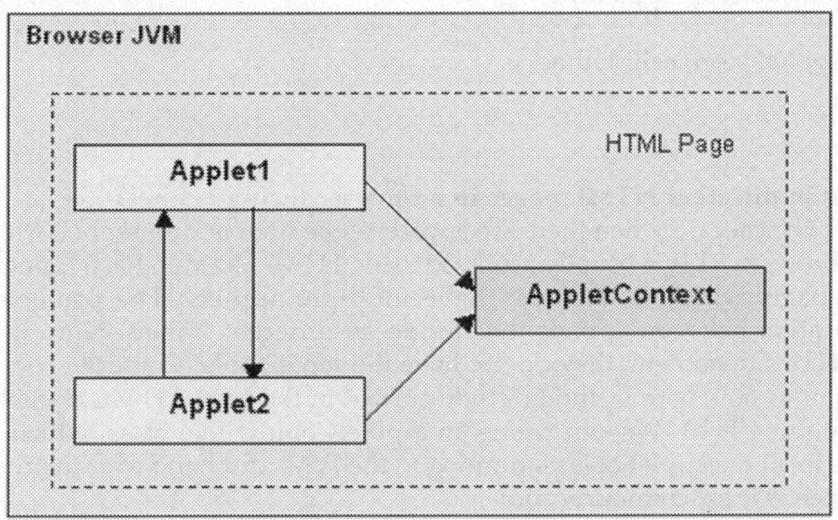

Inter-Applet Commuication between applets on same HTML page

In the above figure, Applet1 can easily communicate with Applet2 and vice versa using the AppletContext. This can be done by using the getApplets() and getApplet(String name) method of the AppletContext. The getApplets()method will return the enumeration of all the Applets in the document and getApplet(String name) method will return the Applet object by passing the applet name, this name is set using the NAME attribute in the Applet HTML tag. Suppose the HTML Page in the above figure contains the following tags,

```
< APPLET CODE="Applet1" NAME="Applet_1" WIDTH="200"
HEIGHT="100">
< / APPLET >
< BR >< BR >
< APPLET CODE="Applet2" NAME="Applet_2" WIDTH ="400" HEIGHT
="100" >
< / APPLET >
```

Applet1 can do the following to communicate with Applet2.

Get the reference to Applet2 using the name, i.e. Applet_2

```
Applet applet2 = getApplet("Applet_2");
((Applet2)applet2).callMethod(); //calling method of Applet2
```

Or, it can enumerate through the list of applets to get the reference, if the name is not known

```
Enumeration allApplets = getApplets();
while(allApplets.hasMoreElements())
{
  Applet app = (Applet) allApplets.nextElement();
  if(app instanceof Applet2)
  {
    ((Applet2)app).callMethod();
  }
}
```

Applets in different HTML pages in a browser (using Frames)

This is a typical case when there are applets embedded in different HTML pages in the same browser. This is usually achieved using HTML Frames. Each frame can have a HTML page, which in turn can have multiple applets. The problem here is communication between applets that reside on different frames. Since they are in different HTMLdocument, they do not have the sameApplet Context.

The point to note here is, though they are not in the same HTML document, they run in the same JVM. We can create an AppletContext like class, which can hold references to all the applet objects running in the JVM, and hand over the reference to whoever needs it for communication.

Let us call this class as AppletRegistry. AppletRegistry will maintain static Hashtable that will keep the map of all the applets. Applets interested in allowing other applets to communicate should registry itself with AppletRegistry while starting. Applets should also take care of unregistering themselves while stopping.

In the figure there are 3 Frames with respective HTML Pages. Each HTML page contains an applet. Each applet registers itself to the AppletRegistry using the register(String, Applet) static method. Here is how the AppletRegistry class looks,

The following figure shows the use of AppletRegistry.

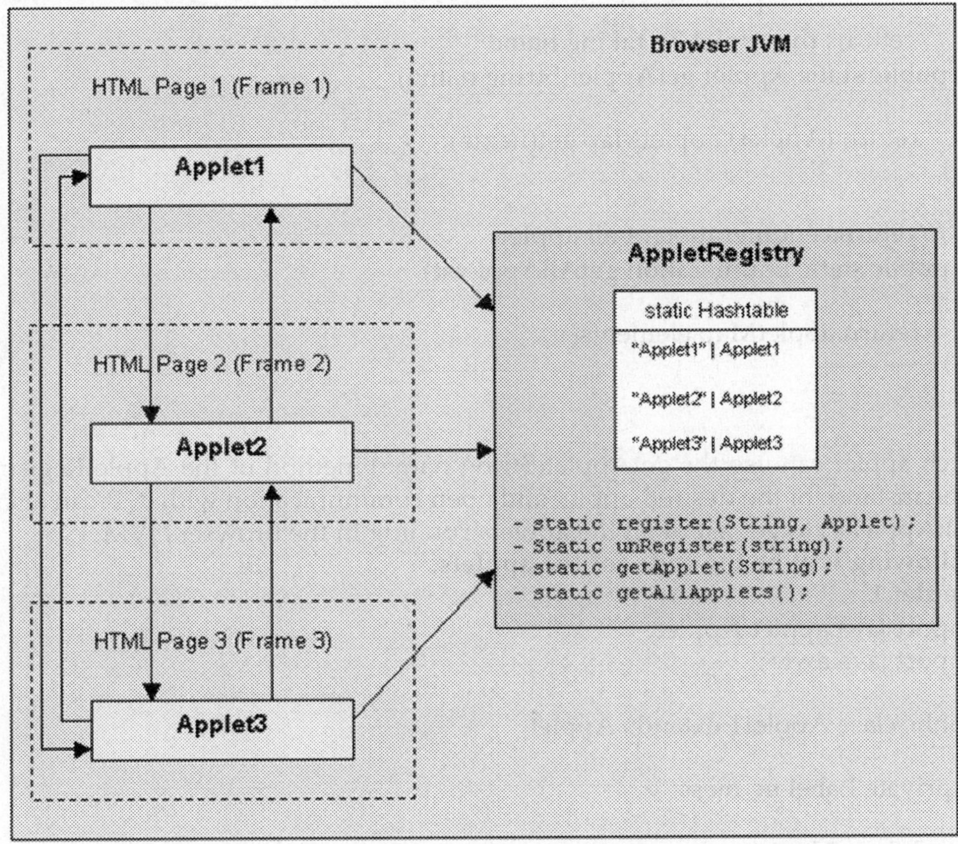

Inter-Applet Communication between applets of different HTML pages

AppletRegistry

```
import java.applet.Applet;
import java.util.*;

public class AppletRegistry extends Applet
{
   //static hashtable maintaining the applet map
   private static Hashtable appletMap = new Hashtable();

   //registers the given applet
   public static void register(String name, Applet applet)
   {
      appletMap.put(name, applet);
   }
```

```
    //unregisters the given applet
    public static void unRegister(String name)
    {
        appletMap.remove(name);
    }

    //returns the applet by taking name
    public static Applet getApplet(String name)
    {
        return (Applet) appletMap.get(name);
    }

    //returns enumeration of all applets
    public static Enumeration getAllApplets()
    {
        return appletMap.elements();
    }
}
```

Each applet can use the getApplet(String name) method of the AppletRegistry to get the instance of the desired applet and open communication with it. It can also use getAllApplets() to get list of all the applets running in the browsers JVM.

Following will be the code for the applets.

Applet 1

```
import java.applet.Applet;
import java.awt.*;

public class Applet1 extends Applet
{
    private Label m_mess;

    public void init()
    {
        setLayout(new BorderLayout());
        m_mess = new Label("Applet1 is running...", Label.CENTER);
        add(BorderLayout.CENTER, m_mess);
    }

    public void start()
    {
        AppletRegistry.register("Applet1", this);
    }

    public void stop()
    {
        AppletRegistry.unRegister("Applet1");
    }

    public void setData(String message)
```

```
    {
        //process message
        m_mess.setText("Welcome " + message);

        //send this message tp Applet3
        //get Applet1 instance
        Applet applet3 = AppletRegistry.getApplet("Applet3");

        //call method
        ((Applet3)applet3).setData(message);
    }
}
```

Applet 2

```
import java.applet.Applet;
import java.awt.*;
import java.awt.event.*;

public class Applet2 extends Applet implements ActionListener
{
    private Label m_mess;
    private TextField m_input;
    private Button m_send;

    public void init()
    {
        //create UI
        setLayout(new BorderLayout());
        Panel p = new Panel(new FlowLayout(FlowLayout.CENTER));
        m_input = new TextField(20);
        m_mess = new Label("Applet2 is running...", Label.CENTER);
        m_send = new Button("Send");
        m_send.addActionListener(this);
        p.add(new Label("Enter your Name : "));
        p.add(m_input);
        p.add(m_send);

        add(BorderLayout.CENTER, p);
        add(BorderLayout.SOUTH, m_mess);
    }

    public void start()
    {
        AppletRegistry.register("Applet2", this);
    }

    public void stop()
    {
```

```
        AppletRegistry.unRegister("Applet2");
    }

    public void actionPerformed(ActionEvent e)
    {
        if(e.getSource().equals(m_send))
        {
            String message = m_input.getText();

            //send this message to Applet1
            //get Applet1 instance
            Applet applet1 = AppletRegistry.getApplet("Applet1");

            //call method
            ((Applet1)applet1).setData(message);
            m_mess.setText("Message Send : " + message);
        }
    }
}
```

Applet 3

```
import ja va.applet.Applet;
import java.awt.*;
public class Applet3 extends Applet
{
    private Label m_mess;

    public void init()
    {
        setLayout(new BorderLayout());
        m_mess = new Label("Applet3 is running...", Label.CENTER);
        add(BorderLayout.CENTER, m_mess);
    }

    public void start()
    {
        AppletRegistry.register("Applet3", this);
    }

    public void stop()
    {
        AppletRegistry.unRegister("Applet3");
    }

    public void setData(String message)
    {
        //process message
        m_mess.setText("How are you " + message);
    }
}
```

In the above code, Applet2 calls the setData() method of Applet1 when the Send button is clicked. Applet1 then receives the message and calls setData() method of Applet3.

The important points that need to be considered in this approach are

- Make sure that the applet is registered before communicating with it; else you can get nullpointer exception while doing the lookup.
- Make sure to unRegister the applet while stopping or destroying.

The following figure shows the applets running.

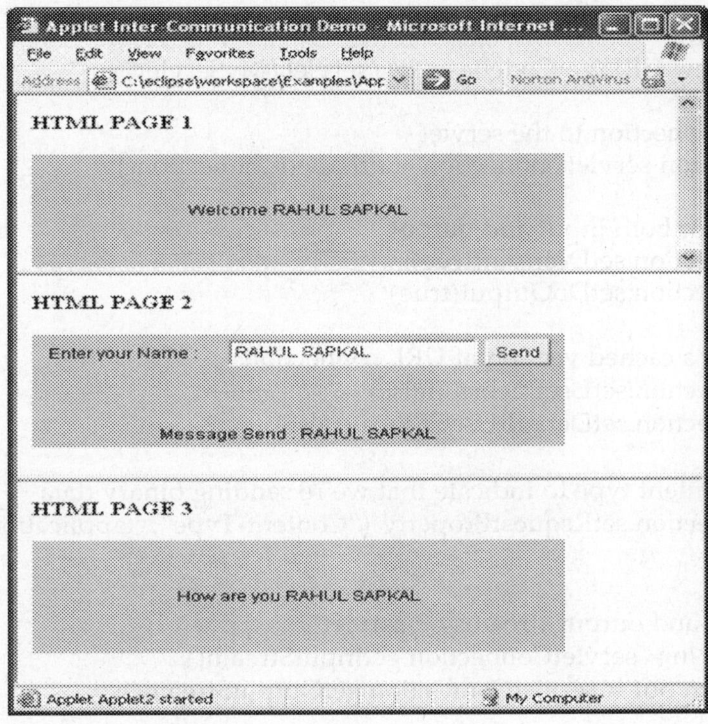

6. APPLET TO SERVLET COMMUNICATION

There are several ways for an applet to communicate with a servlet:

(1) HTTP GET Method: Easy to pass simple parameter types like strings to the servlet.

The applet can send a GET request to the servlet by opening a servlet URL with a query string. For example,

```
// servlet URL with a name paremeter in a query string
URL url = new URL(getCodeBase(), "/servlet/MyServlet?nameparam=John+
Smith");

// open a connection to the servlet
URLConnection servletConnection = url.openConnection();
servletConnection.setDoInput(true); // true, if we get data back

// get the input stream from the servlet
```

```
InputStream in = servletConnection.getInputStream();

// read the response from the servlet
...
```

(2) HTTP POST Method: Useful for passing a large amount of text or binary data to the servlet

The applet can send a POST request to the servlet by opening a servlet URL and then writing the stream of data to the servlet. For example,

```
// servlet URL
URL url = new URL(getCodeBase(), "/servlet/MyServlet");

// open a connection to the servlet
URLConnection servletConnection = url.openConnection();

// prepare for both input and output
servletConnection.setDoInput(true);
servletConnection.setDoOutput(true);

// don't use a cached version of URL connection
servletConnection.setUseCaches (false);
servletConnection.setDefaultUseCaches (false);

// set the content type to indicate that we're sending binary data
servletConnection.setRequestProperty ("Content-Type", "application/octet-
stream");

// get input and output streams on servlet
InputStream in = servletConnection.getInputStream();
OutputStream out = servletConnection.getOutputStream();

// send your data to the servlet
...

// read the response from the servlet
...
```

(3) Object Serialization - PROS: Easy to pass complex data to or from a servlet, CONS: only works if the browser running your applet supports JDK 1.1 or later

The applet can use Java serialization to send and/or receive class objects to/from the servlet. The classes you are passing must implement the java.io.Serializable interface.

The applet invokes the servlet directly and passes the serializable class object in an object stream:

```
// servlet URL
URL url = new URL(getCodeBase(), "/servlet/MyServlet");
```

```
// open a connection to the servlet
URLConnection servletConnection = url.openConnection();

// prepare for both input and output
servletConnection.setDoInput(true);
servletConnection.setDoOutput(true);

// don't use a cached version of URL connection
servletConnection.setUseCaches (false);
servletConnection.setDefaultUseCaches (false);

// set the content type to indicate that we're sending binary data
servletConnection.setRequestProperty ("Content-Type", "application/octet-
stream");

// get input and output streams on servlet
InputStream in = servletConnection.getInputStream();
OutputStream out = servletConnection.getOutputStream() ;

// data to pass to the servlet
DataIn data_in = new DataIn();

// create an output stream
ObjectOutputStream oos = new ObjectOutputStream(out);

// write the serialized data object to the output stream
oos.WriteObject(data_in);
oos.flush();
oos.close();

// create an input stream
ObjectInputStream ois = new ObjectInputStream(in);

// now read the object returned from the servlet
DataOut data_out = (DataOut) ois.ReadObject().
```

(4) Raw Sockets - PROS: Bidirectional communication, CONS: Most firewalls do not allow raw socket connections

(5) RMI - PROS: Object-oriented, works with firewalls, CONS: Most browsers do not support RMI

For example, 1Echo program

EchoApplet.java

```
import java.applet.Applet;
import java.awt.*;
import java.awt.event.*;
import java.io.*;
import java.net.*;

/**
 * Simple demonstration for an Applet <-> Servlet communication.
```

```
*/
public class EchoApplet extends Applet {
        private TextField inputField = new TextField();
        private TextField outputField = new TextField();
        private TextArea exceptionArea = new TextArea();

        /**
         * Setup the GUI.
         */
        public void init() {
                // set new layout
                setLayout(new GridBagLayout());

                // add title
                Label title = new Label("Echo Applet", Label.CENTER);
                title.setFont(new Font("SansSerif", Font.BOLD, 14));
                GridBagConstraints c = new GridBagConstraints();
                c.gridwidth = GridBagConstraints.REMAINDER;
                c.weightx = 1.0;
                c.fill = GridBagConstraints.HORIZONTAL;
                c.insets = new Insets(5, 5, 5, 5);
                add(title, c);

                // add input label, field and send button
                c = new GridBagConstraints();
                c.anchor = GridBagConstraints.EAST;
                add(new Label("Input:", Label.RIGHT), c);
                c = new GridBagConstraints();
                c.fill = GridBagConstraints.HORIZONTAL;
                c.weightx = 1.0;
                add(inputField, c);
                Button sendButton = new Button("Send");
                c = new GridBagConstraints();
                c.gridwidth = GridBagConstraints.REMAINDER;
                add(sendButton, c);
                sendButton.addActionListener(new ActionListener() {
                        public void actionPerformed(ActionEvent e) {
                                onSendData();
                        }
                });

                // add output label and non-editable field
                c = new GridBagConstraints();
                c.anchor = GridBagConstraints.EAST;
                add(new Label("Output:", Label.RIGHT), c);
                c = new GridBagConstraints();
                c.gridwidth = GridBagConstraints.REMAINDER;
                c.fill = GridBagConstraints.HORIZONTAL;
                c.weightx = 1.0;
```

```
        add(outputField, c);
        outputField.setEditable(false);

        // add exception label and non-editable textarea
        c = new GridBagConstraints();
        c.anchor = GridBagConstraints.EAST;
        add(new Label("Exception:", Label.RIGHT), c);
        c = new GridBagConstraints();
        c.gridwidth = GridBagConstraints.REMAINDER;
        c.weighty = 1;
        c.fill = GridBagConstraints.BOTH;
        add(exceptionArea, c);
        exceptionArea.setEditable(false);
}

/**
 * Get a connection to the servlet.
 */
private URLConnection getServletConnection()
        throws MalformedURLException, IOException {

        URL urlServlet = new URL(getCodeBase(), "echo");
        URLConnection con = urlServlet.openConnection();

        con.setDoInput(true);
        con.setDoOutput(true);
        con.setUseCaches(false);
        con.setRequestProperty(
                "Content-Type","application/x-java-serialized-object");

        return con;
}

/**
* Send the inputField data to the servlet and show the result in the outputField.
*/
private void onSendData() {
        try {
                // get input data for sending
                String input = inputField.getText();
                // send data to the servlet
                URLConnection con = getServletConnection();
                OutputStream outstream = con.getOutputStream();
                ObjectOutputStream oos = new ObjectOutputStream(outstream);
                oos.writeObject(input);
                oos.flush();
                oos.close();
```

```
            // receive result from servlet
            InputStream instr = con.getInputStream();
            ObjectInputStream inputFromServlet = new ObjectInputStream(instr);
            String result = (String) inputFromServlet.readObject();
            inputFromServlet.close();
            instr.close();
            // show result
            outputField.setText(result);
        } catch (Exception ex) {
            ex.printStackTrace();
            exceptionArea.setText(ex.toString());
        }
    }
}
```

EchoServlet.java

```
import java.io.*;

import javax.servlet.ServletException;
import javax.servlet.http.*;

/**
 * Simple demonstration for an Applet <-> Servlet communication.
 */
public class EchoServlet extends HttpServlet {
    /**
     * Get a String-object from the applet and send it back.
     */
    public void doPost(
            HttpServletRequest request,
            HttpServletResponse response)
            throws ServletException, IOException {
            try {
                    response.setContentType("application/x-java-serialized-object");

                    // read a String-object from applet
                    // instead of a String-object, you can transmit any object, which
                    // is known to the servlet and to the applet
                    InputStream in = request.getInputStream();
                    ObjectInputStream inputFromApplet = new ObjectInputStream(in);
                    String echo = (String) inputFromApplet.readObject();

                    // echo it to the applet
                    OutputStream outstr = response.getOutputStream();
                    ObjectOutputStream oos = new ObjectOutputStream(outstr);
                    oos.writeObject(echo);
                    oos.flush();
```

```
                oos.close();

            } catch (Exception e) {
                e.printStackTrace();
            }
    }

}
```

Web.xml in tomcat's page

```
<web-app>
<!— General description of your web application —>
<display-name>Echo Servlet</display-name>
<description>Echo Servlet</description>
<!— define servlets and mapping —>
<servlet>
<servlet-name>echo</servlet-name>
<servlet-class>EchoServlet</servlet-class>
</servlet>
<servlet-mapping>
<servlet-name>echo</servlet-name>
<url-pattern>/echo</url-pattern>
</servlet-mapping>
</web-app>
```

Example

Daytimeservlet.java

```
// a servlet that returns the current time to an applet
import java.io.*;
import java.util.*;
import javax.servlet.*;
import javax.servlet.http.*;

public class daytimeservlet extends HttpServlet
{
    public void doGet(HttpServletRequest req, HttpServletResponse res) throws
ServletException, IOException
    {

        res.setContentType("text/plain");
        PrintWriter out = res.getWriter();
        out.println(new java.util.Date().toString());
    }

    public void doPost(HttpServletRequest req, HttpServletResponse res) throws
ServletException, IOException
    {
        doGet(req,res);
    }
}
```

```
/*
 * daytimeapplet.java
 /

import javax.swing.*;
import java.applet.Applet;
import java.awt.*;
import java.awt.event.*;
import java.io.*;
import java.util.*;
import java.net.*;

public class daytimeapplet extends java.applet. Applet implements ActionListener
{

    TextField dt = new TextField();
    Label dl = new Label("The Daytime : ",Label.RIGHT);
    Button db = new Button("Get the Time");

    /** Creates new daytimeapplet */

    public void init()
    {
       setLayout(new BorderLayout());
       add("West",dl);
       add("Center",dt);
       db.addActionListener(this);
       add("South",db);
       //pack();
       setVisible(true);
    }

     public void actionPerformed(java.awt.event.ActionEvent p1) {
        dt.setText(getdate());
     }

     private String getdate()
     {
        String ts="";
        String argstring="",urlString="";

        try{

                          urlString = getParameter("servletURL");
           //Reading Servlet Data through URL
           URL url = new URL(urlString);
```

```
        URLConnection con = url.openConnection();
        con.setUseCaches(false);
        InputStream in = con.getInputStream();

        //Converting the byte data into String and Displaying in the Applet
        BufferedReader result = new BufferedReader(new InputStreamReader(in));
        ts = result.readLine();
        in.close();
        return ts;

      }catch(Exception e)
      {
       System.out.println("Error\n"+e);
      }
      return "Security Imposed. Unable to read the time from the server";
    }

  }
```

Daytimeapplet.html

```
<HTML>
<HEAD>
  <TITLE>Applet Servlet Communication Page</TITLE>
</HEAD>
<BODY>

<H3><HR WIDTH="100%">Applet Servlet Communication Page<HR
WIDTH="100%"></H3>

<APPLET codebase="http://gowri:8080/test" code="daytimeapplet.class"
width=350 height=200>
<param name="servletURL" value="http://gowri:8080/test/servlet/
daytimeservlet">

</APPLET>
</BODY>
</HTML>
```

7. SERVLET TO APPLET COMMUNICATION

We are providing you the code that will display the message sent from the servlet to applet.

(1) Here is the code of 'ServletExample.java'

```
import java.io.*;
import javax.servlet.*;
import javax.servlet.http.*;

public class ServletExample extends HttpServlet{
public void service(HttpServletRequest request,
```

```
HttpServletResponse response) throws ServletException , IOException{

ObjectOutputStream output = null;
try{
output = new ObjectOutputStream(response.getOutputStream());
String str = new String("Hello World");
output.writeObject(str);
output.flush();
output.close();
System.out.println("Message is===== " + str);
}
catch( Exception e){
e. printStackTrace();
}
}
}
```

(2) Now call the servlet with the 'AppletCallingServlet.java'

```
import java.io.*;
import java.awt.*;
import java.net.*;
import java.applet.*;

public class AppletCallingServlet extends Applet{
URL url = null;
URLConnection servletConnection = null;
public void init()
{
try{
url = new URL("http://localhost:8080/examples/ServletExample";;);
servletConnection = url.openConnection();
servletConnection.setDoInput(true);
servletConnection.setDoOutput(true);
servletConnection.setUseCaches(false);
servletConnection.setDefaultUseCaches(false);
servletConnection.setRequestProperty("Content-Type","application/octet-stream");
}
catch(Exception e){
e.printStackTrace();
}
}
public void paint(Graphics g)
{
try{
ObjectInputStream input = new
ObjectInputStream(servletConnection.getInputStream());
```

```
g.drawString("Applet Servlet Communication",50,50);
String str = new String();
str = (String)input.readObject();
g.drawString(" Message sent from server: " + str,50,100);

input.close();
}
catch( Exception e)
{
e.printStackTrace();
}
}
}
```

(3) Call this applet with the html file.

```
<html>
<body>
<h1>Java Applet Demo</h1>
<applet code=AppletCallingServlet.class width=500 height=500>
</applet>
</body>
</html>
```

8. JAVA MEDIA FRAMEWORK

JMF is a framework for handling streaming media in Java programs. JMF is an optional package of Java 2 standard platform. JMF provides a unified architecture and messaging protocol for managing the acquisition, processing and delivery of time-based media. JMF enables Java programs to

(i) Present (playback) multimedia contents,
(ii) capture audio through microphone and video through camera,
(iii) do real-time streaming of media over the Internet,
(iv) process media (such as changing media format, adding special effects),
(v) store media into a file.

Features of JMF

JMF supports many popular media formats such as JPEG, MPEG-1, MPEG-2, QuickTime, AVI, WAV, MP3, GSM, G723, H263, and MIDI. JMF supports popular media access protocols such as file, HTTP, HTTPS, FTP, RTP, and RTSP.

JMF uses a well-defined event reporting mechanism that follows the "Observer" design pattern. JMF uses the "Factory" design pattern that simplifies the creation of JMF objects. The JMF support the reception and transmission of media streams using Real-time Transport Protocol (RTP) and JMF supports management of RTP sessions.

JMF scales across different media data types, protocols and delivery mechanisms. JMF provides a plug-in architecture that allows JMF to be customized and extended. Technology providers can extend JMF to support additional media formats. High performance custom implementation of media players, or codecs possibly using hardware accelerators can be defined and integrated with the JMF.

Criticisms on JMF

Multimedia processing and presentation is compute-intensive. Therefore most of the existing media players and processors for desktop computers are implemented using native code for performing computationally intensive tasks like media encoding, decoding, and rendering.

The general criticism on Java-based applications and therefore on JMF is that they lack performance as compared to native codes. The answer to this criticism is as follows.

Why JMF?

The main drawback of native implementations of media players is that they are platform dependent. Hence they are not portable across platforms. This directly means applications using platform-dependent media players and processors are unsuitable for web-deployment. JMF provides a platform-neutral framework for handling multimedia.

The JMF API provides an abstraction that hides these implementation details from the developer. For example, a particular JMF Player implementation might choose to leverage an operating system's capabilities by using native methods. Indeed Sun's implementation of JMF has different versions each one tailored for one platform.

The JMF Model

JMF adopts the same model that is used by the consumer electronics industry in handling the media. According to the JMF model, the life cycle of the media starts from a media source, and ends in a media sink. In between the media is handled by media handlers.

The media source can be a (i) a capture device, or (ii) a media file stored locally or remotely on the network or (iii) a real-time media stream available on the network. The media handlers process the media which may involve demultiplexing or multiplexing or encoding or decoding. The media processing can be implemented partly in hardware but mostly it is done by software. The media sink or destination can be rendering devices, or storage files or media streams.

For audio the capture device is a microphone along with a sound card. For images and video, the capture device is a PC add-on digital camera. We typically use CRT monitor for rendering images or video and speakers for rendering audio.

The JMF Time Model

Media is any form of data that changes meaningfully with respect to time. Therefore programs that handle multimedia contents should have a sense for time and its progression. To meet this requirement JMF defines the class Time, and interfaces Timebase, Clock and Duration.

The time is represented in JMF using the class Time. A Time object represents a particular instant of time in the time axis. The interface TimeBase is an uncontrolled source for time. The interface clock represents a controlled source of time. A clock can be stopped, started and its rate can be adjusted. The interface Duration represents time duration.

Representing Media

All multimedia contents are invariably stored in a compressed form using one of the various standard formats. Each format basically defines the method used to encode the media. Therefore, we need a class to define the format of the multimedia contents we are handling.

To this end JMF defines the class Format that specifies the common attributes of the media Format. The class Format is further specialized into the classes AudioFormat and VideoFormat.

Specifying the Source of Media

The next most important support an API should offer is the ability to specify the media data source, and to control the access of data from it. We all know that Java provides us with the class java.net.URL to identify any resource in the Internet. Using an URL object we can indeed specify the media source.

JMF provides another class called MediaLocator to locate a media source. You can construct a MediaLocator using a "Locator string" which identifies the source or using an URL.

The source of the media can be of varying nature. The JMF class "DataSource" abstracts a source of media and offers a simple connect-protocol to access the media data.

Specifying the Media Destination

A DataSink abstracts the location of the media destination and provides a simple protocol for rendering media into destination. A DataSink can read the media from a DataSource and render the media to a file or a stream.

Media Playback

JMF supports playback of media by defining the Player interface. The Player interface extends the interfaces "Controller" and MediaHandler.

Player Lifecycle

The Player (Controller) defines various states that represent different stages that are involved in the construction of a media player. When a Player is first constructed, it has no knowledge of the media it has to handle. Therefore, a Controller cannot identify the resources it needs to handle the media. This state of the Player is called the Unrealized state. The Controller then enters into the Realizing state.

In the Realizing state, the Player identifies and locates all the resources it needs to handle the media. Once the process of realization is over, the Player enters the Realized state. Next the Player gets into the Prefetching state wherein it prefetches the media needed for the presentation. A Player that has prefetched the media is in the Prefetched state. After that the Player can be started using the start() method. A started Player can be stopped and its resources can be deallocated.

The Controller posts a number of ControllerEvents to advertise its state transitions and the error conditions. To listen to these events you can define a class implementing the ControllerListener interface. Then instantiate an object belonging to that class implementing the ControllerListener. The instantiated object should be registered with the Controller whose events we want to listen.

The interface Player defines additional methods that are required for the playback of media. Player provides a GUI for the user to control the media playback. Player may also provide a visual component associated with the media presentation. Player provides a way to manage a group of Controllers. Managing multiple Controllers is important in the case of synchronized media presentation.

Media Processing

Typical stages that are involved in the media processing are: (i) Demultiplexing, (ii) transcoding, (iii) multiplexing, and (iv) rendering. In the demultiplexing stage different streams (say audio and video) of the media are demultiplexed or extracted from a composite stream. Transcoding involves a change in the encoding format of the media. During multiplexing different streams of the media are multiplexed into a single stream. Rendering involves the presentation of the media.

PlugIn

JMF defines the interface PlugIn to represent a media-processing stage. PlugIns are of following types. They are: (i) Multiplexer, (ii) Demultiplexer, (iii) Codec, (iv) Effect and (v) Renderer.

Processor

The interface Processor abstracts a media processing class. It extends Player and allows programmatic control over the media processing stages. Processor introduces a state called Configuring wherein you can set the Codec and renderer PlugIns. The output of a Processor can be taken and can be handed over to (i) another Processor for further processing or (ii) to a Player for playback or (iii) to a DataSink for media transmission or storage.

Manager

The class Manager acts as a factory for creating the JMF objects such as DataSource, DataSink, Player, Processor, cloneableDataSource, and MergedDataSource.

Media Capture

Capture devices include cameras and sound cards. All capture devices should be registered with a registry called CaptureDeviceManager in order to use them with the JMF. You can query the CaptureDeviceManager for a list of all capture devices that support a given output format.

JMF Controls

JMF API has many interfaces that allows us to control the attributes of JMF objects such as Players, Processors, DataSources, etc. For example, the GainControl object can be used to control the audio volume of a Player.

JMF RTP API

JMF supports a protocol called Real-time Transport Protocol (RTP) for media transmission and reception [rtp]. RTP is a transport layer protocol and is typically used above the UDP layer. RTP packets has (i) time stamping to indicate the time instant at which the media carried by the packet has to be played, (ii) sequence number that can be used for the ordered delivery of the packets, (iii) identification of the media source and (iv) payload media format identification.

A RTP session consists of a set of applications exchanging media using the RTP. Each of these applications is called a participant. Every participant uses an object called RTPManager to co-ordinate the RTP session on its behalf. Media streams exchanged in a RTP session are called RTPStreams. The RTPStreams can be of two types, sendStream and ReceiveStream.

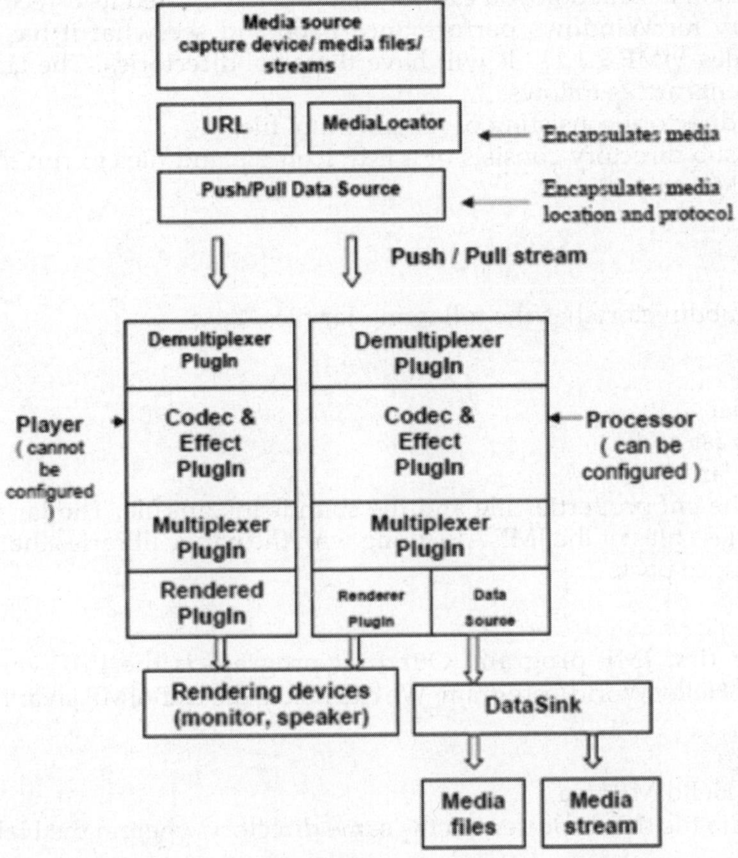

JMF Architecture

The JMF RTP API allows us to construct RTPManagers using which we can send and receive RTPStreams among the participating JMF applications.

Installing JMF

The following are the different versions of JMF 2.1.1 that are currently available[jmf]

JMF 2.1.1 FCS Cross-Platform

This is an implementation of JMF with only Java byte codes and without any native codes.

(2) JMF 2.1.1 FCS with the Windows Performance Pack which is optimized for Window's platform.

JMF 2.1.1 FCS with the Solaris Sparc Performance pack

JMF 2.1.1 FCS for Linux from Blackdown

Decide the version of the JMF you wish to install. The version depends primarily on whether you want to work on Windows, or Solaris or Linux platform, etc. You can download the corresponding version of JMF from the following JMF website and install it.

http://java.sun.com/products/java-media/jmf/index.jsp

After successful installation you can see that JMF has created its directory. The JMF home directory for Windows performance pack and see what it has is typically c:\Program Files \JMF 2.1.1\. It will have three subdirectories. The subdirectories and their contents are as follows.

(1) doc sub directory consisting of readme.html file.

(2) The bin sub directory consists of a JMF Icon file and files to run the following applications. JMF customizer

JMFinit

JMFRegistry

JMStudio.

(3) The lib subdirectory has the following Jar files,

jmf.jar

sound.jar

customizer.jar

mediaplayer.jar and

multiplayer.jar.

It also has the jmf.properties file and the soundbank.gm file. The Jar files contain the compiled class files of the JMF API along with the native libraries that are specific to the performance pack.

HelloJMF.java

Let us try our first JMF program. Our first program is the JMF version of the indispensable "Hello, World" program. We call this as the HelloJMF.java. The program is given below.

```
/* Program HelloJMF.java
Plays an audio file that is present in the same directory wherein the HelloJMF.class file resides.
*/
import javax.swing.*;
import java.awt.*;
import javax.media.*;
import java.awt.event.*;
import java.net.*;
public class HelloJMF {
JFrame frame = new JFrame(" Hello JMF Player");
static Player helloJMFPlayer = null;
public HelloJMF(){
try{ // method using URL
URL url=new URL("file", null,"hello.wav");
helloJMFPlayer = Manager.createRealizedPlayer(url);
} catch(Exception e) {
System.out.println(" Unable to create the audioPlayer :" + e );
}
Component control = helloJMFPlayer.getControlPanelComponent();
frame.getContentPane().add(control, BorderLayout.CENTER);
frame.addWindowListener(new WindowAdapter() {
public void windowClosing(WindowEvent we) {
HelloJMF.stop();
```

```
System.exit(0);
}
});
frame.pack();
frame.setSize(new Dimension(200,50));
frame.setVisible(true);
helloJMFPlayer.start();
}
public static void stop(){
helloJMFPlayer.stop();
helloJMFPlayer.close();
}
public static void main( String args[]){
HelloJMF helloJMF = new HelloJMF();
}
}
```

Here is a brief explanation of the HelloJMF.java. The constructor of HelloJMF.java constructs an URL (Uniform Resource Locator) object representing the file called Hello.wav. The file Hello.wav is expected to be present in the same directory as that of HelloJMF.class. We use the following constructor of the java.net.URL class to construct the URL object.

```
public URL(String protocol, // the protocol to use
String host, // the host name to connect to
String file) // the specified file name on that host
throws MalformedURLException
```

The protocol used to access the hello.wav resource is the "file" protocol. The host name can be null since for the case of file protocol, the host is the local host. With this URL we then construct a media Player object using the utility class Manager.

A Simple Audio Player

In this section, we will walk through the first exercise of creating a simple audio player. This example introduces you to the Manager class and the Player interface, which are two of the major pieces in building most any JMF-based application. The functional goal of this example is to play a local audio file through a command-line interface. We will walk through the source code and review what is happening in each line. After completing this section, you will have a demo application on which you can play any audio file type supported by JMF, including MP3, WAV, and AU, among many others.

Refer to the SimpleAudioPlayer.java file in the source code distribution to follow this exercise.

Importing the necessary classes

The first few lines of the SimpleAudioPlayer class include the following calls, which import all necessary classes:

```
import javax.media.*;
import java.io.File;
import java.io.IOException;
import java.net.URL;
import java.net.MalformedURLException;
```

The javax.media package is one of the many packages defined by JMF.

javax.media is the core package, containing the definitions of the Manager class and the Player interface, among others. We will focus on the Manager class and the Player interface in this section, and deal with some of the other javax.media classes in later sections. In addition to the import javax.media statement, the above code fragment includes several import statements that create the input to our media player.

The Player Interface

In the next code fragment, the public class SimpleAudioPlayer and the Player instance variable are defined:

```
public class SimpleAudioPlayer {
private Player audioPlayer = null;
```

The term Player may sound quite familiar, because it is based on our common use of audio- and video-based media players. In fact, instances of this interface act much like their real-life counterparts. Players expose methods that relate to the functions of a physical media player such as a stereo system or VCR. For example, a JMF media Player has the ability to start and stop a stream of media. We will use the start and stop functionality of the Player throughout this section.

Creating a Player Over a File

JMF makes it quite simple to obtain a Player instance for a given media file. The Manager class acts as a factory for creating many of the specific interface types exposed in JMF, including the Player interface. Therefore, the Manager class is responsible for creating our Player instance, as shown below:

```
public SimpleAudioPlayer(URL url) throws IOException,
NoPlayerException,
CannotRealizeException {
audioPlayer = Manager.createRealizedPlayer(url);
}
public SimpleAudioPlayer(File file) throws IOException,
NoPlayerException,
CannotRealizeException {
this(file.toURL());
}
```

If you are following along with the source for this section, you may have noticed that the Manager class contains other methods for creating Player instances. We will explore some of these methods — such as passing in instances of a DataSource or MediaLocator—in a later section.

Player States

JMF defines a number of different states that a Player instance may be in. These states are as follows:

- Prefetched
- Prefetching

- Realized
- Realizing
- Started
- Unrealized

Working with states

Because working with media is often quite resource intensive, many of the methods exposed by JMF objects are non-blocking and allow for asynchronous notification of state changes through a series of event listeners. For example, a Player must go through both the *Prefetched* and *Realized* states before it may be started. Because these state changes can take some time to complete, a JMF media application can assign one thread to the initial creation of a Player instance, then move on to other operations. When the Player is ready, it will notify the application of its state changes.

In a simple application such as ours, this type of versatility is not so important. For this reason, the Manager class also exposes utility methods for creating *Realized* players. Calling a createRealizedPlayer() method causes the calling thread to block until the player reaches the *Realized* state. To call a non-blocking player-creation method, we use one of the createPlayer() methods on the Manager class. The following line of code creates a *Realized* player, which we need in our example application: audioPlayer = Manager.createRealizedPlayer(url);

Starting and Stopping the Player

Setting up a Player instance to be started or stopped is as simple as calling the easily recognized methods on the Player, as shown here:

```
public void play() {
audioPlayer.start();
}
public void stop() {
audioPlayer.stop();
audioPlayer.close();
}
```

Calling the play() method on the SimpleAudioPlayer class simply delegates the call to the start() method on the Player instance. After calling this method, you should hear the audio file played through the local speakers. Likewise, the stop() method delegates to the player to both stop and close the Player instance. Closing the Player instance frees any resources that were used for reading or playing the media file. Because this is a simple example, closing the Player is an acceptable way of ending a session. In a real application, however, you should carefully consider whether you want to get rid of the Player before you close it. Once you have closed the player, you will have to create a new Player instance (and wait for it to go through all of its state changes) before you can play your media again.

Creating a SimpleAudioPlayer

Finally, this media player application contains a main() method, which lets it be invoked from the command line by passing in the file name. In the main() method, we make the following call, which creates the SimpleAudioPlayer:

```
File audioFile = new File(args[0]);
SimpleAudioPlayer player = new SimpleAudioPlayer(audioFile);
```

The only other thing we have to do before we can play our audio file is to call the play() method on the created audio player, as shown here:

```
player.play();
```

To stop and clean up the audio player, we make the following call, also found in the main() method:

```
player.stop();
```

Compiling and running the SimpleAudioPlayer

Compile the example application by typing javac SimpleAudioPlayer.java at a command prompt. This creates the SimpleAudioPlayer.class file in the working directory. Then run the example application by typing the following at a command prompt:

java SimpleAudioPlayer *audioFile*

Replace *audioFile* with the file name of an audio file on your local system. Any relative file names will be resolved relative to the current working directory. You should see some messages indicating the file that is being played. To stop playing, press the Enter key. If compilation failed, check to make sure that the JMF jar files are included in the current CLASSPATH environment variable.

JMF User Interface Components

Playing video

In the previous section, we walked through the steps of setting up an application that lets you play audio files through a command-line interface. One of the great features of JMF is that you do not need to know anything about the media file types in order to configure a media player; everything is handled internally. For instance, in our previous example, we did not need to tell the application to create a Player specifically for an MP3 file, since the MP3 setup was handled for us. As you will see in this section, the same holds true for handling video files. JMF handles all of the details of interfacing with the media file types.

The main difference in dealing with video media rather than audio is that we must create a visual representation of a screen to be able to display the video. Luckily, JMF handles many of these details for us. We will create a Player instance much like we did in the previous example, and obtain many of the visual components to create our visual media viewer directly from JMF objects.

Getting the GUI Components

The Player interface exposes methods to obtain references to selected visual components. In the MediaPlayerFrame we use the following components:

- player.getVisualComponent() is the visual component responsible for displaying any video media.
- player.getControlPanelComponent() is a visual component for handling time-based operations (that is, start, stop, rewind) as well as containing some useful information on the media stream.
- player.getGainControl().getControlComponent() is a visual component for handling volume (gain) operations. The getGainControl() method returns a GainControl instance, which may be used to change gain levels programmatically.

Working With Visual Components

All of the above interface methods return an instance of the java.awt.Component class. Each instance is a visual component that may be added to our frame. These components are tied directly to the Player, so any manipulation of visual elements on these components will cause a corresponding change to the media displayed by the Player. It is important that we ensure each of these components is not null before we add it to our frame. Because not every type of media player contains every type of visual component, we should only add components that are relevant for the type of player we have. For instance, an audio player generally does not have a visual component, so getVisualComponent() returns null. You would not want to add a visual component to the audio player frame.

Obtaining Media-Specific Controls

A Player instance may also expose other controls through its getControl() and getControls() methods—getControls() returns a collection of Control objects, whereas the getControl() method looks for a specific Control. Different types of players may choose to expose controls for operations specific to a given media type or to the transport mechanism used to obtain that media. If you were to write an application that handled only certain media types, you could count on certain Control objects being available through the Player instance. Because our player is very abstract and designed to work with many different media types, we simply expose all the Control objects to the user. If we find any extra controls, we can use the getControlComponent() method to add their corresponding visual component to a tabbed pane. This way, the user will be able to view any of these components through the player. The following code fragment exposes all the control objects to the user:

```
Control[] controls = player.getControls();
for (int i = 0; i < controls.length; i++) {
if (controls[i].getControlComponent() != null) {
tabPane.add(controls[i].getControlComponent());
}
}
```

For a real application to do something useful with a Control instance (besides being able to display its visual component), the application would need to know the specific type of the Control and cast it to that type. After that point, the application could use the control to manipulate the media programmatically. For example, if you knew the media you were working with always exposed a Control of type javax.media.control.QualityControl, you could cast to the QualityControl interface and then change any quality settings by calling any of the methods on the QualityControl interface. Using a MediaLocator the last big difference between our new GUI-based media player and our first simple player is that we will use a MediaLocator object rather than a URL to create the Player instance, as shown below:

```
public void setMediaLocator(MediaLocator locator) throws IOException,
NoPlayerException, CannotRealizeException {
setPlayer(Manager.createRealizedPlayer(locator));
}
```

We will discuss the reason for this change in a later section. For now, think of a MediaLocator object as being very similar to a URL, in that both describe a resource location on a network. In fact, you may create a MediaLocator from a URL, and you may get a URL from a MediaLocator. Our new media player creates a MediaLocator instance from a URL and uses that to create a Player over the file.

Compiling and Running the MediaPlayerFrame

Compile the example application by typing javac MediaPlayerFrame.java at a command prompt. This creates a file named MediaPlayerFrame.class in the working directory. To run the example application, type the following at a command prompt:

java MediaPlayerFrame *mediaFile*

You should replace *mediaFile* with the file name of a media file on your local system (either audio or video will do). Any relative file names will be resolved relative to the current working directory. You should see a window that displays GUI controls for manipulating your media file. For a list of audio and video file formats acceptable for use in JMF, see Resources. If the initial compilation failed, check to make sure that the JMF jar files are included in the current CLASSPATH environment variable.

MediaPlayerFrame in Action

Earlier in this section you saw a screenshot of a video player playing an MPEG video file. The control panel of the media player is attached to the application frame. The application frame is made visible and the player is started. When the user closes the application frame, the anonymous WindowAdaptor Object uses the WindowClosing event to stop the media player and exit.

Type this program in a file called HelloJMF.java or you may cut and copy this program listing into a file. We hope you are familiar with compiling and executing Java programs. If you use any IDE for program development you should consult its manual for how to edit, compile, execute and debug programs. In case you use the JDK for program development use the following command

javac HelloJMF.java
to compile HelloJMF.java

In order to test this program create a wav file named hello.wav and store it in same directory where the helloJMF.class is present.

To execute the program use the command java HelloJMF from the directory which holds both the HelloJMF.class and the Hello.wav file. On execution of HelloJMF.class a window entitled "Hello JMF Player" is created and you should hear the hello.wav. If you encounter any problem, it may be because JMF is not installed properly. Check whether you can invoke and use JMStudio to play Hello.wav.

9 Application in Distributed Environment

INTRODUCTION TO JAVA RMI

Remote Method Invocation (RMI) is a part of Java Development Kit. It allows us to develop distributed applications. Distributed systems require computations that are running in different address spaces, practically on different machines, must be also to communicate between one machine to another. Java RMI facilitates such a communication specifically for java applications. The RMI is platform independent because Java is platform independent. The RMI can communicate only from one Java Virtual Machine to another. So, the user need not learn any other language to develop the distributed application by using RMI. Using RMI we can write reliable distributed applications that are as simple as possible.

In RMI the application is divided into objects. The objects communicate with each other through an interface. This interface is used to access the remote object and its methods. RMI passes objects by their true type, as a result the behavior of those objects is not changed when they are sent to another virtual machine. To develop the distributed application using RMI, we have to follow the steps given below.

- Define the Interfaces
- Implementing these interfaces
- Compile the interfaces and their implementations with the Java compiler
- Compile the server implementation with RMI compiler
- Run the application.

There are many ways of developing the distributed applications. We can use Sockets, Remote Procedure Call (RPC), Microsoft's Distributed Component Object Model (DCOM), Common Object Request Broker Architecture (CORBA) and RMI. Each technology has its own features. First we will look into the need for using a distributed application.

1. NEED FOR DISTRIBUTED APPLICATION

The traditional applications are monolithic in nature. If we need to make any extension or enhancement in this application, we have to recompile and integrate the whole application after making the change since each method or function depends on another. Hence, we use distributed application in which the application is divided into parts called components or objects. Each object (or component) may have the interface to refer components. We can make changes to a particular object without affecting other objects. Also, it is sufficient if we compile the object that has been changed.

In a distributed application we can add or remove pre-defined components from other distributed applications. This will not affect the structure of any of these applications. For example, let us consider two objects: One for performing addition and another for performing subtraction.

If we develop an application to perform both addition and subtraction, then we can use both the objects in our application. If an application supports only one of these operations, then we can use only that particular object.

When we use the network to develop distributed applications, we can keep our objects separately. The Internet Protocol (IP) address and interface uniquely determine the object. The interface is registered in the local registry. The objects obtain the remote object's interface from the registry. The two objects communicate through the Object Request Broker (ORB). When we develop the distributed applications using RMI, then the connections will be provided by RMI system. The ORB is not needed when we use the RMI to develop application. Hence it reduces the cost. The communication between 2 objects that are placed in different machines is given in Figure 9.1.

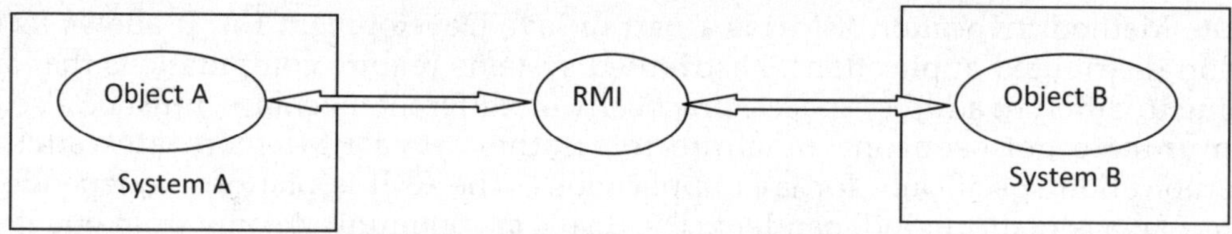

Figure 9.1 Communication between two objects

In Figure 9.1 **object B** wants to communicate with **object A.** Hence it has to obtain the handle (full name) of the object A. If the object A is registered in the RMI registry or any ORB registry as **objectA,** then the handle of object A is 80.0.0.22/objectA. Object B can communicate with the object B through this handle.

The most common distributed applications used nowadays consist of two parts: the server part and the client part. The server is placed in a machine that is connected to a network. The client can be placed in a different machine that is also connected to the same network. The client can invoke the server's objects or methods whenever needed. Using middleware-technologies like the RMI facilitate this communication. Usage of distributed application avoids the replication of objects.

Most of the distributed objects are in executable form, i.e. in binary form. Once the distributed application is developed we can use it at anytime and anywhere.

Advantages of distributed applications are:

- Component reuse—Once the object is created for a certain application, it can be used by other developers for their applications also.
- Less time and resources—We can use the existing components for our application and we can save our time and resources.
- Less complexity—The application is divided into objects or components, hence the complexity of the entire project is reduced.
- Less storage space—Since multiple applications use the same set of objects, replication of the objects is avoided. Hence, the storage requirement is less.

2. OVERVIEW OF RMI

RMI applications are developed as two separate programs: A server and a client. Server application creates a number of remote objects, makes references to those remote objects. Client application gets a remote reference to one or more remote objects in the server and then invokes methods on them. RMI provides the mechanism by which the server and the client communicate and pass information back and forth.

RMI distributed application uses the registry to obtain references to a remote object. The server creates the objects, registers it in the local registry by the object's name. The client looks up the remote object by its name in the server's registry and then invokes a method of the server object. The RMI system also uses the existing web server to load Java class bytecodes (objects) from server to client and from client to server, when needed. RMI can load class bytecodes using any Uniform Resource Locator (URL) protocol. These are HyperText Transfer Protocol (HTTP), File transfer Protocol (FTP) etc., which are supported by java system.

RMI architecture consists of four layers and each layer can perform specific functions:

- Application layer: It has contained the actual object definition.
- Proxy layer: It consists of two parts, namely Stub and Skeleton. These are used for marshalling and unmarshalling the data that is transferred through the network. Marshalling is the process by which we can convert the Java bytecodes into a stream of bytes, and unmarshaling is the reverse process of it. Stub is a proxy for the server. It is placed on the client side of the application, whereas the skeleton is placed on the server side.
- Remote reference layer: It gets the stream of bytes from the transport layer and sends it to the proxy layer.
- Transport layer: This layer is responsible for handling the actual machine to machine communication.

The following section deals with the other middleware technologies that act as alternatives to RMI.

2.1 Alternative to RMI

When designing and implementing the distributed application with the RMI, we can establish communication only between one JVM (java object) and another JVM (java object). We have other technologies that provide features like language neutrality. These are Sockets, Remote Procedure Call, DCOM, and CORBA.

a. Sockets

Sockets are used to establish communication between machines, sometimes used to establish communication between two processes in the same machine. Socket is a channel through which applications can connect with each other and communicate. The most straightforward way to communicate between application components is to use sockets directly, i.e. the developer can write data to the socket and read from the socket. This is called the socket programming.

The socket programming is not well suited to handle complex data types, especially when application components reside on different types of machines. It is not suited when the components are developed in different languages, because the Application Programming Interface (API) for socket programming is low level.

Advantage: The direct socket programming can result in very efficient applications

Limitation: This technique is unsuitable for developing complex applications. Using RMI we can avoid this limitation.

b. Remote Procedure Call (RPC)

RPC provides a function-oriented interface to socket level communications. Using RPC, the developer defines the function like C language function. The function actually uses sockets to communicate with remote server, which executes the function and returns the result again using sockets.

RPC makes a network connection to a remote computer using sockets. After the socket connection is made with the server computer, we can invoke a function on the remote machine. The parameters from client to the server and the results from the server to the client are passed through sockets. If we write pure socket-based client or server, we must marshal the parameters. If we write using the RPC, it marshals the parameter automatically.

When using the remote objects we have to run special registry service. This registry is a program that listens for the client requests on a designated port. The combination of hostname and port address will form a socket address. Then by using this socket address the connection is made. Once the connection is established, the client can ask the remote registry to give a reference to the remote object. If the client gets the reference to the object, then it can invoke the methods in the remote object.

Advantages:

- The function-oriented interface is easier to use than socket programming.
- RPC is also powerful for many client server applications.

c. Common Object Request Broker Architecture (CORBA)

CORBA is a middleware technology developed by a consortium called OMG. Fundamental to the Common Object Request Broker Architecture is the Object Request Broker (ORB). An ORB is a software component used to communicate between objects. The Internet Inter-ORB protocol (IIOP) enables the objects to interact across a network, using ORB.IIOP forms an object bus that links the objects and ORBs into remote systems. The ORB provides a number of capabilities; one is to locate a remote object, given an interface. Another service provided by the ORB is the marshaling of parameters and return values to and from remote method invocations.

Another fundamental piece of the CORBA architecture is the usage of Interface Definition Language (IDL). IDL specifies interfaces between CORBA objects. CORBA is language independent, because interfaces described in IDL can be mapped to any programming language. A client written in C++ can communicate with a server written in Java, which can communicate with another server written in COBOL, and so no.

IDL is not an implementation language like C++, Java, etc. Its only purpose is to define interface. The implementation of the interface is written in some other programming language.

System services and facilities are also important elements of CORBA that provide general-purpose functionality to objects travelling on CORBA's IIOP bus.

The advantages of CORBA are:

- It is much easier to use than sockets
- It enables us to access to software running on different platforms
- Language neutrality

Limitation: We have to use ORB to connect two objects. We have to pay money for buying the ORB. Using RMI we can avoid this limitation.

d. Distributed Component Object Model (DCOM)

Distributed Component Object Model is an extension of the Component Object Model (COM). There are basic types of COM services:

- Inproc server—It is implemented as Dynamic Link Libraries (DLL) file. It resides in the same address space as that of the client process.
- Outproc server—It is implemented as .exc file. It resides on the different address spaces in the same machine. COM can access both Inproc and Outproc servers, because both reside on single machine.
- Remote server—It resides on a remote machine. DCOM is used access this server.

COM is used to develop the distributed application in the same machine. Hence it can communicate between components in the same machine, and also communicate between two processes in the same machine. DCOM places COM applications on the network.

While developing an application using DCOM technology, the application is divided into components. A component is a binary object. All the components are integrated to form an application. Any change can be made to the application by replacing the existing component by a new component.

All the Components support interfaces. The Components offer their service through methods. To call a method of a component, a pointer to the interface is obtained and the method is called using the pointer.

Whenever a component is invoked, its reference count is incremented by one, and whenever the client has finished using the component, the reference counts decrements by one. If the reference count reaches zero, the component is discarded from memory and the memory is freed. A ping message is sent at regular intervals from the client to the server. If the client crashes, the server will not receive the ping message. If three pings are not received continuously, the server assumes that the client has crashed and frees all the resources associated with it.

Advantage: It is language independent; hence DCOM is a powerful technology to develop the distributed applications.

Limitation: It can communicate only between applications developed in windows environment. Using RMI we can avoid this limitation.

2.2 Defining Interfaces

Defining interfaces is the first step to writing RMI programs. In RMI programming, any class that exports objects must implement an interface that defines the methods that can be accessed via a remote application (i.e. client). The class may have methods that are not defined in this interface. Then the client cannot access these methods. It is also possible for a single class to implement many remote interfaces.

Each remote object is identified by its remote interface. The remote interface is an interface that declares a set of methods that may be invoked from a remote Java virtual machine. A client gets a handle to the interface describing the remote method of an object. This interface must extend from java.rmi. Remote. The java.rmi. Remote serves to identify all remote interfaces. All the remote objects must directly or indirectly implement this interface. Only the remote interfaces can be invoked via RMI. Local interfaces cannot be called in this manner.

Case Study

CBS Publishers is one of the growing publishers in India. The head office is placed in Chennai. It has 10 branches, placed in New Delhi. In order to calculate the monthly salary of the employees, the accounts department has to keep track of the attendance of its employees in all its branches. Hence it needs a distributed application that performs the following operations:

- The server should run continuously.
- The client has to get the name and password of the employees.
- It has to check whether the employee name and his password are valid or not.
- It has to display the entry time and exit of the employee.
- It has to store the attendance of the employees in a file.

The application is developed under the package named attendance. The server is installed in the head office and clients are installed in all its branches. The following are the names of the classes and the interface used in the application:

EmpInt is the interface

EmpServer is the server

EmpApp is the client. The client side has an applet and an HTML file. Both of them have the same primary name and differ only in the extension.

The head office keeps a list of the employee names and their corresponding passwords in a file named **EmpRegister.** This is helpful for them to identify each employee uniquely. The attendance of the employees is registered in a file named EAttendance.

To develop the application, first we have to define the remote interface as given Snippet 9.1. The interface definition is part of the package named attendance.

Snippet 9.1

1. Create a directory named examples.
2. Create a directory named attendance inside the examples directory.
3. Open Notepad application and type the following code:

```
// Remote Interface Definition

1. package  attendance;
2. import java.rmi.*;
3. public interface EmpInt extends Remote{
4. public string (String ename, String key) thrown RemoteException;
```

Note : Do not enter the line number while typing the program in the system, line numbers are given only for explanation purpose.

We can also use any other editor to type the code

```
5. }
```

First line of the code, **package attendance;** shows that the software is developed in the package **attendance.**

Second line of the code, **import java.rmi.*;** shows that the interface can extend the classes in the **java.rmi** package.

Third line of the code, public interface EmpInt extends Remote { shows the declaration of the interface named EmpInt that extends the java.rim.Remote class.

Fourth line of the code public string enter (String ename, String key) throws Remote Exception; shows that the EmpInt interface has only method, namely enter (). As a member of a remote interface, the enter () method is a remote method. Therefore, the method must be defined as being capable of throwing a java.rmi.RemoteException. This exception is thrown by the RMI system during a remote method call to indicate if either a communication failure or a protocol error occurs. It takes two arguments. Client must call only this method, which is declared in the interface remotely. In interface definition the enter () is the user defined method, it returns a String type value. We can also declare the system methods like getText (), purText (), etc. These methods should be implemented in the server implementation. The client can access only these methods remotely. (The implementation of interface is dealt with in the second chapter.)

EmpInt is the remote interface. The **enter** () method is declared within this interface. It has to be implemented by the server. The client can access the method through the proxy of the server, namely the server.

Notice that a remote interface must extend the java.rmi.Remote interface either directly or indirectly.

Implementation of Interfaces

Implementation of interfaces may be described in the following topics:

- Server implementation
- Client implementation
- Packages used in RMI

Server Implementation

Now that the interface EmpInt has been defined, Perfect Solutions Limited has decided to do the following operations in the server side.

- To check whether the employee name and his password are valid or not.
- To store the attendance of the employees in a file.

In general, the server implementation class of a remote interface should do the following:

- Declare the remote interfaces being implemented
- Define the constructor for the remote object
- Provide an implementation for each remote method in the remote interfaces

The server needs to create a remote object and register it with the local registry. This procedure can be encapsulated in **main** () method in the remote object implementation class itself, or it can be included in another class entirely. The procedure should

- Create one or more instances of REMOTE OBJECT
- Register at least one of the remote objects with the RMI registry (or another naming service) in order to be located by prospective clients.

In the case of Perfect Solutions Limited, the server declares the interface **EmpInt**, defines a constructor without any input parameters and provides the implementation for the **enter** ( ) method.

Declare the Remote Interfaces Being Implemented

The server implementation **EmpServer** is also a part of the **attendance** package as the interface EmpInt. Therefore, it includes the same package statement as the interface (Snippet 9.3).

Example

1. Open notepad application and type the following code:

```
package attendance;
import java.rmi.*;
import java.rmi.server. UnicastRemoteobject;
```

2. Save the file as EmpServer.java in the attendance directory.

Note: Do not try to compile or run the examples from Snippet 9.1 to Snippet 9.12 as these are not complete programs. These examples are given only to illustrate the concepts.

The **EmpServer** class implements the **EmpInt** interface and extend **java.rmi.server.UnicastRemoteObject**. There are two choices when designing a remote object. The first choice is to extend the **UnicastRemoteObject** and the other choice is to export the object using one of the member functions of **UnicastRemoteObject.** We have opted to use the first choice in Snippet 9.1.

Snippet 9.4

Add the following lines to the EmpServer.java file.

```
public class EmpServer extends UnicastRemoteObject implements EmpInt {
```

The server maintains a Vector ( ) of employees of the company in the variable named **employees**. This Vector stores the employee name and the entry time. There is another Vector named **Key** that contains a list of valid employee names and passwords. This Vector is read from the file **EmpRegister** when the server first starts up. (The procedure for creating the EmpRegister file will be explained in the fourth chapter). The class also defines a variable of the type RandomAccessFile, which will be later assigned to the file **EAttendance.** Maintaining the entry and exit details in the server helps the head office to have control over the whole application. These declarations are shown in Snippet 9.5.

Add the following lines to the EmpServer.java file.

```
vector employees = new vector();
vector key = null;
RandomAccessFile EAttendance = null;
```

Define the Constructor for the Remote Object

The first method in the class is the constructor **EmpServer** ( ). The constructor does not take any arguments. It throws the **RemoteException.** Each and every remote method in a remote object should throw the **RemoteException** so that the client will be informed if there is any problem in passing the parameters or in the connection between the client and the server. The constructor opens the file **EAttendance** and points to the end of the file to keep it ready for appending new records. The file **EAttendance** should be present in the **examples** directory in which the attendance package is present. If the file is not present in this directory, then the constructor creates a new one in that name. The constructor also calls the local method **readPasswords( )** to populate the Vector key with the contents of the file **EAttendance.**

Note: Any object can access a remote method of a remote object, whereas the local method of a remote object can be accessed only within the object.

Example

1. Add the following lines to the EmpServer.java file.

```
public EmpServer () throws RemoteException {
      try {
      EAttendance = new RandomAccessFile ("EAttendance","rw");
                          EAttendance.seek (EAttendance.length() );
      } catch (IOException e) {
            System.out.println("EAttendance: " + e);
      }
      System.out.println("EmpServer starting at " + new
                                               Data(). toString());
      readPasswords();
}
```

The **readPasswords** () method (Example below) has a try and two catch clauses in order to account for the case where the file EAttendance is not found or when there is an IO error while reading from the file. It adds an object of the **EmpKey** class to the **key** Vector.

Example

1. Add the following lines to the EmpServer.java file.

```
void readPasswords() {
      String instry;
      Key = new Vector ();
      BufferedReader EmpRegister = null;
      try {
            EmpRegister = new BufferedReader (new)
                          FileReader("attendance\\ EmpRegister"));
            while(true) {
            if(( instry = EmpRegister.readLine())==null){
                  EmpRegister=null;
                        return;
                  }
                  key. addElement(new
                              Empkey (instry.substring (o,instrg.indexof
                              (' ' )), instrg.substring(instrg.indexof
                              (' ')+1)));
            }
      } catch (FileNotFoundException e) {
      System.out.println("EmpRegister not found: "+e);
      } catch (IOException  e) {
      EmpRegister = null;
      }
}
```

The **EmpKey** class declares variables for the employee name and the password. Its constructor takes in the employee name and the password as arguments and sets them to an object of **EmpKey**. The **isEqual** ( ) method takes in an object **pw** of **EmpKey** type and checks whether the name and password of the employee we have entered is found in the **EmpRegister** file. This method helps the head office to maintain a list of the employee that will be used to check the validity of an entry when an employee enters or exits.

1. Open the Notepad application and type the following code:

```
package attendance;
class EmpKey {
        string ename;
        string key;
        Empkey (string ename; string key; ) {
                this.ename = ename;
                this. Key = key;
        }
        public Boolean isEqual (Empkey pw) {
                if (pw.ename.equals(this.ename)&&pw.key.equals.key))
                        return true;
                else
                        return false;
                }
        }
```

2. Save the file as Empkey.java in the attendance directory.
3. Change the current directory to the examples directory.
4. Compile the file using the following command in the MSDOS prompt;

```
javac –d.attendance.Empkey.java
```

The Empkey.Class file will be created in the attendance directory.

Provide an Implementation for Each Remote Method

The remote method **enter** ( ) takes two arguments, namely the employee name and the password that is entered in the client side. It checks if the employee information that is entered is valid by using the **isValidID** () method. If the entered employee information is invalid, then a corresponding message is returned to the client. If the information is valid, then the **enter** () method calls the **ontheClock**() method to check whether that name is in the employees Vector. If it exists, then the method calls the **removeEmployee** () method to remove the entry from the Vector and update the **EAttendance** file. If it does not exist, then the **enter** () method calls the **addEmployee** () method to add the entry to the Vector.

Example

1. Add the following lines to the Empserver.java file.

```
public string enter string empName, String key ) {
      if (isValidID(new EmpKey (empName, key))) {
              if (onTheclock (empName))
                      return removeEmployee (empName);
              else
```

```
                return addEmployee (empName);
        }
        Return "Sorry, Invalid employee name or password";
}
```

The **isValidID** () method also takes in the same arguments as the **enter** ( ) method. It checks for the validity of the employee information that is entered, by checking it against the **key** Vector that was populated using the **readPasswords** ( ) method. This helps the head office of Perfect Solutions in ensuring the validity of the data entered.

Example

1. Add the following lines to the Empserver.java file.

```
private boolean isValidID (Empkey pw) {
        Empkey k1;
        for (int I = 0; I < key.size(); i++) {
                k1 = (Empkey) key.elementAt(i);
                if (k1.isEqual (pw) )
                        return true;
        }
        Return false;
}
```

The **onTheClock** () method accepts the employee name as the input parameter and checks whether it is present in the **employees** Vector. If it is present, the method returns true, else it returns false.

Example

1. Add the following lines to the Empserver.java file.

```
Private boolean onTheClock (String empName) {
        for (int I = 0; I < employees.size(); i++)
        if (((EmpRec) employee.elementAt(i)).ename.equals(empName))
                return false;
}
```

The **addEmployee** () accepts the employee name as the input parameter. It creates a new object **emp** of **EmpRec** class and returns a String containing the employee name and his entry time. This information will be printed on the client side so that the employee can confirm the registration of his entry.

Example

1. Add the following lines to the Empserver.java file.

```
private String addEmployee (String empName) {
        EmpRec emp;
        employees. addElement(emp = new EmpRec(empName));
        return "Employee name: "+empName + "Entry time: " +
                                emp.EntryTime.toString();
}
```

The class **EmpRec** is in the package **attendance.** It declares the variables **ename, EntryTime** and **ExitTime** which correspond to the employee name, his entry time and his exit time respectively. The constructor **EmpRec** () takes in the employee name as the input parameter. It then allocates the name to the current object and sets the system time as the entry time. The code for the class EmpRec is given below. Since the entry time is taken from the system, the employee cannot modify the entry time. This prevents any misuse regarding the time of entry. The same holds true with respect to exit time.

Example

1. Open the Notepad application and type the following code:

```
package  attendance;
import java. Util.*;

class EmpRec {
                String ename;
                Data EntryTime;
                Data ExitTime;

                EmpRec (String ename) {
                        this . ename = Data();
                }
        }
```

2. Change the current directory to examples directory.
3. Compile the file using the following command in the MSDOS prompt:

```
java c –d . attendance. EmpRec.java
```

The EmpRec. Class file will be generated in the attendance directory.

The **removeEmployee** () method is used to remove an entry from the **employees** Vector. It gets the employee name to be deleted as an input parameter. Then it finds the position of **empName** in the **employees** Vector and removes it from the Vector. It has a **try** and **catch** clause to write the employee name, his entry time and his exit time in the **EAttendance** file. It also returns the same to the client. The head office uses the **EAttendance** file to get the attendance details of the employees at the end of each month.

Example

1. Add the following lines to the Empserver.java file.

```
private String  removeEmployee (String empName) {
        EmpRec recl;
        for (int I = 0; I < employees. Size(); i++) {
                recl = (EmpRec) employees. elementEAt(i);
                if(recl.ename.equals (empName)) {
                        recl.ExitTime = new Data();
                employees.removeElementAt (i);
                try {
                        EAttendance. writeChars(recl.ename + "  "  " +
```

```
                        recl. EntryTime. toString() +  "  " +
                        recl. ExitTime. toString() + " \n ")
                } catch (IOException e ){
                        System.out.println(e) ;
                }
                return "Employee name: " + recl.ename +
                        "\nEntry time: " =
                        recl. EntryTime.toString () + "\nExit
                        time : " + recl. ExitTime.toString ();
            }
        }
        return "Employee not found";
    }
```

2.3 Passing Objects in RMI

Arguments to or return values from remote methods can be of almost any type, including local objects, remote objects, primitive types or serializable objects that implement **the java.io.Serializable** interface.

A few object types do not meet any of the above criteria and thus cannot be passed to or returned from a remove method. Most of these objects, such as a file descriptor, encapsulate information that makes sense only within a single address space. Many of the core classes, including those in the packages java.lang and java.util, implement the **serializable** interface.

The rules governing how arguments and return values are passed are as follows:

- Remote objects are essentially passed by reference. A remote object reference is a stub, which is a client-side proxy. The stub implements the complete set of remote interfaces that the remote object implements.
- Local objects are passed by value, using object serialization. By default all fields are copied, except those that are marked **static** or **transient.** Default serialization behavior can be overridden on a class-by-class basis.

Passing an object by reference (as is done with remote objects) means that any changes made to the state of the object by remote method calls are reflected in the original remote object. When passing a remote object, only those interfaces that are remote interfaces are available to the receiver; any methods defined in the implementation class or defined in non-remote interfaces implemented by the class are not available to that receiver.

For example, if we pass a reference to an instance of the **EmpInt** class, the receiver would have access only to the e**nter** () method of the **EmpInt** class. The receiver cannot access any other method declared in **EmpServer.**

Objects that are not remote, such as parameters, return values and exceptions, are passed by value in remote method calls. This means that a copy of the object is created in the receving virtual machine. Any changes to this object's state at the receiver are reflected only in the receiver's copy, not in the original instance.

The **main** () method is not remote method, which means that it cannot be called from a different virtual machine. Since the **main** method is declared **static**, the method is not associated with any object. Instead, the method is associated with the class in which it is declared.

The **main** () method creates an object of the **EmpServer** class, registers it with the registry using **Naming.rebind** () and then gives an output that the server is ready for use. The java.rmi.Naming interface is used as a front-end API for binding, or registering, and looking up remote objects in the registry. Once a remote object is registered with the **RMI** registry on the local machine, callers on any machine can look up the remote object by name, obtain its reference, and then invoke remote methods on the object. All servers running on a machine may share the registry, or an individual server process may create and use its own registry, if desired.

The following parameters are the arguments to call Naming.rebind();

- An URL-formatted name associated with the remote object
- A new remote object to associate with the name

The first parameter, which is an URL-formatted java.lang.String, represents the location and the name of the remote object. The location includes the name or IP address of the server machine. If both of these are omitted from the URL, it defaults to the IP address of the local machine. In addition, we need not specify a protocol in the URL. For example, supplying **EmpServer** as the name in the **Naming.rebind**() calls is allowed. Optionally, a port number may be supplied in the URL. For example, the name //**host:1023/ES** is legal, where **host** refers to the name or the IP address of the server machine. If the port is omitted, it defaults to 1099. A port number is necessary only if a server creates a registry on a port other than the default 1099. The default port is useful because it provides a well-known place to look for the remote objects that offer services on a particular machine.

The RMI runtime substitutes a reference to the stub for the remote object reference specified by the argument. Remote implementation objects, such as instances of **EmpServer,** never leave the virtual machine where they are created. So, when a client performs a lookup in a server's remote object registry, a reference to the stub is returned. As discussed earlier, remote objects in such cases are passed by reference rather than by value.

An application can bind, unbind or rebind remote object references only with a registry running on the same machine. This restriction prevents a remote client from removing or overwriting any of the entries in a server's registry. The advantage of rebind method over the bind method is that any existing binding for the same name is replaced by rebind, whereas bind gives an error if a binding for the same name exists already.

Once the server has registered with the local RMI registry, it prints out a message indicating that it is ready to start handling calls and then the main method exits. It is not necessary to have a thread wait to keep the server alive. As long as there is a reference to the server object in another virtual machine, local or remote, the server object will not be shut down, or garbage collected. The server object is reachable from a remote client because the program binds a reference to the server object in the registry. The RMI system takes care of retaining the server object in its run state. Thus, the server object Empserver is now available to accept calls. It will not be reclaimed until its binding is removed from the registry, and no remote client hold a remote reference to the **Empserver** object.

Example

1. Add the following lines to the Empserver.java file.

```
public static void main(String arge[]) {
    ret {
    EmpServer ES = new EmpServer ();
            Naming.rebind ("EmpServer",ES);
```

```
            Systerm.out.println ("Server  is Ready");
            } catch (Exception e) {
                  Systerm.out.println("EmpServer error: " + e);
            }
      }
```

The complete implementation of EmpInt interface (Server) is given below:

```
package  attendance;

import java.rmi.*;
import java.rmi.server.*;
import java.io.*;
import java.util.*;
public class EmpServer extends UnicastRemoteObject implements EmpInt {
vector employees = new vector();
vector key = null;
RandomAccessFile EAttendance = null;
public  EmpServer () throws RemoteException {
      try {
EAttendance = new RandomAccessFile ("EAttendance", "rw");
            EAttendance.seek (EAttendance.length());
      } catch (IOException e) {
      Systerm.out.println ("EAttendance: " + e);
      }
Systerm.out.println("EmpServer starting at " +
                              new Data (). toString());
readPasswords();
} //END OF EMPSERVER()
void readPasswords() {
String instry:
Key = new Vector();
BufferedReader EmpRegister  = null;

try {
EmpRegister = new BufferedReader (new
            FileReader ("attendance\\EmpRegister"));
      while (true) {
      if ((instrg = EmpRegister.readLine())==null) {
            EmpRegister=null;
            return;
      }
      key.addElement (new
            Empkey(instry.substring(0,instry.indexof
            (' ')),instry.substring (instrg.indexof
            (' ')+1)));
}
} catch (FileNotFoundException e) {
Systerm.out.println("EmpRegister not found: " + e);
} catch (IOException  e)  {
```

```
EmpRegister = null;
}
} // END OF READPASSWORDS()
public String enter (String empName, String key) {
if (isValidID(new Empkey (empName, key))) {
                if (onTheClock(empName))
                return removeEmployee(empName);
        else
        return addEmployee(empName);
}
return "Sorry, Invalid employee name or password";
    } //END OF ENTER ()
    private boolean isValidID (Empkey pw) {
        Empkey k1;
        for (int i= 0; i< key.size(); i++) {
                k1 = (Empkey) key.elementAt(i);
                if(k1.isEqual (pw) )
                        return true;
                }
                return false;
        } // END OF ISVALIDID()
        private String addEmployee (String empName) {
                EmpRec emp;
                Employees.addElement(emp =new EmpRec(empName));
                return "Employee name: " + empName + "Entry time: " +
                                        emp. EntryTime.toString();
        } // END OF ADDEMPLOYEE()
        private string removeEmployee (String empName) {
                EmpRec recl;
                for (int i = 0; I < employees. Size (); i++) {
                        recl = (EmpRec) employees.elementAT(i) ;
                        if(recl.ename.equals(empName)) {
                                recl.ExitTime = new Data();
                                employees. removeElementAT(i);
                                try {
                                EAttendance.writeChars(recl.ename + "  "
                                        + recl.EntryTime.toString() + "  " +
                                        Recl.ExitTime.toString () + " \n");
                                } catch (IOException e ) {
                                        Systerm.out.println(e);
                                }
                                return "Employee name: " + recl.ename +
                                        "\nEntry time: " +
                                recl.EntryTime.toString () + "\nExit time:
                                        " + recl.ExitTime.toString ();

                        }
                }
                return "Employee not found";
    } //END OF REMOVEEMPLOYEE()
```

```
        private  boolean onTheClock (String empName)  {
                for (int  I = 0; I < employees. Size(); i++)
                if ((( EmpRec) employees. elementAt (i)). Ename.equals (empName))
                                return true;
                        return false;
        } // END OF ONTHECLOCK()
        public static void main (String args[] ) {
                try {
                EmpServer ES  =  new EmpServer () ;
                        Naming.rebind  ("EmpServer",ES);
                        Systerm.out.println("Server   is Ready");
                } catch (Exception  e ) {
                Systerm.out.println ("EmpServer  error:  "  + e) :
                }
        }  // END OF MAIN ()
} // END OF THE CLASS EMPSERVER
```

Open the file Empserver. Java in the Notepad application and ensure that the code is as given above. Then follow the steps as below.

Example

1. Change the current directory to examples directory.
2. Compile the file using the following command in the MSDOS prompt:

```
javac  -d . attendance. EmpServer. java
```

The EmpServer.class file will be generated in the attendance directory.

2.4 Client Implementation

After writing the server implementation, we move onto the client implementation. This has to be separate because this part alone should be run in all the branches of Perfect Solutions Limited. For the client implementation, we will use an applet so that the user can interact easily with the application. Any method in the client that calls a remote method should have a **try** and **catch** clause or throw a **RemoteException.** The client is also present in the same package **attendance** as the server. Since it is an applet it should import **java.applet**.* and it extends the **Applet** class.

In the **EmpApp** class we declare two text fields names **tfName** and **tfKey** to get the name and password of the employee. We declare a text area named **taInfo** to print the information regarding the processes executed in the application. The information displayed in this text area ensures the employee of the registration of his entry or exit time. It also shows an exception if there is any problem with the execution of the application. We also declare an object of **EmpKey** class to store the input information from the users. The initial code is shown below:

```
package attendance;
import java.rmi.*;
import java.applet.*;
import java.awt.*;
public class EmpApp extends Applet  {
        private TextField tfName;
        private TextField tfKey;
```

```
private TextArea taInfo;
EmpInt EK;
```

The **init** () method adds two labels, the text fields and the text area to the applet and sets the EchoCharacter as "* " for the password text field. It looks up the registry to get an interface of the server or the remote object, using the **Naming.lookup**() method. With the help of this method, the client running in the branch office of Perfect Solutions Limited gets a reference of the server running in the head office. It is enclosed in a try catch clause so that any error during connection to the server can be caught and displayed to the user. For example, if the server is not ready, the **NotBoundException** is displayed in the text area **taInfo**. The code for the init ( ) method is shown below:

```
public void init() {
        Label 11=new Lable ("Enter the name:");
        add (11);
        add(tfName  = new TextField(20));
        Label  12=new Label ("Enter the password: ");
        add(12);
        add(tfkey  =  new Textfield(20));
        tfKey.setEchocharacter ('*');
        add(taInfo  =  new TextArea(5,80));
        add(new Button("Click to Enter/Exit"));
        taInfo. setEditable(false);
        taInfo.setText("Connecting to the Server\n");
        try {
        EK  =  (EmpInt) Naming. Lookup ("EmpServer");
        } catch  (Exception  e) {
        System.out.println("Error: " + e.toString());
                taInfo.append ("Error:" + e.toString());
        }
}
```

In the above code we have assumed that the client part and the server part reside in the same machine. Suppose they reside in different machines and the server resides in a machine whose IP address is 80.0.0.100, then the statement:

```
EK = (EmpInt)Naming.lookup ("EmpServer");
```

can be rewritten as follows:

```
EK = (EmpInt) Naming.looup ("rmi://80.0.0.100/EmpServer");
```

If we wish to specify the IP address during runtime, the statement can still be modified as follows:

```
EK = (EmpInt) Naming.looup ("rmi://"+ args[0] + "/ EmpServer");
```

Where args[0] represents the IP address of the machine.

The **Naming.lookup** ( ) is a static method that searches for the server using the server name registered with the RMI registry. It takes in a single parameter of String type representing the URL-formatted name for the remote object and returns a reference (i.e. a stub) for the remote object associated with the specified name. If the name is not currently bound to the registry, then it throws the NotBoundException. If the registry could not be contacted, then the lookup method throws a RemoteException. If the operation is not permitted, then the lookup method throws the AccessException. Once the connection is established, the **enter**() method can be invoked. The whole process is depicted in Fig. 9.2.

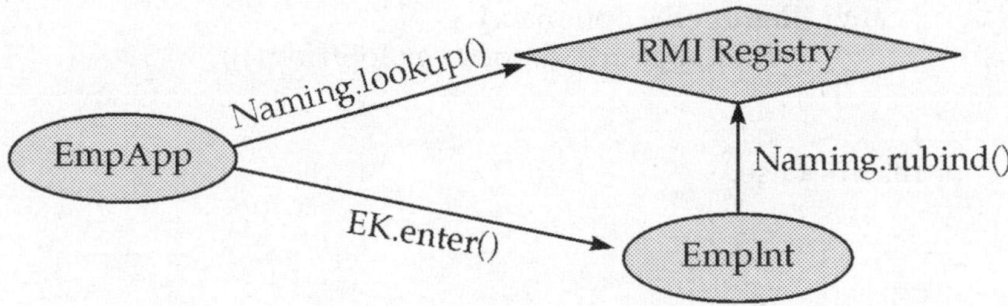

Fig. 9.2 Process of RMI

The complete client side code is given in example below.

Example

1. Open the Notepad application and type the following code:

```
package attendance ;
import java.rmi.*;
import java.applet.*;
import java.awti.*;
public class EmpApp extends Applet {
    private TextField tfName;
    private TextField tfKey;
    private TextArea taInfo;
    EmpInt EK;
    public void init( ) {
    Label 11=new Label ("Enter the name:");
        add(11);
        add(tfName = new Textfield(20));
        Label 12=new Label (Enter the password:");
        add(12);
        add(tfKey = new TextField(20));
        tfKey.setEchoCharacter('*');
        add(taInfo = new TextArea (5,80));
        add(new Button("click to Enter/Exit"));
        taInfo.setEditable(false);
        taINfo.setText ("Connecting to the Server\n");
        try {
        EK = (EmpInt) Naming.looup("EmpServer");
        } catch (Exception e) {
            System.out.println ("Error: " + e.toString());
            taInfo.append ("Error:" + e.toString());
        }
```

```
    }
public boolean handleEvent (Event  e) {
if (e.target instanceof  Button &&  e.id==Event.ACTION_EVENT) {
        try {
                taInfo.append(EK.enter(tfName.getText(),
                                tfKey.getText())+"\n");
                return true;
        } catch (RemoteException ex) {
                taInfo.append("Error:"+ex.toString());
                }
        }
        return  false;
    }
{
```

2. Save the file as EmpApp.java in the attendance directory.
3. Change the current directory to examples directory.
4. Compile the file using the following command in the MSDOS prompt:

The EmpApp.class file will be generated in the attendance directory.

To execute the above applet we need an HTML file. The HTML code used to execute the applet is given in example below.

Example

1. Open the Notepad application and type the following code:

```
<HTML>
<HEAD
<TITLE> Employee Attendance Applet</TITLE
</HEAD>
<BODY>
<H1>please enter your name and password to enter or exit.</H!>
<APPLET     code=attendance.EmpApp.class
        width=600
        height =250  >
</APPLET>
</BODY>
</HTML>
```

2. Save the file as EmpApp.html in examples directory.

2.5 Packages Used in RMI

The following are four of the important packages used in Java RMI:

- java.rmi
- java.rmi.dgc
- java.rmi.registry
- java.rmi.server

This section gives a brief picture about the classes and the interfaces available in each of these packages.

(a) java.rmi

Any object that can be invoked using RMI must implement the **Remote** interface in the **java.rmi** package. When such an object is invoked, its arguments are marshalled and sent from the local virtual machine to the remote one, where the arguments are unmarshalled. When the method terminates, the results are marshaled from the remote machine and sent to the caller's virtual machine. If the method invocation results in an exception beging thrown, the exception is indicated to the caller.

Note : Process of converting input parameters to a format that can be transmitted across the network is called marshalling.

A reverse process of marshalling in which the data is converted from a format portable across the network to the format used by methods is called unmarshalling.

Input parameters and objects are always in bytecodes (in Java), but the network always transmits them as a stream of bits.

Classes in java.rmi

The **Naming** class extends the Object class. The **Naming** class provides methods for storing and obtaining references to remote objects in the remote object registry. One of the arguments to the methods of the Naming class URL formatted **java.lang.String.** The **bind** and **rebind** methods associate a name with a remote object using this class.

The **MarshalledObject** class extends the **Object** class and implements the **Serializable** interface. A **MarshalledObject** contains a byte stream will the serialized representation of an object given to its constructor. The get method returns a new copy of the original object, as deserialized from the contained byte stream. The contained object is serialized and deserialized with the same serialization semantics used for marshalling and unmarshalling parameters and return values of RMI calls. **MarshalledObject** facilitates passing objects in RMI calls that are not automatically deserialized immediately by the remote peer.

The **RMISecurityManager** class extends the **SecurityManager** class.

RMISecurityManager provides an example security manager for use by RMI applications that use downloaded code. RMI's class loader will not download any classes from remote locations if no security manager has been set. **RMISecurityManager** does not apply to applets, which run under the protection of their browser's security manager.

Interfaces in java.rmi

The **Remote** interface serves to identify interfaces whose methods may be invoked from a remote virtual machine. Remote objects must directly or indirectly implement this interface. Only those methods specified in a remote interface, an interface that extends **java.rmi.Remote** are available remotely. Implementation classes can implement any number of remote interfaces and can extend other remote implementation classes.

(b) java.rmi.dgc

The **DGC** abstraction is used for the server side of the distributed garbage collection algorithm. When an object returned from a server is no more in use by the client, the server garbage-collects the remote object. The **java.rmi.dgc** package provides classes and interface for this RMI Distributed Garbage-Collection (DGC).

Classes in java.rmi.dgc

The **VMID** class extends the **Object** class and implements the **Serializable** interface. A Virtual Machine ID (VMID) is an identifier that is unique across all Java virtual machines. VMIDs are used by the distributed garbage collector to identify client VMs.

The **Lease** class extends the **Object** class and implements the **Serializable** interface. A lease contains a unique VM identifier and lease duration. A lease object is used to request and grant leases to remote object references. The **getValue**( ) method of the class returns the lease duration and the **getVMID** ( ) method returns the client **VMID** associated with the lease.

Interfaces in java.rmi.dgc

The **DGC** interface extends the **Remote** interface. This interface contains two methods, namely **dirty** and **clean.** A dirty call is made when a remote reference is unmarshalled in a client (indicated by its VMID). A corresponding clean call is made when no more references to the remote reference exist in the client.

(c) java.rmi.registry

The **java.rmi.registry** package provides a class and two interfaces for the **RMI** registry. A registry is a remote object that maps names to remote objects. A server registers its remote objects with the registry so that they can be invoked by other objects searching the registry. When an object wants to invoke a method on a remote object, it must first lookup the remote object using its name. The registry returns to the calling object a reference to the remote object, using which a remote method can be invoked.

Classes in java.rmi.registry

The **LocateRegistry** class extends the **Object** class. **LocateRegistry** is used to obtain a reference to a bootstrap remote object registry on a particular machine (including the local machine), or to create a remote object registry that accepts calls on a specific port. The **getregistry** ( ) method of this class creates a local reference to the remote registry, even if no registry is running on the remote machine.

Interfaces in java.rmi.registry

The **Registry** interface extends the **Remote** class. The remote object Registry interface helps to obtain reference to remote objects. It is implemented by RMI's **rmiregistry** that provides methods for storing and retrieving remote object reference. The most important methods used by the server for registering with the server, namely **bind** ( ), **unbind** ( ) and **rebind** ( ), are present in this interface. The **lookup**( ) method of this interface is used by the client to search for the server in the RMI registry. The interface also contains the **list**( ) method which returns an array of names bounds in the registry.

RegistryHandler is a deprecated interface used internally by the RMI runtime in previous implementation versions. It should never be accessed by application code.

java.rmi.server

The **java.rmi.server** package provides classes and interfaces for supporting the server side of RMI. A group of classes are used by the stubs and skeletons generated by the rmic atub compiler (we will learn more about this compiler in the next chapter). Another group of classes implements the RMI Transport protocol and HTTP tunneling.

(d) java.rmi.server

The **ObjID** class extends the **Object** class and implements the **Serializable** interface. An **ObjID** is used to identify remote objects uniquely in a VM over time. Each identifier contains an object number and an address space identifier that is unique with respect to a specific machine. An object identifier is assigned to a remote object when it is exported.

The **RemoteObject** class extends the **Object** class and implements the two interfaces Remote and **Serializable**. The **RemoteObject** class implements the java.lang.Object behavior for remote objects. **RemoteObject** provides the remote semantics of Object by implementing methods for hashCode, equals, and toString.

The **RemoteServer** class extends the **RemoteObject** class and the common superclass to server implementations. It provides the framework to support a wide range of remote reference semantics.

The **RemoteStub** class also extends the **RemoteObject** class and is the common superclass to client stubs. It provides the framework to support a wide range of remote reference semantics.

The **RMIClassLoader** class that extends the **Object** class provides static methods for loading classes from a network location and obtaining the location from which an existing class can be loaded. These methods are used by the RMI runtime when marshalling and unmarshlling classes of parameters and return values.

The **RMISocketFactory** class extends the Object class and implements the **RMIClientSocketFactory** and **RMISocketFactory** interfaces. An instance of the **RMISocketFactory** class is used by the RMI runtime in order to obtain client and server sockets for RMI calls. An application may use the setSocketFactory method of this class to request that the RMI runtime use its socket factory instance instead of the default implementation.

The UID class extends the Object class and implements the Serializable interface. The UID class is an abstraction for creating identifies that are unique with respect to the machine on which it is generated.

The **UnicastRemoteObject** is a very important class that extends the **RemoteServer** class. The UnicastRemoteObject class defines a non-replicated remote object whose reference are valid only while the server process is alive. The UnicastRemoteObject uses the TCP streams to provide support for active object references from one end to the other. Objects that require remote behavior should extend RemoteObject through UnicastRemoteObject.

The **LogStream** is a deprecated class that extends the **PrintStream** class. The LogStream class provides a mechanism for logging errors that are useful for monitoring a system.

The **operation** is a deprecated class that extends the **Object** class. An **Operatio**n contains a description of a Java method.

Interfaces in java.rmi.server

The **RemoteRef** interface extends the **Externalizable** interface. RemoteRef represents the handle for a remote object. A Remotestub uses a remote reference to carry out a remote method invocation to a remote object.

The **ServerRef** interface extends the **RemoteRef** interface and represents the server-side handle for a remote object implementation.

An instance of the **RMIClientSocketFactory** interface is used by the **RMI** runtime to acquire client sockets for RMI calls. A remote object can be associated with a **RMIClientSocketFactory** when it is created/exported through the constructors or **exportObject** methods of java.rmi.server.UnicastRemoteObject.

A **RMIFailureHandler** interface can be registered using the setFailureHandler method of the **RMISocketFactory** class. The only method of the handler, namely failure(), is invoked when the RMI runtime is unable to create a ServerSocket through the **RMISocketFactory,** to listen for incoming calls. This method returns a boolean indicating whether or not the runtime should attempt to re-create the **ServerSocket**

An instance of the **RMIServerSocketFactory** interface is used by the RMI runtime in order to obtain server sockets for RMI calls. A remote object can be associated with an **RMIServerSocketFactory** when it is created / exported via the constructors or exportObject methods of java.rmi.server.UnicastRemoteObject. An **RMIServerSocketFactory** instance associated with a remote object is used to obtain the **ServerSocket** used to accept incoming calls from clients.

A remote object implementation should implement the **Unreferenced** interface in order to receive notification when there are no more clients that reference that remote object.

The **LoaderHandler** is a deprecated interface used internally by the RMI runtime in previous implementation versions. It should never be accessed by application code.

The **RemoteCall** interface is an abstraction used solely by the RMI runtime to carry out a call to a remote object. The **RemoteCall** interface is deprecated in JDK 1.2 since it is only used by deprecated methods of the **RemoteRef interface**.

The **Skeleton** is a deprecated interface used solely by the RMI implementation. Every skeleton class generated by the rmic stub compiler in the version 1.1 implements this interface. A skeleton for a remote object is a server-side entity that dispatches calls to the actual remote object implementation. In JDK 1.2 and higher version skeletons are not needed for remote method calls.

Summary

- The server needs to create and register objects in the local RMI registry.
- The Server should provide an implementation for each remote method defined in the interface.
- In remote method invocations, the arguments are passed by reference if they are remote objects and are passed by value if they are local.
- Each remote method must throw the RemoteException.
- The client searches for the server name in the registry by using the Naming.lookup( ) method.
- Four of the important package in RMI are java.rmi, java.rmi.dgc, java.rmi.registry and java.rmi.server.

3. ARCHITECTURE OF RMI

In this chapter we are going to learn about the RMI layers, namely the application layer, the Stub and Skeleton layer (Proxy layer), the Remote reference layer and the Transport layer. Each of these layers performs specific functions during the execution of a distributed application. We will learn how the RMI supports garbage collection and the registry system. Finally, we are going to understand how the RMI system works.

The complete RMI system is organized as a four-layer model. Each layer can perform specific functions, like establish the connection, marshal and unmarshal the parameters, transmitting the object, etc. The four layers are:

- Layer 1 is the **Application layer**, it is the actual object definition.
- Layer 2 is the **Proxy layer**, it consists of two parts; Stub and Skeleton. These are used for marshalling and unmarshalling the data that is transferred through the network. We create the stub and skeleton classes using the rmic (RMI compiler).
- Layer 3 is the **Remote Reference layer**, it can get the stream of bytes from the transport layer and gives it to proxy layer.
- Layer 4 is the **Transport layer**, it is responsible for handling the actual machine to machine communication. We can see details about these layers in this chapter.

The garbage collection is one of the advantages of the Java language. In this chapter we are going to see how **RMI** supports garbage collection. The registry simply keeps track of the addresses of remote objects that are being exported by the applications. All the distributed systems use the registry to keep track of the names of the remote objects.

3.1 RMI Layers

We already know that, RMI communicates only between one Java Virtual Machine to another Java Virtual Machine. When the client wants to access the remote object that resides in the server, it cannot communicate directly with the server. The other layers should support the communication. Now we can see the layers and their functions.

a. Application Layer

The application layer is the actual implementation of the client and server application. Here, the high level calls are made in order to access and export remote objects. The client can access the remote method through an interface that extends java.rmi.Remote. When we want to define a set of methods that can be remotely called, they must be declared in one or more interfaces that should extend java.rmi.Remote.

Once the methods described in the remote interfaces have been implemented, the object must be exported. This can be done implicitly if the object extends the unicastRemoteObject class of the java.rmi.server package. Then, the application will register itself with a naming server or registry. This is used by the client to obtain a reference of the objects. On the client side, a client simply requests a remote object from either a registry or remote object whose reference has already been obtained. After getting the reference, the proxy layer has to perform the important role of communication between the client and the server.

Note: In our application, the EmpClient and the EmpServer from the application layer.

b. Proxy Layer

To understand how RMI works, we must appreciate the roles of stubs and skeletons. The stub and skeleton are created using the RMI compiler (RMIC). These are simply class files that represent the client and server side of a remote object. In our employee attendance application the client EmpClient has to access the remote method enter( ), that is in the server side program EmpServer. The stub is placed in the client side while the skeleton is placed in the server side as in Fig. 9.3.

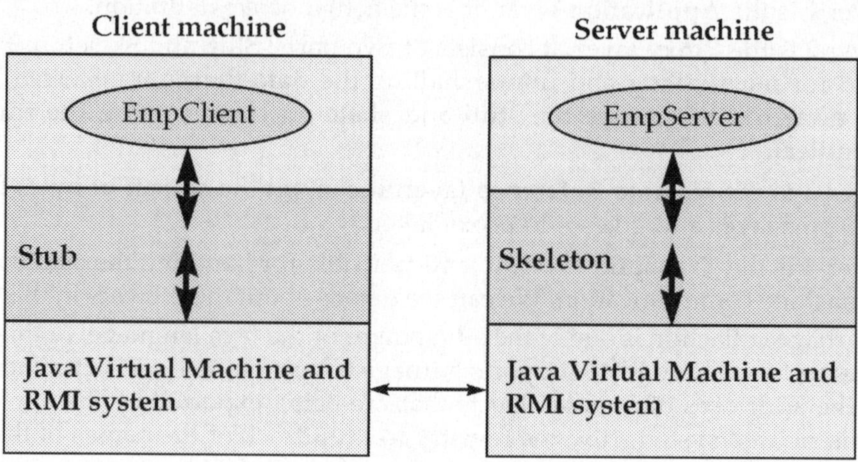

Fig. 9.3 Stub and skeleton

The Stub Class

The stub is client side proxy of the remote object. It has three primary responsibilities:

- It presents the same remote interfaces as the object EmpServer. Therefore, from the perspective of the EmpClient, the stub is equivalent to the remote object.
- It works with Java Virtual Machine and RMI system on client machine to serialize any arguments to a remote method call and sends this information to server machine.
- The stub receives any result from the remote method and return it to the client.

The Skeleton Class

The skeleton is a server side proxy of the remote object. It has three primary responsibilities:

- It receives the remote method call and any associated arguments. It works with the JVM and RMI system on the Server machine to deserialize any arguments for this remote method call.
- It invokes the appropriate method in the EmpServer, using these arguments.
- It receives any return value from this method call and works with the JVM and RMI system on Server machine to serialize this return value and sends the information back to EmpClient.

How to Create Stub and Skeleton Classes

We have compiled all the Java files using the javac command (in Chapter 2) and hence class files for all the Java files are available. These files have intermediate code called byte codes. We can create stub and skeleton classes by using the command rmic. This is illustrated in Fig. 9.4. RMIC stands for RMI compiler. In order to create stub and skeleton files we have to type the following command in the **examples** directory as:

rmic -d . attendance.EmpServer

After this command is executed the Stub and Skeleton classes will be created in the **attendance** directory, with the names: EmpServer_stub.class and Empserver_skeleton.class.

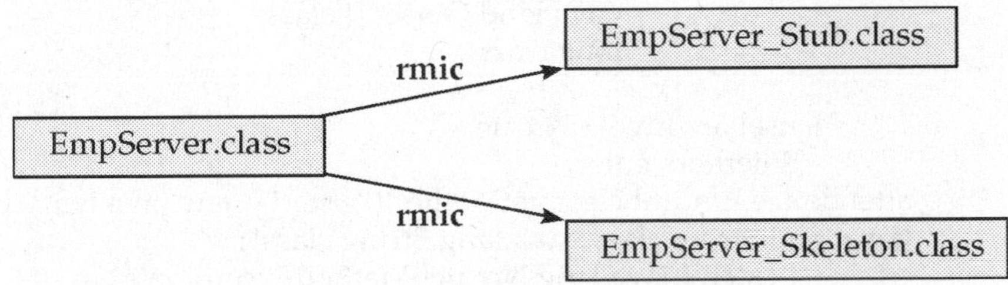

Fig. 9.4 Stub and skeleton class

While creating these classes, the compiler generates the intermediate Java files as EmpServer_stub.java and EmpClient_Client.java. We cannot view these files as they are temporary files and rmic (RMI Compiler) will delete these files. If we want to see these files we have to compile the EmpServer with the following command:

rmic -d . –keepgenerated attendance.EmpServer

After executing this command EmpServer_stub.java and EmpServer_skeleton.java are obtained.

Note: The files that are created by rmic and with the extension as.java are not required. The application can run successfully without these files. The following coding gives us a clear idea of how the system works.

EmpServer_stub.java

```
//stub class generated by rmic, do not edit.
//contents subject to change without notice.
package attendance;
/*EmpServer is an implementation of the interface EmpInt
in the package attendance.*/
public final class EmpServer_stub
        extends java.rmi.server.RemoteStub
        implements attendance.EmpInt, java.rmi.Remote
{
        private static final java.rmi.server.operation[] operation = {
                new java,rmi.server.operation("java.lang.String
enter (java.lang.String,  java.lang.String)")
        };
private static final long interfaceHash  =3866617036126357469L;
private static final long serialversionUID  = 2;

private static boolean useNewInvoke;
private static java.lang.reflect.Method smethod_enter_0;

static {
        try {
                java.rmi.server.RemoteRef.class.getMethod("invoke",
                        new java.lang.Class[]  {
                                java.rmi..Remote.class,
                                java.lang.reflect.Method.class,
```

```
                              java.land.Object [].class,
                              long.class
                         });
                    useNewInvoke =true;
                    $method_enter_0 =
          attendance.EmpInt.class.getMethod("enter", new java.lang.class[]
          (java.lang.String.class, java.lang.String.class});
                    } catch ((java.lang.NoSuchMethodException e) {
                              useNewInvoke = false;
                    }
          }
          //constructors
          public EmpServer_Stub( ) {
                    super ();
          }
          public Empserver_stub(java.rmi.server.RemoteRef ref) {
                    super (ref);
          }

          //methods from remote interfaces

          //implementation of enter(String, string)
          public  java.lang.String enter( java.lang.String $param_string_1,
java.lang.String $param_string_2)
                    throws java.rmi.RemoteException
          {
                    try {
                              if (useNewInvoke) {
                                        object $result = ref.invoke(this,
$method_enter_0, new java.lang.object[] ($param_string_1,
$param_String_2},  -6656453098156255660L);
                    return  ((java.lang.String) $result);
          } else {
                    Java.rmi.server.Remotecall call =
ref.newCall( (Java.rmi.server.Remoteobject) this, operations, 0,
interfaceHash);
                    try {
                              java.io.objectoutput out =
call.getOutputstream( );
                              out.writeobject($psram_srring_1);
                              out.writeobject($psram_srring_2);
                    } catch (java.io.IOException e) {
                              Throw new java.rmi.MarshalException ("error
marshalling arguments", e):
                    }
                    ref.invoke(call);
                    java.lang.string $result;
                    try {
                              java.io.objectInput  in =
```

```
        call.getInputStream( );
                                $result  = (java.lang.String)
        in.readobject( );
                        } catch (java.io.IOException  e) {
                                throw  new
        java.rmi.UnmarshalException ("error unmarshalling return", e) ;
                        } catch (java.lang.clssNotFoundException  e) {
                                throw  new
        java.rmi.UnmarshalException ("error unmarshalling return", e) ;
                        } finally {
                                ref.done(call);
                        }
                        return $result;
                }
        } catch (java.lang.RuntimeException  e) {
                                throw  e;
                } catch (java.rmi.RemoteException  e) {
                                throw  e;
                } catch (javalang.Exception  e) {
                                throw new java.rmi.UnexpectedException("undeclared
        checked exception", e);
                }
        }
}
```

The EmpServer_skeleton.Java is as follows

```
        // skeleton class generated by rmic,  do not edit.
        //  contents subject to change without notice.

        package attendance;

        public final class EmpServer_skel
                implements "server.skeleton
        {
                private static final java. rmi.server.operation[] operations  = {
                        new  java. rmi.server.operation("java.lang.String
        enter(java.lang.String, java.lang.String) ")
                };
                private static final long interfaceHash = 3866617036126357469L;
                public java. rmi.server.operation[] getOperations( ) {
                   return (java. rmi.server.operation []) operations.clone( );
                }
                public void dispatch (java.rmi.Remote obj,
        java.rmi.server.RemoteCall call, int opnum, long hash)
                throws java.lang.Exception
                {
                        If (opnum <0) {
                                If (hash == -6656453098156255660L) {
                        Opnum  = 0;
```

```
                } else {
                throw new java.rmi.UnmarshalException("invalid
        method hash");
                }
        } else {
                If (hash ! = interfaceHash)
                                throw new
        java.rmi.server.SkeletonMismatchException("interface hash mismatch");
                }
        attendance.EmpServer server = (attendance.EmpServer) obj;
                switch (opnum) {
                case 0://enter(String,String)
                {
                        java.lang.String $param_String_1;
                        java.lang.String $param_String_2;
                        try {
                        java.io.objectInput in = call get InputStreem ( );
                                $param_String_1 = (java.lang.String)
                in.readObject( );
                                $param_String2 = (java.lang.String)
                in.readObject( );
                        } catch (java.io.IOException e) {
                        throw new java.rmi.UnmarshalException("error
                        unmarshalling arguments", e);
                        } catch (java.lang.ClassNotFoundException e) {
                        throw new java.rmi.UnmarshalException("error
                unmarshalling arguments", e);
                        } finally {
                                call.releaseInputStream( );
                        }
                        java.lang.String $result =
                        server.enter($param_String_1, $param_String_2);
                                try {
                                        java.io.objectOutput out =
                        call.getResultStream(true);
                                        out.writeObject($result);
                                } catch (java.io.IOException e) {
                                throw new java.rmi.marshalException("error
                marshalling return", e);
                                }
                        break;
                }
                default:
                throw new java.rmi.UnmarshalException ("invalid method
                number");
                }
        }
}
```

The main function of proxy layer is marshalling and unmarshalling the parameters. The other options of rmic are given in Table 9.1.

Usage of rmic is:

rmic <options> <class names>

Table 9.1: rmic Options

Option	Description
-Keep	When generating the stub and skeleton RMI compiler can generate the intermediate code. If we want to see these codes we have to use this option. Now it will not delete the intermediate code.
-keepgenerated	Same as "-keep"
-v1.1	To create stubs/skeletons for Java Development kit 1.1 stub protocol version
-v1.2	To create stubs for JDK 1.2 stub protocol version only
-g	To generate debugging information
-depend	Recompile out-of-date files recursively
-nowarn	Generate no warnings
-verbose	Output messages about what the compiler is doing
-classpath <path>	Specify where to find input source and class files
-d <directory>	Specify where to place generated class files
-J<runtime flag>	Pass argument to the java interpreter

c. Remote Reference Layer

The remote reference layer is effectively between stub and skeleton classes and the transport layer handles the actual communication protocols. When transmitting the parameter or objects through the network it should be in the form of a stream. The Java Virtual Machine works with the Java byte codes. It can get the stream-oriented data from the transport layer and give it to the proxy layer and vice versa.

This layer is used to

- Handle the replicated objects. Once this feature is incorporated into the RMI system, the replicated objects will allow simple dispatch to many programs that are exporting substantially the same Remote objects.
- It is responsible for establishing persistence and strategies for recovery of lost connections.

d. The Transport Layer

The transport layer's responsibilities are as follows:

- It is responsible for handling the actual machine to machine communication, the default communication will take place through a standard TCP/IP (Transfer Control Protocol/ Internet Protocol).

- It creates a stream that is accessed by the remote reference layer to send and receive data to and from other machines.
- It sets up the connections to remote machines.
- It manages the connections.
- It monitors the connections to make sure that they (remote machines) are live.
- It listens for connections from other machines.

The transport layer can be modified to handle encrypted streams, compression algorithms, and a number of other security enhancements. Hence the transport layer is independent of the other three layers, an RMI application does not need to know the specifics of any changes made to the transport layer.

3.2 RMI Registry

RMI registry is a simple server that enables an application to lookup objects that are exported for remote invocation. It is also called **bootstrap registry**. The registry simply keeps track of the address of remote objects that are being exported by their application. In any distributed system all the objects are assigned unique names that are used to identify the object. Let us learn how to register the object into the registry and how to get the object or reference to the object from the registry.

a. Register the Object into the Registry

In the RMI system, certain methods can be called from the rmi.registry. Registry interface, or from the rmi.Naming class. It allows the application to add, remove and access remote objects in the registry's table of objects and associated names. In the application created for CBS Publishers, the server was called from the registry using java.rmi.Naming as follows:

```
EmpServer ES = newEmpServer( );  //creates the object ES
Naming.rebind("EmpServer",ES);  // registering the object
```

When we use it in the network we have to rewrite the code as given in Example, so that it can be registered as a remote object.

Example

If the IP of the system in which the resides is 80.0.0.150, then the code will be as follows:

```
//creates the object ES
EmpServer ES  = new EmpServer( );

// registering the object
Naming.rebind ("rmi://80.0.0.150/EmpServer",ES);
```

A server registers the object with registry by calling bind ( ) or rebind ( ) method. The method can take two parameters that uniquely identify the object. It also identifies a reference to an instance of the object that is being exported. If we use the method bind( ) and the object's name already exists in the registry, then the result is an AlreadyboundException being thrown, but the method rebind( ) replaces the binding to point to the new object.

Note: All the objects must have unique name.

After compiling all the Java programs by java command and the server class file by rmic command, the application is now ready to run. (The process of running the application will be discussed in Chapter 4).

RMI registry application should run continuously as the background process. If the rmiregistry is run, a blank screen will be displayed. Before running the application we have to know where to locate the files and then a new instant has to be opened. The client and stub classes have to be placed in one machine while all other files that are in the examples directory have to be placed in another machine. If we run our application without running the rmiregistry, it will thrown an exception as given in the following Example.

Example

Java.rmi.connectException: Connection refused to host; nested exception is:
Java.net.ConnectException: Connect refused: no further information.

b. Searching for Object in Registry

If the application is located in a single machine, the communication between the client and server is performed easily. For a distributed application, client and server should be placed on different machines. The stub is placed on the client side, the skeleton and remote interface files are placed in the server side. If we keep the files on different machine, the client can access the registry through URL (Uniform Resource Locator). A standard HTTP URL is

http://80.0.0150:80/EmpApp.html

- HTTP is the protocol used to transfer the messages.
- 80.0.0.150- is the Internet Protocol Address of the machine (machine name) that has the EmpApp.html file.
- EmpApp.html is the file we want to execute in the web.
- 80 is the port number, it is optional.
- The first":" is used to mark the end of the protocol identifier, the "//"is used to note the start of the machine name, and the " : " following the machine name is used to attach the port number.

The URL convention used by the Software Systems Limited is as follows, because the application is developed using RMI, it uses rmi protocol. So, the URL will be as follows:

rmi:// 80.0.0.150/EmpServer

The registry is searched when executing the following statement in EmpApp.java file

```
// If application in local machine
EK = (EmInt) Naming.lookup ("EmpServer");
// If application in remote machine
EK=(EmInt) Naming.lookup ("rmi:// 80.0.0150/EmpServer");
```

The client can access the remote object as soon as it gets the reference.

3.3 RMI Flow

When the client needs to communicate with the remote objects or methods, the RMI system follows certain steps.

First two steps of application flow are given in the Fig. 9.4. The steps taken by the system are:

Step 1: The server creates the remote object.

Step 2: Server registers the remote object with the RMI registry.

Object creation is implemented in the server program, with following statement:

EmpServer ES = new EmpServer( );

The object is registered with the following statement:

Naming.rebind ("EmpServer:, ES);

ES is the object name, it is registered in the RMI registry. The next two steps (3 and 4) are illustrated in Fig. 9.5.

Step 3: Client gives a request for the object to the RMI registry.

Step 4: The RMI registry returns the remote interface to the stub.

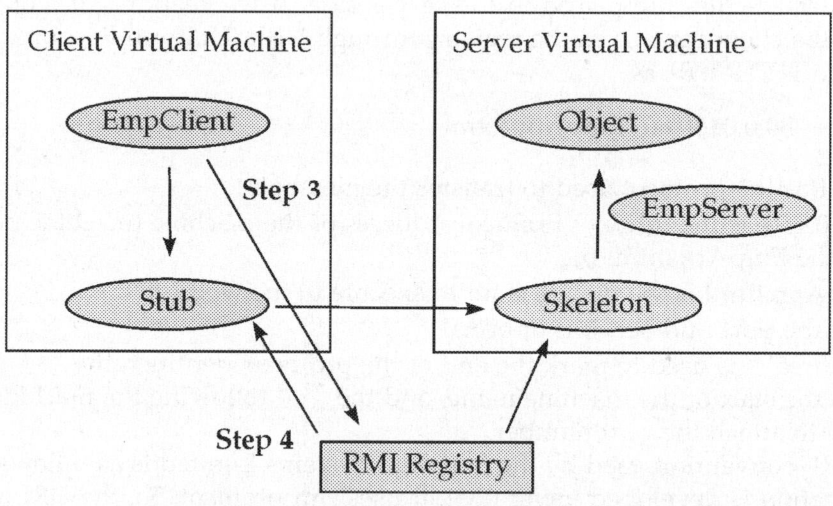

Fig. 9.5 Object registering in RMI registry

To request an object, the client has to lookup the server's name in the registry as follows.

EK = (EmpInt) Naming.looup("EmpServer");

The registry gives the remote reference (EK) through the stub. The last three steps (5 to 7) are illustrated in Fig. 9.6.

Step 5: The client invokes the stub method. We know that the stub is the proxy for the server.

Step 6: Stub passes the request to the skeleton.

Step 7: Skeleton invokes the remove object's methods.

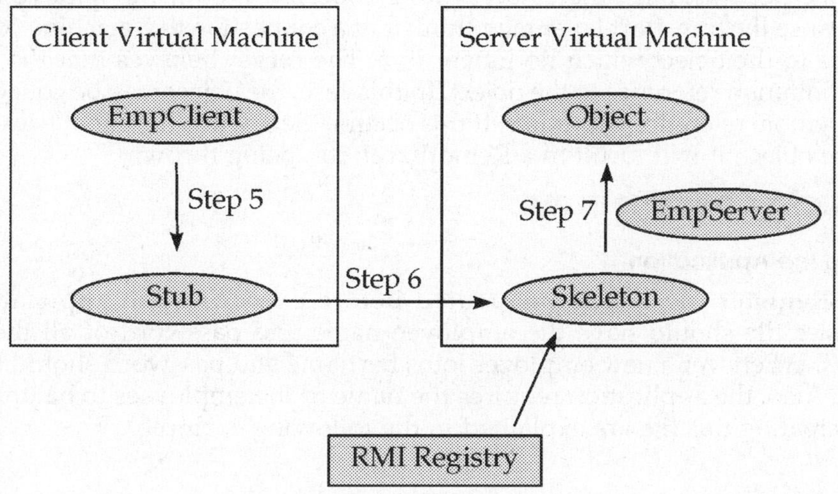

Fig. 9.6 RMI steps 5–7

```
//This code is from EmpApp.java
taInfo.append( EK.enter (tfName.getText( ), tfkey.getText ( ) )+"\n");
```

In this statement the client invokes the remote method enter () through the remote interface EK. Only this step is transparent to the programmer. The last two steps are performed internally by the RMI system. When running the application, the network may fail or the client might remove its reference. In such a situation the process of garbage collection is performed.

3.4 Garbage Collection

One of the advantages of using Java is memory management. In a distributed system, it is desirable to automatically delete remove objects that are no longer referenced by any client. This is called garbage collection. As a result, the programmer need not to keep track of the remote objects life cycle, hence that RMI uses a reference-counting garbage collection algorithm.

To accomplish reference-counting garbage collection, the RMI runtime keeps track of all live references within each Java virtual machine. When a client first receives a reference to a remove object, a "referenced" message is sent to the server that exports the object. Every subsequent reference within the client's local machine causes the reference counter to be incremented by one. A local reference is finalized; the reference count will be reduced by one. When the reference counter reaches zero, an "unreferenced" message is sent to the server. Once the server has no more live references to an object and there are no local references, object is free to be finalized and garbage collected. As long as a local reference to a remote object exists, it cannot be garbage-collected.

If a remote object implements the java.rmi.server.Unreferenced interface, it must implement a body for the unreferenced. The unreferenced method will be called, once the object is no longer referenced. This allows us to write a code that will be called whenever object becomes unreferenced. The method "unreferenced" may be called more than once as the set of references are removed. The object will not be collected until all references are removed, including both local and remote references.

When the network fails, the object will be collected even if the remove reference exists, because the transport layer may think that a connection to a machine containing a reference to the object which no longer live. The server believes that the client no longer maintains a reference to the object. In this case, the object may be collected even though a remote reference still exists. If this occurs, then, when the client tries to access the remote object, it will result in a RemotException being thrown.

4. ADVANCED CONCEPTS IN RMI

a. Running the Application

The **EmpRegister** file has to be created before executing the application. The EmpRegister file should have the employee name and password of all the present employees. Whenever a new employee joins his name and password should be added to the list. Also, the application requires the name of the employees to be unique. The steps for creating this file are explained in the following example.

Example

1. Open the Notepad application.
2. Type the name and password for the employee, separated by a space.
3. Press enter.
4. Repeat steps 2 and 3 for each employ
5. Save the file as **EmpRegister** in the **attendance** directory.

> **Note:** Do not press the enter-key after the last employee record because it will raise an exception while running the server.

Now we are ready to run the application. First we have to run the registry, followed by the server and then the client. The steps are explained in the following example.

Example

1. Open an instance of the MSDOS prompt.
2. Run the registry by giving the following command at the prompt:

   rmiregistry

3. Minimize the instance.
4. Open another instance of MSDOS prompt.
5. Set the CLASSPATH to indicate that the class files are in the examples directory. Suppose the examples directory is in C:\, then we give:

   Set CLASSPATH=C:\examples

6. Change the current directory to examples and give the following command to run the server:

   Java attendance.EmpServer

7. Minimize the instance.
8. Open a third instance of the MSDOS prompt.
9. Change the current directory to examples and give the following command to run the client:

   appletviewer EmpApp.html

An applet will be opened and the connection with the server will be established.

10. Enter the employee name and password, say,

Name : **Krishna**
Password: **KK**

11. Click the "Click to Enter/Exit" button.
    The current time will be registered as the entry time for the employee named "Krishna".
12. Repeat steps 10 and 11 for the employee "ponraj".
13. Repeat steps 10 and 11 for the employee "Krishna".

Since the employee has already registered his entry time, the current time will be taken as his exit time.

14. Close the client.
15. Close the server and the registry.

In the **examples** directory, a file named **EAttendance** will be created and an entry will be made in it as follows:

```
Krishna Tue Aug 03 11:21:21 GMT+05 : 30
1999 Tue Aug 03 11:27: 52 GMT+05:30 2013
```

If the file exists already in the specified directory, then the above entry will be appended to its contents.

Note : Do not close the instances for the registry and the server before closing the client.

Closing the client does not end the application. In case an employee's details with respect to his/her entry time are registered and the instance is close, we can open another instance and register his exit time. This is illustrated in the following example.

Example

1. Repeat the steps 1 to 14 of the above example.
2. Open another instance of the client and enter the detains of the employee named "ponraj".

The current time will be considered as his exit time due to two factors:

- The employee, "ponraj" has already registered his entry time.
- We have closed only the client and the server which stores the data is not closed.

3. Close the client, server and the registry.

The following entry will be made in the **EAttendance** file.

```
Ponraj TueAu g 03 11:25:56 GMT+05:30
1999 T ue A ug 03 11:33:21 GMT +05:30 2013
```

b. Security

The Java security model considers applications in the local system to be safe. But, applets are are not considered to be safe. Most of the concerns over security are closely associated with network applications. This is because the code loaded from networks can be utterly malicious, or someone could be listening in on our conversations as they are transmitted over the network.

In order to avoid these risks the security manager must be set. In the case of an applet, the **AppletSecurityManager** is automatically set. But for stand-alone applications, the default security manager will prevent the system from loading classes from anywhere except but local file system. The **RMISecurityManager** extends the **java.lang.SecurityManager**.I which is used as security manager and is set using the command:

```
System.setSecurityManager(new RMISecurityManager());
```

It must be set before any remote stub classes are loaded over the network. The **RMISecurityManager** allows the stub classes only to load the necessary class files over the network. If **java.rmi.useCodebaseOnly** is set, then network loading of stub classes will be prevented, even if **RMISecurityManager** is set. Using the security manager is necessary only while using a firewall. Perfect Soluctions Limited does not use the firewall. Therefore, the security manager is not set in their application.

While passing parameters and return values to and from the methods in an application, we use the concept of serialization. Therefore, let us try to understand how object are serialized.

c. Serialization

In the RMI architecture, it is imperative to have some method of sending objects over network connections. Even if we have applications that pass only simple data types over the network, the data for stub objects need to be sent across the network. This is where serialization comes into play. Through object serialization an object can be duplicated by writing the values of its fields to a stream and recreating a copy of the object from these field values.

Recreation of an object is a copy of the object. Therefore, if we need to pass an object by reference, then passing by value, we can use a Remote object. A Remote object either extends **Java.rmi.server.UnicastRemoteObject** or is passed to**UnicastRemoteObject.exportObject(Remote).** In both these cases, the object implements the **Remote** interface. The seralizable classes in the standard Java class libraries are listed in Table 9.2 on the next page.

The following classes are not serializable:

- Classes that contain only or transient fields
- Classes that are used to represent objects or resources that are specific to a local Virtual Machine.

A serializable object must:

- Implement a **Serializable** interface (or)
- Implement a subclass of **Serializable** interface (or)
- Extend an object that implement the **Serializable** interface

The serializable interface is used as a mark for objects that should be permitted serialization. It does not define any fields or methods. Since transient and static objects are not serialized by default serialization, any such objects is reconstructed from the stream that it was sent to and the corresponding fields are set to their initial values. This provides data security.

Table 9.2

Serializable Classes	Packages	Serializable Classes	Packages
BitSet	java.util	GridLayout	java.awt
Boolean	java.lang	HashTable	java.util
Border Layout	java.awt	Image	ava.awt
CardLayout	java.awt	InetAddress	java.net
Character	java.lang	Insets	java.awt
CheckboxGroup	java.awt	MediaTracker	java.awt
Color	java.awt	MenuComponent	java.awt
Component	java.awt	Number	java.lang
Cursor	java.awt	Point	java.awt
Date	java.util	Polygon	java.awt
Dimension	java.awt	Random	java.until
Event	java.awt	Rectangle	java.awt
File	java.io	String	java.lang
FlowLayout	java.awt	StringBuffer	java.lang
Font	java.awt	SystemColor	java.awt
FontMetrics	java.awt	Throwable	java.lang
GridBagConstraints	java.awt	Vector	java.until
GridBagLayout	java.awt	Window	java.awt

d. Object Streams

An object stream, or a subclass of a stream that is capable of carrying objects, transmits a representation of its structure and related data. An object stream contains header information that is used to rebuild objects that it contains, along with the data for these objects. In order to transmit an object over an object stream, the system transmitting it, writes the type of that object and the values of each of its fields to the stream. To rebuild an object, the receiver does the following:

- Reads the type of the object from the stream
- Constructs a new objects of that type
- Reads the value of the fields for the object from the stream

Note : A stream is a series of zeros and ones.

For these steps, the run-time environment on the receiving side must have access to the bytecode for the class of the object. When the object is being recreated, the bytecode is loaded from the local system, if it is available. If it cannot be found locally, the **AppletClassLoader** or the **RMIClassLoader** attempts to load the class file. If the bytecode is inaccessible through both these mechanisms, a **ClassNotFoundException** is thrown and deserialization fails. All this happens automatically while using RMI.

Serialization is built upon stream objects. Stream objects implement either the **ObjectInput** or the **ObjectOutput** interfaces. DataInput interface includes methods for the input of primitive types. **ObjectInput** interfaces extends the DataInput interface to include objects, arrays and Strings. DataOutput interface includes methods for output of primitive types. **ObjectOutput** interface extends the DataOutput interface to include objects, arrays, and Strings.

Defining the Read/Write routines

If a class implements the Serlizable interface, default serialization takes place. But we can also have custom serialization methods. First let us understand how an object is sent down an object stream using default serialization. Let us consider the **EmpRec** class. Assume that the class is defined as follows:

```
Class EmpRec implements java.io.Serializable {
        String ename;
        EmpRec (String ename) {
                this .ename = ename;
        }
}
```

To send the EmpRec object to an object stream we have to use the following code:

```
EmpRec anObject  =  new EmpRec("Sudha");
FileOutputStream  fos  =  new FileOutputStream("temporary.tmp");
objectoutputstream outstr = new objectoutputstream(fos);
outstr.writeobject(anObject);
```

Using the default serialization methods, the system (here, the client) sends the values of **ename, EntryTime** and **ExitTime** to the stream along with header and version information that is used to identify the class type. The receiver (here, the server) can then rebuilt the object, assuming there is local access to the **EmpRec** bytecode, with the following code:

```
EmpRec anObject ;
FileIntputStream  fis  = new  FileIntputStream("temp.tmp");
objectinputstream instr = objectinputstream(fis);
anObject  = (EmpRec) instr.readobject( );
```

Now let us create the custom serialization methods. The following are the rules to create custom serialization:

- The object must implement **java.io.Serializable** (Rule 1)
- The object must define a constructor with no arguments (Rules 2)
- The object must define its own **readObject**( ) and **writeObject**( ) methods (Rules 3)

Using custom serialization methods we can rewrite Example 4.3 as follows:

```
class  EmpRec implements java.io.serializable (  / / By Rule  1
        string ename;
        public EmpRec( ) {
        }
                                        / /By Rule 2
        public EmpRec(string ename) {
                this.ename = ename;
        }
```

```
void writeObject (objectOutputStream outStr) throws
                                        java.io.IOException{
        outStr.writeChars(ename);
}
                                                // By Rule 3
void  readObject(objectinputStream inStr) throws
                                        java.io.IOException{
        ename = inStr.readString( );
    }
}
                                                //By Rule 3
```

The object reconstructed from the above code will not differ from an object reconstructed from default serialization. The advantage of writing these **readObject( )** and **writeObject( )** methods is that we can also handle special processing of data. While serializing an object that contains other objects, we must make sure that all of them are serializable. When the object is being serialized, a graph representing the object and the object references that it contains is constructed. When the object is written to an object stream, this graph is traversed and the data for all of the objects in the graph are serialized in the same manner. If we want to allow the reading and writing of some data provided that it satisfies some conditions, then the **readObject( )** and **writeObject( )** methods can throw a **NotSerializableException ( )** if access to these fields is not permitted. This prevents any further attempts to serialize or deserialize the object.

e. Externalized Objects

If we serialize an object that extends another object, the system automatically serializes and reconstitutes the parent object. The readObject and writeObject methods are concerned only with the classes to which they belong. Sometimes it is beneficial to handle the serialization of the entire object graph manually. In such cases, the object must extend the **Externalizable** interface which actually extends the **Serializable** interface.

The major difference between Externalizable and Serializable objects is that an Externalizable object does not have default serialization. An Externalizable object must define **readExternal( )** and **writeExternal( )** methods to handle serialization of the objects, traversal of the object graph and the serialization of the base classes (if any).

During serialization we use different protocols. The following paragraphs give an overview of these protocols.

f. Protocols

The **RMI protocol** makes use of two other protocols for its on-the-wire format: **Java Object Serialization Protocol** and **HTTP**. The Object Serialization protocol is used to marshal call and return data. The HTTP protocol is used to "POST" a remote method invocation and obtain return data when circumstances warrant.

Java Object Serialization Protocol

The **Java Object Serialization Protocol** is used to format the call and return data in RMI calls. Each method invocation's **CallData** contains the following:

- An **ObjectIdentifier** which is the target of the call
- An **Operation** which is a number representing the method to be invoked

- A **Hash** which is a number that verifies that client stub and remote object skeleton use the same stub protocol
- A list of zero or more **Arguments** for the call

RMI overrides the **annotateClass** of the **ObjectOutputStream** and **resolveClass** of the **ObjectInputStream**. Each class is annotated with the codebase URL that represents the location from which the class can be loaded. In the **annotateClass** method, the classloader that loaded the class is queried for its codebase URL. If the classloader is not null and the classloader has a not null codebase, then the codebase is written to the stream using the **ObjectOutputStream.writeObject** method; otherwise a null is written to the stream using the writeObject method.

The class annotation is resolved during deserialization using the **ObjectInputStream.resolveClass** method. The **resolveClass** method first reads the annotation via the **ObjectInputStream.readObject** method. If the annotation, a codebase URL, is not null, then it obtains the classloader for that URL and attempts to load the class. The class is loaded by using a **java.net.URLConnection** to fetch the class bytes.

HTTP Protocol

In RMI, the HTTP protocol is used to invoke remote methods through a firewall. The URL specified in the post header can be one on the following:

http://<host>:<port>/
http://<host>:80/cgi-bin/java- rmi?forward=<port>

The first URL is used for direct communication with an RMI server on the specific **host** and **port**. The second URL form is used to invoke a cgi script on the server, which forwards the invocation to the server on the specified **port.** CGI stands for Common Gateway Interface. It is used to connect dissimilar networks.

Multiplexing Protocol

Let us consider two endpoints trying to open multiple full duplex connections to the other endpoint. Suppose, only one of the endpoints is able to open such a bi-directional connection. In such a situation, the **multiplexing protocol** provides a model for both the endpoints to open multiple full duplex connections. RMI uses this simple multiplexing protocol to allow a client to connect to an RMI server object in some situations where that is otherwise not possible. Using the full duplex connection, simultaneous two-way communication is possible.

g. Exceptions in RMI

In Chapter 2, we have learnt about the important packages in RMI. In the following paragraphs, we will be learning about the exceptions in these packages.

Exception in java.rmi

The **NotBoundException** class extends the **Exception** class. A NotBoundException is thrown if an attempt is made to **lookup** or **unbind** a name that has no associated binding in the registry.

The **RemoteException** class extends the **IOException** class. A RemoteException is the common superclass for a number of communication-related exception that may occur during the exception of a remote method call. Each method of a remote interface must list RemoteException in its thrown clause. The following exceptions extends the RemoteException class.

An AccessException is thrown by certain methods of the **java.rmi.Naming** class to indicate that the caller does not have permission to perform the action requested by the method call.

An **AlreadyBoundException** is thrown if an attempt is made to bind an object in the registry to a name that already has an associated binding. To avoid this exception we can use the **rebind** method instead of the **bind** method.

A **ConnectException** is thrown if a connection is refused to the remote host for a remote method call.

A **ConnectIOException** is thrown if an IOException occurs while making a connection to the remote host for a remote method call.

A **MarshalException** is thrown if a **java.io.IOException** occurs while marshalling the remote call header, arguments or return value for a remote method call. A MarshalException is also thrown if the receiver does not support the protocol version of the sender.

A **NoSuchObjectException** is thrown if we attempt to invoke a method on a remote object that does not exist. A NoSuchObjectException is also thrown by the method **java.rmi.server.RemoteObject.toStub** and by the **unexportObject** method of **java.rmi.server.Unicast RemoteObject.**

A **ServerError** is thrown as a result of a remote method call if the execution of the remote method on the server machine throws a **java.lang.Error**. The ServerError contains a nested exception which is the **java.lang.Error** that occurred during a remote method exception.

A **ServerException** is thrown as a result of a remote method call if the exception of the remote method on the server machine thrown a RemoteException.

A **StubNotFoundException** is thrown if the client of a remote method call receives a checked exception that is not among the checked exception types declared in the throws clause of the method in the remote interface.

An **UnexpectedException** is thrown if the client of a remote method call receives a checked exception that is not among the checked exception types declared in the throws clause of the method in the remote interface.

An **UnknownHostException** is thrown if a java.net.UnknownHostException occurs while creating a connection to the remote host for a remote method call.

An **UnmarshalException** is thrown while unmarshalling the parameters or results of a remote method call if any of the following conditions occur:

- if an exception occurs while unmarshaling the call header
- if the protocol for the return value is invalid
- if a **java.io.IOException** occurs while unmarshalling parameters on the server side
- if a **java.io.IOException** occurs while the return value on the client side
- if a **java.lang.ClassNotFoundException** occurs during unmarshalling parameters or return values
- if the method hash is invalid (i.e. missing method).
- If there is a failure to create a remote reference object for a remote object's stub when it is unmarshalled.

A ServerRuntimeException is a deprecated exception that is thrown as a result of a remote method call if the execution of the remote method on the server machine throws a **java.lang.RuntimeException**. A ServerRuntimeException is not thrown from servers executing in JDK 1.2 or later versions.

The **RMISecurityException** class is a deprecated class that extends the **SecurityException** class. A RMISecurityException signals that a security exception has occurred during the exception of one of the methods in **java.rmi.RMISecurityManager.**

Exception in java.rmi.server

The **ExportException** extends the **RemoteException**. An ExportException is a RemoteException thrown if an attempt to export a remote object fails.

The **SocketSecurityException** extends the **ExportException.** A SocketSecurityException is thrown during a remote object export if the code exporting the remote object does not have permission to create an instance of the **ServerSocket** class of the **java.net** package on the port number specified.

The **ServerCloneException** extends the **CloneNotSupportedException**. A serverCloneException is thrown if a remote exception occurs during the cloning of a UnicastRemoteObject.

The **ServerNotActiveException** extends the **Exception.** If the **getClientHost**( ) method of the RemoteServer class is called outside of servicing a remote method call, the ServerNotActive3Exception is thrown.

The **SkeletonMismatchException** is a deprecated exception that extends the **RemoteException.** This exception is thrown when a call is received that does not match the available skeleton. It indicates either that the remote method names or signatures in this interface have changed or that the stub class used to make the call and the skeleton receiving the call were not generated by the same version of the stub compiler (**rmic**).

The SkeletonNotFoundException extends the **RemoteException**. A SkeletonNotFoundException is thrown if the skeleton corresponding to the remote object being exported is not found. Since skeletons are not required in JDK 1.2, this exception is deprecated.

Summary

- We must run the registry before running the application.
- The RMISecurityManager is used to set the security policies used to govern the behavior of the stub and skeleton classes.
- Object serialization is used to duplicate an object by writing the values of its fields to a stream and recreating a copy of the object from these field values.
- A remote object should implement the Remote interface.
- If a class implements a serializable interface default serialization takes place.
- Custom serialization methods must define its own readObject( ) and writeObject( ) methods.
- Since Externalizable objects do not have default serialization, they must define readExternal() and writeExternal( ) methods to handle serialization.
- The RMI protocol uses the Java Object Serialization protocol and HTTP for its on-the-wire format.
- The important exceptions in RMI are found in java.rmi and java.rmi.server packages.

Interview Questions

Language Fundamentals

1. **Which four options describe the correct default values for array elements of the types indicated?**

   1. int -> 0
   2. String -> "null"
   3. Dog -> null
   4. char -> '\u0000'
   5. float -> 0.0f
   6. boolean -> true

   A. 1, 2, 3, 4 **B.** 1, 3, 4, 5
   C. 2, 4, 5, 6 **D.** 3, 4, 5, 6

Ans. B. 1, 3, 4, 5

Explanation

(1), (3), (4), (5) are the correct statements.

(2) is wrong because the default value for a String (and any other object reference) is null, with no quotes.

(6) is wrong because the default value for boolean elements is false.

2. **Which one of these lists contains only Java programming language keywords?**

   A. class, if, void, long, Int, continue
   B. goto, instanceof, native, finally, default, throws
   C. try, virtual, throw, final, volatile, transient
   D. strictfp, constant, super, implements, do
   E. byte, break, assert, switch, include

Ans. B. goto, instanceof, native, finally, default, throws

Explanation

All the words in option B are among the 49 Java keywords. Although goto reserved as a keyword in Java, goto is not used and has no function.

Option A is wrong because the keyword for the primitive int starts with a lowercase i.

Option C is wrong because "virtual" is a keyword in C++, but not Java.

Option D is wrong because "constant" is not a keyword. Constants in Java are marked static and final.

Option E is wrong because "include" is a keyword in C, but not in Java.

3. **Which will legally declare, construct, and initialize an array?**

   A. int [] myList = {"1", "2", "3"};
   B. int [] myList = (5, 8, 2);
   C. int myList [] [] = {4,9,7,0};
   D. int myList [] = {4, 3, 7};

Ans. D. int myList [] = {4, 3, 7};

Explanation

The only legal array declaration and assignment statement is Option D

Option A is wrong because it initializes an int array with String literals.

Option B is wrong because it use something other than curly braces for the initialization.

Option C is wrong because it provides initial values for only one dimension, although the declared array is a two-dimensional array.

4. **Which is a reserved word in the Java programming language?**

   A. Method **B.** Native
   C. Subclasses **D.** Reference
   E. Array

Ans. B. Native

Explanation

The word "native" is a valid keyword, used to modify a method declaration.

Options A, D and E are not keywords. Option C is wrong because the keyword for subclassing in Java extends, not 'subclasses'.

5. **Which is a valid keyword in Java?**

   A. Interface **B.** String
   C. Float **D.** Unsigned

Ans. A. Interface

Explanation

Interface is a valid keyword.

Option B is wrong because although "String" is a class type in Java, "string" is not a keyword.

Option C is wrong because "Float" is a class type. The keyword for the Java primitive is float.

Option D is wrong because "unsigned" is a keyword in C/C++ but not in Java.

6. **Which three are legal array declarations?**
   1. int [] myScores [];
   2. char [] myChars;
   3. int [6] myScores;
   4. Dog myDogs [];
   5. Dog myDogs [7];

   A. 1, 2, 4 **B.** 2, 4, 5
   C. 2, 3, 4 **D.** All are correct.

Ans. A. 1, 2, 4

Explanation

(1), (2), and (4) are legal array declarations. With an array declaration, you can place the brackets to the right or left of the identifier. Option A looks strange, but it's perfectly legal to split the brackets in a multidimensional array, and place them on both sides of the identifier. Although coding this way would only annoy your fellow programmers, for the exam, you need to know it's legal.

(3) and (5) are wrong because you can't declare an array with a size. The size is only needed when the array is actually instantiated (and the JVM needs to know how much space to allocate for the array, based on the type of array and the size).

7. **public interface Foo**

```
{
    int k = 4; /* Line 3 */
}
```

   Which three piece of codes are equivalent to line 3?
   1. final int k = 4;
   2. public int k = 4;
   3. static int k = 4;
   4. abstract int k = 4;
   5. volatile int k = 4;
   6. protected int k = 4;

   A. 1, 2 and 3 **B.** 2, 3 and 4
   C. 3, 4 and 5 **D.** 4, 5 and 6

Ans. A. 1, 2 and 3

Explanation

(1), (2) and (3) are correct. Interfaces can have constants, which are always implicitly public, static, and final. Interface constant declarations of public, static, and final are optional in any combination.

8. **Which one of the following will declare an array and initialize it with five numbers?**

   A. Array a = new Array(5);
   B. int [] a = {23,22,21,20,19};
   C. int a [] = new int[5];
   D. int [5] array;

Ans. B. int [] a = {23,22,21,20,19};

Explanation

Option B is the legal way to declare and initialize an array with five elements.

Option A is wrong because it shows an example of instantiating a class named Array, passing the integer value 5 to the object's constructor. If you don't see the brackets, you can be certain there is no actual array object! In other words, an Array object (instance of class Array) is not the same as an array object.

Option C is wrong because it shows a legal array declaration, but with no initialization.

Option D is wrong (and will not compile) because it declares an array with a size. Arrays must never be given a size when declared.

9. **Which three are valid declarations of a char?**
   1. char c1 = 064770;
   2. char c2 = 'face';
   3. char c3 = 0xbeef;
   4. char c4 = \u0022;
   5. char c5 = '\iface';
   6. char c6 = '\uface';

   A. 1, 2, 4 **B.** 1, 3, 6
   C. 3, 5 **D.** 5 only

Ans. B. 1, 3, 6

Explanation

(1), (3), and (6) are correct. char c1 = 064770; is an octal representation of the integer value 27128, which is legal because it fits into an unsigned 16-bit integer. char c3 = 0xbeef; is a hexadecimal representation of the integer value 48879, which fits into an unsigned 16-bit integer. char c6 = '\uface'; is a Unicode representation of a character.

char c2 = 'face'; is wrong because you can't put more than one character in a char literal. The only other acceptable char literal that can go between single quotes is a Unicode value, and Unicode literals must always start with a '\u'.

char c4 = \u0022; is wrong because the single quotes are missing.

char c5 = '\iface'; is wrong because it appears to be a Unicode representation (notice the backslash), but starts with '\i' rather than '\u'.

10. Which is the valid declarations within an interface definition?

A. public double methoda();
B. public final double methoda();
C. static void methoda(double d1);
D. protected void methoda(double d1);

Ans. A. public double methoda();

Explanation

Option A is correct. A public access modifier is acceptable. The method prototypes in an interface are all abstract by virtue of their declaration, and should not be declared abstract.
Option B is wrong. The final modifier means that this method cannot be constructed in a subclass. A final method cannot be abstract.
Option C is wrong. Static is concerned with the class and not an instance.
Option D is wrong. Protected is not permitted when declaring a method of an interface. See information below.
Member declarations in an interface disallow the use of some declaration modifiers; you cannot use transient, volatile, or synchronized in a member declaration in an interface. Also, you may not use the private and protected specifiers when declaring members of an interface.

11. Which one is a valid declaration of a boolean?

A. boolean b1 = 0;
B. boolean b2 = false;
C. boolean b3 = false;
D. boolean b4 = Boolean.false();
E. boolean b5 = no;

Ans. C. boolean b3 = false;

Explanation

A boolean can only be assigned the literal true or false.

12. Which three are valid declarations of a float?

1. float f1 = –343;
2. float f2 = 3.14;
3. float f3 = 0x12345;
4. float f4 = 42e7;
5. float f5 = 2001.0D;
6. float f6 = 2.81F;

A. 1, 2, 4 **B.** 2, 3, 5
C. 1, 3, 6 **D.** 2, 4, 6

Ans. C. 1, 3, 6

Explanation

(1) and (3) are integer literals (32 bits), and integers can be legally assigned to floats (also 32 bits). (6) is correct because (F) is appended to the literal, declaring it as a float rather than a double (the default for floating point literals).
(2), (4),and (5) are all doubles.

13. Which is a valid declarations of a String?

A. String s1 = null;
B. String s2 = 'null';
C. String s3 = (String) 'abc';
D. String s4 = (String) '\ufeed';

Ans. A. String s1 = null;

Explanation

Option A sets the String reference to null.
Option B is wrong because null cannot be in single quotes.
Option C is wrong because there are multiple characters between the single quotes ('abc').
Option D is wrong because you can't cast a char (primitive) to a String (object).

14. What is the numerical range of a char?

A. –128 to 127 **B.** –(215) to (215) –1
C. 0 to 32767 **D.** 0 to 65535

Ans. D. 0 to 65535

Explanation

A char is really a 16-bit integer behind the scenes, so it supports 216 (from 0 to 65535) values.

15. What will be the output of the program?

```
public class CommandArgsThree
{
    public static void main(String [] args)
    {
        String [][] argCopy = new String[2][2];
        int x;
        argCopy[0] = args;
        x = argCopy[0].length;
        for (int y = 0; y < x; y++)
        {
        System.out.print(" " + argCopy[0][y]);
        }
    }
}
```

and the command-line invocation is
> java CommandArgsThree 1 2 3

A. 0 0 **B.** 1 2
C. 0 0 0 **D.** 1 2 3

Ans. D. 1 2 3

Explanation

In argCopy[0] = args;, the reference variable argCopy[0], which was referring to an array with two elements, is reassigned to an array (args) with three elements.

16. What will be the output of the program?

```
public class CommandArgs
{
    public static void main(String [] args)
    {
        String s1 = args[1];
        String s2 = args[2];
        String s3 = args[3];
        String s4 = args[4];
        System.out.print(" args[2] = " + s2);
    }
}
```

and the command-line invocation is

> java CommandArgs 1 2 3 4

A. args[2] = 2

B. args[2] = 3

C. args[2] = null

D. An exception is thrown at runtime.

Ans. D. An exception is thrown at runtime

Explanation

An exception is thrown because in the code String s4 = args[4];, the array index (the fifth element) is out of bounds. The exception thrown is cleverly named ArrayIndexOutOfBoundsException.

17. public class F0091

```
{
    public void main( String[] args )
    {
        System.out.println( "Hello" + args[0] );
    }
}
```

What will be the output of the program, if this code is executed with the command line:

> java F0091 world

A. Hello

B. Hello Foo91

C. Hello world

D. The code does not run.

Ans. D. The code does not run

Explanation

Option D is correct. A runtime error will occur owing to the main method of the code fragment not being declared static:

Exception in thread "main" java.lang.NoSuchMethodError: main

The Java Language Specification clearly states: "The main method must be declared public, static, and void. It must accept a single argument that is an array of strings."

18. What will be the output of the program?

```
public class TestDogs
{
    public static void main(String [] args)
    {
        Dog [][] theDogs = new Dog[3][];
        System.out.println(theDogs[2][0].toString());
    }
}
class Dog { }
```

A. null

B. theDogs

C. Compilation fails

D. An exception is thrown at runtime

Ans. D. An exception is thrown at runtime

Explanation

The second dimension of the array referenced by theDogs has not been initialized. Attempting to access an uninitialized object element (System.out.println(theDogs[2][0].toString());) raises a NullPointerException.

19. What will be the output of the program ?

```
public class Test
{
    public static void main(String [] args)
    {
        signed int x = 10;
        for (int y=0; y<5; y++, x—)
            System.out.print(x + ", ");
    }
}
```

A. 10, 9, 8, 7, 6

B. 9, 8, 7, 6, 5

C. Compilation fails.

D. An exception is thrown at runtime.

Ans. C. Compilation fails

Explanation

The word "signed" is not a valid modifier keyword in the Java language. All number primitives in Java are signed. Hence the compilation will fail.

20. What will be the output of the program?

```
public class CommandArgsTwo
{
    public static void main(String [] argh)
    {
        int x;
        x = argh.length;
        for (int y = 1; y <= x; y++)
        {
            System.out.print(" " + argh[y]);
        }
    }
}
```

and the command-line invocation is

> java CommandArgsTwo 1 2 3

A. 0 1 2

B. 1 2 3

C. 0 0 0

D. An exception is thrown at runtime

Ans. D. An exception is thrown at runtime

Explanation

An exception is thrown because at some point in (System.out.print(" " + argh[y]);), the value of x will be equal to y, resulting in an attempt to access an index out of bounds for the array. Remember that you can access only as far as length - 1, so loop logical tests should use x < someArray.length as opposed to x < = someArray.length.

21. In the given program, how many lines of output will be produced?

```
public class Test
{
    public static void main(String [] args)
    {
    int [] [] [] x = new int [3] [] [];
    int i, j;
    x[0] = new int[4][];
    x[1] = new int[2][];
    x[2] = new int[5][];
    for (i = 0; i < x.length; i++)
    {
        for (j = 0; j < x[i].length; j++)
        {
            x[i][j] = new int [i + j + 1];
            System.out.println("size = " +
x[i][j].length);
        }
    }
    }
}
```

A. 7 **B.** 9

C. 11 **D.** 13

E. Compilation fails

Ans. C. 11

Explanation

The loops use the array sizes (length).

It produces 11 lines of output as given below.

```
D:\Java>javac Test.java
D:\Java>java Test
size = 1
size = 2
size = 3
size = 4
size = 2
size = 3
size = 3
size = 4
size = 5
size = 6
size = 7
```

Therefore, 11 is the answer.

22. What will be the output of the program?

```
public class X
{
    public static void main(String [] args)
    {
        String names [] = new String[5];
        for (int x=0; x < args.length; x++)
            names[x] = args[x];
        System.out.println(names[2]);
    }
}
```

and the command line invocation is

> java X a b

A. Names

B. Null

C. Compilation fails

D. An exception is thrown at runtime

Ans. B. Null

Explanation

The names array is initialized with five null elements. Then elements 0 and 1 are assigned the String values "a" and "b" respectively (the command-line arguments passed to main). Elements of names array 2, 3, and 4 remain unassigned, so they have a value of null.

Declarations and Access Control

23. You want subclasses in any package to have access to members of a superclass. Which is the most restrictive access that accomplishes this objective?

A. Public **B.** Private

C. Protected **D.** Transient

Ans. C. Protected

Explanation

Access modifiers dictate which classes, not which instances, may access features.

Methods and variables are collectively known as members. Method and variable members are given access control in exactly the same way.

Private makes a member accessible only from within its own class.

Protected makes a member accessible only to classes in the same package or subclass of the class.

Default access is very similar to protected (make sure you spot the difference) default access makes a member accessible only to classes in the same package.

Public means that all other classes regardless of the package that they belong to, can access the member (assuming the class itself is visible).

Final makes it impossible to extend a class, when applied to a method it prevents a method from being overridden in a subclass, when applied to a variable it makes it impossible to reinitialise a variable once it has been initialised.

Abstract declares a method that has not been implemented.

Transient indicates that a variable is not part of the persistent state of an object.

Volatile indicates that a thread must reconcile its working copy of the field with the master copy every time it accesses the variable.

After examining the above it should be obvious that the access modifier that provides the most restrictions for methods to be accessed from the subclasses of the class from another package is C - protected. A is also a contender but C is more restrictive, B would be the answer if the constraint was the "same package" instead of "any package", in other words, the subclasses clause in the question eliminates default.

24.
```
public class Outer
{
    public void someOuterMethod()
    {
        //Line 5
    }
    public class Inner { }

    public static void main(String[] argv)
    {
        Outer ot = new Outer();
        //Line 10
    }
}
```

Which of the following code fragments inserted, will allow to compile?

A. new Inner(); //At line 5

B. new Inner(); //At line 10

C. new ot.Inner(); //At line 10

D. new Outer.Inner(); //At line 10

Ans. A. new Inner(); //At line 5

Explanation

Option A compiles without problem.

Option B gives error—non-static variable cannot be referenced from a static context.

Option C package does not exist.

Option D gives error—non-static variable cannot be referenced from a static context.

25.
```
interface Base
{
    boolean m1 ();
    byte m2(short s);
}
```

which two code fragments will compile?

```
interface Base2 implements Base {}
abstract class Class2 extends Base
{ public boolean m1(){ return true; }}
abstract class Class2 implements Base {}
abstract class Class2 implements Base
{ public boolean m1(){ return (7 > 4); }}
abstract class Class2 implements Base
{ protected boolean m1(){ return (5 > 7) }}
```

A. 1 and 2 **B.** 2 and 3

C. 3 and 4 **D.** 1 and 5

Ans. C. 3 and 4

Explanation

(3) is correct because an abstract class doesn't have to implement any or all of its interface's methods. (4) is correct because the method is correctly implemented ((7 > 4) is a boolean).

(1) is incorrect because interfaces don't implement anything. (2) is incorrect because classes don't extend interfaces. (5) is incorrect because interface methods are implicitly public, so the methods being implemented must be public.

26. Which three forms part of correct array declarations?

1. public int a [ ]
2. static int [ ] a
3. public [ ] int a
4. private int a [3]
5. private int [3] a [ ]
6. public final int [ ] a

A. 1, 3, 4 **B.** 2, 4, 5
C. 1, 2, 6 **D.** 2, 5, 6

Ans. C. 1, 2, 6

Explanation

(1), (2) and (6) are valid array declarations.
Option (3) is not a correct array declaration. The compiler complains with: illegal start of type. The brackets are in the wrong place. The following would work: public int[ ] a
Option (4) is not a correct array declaration. The compiler complains with: ']' expected. A closing bracket is expected in place of the 3. The following works: private int a []
Option (5) is not a correct array declaration. The compiler complains with 2 errors:
']' expected. A closing bracket is expected in place of the 3 and
<identifier> expected A variable name is expected after a[ ].

27. public class Test { }
What is the prototype of the default constructor?

A. Test( ) **B.** Test(void)
C. public Test( ) **D.** public Test(void)

Ans. C. public Test( )

Explanation

Options A and B are wrong because they use the default access modifier and the access modifier for the class is public (remember, the default constructor has the same access modifier as the class).
Option D is wrong. The void makes the compiler think that this is a method specification—in fact if it were a method specification, the compiler would spit it out.

28. What is the most restrictive access modifier that will allow members of one class to have access to members of another class in the same package?

A. Public **B.** Abstract
C. Protected **D.** Synchronized
E. Default access

Ans. E. Default access

Explanation

Default access is the "package oriented" access modifier.
Options A and C are wrong because public and protected are less restrictive. Options B and D are wrong because abstract and synchronized are not access modifiers.

29. Which of the following is/are legal method declarations?

1. protected abstract void m1();
2. static final void m1(){}
3. synchronized public final void m1() {}
4. private native void m1();

A. 1 and 3
B. 2 and 4
C. 1 only
D. All of them are legal declarations.

Ans. D. All of them are legal declarations

All the given statements are legal declarations.

30. Which cause a compiler error?

A. int[ ] scores = {3, 5, 7};
B. int [ ][ ] scores = {2,7,6}, {9,3,45};
C. String cats[ ] = {"Fluffy", "Spot", "Zeus"};
D. boolean results[ ] = new boolean [] {true, false, true};
E. Integer results[ ] = {new Integer(3), new Integer(5), new Integer(8)};

Ans. B. int [ ][ ] scores = {2,7,6}, {9,3,45};

Explanation

Option B generates a compiler error: <identifier> expected. The compiler thinks you are trying to create two arrays because there are two array initialisers to the right of the equals, whereas your intention was to create one 3 x 3 two-dimensional array.
To correct the problem and make option B compile you need to add an extra pair of curly brackets:
int [ ] [ ] scores = { {2,7,6}, {9,3,45} };

31. Which three are valid method signatures in an interface?
1. private int getArea();
2. public float getVol(float x);
3. public void main(String [] args);
4. public static void main(String [] args);
5. boolean setFlag(Boolean [] test);

A. 1 and 2 **B.** 2, 3 and 5
C. 3, 4, and 5 **D.** 2 and 4

Ans. B. 2, 3 and 5

Explanation

(2), (3), and (5). These are all valid interface method signatures.
(1) is incorrect because an interface method must be public; if it is not explicitly declared public, it will be made public implicitly. (4) is incorrect because interface methods cannot be static.

32. You want a class to have access to members of another class in the same package. Which is the most restrictive access that accomplishes this objective?

A. Public **B.** Private
C. Protected **D.** Default access

Ans. D. Default access

Explanation

The only two real contenders are C and D. Protected access option C makes a member accessible only to classes in the same package or subclass of the class. While default access option D makes a member accessible only to classes in the same package.

33. What is the widest valid returnType for methodA in line 3?

```
public class ReturnIt
{
  returnType methodA(byte x, double y) /
  * Line 3 */
  {
    return (long)x / y * 2;
  }
}
```

A. Int **B.** Byte
C. Long **D.** Double

Ans. D. Double

Explanation

However, A, B and C are all wrong. Each of these would result in a narrowing conversion. Whereas we want a widening conversion, therefore the only correct answer is D. Don't be put off by the long cast, this applies only to the variable x and not the rest of the expression. It is the variable y (of type double) that forces the widening conversion to double.

Java's widening conversions are:
- From a byte to a short, an int, a long, a float, or a double.
- From a short, an int, a long, a float, or a double
- From a char to an int, a long, a float, or a double.
- From an int to a long, a float, or a double.
- From a long to a float, or a double.
- From a float to a double.

34. class A

```
{
  protected int method1(int a, int b)
  {
    return 0;
  }
}
```

Which is valid in a class that extends class A?

A. public int method1(int a, int b) {return 0; }
B. private int method1(int a, int b) { return 0; }
C. public short method1(int a, int b) { return 0; }
D. static protected int method1(int a, int b) { return 0; }

Ans. A. public int method1(int a, int b) {return 0; }

Explanation

Option A is correct—because the class that extends A is just simply overriding method1.
Option B is wrong—because it can't override as there are less access privileges in the subclass method1.Option C is wrong—because to override it, the return type needs to be an integer. The different return type means that the method is not overriding but the same argument list means that the method is not overloading. Conflict—compile time error.
Option D is wrong—because you can't override a method and make it a class method, i.e. using static.

35. Which one creates an instance of an array?

A. int[ ] ia = new int[15];
B. float fa = new float[20];
C. char[ ] ca = "Some String";
D. int ia[ ] [ ] = { 4, 5, 6 }, { 1,2,3 };

Ans. A. int[ ] ia = new int[15];

Explanation

Option A is correct. It uses correct array declaration and correct array construction.

Option B is incorrect. It generates a compiler error: Incompatible types because the array variable declaration is not correct. The array construction expects a reference type, but it is supplied with a primitive type in the declaration. Option C is incorrect. It generates a compiler error: incompatible types because a string literal is not assignable to a character type variable. Option D is wrong, it generates a compiler error <identifier> expected. The compiler thinks that you are trying to create two arrays because there are two array initialisers to the right of the equals, whereas your intention was to create a 3 x 3 two-dimensional array.

36. Which two of the following are legal declarations for nonnested classes and interfaces?

1. final abstract class Test {}
2. public static interface Test {}
3. final public class Test {}
4. protected abstract class Test {}
5. protected interface Test {}
6. abstract public class Test {}

A. 1 and 4 **B.** 2 and 5
C. 3 and 6 **D.** 4 and 6

Ans. C. 3 and 6

Explanation

(3) and (6). Both are legal class declarations.

(1) is wrong because a class cannot be abstract and final—there would be no way to use such a class. (2) is wrong because interfaces and classes cannot be marked as static. (4) and (5) are wrong because classes and interfaces cannot be marked as protected.

37. Which of the following class level (nonlocal) variable declarations will not compile?

A. protected int a;

B. transient int b = 3;

C. private synchronized int e;

D. volatile int d;

Ans. C. private synchronized int e;

Explanation

Option C will not compile; the synchronized modifier applies only to methods. Options A and B will compile because protected and transient are legal variable modifiers. Option D will compile because volatile is a proper variable modifier.

38. Which two cause a compiler error?

1. float[ ] f = new float(3);
2. float f2[ ] = new float[ ];
3. float[ ]f1 = new float[3];
4. float f3[ ] = new float[3];
5. float f5[ ] = {1.0f, 2.0f, 2.0f};

A. 2, 4 **B.** 3, 5
C. 4, 5 **D.** 1, 2

Ans. D. 1, 2

Explanation

(1) causes two compiler errors ('[' expected and illegal start of expression) because the wrong type of bracket is used, ( ) instead of [ ]. The following is the correct syntax: float[ ] f = new float[3];

(2) causes a compiler error ('{' expected) because the array constructor does not specify the number of elements in the array. The following is the correct syntax: float f2[ ] = new float[3];

(3), (4), and (5) compile without error.

39. Given a method in a protected class, what access modifier do you use to restrict access to that method to only the other members of the same class?

A. Final **B.** Static
C. Private **D.** Protected
E. Volatile

Ans. C. Private

Explanation

The private access modifier limits access to members of the same class.

Options A, B, D, and E are wrong because protected are the wrong access modifiers, and final, static, and volatile are modifiers but not access modifiers.

40. Which is a valid declaration within an interface?

A. public static short stop = 23;

B. protected short stop = 23;

C. transient short stop = 23;

D. final void madness(short stop);

Ans. A. public static short stop = 23;

Explanation

(A) is valid interface declarations.

(B) and (C) are incorrect because interface variables cannot be either protected or transient.

(D) is incorrect because interface methods cannot be final or static.

Find the Output

41. What will be the output of the program?

```
class A
{
    final public int GetResult(int a, int b) {
return 0; }
}
class B extends A
{
    public int GetResult(int a, int b) {return
1; }
}
public class Test
{
    public static void main(String args[])
    {
        B b = new B();
            System.out.println("x = " +
b.GetResult(0, 1));
    }
}
```

A. x = 0
B. x = 1
C. Compilation fails.
D. An exception is thrown at runtime.

Ans. C. Compilation fails.

Explanation

The code doesn't compile because the method GetResult() in class A is final and so cannot be overridden.

42. What will be the output of the program?

```
public class Test
{
    public static void main(String args[])
    {
        class Foo
        {
            public int i = 3;
        }
        Object o = (Object)new Foo();
        Foo foo = (Foo)o;
        System.out.println("i = " + foo.i);
    }
}
```

A. i = 3
B. Compilation fails.
C. i = 5
D. A ClassCastException will occur.

Ans. A. i = 3

Explanation

No answer description available for this question. Let us discuss.

43. What will be the output of the program?

```
public class A
{
    void A() /* Line 3 */
    {
        System.out.println("Class A");
    }
    public static void main(String[] args)
    {
        new A();
    }
}
```

A. Class A
B. Compilation fails.
C. An exception is thrown at line 3.
D. The code executes with no output.

Ans. D. The code executes with no output

Explanation

Option D is correct. The specification at line 3 is for a method and not a constructor and this method is never called therefore there is no output. The constructor that is called is the default constructor.

44. What will be the output of the program?

```
class Super
{
    public int i = 0;

    public Super(String text) /* Line 4 */
    {
        i = 1;
    }
}

class Sub extends Super
{
    public Sub(String text)
    {
        i = 2;
    }
    public static void main(String args[])
    {
```

```
        Sub sub = new Sub("Hello");
        System.out.println(sub.i);
    }
}
```

A. 0 **B.** 1

C. 2 **D.** Compilation fails.

Ans. D. Compilation fails

Explanation

A default no-args constructor is not created because there is a constructor supplied that has an argument, line 4. Therefore, the sub-class constructor must explicitly make a call to the superclass constructor:

```
public Sub(String text)
{
    super(text); // this must be the first line constructor
    i = 2;
}
```

45. What will be the output of the program?

```
public class Test
{
    public int aMethod()
    {
        static int i = 0;
        i++;
        return i;
    }
    public static void main(String args[])
    {
        Test test = new Test();
        test.aMethod();
        int j = test.aMethod();
        System.out.println(j);
    }
}
```

A. 0 **B.** 1

C. 2 **D.** Compilation fails.

Ans. D. Compilation fails

Explanation

Compilation failed because static was an illegal start of expression—method variables do not have a modifier (they are always considered local).

46. What will be the output of the program?

```
interface Count
{
    short counter = 0;
    void countUp();
}
public class TestCount implements Count
{
    public static void main(String [] args)
    {
        TestCount t = new TestCount();
        t.countUp();
    }
    public void countUp()
    {
        for (int x = 6; x>counter; x—, ++counter) /* Line 14 */
        {
            System.out.print(" " + counter);
        }
    }
}
```

A. 0 1 2 **B.** 1 2 3

C. 0 1 2 3 **D.** 1 2 3 4

E. Compilation fails

Ans. E. Compilation fails

Explanation

The code will not compile because the variable counter is an interface variable that is by default final static. The compiler will complain at line 14 when the code attempts to increment counter.

47. What will be the output of the program?

```
class Base
{
    Base()
    {
        System.out.print("Base");
    }
}
public class Alpha extends Base
{
    public static void main(String[] args)
    {
        new Alpha(); /* Line 12 */
        new Base(); /* Line 13 */
    }
}
```

A. Base

B. BaseBase

C. Compilation fails

D. The code runs with no output

Ans. B. BaseBase

Explanation

Option B is correct. It would be correct if the code had compiled, and the subclass Alpha had been saved in its own file. In this case Java supplies an implicit call from the subclass constructor to the no-args constructor of the superclass, therefore line 12 causes Base to be output. Line 13 also causes Base to be output.

Option A is wrong. It would be correct if either the main class or the subclass had not been instantiated.

Option C is wrong. The code compiles.

Option D is wrong. There is output.

48. What will be the output of the program?

```
import java.util.*;
public class NewTreeSet2 extends NewTreeSet
{
    public static void main(String [] args)
    {
        NewTreeSet2 t = new NewTreeSet2();
        t.count();
    }
}
protected class NewTreeSet
{
    void count()
    {
        for (int x = 0; x < 7; x++,x++ )
        {
            System.out.print(" " + x);
        }
    }
}
```

A. 0 2 4

B. 0 2 4 6

C. Compilation fails at line 2

D. Compilation fails at line 10

Ans. D. Compilation fails at line 10

Explanation

Nonnested classes cannot be marked protected (or final for that matter), so the compiler will fail at protected class NewTreeSet.

49. What will be the output of the program?

```
public class ArrayTest
{
    public static void main(String[ ] args)
    {
        float f1[ ], f2[ ];
        f1 = new float[10];
        f2 = f1;
        System.out.println("f2[0] = " + f2[0]);
    }
}
```

A. It prints f2[0] = 0.0

B. It prints f2[0] = NaN

C. An error at f2 = f1; causes compile to fail.

D. It prints the garbage value.

Ans. A. It prints f2[0] = 0.0

Explanation

Option A is correct. When you create an array (f1 = new float[10];) the elements are initialises to the default values for the primitive data type (float in this case - 0.0), so f1 will contain 10 elements each with a value of 0.0. f2 has been declared but has not been initialised, it has the ability to reference or point to an array but as yet does not point to any array. f2 = f1; copies the reference (pointer/memory address) of f1 into f2 so now f2 points at the array pointed to by f1.

This means that the values returned by f2 are the values returned by f1. Changes to f1 are also changes to f2 because both f1 and f2 point to the same array.

50. What will be the output of the program?

```
class Super
{
    public Integer getLength()
    {
        return new Integer(4);
    }
}

public class Sub extends Super
{
    public Long getLength()
    {
        return new Long(5);
    }

    public static void main(String[] args)
    {
        Super sooper = new Super();
        Sub sub = new Sub();
        System.out.println(
            sooper.getLength().toString() + "," +
sub.getLength().toString() );
    }
}
```

A. 4, 4

B. 4, 5

C. 5, 4

D. Compilation fails.

Ans. D. Compilation fails.

Explanation

Option D is correct, compilation fails: The return type of getLength( ) in the superclass is an object of reference type Integer and the return type in the sub class is an object of reference type Long. In other words, it is not an override because of the change in the return type and it is also not an overload because the argument list has not changed.

Choose the Correct Statement

51. interface DoMath

```
{
    double getArea(int rad);
}
interface MathPlus
{
    double getVol(int b, int h);
}
/* Missing Statements ? */
```

Which two code fragments inserted at end of the program, will allow to compile?

1. class AllMath extends DoMath { double getArea(int r); }
2. interface AllMath implements MathPlus { double getVol(int x, int y); }
3. interface AllMath extends DoMath { float getAvg(int h, int l); }
4. class AllMath implements MathPlus { double getArea(int rad); }
5. abstract class AllMath implements DoMath, MathPlus { public double getArea(int rad) { return rad * rad * 3.14; } }

A. 1 only **B.** 2 only
C. 3 and 5 **D.** 1 and 4

Ans. C. 3 and 5

Explanation

(3) are (5) are correct because interfaces and abstract classes do not need to fully implement the interfaces they extend or implement (respectively).

(1) is incorrect because a class cannot extend an interface. (2) is incorrect because an interface cannot implement anything. (4) is incorrect because the method being implemented is from the wrong interface.

52. Which two statements are true for any concrete class implementing the java.lang.Runnable interface?

1. You can extend the Runnable interface as long as you override the public run() method.
2. The class must contain a method called run() from which all code for that thread will be initiated.
3. The class must contain an empty public void method named run().
4. The class must contain a public void method named runnable().
5. The class definition must include the words implements Threads and contain a method called run().
6. The mandatory method must be public, with a return type of void, must be called run(), and cannot take any arguments.

A. 1 and 3 **B.** 2 and 4
C. 1 and 5 **D.** 2 and 6

Ans. D. 2 and 6

Explanation

(2) and (6). When a thread's run() method completes, the thread will die. The run() method must be declared public void and not take any arguments.

(1) is incorrect because classes can never extend interfaces. (3) is incorrect because the run() method is typically not empty; if it were, the thread would do nothing. (4) is incorrect because the mandatory method is run(). (5) is incorrect because the class implements Runnable.

53. /* Missing statements ? */

```
public class NewTreeSet extends java.util.TreeSet
{
    public static void main(String [] args)
    {
        java.util.TreeSet t = new java.util.TreeSet();
        t.clear();
    }
    public void clear()
    {
        TreeMap m = new TreeMap();
        m.clear();
    }
}
```

Which two statements, added independently at beginning of the program, allow the code to compile?

1. No statement is required
2. import java.util.*;
3. import.java.util.Tree*;
4. import java.util.TreeSet;
5. import java.util.TreeMap;

A. 1 only **B.** 2 and 5
C. 3 and 4 **D.** 3 and 5

Ans. B. 2 and 5

Explanation

(2) and (5). TreeMap is the only class that must be imported. TreeSet does not need an import statement because it is described with a fully qualified name.

(1) is incorrect because TreeMap must be imported. (3) is incorrect syntax for an import statement. (4) is incorrect because it will not import TreeMap, which is required.

54. Which three statements are true?

1. The default constructor initialises method variables.
2. The default constructor has the same access as its class.
3. The default constructor invokes the no-arg constructor of the superclass.
4. If a class lacks a no-arg constructor, the compiler always creates a default constructor.
5. The compiler creates a default constructor only when there are no other constructors for the class.

A. 1, 2 and 4 **B.** 2, 3 and 5
C. 3, 4 and 5 **D.** 1, 2 and 3

Ans. B. 2, 3 and 5

Explanation

(2) sounds correct as in the example below

```
class CoffeeCup {
  private int innerCoffee;
  public CoffeeCup() {
  }

  public void add(int amount) {
  innerCoffee += amount;
  }
  //...
}
```

The compiler gives default constructors the same access level as their class. In the example above, class CoffeeCup is public, so the default constructor is public. If CoffeeCup had been given package access, the default constructor would be given package access as well.

(3) is correct. The Java compiler generates at least one instance initialisation method for every class it compiles. In the Java class file, the instance initialisation method is named "<init>." For each constructor in the source code of a class, the Java compiler generates one <init>() method. If the class declares no constructors explicitly, the compiler generates a default no-arg constructor that just invokes the superclass's no-arg constructor. As with any other constructor, the compiler creates an <init>() method in the class file that corresponds to this default constructor.

(5) is correct. The compiler creates a default constructor if you do not declare any constructors in your class.

55. package testpkg.p1;

```
  public class ParentUtil
  {
    public int x = 420;
    protected int doStuff() { return x; }
  }

package testpkg.p2;
import testpkg.p1.ParentUtil;
public class ChildUtil extends ParentUtil
{
  public static void main(String [] args)
  {
    new ChildUtil().callStuff();
  }
  void callStuff()
  {
        System.out.print("this " +
this.doStuff() ); /* Line 18 */
    ParentUtil p = new ParentUtil();
      System.out.print(" parent " +
p.doStuff() ); /* Line 20 */
  }
}
```

which statement is true?

A. The code compiles and runs, with output this 420 parent 420.

B. If line 18 is removed, the code will compile and run.

C. If line 20 is removed, the code will compile and run.

D. An exception is thrown at runtime.

Ans. C. If line 20 is removed, the code will compile and run.

Explanation

The ParentUtil instance p cannot be used to access the doStuff() method. Because doStuff() has protected access, and the ChildUtil class is not in the same package as the ParentUtil class, doStuff() can be accessed only by instances of the ChildUtil class (a subclass of ParentUtil). Options A, B, D, and E are incorrect because of the access rules described previously.

Operators and Assignments

56. What will be the output of the program?

```
class PassA
{
    public static void main(String [] args)
    {
        PassA p = new PassA();
        p.start();
    }

    void start()
    {
        long [] a1 = {3,4,5};
        long [] a2 = fix(a1);
        System.out.print(a1[0] + a1[1] + a1[2]
+ " ");
            System.out.println(a2[0] + a2[1] +
a2[2]);
    }

    long [] fix(long [] a3)
    {
        a3[1] = 7;
        return a3;
    }
}
```

A. 12 15 **B.** 15 15

C. 3 4 5 3 7 5 **D.** 3 7 5 3 7 5

Ans. B. 15 15

Explanation

Output: 15 15

The reference variables a1 and a3 refer to the same long array object. When the [1] element is updated in the fix() method, it is updating the array referred to by a1. The reference variable a2 refers to the same array object.

So Output: 3+7+5+" "3+7+5

Output: 15 15 because numeric values will be added.

57. What will be the output of the program?

```
class Test
{
    public static void main(String [] args)
    {
        Test p = new Test();
        p.start();
    }

    void start()
    {
        boolean b1 = false;
        boolean b2 = fix(b1);
        System.out.println(b1 + " " + b2);
    }

    boolean fix(boolean b1)
    {
        b1 = true;
        return b1;
    }
}
```

A. true true **B.** false true

C. true false **D.** false false

Ans. B. false true

Explanation

The boolean b1 in the fix() method is a different boolean than the b1 in the start() method. The b1 in the start() method is not updated by the fix() method.

58. What will be the output of the program?

```
class PassS
{
    public static void main(String [] args)
    {
        PassS p = new PassS();
        p.start();
    }
```

```
void start()
{
    String s1 = "slip";
    String s2 = fix(s1);
    System.out.println(s1 + " " + s2);
}

String fix(String s1)
{
    s1 = s1 + "stream";
    System.out.print(s1 + " ");
    return "stream";
}
}
```

A. slip stream
B. slipstream stream
C. stream slip stream
D. slipstream slip stream

Ans. D. slipstream slip stream

Explanation

When the fix() method is first entered, start()'s s1 and fix()'s s1 reference variables both refer to the same String object (with a value of "slip"). Fix()'s s1 is reassigned to a new object that is created when the concatenation occurs (this second String object has a value of "slipstream"). When the program returns to start(), another String object is created, referred to by s2 and with a value of "stream".

59. What will be the output of the program?

```
class BitShift
{
    public static void main(String [] args)
    {
        int x = 0x80000000;
        System.out.print(x + " and ");
        x = x >>> 31;
        System.out.println(x);
    }
}
```

A. –2147483648 and 1
B. 0x80000000 and 0x00000001
C. –2147483648 and –1
D. 1 and –2147483648

Ans. A. –2147483648 and 1

Explanation

Option A is correct. The >>> operator moves all bits to the right, zero filling the left bits. The bit transformation looks like this:

Before: 1000 0000 0000 0000 0000 0000 0000 0000
After: 0000 0000 0000 0000 0000 0000 0000 0001

Option C is incorrect because the >>> operator zero fills the left bits, which in this case changes the sign of x, as shown.
Option B is incorrect because the output method print() always displays integers in base 10
Option D is incorrect because this is the reverse order of the two output numbers.

60. What will be the output of the program?

```
class Equals
{
    public static void main(String [] args)
    {
        int x = 100;
        double y = 100.1;
        boolean b = (x = y); /* Line 7 */
        System.out.println(b);
    }
}
```

A. True
B. False
C. Compilation fails
D. An exception is thrown at runtime

Ans. C. Compilation fails

Explanation

The code will not compile because in line 7, the line will work only if we use (x==y) in the line. The == operator compares values to produce a boolean, whereas the = operator assigns a value to variables.

Options A, B, and D are incorrect because the code does not get as far as compiling. If we corrected this code, the output would be false.

61. What will be the output of the program?

```
class Test
{
    public static void main(String [] args)
    {
        int x=20;
        String sup = (x < 15) ? "small" : (x <
22)? "tiny" : "huge";
        System.out.println(sup);
    }
}
```

A. Small **B.** Tiny
C. Huge **D.** Compilation fails

Ans. B. tiny

Explanation

This is an example of a nested ternary operator. The second evaluation (x < 22) is true, so the "tiny" value is assigned to sup.

62. What will be the output of the program?

```
class Test
{
   public static void main(String [] args)
   {
      int x= 0;
      int y= 0;
      for (int z = 0; z < 5; z++)
      {
         if (( ++x > 2 ) && (++y > 2))
         {
            x++;
         }
      }
      System.out.println(x + " " + y);
   }
}
```

A. 5 2 **B.** 5 3
C. 6 3 **D.** 6 4

Ans. C. 6 3

Explanation

In the first two iterations x is incremented once and y is not because of the short circuit && operator. In the third and forth iterations x and y are each incremented, and in the fifth iteration x is doubly incremented and y is incremented.

63. What will be the output of the program?

```
class Test
{
   public static void main(String [] args)
   {
      int x= 0;
      int y= 0;
      for (int z = 0; z < 5; z++)
      {
         if (( ++x > 2 ) || (++y > 2))
         {
            x++;
         }
      }
   System.out.println(x + " " + y);
   }
}
```

A. 5 3 **B.** 8 2
C. 8 3 **D.** 8 5

Ans. B. 8 2

Explanation

The first two iterations of the for loop both x and y are incremented. On the third iteration x is incremented, and for the first time becomes greater than 2. The short circuit or operator || keeps y from ever being incremented again and x is incremented twice on each of the last three iterations.

64. What will be the output of the program?

```
class Bitwise
{
   public static void main(String [] args)
   {
      int x = 11 & 9;
      int y = x ^ 3;
      System.out.println( y | 12 );
   }
}
```

A. 0 **B.** 7
C. 8 **D.** 14

Ans. D. 14

Explanation

The & operator produces a 1 bit when both bits are 1. The result of the & operation is 9. The ^ operator produces a 1 bit when exactly one bit is 1; the result of this operation is 10. The | operator produces a 1 bit when at least one bit is 1; the result of this operation is 14.

65. What will be the output of the program?

```
class SSBool
{
   public static void main(String [] args)
   {
      boolean b1 = true;
      boolean b2 = false;
      boolean b3 = true;
      if ( b1 & b2 | b2 & b3 | b2 ) /* Line 8 */
         System.out.print("ok ");
      if ( b1 & b2 | b2 & b3 | b2 | b1 ) /*Line
10*/
         System.out.println("dokey");
   }
}
```

A. ok
B. dokey
C. ok dokey
D. No output is produced
E. Compilation error

Ans. B. dokey

Explanation

The & operator has a higher precedence than the | operator so that on line 8 b1 and b2 are evaluated together as are b2 & b3. The final b1 in line 10 is what causes that if test to be true. Hence it prints "dokey".

66. What will be the output of the program?

```
class SC2
{
    public static void main(String [] args)
    {
        SC2 s = new SC2();
        s.start();
    }

    void start()
    {
        int a = 3;
        int b = 4;
        System.out.print(" " + 7 + 2 + " ");
        System.out.print(a + b);
        System.out.print(" " + a + b + " ");
        System.out.print(foo() + a + b + " ");
        System.out.println(a + b + foo());
    }

    String foo()
    {
        return "foo";
    }
}
```

A. 9 7 7 foo 7 7foo

B. 72 34 34 foo34 34foo

C. 9 7 7 foo34 34foo

D. 72 7 34 foo34 7foo

Ans. D. 72 7 34 foo34 7foo

Explanation

Because all of these expressions use the + operator, there is no precedence to worry about and all of the expressions will be evaluated from left to right. If either operand being evaluated is a String, the + operator will concatenate the two operands; if both operands are numeric, the + operator will add the two operands.

67. What will be the output of the program?

```
class Test
{
    static int s;
    public static void main(String [] args)
    {
        Test p = new Test();
        p.start();
        System.out.println(s);
    }

    void start()
    {
        int x = 7;
        twice(x);
        System.out.print(x + " ");
    }

    void twice(int x)
    {
        x = x*2;
        s = x;
    }
}
```

A. 7 7

B. 7 14

C. 14 0

D. 14 14

Ans. B. 7 14

Explanation

The int x in the twice() method is not the same int x as in the start() method. Start()'s x is not affected by the twice() method. The instance variable s is updated by twice()'s x, which is 14.

68. What will be the output of the program?

```
class Two
{
    byte x;
}

class PassO
{
    public static void main(String [] args)
    {
        PassO p = new PassO();
        p.start();
    }

    void start()
    {
        Two t = new Two();
        System.out.print(t.x + " ");
        Two t2 = fix(t);
        System.out.println(t.x + " " + t2.x);
    }

    Two fix(Two tt)
    {
        tt.x = 42;
```

```
        return tt;
    }
}
```

A. null null 42 B. 0 0 42
C. 0 42 42 D. 0 0 0

Ans. C. 0 42 42

Explanation

In the fix() method, the reference variable tt refers to the same object (class Two) as the t reference variable. Updating tt.x in the fix() method updates t.x (they are one in the same object). Remember also that the instance variable x in the Two class is initialized to 0.

69. What will be the output of the program?

```
class BoolArray
{
    boolean [] b = new boolean[3];
    int count = 0;

    void set(boolean [] x, int i)
    {
        x[i] = true;
        ++count;
    }

    public static void main(String [] args)
    {
        BoolArray ba = new BoolArray();
        ba.set(ba.b, 0);
        ba.set(ba.b, 2);
        ba.test();
    }

    void test()
    {
        if ( b[0] && b[1] | b[2] )
            count++;
        if ( b[1] && b[(++count - 2)] )
            count += 7;
        System.out.println("count = " + count);
    }
}
```

A. count = 0 B. count = 2
C. count = 3 D. count = 4

Ans. C. count = 3

Explanation

The reference variables b and x both refer to the same boolean array. count is incremented for each call to the set() method, and once again when the first if test is true. Because of the && short circuit operator, count is not incremented during the second if test.

70. What will be the output of the program?

```
public class Test
{
    public static void leftshift(int i, int j)
    {
        i <<= j;
    }
    public static void main(String args[])
    {
        int i = 4, j = 2;
        leftshift(i, j);
        System.out.println(i);
    }
}
```

A. 2 B. 4
C. 8 D. 16

Ans. B. 4

Explanation

Java only ever passes arguments to a method by value (i.e. a copy of the variable) and never by reference. Therefore, the value of the variable i remains unchanged in the main method.

If you are clever you will spot that 16 is 4 multiplied by 2 twice, (4 * 2 * 2) = 16. If you had 16 left shifted by three bits then 16 * 2 * 2 * 2 = 128. If you had 128 right shifted by 2 bits, then 128 / 2 / 2 = 32. Keeping these points in mind, you don't have to go converting to binary to do the left and right bit shifts.

Choose the Correct Statements

71. import java.awt.*;

```
class Ticker extends Component
{
    public static void main (String [] args)
    {
        Ticker t = new Ticker();
        /* Missing Statements ? */
    }
}
```

which two of the following statements, inserted independently, could legally be inserted into missing section of this code?

boolean test = (Component instanceof t);
boolean test = (t instanceof Ticker);
boolean test = t.instanceof(Ticker);
boolean test = (t instanceof Component);

A. 1 and 4 **B.** 2 and 3
C. 1 and 3 **D.** 2 and 4

Ans. D. 2 and 4

Explanation

(2) is correct because class type Ticker is part of the class hierarchy of t; therefore it is a legal use of the instanceof operator. (4) is also correct because Component is part of the hierarchy of t, because Ticker extends Component.

(1) is incorrect because the syntax is wrong. A variable (or null) always appears before the instanceof operator, and a type appears after it.

(3) is incorrect because the statement is used as a method (t.instanceof(Ticker);), which is illegal.

72. Which of the following are legal lines of code?

1. int w = (int)888.8;
2. byte x = (byte)1000L;
3. long y = (byte)100;
4. byte z = (byte)100L;

A. 1 and 2
B. 2 and 3
C. 3 and 4
D. All statements are correct.

Ans. D. All statements are correct

Explanation

Statements (1), (2), (3), and (4) are correct. (1) is correct because when a floating-point number (a double in this case) is cast to an int, it simply loses the digits after the decimal.

(2) and (4) are correct because a long can be cast into a byte. If the long is over 127, it loses its most significant (leftmost) bits.

(3) actually works, even though a cast is not necessary, because a long can store a byte.

73. Which two statements are equivalent?

16*4
16>>2
16/2^2
16>>>2

A. 1 and 2 **B.** 2 and 4
C. 3 and 4 **D.** 1 and 3

Ans. B. 2 and 4

Explanation

(2) is correct. 16 >> 2 = 4
(4) is correct. 16 >>> 2 = 4
(1) is wrong. 16 * 4 = 64
(3) is wrong. 16/2 ^ 2 = 10

74. Which two statements are equivalent?

1. 3/2
2. 3<2
3. 3*4
4. 3<<2

A. 1 and 2 **B.** 2 and 3
C. 3 and 4 **D.** 1 and 4

Ans. C. 3 and 4

Explanation

(1) is wrong. 3/2 = 1 (integer arithmetic).
(2) is wrong. 3 < 2 = false.
(3) is correct. 3 * 4 = 12.
(4) is correct. 3 <<2= 12. In binary 3 is 11, now shift the bits two places to the left and we get 1100 which is 12 in binary (3*2*2).

75. import java.awt.Button;

```
class CompareReference
{
    public static void main(String [] args)
    {
        float f = 42.0f;
        float [] f1 = new float[2];
        float [] f2 = new float[2];
        float [] f3 = f1;
        long x = 42;
        f1[0] = 42.0f;
    }
}
```

which three statements are true?

1. f1 == f2
2. f1 == f3
3. f2 == f1[1]
4. x == f1[0]
5. f == f1[0]

A. 1, 2 and 3 **B.** 2, 4 and 5
C. 3, 4 and 5 **D.** 1, 4 and 5

Ans. B. 2, 4 and 5

Explanation

(2) is correct because the reference variables f1 and f3 refer to the same array object.

(4) is correct because it is legal to compare integer and floating-point types.

(5) is correct because it is legal to compare a variable with an array element.

(3) is incorrect because f2 is an array object and f1[1] is an array element.

76. Which two are equal?

1. 32/4
2. (8 >> 2) << 4
3. 2^5
4. 128 >>> 2
5. 2 >> 5

A. 1 and 2 **B.** 2 and 4

C. 1 and 3 **D.** 2 and 3

Ans. B. 2 and 4

Explanation

(2) and (4) are correct. (2) and (4) both evaluate to 32. (2) is shifting bits right, then left using the signed bit shifters >> and <<. (4) is shifting bits using the unsigned operator >>>, but since the beginning number is positive, the sign is maintained.

(1) evaluates to 8, (3) looks like 2 to the 5th power, but ^ is the Exclusive OR operator so (3) evaluates to 7. (5) evaluates to 0 (2 >> 5 is not 2 to the 5th).

Flow Control

77. public void foo(boolean a, boolean b)

```
{
    if( a )
    {
        System.out.println("A"); /* Line 5 */
    }
    else if(a && b) /* Line 7 */
    {
        System.out.println( "A && B");
    }
    else /* Line 11 */
    {
        if ( !b )
        {
            System.out.println( "notB") ;
        }
        else
        {
            System.out.println( "ELSE" ) ;
        }
    }
}
```

A. If a is true and b is true, then the output is "A && B"

B. If a is true and b is false, then the output is "notB"

C. If a is false and b is true, then the output is "ELSE"

D. If a is false and b is false, then the output is "ELSE"

Ans. C. If a is false and b is true, then the output is "ELSE"

Explanation

Option C is correct. The output is "ELSE". Only when a is false do the output lines after 11 get some chance of executing.

Option A is wrong. The output is "A". When a is true, irrespective of the value of b, only the line 5 output will be executed. The condition at line 7 will never be evaluated (when a is true, it will always be trapped by the line 12 condition) therefore the output will never be "A && B".

Option B is wrong. The output is "A". When a is true, irrespective of the value of b, only the line 5 output will be executed.

Option D is wrong. The output is "notB".

78. switch(x)

```
{
    default:
        System.out.println("Hello");
}
```

Which two are acceptable types for x?

byte
long
char
float
Short
Long

A. 1 and 3 **B.** 2 and 4

C. 3 and 5 **D.** 4 and 6

Ans. A. 1 and 3

Explanation

Switch statements are based on integer expressions and since both bytes and chars can implicitly be widened to an integer, these can also be used. Also shorts can be used. Short and long are wrapper classes and reference types can not be used as variables.

79. public void test(int x)

```
{
    int odd = 1;
    if(odd) /* Line 4 */
```

```
    {
        System.out.println("odd");
    }
    else
    {
        System.out.println("even");
    }
}
```

Which statement is true?

A. Compilation fails.
B. "odd" will always be output.
C. "even" will always be output.
D. "odd" will be output for odd values of x, and "even" for even values.

Ans. A. Compilation fails.

Explanation

The compiler will complain because of incompatible types (line 4), the if expects a boolean but it gets an integer.

80. public class While

```
{
    public void loop()
    {
        int x= 0;
        while ( 1 ) /* Line 6 */
        {
            System.out.print("x plus one is " +
(x + 1)); /* Line 8 */
        }
    }
}
```

Which statement is true?

A. There is a syntax error on line 1.
B. There are syntax errors on lines 1 and 6.
C. There are syntax errors on lines 1, 6, and 8.
D. There is a syntax error on line 6.

Ans. D. There is a syntax error on line 6.

Explanation

Using the integer 1 in the while statement, or any other looping or conditional construct for that matter, will result in a compiler error. This is old C Program syntax, not valid Java.

A, B and C are incorrect because line 1 is valid (Java is case sensitive so While is a valid class name). Line 8 is also valid because an equation may be placed in a String operation as shown.

Choose Output

81. What will be the output of the program?

```
int i = l, j = -1;
switch (i)
{
    case 0, 1: j = 1; /* Line 4 */
    case 2: j = 2;
    default: j = 0;
}
System.out.println("j = " + j);
```

A. j = –1 **B.** j = 0
C. j = 1 **D.** Compilation fails.

Ans. D. Compilation fails.

Explanation

The case statement takes only a single argument. The case statement on line 4 is given two arguments so the compiler complains.

82. What will be the output of the program?

```
int i = 1, j = 10;
do
{
    if(i > j)
    {
        break;
    }
    j—;
} while (++i < 5);
System.out.println("i = " + i + " and j = " +
j);
```

A. i = 6 and j = 5 **B.** i = 5 and j = 5
C. i = 6 and j = 4 **D.** i = 5 and j = 6

Ans. D. i = 5 and j = 6

Explanation

This loop is a do-while loop, which always executes the code block within the block at least once, due to the testing condition being at the end of the loop, rather than at the beginning. This particular loop is exited prematurely if i becomes greater than j.

The order is, test i against j, if bigger, it breaks from the loop, decrements j by one, and then tests the loop condition, where a pre-incremented by one i is tested for being lower than 5. The test is at the end of the loop, so i can reach the value of 5 before it fails. So it goes, start:

```
1, 10
2, 9
3, 8
4, 7
5, 6 loop condition fails.
```

83. What will be the output of the program?

```
public class Switch2
{
    final static short x = 2;
    public static int y = 0;
    public static void main(String [] args)
    {
        for (int z=0; z < 3; z++)
        {
            switch (z)
            {
                case x: System.out.print("0 ");
                case x-1: System.out.print("1 ");
                case x-2: System.out.print("2 ");
            }
        }
    }
}
```

A. 0 1 2   B. 0 1 2 1 2 2

C. 2 1 0 1 0 0   D. 2 1 2 0 1 2

Ans. D. 2 1 2 0 1 2

Explanation

The case expressions are all legal because x is marked final, which means the expressions can be evaluated at compile time. In the first iteration of the for loop case x-2 matches, so 2 is printed. In the second iteration, x-1 is matched so 1 and 2 are printed (remember, once a match is found all remaining statements are executed until a break statement is encountered). In the third iteration, x is matched. So 0 1 and 2 are printed.

84. What will be the output of the program?

```
public class SwitchTest
{
    public static void main(String[] args)
    {
        System.out.println("value =" + switchIt(4));
    }
    public static int switchIt(int x)
    {
        int j = 1;
        switch (x)
        {
            case l: j++;
            case 2: j++;
            case 3: j++;
            case 4: j++;
            case 5: j++;
            default: j++;
        }
        return j + x;
    }
}
```

A. value = 2   B. value = 4

C. value = 6   D. value = 8

Ans. D. value = 8

Explanation

Because there is no break statement, once the desired result is found, the program continues though each of the remaining options.

85. What will be the output of the program?

```
public class If2
{
    static boolean b1, b2;
    public static void main(String [] args)
    {
        int x = 0;
        if ( !b1 ) /* Line 7 */
        {
            if ( !b2 ) /* Line 9 */
            {
                b1 = true;
                x++;
                if ( 5 > 6 )
                {
                    x++;
                }
                if ( !b1 )
                    x = x + 10;
                else if ( b2 = true ) /* Line 19 */
                    x = x + 100;
                else if ( b1 | b2 ) /* Line 21 */
                    x = x + 1000;
            }
        }
        System.out.println(x);
    }
}
```

A. 0   B. 1

C. 101   D. 111

Ans. C. 101

Explanation

As instance variables, b1 and b2 are initialized to false. The if tests on lines 7 and 9 are successful, so b1 is set to true and x is incremented. The next if test to succeed is on line 19 (note that the code is not testing to see if b2 is true, it is setting b2 to be true). Since line 19 was successful, subsequent else-if's (line 21) will be skipped.

86. What will be the output of the program?

```
public class Switch2
{
    final static short x = 2;
    public static int y = 0;
    public static void main(String [] args)
    {
        for (int z=0; z < 3; z++)
        {
            switch (z)
            {
                case y: System.out.print("0 "); /*
Line 11 */
                case x-1: System.out.print("1 "); /
* Line 12 */
                case x: System.out.print("2 "); /*
Line 13 */
            }
        }
    }
}
```

A. 0 1 2

B. 0 1 2 1 2 2

C. Compilation fails at line 11.

D. Compilation fails at line 12.

Ans. C. Compilation fails at line 11.

Explanation

Case expressions must be constant expressions. Since x is marked final, lines 12 and 13 are legal; however, y is not a final so the compiler will fail at line 11.

87. What will be the output of the program?

```
public class If1
{
    static boolean b;
    public static void main(String [] args)
    {
        short hand = 42;
        if ( hand < 50 & !b ) /* Line 7 */
            hand++;
        if ( hand > 50 );    /* Line 9 */
        else if ( hand > 40 )
        {
            hand += 7;
            hand++;
        }
        else
            —hand;
        System.out.println(hand);
    }
}
```

A. 41 B. 42

C. 50 D. 51

Ans. D. 51

Explanation

In Java, boolean instance variables are initialized to false, so the if test on line 7 is true and hand is incremented. Line 9 is legal syntax, a do nothing statement. The else-if is true so hand has 7 added to it and is then incremented.

88. What will be the output of the program?

```
public class Test
{
    public static void main(String [] args)
    {
        int I = 1;
        do while ( I < 1 )
        System.out.print("I is " + I);
        while ( I > 1 ) ;
    }
}
```

A. I is 1

B. I is 1 I is 1

C. No output is produced.

D. Compilation error

Ans. C. No output is produced.

Explanation

There are two different looping constructs in this problem. The first is a do-while loop and the second is a while loop, nested inside the do-while. The body of the do-while is only a single statement—brackets are not needed. You are assured that the while expression will be evaluated at least once, followed by an evaluation of the do-while expression. Both expressions are false and no output is produced.

89. What will be the output of the program?

```
int x = l, y = 6;
while (y—)
{
    x++;
}
System.out.println("x = " + x +" y = " + y);
```

A. x = 6 y = 0 **B.** x = 7 y = 0

C. x = 6 y = –1 **D.** Compilation fails.

Ans. D. Compilation fails.

Explanation

Compilation fails because the while loop demands a boolean argument for its looping condition, but in the code, it's given an int argument.

```
while(true) { //insert code here }
```

90. What will be the output of the program?

```
int I = 0;
outer:
while (true)
{
    I++;
    inner:
    for (int j = 0; j < 10; j++)
    {
        I += j;
        if (j == 3)
            continue inner;
        break outer;
    }
    continue outer;
}
System.out.println(I);
```

A. 1 **B.** 2

C. 3 **D.** 4

Ans. A. 1

Explanation

The program flows as follows: I will be incremented after the while loop is entered, then I will be incremented (by zero) when the for loop is entered. The if statement evaluates to false, and the continue statement is never reached. The break statement tells the JVM to break out of the outer loop, at which point I is printed and the fragment is done.

91. What will be the output of the program?

```
for (int i = 0; i < 4; i += 2)
{
    System.out.print(i + " ");
}
System.out.println(i); /* Line 5 */
```

A. 0 2 4 **B.** 0 2 4 5

C. 0 1 2 3 4 **D.** Compilation fails

Ans. D. Compilation fails.

Explanation

Compilation fails on the line 5—System.out.println(i); as the variable i has only been declared within the for loop. It is not a recognised variable outside the code block of loop.

92. What will be the output of the program?

```
int x = 3;
int y = 1;
if (x = y) /* Line 3 */
{
    System.out.println("x =" + x);
}
```

A. x = 1

B. x = 3

C. Compilation fails.

D. The code runs with no output.

Ans. C. Compilation fails.

Explanation

Line 3 uses an assignment as opposed to comparison. Because of this, the if statement receives an integer value instead of a boolean. And so the compilation fails.

93. What will be the output of the program?

```
Float f = new Float("12");
switch (f)
{
    case 12: System.out.println("Twelve");
    case 0: System.out.println("Zero");
    default: System.out.println("Default");
}
```

A. Zero **B.** Twelve

C. Default **D.** Compilation fails

Ans. D. Compilation fails

Explanation

The switch statement can only be supported by integers or variables more "narrow" than an integer, i.e. byte, char, short. Here a Float wrapper object is used and so the compilation fails.

94. What will be the output of the program?

```
int i = O;
while(1)
{
    if(i == 4)
    {
        break;
    }
    ++i;
}
System.out.println("i = " + i);
```

A. i = 0 **B.** i = 3
C. i = 4 **D.** Compilation fails.

Ans. D. Compilation fails.

Explanation

Compilation fails because the argument of the while loop, the condition, must be of primitive type boolean. In Java, 1 does not represent the true state of a boolean, rather it is seen as an integer.

95. What will be the output of the program?

```
public class Delta
{
  static boolean foo(char c)
  {
    System.out.print(c);
    return true;
  }
  public static void main( String[] argv )
  {
    int i = 0;
        for (foo('A'); foo('B') && (i < 2);
foo('C'))
    {
      i++;
      foo('D');
    }
  }
}
```

A. ABDCBDCB
B. ABCDABCD
C. Compilation fails.
D. An exception is thrown at runtime.

Ans. A. ABDCBDCB

Explanation

'A' is only printed once at the very start as it is in the initialisation section of the for loop. The loop will only initialise that once.
'B' is printed as it is part of the test carried out in order to run the loop.
'D' is printed as it is in the loop.
'C' is printed as it is in the increment section of the loop and will 'increment' only at the end of each loop. Here ends the first loop. Again 'B' is printed as part of the loop test.
'D' is printed as it is in the loop.
'C' is printed as it 'increments' at the end of each loop.
Again 'B' is printed as part of the loop test. At this point the test fails because the other part of the test (i < 2) is no longer true. i has been increased in value by 1 for each loop with the line: i++;
This results in a printout of ABDCBDCB

96. What will be the output of the program?

```
for(int i = 0; i < 3; i++)
{
  switch(i)
  {
    case 0: break;
    case 1: System.out.print("one ");
    case 2: System.out.print("two ");
    case 3: System.out.print("three ");
  }
}
System.out.println("done");
```

A. done
B. one two done
C. one two three done
D. one two three two three done

Ans. D. one two three two three done

Explanation

The variable i will have the values 0, 1 and 2.
When i is 0, nothing will be printed because of the break in case 0.
When i is 1, "one two three" will be output because case 1, case 2 and case 3 will be executed (they don't have break statements). When i is 2, "two three" will be output because case 2 and case 3 will be executed (again no break statements). Finally, when the for loop finishes "done" will be output.

97. What will be the output of the program?

```
public class Test
{
  public static void main(String args[])
  {
    int i = 1, j = 0;
    switch(i)
    {
      case 2: j += 6;
      case 4: j += 1;
      default: j += 2;
      case 0: j += 4;
    }
    System.out.println("j = " + j);
  }
}
```

A. 0 **B.** 2
C. 4 **D.** 6

Ans. D. 6

Explanation

Because there is no break statement, the program gets to the default case and adds 2 to j, then goes to case 0 and adds 4 to the new j. The result is j = 6.

98. What will be the output of the program?

```
boolean bool = true;
if(bool = false) /* Line 2 */
{
    System.out.println("a");
}
else if(bool) /* Line 6 */
{
    System.out.println("b");
}
else if(!bool) /* Line 10 */
{
    System.out.println("c"); /* Line 12 */
}
else
{
    System.out.println("d");
}
```

A. a **B.** b

C. c **D.** d

Ans. C. c

Explanation

Look closely at line 2, whether this is an equality check (==) or an assignment (=). The condition at line 2 evaluates to false and also assigns false to bool. bool is now false so the condition at line 6 is not true. The condition at line 10 checks to see if bool is not true (if !(bool == true)), it isn't so line 12 is executed.

99. What will be the output of the program?

```
public class Switch2
{
    final static short x = 2;
    public static int y = 0;
    public static void main(String [] args)
    {
        for (int z=0; z < 4; z++)
        {
            switch (z)
            {
                case x: System.out.print("0 ");
                default: System.out.print("def ");
                case x-1: System.out.print("1 ");
                    break;
                case x-2: System.out.print("2 ");
            }
        }
    }
}
```

A. 0 def 1

B. 2 1 0 def 1

C. 2 1 0 def def

D. 2 1 0 def 1 def 1

Ans. D. 2 1 0 def 1 def 1

Explanation

When z == 0 , case x-2 is matched. When z == 1, case x-1 is matched and then the break occurs. When z == 2, case x, then default, then x-1 are all matched. When z == 3, default, then x-1 are matched. The rules for default are that it will fall through from above like any other case (for instance when z == 2), and that it will match when no other cases match (for instance when z==3).

100. What will be the output of the program?

```
int i = 0, j = 5;
tp: for (;;)
    {
        i++;
        for (;;)
        {
            if(i > —j)
            {
                break tp;
            }
        }
        System.out.println("i =" + i + ", j = " + j);
```

A. i = 1, j = 0

B. i = 1, j = 4

C. i = 3, j = 4

D. Compilation fails.

Ans. D. Compilation fails.

Explanation

If you examine the code carefully you will notice a missing curly bracket at the end of the code, this would cause the code to fail.

Objects and Collections

101. Suppose that you would like to create an instance of a new Map that has an iteration order that is the same as the iteration order of an existing instance of a Map. Which concrete implementation of the Map interface should be used for the new instance?

A. TreeMap

B. HashMap

C. LinkedHashMap

D. The answer depends on the implementation of the existing instance.

Ans. C. LinkedHashMap

Explanation

The iteration order of a Collection is the order in which an iterator moves through the elements of the Collection. The iteration order of a LinkedHashMap is determined by the order in which elements are inserted.

When a new LinkedHashMap is created by passing a reference to an existing Collection to the constructor of a LinkedHashMap, the Collection.addAll method will ultimately be invoked.

The addAll method uses an iterator to the existing Collection to iterate through the elements of the existing Collection and add each to the instance of the new LinkedHashMap.

Since the iteration order of the LinkedHashMap is determined by the order of insertion, the iteration order of the new LinkedHashMap, must be the same as the iteration order of the old Collection.

102. Which class does not override the equals() and hashCode() methods, inheriting them directly from class Object?

A. java.lang.String

B. java.lang.Double

C. java.lang.StringBuffer

D. java.lang.Character

Ans. C. java.lang.StringBuffer

Explanation

java.lang.StringBuffer is the only class in the list that uses the default methods provided by class Object.

103. Which collection class allows you to grow or shrink its size and provides indexed access to its elements, but whose methods are not synchronized?

A. java.util.HashSet

B. java.util.LinkedHashSet

C. java.util.List

D. java.util.ArrayList

Ans. D. java.util.ArrayList

Explanation

All of the collection classes allow you to grow or shrink the size of your collection. ArrayList provides an index to its elements. The newer collection classes tend not to have synchronized methods. Vector is an older implementation of ArrayList functionality and has synchronized methods; it is slower than ArrayList.

104. You need to store elements in a collection that guarantees that no duplicates are stored and all elements can be accessed in natural order. Which interface provides that capability?

A. java.util.Map

B. java.util.Set

C. java.util.List

D. java.util.Collection

Ans. B. java.util.Set

Explanation

Option B is correct. A set is a collection that contains no duplicate elements. The iterator returns the elements in no particular order (unless this set is an instance of some class that provides a guarantee). A map cannot contain duplicate keys but it may contain duplicate values. List and Collection allow duplicate elements.

Option A is wrong. A map is an object that maps keys to values. A map cannot contain duplicate keys; each key can map to at most one value. The Map interface provides three collection views, which allow a map's contents to be viewed as a set of keys, collection of values, or set of key-value mappings. The order of a map is defined as the order in which the iterators on the map's collection views return their elements. Some map implementations, like the TreeMap class, make specific guarantees as to their order (ascending key order); others, like the HashMap class, do not (does not guarantee that the order will remain constant over time).

Option C is wrong. A list is an ordered collection (also known as a sequence). The user of this interface has precise control over where in the list each element is inserted. The user can access elements by their integer index (position in the list), and search for elements in the list. Unlike sets, lists typically allow duplicate elements.

Option D is wrong. A collection is an ordered collection (also known as a sequence). The user of this interface has precise control over where in the list each element is inserted. The user can access elements by their integer index (position in the list), and search for elements in the list. Unlike sets, lists typically allow duplicate elements.

105. Which interface does java.util.Hashtable implement?

A. Java.util.Map
B. Java.util.List
C. Java.util.HashTable
D. Java.util.Collection

Ans. A. Java.util.Map

Explanation

Hash table based implementation of the Map interface.

106. Which interface provides the capability to store objects using a key-value pair?

A. Java.util.Map **B.** Java.util.Set
C. Java.util.List **D.** ava.util.Collection

Ans. A. Java.util.Map

Explanation

An object that maps keys to values. A map cannot contain duplicate keys; each key can map to at most one value.

107. Which collection class allows you to associate its elements with key values, and allows you to retrieve objects in FIFO (first-in, first-out) sequence?

A. java.util.ArrayList
B. java.util.LinkedHashMap
C. java.util.HashMap
D. java.util.TreeMap

Ans. B. java.util.LinkedHashMap

Explanation

LinkedHashMap is the collection class used for caching purposes. FIFO is another way to indicate caching behavior. To retrieve LinkedHashMap elements in cached order, use the values() method and iterate over the resultant collection.

108. Which collection class allows you to access its elements by associating a key with an element's value, and provides synchronization?

A. java.util.SortedMap
B. java.util.TreeMap
C. java.util.TreeSet
D. java.util.Hashtable

Ans. D. java.util.Hashtable

Explanation

Hashtable is the only class listed that provides synchronized methods. If you need synchronization great; otherwise, use HashMap, it is faster.

109. Which is valid declaration of a float?

A. float f = 1F; **B.** float f = 1.0;
C. float f = "1"; **D.** float f = 1.0d;

Ans. A. float f = 1F;

Explanation

Option A is valid declaration of float.

Option B is incorrect because any literal number with a decimal point you declare the computer will implicitly cast to double unless you include "F or f"

Option C is incorrect because it is a String.

Option D is incorrect because "d" tells the computer it is a double therefore, you are trying to put a double value into a float variables, i.e there might be a loss of precision.

110. /* Missing Statement ? */

```
public class foo
{
public static void main(String[]args)throws
Exception
    {
        java.io.PrintWriter out = new
java.io.PrintWriter();
                                            new
java.io.OutputStreamWriter(System.out,true);
        out.println("Hello");
    }
}
```

What line of code should replace the missing statement to make this program compile?

A. No statement required.

B. import java.io.*;

C. include java.io.*;

D. import java.io.PrintWriter;

Ans. A. No statement required.

Explanation

The usual method for using/importing the java packages/classes is by using an import statement at the top of your code. However, it is possible to explicitly import the specific class that you want to use as you use it which is shown in the code above. The disadvantage of this, however, is that every time you create a new object you will have to use the class path in the case "java.io" then the class name in the long run leading to a lot more typing.

111. What is the numerical range of char?

A. 0 to 32767 **B.** 0 to 65535
C. –256 to 255 **D.** –32768 to 32767

Ans. B. 0 to 65535

Explanation

The char type is integral but unsigned. The range of a variable of type char is from 0 to 216-1 or 0 to 65535. Java characters are Unicode, which is a 16-bit encoding capable of representing a wide range of international characters. If the most significant nine bits of a char are 0, then the encoding is the same as seven-bit ASCII.

112. Which of the following are Java reserved words?

1. run
2. import
3. default
4. implement

A. 1 and 2 **B.** 2 and 3
C. 3 and 4 **D.** 2 and 4

Ans. B. 2 and 3

Explanation

(2) - This is a Java keyword
(3) - This is a Java keyword
(1) - Is incorrect because although it is a method of Thread/Runnable it is not a keyword
(4) - This is not a Java keyword, the keyword is implements

Choose the Output

113. What will be the output of the program?

```
public class Test
{
    public static void main (String[] args)
    {
        String foo = args[1];
        String bar = args[2];
        String baz = args[3];
        System.out.println("baz = " + baz); /*
Line 8 */
    }
}
```

And the command line invocation:

> java Test red green blue

A. baz = **B.** baz = null
C. baz = blue **D.** Runtime Exception

Ans. D. Runtime Exception

Explanation

When running the program you entered 3 arguments "red", "green" and "blue". When dealing with arrays in java you must remember ALL ARRAYS IN JAVA ARE ZERO BASED, therefore args[0] becomes "red", args[1] becomes "green" and args[2] becomes "blue".

When the program entcounters line 8 above at runtime, it looks for args[3] which has never been created therefore you get an ArrayIndexOutOfBoundsException at runtime.

114. What will be the output of the program?

```
public class Test
{
    public static void main (String args[])
    {
        String str = NULL;
        System.out.println(str);
    }
}
```

A. NULL
B. Compile Error
C. Code runs but no output
D. Runtime Exception

Ans. B. Compile Error

Explanation

Option B is correct because to set the value of a String variable to null, you must use "null" and not "NULL".

115. What will be the output of the program?

```
package foo;
import java.util.Vector; /* Line 2 */
private class MyVector extends Vector
{
    int i = 1; /* Line 5 */
    public MyVector()
    {
        i = 2;
    }
}
public class MyNewVector extends MyVector
```

```
{
    public MyNewVector ()
    {
        i = 4; /* Line 15 */
    }
    public static void main (String args [])
    {
        MyVector v = new MyNewVector(); /
* Line 19 */
    }
}
```

A. Compilation will succeed.
B. Compilation will fail at line 3.
C. Compilation will fail at line 5.
D. Compilation will fail at line 15.

Ans. B. Compilation will fail at line 3.

Explanation

Option B is correct. The compiler complains with the error "modifier private not allowed here". The class is created private and is being used by another class on line 19.

116. What will be the output of the program?

```
public class Test
{
    private static int[] x;
    public static void main(String[] args)
    {
        System.out.println(x[0]);
    }
}
```

A. 0
B. null
C. Compile Error
D. NullPointerException at runtime

Ans. D. NullPointerException at runtime

Explanation

In the above code the array reference variable x has been declared but it has not been instantiated, i.e. the new statement is missing, for example:
private static int[]x = new int[5];
private static int[x] declares a static, i.e. class level array.
The "new" keyword is the word that actually creates said array.
int[5] in association with the new sets the size of the array. So since the above code contains no new or size decalarations when you try and access x[0] you are trying to access a member of an array that has been declared but not initialized, hence you get a NullPointerException at runtime.

117. What will be the output of the program?

```
import java.util.*;
class I
{
    public static void main (String[] args)
    {
        Object i = new ArrayList().iterator();
            System.out.print((i instanceof
List)+",");
            System.out.print((i instanceof
Iterator)+",");
            System.out.print(i instanceof
ListIterator);
    }
}
```

A. Prints: false, false, false
B. Prints: false, false, true
C. Prints: false, true, false
D. Prints: false, true, true

Ans. C. Prints: false, true, false

Explanation

The iterator() method returns an iterator over the elements in the list in proper sequence, it doesn't return a List or a ListIterator object.
A ListIterator can be obtained by invoking the ListIterator method.

118. What will be the output of the program?

```
public class Test
{
    private static float[] f = new float[2];
    public static void main (String[] args)
    {
        System.out.println("f[0] = " + f[0]);
    }
}
```

A. f[0] = 0 **B.** f[0] = 0.0
C. Compile Error **D.** Runtime Exception

Ans. B. f[0] = 0.0

Explanation

The choices are between options A and B, what this question is really testing is your knowledge of default values of an initialized array. This is an array type float, i.e. it is a type that uses decimal point numbers, therefore its initial value will be 0.0 and not 0.

119. What will be the output of the program?

```
import java.util.*;
class H
{
    public static void main (String[] args)
```

```
{
  Object x = new Vector().elements();
        System.out.print((x instanceof
Enumeration)+",");
        System.out.print((x instanceof
Iterator)+",");
        System.out.print(x instanceof
ListIterator);
  }
}
```

A. Prints: false,false,false
B. Prints: false,false,true
C. Prints: false,true,false
D. Prints: true,false,false

Ans. D. Prints: true,false,false

Explanation

The Vector.elements method returns an Enumeration over the elements of the vector. Vector implements the List interface and extends AbstractList so it is also possible to get an Iterator over a Vector by invoking the iterator or listIterator method.

120. What will be the output of the program?

```
TreeSet map = new TreeSet();
map.add("one");
map.add("two");
map.add("three");
map.add("four");
map.add("one");
Iterator it = map.iterator();
while (it.hasNext() )
{
  System.out.print( it.next() + " " );
}
```

A. one two three four
B. four three two one
C. four one three two
D. one two three four one

Ans. C. four one three two

Explanation

TreeSet assures no duplicate entries; also, when it is accessed it will return elements in natural order, which typically means alphabetical.

121. What will be the output of the program?

```
public static void main(String[] args)
{
   Object obj = new Object()
   {
     public int hashCode()
     {
        return 42;
     }
   };
   System.out.println(obj.hashCode());
}
```

A. 42
B. Runtime Exception
C. Compile Error at line 2
D. Compile Error at line 5

Ans. A. 42

Explanation

This code is an example of an anonymous inner class. They can be declared to extend another class or implement a single interface. Since they have no name you cannot use the "new" keyword on them.

In this case the annoynous class is extending the Object class. Within the {} you place the methods you want for that class. After this class has been declared its methods can be used by that object in the usual way, e.g. objectname.annoymousClassMethod()

Point to the Correct Statements

122.

```
class Test1
{
  public int value;
  public int hashCode() { return 42; }
}
class Test2
{
  public int value;
      public int hashcode() { return
(int)(value^5); }
}
```

Which statement is true?

A. class Test1 will not compile.
B. The Test1 hashCode() method is more efficient than the Test2 hashCode() method.
C. The Test1 hashCode() method is less efficient than the Test2 hashCode() method.
D. class Test2 will not compile.

Ans. C. The Test1 hashCode() method is less efficient than the Test2 hashCode() method.

Explanation

The so-called "hashing algorithm" implemented by class Test1 will always return the same value, 42, which is legal but which will place all of the hash table entries into a single bucket, the most inefficient setup possible.

Options A and D are incorrect because these classes are legal.

Option B is incorrect based on the logic described above.

123. Which statement is true for the class java.util.HashSet?

A. The elements in the collection are ordered.

B. The collection is guaranteed to be immutable.

C. The elements in the collection are guaranteed to be unique.

D. The elements in the collection are accessed using a unique key.

Ans. C. The elements in the collection are guaranteed to be unique.

Explanation

Option C is correct. HashSet implements the Set interface and the Set interface specifies collection that contains no duplicate elements.

Option A is wrong. HashSet makes no guarantees as to the iteration order of the set; in particular, it does not guarantee that the order will remain constant over time.

Option B is wrong. The set can be modified.

Option D is wrong. This is a Set and not a Map.

124. Which of the following statements about the hashcode() method are incorrect?

1. The value returned by hashcode() is used in some collection classes to help locate objects.
2. The hashcode() method is required to return a positive int value.
3. The hashcode() method in the String class is the one inherited from Object.
4. Two new empty String objects will produce identical hashcodes.

A. 1 and 2 **B.** 2 and 3
C. 3 and 4 **D.** 1 and 4

Ans. B. 2 and 3

Explanation

(2) is an incorrect statement because there is no such requirement.

(3) is an incorrect statement and therefore a correct answer because the hashcode for a string is computed from the characters in the string.

125. What two statements are true about properly overridden hashCode() and equals() methods?

1. hashCode() doesn't have to be overridden if equals() is.
2. equals() doesn't have to be overridden if hashCode() is.
3. hashCode() can always return the same value, regardless of the object that invoked it.
4. equals() can be true even if it's comparing different objects.

A. 1 and 2 **B.** 2 and 3
C. 3 and 4 **D.** 1 and 3

Ans. C. 3 and 4

Explanation

(3) and (4) are correct.

(1) and (2) are incorrect because by contract hashCode() and equals() can't be overridden unless both are overridden.

126. Which two statements are true about comparing two instances of the same class, given that the equals() and hashCode() methods have been properly overridden?

1. If the equals() method returns true, the hashCode() comparison == must return true.
2. If the equals() method returns false, the hashCode() comparison != must return true.
3. If the hashCode() comparison == returns true, the equals() method must return true.
4. If the hashCode() comparison == returns true, the equals() method might return true.

A. 1 and 4 **B.** 2 and 3
C. 3 and 4 **D.** 1 and 3

Ans. A. 1 and 4

Explanation

(1) is a restatement of the equals() and hashCode() contract. (4) is true because if the hashCode() comparison returns ==, the two objects might or might not be equal.

(2) and (3) are incorrect because the hashCode() method is very flexible in its return values, and often two dissimilar objects can return the same hash code value.

127. x = 0;
if (x1.hashCode() != x2.hashCode()) x = x + 1;
if (x3.equals(x4)) x = x + 10;
if (!x5.equals(x6)) x = x + 100;
if (x7.hashCode() == x8.hashCode()) x = x + 1000;
System.out.println("x = " + x);
and assuming that the equals() and hashCode() methods are property implemented, if the output is "x = 1111", which of the following statements will always be true?

A. x2.equals(x1)
B. x3.hashCode() == x4.hashCode()
C. x5.hashCode() != x6.hashCode()
D. x8.equals(x7)

Ans. B. x3.hashCode() == x4.hashCode()

Explanation

By contract, if two objects are equivalent according to the equals() method, then the hashCode() method must evaluate them to be ==.
Option A is incorrect because if the hashCode() values are not equal, the two objects must not be equal.
Option C is incorrect because if equals() is not true, there is no guarantee of any result from hashCode().
Option D is incorrect because hashCode() will often return == even if the two objects do not evaluate to equals() being true.

128. Which statement is true for the class java.util.ArrayList?

A. The elements in the collection are ordered.
B. The collection is guaranteed to be immutable.
C. The elements in the collection are guaranteed to be unique.
D. The elements in the collection are accessed using a unique key.

Ans. A. The elements in the collection are ordered.

Explanation

Yes, always the elements in the collection are ordered.

Exceptions

129. What will be the output of the program?

```
public class Foo
{
    public static void main(String[] args)
    {
        try
        {
            return;
        }
        finally
        {
            System.out.println( "Finally" );
        }
    }
}
```

A. Finally
B. Compilation fails.
C. The code runs with no output.
D. An exception is thrown at runtime.

Ans. A. Finally

Explanation

If you put a finally block after a try and its associated catch blocks, then once execution enters the try block, the code in that finally block will definitely be executed except in the following circumstances:

An exception arising in the finally block itself.
The death of the thread.
The use of System.exit()
Turning off the power to the CPU.

I suppose the last three could be classified as VM shutdown.

130. What will be the output of the program?

```
try
{
    int x = 0;
    int y = 5 / x;
}
catch (Exception e)
{
    System.out.println("Exception");
}
catch (ArithmeticException ae)
{
    System.out.println(" Arithmetic Exception");
```

```
}
System.out.println("finished");
```

A. Finished

B. Exception

C. Compilation fails.

D. Arithmetic Exception

Ans. C. Compilation fails.

Explanation

Compilation fails because ArithmeticException has already been caught. ArithmeticException is a subclass of java.lang.Exception, by time the ArithmeticException has been specified it has already been caught by the Exception class.

If ArithmeticException appears before Exception, then the file will compile. When catching exceptions, the more specific exceptions must be listed before the more general (the subclasses must be caught before the superclasses).

131. What will be the output of the program?

```
public class X
{
    public static void main(String [] args)
    {
        try
        {
            badMethod();
            System.out.print("A");
        }
        catch (Exception ex)
        {
            System.out.print("B");
        }
        finally
        {
            System.out.print("C");
        }
        System.out.print("D");
    }
    public static void badMethod()
    {
        throw new Error(); /* Line 22 */
    }
}
```

A. ABCD

B. Compilation fails.

C. C is printed before exiting with an error message.

D. BC is printed before exiting with an error message.

Ans. C. C is printed before exiting with an error message.

Explanation

Error is thrown but not recognised line(22) because the only catch attempts to catch an Exception and Exception is not a superclass of Error. Therefore, only the code in the finally statement can be run before exiting with a runtime error (Exception in thread "main" java.lang.Error).

132. What will be the output of the program?

```
public class X
{
    public static void main(String [] args)
    {
        try
        {
            badMethod();
            System.out.print("A");
        }
        catch (RuntimeException ex) /* Line 10 */
        {
            System.out.print("B");
        }
        catch (Exception ex1)
        {
            System.out.print("C");
        }
        finally
        {
            System.out.print("D");
        }
        System.out.print("E");
    }
    public static void badMethod()
    {
        throw new RuntimeException();
    }
}
```

A. BD **B.** BCD

C. BDE **D.** BCDE

Ans. C. BDE

Explanation

A Runtime exception is thrown and caught in the catch statement on line 10. All the code after the finally statement is run because the exception has been caught.

133. What will be the output of the program?

```
public class RTExcept
{
  public static void throwit ()
  {
    System.out.print("throwit ");
    throw new RuntimeException();
  }
  public static void main(String [] args)
  {
    try
    {
      System.out.print("hello ");
      throwit();
    }
    catch (Exception re )
    {
      System.out.print("caught ");
    }
    finally
    {
      System.out.print("finally ");
    }
    System.out.println("after ");
  }
}
```

A. hello throwit caught
B. Compilation fails
C. hello throwit RuntimeException caught after
D. hello throwit caught finally after

Ans. D. hello throwit caught finally after

Explanation

The main() method properly catches and handles the RuntimeException in the catch block, finally runs (as it always does), and then the code returns to normal.

A, B and C are incorrect based on the program logic described above. Remember that properly handled exceptions do not cause the program to stop executing.

134. What will be the output of the program?

```
public class Test
{
public static void aMethod() throws Exception
  {
  try /* Line 5 */
  {
    throw new Exception(); /* Line 7 */
  }
  finally /* Line 9 */
  {
    System.out.print("finally "); /* Line
11 */
  }
 }
 public static void main(String args[])
 {
  try
  {
    aMethod();
  }
  catch (Exception e) /* Line 20 */
  {
    System.out.print("exception ");
  }
  System.out.print("finished"); /*
Line 24 */
 }
}
```

A. finally
B. exception finished
C. finally exception finished
D. Compilation fails

Ans. C. finally exception finished

Explanation

This is what happens:

(1) The execution of the try block (line 5) completes abruptly because of the throw statement (line 7).

(2) The exception cannot be assigned to the parameter of any catch clause of the try statement, therefore the finally block is executed (line 9) and "finally" is output (line 11).

(3) The finally block completes normally, and then the try statement completes abruptly because of the throw statement (line 7).

(4) The exception is propagated up the call stack and is caught by the catch in the main method (line 20). This prints "exception".

(5) Lastly program execution continues, because the exception has been caught, and "finished" is output (line 24).

135. What will be the output of the program?

```
public class X
{
  public static void main(String [] args)
  {
  try
```

```
        {
            badMethod();
            System.out.print("A");
        }
        catch (Exception ex)
        {
            System.out.print("B");
        }
        finally
        {
            System.out.print("C");
        }
        System.out.print("D");
    }
    public static void badMethod() {}
}
```

A. AC **B.** BC
C. ACD **D.** ABCD

Ans. C. ACD

Explanation

There is no exception thrown, so all the code with the exception of the catch statement block is run.

136. What will be the output of the program?

```
public class X
{
    public static void main(String [] args)
    {
        try
        {
            badMethod(); /* Line 7 */
            System.out.print("A");
        }
        catch (Exception ex) /* Line 10 */
        {
            System.out.print("B"); /* Line 12 */
        }
        finally /* Line 14 */
        {
            System.out.print("C"); /* Line 16 */
        }
        System.out.print("D"); /* Line 18 */
    }
    public static void badMethod()
    {
        throw new RuntimeException();
    }
}
```

A. AB **B.** BC
C. ABC **D.** BCD

Ans. D. BCD

Explanation

(1) A RuntimeException is thrown, this is a subclass of exception.
(2) The exception causes the try to complete abruptly (line 7), therefore line 8 is never executed.
(3) The exception is caught (line 10) and "B" is output (line 12).
(4) The finally block (line 14) is always executed and "C" is output (line 16).
(5) The exception was caught, so the program continues with line 18 and outputs "D".

137. What will be the output of the program?

```
public class MyProgram
{
    public static void main(String args[])
    {
        try
        {
            System.out.print("Hello world ");
        }
        finally
        {
            System.out.println("Finally executing ");
        }
    }
}
```

A. Nothing. The program will not compile because no exceptions are specified.
B. Nothing. The program will not compile because no catch clauses are specified.
C. Hello world
D. Hello world Finally executing

Ans. D. Hello world Finally executing

Explanation

Finally clauses are always executed. The program will first execute the try block, printing Hello world, and will then execute the finally block, printing Finally executing.

Options A, B, and C are incorrect based on the program logic described above. Remember that either a catch or a finally statement must follow a try. Since the finally is present, the catch is not required.

138. What will be the output of the program?

```
class Exc0 extends Exception { }
class Exc1 extends Exc0 { } /* Line 2 */
public class Test
```

```
{
    public static void main(String args[])
    {
        try
        {
            throw new Exc1(); /* Line 9 */
        }
        catch (Exc0 e0) /* Line 11 */
        {
            System.out.println("Ex0 caught");
        }
        catch (Exception e)
        {
            System.out.println("exception caught");
        }
    }
}
```

A. Ex0 caught
B. exception caught
C. Compilation fails because of an error at line 2.
D. Compilation fails because of an error at line 9.

Ans. A. Ex0 caught

Explanation

An exception Exc1 is thrown and is caught by the catch statement on line 11. The code is executed in this block. There is no finally block of code to execute.

139. import java.io.*;

```
public class MyProgram
{
    public static void main(String args[])
    {
        FileOutputStream out = null;
        try
        {
            out = new FileOutputStream("test.txt");
            out.write(122);
        }
        catch(IOException io)
        {
            System.out.println("IO Error.");
        }
        finally
        {
            out.close();
        }
    }
}
```

and given that all methods of class FileOutputStream, including close(), throw an IOException, which of these is true?

A. This program will compile successfully.
B. This program fails to compile due to an error at line 4.
C. This program fails to compile due to an error at line 6.
D. This program fails to compile due to an error at line 13.

Ans. D. This program fails to compile due to an error at line 13.

Explanation

Any method (in this case, the main() method) that throws a checked exception (in this case, out.close()) must be called within a try clause, or the method must declare that it throws the exception. Either main() must declare that it throws an exception, or the call to out.close() in the finally block must fall inside a (in this case nested) try-catch block.

140. public class MyProgram

```
{
    public static void throwit()
    {
        throw new RuntimeException();
    }
    public static void main(String args[])
    {
        try
        {
            System.out.println("Hello world ");
            throwit();
            System.out.println("Done with try block ");
        }
        finally
        {
            System.out.println("Finally executing ");
        }
    }
}
```

Which answer most closely indicates the behavior of the program?

A. The program will not compile.
B. The program will print Hello world, then print that a RuntimeException

has occurred, then print Done with try block, and then print Finally executing.

C. The program print Hello world, then print that a RuntimeException has occurred, and then print Finally executing.

D. The program will print Hello world, then print Finally executing, then print that a RuntimeException has occurred.

Ans. D. The program will print Hello world, then print Finally executing, then print that a RuntimeException has occurred.

Explanation

Once the program throws a RuntimeException (in the throwit() method) that is not caught, the finally block will be executed and the program will be terminated. If a method does not handle an exception, the finally block is executed before the exception is propagated.

141. public class ExceptionTest

```
{
   class TestException extends Exception {}
        public void runTest() throws
TestException {}
   public void test() /* Point X */
   {
      runTest();
   }
}
```

At Point X on line 5, which code is necessary to make the code compile?

A. No code is necessary.

B. throws Exception

C. catch (Exception e)

D. throws RuntimeException

Ans. B. throws Exception

Explanation

Option B is correct. This works because it DOES throw an exception if an error occurs.

Option A is wrong. If you compile the code as given the compiler will complain:

"unreported exception must be caught or declared to be thrown". The class extends Exception so we are forced to test for exceptions.

Option C is wrong. The catch statement belongs in a method body, not a method specification.

Option D is wrong. TestException is a subclass of Exception, therefore the test method, in this example, must throw TestException or some other class further up the Exception tree. Throwing RuntimeException is just not on as this belongs in the java.lang.RuntimeException branch (it is not a superclass of TestException). The compiler complains with the same error as in A above.

142. System.out.print("Start ");

```
try
{
   System.out.print("Hello world");
   throw new FileNotFoundException();
}
System.out.print(" Catch Here "); /* Line 7
*/
catch(EOFException e)
{
   System.out.print("End of file exception");
}
catch(FileNotFoundException e)
{
   System.out.print("File not found");
}
```

and given that EOFException and FileNotFoundException are both subclasses of IOException, and further assuming this block of code is placed into a class, which statement is most true concerning this code?

A. The code will not compile.

B. Code output: Start Hello world File Not Found.

C. Code output: Start Hello world End of file exception.

D. Code output: Start Hello world Catch Here File not found.

Ans. A. The code will not compile.

Explanation

Line 7 will cause a compiler error. The only legal statements after try blocks are either catch or finally statements.

Options B, C, and D are incorrect based on the program logic described above. If line 7 was removed, the code would compile and the correct answer would be option B.

143. Which statement is true?

A. catch(X x) can catch subclasses of X where X is a subclass of Exception.

B. The Error class is a RuntimeException.

C. Any statement that can throw an Error must be enclosed in a try block.

D. Any statement that can throw an Exception must be enclosed in a try block.

Ans. A. catch(X x) can catch subclasses of X where X is a subclass of Exception.

Explanation

Option A is correct. If the class specified in the catch clause does have subclasses, any exception object that subclasses the specified class will be caught as well.

Option B is wrong. The error class is a subclass of Throwable and not Runtime Exception.

Option C is wrong. You do not catch this class of error.

Option D is wrong. An exception can be thrown to the next method higher up the call stack.

144. Which four can be thrown using the throw statement?

1. Error
2. Event
3. Object
4. Throwable
5. Exception
6. RuntimeException

A. 1, 2, 3 and 4 **B.** 2, 3, 4 and 5
C. 1, 4, 5 and 6 **D.** 2, 4, 5 and 6

Ans. C. 1, 4, 5 and 6

Explanation

The (1), (4), (5) and (6) are the only four that can be thrown.

An Error is a subclass of Throwable that indicates serious problems that a reasonable application should not try to catch.

The Throwable class is the superclass of all errors and exceptions in the Java language.

The class Exception and its subclasses are a form of Throwable that indicates conditions that a reasonable application might want to catch (checked exceptions).

RuntimeException is the superclass of those exceptions that can be thrown during the normal operation of the Java Virtual Machine.

145. Which statement is true?

A. A try statement must have at least one corresponding to catch block.

B. Multiple catch statements can catch the same class of exception more than once.

C. An error that might be thrown in a method must be declared as thrown by that method, or be handled within that method.

D. Except in case of VM shutdown, if a try block starts to execute, a corresponding finally block will always start to execute.

Ans. D. Except in case of VM shutdown, if a try block starts to execute, a corresponding finally block will always start to execute.

Explanation

A is wrong. A try statement can exist without catch, but it must have a finally statement.

B is wrong. A try statement executes a block. If a value is thrown and the try statement has one or more catch clauses that can catch it, then control will be transferred to the first such catch clause. If that catch block completes normally, then the try statement completes normally.

C is wrong. Exceptions of type Error and RuntimeException do not have to be caught, only checked exceptions (java.lang.Exception) have to be caught. However, speaking of Exceptions, Exceptions do not have to be handled in the same method as the throw statement. They can be passed to another method.

If you put a finally block after a try and its associated catch blocks, then once execution enters the try block, the code in that finally block will definitely be executed except in the following circumstances:

An exception arising in the finally block itself.

The death of the thread.

The use of System.exit()

Turning off the power to the CPU.

I suppose the last three could be classified as VM shutdown.

Inner Class

146. Which is true about an anonymous inner class?

A. It can extend exactly one class and implement exactly one interface.

B. It can extend exactly one class and can implement multiple interfaces.

C. It can extend exactly one class or implement exactly one interface.

D. It can implement multiple interfaces regardless of whether it also extends a class.

Ans. C. It can extend exactly one class or implement exactly one interface.

Explanation

Option C is correct because the syntax of an anonymous inner class allows for only one named type after the new, and that type must be either a single interface (in which case the anonymous class implements that one interface) or a single class (in which case the anonymous class extends that one class).

Options A, B, D, and E are all incorrect because they don't follow the syntax rules described in the response for answer option C.

147. class Boo

```
{
    Boo(String s) { }
    Boo() { }
}
class Bar extends Boo
{
    Bar() { }
    Bar(String s) {super(s);}
    void zoo()
    {
    // insert code here
    }
}
```

Which one create an anonymous inner class from within class Bar?

A. Boo f = new Boo(24) { };

B. Boo f = new Bar() { };

C. Bar f = new Boo(String s) { };

D. Boo f = new Boo.Bar(String s) { };

Ans. B. Boo f = new Bar() { };

Explanation

Option B is correct because anonymous inner classes are no different from any other class when it comes to polymorphism. That means you are always allowed to declare a reference variable of the superclass type and have that reference variable refer to an instance of a subclass type, which in this case is an anonymous subclass of Bar. Since Bar is a subclass of Boo, it all works.

Option A is incorrect because it passes an int to the Boo constructor, and there is no matching constructor in the Boo class.

Option C is incorrect because it violates the rules of polymorphism—you cannot refer to a superclass type using a reference variable declared as the subclass type. The superclass is not guaranteed to have everything the subclass has.

Option D uses incorrect syntax.

148. Which is true about a method-local inner class?

A. It must be marked final.

B. It can be marked abstract.

C. It can be marked public.

D. It can be marked static.

Ans. B. It can be marked abstract.

Explanation

Option B is correct because a method—local inner class can be abstract, although it means a subclass of the inner class must be created if the abstract class is to be used (so an abstract method—local inner class is probably not useful).

Option A is incorrect because a method—local inner class does not have to be declared final (although it is legal to do so).

C and D are incorrect because a method—local inner class cannot be made public (remember-you cannot mark any local variables as public), or static.

149. Which statement is true about a static nested class?

A. You must have a reference to an instance of the enclosing class in order to instantiate it.

B. It does not have access to nonstatic members of the enclosing class.

C. Its variables and methods must be static.

D. It must extend the enclosing class.

Ans. B. It does not have access to nonstatic members of the enclosing class.

Explanation

Option B is correct because a static nested class is not tied to an instance of the enclosing class, and thus can't access the nonstatic members of the class (just as a static method can't access nonstatic members of a class).

Option A is incorrect because static nested classes do not need (and can't use) a reference to an instance of the enclosing class.

Option C is incorrect because static nested classes can declare and define nonstatic members.

Option D is wrong because it just is. There's no rule that says an inner or nested class has to extend anything.

150. Which constructs an anonymous inner class instance?

A. Runnable r = new Runnable() { };

B. Runnable r = new Runnable(public void run() { });

C. Runnable r = new Runnable { public void run(){}};

D. System.out.println(new Runnable() {public void run() { }});

Ans. D. System.out.println(new Runnable() {public void run() { }});

Explanation

D is correct. It defines an anonymous inner class instance, which also means it creates an instance of that new anonymous class at the same time. The anonymous class is an implementer of the Runnable interface, so it must override the run() method of Runnable.

A is incorrect because it doesn't override the run() method, so it violates the rules of interface implementation.

B and C use incorrect syntax.

151. class Foo

```
class Foo
{
    class Bar{ }
}
class Test
{
    public static void main (String [] args)
    {
        Foo f = new Foo();
        /* Line 10: Missing statement ? */
    }
}
```

Which statement, inserted at line 10, creates an instance of Bar?

A. Foo.Bar b = new Foo.Bar();

B. Foo.Bar b = f.new Bar();

C. Bar b = new f.Bar();

D. Bar b = f.new Bar();

Ans. B. Foo.Bar b = f.new Bar();

Explanation

Option B is correct because the syntax is correct- using both names (the enclosing class and the inner class) in the reference declaration, then using a reference to the enclosing class to invoke new on the inner class.

Options A, C and D all use incorrect syntax. A is incorrect because it doesn't use a reference to the enclosing class, and also because it includes both names in the new.

C is incorrect because it doesn't use the enclosing class name in the reference variable declaration, and because the new syntax is wrong.

D is incorrect because it doesn't use the enclosing class name in the reference variable declaration.

152. public class MyOuter

```
public class MyOuter
{
    public static class MyInner
    {
        public static void foo() { }
    }
}
```

Which statement, if placed in a class other than MyOuter or MyInner, instantiates an instance of the nested class?

A. MyOuter.MyInner m = new MyOuter.MyInner();

B. MyOuter.MyInner mi = new MyInner();

C. MyOuter m = new MyOuter(); MyOuter.MyInner mi = m.new MyOuter.MyInner();

D. MyInner mi = new MyOuter.MyInner();

Ans. A. MyOuter.MyInner m = new MyOuter.MyInner();

Explanation

MyInner is a static nested class, so it must be instantiated using the fully-scoped name of MyOuter.MyInner.

Option B is incorrect because it doesn't use the enclosing name in the new.

Option C is incorrect because it uses incorrect syntax. When you instantiate a nested class by invoking new on an instance of the enclosing class, you do not use the enclosing name. The difference between options A and C is that option C is calling new on an instance of the enclosing class rather than just new by itself.

Option D is incorrect because it doesn't use the enclosing class name in the variable declaration.

153. What will be the output of the program?

```
public class Foo
{
    Foo()
    {
        System.out.print("foo");
    }

    class Bar
    {
        Bar()
        {
            System.out.print("bar");
        }
        public void go()
```

```
    {
        System.out.print("hi");
    }
} /* class Bar ends */

    public static void main (String [] args)
    {
        Foo f = new Foo();
        f.makeBar();
    }
    void makeBar()
    {
        (new Bar() {}).go();
    }
}/* class Foo ends */
```

A. Compilation fails
B. An error occurs at runtime
C. It prints "foobarhi"
D. It prints "barhi"

Ans. C. It prints "foobarhi"

Explanation

Option C is correct because first the Foo instance is created, which means the Foo constructor runs and prints "foo". Next, the makeBar() method is invoked which creates a Bar, which means the Bar constructor runs and prints "bar", and finally the go() method is invoked on the new Bar instance, which means the go() method prints "hi".

154. What will be the output of the program?

```
public class HorseTest
{
    public static void main (String [] args)
    {
        class Horse
        {
            public String name; /* Line 7 */
            public Horse(String s)
            {
                name = s;
            }
        } /* class Horse ends */
        Object obj = new Horse("Zippo"); /*
Line 13 */
        Horse h = (Horse) obj; /* Line 14 */
        System.out.println(h.name);
    }
} /* class HorseTest ends */
```

A. An exception occurs at runtime at line 10.
B. It prints "Zippo".
C. Compilation fails because of an error on line 7.
D. Compilation fails because of an error on line 13.

Ans. B. It prints "Zippo".

Explanation

The code in the HorseTest class is perfectly legal. Line 13 creates an instance of the method—local inner class Horse, using a reference variable declared as type Object. Line 14 casts the Horse object to a Horse reference variable, which allows line 15 to compile. If line 14 were removed, the HorseTest code would not compile, because class Object does not have a name variable.

155. What will be the output of the program?

```
public class TestObj
{
    public static void main (String [] args)
    {
        Object o = new Object() /* Line 5 */
        {
            public boolean equals(Object obj)
            {
                return true;
            }
        }    /* Line 11 */

        System.out.println(o.equals("Fred"));
    }
}
```

A. It prints "true".
B. It prints "Fred".
C. An exception occurs at runtime.
D. Compilation fails

Ans. D. Compilation fails

Explanation

This code would be legal if line 11 ended with a semicolon. Remember that line 5 is a statement that doesn't end until line 11, and a statement needs a closing semicolon!

156. What will be the output of the program?

```
public abstract class AbstractTest
{
    public int getNum()
    {
        return 45;
    }
    public abstract class Bar
    {
        public int getNum()
        {
```

```
        return 38;
    }
}
public static void main (String [] args)
{
    AbstractTest t = new AbstractTest()
    {
        public int getNum()
        {
            return 22;
        }
    };
    AbstractTest.Bar f = t.new Bar()
    {
        public int getNum()
        {
            return 57;
        }
    };

    System.out.println(f.getNum() + " " +
t.getNum());
  }
}
```

A. 57 22

B. 45 38

C. 45 57

D. An exception occurs at runtime.

Ans. A. 57 22

Explanation

You can define an inner class as abstract, which means you can instantiate only concrete subclasses of the abstract inner class. The object referenced by the variable t is an instance of an anonymous subclass of AbstractTest, and the anonymous class overrides the getNum() method to return 22. The variable referenced by f is an instance of an anonymous subclass of Bar, and the anonymous Bar subclass also overrides the getNum() method (to return 57). Remember that to instantiate a Bar instance, we need an instance of the enclosing AbstractTest class to tie to the new Bar inner class instance. AbstractTest can't be instantiated because it's abstract, so we created an anonymous subclass (non-abstract) and then used the instance of that anonymous subclass to tie to the new Bar subclass instance.

Threads

157. What is the name of the method used to start a thread execution?

A. init(); **B.** start();

C. run(); **D.** resume();

Ans. B. start();

Explanation

Option B is correct. The start() method causes this thread to begin execution; the Java Virtual Machine calls the run method of this thread.

Option A is wrong. There is no init() method in the Thread class.

Option C is wrong. The run() method of a thread is like the main() method to an application. Starting the thread causes the object's run method to be called in that separately executing thread.

Option D is wrong. The resume() method is deprecated. It resumes a suspended thread.

158. Which two are valid constructors for Thread?

1. Thread(Runnable r, String name)
2. Thread()
3. Thread(int priority)
4. Thread(Runnable r, ThreadGroup g)
5. Thread(Runnable r, int priority)

A. 1 and 3 **B.** 2 and 4

C. 1 and 2 **D.** 2 and 5

Ans. C. 1 and 2

Explanation

(1) and (2) are both valid constructors for Thread. (3), (4), and (5) are not legal Thread constructors, although (4) is close. If you reverse the arguments in (4), you'd have a valid constructor.

159. Which three are methods of the Object class?

1. notify();
2. notifyAll();
3. isInterrupted();
4. synchronized();
5. interrupt();
6. wait(long msecs);
7. sleep(long msecs);
8. yield();

A. 1, 2, 4 **B.** 2, 4, 5

C. 1, 2, 6 **D.** 2, 3, 4

Ans. C. 1, 2, 6

Explanation

(1), (2), and (6) are correct. They are all related to the list of threads waiting on the specified object. (3), (5), (7), and (8) are incorrect answers. The methods isInterrupted() and interrupt() are instance methods of Thread.

The methods sleep() and yield() are static methods of Thread.

D is incorrect because synchronized is a keyword and the synchronized() construct is part of the Java language.

160. Class X implements Runnable

```
{
    public static void main(String args[])
    {
        /* Missing code? */
    }
    public void run() {}
}
```

Which of the following line of code is suitable to start a thread?

A. Thread t = new Thread(X);

B. Thread t = new Thread(X); t.start();

C. X run = new X(); Thread t = new Thread(run); t.start();

D. Thread t = new Thread(); x.run();

Ans. C. X run = new X(); Thread t = new Thread(run); t.start();

Explanation

Option C is suitable to start a thread.

161. Which cannot directly cause a thread to stop executing?

A. Calling the SetPriority() method on a Thread object.

B. Calling the wait() method on an object.

C. Calling notify() method on an object.

D. Calling read() method on an InputStream.

162. Which two of the following methods are defined in class Thread?

1. start()
2. wait()
3. notify()
4. run()
5. terminate()

A. 1 and 4 **B.** 2 and 3

C. 3 and 4 **D.** 2 and 4

Ans. A. 1 and 4

Explanation

(1) and (4). Only start() and run() are defined by the Thread class.

(2) and (3) are incorrect because they are methods of the Object class. (5) is incorrect because there's no such method in any thread-related class.

163. Which three guarantee that a thread will leave the running state?

1. yield()
2. wait()
3. notify()
4. notifyAll()
5. sleep(1000)
6. aLiveThread.join()
7. Thread.killThread()

A. 1, 2 and 4 **B.** 2, 5 and 6

C. 3, 4 and 7 **D.** 4, 5 and 7

Ans. B. 2, 5 and 6

Explanation

(2) is correct because wait() always causes the current thread to go into the object's wait pool.

(5) is correct because sleep() will always pause the currently running thread for at least the duration specified in the sleep argument (unless an interrupted exception is thrown).

(6) is correct because assuming that the thread you're calling join() on is alive, the thread calling join() will immediately block until the thread you're calling join() on is no longer alive.

(1) is wrong, but tempting. The yield() method is not guaranteed to cause a thread to leave the running state, although if there are runnable threads of the same priority as the currently running thread, then the current thread will probably leave the running state.

(3) and (4) are incorrect because they don't cause the thread invoking them to leave the running state.

(7) is wrong because there's no such method.

164. Which of the following will directly stop the execution of a Thread?

A. wait()

B. notify()

C. notifyall()

D. exits synchronized code

Ans. A. wait()

Explanation

Option A is correct. wait() causes the current thread to wait until another thread invokes the notify() method or the notifyAll() method for this object.

Option C is wrong. notifyAll() - wakes up all threads that are waiting on this object's monitor.

Option D is wrong. Typically, releasing a lock means the thread holding the lock (in other words, the thread currently in the synchronized method) exits the synchronized method. At that point, the lock is free until some other thread enters a synchronized method on that object. Does entering/exiting synchronized code mean that the thread execution stops? Not necessarily because the thread can still run code that is not synchronized. I think the word directly in the question gives us a clue. Exiting synchronized code does not directly stop the execution of a thread.

165. Which method must be defined by a class implementing the java.lang.Runnable interface?

A. void run()

B. public void run()

C. public void start()

D. void run(int priority)

Ans. B. public void run()

Explanation

Option B is correct because in an interface all methods are abstract by default, therefore they must be overridden by the implementing class. The Runnable interface only contains 1 method, the void run() method, therefore it must be implemented.

Options A and D are incorrect because they are narrowing the access privileges, i.e. package(default) access is narrower than public access.

Option C is not method in the Runnable interface, therefore it is incorrect.

166. Which will contain the body of the thread?

A. run(); **B.** start();

C. stop(); **D.** main();

Ans. A. run();

Explanation

Option A is correct. The run() method to a thread is like the main() method to an application. Starting the thread causes the object's run method to be called in that separately executing thread.

Option B is wrong. The start() method causes this thread to begin execution; the Java Virtual Machine calls the run method of this thread.

Option C is wrong. The stop() method is deprecated. It forces the thread to stop executing.

Option D is wrong. It is the main entry point for an application.m object.

167. Which method registers a thread in a thread scheduler?

A. run(); **B.** construct();

C. start(); **D.** register();

Ans. C. start();

Explanation

Option C is correct. The start() method causes this thread to begin execution; the Java Virtual Machine calls the run method of this thread.

Option A is wrong. The run() method of a thread is like the main() method to an application. Starting the thread causes the object's run method to be called in that separately executing thread.

Option B is wrong. There is no construct() method in the Thread class.

Option D is wrong. There is no register() method in the Thread class.

168. Assume the following method is properly synchronized and called from a thread A on an object B:

wait(2000);

After calling this method, when will the thread A become a candidate to get another turn at the CPU?

A. After thread A is notified, or after two seconds.

B. After the lock on B is released, or after two seconds.

C. Two seconds after thread A is notified.

D. Two seconds after lock B is released.

Ans. A. After thread A is notified, or after two seconds.

Explanation

Option A. Either of the two events (notification or wait time expiration) will make the thread become a candidate for running again.

Option B is incorrect because a waiting thread will not return to runnable when the lock is released, unless a notification occurs.

Option C is incorrect because the thread will become a candidate immediately after notification, not two seconds afterwards.

Option D is also incorrect because a thread will not come out of a waiting pool just because a lock has been released.

169. Which of the following will not directly cause a thread to stop?

A. notify()
B. wait()
C. InputStream access
D. sleep()

Ans. A. notify()

Explanation

Option A is correct. notify()—wakes up a single thread that is waiting on this object's monitor.

Option B is wrong. wait() causes the current thread to wait until another thread invokes the notify() method or the notifyAll() method for this object.

Option C is wrong. Methods of the InputStream class block until input data is available, the end of the stream is detected, or an exception is thrown. Blocking means that a thread may stop until certain conditions are met.

Option D is wrong. sleep()—Causes the currently executing thread to sleep (temporarily cease execution) for a specified number of milliseconds. The thread does not lose ownership of any monitors.

170. Which class or interface defines the wait(), notify(), and notifyAll() methods?

A. Object **B.** Thread
C. Runnable **D.** Class

Ans. A. Object

Explanation

The Object class defines these thread-specific methods.

Options B, C, and D are incorrect because they do not define these methods. And yes, the Java API does define a class called Class, though you do not need to know it for the exam.

171. public class MyRunnable implements Runnable

```
{
    public void run()
    {
        // some code here
    }
}
```

Which of these will create and start this thread?

A. new Runnable(MyRunnable).start();
B. new Thread(MyRunnable).run();
C. new Thread(new MyRunnable()).start();
D. new MyRunnable().start();

Ans. C. new Thread(new MyRunnable()).start();

Explanation

Because the class implements Runnable, an instance of it has to be passed to the Thread constructor, and then the instance of the Thread has to be started.

A is incorrect. There is no constructor like this for Runnable because Runnable is an interface, and it is illegal to pass a class or interface name to any constructor.

B is incorrect for the same reason; you can't pass a class or interface name to any constructor.

D is incorrect because MyRunnable doesn't have a start() method, and the only start() method that can start a thread of execution is the start() in the Thread class.

172. What will be the output of the program?

```
class MyThread extends Thread
{
    MyThread()
    {
        System.out.print(" MyThread");
    }
    public void run()
    {
        System.out.print(" bar");
    }
    public void run(String s)
    {
        System.out.println(" baz");
    }
}
public class TestThreads
{
    public static void main (String [] args)
    {
        Thread t = new MyThread()
        {
            public void run()
            {
                System.out.println(" foo");
            }
        };
        t.start();
    }
}
```

A. foo **B.** MyThread foo
C. MyThread bar **D.** foo bar

Ans. B. MyThread foo

Explanation

Option B is correct because in the first line of main we're constructing an instance of an anonymous inner class extending from MyThread. So the MyThread constructor runs and prints "MyThread". The next statement in main invokes start() on the new thread instance, which causes the overridden run() method (the run() method defined in the anonymous inner class) to be invoked, which prints "foo".

173. What will be the output of the program?

```
class MyThread extends Thread
{
    public static void main(String [] args)
    {
        MyThread t = new MyThread();
        t.start();
        System.out.print("one. ");
        t.start();
        System.out.print("two. ");
    }
    public void run()
    {
        System.out.print("Thread ");
    }
}
```

A. Compilation fails

B. An exception occurs at runtime.

C. It prints "Thread one. Thread two."

D. The output cannot be determined.

Ans. B. An exception occurs at runtime.

Explanation

When the start() method is attempted a second time on a single Thread object, the method will throw an IllegalThreadStateException (you will not need to know this exception name for the exam). Even if the thread has finished running, it is still illegal to call start() again.

174. What will be the output of the program?

```
class MyThread extends Thread
{
    MyThread() {}
    MyThread(Runnable r) {super(r); }
    public void run()
    {
        System.out.print("Inside Thread ");
    }
}
class MyRunnable implements Runnable
{
    public void run()
    {
        System.out.print(" Inside Runnable");
    }
}
class Test
{
    public static void main(String[] args)
    {
        new MyThread().start();
        new MyThread(new
MyRunnable()).start();
    }
}
```

A. Prints "Inside Thread Inside Thread"

B. Prints "Inside Thread Inside Runnable"

C. Does not compile

D. Throws exception at runtime

Ans. A. Prints "Inside Thread Inside Thread"

Explanation

If a Runnable object is passed to the Thread constructor, then the run method of the Thread class will invoke the run method of the Runnable object.

In this case, however, the run method in the Thread class is overridden by the run method in MyThread class. Therefore, the run() method in MyRunnable is never invoked.

Both times, the run() method in MyThread is invoked instead.

175. What will be the output of the program?

```
class s1 implements Runnable
{
    int x = 0, y = 0;
    int addX() {x++; return x;}
    int addY() {y++; return y;}
    public void run() {
    for(int i = 0; i < 10; i++)
        System.out.println(addX() + " " +
addY());
    }
    public static void main(String args[])
    {
        s1 run1 = new s1();
        s1 run2 = new s1();
        Thread t1 = new Thread(run1);
        Thread t2 = new Thread(run2);
        t1.start();
        t2.start();
    }
}
```

A. Compile time Error: There is no start() method
B. Will print in this order: 1 1 2 2 3 3 4 4 5 5...
C. Will print but not exactly in an order (e.g. 1 1 2 2 1 1 3 3...)
D. Will print in this order: 1 2 3 4 5 6... 1 2 3 4 5 6...

Ans. C. Will print but not exactly in an order (e.g. 1 1 2 2 1 1 3 3...)

Explanation

Both threads are operating on different sets of instance variables. If you modify the code of the run() method to print the thread name, it will help to clarify the output:

```
public void run()
{
for(int i = 0; i < 10; i++)

System.out.println(
Thread.currentThread().getName() + ": " +
addX() + " " + addY()
);

}
```

176. What will be the output of the program?

```
public class Q126 implements Runnable
{
  private int x;
  private int y;

  public static void main(String [] args)
  {
    Q126 that = new Q126();
     (new Thread(that)).start( ); /* Line 8
*/
     (new Thread(that)).start( ); /* Line 9
*/
  }
 public synchronized void run( ) /* Line 11 */
  {
    for (;;) /* Line 13 */
    {
      x++;
      y++;
      System.out.println("x = " + x + "y =
" + y);
    }
  }
}
```

A. An error at line 11 causes compilation to fail
B. Errors at lines 8 and 9 cause compilation to fail.
C. The program prints pairs of values for x and y that might not always be the same on the same line (for example, "x=2, y=1")
D. The program prints pairs of values for x and y that are always the same on the same line (for example, "x=1, y=1"). In addition, each value appears once (for example, "x=1, y=1" followed by "x=2, y=2")

Ans. D. The program prints pairs of values for x and y that are always the same on the same line (for example, "x=1, y=1"). In addition, each value appears once (for example, "x=1, y=1" followed by "x=2, y=2")

Explanation

The synchronized code is the key to answering this question. Because x and y are both incremented inside the synchronized method, they are always incremented together. Also keep in mind that the two threads share the same reference to the Q 126 object.

Also note that because of the infinite loop at line 13, only one thread ever gets to execute.

177. What will be the output of the program?

```
class s1 extends Thread
{
  public void run()
  {
    for(int i = 0; i < 3; i++)
    {
      System.out.println("A");
      System.out.println("B");
    }
  }
}
class Test120 extends Thread
{
  public void run()
  {
    for(int i = 0; i < 3; i++)
    {
      System.out.println("C");
      System.out.println("D");
    }
```

```
    }
    public static void main(String args[])
    {
        s1 t1 = new s1();
        Test120 t2 = new Test120();
        t1.start();
        t2.start();
    }
}
```

A. Compile time Error There is no start() method

B. Will print in this order AB CD AB...

C. Will print but not be able to predict the Order

D. Will print in this order ABCD...ABCD...

Ans. C. Will print but not be able to predict the Order

Explanation

We cannot predict the order in which threads are going to run.

178. What will be the output of the program?

```
class s implements Runnable
{
    int x, y;
    public void run()
    {
        for(int i = 0; i < 1000; i++)
            synchronized(this)
            {
                x = 12;
                y = 12;
            }
        System.out.print(x + " " + y + " ");
    }
    public static void main(String args[])
    {
        s run = new s();
        Thread t1 = new Thread(run);
        Thread t2 = new Thread(run);
        t1.start();
        t2.start();
    }
}
```

A. DeadLock

B. It prints 12 12 12 12

C. Compilation Error

D. Cannot determine output.

Ans. B. It prints 12 12 12 12

Explanation

The program will execute without any problems and print 12 12 12 12.

179. What will be the output of the program?

```
public class ThreadDemo
{
    private int count = 1;
    public synchronized void doSomething()
    {
        for (int i = 0; i < 10; i++)
            System.out.println(count++);
    }
    public static void main(String[] args)
    {
        ThreadDemo demo = new
ThreadDemo();
        Thread a1 = new A(demo);
        Thread a2 = new A(demo);
        a1.start();
        a2.start();
    }
}
class A extends Thread
{
    ThreadDemo demo;
    public A(ThreadDemo td)
    {
        demo = td;
    }
    public void run()
    {
        demo.doSomething();
    }
}
```

A. It will print the numbers 0 to 19 sequentially

B. It will print the numbers 1 to 20 sequentially

C. It will print the numbers 1 to 20, but the order cannot be determined

D. The code will not compile.

Ans. B. It will print the numbers 1 to 20 sequentially

Explanation

You have two different threads that share one reference to a common object.

The updating and output takes place inside synchronized code.

One thread will run to completion printing the numbers 1–10.

180. What will be the output of the program?

```
public class WaitTest
{
    public static void main(String [] args)
    {
        System.out.print("1 ");
        synchronized(args)
        {
            System.out.print("2 ");
            try
            {
                args.wait(); /* Line 11 */
            }
            catch(InterruptedException e){ }
        }
        System.out.print("3 ");
    }
}
```

A. It fails to compile because the IllegalMonitorStateException of wait() is not dealt within line 11.

B. 1 2 3

C. 1 3

D. 1 2

Ans. D. 1 2

Explanation

1 and 2 will be printed, but there will be no return from the wait call because no other thread will notify the main thread, so 3 will never be printed. The program is essentially frozen at line 11.

A is incorrect; IllegalMonitorStateException is an unchecked exception so it doesn't have to be dealt with explicitly.

B and C are incorrect; 3 will never be printed, since this program will never terminate because it will wait forever.

181. What will be the output of the program?

```
public class SyncTest
{
    public static void main (String [] args)
    {
        Thread t = new Thread()
        {
            Foo f = new Foo();
            public void run()
            {
                f.increase(20);
            }
        };
        t.start();
    }
}
class Foo
{
    private int data = 23;
    public void increase(int amt)
    {
        int x = data;
        data = x + amt;
    }
}
```

and assuming that data must be protected from corruption, what—if anything—can you add to the preceding code to ensure the integrity of data?

A. Synchronize the run method.

B. Wrap a synchronize(this) around the call to f.increase().

C. The existing code will cause a runtime exception.

D. Synchronize the increase() method

Ans. D. Synchronize the increase() method

Explanation

Option D is correct because synchronizing the code that actually does the increase will protect the code from being accessed by more than one thread at a time.

Option A is incorrect because synchronizing the run() method would stop other threads from running the run() method (a bad idea) but still would not prevent other threads with other runnables from accessing the increase() method.

Option B is incorrect for virtually the same reason as A—synchronizing the code that calls the increase() method does not prevent other code from calling the increase() method.

182. What will be the output of the program?

```
class Happy extends Thread
{
    final StringBuffer sb1 = new StringBuffer();
    final StringBuffer sb2 = new StringBuffer();

    public static void main(String args[])
    {
        final Happy h = new Happy();

        new Thread()
```

```
    {
        public void run()
        {
            synchronized(this)
            {
                h.sb1.append("A");
                h.sb2.append("B");
                System.out.println(h.sb1);
                System.out.println(h.sb2);
            }
        }
    }.start();

    new Thread()
    {
        public void run()
        {
            synchronized(this)
            {
                h.sb1.append("D");
                h.sb2.append("C");
                System.out.println(h.sb2);
                System.out.println(h.sb1);
            }
        }
    }.start();
  }
}
```

A. ABBCAD

B. ABCBCAD

C. CDADACB

D. Output determined by the underlying platform.

Ans. D. Output determined by the underlying platform.

Explanation

Can you guarantee the order in which threads are going to run? No, you can't. So how do you know what the output will be? The output cannot be determined.

183. class Test

```
{
    public static void main(String [] args)
    {
        printAll(args);
    }

    public static void printAll(String[] lines)
    {
        for(int i = 0; i < lines.length; i++)
        {
            System.out.println(lines[i]);
            Thread.currentThread().sleep(1000);
        }
    }
}
```

the static method Thread.currentThread() returns a reference to the currently executing Thread object. What is the result of this code?

A. Each String in the array lines will output, with a 1-second pause.

B. Each String in the array lines will output, with no pause in between because this method is not executed in a Thread.

C. Each String in the array lines will output, and there is no guarantee there will be a pause because currentThread() may not retrieve this thread.

D. This code will not compile.

Ans. D. This code will not compile.

Explanation

D. The sleep() method must be enclosed in a try/catch block, or the method printAll() must declare it throws the InterruptedException.

A is incorrect, but it would be correct if the InterruptedException was dealt with.

B is incorrect, but it would still be incorrect if the InterruptedException was dealt with because all Java code, including the main() method, runs in threads.

C is incorrect. The sleep() method is static, so even if it is called on an instance, it still always affects the currently executing thread.

184. What will be the output of the program?

```
class MyThread extends Thread
{
    public static void main(String [] args)
    {
        MyThread t = new MyThread(); /*
Line 5 */
        t.run(); /* Line 6 */
    }

    public void run()
    {
        for(int i=1; i < 3; ++i)
        {
            System.out.print(i + "..");
        }
    }
}
```

A. This code will not compile due to line 5.
B. This code will not compile due to line 6.
C. 1..2..
D. 1..2..3..

Ans. C. 1..2..

Explanation

Line 6 calls the run() method, so the run() method executes as a normal method and it prints "1..2.."
A is incorrect because line 5 is the proper way to create an object.
B is incorrect because it is legal to call the run() method, even though this will not start a true thread of execution. The code after line 6 will not execute until the run() method is complete.
D is incorrect because the for loop only does two iterations.

185. What will be the output of the program?

```
class Test116
{
static final StringBuffer sb1 = new
StringBuffer();
static final StringBuffer sb2 = new
StringBuffer();
public static void main(String args[])
{
   new Thread()
   {
      public void run()
      {
         synchronized(sb1)
         {
            sb1.append("A");
            sb2.append("B");
         }
      }
   }.start();

   new Thread()
   {
      public void run()
      {
         synchronized(sb1)
         {
            sb1.append("C");
            sb2.append("D");
         }
      }
   }.start(); /* Line 28 */

   System.out.println (sb1 + " " + sb2);
   }
}
```

A. main() will finish before starting threads.
B. main() will finish in the middle of one thread.
C. main() will finish after one thread.
D. Cannot be determined.

Ans. D. Cannot be determined.

Explanation

Can you guarantee the order in which threads are going to run? No, you can't. So how do you know what the output will be? The output cannot be determined.
add this code after line 28:

```
try { Thread.sleep(5000); }
catch(InterruptedException e) { }
```

and you have some chance of predicting the outcome.

186. What will be the output of the program?

```
public class ThreadTest extends Thread
{
   public void run()
   {
      System.out.println("In run");
      yield();
      System.out.println("Leaving run");
   }
   public static void main(String []argv)
   {
      (new ThreadTest()).start();
   }
}
```

A. The code fails to compile in the main() method
B. The code fails to compile in the run() method
C. Only the text "In run" will be displayed
D. The text "In run" followed by "Leaving run" will be displayed

Ans. D. The text "In run" followed by "Leaving run" will be displayed

187. What will be the output of the program?

```
public class Test107 implements Runnable
{
   private int x;
   private int y;
   public static void main(String args[])
   {
      Test107 that = new Test107();
      (new Thread(that)).start();
      (new Thread(that)).start();
   }
```

```
public synchronized void run()
{
    for(int i = 0; i < 10; i++)
    {
        x++;
        y++;
        System.out.println("x = " + x + ", y
= " + y); /* Line 17 */
    }
}
}
```

A. Compilation error.

B. Will print in this order: x = 1 y = 1 x = 2 y = 2 x = 3 y = 3 x = 4 y = 4 x = 5 y = 5... but the output will be produced by both threads running simultaneously.

C. Will print in this order: x = 1 y = 1 x = 2 y = 2 x = 3 y = 3 x = 4 y = 4 x = 5 y = 5... but the output will be produced by first one thread then the other. This is guaranteed by the synchronised code.

D. Will print in this order x = 1 y = 2 x = 3 y = 4 x = 5 y = 6 x = 7 y = 8...

Ans. C. Will print in this order: x = 1 y = 1 x = 2 y = 2 x = 3 y = 3 x = 4 y = 4 x = 5 y = 5... but the output will be produced by first one thread then the other. This is guaranteed by the synchronised code.

Explanation

Both threads are operating on the same instance variables. Because the code is synchronized, the first thread will complete before the second thread begins. Modify line 17 to print the thread names:

System.out.println(Thread.currentThread().getName() + " x = " + x + ", y = " + y);

188. What will be the output of the program?

```
public class Test
{
    public static void main (String [] args)
    {
        final Foo f = new Foo();
            Thread t = new Thread(new
Runnable()
        {
            public void run()
            {
                f.doStuff();
            }
        });
        Thread g = new Thread()
        {
            public void run()
            {
                f.doStuff();
            }
        };
        t.start();
        g.start();
    }
}
class Foo
{
    int x = 5;
    public void doStuff()
    {
        if (x < 10)
        {
            // nothing to do
            try
            {
                wait();
            } catch(InterruptedException ex) {
        }
        }
        else
        {
            System.out.println("x is " + x++);
            if (x >= 10)
            {
                notify();
            }
        }
    }
}
```

A. The code will not compile because of an error on notify(); of class Foo.

B. The code will not compile because of some other error in class Test.

C. An exception occurs at runtime.

D. It prints "x is 5 x is 6".

Ans. C. An exception occurs at runtime.

Explanation

C is correct because the thread does not own the lock of the object it invokes wait() on. If the method were synchronized, the code would run without exception.

A, B are incorrect because the code compiles without errors.

D is incorrect because the exception is thrown before there is any output.

189. What will be the output of the program?

```
class MyThread extends Thread
{
  public static void main(String [] args)
  {
    MyThread t = new MyThread();
    Thread x = new Thread(t);
    x.start(); /* Line 7 */
  }
  public void run()
  {
    for(int i = 0; i < 3; ++i)
    {
      System.out.print(i + "..");
    }
  }
}
```

A. Compilation fails.

B. 1..2..3..

C. 0..1..2..3..

D. 0..1..2..

Ans. D. 0..1..2

Explanation

The thread MyThread will start and loop three times (from 0 to 2).

Option A is incorrect because the Thread class implements the Runnable interface; therefore, in line 7, Thread can take an object of type Thread as an argument in the constructor.

Options B and C are incorrect because the variable i in the for loop starts with a value of 0 and ends with a value of 2.

190. Which statement is true?

A. A static method cannot be synchronized.

B. If a class has synchronized code, multiple threads can still access the nonsynchronized code.

C. Variables can be protected from concurrent access problems by marking them with the synchronized keyword.

D. When a thread sleeps, it releases its locks.

Ans. B. If a class has synchronized code, multiple threads can still access the nonsynchronized code.

Explanation

B is correct because multiple threads are allowed to enter nonsynchronized code, even within a class that has some synchronized methods.

A is incorrect because static methods can be synchronized; they synchronize on the lock on the instance of class java.lang.Class that represents the class type.

C is incorrect because only methods—not variables—can be marked synchronized.

D is incorrect because a sleeping thread still maintains its locks.

191. Which two can be used to create a new Thread?

1. Extend java.lang.Thread and override the run() method.
2. Extend java.lang.Runnable and override the start() method.
3. Implement java.lang.Thread and implement the run() method.
4. Implement java.lang.Runnable and implement the run() method.
5. Implement java.lang.Thread and implement the start() method.

A. 1 and 2 **B.** 2 and 3

C. 1 and 4 **D.** 3 and 4

Ans. C. 1 and 4

Explanation

There are two ways of creating a thread; extend (sub-class) the Thread class and implement the Runnable interface. For both of these ways you must implement (override and not overload) the public void run() method.

(1) is correct: Extending the Thread class and overriding its run method is a valid procedure.

(4) is correct: You must implement interfaces, and runnable is an interface and you must also include the run method.

(2) is wrong: Runnable is an interface which implements not Extends. Gives the error (no interface expected here)

(3) is wrong: You cannot implement java.lang.Thread (this is a Class). (Implements Thread, gives the error: Interface expected). Implements expects an interface.

(5) is wrong: You cannot implement java.lang.Thread (this is a class). You Extend classes, and implement interfaces. (Implements Thread, gives the error: Interface expected)

192. Which statement is true?

A. If only one thread is blocked in the wait method of an object, and another thread executes the modify on that same object, then the first thread immediately resumes execution.

B. If a thread is blocked in the wait method of an object, and another thread executes the notify method on the same object, it is still possible that the first thread might never resume execution.

C. If a thread is blocked in the wait method of an object, and another thread executes the notify method on the same object, then the first thread definitely resumes execution as a direct and sole consequence of the notify call.

D. If two threads are blocked in the wait method of one object, and another thread executes the notify method on the same object, then the first thread that executed the wait call first definitely resumes execution as a direct and sole consequence of the notify call.

Ans. B. If a thread is blocked in the wait method of an object, and another thread executes the notify method on the same object, it is still possible that the first thread might never resume execution.

Explanation

Option B is correct: The notify method only wakes the thread. It does not guarantee that the thread will run.

Option A is incorrect: Just because another thread activates the modify method in A, this does not mean that the thread will automatically resume execution.

Option C is incorrect: This is incorrect because as said in Answer B notify only wakes the thread but further to this, once it is awake it goes back into the stack and awaits execution, therefore it is not a "direct and sole consequence of the notify call"

Option D is incorrect: The notify method wakes one waiting thread up. If there are more than one sleeping threads, then the choice as to which thread to wake is made by the machine rather than you, therefore you cannot guarantee that the notified thread will be the first waiting thread.

193. Which two statements are true?

1. Deadlock will not occur if wait()/ notify() is used
2. A thread will resume execution as soon as its sleep duration expires.
3. Synchronization can prevent two objects from being accessed by the same thread.
4. The wait() method is overloaded to accept a duration.
5. The notify() method is overloaded to accept a duration.
6. Both wait() and notify() must be called from a synchronized context.

A. 1 and 2 **B.** 3 and 5
C. 4 and 6 **D.** 1 and 3

Ans. C. 4 and 6

Explanation

Statements (4) and (6) are correct. (4) is correct because the wait() method is overloaded to accept a wait duration in milliseconds. If the thread has not been notified by the time the wait duration has elapsed, then the thread will move back to runnable even without having been notified.

(6) is correct because wait()/notify()/notifyAll() must all be called from within a synchronized, context. A thread must own the lock on the object its invoking wait()/notify()/notifyAll() on.

(1) is incorrect because wait()/notify() will not prevent deadlock.

(2) is incorrect because a sleeping thread will return to runnable when it wakes up, but it might not necessarily resume execution right away. To resume executing, the newly awakened thread must still be moved from runnable to running by the scheduler.

(3) is incorrect because synchronization prevents two or more threads from accessing the same object.

(5) is incorrect because notify() is not overloaded to accept a duration.

194. The following block of code creates a Thread using a Runnable target:

Runnable target = new MyRunnable();
Thread myThread = new Thread(target);

Which of the following classes can be used to create the target, so that the preceding code compiles correctly?

A. public class MyRunnable extends Runnable{public void run(){}}

B. public class MyRunnable extends Object{public void run(){}}

C. public class MyRunnable implements Runnable{public void run(){}}

D. public class MyRunnable implements Runnable{void run(){}}

Ans. C. public class MyRunnable implements Runnable{public void run(){}}

Explanation

The class correctly implements the Runnable interface with a legal public void run() method.

Option A is incorrect because interfaces are not extended; they are implemented.

Option B is incorrect because even though the class would compile and it has a valid public void run() method, it does not implement the Runnable interface, so the compiler would complain when creating a Thread with an instance of it.

Option D is incorrect because the run() method must be public.

195. Which statement is true?

A. The notifyAll() method must be called from a synchronized context.

B. To call wait(), an object must own the lock on the thread.

C. The notify() method is defined in class java.lang.Thread.

D. The notify() method causes a thread to immediately release its locks.

Ans. A. The notifyAll() method must be called from a synchronized context.

Explanation

Option A is correct because the notifyAll() method (along with wait() and notify()) must always be called from within a synchronized context.

Option B is incorrect because to call wait(), the thread must own the lock on the object that wait() is being invoked on, not the other way around.

Option C is wrong because notify() is defined in java.lang.Object.

Option D is wrong because notify() will not cause a thread to release its locks. The thread can only release its locks by exiting the synchronized code.

Garbage Collections

196. void start() {

```
    A a = new A();
    B b = new B();
    a.s(b);
    b = null; /* Line 5 */
    a = null; /* Line 6 */
    System.out.println("start completed"); /
* Line 7 */
}
```

When is the B object, created in line 3, eligible for garbage collection?

A. after line 5

B. after line 6

C. after line 7

D. There is no way to be absolutely certain.

Ans. D. There is no way to be absolutely certain.

197. class HappyGarbage01

```
{
    public static void main(String args[])
    {
     HappyGarbage01    h    =    new
HappyGarbage01();
       h.methodA(); /* Line 6 */
    }
    Object methodA()
    {
        Object obj1 = new Object();
        Object [] obj2 = new Object[1];
        obj2[0] = obj1;
        obj1 = null;
        return obj2[0];
    }
}
```

Where will be the most chance of the garbage collector being invoked?

A. After line 9

B. After line 10

C. After line 11

D. Garbage collector never invoked in methodA()

Ans. D. Garbage collector never invoked in methodA()

Explanation

Option D is correct. Garbage collection takes place after the method has returned its reference to the object. The method returns to line 6, there is no reference to store the return value. so garbage collection takes place after line 6.

Option A is wrong. Because the reference to obj1 is stored in obj2[0]. The Object obj1 still exists on the heap and can be accessed by an active thread through the reference stored in obj2[0].

Option B is wrong. Because it is only one of the references to the object obj1, the other reference is maintained in obj2[0].
Option C is wrong. The garbage collector will not be called here because a reference to the object is being maintained and returned in obj2[0].

198. class Bar { }

```
class Test
{
    Bar doBar()
    {
        Bar b = new Bar(); /* Line 6 */
        return b; /* Line 7 */
    }
    public static void main (String args[])
    {
        Test t = new Test();  /* Line 11 */
        Bar newBar = t.doBar();  /* Line 12 */
        System.out.println("newBar");
        newBar = new Bar(); /* Line 14 */
          System.out.println("finishing"); /*
Line 15 */
    }
}
```

At what point is the Bar object, created on line 6, eligible for garbage collection?

A. after line 12
B. after line 14
C. after line 7, when doBar() completes
D. after line 15, when main() completes

Ans. B. after line 14

Explanation

Option B is correct. All references to the Bar object created on line 6 are destroyed when a new reference to a new Bar object is assigned to the variable newBar on line 14. Therefore the Bar object, created on line 6, is eligible for garbage collection after line 14.
Option A is wrong. This actually protects the object from garbage collection.
Option C is wrong. Because the reference in the doBar() method is returned on line 7 and is stored in newBar on line 12. This preserver the object created on line 6.
Option D is wrong. Not applicable because the object is eligible for garbage collection after line 14

199. class Test

```
{
    private Demo d;
    void start()
    {
        d = new Demo();
        this.takeDemo(d); /* Line 7 */
    } /* Line 8 */
    void takeDemo(Demo demo)
    {
        demo = null;
        demo = new Demo();
    }
}
```

When is the Demo object eligible for garbage collection?

A. After line 7
B. After line 8
C. After the start() method completes
D. When the instance running this code is made eligible for garbage collection.

Ans. D. When the instance running this code is made eligible for garbage collection.

Explanation

Option D is correct. By a process of elimination.
Option A is wrong. The variable d is a member of the Test class and is never directly set to null.
Option B is wrong. A copy of the variable d is set to null and not the actual variable d.
Option C is wrong. The variable d exists outside the start() method (it is a class member). So, when the start() method finishes, the variable d still holds a reference.

200. public class X

```
{
    public static void main(String [] args)
    {
        X x = new X();
        X x2 = m1(x); /* Line 6 */
        X x4 = new X();
        x2 = x4; /* Line 8 */
        doComplexStuff();
    }
    static X m1(X mx)
    {
        mx = new X();
        return mx;
    }
}
```

After line 8 runs, how many objects are eligible for garbage collection?

A. 0
B. 1
C. 2
D. 3

Ans. B. 1

Explanation

By the time line 8 has run, the only object without a reference is the one generated as a result of line 6. Remember that "Java is pass by value," so the reference variable x is not affected by the m1() method.

201. public Object m()

```
{
    Object o = new Float(3.14F);
    Object [] oa = new Object[l];
    oa[0] = o; /* Line 5 */
    o = null;  /* Line 6 */
    oa[0] = null; /* Line 7 */
    return o; /* Line 8 */
}
```

When is the Float object, created in line 3, eligible for garbage collection?

A. Just after line 5
B. Just after line 6
C. Just after line 7
D. Just after line 8

Ans. C. Just after line 7

Explanation

Option A is wrong. This simply copies the object reference into the array.

Option B is wrong. The reference o is set to null, but, oa[0] still maintains the reference to the Float object.

Option C is correct. The thread of execution will then not have access to the object.

202. class X2

```
{
    public X2 x;
    public static void main(String [] args)
    {
        X2 x2 = new X2();  /* Line 6 */
        X2 x3 = new X2();  /* Line 7 */
        x2.x = x3;
        x3.x = x2;
        x2 = new X2();
        x3 = x2; /* Line 11 */
        doComplexStuff();
    }
}
```

After line 11 runs, how many objects are eligible for garbage collection?

A. 0
B. 1
C. 2
D. 3

Ans. C. 2

Explanation

This is an example of the islands of isolated objects. By the time line 11 has run, the objects instantiated in lines 6 and 7 are referring to each other, but no live thread can reach either of them.

203. What allows the programmer to destroy an object x?

A. x.delete()
B. x.finalize()
C. Runtime.getRuntime().gc()
D. Only the garbage collection system can destroy an object.

Ans. D. Only the garbage collection system can destroy an object.

Explanation

Option D is correct. When an object is no longer referenced, it may be reclaimed by the garbage collector. If an object declares a finalizer, the finalizer is executed before the object is reclaimed to give the object a last chance to clean up resources that would not otherwise be released. When a class is no longer needed, it may be unloaded.

Option A is wrong. I found 4 delete() methods in all of the Java class structure. They are:

delete()—Method in class java.io.File: Deletes the file or directory denoted by this abstract pathname.

delete(int, int)—Method in class java.lang.StringBuffer: Removes the characters in a substring of this StringBuffer.

delete(int, int)—Method in interface javax.accessibility.AccessibleEditableText: Deletes the text between two indices.

delete(int, int)—Method in class: javax.swing.text.JTextComponent.AccessibleJTextComponent; Deletes the text between two indices.

None of these destroy the object to which they belong.

Option B is wrong. I found 19 finalize() methods. The most interesting, from this question point of view, was the finalize() method in class java.lang.Object which is called by the garbage collector on an object when garbage collection determines that there are no more references to the object. This method does not destroy the object to which it belongs.

Option C is wrong. But it is interesting. The Runtime class has many methods, two of which are:

getRuntime() - Returns the runtime object associated with the current Java application.

gc() - Runs the garbage collector. Calling this method suggests that the Java virtual machine expend effort toward recycling unused objects in order to make the memory they currently occupy available for quick reuse. When control returns from the method call, the virtual machine has made its best effort to recycle all discarded objects. Interesting as this is, it doesn't destroy the object.

204. Which statement is true?

A. Programs will not run out of memory.

B. Objects that will never again be used are eligible for garbage collection.

C. Objects that are referred to by other objects will never be garbage collected.

D. Objects that can be reached from a live thread will never be garbage collected.

Ans. D. Objects that can be reached from a live thread will never be garbage collected.

Explanation

Option D is correct.

Option C is wrong. See the note above on Islands of Isolation. (An object is eligible for garbage collection when no live thread can access it—even though there might be references to it.)

Option B is wrong. "Never again be used" does not mean that there are no more references to the object.

Option A is wrong. Even though Java applications can run out of memory, the another answer supplied that is more right.

205. Which statement is true?

A. All objects that are eligible for garbage collection will be garbage collected by the garbage collector.

B. Objects with at least one reference will never be garbage collected.

C. Objects from a class with the finalize() method overridden will never be garbage collected.

D. Objects instantiated within anonymous inner classes are placed in the garbage collectible heap.

Ans. D. Objects instantiated within anonymous inner classes are placed in the garbage collectible heap.

Explanation

All objects are placed in the garbage collectible heap.

Option A is incorrect because the garbage collector makes no guarantees.

Option B is incorrect because islands of isolated objects can exist.

Option C is incorrect because finalize() has no such mystical powers.

206. Which statement is true?

A. Memory is reclaimed by calling Runtime.gc().

B. Objects are not collected if they are accessible from live threads.

C. An OutOfMemory error is only thrown if a single block of memory cannot be found that is large enough for a particular requirement.

D. Objects that have finalize() methods always have their finalize() methods called before the program ends.

Ans. B. Objects are not collected if they are accessible from live threads.

Explanation

Option B is correct. If an object can be accessed from a live thread, it can't be garbage collected.

Option A is wrong. Runtime.gc() asks the garbage collector to run, but the garbage collector never makes any guarantees about when it will run or what unreachable objects it will free from memory.

Option C is wrong. The garbage collector runs immediately the system is out of memory before an OutOfMemoryException is thrown by the JVM.

Option D is wrong. If this were the case then the garbage collector would actively hang onto objects until a program finishes—this goes against the purpose of the garbage collector.

207. Which statement is true?

A. Calling Runtime.gc() will cause eligible objects to be garbage collected.

B. The garbage collector uses a mark and sweep algorithm.

C. If an object can be accessed from a live thread, it can't be garbage collected.

D. If object 1 refers to object 2, then object 2 can't be garbage collected.

Ans. C. If an object can be accessed from a live thread, it can't be garbage collected.

Explanation

This is a great way to think about when objects can be garbage collected.

Options A and B assume guarantees that the garbage collector never makes.

Option D is wrong because of the now famous islands of isolation scenario.

Assertions

208. What will be the output of the program?

```
public class Test
{
  public static void main(String[] args)
  {
    int x = 0;
      assert (x > 0) ? "assertion failed" :
"assertion passed" ;
    System.out.println("finished");
  }
}
```

A. finished

B. Compilation fails.

C. An AssertionError is thrown and finished is output.

D. An AssertionError is thrown with the message "assertion failed."

Ans. B. Compilation fails.

Explanation

Compilation fails. You can't use the Assert statement in a similar way to the ternary operator. Don't confuse.

209. public class Test

```
{
  public void foo()
  {
    assert false; /* Line 5 */
    assert false; /* Line 6 */
  }
  public void bar()
  {
    while(true)
    {
      assert false; /* Line 12 */
    }
    assert false;  /* Line 14 */
  }
}
```

What causes compilation to fail?

A. Line 5

B. Line 6

C. Line 12

D. Line 14

Ans. D. Line 14

Explanation

Option D is correct. Compilation fails because of an unreachable statement at line 14. It is a compile-time error if a statement cannot be executed because it is unreachable. Now the question is, why is line 20 unreachable. If it is because of the assert, then surely line 6 would also be unreachable. The answer must be something other than assert.

Examine the following:

A while statement can complete normally if and only if at least one of the following is true:

- The while statement is reachable and the condition expression is not a constant expression with value true.

- There is a reachable break statement that exits the while statement.

The while statement at line 11 is infinite and there is no break statement therefore line 14 is unreachable. You can test this with the following code:

```
public class Test80
{
  public void foo()
  {
    assert false;
    assert false;
  }
  public void bar()
  {
    while(true)
    {
      assert false;
```

```
        break;
    }
    assert false;
  }
}
```

210. What will be the output of the program?

```
public class Test
{
    public static int y;
    public static void foo(int x)
    {
        System.out.print("foo ");
        y = x;
    }
    public static int bar(int z)
    {
        System.out.print("bar ");
        return y = z;
    }
    public static void main(String [] args )
    {
        int t = 0;
        assert t > 0 : bar(7);
        assert t > 1 : foo(8); /* Line 18 */
        System.out.println("done ");
    }
}
```

A. bar
B. bar done
C. foo done
D. Compilation fails

Ans. D. Compilation fails

Explanation

The foo() method returns void. It is a perfectly acceptable method, but because it returns void it cannot be used in an assert statement, so line 18 will not compile.

211. What will be the output of the program (when you run with the -ea option) ?

```
public class Test
{
    public static void main(String[] args)
    {
        int x = 0;
         assert (x > 0) : "assertion failed"; /*
Line 6 */
        System.out.println("finished");
    }
}
```

A. finished
B. Compilation fails.
C. An AssertionError is thrown.
D. An AssertionError is thrown and finished is output.

Ans. C. An AssertionError is thrown.

Explanation

An assertion Error is thrown as normal giving the output "assertion failed". The word "finished" is not printed (ensure you run with the -ea option)

Assertion failures are generally labeled in the stack trace with the file and line number from which they were thrown, and also in this case with the error's detail message "assertion failed". The detail message is supplied by the assert statement in line 6.

212. public class Test2

```
{
    public static int x;
    public static int foo(int y)
    {
        return y * 2;
    }
    public static void main(String [] args)
    {
        int z = 5;
        assert z > 0; /* Line 11 */
        assert z > 2: foo(z); /* Line 12 */
        if ( z < 7 )
            assert z > 4; /* Line 14 */

        switch (z)
        {
            case 4: System.out.println("4 ");
            case 5: System.out.println("5 ");
            default: assert z < 10;
        }

        if ( z < 10 )
            assert z > 4: z++; /* Line 22 */
        System.out.println(z);
    }
}
```

Which line is an example of an inappropriate use of assertions?

A. Line 11 **B.** Line 12
C. Line 14 **D.** Line 22

Ans. D. Line 22

Explanation

Assert statements should not cause side effects. Line 22 changes the value of z if the assert statement is false.

Option A is fine; a second expression in an assert statement is not required.

Option B is fine because it is perfectly acceptable to call a method with the second expression of an assert statement.

Option C is fine because it is proper to call an assert statement conditionally.

213. Which of the following statements is true?

A. If assertions are compiled into a source file, and if no flags are included at runtime, assertions will execute by default.

B. As of Java version 1.4, assertion statements are compiled by default.

C. With the proper use of runtime arguments, it is possible to instruct the VM to disable assertions for a certain class, and to enable assertions for a certain package, at the same time.

D. When evaluating command-line arguments, the VM gives -ea flags precedence over -da flags.

Ans. C. With the proper use of runtime arguments, it is possible to instruct the VM to disable assertions for a certain class, and to enable assertions for a certain package, at the same time.

Explanation

Option C is true because multiple VM flags can be used on a single invocation of a Java program.

Option A is incorrect because at runtime assertions are ignored by default.

Option B is incorrect because as of Java 1.4 you must add the argument—source 1.4 to the command line if you want the compiler to compile assertion statements.

Option D is incorrect because the VM evaluates all assertion flags left to right.

214. Which statement is true?

A. Assertions can be enabled or disabled on a class-by-class basis.

B. Conditional compilation is used to allow tested classes to run at full speed.

C. Assertions are appropriate for checking the validity of arguments in a method.

D. The programmer can choose to execute a return statement or to throw an exception if an assertion fails.

Ans. A. Assertions can be enabled or disabled on a class-by-class basis.

Explanation

Option A is correct. The assertion status can be set for a named top-level class and any nested classes contained therein. This setting takes precedence over the class loader's default assertion status, and over any applicable per-package default. If the named class is not a top-level class, the change of status will have no effect on the actual assertion status of any class.

Option B is wrong. Is there such a thing as conditional compilation in Java?

Option C is wrong. For private methods—yes. But do not use assertions to check the parameters of a public method. An assert is inappropriate in public methods because the method guarantees that it will always enforce the argument checks. A public method must check its arguments whether or not assertions are enabled. Further, the assert construct does not throw an exception of the specified type. It can throw only an AssertionError.

Option D is wrong. Because you're never supposed to handle an assertion failure. That means don't catch it with a catch clause and attempt to recover.

215. Which statement is true about assertions in the Java programming language?

A. Assertion expressions should not contain side effects.

B. Assertion expression values can be any primitive type.

C. Assertions should be used for enforcing preconditions on public methods.

D. An AssertionError thrown as a result of a failed assertion should always be handled by the enclosing method.

Ans. A. Assertion expressions should not contain side effects.

Explanation

Option A is correct. Because assertions may be disabled, programs must not assume that the boolean expressions contained in assertions will be evaluated. Thus these expressions should be free of side effects. That is, evaluating such an expression should not affect any state that is visible after the evaluation is complete. Although it is not illegal for a boolean expression contained in an assertion to have a side effect, it is generally inappropriate, as it could cause program behaviour to vary depending on whether assertions are enabled or disabled.

Assertion checking may be disabled for increased performance. Typically, assertion checking is enabled during program development and testing and disabled for deployment.

Option B is wrong. Because you assert that something is "true". True is Boolean. So, an expression must evaluate to Boolean, not int or byte or anything else. Use the same rules for an assertion expression that you would use for a while condition.

Option C is wrong. Usually, enforcing a precondition on a public method is done by condition-checking code that you write yourself, to give you specific exceptions.

Option D is wrong. "You're never supposed to handle an assertion failure"

Not all legal uses of assertions are considered appropriate. As with so much of Java, you can abuse the intended use for assertions, despite the best efforts of Sun's Java engineers to discourage you. For example, you're never supposed to handle an assertion failure. That means don't catch it with a catch clause and attempt to recover. Legally, however, AssertionError is a subclass of Throwable, so it can be caught. But just don't do it! If you're going to try to recover from something, it should be an exception. To discourage you from trying to substitute an assertion for an exception, the AssertionError doesn't provide access to the object that generated it. All you get is the String message.

216. Which of the following statements is true?

A. It is sometimes good practice to throw an AssertionError explicitly.

B. Private getter() and setter() methods should not use assertions to verify arguments.

C. If an AssertionError is thrown in a try-catch block, the finally block will be bypassed.

D. It is proper to handle assertion statement failures using a catch (AssertionException ae) block.

Ans. A. It is sometimes good practice to throw an AssertionError explicitly.

Explanation

Option A is correct because it is sometimes advisable to thrown an assertion error even if assertions have been disabled.

Option B is incorrect because it is considered appropriate to check argument values in private methods using assertions.

Option C is incorrect; finally is never bypassed.

Option D is incorrect because AssertionErrors should never be handled.

217. Which of the following statements is true?

A. In an assert statement, the expression after the colon (:) can be any Java expression.

B. If a switch block has no default, adding an assert default is considered appropriate.

C. In an assert statement, if the expression after the colon (:) does not have a value, the assert's error message will be empty.

D. It is appropriate to handle assertion failures using a catch clause.

Ans. B. If a switch block has no default, adding an assert default is considered appropriate.

Explanation

Adding an assertion statement to a switch statement that previously had no default case is considered an excellent use of the assert mechanism.

Option A is incorrect because only Java expressions that return a value can be used. For instance, a method that returns void is illegal.

Option C is incorrect because the expression after the colon must have a value.

Option D is incorrect because assertions throw errors and not exceptions, and assertion errors do cause program termination and should not be handled.

218. Which three statements are true?

1. Assertion checking is typically enabled when a program is deployed.
2. It is never appropriate to write code to handle failure of an assert statement.
3. Assertion checking is typically enabled during program development and testing.
4. Assertion checking can be selectively enabled or disabled on a per-package basis, but not on a per-class basis.
5. Assertion checking can be selectively enabled or disabled on both a per-package basis and a per-class basis.

A. 1, 2 and 4 **B.** 2, 3 and 5

C. 3, 4 and 5 **D.** 1, 2 and 5

Ans. B. 2, 3 and 5

Explanation

(1) is wrong. It's just not true.

(2) is correct. You're never supposed to handle an assertion failure.

(3) is correct. Assertions let you test your assumptions during development, but the assertion code—in effect—evaporates when the program is deployed, leaving behind no overhead or debugging code to track down and remove.

(4) is wrong. See the explanation for (5) below.

(5) is correct. Assertion checking can be selectively enabled or disabled on a per-package basis. Note that the package default assertion status determines the assertion status for classes initialized in the future that belong to the named package or any of its "subpackages".

The assertion status can be set for a named top-level class and any nested classes contained therein. This setting takes precedence over the class loader's default assertion status, and over any applicable per-package default. If the named class is not a top-level class, the change of status will have no effect on the actual assertion status of any class.

219. public class Test2

```
{
    public static int x;
    public static int foo(int y)
    {
        return y * 2;
    }
    public static void main(String [] args)
    {
        int z = 5;
        assert z > 0; /* Line 11 */
        assert z > 2: foo(z); /* Line 12 */
        if ( z < 7 )
            assert z > 4; /* Line 14 */

        switch (z)
        {
            case 4: System.out.println("4 ");
            case 5: System.out.println("5 ");
            default: assert z < 10;
        }

        if ( z < 10 )
            assert z > 4: z++; /* Line 22 */
        System.out.println(z);
    }
}
```

Which line is an example of an inappropriate use of assertions?

A. Line 11

B. Line 12

C. Line 14

D. Line 22

Ans. D. Line 22

Explanation

Assert statements should not cause side effects. Line 22 changes the value of z if the assert statement is false.

Option A is fine; a second expression in an assert statement is not required.

Option B is fine because it is perfectly acceptable to call a method with the second expression of an assert statement.

Option C is fine because it is proper to call an assert statement conditionally.

220. public class Test

```
{
    public void foo()
    {
        assert false; /* Line 5 */
        assert false; /* Line 6 */
    }
    public void bar()
    {
        while(true)
        {
            assert false; /* Line 12 */
        }
        assert false; /* Line 14 */
    }
}
```

What causes compilation to fail?

A. Line 5

B. Line 6

C. Line 12

D. Line 14

Ans. D. Line 14

Explanation

Option D is correct. Compilation fails because of an unreachable statement at line 14. It is a compile-time error if a statement cannot be executed because it is unreachable. The question is now, why is line 20 unreachable. If it is because of the assert, then surely line 6 would also be unreachable. The answer must be something other than assert.

Examine the following:

A while statement can complete normally if and only if at least one of the following is true:

- The while statement is reachable and the condition expression is not a constant expression with value true.

-There is a reachable break statement that exits the while statement.

The while statement at line 11 is infinite and there is no break statement, therefore line 14 is unreachable. You can test this with the following code:

```
public class Test80
{
    public void foo()
    {
        assert false;
        assert false;
    }
    public void bar()
    {
        while(true)
        {
            assert false;
            break;
        }
        assert false;
    }
}
```

lang.class

221. What is the value of "d" after this line of code has been executed?

double d = Math.round (2.5 + Math.random());

A. 2

B. 3

C. 4

D. 2.5

Ans. B. 3

Explanation

The Math.random() method returns a number greater than or equal to 0 and less than 1 . Since we can then be sure that the sum of that number and 2.5 will be greater than or equal to 2.5 and less than 3.5, we can be sure that Math.round() will round that number to 3. So option B is the answer.

222. Which of the following would compile without error?

A. int a = Math.abs(-5);

B. int b = Math.abs(5.0);

C. int c = Math.abs(5.5F);

D. int d = Math.abs(5L);

Ans. A. int a = Math.abs(-5);

Explanation

The return value of the Math.abs() method is always the same as the type of the parameter passed into that method.

In the case of A, an integer is passed in and so the result is also an integer which is fine for assignment to "int a"..

The values used in B, C & D respectively are a double, a float and a long. The compiler will complain about a possible loss of precision if we try to assign the results to an "int".

223. Which of the following are valid calls to Math.max?

1. Math.max(1, 4)
2. Math.max(2.3, 5)
3. Math.max(1, 3, 5, 7)
4. Math.max(-1.5, -2.8f)

A. 1, 2 and 4 **B.** 2, 3 and 4

C. 1, 2 and 3 **D.** 3 and 4

Ans. A. 1, 2 and 4

Explanation

(1), (2), and (4) are correct. The max() method is overloaded to take two arguments of type int, long, float, or double.

(3) is incorrect because the max() method only takes two arguments.

224. public class Myfile

```
{
    public static void main (String[] args)
    {
        String biz = args[1];
        String baz = args[2];
        String rip = args[3];
        System.out.println("Arg is " + rip);
    }
}
```

Select how you would start the program to cause it to print: Arg is 2

A. java Myfile 222
B. java Myfile 1 2 2 3 4
C. java Myfile 1 3 2 2
D. java Myfile 0 1 2 3

Ans. C. java Myfile 1 3 2 2

Explanation

Arguments start at array element 0 so the fourth argument must be 2 to produce the correct output.

225. What will be the output of the program?

```
String x = new String("xyz");
String y = "abc";
x = x + y;
```

How many String objects have been created?

A. 2	**B.** 3
C. 4	**D.** 5

Ans. C. 4

Explanation

Line 1 creates two, one referred to by x and the lost String "xyz". Line 2 creates one (for a total of three). Line 3 creates one more (for a total of four), the concatenated String referred to by x with a value of "xyzabc".

226. What will be the output of the program?

```
public class WrapTest
{
  public static void main(String [] args)
  {
    int result = 0;
    short s = 42;
    Long x = new Long("42");
    Long y = new Long(42);
    Short z = new Short("42");
    Short x2 = new Short(s);
    Integer y2 = new Integer("42");
    Integer z2 = new Integer(42);

    if (x == y) /* Line 13 */
      result = 1;
    if (x.equals(y) ) /* Line 15 */
      result = result + 10;
    if (x.equals(z) ) /* Line 17 */
      result = result + 100;
    if (x.equals(x2) ) /* Line 19 */
      result = result + 1000;
    if (x.equals(z2) ) /* Line 21 */
      result = result + 10000;

    System.out.println("result = " + result);
  }
}
```

A. result = 1	**B.** result = 10
C. result = 11	**D.** result = 11010

Ans. B. result = 10

Explanation

Line 13 fails because == compares reference values, not object values. Line 15 succeeds because both String and primitive wrapper constructors resolve to the same value (except for the Character wrapper). Lines 17, 19, and 21 fail because the equals() method fails if the object classes being compared are different and not in the same tree hierarchy.

227. What will be the output of the program?

```
public class BoolTest
{
  public static void main(String [] args)
  {
    int result = 0;

    Boolean b1 = new Boolean("TRUE");
    Boolean b2 = new Boolean("true");
    Boolean b3 = new Boolean("tRuE");
    Boolean b4 = new Boolean("false");

    if (b1 == b2) /* Line 10 */
      result = 1;
    if (b1.equals(b2) ) /* Line 12 */
      result = result + 10;
    if (b2 == b4) /* Line 14 */
      result = result + 100;
    if (b2.equals(b4) ) /* Line 16 */
      result = result + 1000;
    if (b2.equals(b3) ) /* Line 18 */
      result = result + 10000;

    System.out.println("result = " + result);
  }
}
```

A. 0	**B.** 1
C. 10	**D.** 10010

Ans. D. 10010

Explanation

Line 10 fails because b1 and b2 are two different objects. Lines 12 and 18 succeed because the Boolean String constructors are case insensitive. Lines 14 and 16 fail because true is not equal to false.

228. What will be the output of the program?

```
public class ObjComp
{
    public static void main(String [] args )
    {
        int result = 0;
        ObjComp oc = new ObjComp();
        Object o = oc;

        if (o == oc)
            result = 1;
        if (o != oc)
            result = result + 10;
        if (o.equals(oc) )
            result = result + 100;
        if (oc.equals(o) )
            result = result + 1000;

        System.out.println("result = " + result);
    }
}
```

A. 1 **B.** 10
C. 101 **D.** 1101

Ans. D. 1101

Explanation

Even though o and oc are reference variables of different types, they are both referring to the same object. This means that == will resolve to true and that the default equals() method will also resolve to true.

229. What will be the output of the program?

```
public class Example
{
    public static void main(String [] args)
    {
        double values[] = {-2.3, -1.0, 0.25, 4};
        int cnt = 0;
        for (int x=0; x < values.length; x++)
        {
            if (Math.round(values[x] + .5) ==
Math.ceil(values[x]))
            {
                ++cnt;
            }
        }
        System.out.println("same results " +
cnt + " time(s)");
    }
}
```

A. same results 0 time(s)
B. same results 2 time(s)
C. same results 4 time(s)
D. Compilation fails.

Ans. B. same results 2 time(s)

Explanation

Math.round() adds .5 to the argument, then performs a floor(). Since the code adds an additional .5 before round() is called, it's as if we are adding 1 then doing a floor(). The values that start out as integer values will in effect be incremented by 1 on the round() side but not on the ceil() side, and the noninteger values will end up equal.

230. What will be the output of the program?

```
public class Test178
{
    public static void main(String[] args)
    {
        String s = "foo";
        Object o = (Object)s;
        if (s.equals(o))
        {
            System.out.print("AAA");
        }
        else
        {
            System.out.print("BBB");
        }
        if (o.equals(s))
        {
            System.out.print("CCC");
        }
        else
        {
            System.out.print("DDD");
        }
    }
}
```

A. AAACCC **B.** AAADDD
C. BBBCCC **D.** BBBDDD

Ans. A. AAACCC

231. What will be the output of the program?

```
String x = "xyz";
x.toUpperCase(); /* Line 2 */
String y = x.replace('Y', 'y');
y = y + "abc";
System.out.println(y);
```

A. abcXyZ **B.** abcxyz
C. xyzabc **D.** XyZabc

Ans. C. xyzabc

Explanation

Line 2 creates a new String object with the value "XYZ", but this new object is immediately lost because there is no reference to it. Line 3 creates a new String object referenced by y. This new String object has the value "xyz" because there was no "Y" in the String object referred to by x. Line 4 creates a new String object, appends "abc" to the value "xyz", and refers y to the result.

232. What will be the output of the program?

```
int i = (int) Math.random();
```

A. i = 0
B. i = 1
C. value of i is undetermined
D. Statement causes a compile error

Ans. A. i = 0

Explanation

Math.random() returns a double value greater than or equal to 0 and less than 1. Its value is stored to an int but as this is a narrowing conversion, a cast is needed to tell the compiler that you are aware that there may be a loss of precision.

The value after the decimal point is lost when you cast a double to int and you are left with 0.

233. What will be the output of the program?

```
class A
{
    public A(int x){}
}
class B extends A { }
public class test
{
    public static void main (String args [])
    {
        A a = new B();
        System.out.println("complete");
    }
}
```

A. It compiles and runs printing nothing
B. Compiles but fails at runtime
C. Compile Error
D. Prints "complete"

Ans. C. Compile Error

Explanation

No constructor has been defined for class B, therefore it will make a call to the default constructor but since class B extends class A, it will also call the Super() default constructor. Since a constructor has been defined in class A, Java will no longer supply a default constructor for class A, therefore when class B calls class A's default constructor, it will result in a compile error.

234. What will be the output of the program?

```
int i = 1, j = 10;
do
{
    if(i++ > --j) /* Line 4 */
    {
        continue;
    }
} while (i < 5);
System.out.println("i = " + i + "and j = " +
j); /* Line 9 */
```

A. i = 6 and j = 5 **B.** i = 5 and j = 5
C. i = 6 and j = 6 **D.** i = 5 and j = 6

Ans. D. i = 5 and j = 6

Explanation

This question is not testing your knowledge of the continue statement. It is testing your knowledge of the order of evaluation of operands. Basically the prefix and postfix unary operators have a higher order of evaluation than the relational operators. So on line 4 the variable i is incremented and the variable j is decremented before the greater than comparison is made. As the loop executes the comparison on line 4 will be:

if(i > j)

if(2 > 9)

if(3 > 8)

if(4 > 7)

if(5 > 6) at this point i is not less than 5, therefore the loop terminates and line 9 outputs the values of i and j as 5 and 6 respectively.

The continue statement never gets to execute because i never reaches a value that is greater than j.

235. What will be the output of the program?

```
public class ExamQuestion7
{
```

```
static int j;
static void methodA(int i)
{
    boolean b;
    do
    {
        b = i<10 | methodB(4); /* Line 9 */
        b = i<10 || methodB(8); /* Line 10
*/
    }while (!b);
}
static boolean methodB(int i)
{
    j += i;
    return true;
}
public static void main(String[] args)
{
    methodA(0);
    System.out.println( "j = " + j );
}
}
```

A. j = 0
B. j = 4
C. j = 8
D. The code will run with no output

Ans. B. j = 4

Explanation

The lines to watch here are lines 9 and 10. Line 9 features the non-shortcut version of the OR operator so both of its operands will be evaluated and therefore methodB(4) is executed.

However, line 10 has the shortcut version of the OR operator and if the 1st of its operands evaluates to true (which in this case is true), then the 2nd operand isn't evaluated, so methodB(8) never gets called.

The loop is only executed once, b is initialized to false and is assigned true on line 9. Thus j = 4.

236. What will be the output of the program?

```
try
{
    Float f1 = new Float("3.0");
    int x = f1.intValue();
    byte b = f1.byteValue();
    double d = f1.doubleValue();
    System.out.println(x + b + d);
}
catch (NumberFormatException e) /* Line
9 */
{
    System.out.println("bad number"); /*
Line 11 */
}
```

A. 9.0
B. bad number
C. Compilation fails on line 9.
D. Compilation fails on line 11.

Ans. A. 9.0

Explanation

The xxxValue() methods convert any numeric wrapper object's value to any primitive type. When narrowing is necessary, significant bits are dropped and the results are difficult to calculate.

237. What will be the output of the program?

```
class Q207
{
    public static void main(String[] args)
    {
        int i1 = 5;
        int i2 = 6;
        String s1 = "7";
        System.out.println(i1 + i2 + s1); /* Line
8 */
    }
}
```

A. 18 **B.** 117
C. 567 **D.** Compiler error

Ans. B. 117

Explanation

This question is about the + (plus) operator and the overriden + (string cocatenation) operator. The rules that apply when you have a mixed expression of numbers and strings are:

If either operand is a String, the + operator concatenates the operands.

If both operands are numeric, the + operator adds the operands.

The expression on line 6 above can be read as "Add the values i1 and i2 together, then take the sum and convert it to a string and concatenate it with the String from the variable s1". In code, the compiler probably interprets the expression on line 8 above as:

```
System.out.println( new StringBuffer()
    .append(new Integer(i1 + i2).toString())
    .append(s1)
    .toString() );
```

238. What will be the output of the program?

```
public class SqrtExample
```

```
{
    public static void main(String [] args)
    {
        double value = –9.0;
        System.out.println( Math.sqrt(value));
    }
}
```

A. 3.0 **B.** –3.0
C. NaN **D.** Compilation fails.

Ans. C. NaN

Explanation

The sqrt() method returns NaN (not a number) when its argument is less than zero.

239. What will be the output of the program?

```
String s = "ABC";
s.toLowerCase();
s += "def";
System.out.println(s);
```

A. ABC **B.** abc
C. ABCdef **D.** Compile Error

Ans. C. ABCdef

Explanation

String objects are immutable. The object s above is set to "ABC". Now ask yourself if this object is changed and if so, where remember strings are immutable.

Line 2 returns a string object but does not change the original string object s, so after line 2 s is still "ABC".

So what's happening on line 3? Java will treat line 3 like the following:

s = new StringBuffer().append(s).append("def").toString();

This effectively creates a new String object and stores its reference in the variable s, the old String object containing "ABC" is no longer referenced by a live thread and becomes available for garbage collection.

240. What will be the output of the program?

```
public class NFE
{
    public static void main(String [] args)
    {
    String s = "42";
        try
        {
            s = s.concat(".5"); /* Line 8 */
            double d = Double.parseDouble(s);
            s = Double.toString(d);
            int x = (int) Math.ceil(Double.valueOf(s).doubleValue());
            System.out.println(x);
        }
        catch (NumberFormatException e)
        {
            System.out.println("bad number");
        }
    }
}
```

A. 42 **B.** 42.5
C. 43 **D.** bad number

Ans. C. 43

Explanation

All of this code is legal, and line 8 creates a new String with a value of "42.5". Lines 9 and 10 convert the String to a double and then back again. Line 11 is fun—Math.ceil()'s argument expression is evaluated first. We invoke the valueOf() method that returns an anonymous Double object (with a value of 42.5). Then the doubleValue() method is called (invoked on the newly created Double object), and returns a double primitive (there and back again), with a value of (you guessed it) 42.5. The ceil() method converts this to 43.0, which is cast to an int and assigned to x.

241. What will be the output of the program?

```
System.out.println(Math.sqrt(–4D));
```

A. –2 **B.** NaN
C. Compile Error **D.** Runtime Exception

Ans. B. NaN

Explanation

It is not possible in regular mathematics to get a value for the square-root of a negative number, therefore a NaN will be returned because the code is valid.

242. What will be the output of the program?

```
interface Foo141
{
    int k = 0; /* Line 3 */
}
public class Test141 implements Foo141
{
    public static void main(String args[])
    {
        int i;
        Test141 test141 = new Test141();
        i = test141.k; /* Line 11 */
        i = Test141.k;
```

```
        i = Foo141.k;
    }
}
```

A. Compilation fails.

B. Compiles and runs ok.

C. Compiles but throws an Exception at runtime.

D. Compiles but throws a RuntimeException at runtime.

Ans. B. Compiles and runs ok.

Explanation

The variable k on line 3 is an interface constant, it is implicitly public, static, and final. Static variables can be referenced in two ways:

Via a reference to any instance of the class (line 11)

Via the class name (line 12).

243. What will be the output of the program?

```
String a = "newspaper";
a = a.substring(5,7);
char b = a.charAt(1);
a = a + b;
System.out.println(a);
```

A. apa **B.** app

C. apea **D.** apep

Ans. B. app

Explanation

Both substring() and charAt() methods are indexed with a zero-base, and substring() returns a String of length arg2 - arg1.

244. What will be the output of the program?

```
public class StringRef
{
    public static void main(String [] args)
    {
        String s1 = "abc";
        String s2 = "def";
        String s3 = s2;   /* Line 7 */
        s2 = "ghi";
        System.out.println(s1 + s2 + s3);
    }
}
```

A. abcdefghi **B.** abcdefdef

C. abcghidef **D.** abcghighi

Ans. C. abcghidef

Explanation

After line 7 executes, both s2 and s3 refer to a String object that contains the value "def". When line 8 executes, a new String object is created with the value "ghi", to which s2 refers. The reference variable s3 still refers to the (immutable) String object with the value "def".

245. What will be the output of the program?

```
public class Test138
{
    public static void stringReplace (String text)
    {
        text = text.replace ('j' , 'c'); /* Line 5 */
    }
    public static void bufferReplace (StringBuffer text)
    {
        text = text.append ("c");  /* Line 9 */
    }
    public static void main (String args[])
    {
        String textString = new String ("java");
        StringBuffer textBuffer = new StringBuffer ("java"); /* Line 14 */
        stringReplace(textString);
        bufferReplace(textBuffer);
        System.out.println (textString + textBuffer);
    }
}
```

A. java **B.** javac

C. javajavac **D.** Compile error

Ans. C. javajavac

Explanation

A string is immutable, it cannot be changed, that's the reason for the StringBuffer class. The stringReplace method does not change the string declared on line 14, so this remains set to "java". Method parameters are always passed by value—a copy is passed into the method—if the copy changes, the original remains intact, line 5 changes the reference, i.e. text points to a new String object, however, this is lost when the method completes. The textBuffer is a StringBuffer so it can be changed.

This change is carried out on line 9, so "java" becomes "javac", the text reference on line 9 remains unchanged. This gives us the output of "javajavac"

246. What will be the output of the program?

```
class Tree { }
class Pine extends Tree { }
```

```
class Oak extends Tree { }
public class Forest1
{
   public static void main (String [] args)
   {
      Tree tree = new Pine();
      if( tree instanceof Pine )
         System.out.println ("Pine");
      else if( tree instanceof Tree )
         System.out.println ("Tree");
      else if( tree instanceof Oak )
         System.out.println ( "Oak" );
      else
         System.out.println ("Oops ");
   }
}
```

A. Pine **B.** Tree
C. Forest **D.** Oops

Ans. A. Pine

Explanation

The program prints "Pine".

247. What will be the output of the program?

```
String d = "bookkeeper";
d.substring(1,7);
d = "w" + d;
d.append("woo"); /* Line 4 */
System.out.println(d);
```

A. wookkeewoo **B.** wbookkeeper
C. wbookkeewoo **D.** Compilation fails.

Ans. D. Compilation fails.

Explanation

In line 4 the code calls a StringBuffer method, append() on a String object.

248. What will be the output of the program?

```
String a = "ABCD";
String b = a.toLowerCase();
b.replace('a','d');
b.replace('b','c');
System.out.println(b);
```

A. abcd **B.** ABCD
C. dccd **D.** dcba

Ans. A. abcd

Explanation

String objects are immutable, they cannot be changed, in this case we are talking about the replace method which returns a new String object resulting from replacing all occurrences of oldChar in this string with newChar.
b.replace(char oldChar, char newChar);

But since this is only a temporary String it must either be put to use straight away, i.e.
System.out.println(b.replace('a','d'));
Or a new variable must be assigned its value, i.e.
String c = b.replace('a','d');

249. What will be the output of the program?

```
public class ExamQuestion6
{
   static int x;
   boolean catch()
   {
      x++;
      return true;
   }
   public static void main(String[] args)
   {
      x=0;
      if ((catch() | catch()) || catch())
         x++;
      System.out.println(x);
   }
}
```

A. 1 **B.** 2
C. 3 **D.** Compilation fails

Ans. D. Compilation fails

Explanation

Initially this looks like a question about the logical and logical shortcut operators "|" and "||" but on closer inspection it should be noticed that the name of the boolean method in this code is "catch". "catch" is a reserved keyword in the Java language and cannot be used as a method name. Hence compilation will fail.

250. What will be the output of the program?

```
public class Test
{
   public static void main(String[] args)
   {
         final StringBuffer a = new
StringBuffer();
         final StringBuffer b = new
StringBuffer();

      new Thread()
      {
         public void run()
         {
            System.out.print(a.append("A"));
            synchronized(b)
```

```
            {
            System.out.print(b.append("B"));
            }
        }
    }.start();

    new Thread()
    {
        public void run()
        {
            System.out.print(b.append("C"));
            synchronized(a)
            {
            System.out.print(a.append("D"));
            }
        }
    }.start();
  }
}
```

A. ACCBAD
B. ABBCAD
C. CDDACB
D. Indeterminate output

Ans. D. Indeterminate output

Explanation

It gives different output while executing the same compiled code at different times.

```
C:\>javac Test.java
C:\>java Test
ABBCAD
C:\>java Test
ACADCB
C:\>java Test
ACBCBAD
C:\>java Test
ABBCAD
C:\>java Test
ACBCBAD
C:\>java Test
ACBCBAD
C:\>java Test
ABBCAD
```

251. What will be the output of the program?

```
String s = "hello";
Object o = s;
if( o.equals(s) )
{
    System.out.println("A");
}
else
{
    System.out.println("B");
}
if( s.equals(o) )
{
    System.out.println("C");
}
else
{
    System.out.println("D");
}
```

1. A
2. B
3. C
4. D

A. 1 and 3 **B.** 2 and 4
C. 3 and 4 **D.** 1 and 2

Ans. A. 1 and 3

Explanation

No answer description available for this question. Let us discuss.

251. What will be the output of the program (in jdk1.6 or above)?

```
public class BoolTest
{
    public static void main(String [] args)
    {
        Boolean b1 = new Boolean("false");
        boolean b2;
        b2 = b1.booleanValue();
        if (!b2)
        {
            b2 = true;
            System.out.print("x ");
        }
        if (b1 & b2) /* Line 13 */
        {
            System.out.print("y ");
        }
        System.out.println("z");
    }
}
```

A. z **B.** x z
C. y z **D.** Compilation fails.

Ans. B. x z

252. Which statement is true given the following?

Double d = Math.random();

A. 0.0 < d <= 1.0 B. 0.0 <= d < 1.0
C. Compilation fails D. Cannot say

Ans. B. 0.0 <= d < 1.0

Explanation

The Math.random() method returns a double value with a positive sign, greater than or equal to 0.0 and less than 1.0

253. Which two statements are true about wrapper or String classes?

If x and y refer to instances of different wrapper classes, then the fragment x.equals(y) will cause a compiler failure.

If x and y refer to instances of different wrapper classes, then x == y can sometimes be true.

If x and y are String references and if x.equals(y) is true, then x == y is true.

If x, y, and z refer to instances of wrapper classes and x.equals(y) is true, and y.equals(z) is true, then z.equals(x) will always be true.

If x and y are String references and x == y is true, then y.equals(x) will be true.

A. 1 and 2 **B.** 2 and 3
C. 3 and 4 **D.** 4 and 5

Ans. D. 4 and 5

Explanation

Statement (4) describes an example of the equals() method behaving transitively. By the way, x, y, and z will all be the same type of wrapper. Statement (5) is true because x and y are referring to the same String object.

Statement (1) is incorrect—the fragment will compile. Statement (2) is incorrect because x == y means that the two reference variables are referring to the same object. Statement (3) will only be true if x and y refer to the same String. It is possible for x and y to refer to two different String objects with the same value.

254. Which of the following will produce an answer that is closest in value to a double, d, while not being greater than d?

A. (int)Math.min(d);
B. (int)Math.max(d);
C. (int)Math.abs(d);
D. (int)Math.floor(d);

Ans. D. (int)Math.floor(d);

Explanation

The casting to an int is a smokescreen. Use a process of elimination to answer this question:

Option D is the correct answer, it is syntactically correct and will consistently return a value less than d.

Options A and B are wrong because both the min() and max() methods require 2 arguments, whereas here they are passed only one parameter.

Option C is wrong because it could return a value greater than d (if d was negative).

255. What two statements are true about the result obtained from calling Math.random()?

1. The result is less than 0.0.
2. The result is greater than or equal to 0.0.
3. The result is less than 1.0.
4. The result is greater than 1.0.
5. The result is greater than or equal to 1.0.

A. 1 and 2 **B.** 2 and 3
C. 3 and 4 **D.** 4 and 5

Ans. B. 2 and 3

Explanation

(1) and (2) are correct. The result range for random() is 0.0 to < 1.0; 1.0 is not in range.

CORE JAVA

1. What is the most important feature of Java?

Ans. Java is a platform independent language.

2. What do you mean by platform independence?

Ans. Platform independence means that we can write and compile the java code in one platform (e.g. Windows) and can execute the class in any other supported platform, e.g. Linux, Solaris, etc.

3. What is a JVM?

Ans. JVM is Java Virtual Machine which is a run time environment for the compiled java class files.

4. Are JVM's platform independent?

Ans. JVM's are not platform independent. JVM's are platform specific runtime implementation provided by the vendor.

5. What is the difference between a JDK and a JVM?

Ans. JDK is Java Development Kit which is for development purpose and it includes execution environment also. But JVM is purely a runtime environment and hence you will not be able to compile your source files using a JVM.

6. What is a pointer and does Java support pointers?

Ans. Pointer is a reference handle to a memory location. Improper handling of pointers leads to memory leaks and reliability issues, hence Java doesn't support the usage of pointers.

7. What is the base class of all classes?

Ans. java.lang.Object

8. Does Java support multiple inheritance?

Ans. Java doesn't support multiple inheritance.

9. Is Java a pure object oriented language?

Ans. Java uses primitive data types and hence is not a pure object oriented language.

10. Are arrays primitive data types?

Ans. In Java, arrays are objects.

11. What is the difference between Path and Classpath?

Ans. Path and Classpath are operating system level environment variales. Path is used to define where the system can find the executables(.exe) files and classpath is used to specify the location .class files.

12. What are local variables?

Ans. Local varaiables are those which are declared within a block of code like methods. Local variables should be initialised before accessing them.

13. What are instance variables?

Ans. Instance variables are those which are defined at the class level. Instance variables need not be initialized before using them as they are automatically initialized to their default values.

14. How to define a constant variable in Java?

Ans. The variable should be declared as static and final. So only one copy of the variable exists for all instances of the class and the value can't be changed also.
static final int PI = 2.14; is an example for constant.

15. Should a main() method be compulsorily declared in all java classes?

Ans. No not required. main() method should be defined only if the source class is a java application.

16. What is the return type of the main() method?

Ans. Main() method doesn't return anything, hence declared void.

17. Why is the main() method declared static?

Ans. main() method is called by the JVM even before the instantiation of the class, hence it is declared as static.

18. What is the argument of main() method?

Ans. main() method accepts an array of String object as argument.

19. Can a main() method be overloaded?

Ans. Yes. You can have any number of main() methods with different method signature and implementation in the class.

20. Can a main() method be declared final?

Ans. Yes. Any inheriting class will not be able to have its own default main() method.

21. Does the order of public and static declaration matter in main() method?

Ans. No. It doesn't matter but void should always come before main().

22. Can a source file contain more than one class declaration?

Ans. Yes, a single source file can contain any number of Class declarations but only one of the class can be declared as public.

23. What is a package?

Ans. Package is a collection of related classes and interfaces. package declaration should be first statement in a java class.

24. Which package is imported by default?

Ans. java.lang package is imported by default even without a package declaration.

25. Can a class declared as private be accessed outside its package?

Ans. Not possible.

26. Can a class be declared as protected?

Ans. A class can't be declared as protected. Only methods can be declared as protected.

27. What is the access scope of a protected method?

Ans. A protected method can be accessed by the classes within the same package or by the subclasses of the class in any package.

28. What is the purpose of declaring a variable as final?

Ans. A final variable's value can't be changed. Final variables should be initialized before using them.

29. What is the impact of declaring a method as final?

Ans. A method declared as final can't be overridden. A subclass can't have the same method signature with a different implementation.

30. I don't want my class to be inherited by any other class. What should I do?

Ans. You should declared your class as final. But you can't define your class as final, if it is an abstract class. A class declared as final can't be extended by any other class.

31. Can you give a few examples of final classes defined in Java API?

Ans. java.lang.String, java.lang.Math are final classes.

32. How is final different from finally and finalize()?

Ans. Final is a modifier which can be applied to a class or a method or a variable. final class can't be inherited, final method can't be overridden and final variable can't be changed.

Finally is an exception handling code section which gets executed whether an exception is raised or not by the try block code segment.

Finalize() is a method of Object class which will be executed by the JVM just before garbage collecting object to give a final chance for resource releasing activity.

33. Can a class be declared as static?

Ans. We cannot declare top level class as static, but only inner class can be declared static.

```
public class Test
{
    static class InnerClass
    {
        public static void InnerMethod()
            { System.out.println("Static Inner
    Class!"); }
    }
    public static void main(String args[])
    {
        Test.InnerClass.InnerMethod();
    }
}
//output: Static Inner Class!
```

34. When will you define a method as static?

Ans. When a method needs to be accessed even before the creation of the object of the class, then we should declare the method as static.

35. What are the restriction imposed on a static method or a static block of code?

Ans. A static method should not refer to instance variables without creating an instance and cannot use "this" operator to refer the instance.

36. I want to print "Hello" even before main() is executed. How will you acheive that?

Ans. Print the statement inside a static block of code. Static blocks get executed when the class gets loaded into the memory and even before the creation of an object. Hence it will be executed before the main() method. And it will be executed only once.

37. What is the importance of static variable?

Ans. static variables are class level variables where all objects of the class refer to the same variable. If one object changes the value, then the change gets reflected in all the objects.

38. Can we declare a static variable inside a method?

Ans. Static variables are class level variables and they can't be declared inside a method. If declared, the class will not compile.

39. What is an abstract class and what is its purpose?

Ans. A class which doesn't provide complete implementation is defined as an abstract class. Abstract classes enforce abstraction.

40. Can an abstract class be declared final?

Ans. Not possible. An abstract class without being inherited is of no use and hence will result in compile time error.

41. What is the use of an abstract variable?

Ans. Variables can't be declared as abstract. Only classes and methods can be declared as abstract.

42. Can you create an object of an abstract class?

Ans. Not possible. Abstract classes can't be instantiated.

43. Can a abstract class be defined without any abstract methods?

Ans. Yes it's possible. This is basically to avoid instance creation of the class.

44. Class C implements Interface I containing method m1 and m2 declarations. Class C has provided implementation for method m2. Can I create an object of Class C?

Ans. No, not possible. Class C should provide implementation for all the methods in the Interface I. Since class C didn't provide implementation for m1 method, it has to be declared as abstract. Abstract classes can't be instantiated.

45. Can a method inside a interface be declared as final?

Ans. No, not possible. Doing so will result in compilation error. Public and abstract are the only applicable modifiers for method declaration in an interface.

46. Can an interface implement another interface?

Ans. Intefaces doesn't provide implementation, hence a interface cannot implement another interface.

47. Can an interface extend another interface?

Ans. Yes an Interface can inherit another interface, for that matter an interface can extend more than one interface.

48. Can a class extend more than one class?

Ans. Not possible. A class can extend only one class but can implement any number of interfaces.

49. Why is an interface be able to extend more than one interface but a class can't extend more than one class?

Ans. Basically Java doesn't allow multiple inheritance, so a class is restricted to extend only one class. But an interface is a pure abstraction model and doesn't have inheritance hierarchy like classes(do remember that the base class of all classes is Object). So an interface is allowed to extend more than one interface.

50. Can an interface be final?

Ans. Not possible. Doing so will result in compilation error.

51. Can a class be defined inside an interface?

Ans. Yes it's possible.

52. Can an interface be defined inside a class?

Ans. Yes it's possible.

53. What is a marker interface?

Ans. An interface which doesn't have any declaration inside but still enforces a mechanism.

54. Which object oriented concept is achieved by using overloading and overriding?

Ans. Polymorphism.

55. Why does Java not support operator overloading?

Ans. Operator overloading makes the code very difficult to read and maintain. To maintain code simplicity, Java doesn't support operator overloading.

56. Can we define private and protected modifiers for variables in interfaces?

Ans. No.

57. What is externalizable?

Ans. Externalizable is an interface that extends serializable interface. And sends data into streams in compressed format. It has two methods, writeExternal(ObjectOuput out) and readExternal(ObjectInput in)

58. What modifiers are allowed for methods in an interface?

Ans. Only public and abstract modifiers are allowed for methods in interfaces.

59. What is a local, member and a class variable?

Ans. Variables declared within a method are "local" variables.

Variables declared within the class, i.e. not within any methods are "member" variables (global variables).

Variables declared within the class, i.e. not within any methods and are defined as "static" are class variables.

60. What is an abstract method?

Ans. An abstract method is a method whose implementation is deferred to a subclass.

61. What value does read() return when it has reached the end of a file?

Ans. The read() method returns -1 when it has reached the end of a file.

62. Can a byte object be cast to a double value?

Ans. No, an object cannot be cast to a primitive value.

63. What is the difference between a static and a non-static inner class?

Ans. A non-static inner class may have object instances that are associated with instances of the class's outer class. A static inner class does not have any object instances.

64. What is an object's lock and which object's have locks?

Ans. An object's lock is a mechanism that is used by multiple threads to obtain synchronized access to the object. A thread may execute a synchronized method of an object only after it has acquired the object's lock. All objects and classes have locks. A class's lock is acquired on the class's class object.

65. What is the % operator?

Ans. It is referred to as the modulo or remainder operator. It returns the remainder of dividing the first operand by the second operand.

66. When can an object reference be cast to an interface reference?

Ans. An object reference be cast to an interface reference when the object implements the referenced interface.

67. Which class is extended by all other classes?

Ans. The Object class is extended by all other classes.

68. Which non-Unicode letter characters may be used as the first character of an identifier?

Ans. The non-Unicode letter characters $ and _ may appear as the first character of an identifier.

69. What restrictions are placed on method overloading?

Ans. Two methods may not have the same name and argument list but different return types.

70. What is casting?

Ans. There are two types of casting, casting between primitive numeric types and casting between object references. Casting between numeric types is used to convert larger values, such as double values, to smaller values, such as byte values. Casting between object references is used to refer to an object by a compatible class, interface, or array type reference.

71. What is the return type of a program's main() method?

Ans. void.

72. If a variable is declared as private, where may the variable be accessed?

Ans. A private variable may only be accessed within the class in which it is declared.

73. What do you understand by private, protected and public?

Ans. These are accessibility modifiers. Private is the most restrictive, while public is the least restrictive. There is no real difference between protected and the default type (also known as package protected) within the context of the same package, however, the protected keyword allows visibility to a derived class in a different package.

74. What is downcasting ?

Ans. Downcasting is the casting from a general to a more specific type, i.e. casting down the hierarchy.

75. What modifiers may be used with an inner class that is a member of an outer class?

Ans. A (non-local) inner class may be declared as public, protected, private, static, final, or abstract.

76. How many bits are used to represent Unicode, ASCII, UTF-16, and UTF-8 characters?

Ans. Unicode requires 16 bits and ASCII require 7 bits. Although the ASCII character set uses only 7 bits, it is usually represented as 8 bits.

UTF-8 represents characters using 8, 16, and 18 bit patterns.

UTF-16 uses 16-bit and larger bit patterns.

77. What restrictions are placed on the location of a package statement within a source code file?

Ans. A package statement must appear as the first line in a source code file (excluding blank lines and comments).

78. What is a native method?

Ans. A native method is a method that is implemented in a language other than Java.

79. What are order of precedence and associativity, and how are they used?

Ans. Order of precedence determines the order in which operators are evaluated in expressions. Associativity determines whether an expression is evaluated left-to-right or right-to-left.

80. Can an anonymous class be declared as implementing an interface and extending a class?

Ans. An anonymous class may implement an interface or extend a superclass, but may not be declared to do both.

81. What is the range of the char type?

Ans. The range of the char type is 0 to 216 –1 (i.e. 0 to 65535.)

82. What is the range of the short type?

Ans. The range of the short type is –(215) to 215 – 1. (i.e. –32,768 to 32,767)

83. Why isn't there operator overloading?

Ans. Because C++ has proven by example that operator overloading makes code almost impossible to maintain.

84. What does it mean that a method or field is "static"?

Ans. Static variables and methods are instantiated only once per class. In other words, they are class variables, not instance variables. If you change the value of a static variable in a particular object, the value of that variable changes for all instances of that class. Static methods can be referenced with the name of the class rather than the name of a particular object of the class (though that works too). That's how library methods like System.out.println() work. out is a static field in the java.lang.System class.

85. Is null a keyword?

Ans. The null value is not a keyword.

86. Which characters may be used as the second character of an identifier, but not as the first character of an identifier?

Ans. The digits 0 through 9 may not be used as the first character of an identifier but they may be used after the first character of an identifier.

87. Is the ternary operator written x : y ? z or x ? y : z ?

Ans. It is written x ? y : z.

88. How is rounding performed under integer division?

Ans. The fractional part of the result is truncated. This is known as rounding toward zero.

89. If a class is declared without any access modifiers, where may the class be accessed?

Ans. A class that is declared without any access modifiers is said to have package access. This means that the class can only be accessed by other classes and interfaces that are defined within the same package.

90. Does a class inherit the constructors of its superclass?

Ans. A class does not inherit constructors from any of its superclasses.

91. Name the eight primitive Java types.

Ans. The eight primitive types are byte, char, short, int, long, float, double, and boolean.

92. What restrictions are placed on the values of each case of a switch statement?

Ans. During compilation, the values of each case of a switch statement must evaluate to a value that can be promoted to an int value.

93. What is the difference between a while statement and a do while statement?

Ans. A while statement checks at the beginning of a loop to see whether the next loop iteration should occur. A do while statement checks at the end of a loop to see whether the next iteration of a loop should occur. The do whilestatement will always execute the body of a loop at least once.

94. What modifiers can be used with a local inner class?

Ans. A local inner class may be final or abstract.

95. When does the compiler supply a default constructor for a class?

Ans. The compiler supplies a default constructor for a class if no other constructors are provided.

96. If a method is declared as protected, where may the method be accessed?

Ans. A protected method may only be accessed by classes or interfaces of the same package or by subclasses of the class in which it is declared.

97. What are the legal operands of the instanceof operator?

Ans. The left operand is an object reference or null value and the right operand is a class, interface, or array type.

98. Are true and false keywords?

Ans. The values true and false are not keywords.

99. What happens when you add a double value to a String?

Ans. The result is a String object.

100. What is the diffrence between inner class and nested class?

Ans. When a class is defined within a scope of another class, then it becomes inner class. If the access modifier of the inner class is static, then it becomes nested class.

101. Can an abstract class be final?

Ans. An abstract class may not be declared as final.

102. What is numeric promotion?

Ans. Numeric promotion is the conversion of a smaller numeric type to a larger numeric type, so that integer and floating-point operations may take place. In numerical promotion, byte, char, and short values are converted to int values. The int values are also converted to long values, if necessary. The long and float values are converted to double values, as required.

103. What is the difference between a public and a non-public class?

Ans. A public class may be accessed outside of its package. A non-public class may not be accessed outside of its package.

104. To what value is a variable of the boolean type automatically initialized?

Ans. The default value of the boolean type is false.

105. What is the difference between the prefix and postfix forms of the ++ operator?

Ans. The prefix form performs the increment operation and returns the value of the increment operation. The postfix form returns the current value all of the expression and then performs the increment operation on that value.

106. What restrictions are placed on method overriding?

Ans. Overridden methods must have the same name, argument list, and return type. The overriding method may not limit the access of the method it overrides. The overriding method may not throw any exceptions that may not be thrown by the overridden method.

107. What is a Java package and how is it used?

Ans. A Java package is a naming context for classes and interfaces. A package is used to create a separate name space for groups of classes and interfaces. Packages are also used to organize related classes and interfaces into a single API unit and to control accessibility to these classes and interfaces.

108. What modifiers may be used with a top-level class?

Ans. A top-level class may be public, abstract, or final.

109. What is the difference between an if statement and a switch statement?

Ans. The if statement is used to select among two alternatives. It uses a boolean expression to decide which alternative should be executed. The switch statement is used to select among multiple alternatives. It uses an int expression to determine which alternative should be executed.

110. What are the practical benefits, if any, of importing a specific class rather than an entire package (e.g. import java.net.* versus import java.net.Socket)?

Ans. It makes no difference in the generated class files since only the classes that are actually used are referenced by the generated class file. There is another practical benefit to importing single classes, and this arises when two (or more) packages have classes with the same name. Take java.util.Timer and javax.swing.Timer, for example. If I import java.util.* and javax.swing.* and then try to use "Timer", I get an error while compiling (the class name is ambiguous between both packages). Let's say what you really wanted was the javax.swing.Timer class, and the only classes you plan on using in java.util are Collection and HashMap. In this case, some people will prefer to import java.util.Collection and import java.util.HashMap instead of importing java.util.*. This will now allow them to use Timer, Collection, HashMap, and other javax.swing classes without using fully qualified class names in.

111. Can a method be overloaded based on different return type but same argument type ?

Ans. No, because the methods can be called without using their return type in which case there is ambiguity for the compiler.

112. What happens to a static variable that is defined within a method of a class ?

Ans. Can't do it. You'll get a compilation error.

113. How many static initializers can you have?

Ans. As many as you want, but the static initializers and class variable initializers are executed in textual order and may not refer to class variables declared in the class whose declarations appear textually after the use, even though these class variables are in scope.

114. What is the difference between method overriding and overloading?

Ans. Overriding is a method with the same name and arguments as in a parent, whereas overloading is the same method name but different arguments

115. What is constructor chaining and how is it achieved in Java ?

Ans. A child object constructor always first needs to construct its parent (which in turn calls its parent constructor). In Java it is done via an implicit call to the no-args constructor as the first statement.

116. What is the difference between the Boolean & operator and the && operator?

Ans. If an expression involving the Boolean & operator is evaluated, both operands are evaluated. Then the & operator is applied to the operand. When an expression involving the && operator is evaluated, the first operand is evaluated. If the first operand returns a value of true, then the second operand is evaluated. The && operator is then applied to the first and second operands. If the first operand evaluates to false, the evaluation of the second operand is skipped.

117. Which Java operator is right associative?

Ans. The = operator is right associative.

118. Can a double value be cast to a byte?

Ans. Yes, a double value can be cast to a byte.

119. What is the difference between a break statement and a continue statement?

Ans. A break statement results in the termination of the statement to which it applies (switch, for, do, or while). A continue statement is used to end the current loop iteration and return control to the loop statement.

120. Can a for statement loop indefinitely?

Ans. Yes, a for statement can loop indefinitely. For example, consider the following: for(;;);

121. To what value is a variable of the String type automatically initialized?

Ans. The default value of an String type is null.

122. What is the difference between a field variable and a local variable?

Ans. A field variable is a variable that is declared as a member of a class. A local variable is a variable that is declared local to a method.

123. How are this() and super() used with constructors?

Ans. this() is used to invoke a constructor of the same class. super() is used to invoke a superclass constructor.

124. What does it mean that a class or member is final?

Ans. A final class cannot be inherited. A final method cannot be overridden in a subclass. A final field cannot be changed after it's initialized, and it must include an initializer statement where it's declared.

125. What does it mean that a method or class is abstract?

Ans. An abstract class cannot be instantiated. Abstract methods may only be included in abstract classes. However, an abstract class is not required to have any abstract methods, though most of them do. Each subclass of an abstract class must override the abstract methods of its superclasses or it also should be declared abstract.

126. What is a transient variable?

Ans. Transient variable is a variable that may not be serialized.

127. How does Java handle integer overflows and underflows?

Ans. It uses those low order bytes of the result that can fit into the size of the type allowed by the operation.

128. What is the difference between the >> and >>> operators?

Ans. The >> operator carries the sign bit when shifting right. The >>> zero-fills bits that have been shifted out.

129. Is sizeof a keyword?

Ans. The sizeof operator is not a keyword.

Java Basics

1. What is the difference between a constructor and a method?

Ans. A constructor is a member function of a class that is used to create objects of that class. It has the same name as the class itself, has no return type, and is invoked using the new operator.

A method is an ordinary member function of a class. It has its own name, a return type (which may be void), and is invoked using the dot operator.

2. What is the purpose of garbage collection in Java, and when is it used?

Ans. The purpose of garbage collection is to identify and discard objects that are no longer needed by a program so that their resources can be reclaimed and reused.

A Java object is subject to garbage collection when it becomes unreachable to the program in which it is used.

3. Describe synchronization in respect to multithreading.

Ans. With respect to multithreading, synchronization is the capability to control the access of multiple threads to shared resources.

Without synchonization, it is possible for one thread to modify a shared variable while another thread is in the process of using or updating same shared variable. This usually leads to significant errors.

4. What is an abstract class?

Ans. Abstract class must be extended/ subclassed (to be useful). It serves as a template. A class that is abstract may not be instantiated (i.e. you may not call its constructor), abstract class may contain static data.

Any class with an abstract method is automatically abstract itself, and must be declared as such. A class may be declared abstract even if it has no abstract methods. This prevents it from being instantiated.

5. What is the difference between an interface and an abstract class?

Ans. An abstract class can have instance methods that implement a default behavior. An interface can only declare constants and instance methods, but cannot implement default behavior and all methods are implicitly abstract.

An interface has all public members and no implementation. An abstract class is a class which may have the usual flavors of class members (private, protected, etc.), but has some abstract methods.

6. Explain different way of using thread?

Ans. The thread could be implemented by using runnable interface or by inheriting from the Thread class. The former is more advantageous, 'cause when you are going for multiple inheritance, the only interface can help'.

7. What is an Iterator?

Ans. Some of the collection classes provide traversal of their contents via a java.util.Iterator interface. This interface allows you to walk through a collection of objects, operating on each object in turn.

Remember when using iterators that they contain a snapshot of the collection at the time the iterator was obtained; generally it is not advisable to modify the collection itself while traversing an iterator.

8. State the significance of public, private, protected, default modifiers both singly and in combination and state the effect of package relationships on declared items qualified by these modifiers.

Ans. Public: Public class is visible in other packages, field is visible everywhere (class must be public too)

Private: Private variables or methods may be used only by an instance of the same class that declares the variable or method. A private feature may only be accessed by the class that owns the feature.

Protected: It is available to all classes in the same package and also available to all subclasses of the class that owns the protected feature. This access is provided even to subclasses that reside in a different package from the class that owns the protected feature.

What you get by default, i.e. without any access modifier, (i.e. public, private or protected). It means that it is visible to all within a particular package.

9. What is static in java?

Ans. Static means one per class, not one for each object no matter how many instance of a class might exist. This means that you can use them without creating an instance of a class. Static methods are implicitly final, because overriding is done based on the type of the object, and static methods are attached to a class, not an object.

A static method in a superclass can be shadowed by another static method in a subclass, as long as the original method was not declared final. However, you can't override a static method with a nonstatic method. In other words, you can't change a static method into an instance method in a subclass.

10. What is final class?

Ans. A final class can't be extended, i.e. final class, may not be subclassed. A final method can't be overridden when its class is inherited. You can't change value of a final variable (is a constant).

11. What if the main() method is declared as private?

Ans. The program compiles properly but at runtime it will give message. "main() method not public."

12. What if the static modifier is removed from the signature of the main() method?

Ans. Program compiles. But at runtime throws an error "NoSuchMethodError".

13. What if I write static public void instead of public static void?

Ans. Program compiles and runs properly.

14. What if I do not provide the String array as the argument to the method?

Ans. Program compiles but throws a runtime error "NoSuchMethodError".

15. What is the first argument of the String array in main() method?

Ans. The String array is empty. It does not have any element. This is unlike C/C++ where the first element by default is the program name.

16. If I do not provide any arguments on the command line, then the String array of main() method will be empty or null?

Ans. It is empty. But not null.

17. How can one prove that the array is not null but empty using one line of code?

Ans. Print args.length. It will print 0. That means it is empty. But if it would have been null then it would have thrown a NullPointerException on attempting to print args.length.

18. What environment variables do I need to set on my machine in order to be able to run Java programs?

Ans. CLASSPATH and PATH are the two variables.

19. Can an application have multiple classes having main() method?

Ans. Yes it is possible. While starting the application we mention the class name to be run. The JVM will look for the Main method only in the class whose name you have mentioned.

Hence there is not conflict amongst the multiple classes having main() method.

20. Can I have multiple main() methods in the same class?

Ans. No the program fails to compile. The compiler says that the main() method is already defined in the class.

21. Do I need to import java.lang package any time? Why ?

Ans. No. It is by default loaded internally by the JVM.

22. Can I import same package/class twice? Will the JVM load the package twice at runtime?

Ans. One can import the same package or same class multiple times. Neither compiler nor JVM complains about it. And the JVM will internally load the class only once, no matter how many times you import the same class.

23. What are Checked and UnChecked Exception?

Ans. A checked exception is some subclass of Exception (or Exception itself), excluding class RuntimeException and its subclasses. Making an exception checked forces client programmers to deal with the possibility that the exception will be thrown.

For example: IOException thrown by java.io.FileInputStream's read() method.

Unchecked exceptions are RuntimeException and any of its subclasses. Class Error and its subclasses also are unchecked. With an unchecked exception, however, the compiler doesn't force client programmers either to catch the exception or declare it in a throws clause. In fact, client programmers may not even know that the exception could be thrown.

For example:
StringIndexOutOfBoundsException thrown by String's charAt() method.
Checked exceptions must be caught at compile time. Runtime exceptions do not need to be. Errors often cannot be.

24. What is overriding?

Ans. When a class defines a method using the same name, return type, and arguments as a method in its superclass, the method in the class overrides the method in the superclass.

When the method is invoked for an object of the class, it is the new definition of the method that is called, and not the method definition from superclass. Methods may be overridden to be more public, not more private.

25. Are the imports checked for validity at compile time? For example, will the code containing an import such as java.lang.ABCD compile?

Ans. Yes the imports are checked for the semantic validity at compile time. The code containing above line of import will not

compile. It will throw an error saying, can not resolve symbol
symbol : class ABCD
location: package io
import java.io.ABCD;

26. Does importing a package imports the subpackages as well? For example, does importing com.MyTest.* also import com.MyTest.UnitTests.*?

Ans. No you will have to import the subpackages explicitly. Importing com.MyTest.* will import classes in the package MyTest only. It will not import any class in any of its subpackage.

27. What is the difference between declaring a variable and defining a variable?

Ans. In declaration we just mention the type of the variable and its name. We do not initialize it. But defining means declaration + initialization.

For example: String s; is just a declaration while String s = new String ("abcd"); Or String s = "abcd"; are both definitions.

28. What is the default value of an object reference declared as an instance variable?

Ans. The default value will be null unless we define it explicitly.

29. Can a top level class be private or protected?

Ans. No. A top level class cannot be private or protected. It can have either "public" or no modifier. If it does not have a modifier, it is supposed to have a default access.

If a top level class is declared as private the compiler will complain that the "modifier private is not allowed here". This means that a top level class cannot be private. Same is the case with protected.

30. What type of parameter passing does Java support?

Ans. In Java the arguments are always passed by value.

31. Primitive data types are passed by reference or pass by value?

Ans. Primitive data types are passed by value.

32. Objects are passed by value or by reference?

Ans. Java only supports pass by value. With objects, the object reference itself is passed by value and so both the original reference and parameter copy refer to the same object.

33. What is serialization?

Ans. Serialization is a mechanism by which you can save the state of an object by converting it to a byte stream.

34. How do I serialize an object to a file?

Ans. The class whose instances are to be serialized should implement an interface Serializable. Then you pass the instance to the ObjectOutputStream which is connected to a fileoutputstream. This will save the object to a file.

35. Which methods of Serializable interface should I implement?

Ans. The serializable interface is an empty interface, it does not contain any method. So we do not implement any method.

36. How can I customize the seralization process? That is, how can one have a control over the serialization process?

Ans. Yes, it is possible to have control over, serialization process. The class should implement Externalizable interface. This interface contains two methods, namely readExternal and writeExternal.

You should implement these methods and write the logic for customizing the serialization process.

37. What is the common usage of serialization?

Ans. Whenever an object is to be sent over the network, objects need to be serialized. Moreover, if the state of an object is to be saved, objects need to be serilazed.

38. What is externalizable interface?

Ans. Externalizable is an interface which contains two methods, readExternal and writeExternal. These methods give you a control over the serialization mechanism.

Thus if your class implements this interface, you can customize the serialization process by implementing these methods.

39. When you serialize an object, what happens to the object references included in the object?

Ans. The serialization mechanism generates an object graph for serialization. Thus it determines whether the included object references are serializable or not. This is a recursive process.

Thus when an object is serialized, all the included objects are also serialized along with the original object.

40. What one should take care of while serializing the object?

Ans. One should make sure that all the included objects are also serializable. If any of the objects is not serializable, then it throws a NotSerializableException.

41. What happens to the static fields of a class during serialization?

Ans. There are three exceptions in which serialization does not necessarily read and write to the stream. These are

1. Serialization ignores static fields, because they are not part of any particular state.
2. Base class fields are only handled if the base class itself is serializable.
3. Transient fields.

42. Does Java provide any construct to find out the size of an object?

Ans. No, there is no size of operator in Java. So there is no direct way to determine the size of an object directly in Java.

43. What are wrapper classes?

Ans. Java provides specialized classes corresponding to each of the primitive data types. These are called wrapper classes.

For example: Integer, Character, Double, etc.

44. Why do we need wrapper classes?

Ans. It is sometimes easier to deal with primitives as objects. Moreover, most of the collection classes store objects and not primitive data types. And also the wrapper classes provide many utility methods also.

Because of these reasons we need wrapper classes. And since we create instances of these classes, we can store them in any of the collection classes and pass them around as a collection. Also we can pass them around as method parameters where a method expects an object.

45. What are checked exceptions?

Ans. Checked exceptions are those which the Java compiler forces you to catch.

For example: IOException are checked exceptions.

46. What are runtime exceptions?

Ans. Runtime exceptions are those exceptions that are thrown at runtime because of either wrong input data or because of wrong business logic, etc. These are not checked by the compiler at compile time.

47. What is the difference between error and an exception?

Ans. An error is an irrecoverable condition occurring at runtime, such as OutOfMemory error.

These JVM errors and you cannot repair them at runtime. While exceptions are conditions that occur because of bad input, etc. For example: FileNotFoundException will be thrown if the specified file does not exist. Or a NullPointerException will take place if you try using a null reference.

In most of the cases it is possible to recover from an exception (probably by giving user a feedback for entering proper values, etc.).

48. How to create custom exceptions?

Ans. Your class should extend class Exception, or some more specific type thereof.

49. If I want an object of my class to be thrown as an exception object, what should I do?

Ans. The class should extend from Exception class. Or you can extend your class from some more precise exception type also.

50. If my class already extends from some other class what should I do if I want an instance of my class to be thrown as an exception object?

Ans. One cannot do anything in this scenario. Because Java does not allow multiple inheritance and does not provide any exception interface as well.

51. How does an exception permeate through the code?

Ans. An unhandled exception moves up the method stack in search of a matching. When an exception is thrown from a code which is wrapped in a try block followed by one or more catch blocks, a search is made for matching catch block. If a matching type is found, then that block will be invoked. If a matching type is not found then the exception moves up the method stack and reaches the caller method.

The same procedure is repeated if the caller method is included in a try catch block. This process continues until a catch block handling the appropriate type of exception is found. If it does not find such a block, then finally the program terminates.

52. What are the different ways to handle exceptions?

Ans. There are two ways to handle exceptions:

1. By wrapping the desired code in a try block followed by a catch block to catch the exceptions, and
2. List the desired exceptions in the throws clause of the method and let the caller of the method handle those exceptions.

53. Is it necessary that each try block must be followed by a catch block?

Ans. It is not necessary that each try block must be followed by a catch block. It should be followed by either a catch block or a finally block. And whatever exceptions are likely to be thrown should be declared in the throws clause of the method.

54. If I write return at the end of the try block, will the finally block still execute?

Ans. Yes, even if you write return as the last statement in the try block and no exception occurs, the finally block will execute. The finally block will execute and then the control return.

55. If I write System.exit(0); at the end of the try block, will the finally block still execute?

Ans. No. In this case the finally block will not execute because when you say System.exit(0); the control immediately goes out of the program, and thus finally never executes.

56. How are Observer and Observable used?

Ans. Objects that subclass the Observable class maintain a list of observers. When an Observable object is updated, it invokes the update() method of each of its observers to notify the observers that it has changed state. The Observer interface is implemented by objects that observe Observable objects.

57. What is synchronization and why is it important?

Ans. With respect to multithreading, synchronization is the capability to control the access of multiple threads to shared resources.

Without synchronization, it is possible for one thread to modify a shared object while another thread is in the process of using or updating that object's value. This often leads to significant errors.

58. How does Java handle integer overflows and underflows?

Ans. It uses those low order bytes of the result that can fit into the size of the type allowed by the operation.

59. Does garbage collection guarantee that a program will not run out of memory?

Ans. Garbage collection does not guarantee that a program will not run out of memory. It is possible for programs to use up memory resources faster than they are garbage collected. It is also possible for programs to create objects that are not subject to garbage collection.

60. What is the difference between preemptive scheduling and time slicing?

Ans. Under preemptive scheduling, the highest priority task executes until it enters the waiting or dead states or a higher priority task comes into existence.

Under time slicing, a task executes for a predefined slice of time and then reenters the pool of ready tasks. The scheduler then determines which task should execute next, based on priority and other factors.

61. When a thread is created and started, what is its initial state?

Ans. A thread is in the ready state after it has been created and started.

62. What is the purpose of finalization?

Ans. The purpose of finalization is to give an unreachable object, the opportunity to perform any cleanup processing before the object is garbage collected.

63. What is the Locale class?

Ans. The Locale class is used to tailor program output to the conventions of a particular geographic, political, or cultural region.

64. What is the difference between a while statement and a do statement?

Ans. A while statement checks at the beginning of a loop to see whether the next loop iteration should occur.

A do statement checks at the end of a loop to see whether the next iteration of a loop should occur. The do statement will always execute the body of a loop at least once.

65. What is the difference between static and non-static variables?

Ans. A static variable is associated with the class as a whole rather than with specific instances of a class. Non-static variables take on unique values with each object instance.

66. How are this() and super() used with constructors?

Ans. this() is used to invoke a constructor of the same class. super() is used to invoke a superclass constructor.

67. What is daemon thread and which method is used to create the daemon thread?

Ans. Daemon thread is a low priority thread which runs intermittently in the background doing the garbage collection operation for the java runtime, system.setDaemon method is used to create a daemon thread.

68. Can applets communicate with each other?

Ans. At this point of time applets may communicate with other applets running in the same virtual machine. If the applets are of the same class, they can communicate via shared static variables. If the applets are of different classes, then each will need a reference to the same class with static variables. In any case the basic idea is to pass the information back and forth through a static variable.

An applet can also get references to all other applets on the same page using the getApplets() method of java.applet.AppletContext. Once you get the reference to an applet, you can communicate with it by using its public members.

It is conceivable to have applets in different virtual machines that talk to a server somewhere on the Internet and store any data that needs to be serialized there. Then, when another applet needs this data, it could connect to this same server. Implementing this is non-trivial.

69. What are the steps in the JDBC connection?

Ans. While making a JDBC connection we go through the following steps:

Step 1: Register the database driver by using:

Class.forName(\" driver classs for that specific database\");

Step 2: Now create a database connection using:

Connection con =

DriverManager.getConnection(url,username, password);

Step 3: Now create a query using:

Statement stmt =

Connection.Statement(\"select * from TABLE NAME\");

Step 4: Exceute the query:

stmt.exceuteUpdate();

70. How does a try statement determine which catch clause should be used to handle an exception?

Ans. When an exception is thrown within the body of a try statement, the catch clauses of the try statement are examined in the order in which they appear. The first catch clause that is capable of handling the exception is executed. The remaining catch clauses are ignored.

71. Can an unreachable object become reachable again?

Ans. An unreachable object may become reachable again. This can happen when the object's finalize() method is invoked and the object performs an operation which causes it to become accessible to reachable objects.

72. What method must be implemented by all threads?

Ans. All tasks must implement the run() method, whether they are a subclass of Thread or implement the Runnable interface.

73. What are synchronized methods and synchronized statements?

Ans. Synchronized methods are methods that are used to control access to an object. A thread only executes a synchronized method after it has acquired the lock for the method's object or class.

Synchronized statements are similar to synchronized methods. A synchronized statement can only be executed after a thread has acquired the lock for the object or class referenced in the synchronized statement.

74. What is externalizable?

Ans. Externalizable is an interface that extends Serializable Interface. And sends data into Streams in Compressed Format. It has two methods, writeExternal(ObjectOuput out) and readExternal(ObjectInput in).

75. What modifiers are allowed for methods in an Interface?

Ans. Only public and abstract modifiers are allowed for methods in interfaces.

76. What are some alternatives to inheritance?

Ans. Delegation is an alternative to inheritance. Delegation means that you include an instance of another class as an instance variable, and forward messages to the instance. It is often safer than inheritance because it forces you to think about each message you forward, because the instance is of a known class, rather than a new class, and because it doesn't force you to accept all the methods of the superclass, you can provide only the methods that really make sense. On the other hand, it makes you write more code, and it is harder to re-use (because it is not a subclass).

77. What does it mean that a method or field is "static"?

Ans. Static variables and methods are instantiated only once per class. In other words, they are class variables, not instance variables. If you change the value of a static variable in a particular object, the value of that variable changes for all instances of that class.

Static methods can be referenced with the name of the class rather than the name of a particular object of the class (though that works too). That's how library methods like System.out.println() work out is a static field in the java.lang.System class.

78. What is the difference between pre-emptive scheduling and time slicing?

Ans. Under preemptive scheduling, the highest priority task executes until it enters the waiting or dead states or a higher priority task comes into existence. Under time slicing, a task executes for a predefined slice of time and then reenters the pool of ready tasks.

The scheduler then determines which task should execute next, based on priority and other factors.

79. What is the catch or declare rule for method declarations?

Ans. If a checked exception may be thrown within the body of a method, the method must either catch the exception or declare it in its throws clause.

80. Is Empty .java file a valid source file?

Ans. Yes. An empty .java file is a perfectly valid source file.

81. Can a .java file contain more than one java classes?

Ans. Yes. A .java file contain more than one java classes, provided at the most one of them is a public class.

82. Is String a primitive data type in Java?

Ans. No. String is not a primitive data type in Java, even though it is one of the most extensively used object. Strings in Java are instances of String class defined in java.lang package.

83. Is main a keyword in Java?

Ans. No. Main is not a keyword in Java.

84. Is next a keyword in Java?

Ans. No. Next is not a keyword.

85. Is delete a keyword in Java?

Ans. No. Delete is not a keyword in Java. Java does not make use of explicit destructors the way C++ does.

86. Is exit a keyword in Java?

Ans. No. To exit a program explicitly you use exit method in System object.

87. What happens if you do not initialize an instance variable of any of the primitive types in Java?

Ans. Java by default initializes it to the default value for that primitive type. Thus an int will be initialized to 0(zero), a boolean will be initialized to false.

88. What will be the initial value of an object reference which is defined as an instance variable?

Ans. The object references are all initialized to null in Java. However, in order to do anything useful with these references, you must set them to a valid object, else you will get NullPointerExceptions everywhere you try to use such default initialized references.

89. What are the different scopes for Java variables?

Ans. The scope of a Java variable is determined by the context in which the variable is declared. Thus a java variable can have one of the three scopes at any given point of time.

1. **Instance :** These are typical object level variables, they are initialized to default values at the time of creation of object, and remain accessible as long as the object accessible.

2. **Local:** These are the variables that are defined within a method. They remain accessible only during the course of method excecution. When the method finishes execution, these variables fall out of scope.
3. **Static:** These are the class level variables. They are initialized when the class is loaded in JVM for the first time and remain there as long as the class remains loaded. They are not tied to any particular object instance.

90. What is the default value of the local variables?

Ans. The local variables are not initialized to any default value, neither primitives nor object references. If you try to use these variables without initializing them explicitly, the java compiler will not compile the code. It will complain about the local variable not being initialized.

91. How many objects are created in the following piece of code?

Ans. MyClass c1, c2, c3;
c1 = new MyClass ();
c3 = new MyClass ();
Only 2 objects are created, c1 and c3. The reference c2 is only declared and not initialized.

92. Can a public class MyClass be defined in a source file named YourClass.java?

Ans. No. The source file name, if it contains a public class, must be the same as the public class name itself with a .java extension.

93. Can main() method be declared final?

Ans. Yes, the main() method can be declared final, in addition to being public static.

94. What is HashMap and Map?

Ans. Map is an interface and Hashmap is the class that implements Map.

95. What is the difference between HashMap and HashTable?

Ans. The HashMap class is roughly equivalent to Hashtable, except that it is unsynchronized and permits nulls. (HashMap allows null values as key and value, whereas Hashtable doesn't allow).
HashMap does not guarantee that the order of the map will remain constant over time. HashMap is unsynchronized and Hashtable is synchronized.

96. What is the difference between vector and arraylist?

Ans. Vector is synchronized, whereas arraylist is not.

97. What is the difference between Swing and Awt?

Ans. AWT are heavy-weight components. Swings are light-weight components. Hence swing works faster than AWT.

98. What will be the default values of all the elements of an array defined as an instance variable?

Ans. If the array is an array of primitive types, then all the elements of the array will be initialized to the default value corresponding to that primitive type.
For example: All the elements of an array of int will be initialized to 0(zero), while that of boolean type will be initialized to false. Whereas if the array is an array of references (of any type), all the elements will be initialized to null.

Advanced Java

1. What is a transient variable?

Ans. A transient variable is a variable that may not be serialized.

2. Which containers use a border Layout as their default layout?

Ans. The Window, Frame and Dialog classes use a border layout as their default layout.

3. Why do threads block on I/O?

Ans. Threads block on I/O (that is enters the waiting state) so that other threads may execute while the I/O operation is performed.

4. How are Observer and Observable used?

Ans. Objects that subclass the Observable class maintain a list of observers. When an Observable object is updated, it invokes the update() method of each of its observers to notify the observers that it has changed state. The Observer interface is implemented by objects that observe Observable objects.

5. What is synchronization and why is it important?

Ans. With respect to multithreading, synchronization is the capability to control the access of multiple threads to shared resources. Without synchronization, it is possible for one thread to modify a shared object while another thread is in the process of using or updating that object's value. This often leads to significant errors.

6. Can a lock be acquired on a class?

Ans. Yes, a lock can be acquired on a class. This lock is acquired on the class's Class object.

7. What's new with the stop(), suspend() and resume() methods in JDK 1.2?

Ans. The stop(), suspend() and resume() methods have been deprecated in JDK 1.2.

8. Is null a keyword?

Ans. The null is not a keyword.

9. What is the preferred size of a component?

Ans. The preferred size of a component is the minimum component size that will allow the component to display normally.

10. What method is used to specify a container's layout?

Ans. The setLayout() method is used to specify a container's layout.

11. Which containers use a FlowLayout as their default layout?

Ans. The Panel and Applet classes use the FlowLayout as their default layout.

12. What state does a thread enter when it terminates its processing?

Ans. When a thread terminates its processing, it enters the dead state.

13. What is the Collections API?

Ans. The Collections API is a set of classes and interfaces that support operations on collections of objects.

14. Which characters may be used as the second character of an identifier, but not as the first character of an identifier?

Ans. The digits 0 through 9 may not be used as the first character of an identifier but they may be used after the first character of an identifier.

15. What is the List interface?

Ans. The List interface provides support for ordered collections of objects.

16. How does Java handle integer overflows and underflows?

Ans. It uses those low order bytes of the result that can fit into the size of the type allowed by the operation.

17. What is the Vector class?

Ans. The Vector class provides the capability to implement a growable array of objects.

18. What modifiers may be used with an inner class that is a member of an outer class?

Ans. An (non-local) inner class may be declared as public, protected, private, static, final, or abstract.

19. What is an Iterator interface?

Ans. The Iterator interface is used to step through the elements of a Collection.

20. What is the difference between the >> and >>> operators?

Ans. The >> operator carries the sign bit when shifting right. The >>> zero-fills bits that have been shifted out.

21. Which method of the Component class is used to set the position and size of a component?

Ans. setBounds() method is used to set the position and size of a component.

22. What is the difference between yielding and sleeping?

Ans. When a task invokes its yield() method, it returns to the ready state. When a task invokes its sleep() method, it returns to the waiting state.

23. Which java.util classes and interfaces support event handling?

Ans. The EventObject class and the EventListener interface support event processing.

24. Is sizeof a keyword?

Ans. The sizeof operator is not a keyword.

25. What are wrapped classes?

Ans. Wrapped classes are classes that allow primitive types to be accessed as objects.

26. Does garbage collection guarantee that a program will not run out of memory?

Ans. Garbage collection does not guarantee that a program will not run out of memory. It is possible for programs to use up memory resources faster than they are garbage collected. It is also possible for programs to create objects that are not subject to garbage collection.

27. What restrictions are placed on the location of a package statement within a source code file?

Ans. A package statement must appear as the first line in a source code file (excluding blank lines and comments).

28. Can an object's finalize() method be invoked while it is reachable?

Ans. An object's finalize() method cannot be invoked by the garbage collector while the object is still reachable. However, an object's finalize() method may be invoked by other objects.

29. What is the immediate superclass of the Applet class?

Ans. Panel.

30. What is the difference between preemptive scheduling and time slicing?

Ans. Under preemptive scheduling, the highest priority task executes until it enters the waiting or dead states or a higher priority task comes into existence. Under time slicing, a task executes for a predefined slice of time and then reenters the pool of ready tasks.

The scheduler then determines which task should execute next, based on priority and other factors.

31. Name four component subclasses that support painting.

Ans. The Canvas, Frame, Panel, and Applet classes support painting.

32. What value does readLine() return when it has reached the end of a file?

Ans. The readLine() method returns null when it has reached the end of a file.

33. What is the immediate superclass of the Dialog class?

Ans. Window.

34. What is clipping?

Ans. Clipping is the process of confining paint operations to a limited area or shape.

35. What is a native method?

Ans. A native method is a method that is implemented in a language other than Java.

36. Can a for statement loop indefinitely?

Ans. Yes, a for statement can loop indefinitely. For example, consider the following:

```
for(;;) ;
```

37. What are the order of precedence and associativity, and how are they used?

Ans. Order of precedence determines the order in which operators are evaluated in expressions.

Associativity determines whether an expression is evaluated left-to-right or right-to-left.

38. When a thread blocks on I/O, what state does it enter?

Ans. A thread enters the waiting state when it blocks on I/O.

39. To what value is a variable of the String type automatically initialized?

Ans. The default value of an String type is null.

40. What is the catch or declare rule for method declarations?

Ans. If a checked exception may be thrown within the body of a method, the method must either catch the exception or declare it in its throws clause.

41. What is the difference between a MenuItem and a CheckboxMenuItem?

Ans. The CheckboxMenuItem class extends the MenuItem class to support a menu item that may be checked or unchecked.

42. What is a task's priority and how is it used in scheduling?

Ans. A task's priority is an integer value that identifies the relative order in which it should be executed with respect to other tasks. The scheduler attempts to schedule higher priority tasks before lower priority tasks.

43. What class is the top of the AWT event hierarchy?

Ans. The java.awt.AWTEvent class is the highest-level class in the AWT event-class hierarchy.

44. When a thread is created and started, what is its initial state?

Ans. A thread is in the ready state after it has been created and started.

45. Can an anonymous class be declared as implementing an interface and extending a class?

Ans. An anonymous class may implement an interface or extend a superclass, but may not be declared to do both.

46. What is the immediate superclass of Menu?

Ans. MenuItem.

47. What is the purpose of finalization?

Ans. The purpose of finalization is to give an unreachable object the opportunity to perform any cleanup processing before the object is garbage collected.

48. Which class is the immediate superclass of the MenuComponent class?

Ans. Object.

49. What invokes a thread's run() method?

Ans. After a thread is started, via its start() method or that of the Thread class, the JVM invokes the thread's run() method when the thread is initially executed.

50. What is the difference between the Boolean & operator and the && operator?

Ans. If an expression involving the Boolean & operator is evaluated, both operands are evaluated. Then the & operator is applied to the operand. When an expression involving the && operator is evaluated, the first operand is evaluated.

If the first operand returns a value of true, then the second operand is evaluated. The && operator is then applied to the first and second operands. If the first operand evaluates to false, the evaluation of the second operand is skipped.

51. Name subclasses of the Component class.

Ans. Box.Filler, Button, Canvas, Checkbox, Choice, Container, Label, List, Scrollbar, or TextComponent.

52. What is the GregorianCalendar class?

Ans. The GregorianCalendar class provides support for traditional Western calendars.

53. Which container method is used to cause a container to be laid out and redisplayed?

Ans. validate() method is used to cause a container to be laid out and redisplayed.

54. What is the purpose of the Runtime class?

Ans. The purpose of the Runtime class is to provide access to the Java runtime system.

55. How many times may an object's finalize() method be invoked by the garbage collector?

Ans. An object's finalize() method may only be invoked once by the garbage collector.

56. What is the purpose of the finally clause of a try-catch-finally statement?

Ans. The finally clause is used to provide the capability to execute code no matter whether or not an exception is thrown or caught.

57. What is the argument type of a program's main() method?

Ans. A program's main() method takes an argument of the String[] type.

58. Which Java operator is right associative?

Ans. The = operator is right associative.

59. Can a double value be cast to a byte?

Ans. Yes, a double value can be cast to a byte.

60. What must a class do to implement an interface?

Ans. It must provide all of the methods in the interface and identify the interface in its implements clause.

61. What method is invoked to cause an object to begin executing as a separate thread?

Ans. The start() method of the Thread class is invoked to cause an object to begin executing as a separate thread.

62. Name two subclasses of the TextComponent class.

Ans. TextField and TextArea.

63. Which containers may have a MenuBar?

Ans. Frame.

64. How are commas used in the intialization and iteration parts of a for statement?

Ans. Commas are used to separate multiple statements within the initialization and iteration parts of a for statement.

65. What is the purpose of the wait(), notify(), and notifyAll() methods?

Ans. The wait(), notify(), and notifyAll() methods are used to provide an efficient way for threads to wait for a shared resource. When a thread executes an object's wait() method, it enters the waiting state. It only enters the ready state after another thread invokes the object's notify() or notifyAll() methods.

66. What is an abstract method?

Ans. An abstract method is a method whose implementation is deferred to a subclass.

67. How are Java source code files named?

Ans. A Java source code file takes the name of a public class or interface that is defined within the file. A source code file may contain at most one public class or interface. If a public class or interface is defined within a source code file, then the source code file must take the name of the public class or interface.

If no public class or interface is defined within a source code file, then the file must take on a name that is different than its classes and interfaces. Source code files use the .java extension.

68. What is the relationship between the Canvas class and the Graphics class?

Ans. A Canvas object provides access to a Graphics object via its paint() method.

69. What are the high-level thread states?

Ans. The high-level thread states are ready, running, waiting, and dead.

70. What value does read() return when it has reached the end of a file?

Ans. The read() method returns -1 when it has reached the end of a file.

71. Can a Byte object be cast to a double value?

Ans. No. An object cannot be cast to a primitive value.

72. What is the difference between a static and a non-static inner class?

Ans. A non-static inner class may have object instances that are associated with instances of the class's outer class.

A static inner class does not have any object instances.

73. What is the difference between the String and StringBuffer classes?

Ans. String objects are constants. StringBuffer objects are not constants.

74. If a variable is declared as private, where may the variable be accessed?

Ans. A private variable may only be accessed within the class in which it is declared.

75. What is an object's lock and which object's have locks?

Ans. An object's lock is a mechanism that is used by multiple threads to obtain synchronized access to the object. A thread may execute a synchronized method of an object only after it has acquired the object's lock.

All objects and classes have locks. A class's lock is acquired on the class's Class object.

76. What is the Dictionary class?

Ans. The Dictionary class provides the capability to store key-value pairs.

77. How are the elements of a BorderLayout organized?

Ans. The elements of a BorderLayout are organized at the borders (North, South, East, and West) and the center of a container.

78. What is the % operator?

Ans. It is referred to as the modulo or remainder operator. It returns the remainder of dividing the first operand by the second operand.

79. When can an object reference be cast to an interface reference?

Ans. An object reference be cast to an interface reference when the object implements the referenced interface.

80. What is the difference between a Window and a Frame?

Ans. The Frame class extends Window to define a main application window that can have a menu bar.

81. Which class is extended by all other classes?

Ans. The Object class is extended by all other classes.

82. Can an object be garbage collected while it is still reachable?

Ans. A reachable object cannot be garbage collected. Only unreachable objects may be garbage collected.

83. Is the ternary operator written x : y ? z or x ? y : z ?

Ans. It is written x ? y : z.

84. What is the difference between the Font and FontMetrics classes?

Ans. The FontMetrics class is used to define implementation-specific properties, such as ascent and descent, of a Font object.

85. How is rounding performed under integer division?

Ans. The fractional part of the result is truncated. This is known as rounding toward zero.

86. What happens when a thread cannot acquire a lock on an object?

Ans. If a thread attempts to execute a synchronized method or synchronized statement and is unable to acquire an object's lock, it enters the waiting state until the lock becomes available.

87. What is the difference between the Reader/Writer class hierarchy and the InputStream/OutputStream class hierarchy?

Ans. The Reader/Writer class hierarchy is character-oriented, and the InputStream/OutputStream class hierarchy is byte-oriented.

88. What classes of exceptions may be caught by a catch clause?

Ans. A catch clause can catch any exception that may be assigned to the Throwable type. This includes the Error and Exception types.

89. If a class is declared without any access modifiers, where may the class be accessed?

Ans. A class that is declared without any access modifiers is said to have package access. This means that the class can only be accessed by other classes and interfaces that are defined within the same package.

90. What is the SimpleTimeZone class?

Ans. The SimpleTimeZone class provides support for a Gregorian calendar.

91. What is the Map interface?

Ans. The Map interface replaces the JDK 1.1 Dictionary class and is used associate keys with values.

92. Does a class inherit the constructors of its superclass?

Ans. A class does not inherit constructors from any of its superclasses.

93. For which statements does it make sense to use a label?

Ans. The only statements for which it makes sense to use a label are those statements that can enclose a break or continue statement.

94. What is the purpose of the System class?

Ans. The purpose of the System class is to provide access to system resources.

95. Which TextComponent method is used to set a TextComponent to the read-only state?

Ans. setEditable().

96. How are the elements of a CardLayout organized?

Ans. The elements of a CardLayout are stacked, one on top of the other, like a deck of cards.

97. Is &&= a valid Java operator?

Ans. No. It is not a valid java operator.

98. Name the eight primitive Java types.

Ans. The eight primitive types are byte, char, short, int, long, float, double, and boolean.

99. Which class should you use to obtain design information about an object?

Ans. The Class class is used to obtain information about an object's design.

100. What is the relationship between clipping and repainting?

Ans. When a window is repainted by the AWT painting thread, it sets the clipping regions to the area of the window that requires repainting.

101. Is "abc" a primitive value?

Ans. The String literal "abc" is not a primitive value. It is a String object.

102. What is the relationship between an event-listener interface and an event-adapter class?

Ans. An event-listener interface defines the methods that must be implemented by an event handler for a particular kind of event. An event adapter provides a default implementation of an event-listener interface.

103. What restrictions are placed on the values of each case of a switch statement?

Ans. During compilation, the values of each case of a switch statement must evaluate to a value that can be promoted to an int value.

104. What modifiers may be used with an interface declaration?

Ans. An interface may be declared as public or abstract.

105. Is a class a subclass of itself?

Ans. A class is a subclass of itself.

106. What is the highest-level event class of the event-delegation model?

Ans. The java.util.EventObject class is the highest-level class in the event-delegation class hierarchy.

107. What event results from the clicking of a button?

Ans. The ActionEvent event is generated as the result of the clicking of a button.

108. How can a GUI component handle its own events?

Ans. A component can handle its own events by implementing the required event-listener interface and adding itself as its own event listener.

109. How are the elements of a GridBagLayout organized?

Ans. The elements of a GridBagLayout are organized according to a grid. However, the elements are of different sizes and may occupy more than one row or column of the grid. In addition, the rows and columns may have different sizes.

110. What advantage do Java's layout managers provide over traditional windowing systems?

Ans. Java uses layout managers to layout components in a consistent manner across all windowing platforms. Since Java's layout managers aren't tied to absolute sizing and positioning, they are able to accommodate platform-specific differences among windowing systems.

111. What is the Collection interface?

Ans. The Collection interface provides support for the implementation of a mathematical bag—an unordered collection of objects that may contain duplicates.

112. What modifiers can be used with a local inner class?

Ans. A local inner class may be final or abstract.

113. What is the difference between static and non-static variables?

Ans. A static variable is associated with the class as a whole rather than with specific instances of a class.

Non-static variables take unique values with each object instance.

114. What is the difference between the paint() and repaint() methods?

Ans. The paint() method supports painting via a Graphics object. The repaint() method is used to cause paint() to be invoked by the AWT painting thread.

115. What is the purpose of the File class?

Ans. The File class is used to create objects that provide access to the files and directories of a local file system.

116. Can an exception be rethrown?

Ans. Yes, an exception can be rethrown.

117. Which Math method is used to calculate the absolute value of a number?

Ans. The abs() method is used to calculate absolute values.

118. How does multithreading take place on a computer with a single CPU?

Ans. The operating system's task scheduler allocates execution time to multiple tasks. By quickly switching between executing tasks, it creates the impression that tasks execute sequentially.

119. When does the compiler supply a default constructor for a class?

Ans. The compiler supplies a default constructor for a class if no other constructors are provided.

120. When is the finally clause of a try-catch-finally statement executed?

Ans. The finally clause of the try-catch-finally statement is always executed unless the thread of execution terminates or an exception occurs within the execution of the finally clause.

121. Which class is the immediate superclass of the Container class?

Ans. Component.

122. If a method is declared as protected, where may the method be accessed?

Ans. A protected method may only be accessed by classes or interfaces of the same package or by subclasses of the class in which it is declared.

123. How can the Checkbox class be used to create a radio button?

Ans. By associating Checkbox objects with a CheckboxGroup.

124. Which non-Unicode letter characters may be used as the first character of an identifier?

Ans. The non-Unicode letter characters $ and _ may appear as the first character of an identifier.

125. What restrictions are placed on method overloading?

Ans. Two methods may not have the same name and argument list but different return types.

126. What happens when you invoke a thread's interrupt method while it is sleeping or waiting?

Ans. When a task's interrupt() method is executed, the task enters the ready state. The next time the task enters the running state, an InterruptedException is thrown.

127. What is casting?

Ans. There are two types of casting, casting between primitive numeric types and casting between object references.

Casting between numeric types is used to convert larger values, such as double values, to smaller values, such as byte values.

Casting between object references is used to refer to an object by a compatible class, interface, or array type reference.

128. What is the return type of a program's main() method?

Ans. A program's main() method has a void return type.

129. Name container classes.

Ans. Window, Frame, Dialog, FileDialog, Panel, Applet, or ScrollPane.

130. What is the difference between a Choice and a List?

Ans. A Choice is displayed in a compact form that requires you to pull it down to see the list of available choices. Only one item may be selected from a Choice.

A List may be displayed in such a way that several List items are visible. A List supports the selection of one or more List items.

131. What class of exceptions are generated by the Java runtime system?

Ans. The Java runtime system generates RuntimeException and Error exceptions.

132. What class allows you to read objects directly from a stream?

Ans. The ObjectInputStream class supports the reading of objects from input streams.

133. What is the difference between a field variable and a local variable?

Ans. A field variable is a variable that is declared as a member of a class.

A local variable is a variable that is declared local to a method.

134. Under what conditions is an object's finalize() method invoked by the garbage collector?

Ans. The garbage collector invokes an object's finalize() method when it detects that the object has become unreachable.

135. What is the relationship between a method's throws clause and the exceptions that can be thrown during the method's execution?

Ans. A method's throws clause must declare any checked exceptions that are not caught within the body of the method.

136. What is the difference between the JDK 1.02 event model and the event-delegation model introduced with JDK 1.1?

Ans. The JDK 1.02 event model uses an event inheritance or bubbling approach. In this model, components are required to handle their own events. If they do not handle a particular event, the event is inherited by (or bubbled up to) the component's container. The container then either handles the event or it is bubbled up to its container and so on, until the highest-level container has been tried.

In the event-delegation model, specific objects are designated as event handlers for GUI components. These objects implement event-listener interfaces. The event-delegation model is more efficient than the event-inheritance model because it eliminates the processing required to support the bubbling of unhandled events.

137. How is it possible for two String objects with identical values not to be equal under the == operator?

Ans. The == operator compares two objects to determine if they are the same object in memory. It is possible for two String objects to have the same value, but located in different areas of memory.

138. Why are the methods of the Math class static?

Ans. They can be invoked as if they are a mathematical code library.

139. What Checkbox method allows you to tell if a Checkbox is checked?

Ans. getState().

140. What state is a thread when it is executing?

Ans. An executing thread is in the running state.

141. What are the legal operands of the instanceof operator?

Ans. The left operand is an object reference or null value and the right operand is a class, interface, or array type.

142. How are the elements of a GridBagLayout organized?

Ans. The elements of a GridBagLayout are of equal size and are laid out using the squares of a grid.

143. What an I/O filter?

Ans. An I/O filter is an object that reads from one stream and writes to another, usually altering the data in some way as it is passed from one stream to another.

144. If an object is garbage collected, can it become reachable again?

Ans. Once an object is garbage collected, it ceases to exist. It can no longer become reachable again.

145. What is the Set interface?

Ans. The Set interface provides methods for accessing the elements of a finite mathematical set. Sets do not allow duplicate elements.

146. What classes of exceptions may be thrown by a throw statement?

Ans. A throw statement may throw any expression that may be assigned to the Throwable type.

147. What are E and PI?

Ans. E is the base of the natural logarithm and PI is mathematical value pi.

148. Are true and false keywords?

Ans. The values true and false are not keywords.

149. What is a void return type?

Ans. A void return type indicates that a method does not return a value.

150. What is the purpose of the enableEvents() method?

Ans. The enableEvents() method is used to enable an event for a particular object. Normally, an event is enabled when a listener is added to an object for a particular event.

The enableEvents() method is used by objects that handle events by overriding their event-dispatch methods.

151. What is the difference between the File and RandomAccessFile classes?

Ans. The File class encapsulates the files and directories of the local file system.

The RandomAccessFile class provides the methods needed to directly access data contained in any part of a file.

152. What happens when you add a double value to a String?

Ans. The result is a String object.

153. What is your platform's default character encoding?

Ans. If you are running Java on English Windows platforms, it is probably Cp1252. If you are running Java on English Solaris platforms, it is most likely 8859_1.

154. Which package is always imported by default?

Ans. The java.lang package is always imported by default.

155. What interface must an object implement before it can be written to a stream as an object?

Ans. An object must implement the Serializable or Externalizable interface before it can be written to a stream as an object.

156. How are this and super used?

Ans. 'this' is used to refer to the current object instance.

'super' is used to refer to the variables and methods of the superclass of the current object instance.

157. What is a compilation unit?

Ans. A compilation unit is a Java source code file.

158. What interface is extended by AWT event listeners?

Ans. All AWT event listeners extend the java.util.EventListener interface.

159. What restrictions are placed on method overriding?

Ans. Overridden methods must have the same name, argument list, and return type.

The overriding method may not limit the access of the method it overrides.

The overriding method may not throw any exceptions that may not be thrown by the overridden method.

160. How can a dead thread be restarted?

Ans. A dead thread cannot be restarted.

161. What happens if an exception is not caught?

Ans. An uncaught exception results in the uncaughtException() method of the thread's ThreadGroup being invoked, which eventually results in the termination of the program in which it is thrown.

162. What is a layout manager?

Ans. A layout manager is an object that is used to organize components in a container.

163. Which arithmetic operations can result in the throwing of an ArithmeticException?

Ans. Integer / and % can result in the throwing of an ArithmeticException.

164. What are the three ways in which a thread can enter the waiting state?

Ans. A thread can enter the waiting state by invoking its sleep() method, by blocking on I/O, by unsuccessfully attempting to acquire an object's lock, or by invoking an object's wait() method. It can also enter the waiting state by invoking its (deprecated) suspend() method.

165. Can an abstract class be final?

Ans. An abstract class may not be declared as final.

166. What is the ResourceBundle class?

Ans. The ResourceBundle class is used to store locale-specific resources that can be loaded by a program to tailor the program's appearance to the particular locale in which it is being run.

167. What happens if a try-catch-finally statement does not have a catch clause to handle an exception that is thrown within the body of the try statement?

Ans. The exception propagates up to the next higher level try-catch statement (if any) or results in the program's termination.

168. What is numeric promotion?

Ans. Numeric promotion is the conversion of a smaller numeric type to a larger numeric type, so that integer and floating-point operations may take place. In numerical promotion, byte, char, and short values are converted to int values. The int values are also converted to long values, if necessary. The long and float values are converted to double values, as required.

169. What is the difference between a Scrollbar and a ScrollPane?

Ans. A Scrollbar is a Component, but not a Container. A ScrollPane is a Container.

A ScrollPane handles its own events and performs its own scrolling.

170. What is the difference between a public and a non-public class?

Ans. A public class may be accessed outside of its package.

A non-public class may not be accessed outside of its package.

171. To what value is a variable of the boolean type automatically initialized?

Ans. The default value of the boolean type is false.

172. Can try statements be nested?

Ans. Try statements may be tested.

173. What is the difference between the prefix and postfix forms of the ++ operator?

Ans. The prefix form performs the increment operation and returns the value of the increment operation.

The postfix form returns the current value all of the expression and then performs the increment operation on that value.

174. What is the purpose of a statement block?

Ans. A statement block is used to organize a sequence of statements as a single statement group.

175. What is a Java package and how is it used?

Ans. A Java package is a naming context for classes and interfaces. A package is used to create a separate name space for groups of classes and interfaces.

Packages are also used to organize related classes and interfaces into a single API unit and to control accessibility to these classes and interfaces.

176. What modifiers may be used with a top-level class?

Ans. A top-level class may be public, abstract, or final.

177. What are the Object and Class classes used for?

Ans. The Object class is the highest-level class in the Java class hierarchy. The Class class is used to represent the classes and interfaces that are loaded by a Java program.

178. How does a try statement determine which catch clause should be used to handle an exception?

Ans. When an exception is thrown within the body of a try statement, the catch clauses of the try statement are examined in the order in which they appear. The first catch clause that is capable of handling the exception is executed. The remaining catch clauses are ignored.

179. Can an unreachable object become reachable again?

Ans. An unreachable object may become reachable again. This can happen when the object's finalize() method is invoked and the object performs an operation which causes it to become accessible to reachable objects.

180. When is an object subject to garbage collection?

Ans. An object is subject to garbage collection when it becomes unreachable to the program in which it is used.

181. What method must be implemented by all threads?

Ans. All tasks must implement the run() method, whether they are a subclass of Thread or implement the Runnable interface.

182. What methods are used to get and set the text label displayed by a Button object?

Ans. getLabel() and setLabel().

183. Which component subclass is used for drawing and painting?

Ans. Canvas.

184. What are synchronized methods and synchronized statements?

Ans. Synchronized methods are methods that are used to control access to an object. A thread only executes a synchronized method after it has acquired the lock for the method's object or class.

Synchronized statements are similar to synchronized methods. A synchronized statement can only be executed after a thread has acquired the lock for the object or class referenced in the synchronized statement.

185. What are the two basic ways in which classes that can be run as threads may be defined?

Ans. A thread class may be declared as a subclass of Thread, or it may implement the Runnable interface.

186. What are the problems faced by Java programmers who don't use layout managers?

Ans. Without layout managers, Java programmers are faced with determining how their GUI will be displayed across multiple windowing systems and finding a common sizing and positioning that will work within the constraints imposed by each windowing system.

187. What is the difference between an Interface and an Abstract class?

Ans. An abstract class can have instance methods that implement a default behavior. An interface can only declare constants and instance methods, but cannot implement default behavior and all methods are implicitly abstract. An interface has all public members and no implementation. An abstract class is a class which may have the usual flavors of class members (private, protected, etc.), but has some abstract methods.

188. What is the purpose of garbage collection in Java, and when is it used?

Ans. The purpose of garbage collection is to identify and discard objects that are no longer needed by a program so that their resources can be reclaimed and reused. A Java object is subject to garbage collection when it becomes unreachable to the program in which it is used.

189. Describe synchronization in respect to multithreading.

Ans. With respect to multithreading, synchronization is the capability to control the access of multiple threads to shared resources. Without synchonization, it is possible for one thread to modify a shared variable while another thread is in the process of using or updating same shared variable. This usually leads to significant errors.

190. Explain different ways of using thread.

Ans. The thread could be implemented by using runnable interface or by inheriting from the Thread class. The former is more advantageous, 'cause when you are going for multiple inheritance...the only interface can help.

191. What are pass by reference and passby value?

Ans. Pass by reference means the passing the address itself rather than passing the value. Passby value means passing a copy of the value to be passed.

192. What is HashMap and Map?

Ans. Map is interface and Hashmap is class that implements that.

193. What is the difference between HashMap and HashTable?

Ans. The HashMap class is roughly equivalent to Hashtable, except that it is unsynchronized and permits nulls. (HashMap allows null values as key and value, whereas Hashtable does not allow). HashMap does not guarantee that the order of the map will remain constant over time. HashMap is unsynchronized and Hashtable is synchronized.

194. What is the difference between Vector and ArrayList?

Ans. Vector is synchronized, whereas arraylist is not.

195. What is the difference between Swing and AWT?

Ans. AWT are heavyweight componenets. Swings are lightweight components. Hence swing works faster than AWT.

196. What is the difference between a constructor and a method?

Ans. A constructor is a member function of a class that is used to create objects of that class. It has the same name as the class itself, has no return type, and is invoked using the new operator.

A method is an ordinary member function of a class. It has its own name, a return type (which may be void), and is invoked using the dot operator.

197. What is an Iterator?

Ans. Some of the collection classes provide traversal of their contents via a java.util.Iterator interface. This interface allows you to walk through a collection of objects, operating on each object in turn. Remember when using Iterators that they contain a snapshot of the collection at the time the Iterator was obtained; generally it is not advisable to modify the collection itself while traversing an Iterator.

198. State the significance of public, private, protected, default modifiers both singly and in combination and state the effect of package relationships on declared items qualified by these modifiers.

Ans. Public: Public class is visible in other packages, field is visible everywhere (class must be public too).

Private: Private variables or methods may be used only by an instance of the same class that declares the variable or method. A private feature may only be accessed by the class that owns the feature.

Protected: It is available to all classes in the same package and also available to all subclasses of the class that owns the protected feature.This access is provided even to subclasses that reside in a different package from the class that owns the protected feature.

Default: What you get by default, i.e. without any access modifier (i.e. public private or protected). It means that it is visible to all within a particular package.

199. What is an abstract class?

Ans. Abstract class must be extended/ subclassed (to be useful). It serves as a template. A class that is abstract may not be instantiated (i.e. you may not call its constructor), abstract class may contain static data. Any class with an abstract method is automatically abstract itself, and must be declared as such.

A class may be declared abstract even if it has no abstract methods. This prevents it from being instantiated.

200. What is static in java?

Ans. Static means one per class, not one for each object no matter how many instance of a class might exist. This means that you can use them without creating an instance of a class. Static methods are implicitly final, because overriding is done based on the type of the object, and static methods are attached to a class, not an object. A static method in a superclass can be shadowed by another static method in a subclass, as long as the original method was not declared final. However, you can't override a static method with a nonstatic method. In other words, you can't change a static method into an instance method in a subclass.

201. What is final?

Ans. A final class can't be extended, i.e. final class may not be subclassed. A final method can't be overridden when its class is inherited. You can't change value of a final variable (is a constant).

202. What if the main method is declared as private?

Ans. The program compiles properly but at runtime it will give "Main method not public." message.

203. What if the static modifier is removed from the signature of the main method?

Ans. Program compiles. But at runtime throws an error "NoSuchMethodError".

204. What if I write static public void instead of public static void?

Ans. Program compiles and runs properly.

205. What if I do not provide the String array as the argument to the method?

Ans. Program compiles but throws a runtime error "NoSuchMethodError".

206. What is the first argument of the String array in main method?

Ans. The String array is empty. It does not have any element. This is unlike C/C++ where the first element by default is the program name.

207. If I do not provide any argument on the command line, then will the String array of main method be empty or null.

Ans. It is empty. But not null.

208. How can one prove that the array is not null but empty using one line of code?

Ans. Print args.length. It will print 0. That means it is empty. But if it would have been null, then it would have thrown a NullPointerException on attempting to print args.length.

209. What environment variables do I need to set on my machine in order to be able to run Java programs?

Ans. CLASSPATH and PATH are the two variables.

210. Can an application have multiple classes having main method?

Ans. Yes it is possible. While starting the application we mention the class name to be run. The JVM will look for the main method only in the class whose name you have mentioned. Hence there is not conflict amongst the multiple classes having main method.

211. Can I have multiple main methods in the same class?

Ans. No the program fails to compile. The compiler says that the main method is already defined in the class.

212. Do I need to import java.lang package any time? Why ?

Ans. No. It is by default loaded internally by the JVM.

213. Can I import same package/class twice? Will the JVM load the package twice at runtime?

Ans. One can import the same package or same class multiple times. Neither compiler nor JVM complains about it. And the JVM will internally load the class only once no matter how many times you import the same class.

214. What are Checked and UnChecked Exception?

Ans. A checked exception is some subclass of Exception (or Exception itself), excluding class RuntimeException and its subclasses.

Making an exception checked forces client programmers to deal with the possibility that the exception will be thrown, e.g. IOException thrown by java.io.FileInputStream's read() method.

Unchecked exceptions are RuntimeException and any of its subclasses. Class error and its subclasses also are unchecked. With an unchecked exception, however, the compiler doesn't force client programmers either to catch the exception or declare it in a throws clause. In fact, client programmers may not even know that the exception could be thrown. For example, StringIndexOutOfBoundsException thrown by String's charAt() method.

Checked exceptions must be caught at compile time. Runtime exceptions do not need to be. Errors often cannot be.

215. What is Overriding?

Ans. When a class defines a method using the same name, return type, and arguments as a method in its superclass, the method in the class overrides the method in the superclass.

When the method is invoked for an object of the class, it is the new definition of the method that is called, and not the method definition from superclass. Methods may be overridden to be more public, not more private.

216. What are different types of inner classes?

Ans. Nested top-level classes, member classes, Local classes, Anonymous classes.

Nested top-level classes: If you declare a class within a class and specify the static modifier, the compiler treats the class just like any other top-level class.

Any class outside the declaring class accesses the nested class with the declaring class name acting similarly to a package, e.g. outer.inner. Top-level inner classes implicitly have access only to static variables.There can also be inner interfaces. All of these are of the nested top-level variety.

Member classes: Member inner classes are just like other member methods and member variables and access to the member class is restricted, just like methods and variables. This means a public member class acts similarly to a nested top-level class. The primary difference between member classes and nested top-level classes is that member classes have access to the specific instance of the enclosing class.

Local classes: Local classes are like local variables, specific to a block of code. Their visibility is only within the block of their declaration. In order for the class to be useful beyond the declaration block, it would need to implement a more publicly available interface. Because local classes are not members, the modifiers public, protected, private, and static are not usable.

Anonymous classes: Anonymous inner classes extend local inner classes one level further. As anonymous classes have no name, you cannot provide a constructor.

217. Are the imports checked for validity at compile time? For example, will the code containing an import such as java.lang.ABCD compile?

Ans. Yes the imports are checked for the semantic validity at compile time. The code containing above line of import will not compile. It will throw an error saying, cannot resolve symbol
symbol : class ABCD
location: package io
import java.io.ABCD;

218. Does importing a package imports the subpackages as well? For example, does importing com.MyTest.* also import com.MyTest.UnitTests.*?

Ans. No you will have to import the subpackages explicitly. Importing com.MyTest.* will import classes in the package MyTest only. It will not import any class in any of its subpackage.

219. What is the difference between declaring a variable and defining a variable?

Ans. In declaration we just mention the type of the variable and its name. We do not initialize it. But defining means declaration + initialization.

For example, String s; is just a declaration while String s = new String ("abcd"); Or String s = "abcd"; are both definitions.

220. What is the default value of an object reference declared as an instance variable?

Ans. Null unless we define it explicitly.

221. Can a top level class be private or protected?

Ans. No. A top level class cannot be private or protected. It can have either "public" or no modifier. If it does not have a modifier it is supposed to have a default access. If a top level class is declared as private the compiler will complain that the "modifier private is not allowed here". This means that a top level class cannot be private. Same is the case with protected.

222. What type of parameter passing does Java support?

Ans. In Java the arguments are always passed by value.

223. Are primitive data types passed by reference or pass by value?

Ans. Primitive data types are passed by value.

224. Are objects passed by value or by reference?

Ans. Java only supports pass by value. With objects, the object reference itself is passed by value and so both the original reference and parameter copy refer to the same object.

225. What is serialization?

Ans. Serialization is a mechanism by which you can save the state of an object by converting it to a byte stream.

226. How do I serialize an object to a file?

Ans. The class whose instances are to be serialized should implement an interface serializable. Then you pass the instance to the ObjectOutputStream which is connected to a fileoutputstream. This will save the object to a file.

227. Which methods of Serializable interface should I implement?

Ans. The serializable interface is an empty interface, it does not contain any methods. So we do not implement any method.

228. How can I customize the seralization process? That is, how can one have a control over the serialization process?

Ans. Yes it is possible to have control over serialization process. The class should implement Externalizable interface. This interface contains two methods, namely readExternal and writeExternal. You should implement these methods and write the logic for customizing the serialization process.

229. What is the common usage of serialization?

Ans. Whenever an object is to be sent over the network, objects need to be serialized. Moreover, if the state of an object is to be saved, objects need to be serialized.

230. What is Externalizable interface?

Ans. Externalizable is an interface which contains two methods readExternal and writeExternal. These methods give you a control over the serialization mechanism. Thus if your class implements this interface, you can customize the serialization process by implementing these methods.

231. When you serialize an object, what happens to the object references included in the object?

Ans. The serialization mechanism generates an object graph for serialization. Thus it determines whether the included object references are serializable or not. This is a recursive process. Thus when an object is serialized, all the included objects are also serialized alongwith the original object.

232. What one should take care of while serializing the object?

Ans. One should make sure that all the included objects are also serializable. If any of the objects is not serializable, then it throws a NotSerializableException.

233. What happens to the static fields of a class during serialization?

Ans. There are three exceptions in which serialization does not necessarily read and write to the stream. These are

1. Serialization ignores static fields, because they are not part of any particular state.
2. Base class fields are only handled if the base class itself is serializable.
3. Transient fields.

234. Does Java provide any construct to find out the size of an object?

Ans. No there is not size of operator in Java. So there is not direct way to determine the size of an object directly in Java.

235. Give a simplest way to find out the time a method takes for execution without using any profiling tool?

Ans. Read the system time just before the method is invoked and immediately after method returns. Take the time difference, which will give you the time taken by a method for execution.

To put it in code...

```
long start = System.currentTimeMillis ();
method ();
long end = System.currentTimeMillis ();
System.out.println ("Time taken for execution is " + (end - start));
```

Remember that if the time taken for execution is too small, it might show that it is taking zero milliseconds for execution. Try it on a method which is big enough, in the sense the one which is doing considerable amount of processing.

236. What are wrapper classes?

Ans. Java provides specialized classes corresponding to each of the primitive data types. These are called wrapper classes. For example, they are Integer, Character, Double, etc.

237. Why do we need wrapper classes?

Ans. It is sometimes easier to deal with primitives as objects. Moreover, most of the collection classes store objects and not primitive data types. And also the wrapper classes provide many utility methods also. Because of these reasons we need wrapper classes. And since we create instances of these classes we can store them in any of the collection classes and pass them around as a collection. Also we can pass them around as method parameters where a method expects an object.

238. What are checked exceptions?

Ans. Checked exception are those which the Java compiler forces you to catch, e.g. IOException are checked Exceptions.

239. What are runtime exceptions?

Ans. Runtime exceptions are those exceptions that are thrown at runtime because of either wrong input data or because of wrong business logic, etc. These are not checked by the compiler at compile time.

240. What is the difference between error and an exception?

Ans. An error is an irrecoverable condition occurring at runtime. Such as OutOfMemory error. These JVM errors and you cannot repair them at runtime. While exceptions are conditions that occur because of bad input, etc, e.g. FileNotFoundException will be thrown if the specified file does not exist. Or a NullPointerException will take place if you try using a null reference. In most of the cases it is possible to recover from an exception (probably by giving user a feedback for entering proper values, etc.).

241. How to create custom exceptions?

Ans. Your class should extend class Exception, or some more specific type thereof.

242. If I want an object of my class to be thrown as an exception object, what should I do?

Ans. The class should extend from Exception class. Or you can extend your class from some more precise exception type also.

243. If my class already extends from some other class what should I do if I want an instance of my class to be thrown as an exception object?

Ans. One cannot do anything in this scenario. Because Java does not allow multiple inheritance and does not provide any exception interface as well.

244. How does an exception permeate through the code?

Ans. An unhandled exception moves up the method stack in search of a matching when an exception is thrown from a code which is wrapped in a try block followed by one or more catch blocks, a search is made for matching catch block. If a matching type is found then that block will be invoked. If a matching type is not found then the exception moves up the method stack and reaches the caller method. Same procedure is repeated if the caller method is included in a try catch block. This process continues until a catch block handling the appropriate type of exception is found. If it does not find such a block, then finally the program terminates.

245. What are the different ways to handle exceptions?

Ans. There are two ways to handle exceptions,

1. By wrapping the desired code in a try block followed by a catch block to catch the exceptions and
2. List the desired exceptions in the throws clause of the method and let the caller of the method handle those exceptions.

246. What is the basic difference between the two approaches to exception handling.

1. **try catch block and**
2. **specifying the candidate exceptions in the throws clause?**

When should you use which approach?

Ans. In the first approach as a programmer of the method, you yourself are dealing with the exception. This is fine if you are in a best position to decide should be done in case of an exception. Whereas if it is not the responsibility of the method to deal with its own exceptions, then do not use this approach. In this case use the second approach. In the second approach we are forcing the caller of the method to catch the exceptions, that the method is likely to throw. This is often the approach library creators use. They list the exception in the throws clause and we must catch them. You will find the same approach throughout the java libraries we use.

247. Is it necessary that each try block must be followed by a catch block?

Ans. It is not necessary that each try block must be followed by a catch block. It should be followed by either a catch block OR a finally block. And whatever exceptions are likely to be thrown should be declared in the throws clause of the method.

248. If I write return at the end of the try block, will the finally block still execute?

Ans. Yes even if you write return as the last statement in the try block and no exception occurs, the finally block will execute and then the control return.

249. If I write System.exit (0); at the end of the try block, will the finally block still execute?

Ans. No in this case the finally block will not execute because when you say System.exit (0); the control immediately goes out of the program, and thus finally never executes.

250. How are Observer and Observable used?

Ans. Objects that subclass the Observable class maintain a list of observers. When an Observable object is updated, it invokes the update() method of each of its observers to notify the observers that it has changed state. The Observer interface is implemented by objects that observe Observable objects.

251. What is synchronization and why is it important?

Ans. With respect to multithreading, synchronization is the capability to control the access of multiple threads to shared resources. Without synchronization, it is possible for one thread to modify a shared object while another thread is in the process of using or updating that object's value. This often leads to significant errors.

252. How does Java handle integer overflows and underflows?

Ans. It uses those low order bytes of the result that can fit into the size of the type allowed by the operation.

253. Does garbage collection guarantee that a program will not run out of memory?

Ans. Garbage collection does not guarantee that a program will not run out of memory. It is possible for programs to use up memory resources faster than they are garbage collected. It is also possible for programs to create objects that are not subject to garbage collection.

254. What is the difference between pre-emptive scheduling and time slicing?

Ans. Under preemptive scheduling, the highest priority task executes until it enters the waiting or dead states or a higher priority task comes into existence. Under time slicing, a task executes for a predefined slice of time and then reenters the pool of ready tasks. The scheduler then determines which task should execute next, based on priority and other factors.

255. When a thread is created and started, what is its initial state?

Ans. A thread is in the ready state after it has been created and started.

256. What is the purpose of finalization?

Ans. The purpose of finalization is to give an unreachable object the opportunity to perform any cleanup processing before the object is garbage collected.

257. What is the Locale class?

Ans. The Locale class is used to tailor program output to the conventions of a particular geographic, political, or cultural region.

258. What is the difference between a while statement and a do statement?

Ans. A while statement checks at the beginning of a loop to see whether the next loop iteration should occur. A do statement checks at the end of a loop to see whether the next iteration of a loop should occur. The do statement will always execute the body of a loop at least once.

259. What is the difference between static and non-static variables?

Ans. A static variable is associated with the class as a whole rather than with specific instances of a class. Non-static variables take on unique values with each object instance.

260. How are this() and super() used with constructors?

Ans. This() is used to invoke a constructor of the same class. super() is used to invoke a superclass constructor.

261. What are synchronized methods and synchronized statements?

Ans. Synchronized methods are methods that are used to control access to an object. A thread only executes a synchronized method after it has acquired the lock for the method's object or class. Synchronized statements are similar to synchronized methods. A synchronized statement can only be executed after a thread has acquired the lock for the object or class referenced in the synchronized statement.

262. What is daemon thread and which method is used to create the daemon thread?

Ans. Daemon thread is a low priority thread which runs intermittently in the back-ground doing the garbage collection operation for the java runtime system. setDaemon method is used to create a daemon thread.

263. Can applets communicate with each other?

Ans. At this point of time applets may communicate with other applets running in the same virtual machine. If the applets are of the same class, they can communicate via shared static variables. If the applets are of different classes, then each will need a reference to the same class with static variables. In any case the basic idea is to pass the information back and forth through a static variable.

An applet can also get references to all other applets on the same page using the getApplets() method of java.applet.Applet Context. Once you get the reference to an applet, you can communicate with it by using its public members.

It is conceivable to have applets in different virtual machines that talk to a server somewhere on the Internet and store any data that needs to be serialized there. Then, when another applet needs this data, it could connect to this same server. Implementing this is non-trivial.

264. What are the steps in the JDBC connection?

Ans. While making a JDBC connection we go through the following steps :

Step 1: Register the database driver by using:

Class.forName(\" driver classs for that specific database\");

Step 2: Now create a database connection using:

Connection con =

DriverManager.getConnection(url,username, password);

Step 3: Now Create a query using:
Statement stmt = Connection.Statement (\"select * from TABLE NAME\");
Step 4: Exceute the query:
stmt.exceuteUpdate();

265. How does a try statement determine which catch clause should be used to handle an exception?

Ans. When an exception is thrown within the body of a try statement, the catch clauses of the try statement are examined in the order in which they appear. The first catch clause that is capable of handling the exception is executed. The remaining catch clauses are ignored.

266. Can an unreachable object become reachable again?

Ans. An unreachable object may become reachable again. This can happen when the object's finalize() method is invoked and the object performs an operation which causes it to become accessible to reachable objects.

267. What method must be implemented by all threads?

Ans. All tasks must implement the run() method, whether they are a subclass of Thread or implement the Runnable interface.

268. What are synchronized methods and synchronized statements?

Ans. Synchronized methods are methods that are used to control access to an object. A thread only executes a synchronized method after it has acquired the lock for the method's object or class. Synchronized statements are similar to synchronized methods. A synchronized statement can only be executed after a thread has acquired the lock for the object or class referenced in the synchronized statement.

269. What is Externalizable?

Ans. Externalizable is an Interface that extends Serializable Interface. And sends data into Streams in Compressed Format. It has two methods, writeExternal(ObjectOuput out) and readExternal(ObjectInput in).

270. What modifiers are allowed for methods in an Interface?

Ans. Only public and abstract modifiers are allowed for methods in interfaces.

271. What are some alternatives to inheritance?

Ans. Delegation is an alternative to inheritance. Delegation means that you include an instance of another class as an instance variable, and forward messages to the instance. It is often safer than inheritance because it forces you to think about each message you forward, because the instance is of a known class, rather than a new class, and because it doesn't force you to accept all the methods of the superclass: You can provide only the methods that really make sense. On the other hand, it makes you write more code, and it is harder to re-use (because it is not a subclass).

272. What does it mean that a method or field is "static"?

Ans. Static variables and methods are instantiated only once per class. In other words, they are class variables, not instance variables. If you change the value of a static variable in a particular object, the value of that variable changes for all instances of that class.

Static methods can be referenced with the name of the class rather than the name of a particular object of the class (though that works too). That's how library methods like System.out.println() work out is a static field in the java.lang.System class.

273. What is the difference between preemptive scheduling and time slicing?

Ans. Under preemptive scheduling, the highest priority task executes until it enters the waiting or dead states or a higher priority task comes into existence. Under time slicing, a task executes for a predefined slice of time and then reenters the pool of ready tasks. The scheduler then determines which task should execute next, based on priority and other factors.

274. What is the catch or declare rule for method declarations?

Ans. If a checked exception may be thrown within the body of a method, the method must either catch the exception or declare it in its throws clause.

275. Is Empty .java file a valid source file?

Ans. Yes, an empty .java file is a perfectly valid source file.

276. Can a .java file contain more than one java classes?

Ans. Yes, a .java file contain more than one java classes, provided at the most one of them is a public class.

277. Is String a primitive data type in Java?

Ans. No String is not a primitive data type in Java, even though it is one of the most extensively used object. Strings in Java are instances of String class defined in java.lang package.

278. Is main a keyword in Java?

Ans. No, main is not a keyword in Java.

279. Is next a keyword in Java?

Ans. No, next is not a keyword.

280. Is delete a keyword in Java?

Ans. No, delete is not a keyword in Java. Java does not make use of explicit destructors the way C++ does.

281. Is exit a keyword in Java?

Ans. No. To exit a program explicitly you use exit method in System object.

282. What happens if you do not initialize an instance variable of any of the primitive types in Java?

Ans. Java by default initializes it to the default value for that primitive type. Thus an int will be initialized to 0, a boolean will be initialized to false.

283. What will be the initial value of an object reference which is defined as an instance variable?

Ans. The object references are all initialized to null in Java. However, in order to do anything useful with these references, you must set them to a valid object, else you will get NullPointerExceptions everywhere you try to use such default initialized references.

284. What are the different scopes for Java variables?

Ans. The scope of a Java variable is determined by the context in which the variable is declared. Thus a java variable can have one of the three scopes at any given point in time.

1. Instance : These are typical object level variables, they are initialized to default values at the time of creation of object, and remain accessible as long as the object accessible.
2. Local : These are the variables that are defined within a method. They remain accessible only during the course of method excecution. When the method finishes execution, these variables fall out of scope.
3. Static: These are the class level variables. They are initialized when the class is loaded in JVM for the first time and remain there as long as the class remains loaded. They are not tied to any particular object instance.

285. What is the default value of the local variables?

Ans. The local variables are not initialized to any default value, neither primitives nor object references. If you try to use these variables without initializing them explicitly, the java compiler will not compile the code. It will complain about the local varaible not being initialized.

286. How many objects are created in the following piece of code?

```
MyClass c1, c2, c3;
c1 = new MyClass ();
c3 = new MyClass ();
```

Ans. Only 2 objects are created, c1 and c3. The reference c2 is only declared and not initialized.

287. Can a public class MyClass be defined in a source file named YourClass.java?

Ans. No the source file name, if it contains a public class, must be the same as the public class name itself with a .java extension.

288. Can main method be declared final?

Ans. Yes, the main method can be declared final, in addition to being public static.

[Received fromSandesh Sadhale]

289. What will be the output of the following statement?

```
System.out.println ("1" + 3);
```

Ans. It will print 13.

290. What will be the default values of all the elements of an array defined as an instance variable?

Ans. If the array is an array of primitive types, then all the elements of the array will be initialized to the default value corresponding to that primitive type, e.g. all the elements of an array of int will be initialized to 0, while that of boolean type will be initialized to false. Whereas if the array is an array of references (of any type), all the elements will be initialized to null.

291. What is the Collections API?

Ans. The Collections API is a set of classes and interfaces that support operations on collections of objects.

292. What is the List interface?

Ans. The List interface provides support for ordered collections of objects.

293. What is the Vector class?

Ans. The Vector class provides the capability to implement a growable array of objects.

294. What is an Iterator interface?

Ans. The Iterator interface is used to step through the elements of a Collection.

295. Which java.util classes and interfaces support event handling?

Ans. The EventObject class and the EventListener interface support event processing.

296. What is the GregorianCalendar class?

Ans. The GregorianCalendar provides support for traditional Western calendars.

297. What is the Locale class?

Ans. The Locale class is used to tailor program output to the conventions of a particular geographic, political, or cultural region.

298. What is the SimpleTimeZone class?

Ans. The SimpleTimeZone class provides support for a Gregorian calendar.

299. What is the Map interface?

Ans. The Map interface replaces the JDK 1.1 Dictionary class and is used associate keys with values.

300. What is the highest-level event class of the event-delegation model?

Ans. The java.util.EventObject class is the highest-level class in the event-delegation class hierarchy.

301. What is the Collection interface?

Ans. The Collection interface provides support for the implementation of a mathematical bag—an unordered collection of objects that may contain duplicates.

302. What is the Set interface?

Ans. The Set interface provides methods for accessing the elements of a finite mathematical set. Sets do not allow duplicate elements.

303. What is the typical use of Hashtable?

Ans. Whenever a program wants to store a key value pair, one can use Hashtable.

304. I am trying to store an object using a key in a Hashtable. And some other object already exists in that location, then what will happen? The existing object will be overwritten? Or the new object will be stored elsewhere?

Ans. The existing object will be overwritten and thus it will be lost.

305. What is the difference between the size and capacity of a Vector?

Ans. The size is the number of elements actually stored in the vector, while capacity is the maximum number of elements it can store at a given instance of time.

306. Can a vector contain heterogenous objects?

Ans. Yes a Vector can contain heterogenous objects. Because a Vector stores everything in terms of Object.

307. Can a ArrayList contain heterogenous objects?

Ans. Yes a ArrayList can contain heterogenous objects. Because a ArrayList stores everything in terms of Object.

308. What is an enumeration?

Ans. An enumeration is an interface containing methods for accessing the underlying data structure from which the enumeration is obtained. It is a construct which collection classes return when you request a collection of all the objects stored in the collection. It allows sequential access to all the elements stored in the collection.

309. Considering the basic properties of Vector and ArrayList, where will you use Vector and where will you use ArrayList?

Ans. The basic difference between a Vector and an ArrayList is that, vector is synchronized while ArrayList is not. Thus whenever there is a possibility of multiple threads accessing the same instance, one should use Vector. While if no multiple threads are going to access the same instance then use ArrayList. Nonsynchronized data structure will give better performance than the synchronized one.

310. What is a output comment?

Ans. A comment that is sent to the client in the viewable page source.The JSP engine handles an output comment as uninterpreted HTML text, returning the comment in the HTML output sent to the client. You can see the comment by viewing the page source from your Web browser.

311. What is a Hidden Comment?

Ans. A comments that documents the JSP page but is not sent to the client. The JSP engine ignores a hidden comment, and does not process any code within hidden comment tags. A hidden comment is not sent to the client, either in the displayed JSP page or the HTML page source. The hidden comment is useful when you want to hide or "comment out" part of your JSP page.

You can use any characters in the body of the comment except the closing —%> combination. If you need to use —%> in your comment, you can escape it by typing —%\>.

JSP Syntax

<%— comment —%>

Examples

```
<%@ page language="java" %>
<html>
<head><title>A Hidden Comment </
title></head>
<body>
<%— This comment will not be visible to
the colent in the page source —%>
</body>
</html>
```

312. What is an Expression?

Ans. An expression tag contains a scripting language expression that is evaluated, converted to a String, and inserted where the expression appears in the JSP file. Because the value of an expression is converted to a String, you can use an expression within text in a JSP file. Like

```
<%= someexpression %>
<%= (new java.util.Date()).toLocaleString()
%>
```

You cannot use a semicolon to end an expression

313. What is a Declaration?

Ans. A declaration declares one or more variables or methods for use later in the JSP source file.

A declaration must contain at least one complete declarative statement. You can declare any number of variables or methods within one declaration tag, as long as they are separated by semicolons. The declaration must be valid in the scripting language used in the JSP file.

```
<%! somedeclarations %>
<%! int i = 0; %>
<%! int a, b, c; %>
```

314. What is a Scriptlet?

Ans. A scriptlet can contain any number of language statements, variable or method declarations, or expressions that are valid in the page scripting language.Within scriptlet tags, you can

1. Declare variables or methods to use later in the file (see also Declaration).
2. Write expressions valid in the page scripting language (see also Expression).
3. Use any of the JSP implicit objects or any object declared with a <jsp:useBean> tag.

You must write plain text, HTML-encoded text, or other JSP tags outside the scriptlet. Scriptlets are executed at request time, when the JSP engine processes the client request. If the scriptlet produces output, the output is stored in the out object, from which you can display it.

315. What are implicit objects? List them?

Ans. Certain objects that are available for the use in JSP documents without being declared first. These objects are parsed by the JSP engine and inserted into the generated servlet. The implicit objects relisted below:

- request
- response
- pageContext
- session
- application
- out
- config
- page
- exception

316. Difference between forward and sendRedirect?

Ans. When you invoke a forward request, the request is sent to another resource on the server, without the client being informed that a different resource is going to process the request. This process occurs completely within the web container. When a sendRedirtect method is invoked, it causes the web container to return to the browser indicating that a new URL should be requested. Because the browser issues a completely new request any object that are stored as request attributes before the redirect occurs will be lost. This extra round trip a redirect is slower than forward.

317. What are the different scope values for the <jsp:useBean>?

Ans. The different scope values for <jsp:useBean> are

1. page
2. request
3. session
4. application

318. Explain the lifecycle methods in JSP.

Ans. The generated servlet class for a JSP page implements the HttpJspPage interface of the javax.servlet.jsp package. The HttpJspPage interface extends the JspPage interface which inturn extends the Servlet interface of the javax.servlet package. The generated servlet class thus implements all the methods of these three interfaces. The JspPage interface declares only two mehtods—jspInit() and jspDestroy() that must be implemented by all JSP pages regardless of the client-server protocol. However, the JSP specification has provided the HttpJspPage interface specifically for the JSp pages serving HTTP requests. This interface declares one method _jspService().

The jspInit(): The container calls the jspInit() to initialize the servlet instance. It is called before any other method, and is called only once for a servlet instance.

The _jspservice(): The container calls the _jspservice() for each request, passing it the request and the response objects.

The jspDestroy(): The container calls this when it decides to take the instance out of service. It is the last method called in the servlet instance.

319. How do I prevent the output of my JSP or Servlet pages from being cached by the browser?

Ans. You will need to set the appropriate HTTP header attributes to prevent the dynamic content output by the JSP page from being cached by the browser. Just execute the following scriptlet at the beginning of your JSP pages to prevent them from being cached at the browser. You need both the statements to take care of some of the older browser versions.

```
<%
response.setHeader("Cache-Control","no-store"); //HTTP 1.1
response.setHeader("Pragma\",”no-cache"); //HTTP 1.0
response.setDateHeader ("Expires", 0); //prevents caching at the proxy server
%>
```

320. How does JSP handle runtime exceptions?

Ans. You can use the errorPage attribute of the page directive to have uncaught run-time exceptions automatically forwarded to an error processing page. For example: <%@ page errorPage=\"error.jsp\" %> redirects the browser to the JSP page error.jsp if an uncaught exception is encountered during request processing. Within error.jsp, if you indicate that it is an error-processing page, via the directive: <%@ page isErrorPage=\"true\" %> Throwable object describing the exception may be accessed within the error page via the exception implicit object. *Note*: You must always use a relative URL as the value for the errorPage attribute.

321. How can I implement a thread-safe JSP page? What are the advantages and disadvantages of using it?

Ans. You can make your JSPs thread-safe by having them implement the SingleThreadModel interface. This is done by adding the directive <%@ page isThreadSafe="false" %> within your JSP page. With this, instead of a single instance of the servlet generated for your JSP page loaded in memory, you will have N instances of the servlet loaded and initialized, with the service method of each instance effectively synchronized. You can typically control the number of instances (N) that are instantiated for all servlets implementing SingleThreadModel through the admin screen for your JSP engine. More importantly, avoid using the tag for variables. If you do use this tag, then you should set isThreadSafe to true, as mentioned above. Otherwise, all requests to that page will access those variables, causing a nasty race condition. SingleThreadModel is not recommended for normal use. There are many pitfalls, including the example above of not being able to use <%! %>. You should try really hard to make them thread-safe the old-fashioned way: by making them thread-safe.

322. How do I use a scriptlet to initialize a newly instantiated bean?

Ans. A jsp:useBean action may optionally have a body. If the body is specified, its contents will be automatically invoked when the specified bean is instantiated. Typically, the body will contain scriptlets or jsp:setProperty tags to initialize the newly instantiated bean, although you are not restricted to using those alone.

The following example shows the "today" property of the Foo bean initialized to the current date when it is instantiated. Note that here, we make use of a JSP expression within the jsp:setProperty action.

```
<jsp:useBean id="foo"
class="com.Bar.Foo" >
<jsp:setProperty name="foo"
property="today"
value="<%=java.text.DateFormat.getDateInstance().
format(new java.util.Date()) %>" / >
<%— scriptlets calling bean setter methods
go here —%>
</jsp:useBean >
```

323. How can I prevent the word "null" from appearing in my HTML input text fields when I populate them with a resultset that has null values?

Ans. You could make a simple wrapper function, like

```
<%!
String blanknull(String s) {
return (s == null) ? \"\" : s;
}
%>
```

then use it inside your JSP form, like

```
<input type="text" name="lastName"
value="<%=blanknull(lastName)% >" >
```

324. What's a better approach for enabling thread-safe servlets and JSPs, singlethreadmodel interface or synchronization?

Ans. Although the SingleThreadModel technique is easy to use, and works well for low volume sites, it does not scale well. If you anticipate your users to increase in the future, you may be better off implementing explicit synchronization for your shared data. The key, however, is to effectively minimize the amount of code that is synchronzied so that you take maximum advantage of multithreading.

Also, note that SingleThreadModel is pretty resource intensive from the server\'s perspective. The most serious issue, however, is when the number of concurrent requests exhaust the servlet instance pool. In that case, all the unserviced requests are queued until something becomes free which results in poor performance. Since the usage is non-deterministic, it may not help much even if you did add more memory and increased the size of the instance pool.

325. How can I enable session tracking for JSP pages if the browser has disabled cookies?

Ans. We know that session tracking uses cookies by default to associate a session identifier with a unique user. If the browser does not support cookies, or if cookies are disabled, you can still enable session tracking using URL rewriting. URL rewriting essentially includes the session ID within the link itself as a name/value pair. However, for this to be effective, you need to append the session ID for each and every link that is part of your servlet response. Adding the session ID to a link is greatly simplified by means of a couple of methods: response.encodeURL() associates a session ID with a given URL, and if you are using redirection, response.encodeRedirectURL() can be used by giving the redirected URL as input. Both encodeURL() and encodeRedirectedURL() first determine whether cookies are supported by the browser; if so, the input URL is returned unchanged since the session ID will be persisted as a cookie.

Consider the following example, in which two JSP files, say hello1.jsp and hello2.jsp, interact with each other. Basically, we create a new session within hello1.jsp and place an object within this session. The user can then traverse to hello2.jsp by clicking on the link present within the page. Within hello2.jsp, we simply extract the object that was earlier placed in the session and display its contents. Notice that we invoke the encodeURL() within hello1.jsp on the link used to invoke hello2.jsp; if cookies are disabled, the session ID is automatically appended to the URL, allowing hello2.jsp to still retrieve the session object. Try this example first with cookies enabled. Then disable cookie support, restart the brower, and try again. Each time you should see the maintenance of the session across pages. Do note that to get this example to work with cookies disabled at the browser, your JSP engine has to support URL rewriting.

hello1.jsp

```
<%@ page session=\"true\" %>
<%
Integer num = new Integer(100);
session.putValue("num",num);
String url =response.encodeURL
("hello2.jsp");
%>
<a href=\'<%=url%>\'>hello2.jsp</a>
```

hello2.jsp

```
<%@ page session="true" %>
<%
Integer i= (Integer )session.getValue
("num");
out.println("Num value in session is " +
i.intValue());
%>
```

326. What is the difference b/w variable declared inside a declaration part and variable declared in scriplet part?

Ans. Variable declared inside declaration part is treated as a global variable. This means after convertion jsp file into servlet that variable will be in outside of service method or it will be declared as instance variable. And the scope is available to complete jsp and to complete in the converted servlet class, whereas if you declare a variable inside a scriplet that variable will be declared inside a service method and the scope is within the service method.

327. Is there a way to execute a JSP from the commandline or from my own application?

Ans. There is a little tool called JSPExecutor that allows you to do just that. The developers (Hendrik Schreiber <hs@webapp.de> & Peter Rossbach <pr@webapp.de>) aim was not to write a full blown servlet engine, but to provide means to use JSP for generating source code or reports. Therefore most HTTP-specific features (headers, sessions, etc) are not implemented, i.e. no reponseline or header is generated. Nevertheless you can use it to precompile JSP for your website.

328. Explain the lifecycle methods of a Servlet.

Ans. The javax.servlet.Servlet interface defines the three methods known as lifecycle method.

```
public void init(ServletConfig config)
throws ServletException
public void service( ServletRequest req,
ServletResponse res) throws
ServletException, IOException
public void destroy()
```

First the servlet is constructed, then initialized wih the init() method.

Any request from client are handled initially by the service() method before delegating to the doXxx() methods in the case of HttpServlet.

The servlet is removed from service, destroyed with the destroy() method, then garbage collected and finalized.

329. What is the difference between the getRequestDispatcher(String path) method of javax.servlet.ServletRequest interface and javax.servlet.ServletContext interface?

Ans. The getRequestDispatcher(String path) method of javax.servlet.ServletRequest interface accepts parameter the path to the resource to be included or forwarded to, which can be relative to the request of the calling servlet. If the path begins with a "/" it is interpreted as relative to the current context root.

The getRequestDispatcher(String path) method of javax.servlet.ServletContext interface cannot accepts relative paths. All path must start with a "/" and are interpreted as relative to current context root.

330. Explain the directory structure of a web application.

Ans. The directory structure of a web application consists of two parts.

A private directory called WEB-INF

A public resource directory which contains public resource folder.

WEB-INF folder consists of

1. web.xml
2. classes directory
3. lib directory

331. What are the common mechanisms used for session tracking?

Ans. Cookies

SSL sessions

URL- rewriting

332. Explain ServletContext.

Ans. ServletContext interface is a window for a servlet to view its environment. A servlet can use this interface to get information such as initialization parameters for the web application or servlet container's version. Every web application has one and only one ServletContext and is accessible to all active resource of that application.

333. What is preinitialization of a servlet?

Ans. A container does not initialize the servlets as soon as it starts up, it initializes a servlet when it receives a request for that servlet first time. This is called lazy loading. The servlet specification defines the <load-on-startup> element, which can be specified in the deployment descriptor to make the servlet container load and initialize the servlet as soon as it starts up. The process of loading a servlet before any request comes in is called preloading or preinitializing a servlet.

334. What is the difference between doGet() and doPost()?

Ans. A doGet() method is limited with 2k of data to be sent, and doPost() method doesn't have this limitation. A request string for doGet() looks like the following:

http://www.allapplabs.com/svt1?p1=v1&p2=v2&...&pN=vN

doPost() method call doesn't need a long text tail after a servlet name in a request. All parameters are stored in a request itself, not in a request string, and it's impossible to guess the data transmitted to a servlet only looking at a request string.

335. What is the difference between HttpServlet and GenericServlet?

Ans. A GenericServlet has a service() method aimed to handle requests. HttpServlet extends GenericServlet and adds support for doGet(), doPost(), doHead() methods (HTTP 1.0) plus doPut(), doOptions(), doDelete(), doTrace() methods (HTTP 1.1).

Both these classes are abstract.

336. What is the difference between ServletContext and ServletConfig?

Ans. ServletContext: Defines a set of methods that a servlet uses to communicate with its servlet container, for example, to get the MIME type of a file, dispatch requests, or write to a log file.The ServletContext object is contained within the ServletConfig object, which the Web server provides the servlet when the servlet is initialized.

ServletConfig: The object created after a servlet is instantiated and its default constructor is read. It is created to pass initialization information to the servlet.

337. What is JMS?

Ans. JMS is an acronym used for Java Messaging Service. It is Java's answer to creating software using asynchronous messaging. It is one of the official specifications of the J2EE technologies and is a key technology.

338. How JMS is different from RPC?

Ans. In RPC the method invoker waits for the method to finish execution and return the control back to the invoker. Thus it is completely synchronous in nature. While in JMS the message sender just sends the message to the destination and continues it's own processing. The sender does not wait for the receiver to respond. This is asynchronous behavior.

339. What are the advantages of JMS?

Ans. JMS is asynchronous in nature. Thus not all the pieces need to be up all the time for the application to function as a whole. Even if the receiver is down, the MOM will store the messages on its behalf and will send them once it comes back up. Thus at least a part of application can still function as there is no blocking.

340. Are you aware of any major JMS products available in the market?

Ans. IBM's MQ Series is one of the most popular product used as Message Oriented Middleware. Some of the other products are SonicMQ, iBus etc. All the J2EE compliant application servers come built with their own implementation of JMS.

341. What are the different types of messages available in the JMS API?

Ans. Message, TextMessage, BytesMessage, StreamMessage, ObjectMessage, MapMessage are the different messages available in the JMS API.

342. What are the different messaging paradigms JMS supports?

Ans. Publish and Subscribe, i.e. pub/suc and point-to-point, i.e. p2p.

343. What is the difference between topic and queue?

Ans. A topic is typically used for one to many messaging, i.e. it supports publish subscribe model of messaging. While queue is used for one-to-one messaging, i.e. it supports point-to-point Messaging.

344. What is the role of JMS in enterprise solution development?

Ans. JMS is typically used in the following scenarios

1. Enterprise Application Integration: Where a legacy application is integrated with a new application via messaging.
2. B2B or Business to Business: Businesses can interact with each other via messaging because JMS allows organizations to cooperate without tightly coupling their business systems.
3. Geographically dispersed units: JMS can ensure safe exchange of data amongst the geographically dispersed units of an organization.
4. One to many applications: The applications that need to push data in packet to huge number of clients in a one-to-many fashion are good candidates for the use JMS. Typical such applications are Auction Sites, Stock Quote Services, etc.

345. What is the use of Message object?

Ans. Message is a lightweight message having only header and properties and no payload. Thus if the receivers are to be notified about an event, and no data needs to be exchanged then using Message can be very efficient.

346. What is the basic difference between Publish Subscribe model and P2P model?

Ans. Publish Subscribe model is typically used in one-to-many situation. It is unreliable but very fast. P2P model is used in one-to-one situation. It is highly reliable.

347. What is the use of BytesMessage?

Ans. BytesMessage contains an array of primitive bytes in its payload. Thus it can be used for transfer of data between two applications in their native format which may not be compatible with other Message types. It is also useful where JMS is used purely as a transport between two systems and the message payload is opaque to the JMS client. Whenever you store any primitive type, it is converted into its byte representation and then stored in the payload. There is no boundary line between the different data types stored. Thus you can even read a long and short. This would result in erroneous data and hence it is advisable that the payload be read in the same order and using the same type in which it was created by the sender.

348. What is the use of StreamMessage?

Ans. StreamMessage carries a stream of Java primitive types as its payload. It contains some convenient methods for reading the data stored in the payload. However, StreamMessage prevents reading a long value as short, something that is allowed in case of BytesMessage. This is so because the StreamMessage also writes the type information along with the value of the primitive type and enforces a set of strict conversion rules which actually prevents reading of one primitive type as another.

349. What is the use of TextMessage?

Ans. TextMessage contains instance of java.lang.String as its payload. Thus it is very useful for exchanging textual data. It can also be used for exchanging complex character data such as an XML document.

350. What is the use of ObjectMessage?

Ans. ObjectMessage contains a Serializable java object as its payload. Thus it allows exchange of Java objects between applications. This in itself mandates that both the applications be Java applications. The consumer of the message must typecast the object received to its appropriate type. Thus the consumer should beforehand know the actual type of the object sent by the sender. Wrong type casting would result in ClassCastException. Moreover, the class definition of the object set in the payload should be available on both the machine, the sender as well as the consumer. If the class definition is not available in the consumer machine, an attempt to type cast would result in ClassNotFoundException. Some of the MOMs might support dynamic loading of the desired class over the network, but the JMS specification does not mandate this behavior and would be a value added service if provided by your vendor. And relying on any such vendor specific functionality would hamper the portability of your application. Most of the time the class need to be put in the classpath of both, the sender and the consumer, manually by the developer.

351. What is the use of MapMessage?

Ans. A MapMessage carries name-value pair as its payload. Thus its payload is similar to the java.util.Properties object of Java. The values can be Java primitives or their wrappers.

352. What is the difference between BytesMessage and StreamMessage?

Ans. BytesMessage stores the primitive data types by converting them to their byte representation. Thus the message is one contiguous stream of bytes. While the StreamMessage maintains a boundary between the different data types stored because it also stores the type information along with the value of the primitive being stored. BytesMessage allows data to be read using any type. Thus even if your payload contains a long value, you can invoke a method to read a short and it will return you something. It will not give you a semantically correct data but the call will succeed in reading the first two bytes of data. This is strictly prohibited in the StreamMessage. It maintains the type information of the data being stored and enforces strict conversion rules on the data being read.

353. What is point-to-point messaging?

Ans. With point-to-point message passing the sending application/client establishes a named message queue in the JMS broker/server and sends messages to this queue. The receiving client registers with the broker to receive messages posted to this queue. There is a one-to-one relationship between the sending and receiving clients.

354. Can two different JMS services talk to each other? For instance, if A and B are two different JMS providers, can Provider A send messages directly to Provider B? If not, then can a subscriber to Provider A act as a publisher to Provider B?

Ans. The answers are no to the first question and yes to the second. The JMS specification does not require that one JMS provider be able to send messages directly to another provider. However, the specification does require that a JMS client must be able to accept a message created by a different JMS provider, so a message received by a subscriber to Provider A can then be published to Provider B. One caveat is that the publisher to Provider B is not required to handle a JMSReplyTo header that refers to a destination that is specific to Provider A.

355. What is the advantage of persistent message delivery compared to non-persistent delivery?

Ans. If the JMS server experiences a failure, for example, a power outage, any message that it is holding in primary storage potentially could be lost. With persistent storage, the JMS server logs every message to secondary storage. (The logging occurs on the front end, that is, as part of handling the send operation from the message producing client.) The logged message is removed from secondary storage only after it has been successfully delivered to all consuming clients.

356. Give an example of using the publish/subscribe model.

Ans. JMS can be used to broadcast shutdown messages to clients connected to the Weblogic server on a module wise basis. If an application has six modules, each module behaves like a subscriber to a named topic on the server.

357. Why doesn't the JMS API provide end-to-end synchronous message delivery and notification of delivery?

Ans. Some messaging systems provide synchronous delivery to destinations as a mechanism for implementing reliable applications. Some systems provide clients with various forms of delivery notification so that the clients can detect dropped or ignored messages. This is not the model defined by the JMS API.

JMS API messaging provides guaranteed delivery via the once-and-only-once delivery semantics of PERSISTENT messages. In addition, message consumers can ensure reliable processing of messages by using either CLIENT_ACKNOWLEDGE mode or transacted sessions. This achieves reliable delivery with minimum synchronization and is the enterprise messaging model most vendors and developers prefer.

The JMS API does not define a schema of systems messages (such as delivery notifications). If an application requires acknowledgment of message receipt, it can define an application-level acknowledgment message.

358. What are the various message types supported by JMS?

Ans. Stream Messages ? Group of Java Primitives

Map Messages ? Name Value Pairs. Name being a string& Value being a Java primitive

Text Messages ? String messages (since being widely used a separate messaging Type has been supported)

Object Messages ? Group of serializeable java object

Bytes Message ? Stream of uninterrupted bytes

359. How is a Java object message delivered to a non-java Client?

Ans. It is according to the specification that the message sent should be received in the same format. A non-java client cannot receive a message in the form of java object. The provider inbetween handles the conversion of the data type and the message is transferred to the other end.

360. What is MDB and what is the special feature of that?

Ans. MDB is Message driven bean, which very much resembles the Stateless session bean. The incoming and outgoing messages can be handled by the Message driven bean. The ability to communicate asynchronously is the special feature about the Message driven bean.

361. What are the types of messaging?

Ans. There are two kinds of messaging.

Synchronous messaging: Synchronous messaging involves a client that waits for the server to respond to a message.

Asynchronous messaging: Asynchronous messaging involves a client that does not wait for a message from the server. An event is used to trigger a message from a server.

362. What are the core JMS-related objects required for each JMS-enabled application?

Ans. Each JMS-enabled client must establish the following:

- A connection object provided by the JMS server (the message broker).
- Within a connection, one or more sessions, which provide a context for message sending and receiving.
- Within a session, either a queue or topic object representing the destination (the message staging area) within the message broker.
- Within a session, the appropriate sender or publisher or receiver or subscriber object (depending on whether the client is a message producer or consumer and uses a point-to-point or publish/subscribe strategy, respectively).

Within a session, a message object (to send or to receive).

OOPS Interview Questions

1. What are the principle concepts of OOPS?

Ans. There are four principle concepts upon which object oriented design and programming rest. They are:

- Abstraction
- Polymorphism
- Inheritance
- Encapsulation

(i.e. easily remembered as A-PIE).

2. What is abstraction?

Ans. Abstraction refers to the act of representing essential features without including the background details or explanations.

3. What is encapsulation?

Ans. Encapsulation is a technique used for hiding the properties and behaviors of an object and allowing outside access only as appropriate. It prevents other objects from directly altering or accessing the properties or methods of the encapsulated object.

4. What is the difference between abstraction and encapsulation?

Ans.
- Abstraction focuses on the outside view of an object (i.e. the interface) Encapsulation (information hiding) prevents clients from seeing its inside view, where the behavior of the abstraction is implemented.
- Abstraction solves the problem in the design side while encapsulation is the implementation.
- Encapsulation is the deliverables of abstraction. Encapsulation barely talks about grouping up your abstraction to suit the developer needs.

5. What is Inheritance?

Ans.
- Inheritance is the process by which objects of one class acquire the properties of objects of another class.
- A class that is inherited is called a superclass.
- The class that does the inheriting is called a subclass.
- Inheritance is done by using the keyword extends.
- The two most common reasons to use inheritance are:

  o To promote code reuse

  o To use polymorphism

6. What is polymorphism?

Ans. Polymorphism is briefly described as "one interface, many implementations." Polymorphism is a characteristic of being able to assign a different meaning or usage to something in different contexts specifically, to allow an entity such as a variable, a function, or an object to have more than one form.

7. How does Java implement polymorphism?

Ans. (Inheritance, overloading and overriding are used to achieve polymorphism in java).

Polymorphism manifests itself in Java in the form of multiple methods having the same name.

- In some cases, multiple methods have the same name, but different formal argument lists (overloaded methods).
- In other cases, multiple methods have the same name, same return type, and same formal argument list (overridden methods).

8. Explain the different forms of polymorphism.

Ans. There are two types of polymorphism—one is compile time polymorphism and the other is runtime polymorphism. Compile time polymorphism is method overloading. Runtime polymorphism is done using inheritance and interface.

Note: From a practical programming viewpoint, polymorphism manifests itself in three distinct forms in Java:

- Method overloading
- Method overriding through inheritance
- Method overriding through the Java interface

9. What is runtime polymorphism or dynamic method dispatch?

Ans. In Java, runtime polymorphism or dynamic method dispatch is a process in which a call to an overridden method is resolved at runtime rather than at compile-time. In this process, an overridden method is called through the reference variable of a superclass. The determination of the method to be called is based on the object being referred to by the reference variable.

10. What is dynamic binding?

Ans. Binding refers to the linking of a procedure call to the code to be executed in response to the call. Dynamic binding (also known as late binding) means that the code associated with a given procedure call is not known until the time of the call at run-time. It is associated with polymorphism and inheritance.

11. What is method overloading?

Ans. Method overloading means to have two or more methods with the same name in the same class with different arguments. The benefit of method overloading is that it allows you to implement methods that support the same semantic operation but differ by argument number or type.

Note:

- Overloaded methods MUST change the argument list
- Overloaded methods CAN change the return type
- Overloaded methods CAN change the access modifier
- Overloaded methods CAN declare new or broader checked exceptions
- A method can be overloaded in the same class or in a subclass

12. What is method overriding?

Ans. Method overriding occurs when subclass declares a method that has the same type arguments as a method declared by one of its superclass. The key benefit of overriding is the ability to define behavior that's specific to a particular subclass type.

Note:

- The overriding method cannot have a more restrictive access modifier than the method being overridden (Example: You can't override a method marked public and make it protected).
- You cannot override a method marked final.
- You cannot override a method marked static.

13. What are the differences between method overloading and method overriding?

	Overloaded method	*Overridden method*
Arguments	Must change	Must not change
Return type	Can change	Can't change except for covariant returns
Exceptions	Can change	Can reduce or eliminate. Must not throw new or broader checked exceptions
Access	Can change	Must not make more restrictive (can be less restrictive)
Invocation	Reference type determines which overloaded version is selected. Happens at compile time.	Object type determines which method is selected. Happens at runtime.

14. Can overloaded methods be override too?

Ans. Yes, derived classes still can override the overloaded methods. Polymorphism can still happen. Compiler will not binding the method calls since it is overloaded, because it might be overridden now or in the future.

15. Is it possible to override the main method?

Ans. No, because main is a static method. A static method can't be overridden in Java.

16. How to invoke a superclass version of an overridden method?

Ans. To invoke a superclass method that has been overridden in a subclass, you must either call the method directly through a superclass instance, or use the super prefix in the subclass itself. From the point of the view of the subclass, the super prefix provides an explicit reference to the superclass' implementation of the method.

```
// From subclass
super.overriddenMethod();
```

17. What is super?

Ans. super is a keyword which is used to access the method or member variables from the superclass. If a method hides one of the member variables in its superclass, the method can refer to the hidden variable through the use of the super keyword. In the same way, if a method overrides one of the methods in its superclass, the method can invoke the overridden method through the use of the super keyword.

Note:
- You can only go back one level.
- In the constructor, if you use super(), it must be the very first code, and you cannot access any this.xxx variables or methods to compute its parameters.

18. How do you prevent a method from being overridden?

To prevent a specific method from being overridden in a subclass, use the final modifier on the method declaration, which means "this is the final implementation of this method", the end of its inheritance hierarchy.

```
public final void exampleMethod() {
        // Method statements
        }
```

19. What is an interface?

Ans. An interface is a description of a set of methods that conforming implementing classes must have.

Note:
- You can't mark an interface as final.
- Interface variables must be static.
- An Interface cannot extend anything but another interfaces.

20. Can we instantiate an interface?

Ans. You can't instantiate an interface directly, but you can instantiate a class that implements an interface.

21. Can we create an object for an interface?

Ans. Yes, it is always necessary to create an object implementation for an interface. Interfaces cannot be instantiated in their own right, so you must write a class that implements the interface and fulfill all the methods defined in it.

22. Do interfaces have member variables?

Ans. Interfaces may have member variables, but these are implicitly public, static, and final in other words, interfaces can declare only constants, not instance variables that are available to all implementations and may be used as key references for method arguments as for example.

23. What modifiers are allowed for methods in an Interface?

Ans. Only public and abstract modifiers are allowed for methods in interfaces.

24. What is a marker interface?

Ans. Marker interfaces are those which do not declare any required methods, but signify their compatibility with certain operations. The java.io.Serializable interface and Cloneable are typical marker interfaces. These do not contain any methods, but classes must implement this interface in order to be serialized and de-serialized.

25. What is an abstract class?

Ans. Abstract classes are classes that contain one or more abstract methods. An abstract method is a method that is declared, but contains no implementation.

Note:
- If even a single method is abstract, the whole class must be declared abstract.
- Abstract classes may not be instantiated, and require subclasses to provide implementations for the abstract methods.
- You can't mark a class as both abstract and final.

26. Can we instantiate an abstract class?

Ans. An abstract class can never be instantiated. Its sole purpose is to be extended (subclassed).

27. What are the differences between interface and abstract class?

Ans.

Abstract Class	*Interfaces*
An abstract class can provide complete, default code and/or just the details that have to be overridden.	An interface cannot provide any code at all, just the signature.
In case of abstract class, a class may extend only one abstract class.	A Class may implement several interfaces.
An abstract class can have non-abstract methods.	All methods of an interface are abstract.
An abstract class can have instance variables.	An Interface cannot have instance variables.
An abstract class can have any visibility: public, private, protected.	An interface visibility must be public (or) none.
If we add a new method to an abstract class, then we have the option of providing default implementation and therefore all the existing code might work properly.	If we add a new method to an interface then, we have to track down all the implementations of the interface and define implementation for the new method.
An abstract class can contain constructors.	An interface cannot contain constructors.
Abstract classes are fast.	Interfaces are slow as it requires extra indirection to find corresponding method in the actual class.

28. When should I use abstract classes and when should I use interfaces?

Ans. Use interfaces when...

- You see that something in your design will change frequently.
- If various implementations only share method signatures then it is better to use interfaces.
- You need some classes to use some methods which you don't want to be included in the class, then you go for the interface, which makes it easy to just implement and make use of the methods defined in the interface.

Use abstract class when...

- If various implementations are of the same kind and use common behavior or status, then abstract class is better to use.
- When you want to provide a generalized form of abstraction and leave the implementation task with the inheriting subclass.
- Abstract classes are an excellent way to create planned inheritance hierarchies. They're also a good choice for nonleaf classes in class hierarchies.

29. When you declare a method as abstract, can other nonabstract methods access it?

Ans. Yes, other nonabstract methods can access a method that you declare as abstract.

30. Can there be an abstract class with no abstract methods in it?

Ans. Yes, there can be an abstract class without abstract methods.

Java Basic Interview Questions

31. What is Constructor?

- A constructor is a special method whose task is to initialize the object of its class.
- It is special because its name is the same as the class name.
- They do not have return types, not even void and therefore they cannot return values.
- They cannot be inherited, though a derived class can call the base class constructor.
- Constructor is invoked whenever an object of its associated class is created.

32. How does the Java default constructor be provided?

Ans. If a class defined by the code does not have any constructor, compiler will automatically provide one no-parameter-constructor (default-constructor) for the class in the byte code. The access modifier (public/private/etc.) of the default constructor is the same as the class itself.

33. Can constructor be inherited?

Ans. No, constructor cannot be inherited, though a derived class can call the base class constructor.

34. What are the differences between Contructors and Methods?

Constructors		*Methods*
Purpose	Create an instance of a class	Group Java statements
Modifiers	Cannot be abstract, final, native, static, or synchronized	Can be abstract, final, native, static, or synchronized
Return Type	No return type, not even void	void or a valid return type
Name	Same name as the class (first letter is capitalized by convention) — usually a noun	Any name except the class. Method names begin with a lowercase letter by convention — usually the name of an action
this	Refers to another constructor in the same class. If used, it must be the first line of the constructor	Refers to an instance of the owning class. Cannot be used by static methods.
super	Calls the constructor of the parent class. If used, must be the first line of the constructor	Calls an overridden method in the parent class
Inheritance	Constructors are not inherited	Methods are inherited

35. How are this() and super() used with constructors?

- Constructors use this to refer to another constructor in the same class with a different parameter list.
- Constructors use super to invoke the superclass's constructor. If a constructor uses super, it must use it in the first line; otherwise, the compiler will complain.

36. What are the differences between Class Methods and Instance Methods?

Class methods	*Instance methods*
Class methods can be called without creating an instance of the class	Instance methods, on the other hand, require an instance of the class to exist before they can be called, so an instance of a class needs to be created by using the new keyword.
Class methods can only operate on class members and not on instance members as class methods are unaware of instance members.	Instance methods operate on specific instances of classes. Instance methods of the class can also not be called from within a class method unless they are being called on an instance of that class.
Class methods are methods which are declared as static.	Instance methods are not declared as static.

37. How are this() and super() used with constructors?

Ans.
- Constructors use this to refer to another constructor in the same class with a different parameter list.
- Constructors use super to invoke the superclass's constructor. If a constructor uses super, it must use it in the first line; otherwise, the compiler will complain.

38. What are Access Specifiers?

Ans. One of the techniques in object-oriented programming is encapsulation. It concerns the hiding of data in a class and making this class available only through methods. Java allows you to control access to classes, methods, and fields via so-called access specifiers.

39. What are access specifiers available in Java?

Ans. Java offers four access specifiers, listed below in decreasing accessibility:

- Public: Public classes, methods, and fields can be accessed from everywhere.
- Protected: Protected methods and fields can only be accessed within the same class to which the methods and fields belong, within its subclasses, and within classes of the same package.
- Default(no specifier): If you do not set access to specific level, then such a class, method, or field will be accessible from inside the same package to which the class, method, or field belongs, but not from outside this package.
- Private: Private methods and fields can only be accessed within the same class to which the methods and fields belong. private methods and fields are not visible within subclasses and are not inherited by subclasses.

Situation	Public	Protected	Default	Private
accessible to class from the same package?	Yes	Yes	Yes	No
Accessible to class from different package?	Yes	No, unless it is a subclass	No	No

40. What is final modifier?

Ans. The final modifier keyword makes that the programmer cannot change the value anymore. The actual meaning depends on whether it is applied to a class, a variable, or a method.

- final classes: A final class cannot have subclasses.
- final variables: A final variable cannot be changed once it is initialized.
- final methods: A final method cannot be overridden by subclasses.

41. What are the uses of final method?

Ans. There are two reasons for marking a method as final:

- Disallowing subclasses to change the meaning of the method.
- Increasing efficiency by allowing the compiler to turn calls to the method into inline Java code.

42. What is static block?

Ans. Static block which executed exactly once when the class is first loaded into JVM. Before going to the main method the static block will execute.

43. What are static variables?

Ans. Variables that have only one copy per class are known as static variables. They are not attached to a particular instance of a class but rather belong to a class as a whole. They are declared by using the static keyword as a modifier. For example,

static type varIdentifier;

where, the name of the variable is varIdentifier and its data type is specified by type.

Note: Static variables that are not explicitly initialized in the code are automatically initialized with a default value. The default value depends on the data type of the variables.

44. What is the difference between static and non-static variables?

Ans. A static variable is associated with the class as a whole rather than with specific instances of a class. Non-static variables take on unique values with each object instance.

45. What are static methods?

Ans. Methods declared with the keyword static as modifier are called static methods or class methods. They are so called because they affect a class as a whole, not a particular instance of the class. Static methods are always invoked without reference to a particular instance of a class.

Note: The use of a static method suffers from the following restrictions:

- A static method can only call other static methods.
- A static method must only access static data.
- A static method cannot reference to the current object using keywords super or this.

46. What is an Iterator ?

- The Iterator interface is used to step through the elements of a Collection.
- Iterators let you process each element of a Collection.
- Iterators are a generic way to go through all the elements of a Collection no matter how it is organized.
- Iterator is an Interface implemented a different way for every Collection.

47. How do you traverse through a collection using its Iterator?

Ans. To use an iterator to traverse through the contents of a collection, follow these steps:

- Obtain an iterator to the start of the collection by calling the collectionâ•™s iterator() method.
- Set up a loop that makes a call to hasNext(). Have the loop iterate as long as hasNext() returns true.
- Within the loop, obtain each element by calling next().

48. How do you remove elements during Iteration?

Ans. Iterator also has a method remove() when remove is called, the current element in the iteration is deleted.

49. What is the difference between Enumeration and Iterator?

Ans.

Enumeration	*Iterator*
Enumeration doesn't have a remove() method	Iterator has a remove() method
Enumeration acts as Read-only interface, because it has the methods only to traverse and fetch the objects	Can be abstract, final, native, static, or synchronized

Note: So Enumeration is used whenever we want to make Collection objects as Read-only.

50. How is ListIterator?

Ans. ListIterator is just like Iterator, except it allows us to access the collection in either the forward or backward direction and lets us modify an element

51. What is the List interface?

Ans.
- The List interface provides support for ordered collections of objects.
- Lists may contain duplicate elements.

52. What are the main implementations of the List interface ?

Ans. The main implementations of the List interface are as follows :

- ArrayList: Resizable-array implementation of the List interface. The best all-around implementation of the List interface.

- Vector: Synchronized resizable-array implementation of the List interface with additional "legacy methods."
- LinkedList: Doubly-linked list implementation of the List interface. May provide better performance than the ArrayList implementation if elements are frequently inserted or deleted within the list. Useful for queues and double-ended queues (deques).

53. What are the advantages of ArrayList over arrays ?

Ans. Some of the advantages of ArrayList over arrays are:

- It can grow dynamically
- It provides more powerful insertion and search mechanisms than arrays.

54. What is the difference between ArrayList and Vector?

ArrayList	*Vector*
ArrayList is NOT synchronized by default.	Vector List is synchronized by default.
ArrayList can use only Iterator to access the elements.	Vector list can use Iterator and Enumeration interface to access the elements.
The ArrayList increases its array size by 50 percent if it runs out of room.	A vector defaults to doubling the size of its array if it runs out of room
ArrayList has no default size.	While vector has a default size of 10.

55. How to obtain Array from an ArrayList ?

Ans. Array can be obtained from an ArrayList using toArray() method on ArrayList.

```
List arrayList = new ArrayList();
arrayList.add(a);
ObjectÂ  a[] = arrayList.toArray();
```

56. Why insertion and deletion in ArrayList is slow compared to LinkedList ?

Ans.
- ArrayList internally uses and array to store the elements, when that array gets filled by inserting elements a new array of roughly 1.5 times the size of the original array is created and all the data of old array is copied to new array.
- During deletion, all elements present in the array after the deleted elements have to be moved one step back to fill the space created by deletion. In linked list data is stored in nodes that have reference to the previous node and the next node so adding element is simple as creating the node an updating the next pointer on the last node and the previous pointer on the new node. Deletion in linked list is fast because it involves only updating the next pointer in the node before the deleted node and updating the previous pointer in the node after the deleted node.

57. Why are Iterators returned by ArrayList called Fail Fast ?

Ans. Because, if list is structurally modified at any time after the iterator is created, in any way except through the iterator's own remove or add methods, the iterator will throw a ConcurrentModificationException. Thus, in the face of concurrent modification, the iterator fails quickly and cleanly, rather than risking arbitrary, non-deterministic behavior at an undetermined time in the future.

58. How do you decide when to use ArrayList and when to use LinkedList?

Ans. If you need to support random access, without inserting or removing elements from any place other than the end, then ArrayList offers the optimal collection. If, however, you need to frequently add and remove elements from the middle of the list and only access the list elements sequentially, then LinkedList offers the better implementation.

59. What is the Set interface ?

Ans.
- The Set interface provides methods for accessing the elements of a finite mathematical set
- Sets do not allow duplicate elements
- Contains no methods other than those inherited from Collection
- It adds the restriction that duplicate elements are prohibited
- Two Set objects are equal if they contain the same elements

60. What are the main implementations of the Set interface ?

Ans. The main implementations of the set interface are as follows:

- HashSet
- TreeSet
- LinkedHashSet
- EnumSet

61. What is a HashSet ?

Ans. • A HashSet is an unsorted, unordered Set.

- It uses the hashcode of the object being inserted (so the more efficient your hashcode() implementation the better access performance you'll get).
- Use this class when you want a collection with no duplicates and you don't care about order when you iterate through it.

62. What is a TreeSet ?

Ans. TreeSet is a Set implementation that keeps the elements in sorted order. The elements are sorted according to the natural order of elements or by the comparator provided at creation time.

63. What is an EnumSet ?

Ans. An EnumSet is a specialized set for use with enum types, all of the elements in the EnumSet type that is specified, explicitly or implicitly, when the set is created.

64. What is the difference between HashSet and TreeSet?

HashSet	*TreeSet*
HashSet is under set interface, i.e. it does not guarantee for either sorted order or sequence order.	TreeSet is under set, i.e. it provides elements in a sorted order (ascending order).
We can add any type of elements to hash set.	We can add only similar types of elements to tree set.

65. What is a Map?

Ans. • A map is an object that stores associations between keys and values (key/value pairs).

- Given a key, you can find its value. Both keys and values are objects.
- The keys must be unique, but the values may be duplicated.
- Some maps can accept a null key and null values, others cannot.

66. What are the main implementations of the Map interface ?

Ans. The main implementations of the List interface are as follows:

- HashMap
- HashTable
- TreeMap
- EnumMap

67. What is a TreeMap ?

Ans. TreeMap actually implements the SortedMap interface which extends the Map interface. In a TreeMap the data will be sorted in ascending order of keys according to the natural order for the key's class, or by the comparator provided at creation time. TreeMap is based on the Red-Black tree data structure.

68. How do you decide when to use HashMap and when to use TreeMap ?

Ans. For inserting, deleting, and locating elements in a Map, the HashMap offers the best alternative. If, however, you need to traverse the keys in a sorted order, then TreeMap is your better alternative. Depending upon the size of your collection, it may be faster to add elements to a HashMap, then convert the map to a TreeMap for sorted key traversal.

69. What is the difference between HashMap and Hashtable ?

HashMap	*Hashtable*
HashMap lets you have null values as well as one null key.	HashTable does not allows null values as key and value.
The iterator in the HashMap is fail-safe (If you change the map while iterating, you'll know).	The enumerator for the Hashtable is not fail-safe.
HashMap is unsynchronized.	Hashtable is synchronized.

Note: Only one NULL is allowed as a key in HashMap. HashMap does not allow multiple keys to be NULL. Nevertheless, it can have multiple NULL values.

70. How does a Hashtable internally maintain the key-value pairs?

Ans. TreeMap actually implements the SortedMap interface which extends the Map interface. In a TreeMap the data will be sorted in ascending order of keys according to the natural order for the key's class, or by the comparator provided at creation time. TreeMap is based on the Red-Black tree data structure.

71. What are the different collection views that maps provide?

Ans. Maps provide three collection views.

- Key Set—allow a map's contents to be viewed as a set of keys.

- Values Collection—allow a map's contents to be viewed as a set of values.
- Entry Set—allow a map's contents to be viewed as a set of key-value mappings.

72. What is a KeySet View ?

Ans. KeySet is a set returned by the keySet() method of the Map Interface. It is a set that contains all the keys present in the Map.

73. What is values collection view?

Ans. Values collection view is a collection returned by the values() method of the Map interface. It contains all the objects present as values in the map.

74. What is an EntrySet View ?

Ans. Entry set view is a set that is returned by the entrySet() method in the map and contains Objects of type Map. Entry each of which has both Key and Value.

75. How do you sort an ArrayList (or any list) of user-defined objects ?

Ans. Create an implementation of the java.lang.Comparable interface that knows how to order your objects and pass it to java.util.Collections.sort(List, Comparator).

76. What is the Comparable interface ?

Ans. The Comparable interface is used to sort collections and arrays of objects using the Collections.sort() and java.utils.Arrays.sort() methods respectively. The objects of the class implementing the Comparable interface can be ordered.

The Comparable interface in the generic form is written as follows:

interface Comparable<T>

where T is the name of the type parameter.

All classes implementing the Comparable interface must implement the compareTo() method that has the return type as an integer. The signature of the compareTo() method is as follows:

int i = object1.compareTo(object2)

- If object1 < object2: The value of i returned will be negative.
- If object1 > object2: The value of i returned will be positive.
- If object1 = object2: The value of i returned will be zero.

77. What are the differences between the Comparable and Comparator interfaces ?

Comparable	*Comparator*
It uses the compareTo() method.	int objectOne.compare To(objectTwo).
It uses the compare() method.	int compare(ObjOne, ObjTwo)
It is necessary to modify the class whose instance is going to be sorted.	A separate class can be created in order to sort the instances.
Only one sort sequence can be created.	Many sort sequences can be created.
It is frequently used by the API classes.	It used by third-party classes to sort instances.

Interview Questions

1. What is the difference between JDK, JRE and JVM?

Ans. JVM

JVM is an acronym for Java Virtual Machine, it is an abstract machine which provides the runtime environment in which Java bytecode can be executed.

JVMs are available for many hardware and software platforms (so JVM is platform dependent).

JRE

JRE stands for Java Runtime Environment. It is the implementation of JVM and physically exists.

JDK

JDK is an acronym for Java Development Kit. It physically exists. It contains JRE + development tools.

2. How many types of memory areas are allocated by JVM?

Ans. Five types:

1. Class(Method) Area
2. Heap
3. Stack
4. Program Counter Register
5. Native Method Stack

3. What is JIT compiler?

Ans. Just-In-Time(JIT) compiler: It is used to improve the performance. JIT compiles parts of the byte code that have similar functionality at the same time, and hence reduces the amount of time needed for compilation. Here the term "compiler" refers to a translator from the instruction set of a Java Virtual Machine (JVM) to the instruction set of a specific CPU.

4. What is platform?

Ans. A platform is basically the hardware or software environment in which a program runs. There are two types of platforms: software-based and hardware-based. Java provides software-based platform.

5. What is the main difference between Java platform and other platforms?

Ans. The Java platform differs from most other platforms in the sense that it's a software-based platform that runs on top of other hardware-based platforms. It has two components:

1. Runtime Environment
2. API(Application Programming Interface)

6. What gives Java its 'write once and run anywhere' nature?

Ans. The bytecode. Java is compiled to be a byte code which is the intermediate language between source code and machine code. This byte code is not platorm specific and hence can be fed to any platform.

7. What is classloader?

Ans. The classloader is a subsystem of JVM that is used to load classes and interfaces.There are many types of classloaders, e.g. Bootstrap classloader, Extension classloader, System classloader, Plugin classloader, etc.

8. Is Empty .java file name a valid source file name?

Ans. Yes, save your java file by .java only, compile it by javac .java and run by java your classname. Let's take a simple example:

```
//save by .java only

class A{
public static void main(String args[]){
System.out.println("Hello java");
}
}

//compile by javac .java
//run by    java A
```

compile it by javac .java

run it by java A

9. Is delete, next, main, exit or null keyword in Java?

Ans. No.

10. If I do not provide any argument on the command line, then the String array of Main method will be empty or null?

Ans. It is empty. But not null.

11. What if I write static public void instead of public static void?

Ans. Program compiles and runs properly.

12. What is the default value of the local variables?

Ans. The local variables are not initialized to any default value, neither primitives nor object references.

13. What is the difference between object oriented programming language and object based programming language?

Ans. Object based programming languages follow all the features of OOPs except inheritance. Examples of object based programming languages are JavaScript, VBScript, etc.

14. What will be the initial value of an object reference which is defined as an instance variable?

Ans. The object references are all initialized to null in Java.

15. What is constructor?

Ans. Constructor is just like a method that is used to initialize the state of an object. It is invoked at the time of object creation.

16. What is the purpose of default constructor?

Ans. The default constructor provides the default values to the objects. The java compiler creates a default constructor only if there is no constructor in the class.more details…

17. Does constructor return any value?

Ans. Yes, that is current instance (you cannot use return type yet it returns a value).

18. Is constructor inherited?

Ans. No, constructor is not inherited.

19. Can you make a constructor final?

Ans. No, constructor can't be final.

20. What is static variable?

Ans.
- Static variable is used to refer the common property of all objects (that is not unique for each object), e.g. company name of employees, college name of students, etc.
- Static variable gets memory only once in class area at the time of class loading.

21. What is static method?

Ans.
- A static method belongs to the class rather than object of a class.
- A static method can be invoked without the need for creating an instance of a class.
- Static method can access static data member and can change the value of it.

22. Why main method is static?

Ans. Because object is not required to call static method if it were non-static method, jvm creates object first, then call main() method that will lead to the problem of extra memory allocation.more details...

23. What is static block?

Ans.
- It is used to initialize the static data member.
- It is excuted before main method at the time of classloading.

24. Can we execute a program without main() method?

Ans. Yes, one of the way is, by static block.

25. What if the static modifier is removed from the signature of the main method?

Ans. Program compiles. But at runtime throws an error "NoSuchMethodError".

26. What is the difference between static (class) method and instance method?

(1) A method that is declared as static is known as static method. A method, i.e. not declared as static is known as instance method.

(2) Object is not required to call static method. Object is required to call instance methods.

(3) Non-static (instance) members cannot be accessed in static context (static method, static block and static nested class) directly. Static and non-static variables both can be accessed in instance methods.

(4) For example: public static int cube(int n){ return n*n*n;} For example, public void msg(){...}.

27. What is this in Java?

Ans. It is a keyword that refers to the current object.

28. What is inheritance?

Ans. Inheritance is a mechanism in which one object acquires all the properties and behaviour of another object of another class. It represents IS-A relationship. It is used for Code Reusability and Method Overriding.more details...

29. Which class is the superclass for every class?

Ans. Object class.

30. Why multiple inheritance is not supported in Java?

Ans. To reduce the complexity and simplify the language, multiple inheritance is not supported in Java in case of class.

31. What is composition?

Ans. Holding the reference of the other class within some other class is known as composition.

32. What is the difference between aggregation and composition?

Ans. Aggregation represents weak relationship, whereas composition represents strong relationship. For example, bike has an indicator (aggregation) but bike has an engine (composition).

33. Why Java does not support pointers?

Ans. Pointer is a variable that refers to the memory address. They are not used in Java because they are unsafe (unsecured) and complex to understand.

34. What is super in Java?

Ans. It is a keyword that refers to the immediate parent class object.

35. Can you use this() and super() both in a constructor?

Ans. No. Because super() or this() must be the first statement.

36. What is object cloning?

Ans. The object cloning is used to create the exact copy of an object.

37. What is method overloading?

Ans. If a class have multiple methods by the same name but different parameters, it is known as Method Overloading. It increases the readability of the program.more details...

38. Why method overloading is not possible by changing the return type in Java?

Ans. Because of ambiguity.

39. Can we overload main() method?

Ans. Yes, of course! You can have many main() methods in a class by overloading the main method.

40. What is method overriding?

Ans. If a subclass provides a specific implementation of a method that is already provided by its parent class, it is known as Method Overriding. It is used for runtime polymorphism and to provide the specific implementation of the method.more details...

41. Can we override static method?

Ans. No, you can't override the static method because they are the part of class, not object.

42. Why we cannot override static method?

Ans. It is because the static method is the part of class and it is bound with class, whereas instance method is bound with object and static gets memory in class area and instance gets memory in heap.

43. Can we override the overloaded method?

Ans. Yes.

44. What is the difference between method overloading and overriding?

(1) Method overloading increases the readability of the program. Method overriding provides the specific implementation of the method that is already provided by its superclass.

(2) Method overlaoding occurs within the class. Method overriding occurs in two classes that have IS-A relationship.

(3) In this case, parameter must be different. In this case, parameter must be the same.

45. Can you have virtual functions in Java?

Ans. Yes, all functions in Java are virtual by default.

46. What is covariant return type?

Ans. Now, since java5, it is possible to override any method by changing the return type if the return type of the subclass overriding method is subclass type. It is known as covariant return type.

47. What is final variable?

Ans. If you make any variable as final, you cannot change the value of final variable (it will be constant).

48. What is final method?

Ans. Final methods can't be overriden.

49. What is final class?

Ans. Final class can't be inherited.

50. What is blank final variable?

Ans. A final variable, not initalized at the time of declaration, is known as blank final variable.

51. Can we initialize blank final variable?

Ans. Yes, only in constructor if it is non-static. If it is static blank final variable, it can be initialized only in the static block.

52. Can you declare the main method as final?

Ans. Yes, such as, public static final void main(String[] args){}.

53. What is runtime polymorphism?

Ans. Runtime polymorphism or dynamic method dispatch is a process in which a call to an overridden method is resolved at runtime rather than at compile-time.

In this process, an overridden method is called through the reference variable of a superclass. The determination of the method to be called is based on the object being referred to by the reference variable.

54. Can you achieve runtime polymorphism by data members?

Ans. No

55. What is the difference between static binding and dynamic binding?

Ans. In case of static binding type of object, it is determined at compile time, whereas in dynamic binding type of object, it is determined at runtime.

56. What is abstraction?

Ans. Abstraction is a process of hiding the implementation details and showing only functionality to the user.

Abstraction lets you focus on what the object does instead of how it does it.

57. What is the difference between abstraction and encapsulation?

Ans. Abstraction hides the implementation details, whereas encapsulation hides the data.

Abstraction lets you focus on what the object does instead of how it does it.

58. What is abstract class?

Ans. A class that is declared as abstract is known as abstract class. It needs to be extended and its method implemented. It cannot be instantiated.

59. Can there be any abstract method without abstract class?

Ans. No, if there is any abstract method in a class, that class must be abstract.

60. Can you use abstract and final both with a method?

Ans. No, because abstract method needs to be overridden, whereas you can't override final method.

61. Is it possible to instantiate the abstract class?

Ans. No, abstract class can never be instantiated.

62. What is interface?

Ans. Interface is a blueprint of a class that have static constants and abstract methods. It can be used to achieve fully abstraction and multiple inheritance.

63. Can you declare an interface method static?

Ans. No, because methods of an interface is abstract by default, and static and abstract keywords can't be used together.

64. Can an interface be final?

Ans. No, because its implementation is provided by another class.

65. What is marker interface?

Ans. An interface that have no data member and method is known as a marker interface. For example, Serializable, Cloneable, etc.

66. What is the difference between abstract class and interface?

(1) An abstract class can have method body (non-abstract methods). Interfaces have only abstract methods.

(2) An abstract class can have instance variables. An interface cannot have instance variables.

(3) An abstract class can have constructor. Interface cannot have constructor.

(4) An abstract class can have static methods. Interface cannot have static methods.

(5) You can extend one abstract class.You can implement multiple interfaces.

67. Can we define private and protected modifiers for variables in interfaces?

Ans. No, they are implicitly public.

68. When can an object reference be cast to an interface reference?

Ans. An object reference can be cast to an interface reference when the object implements the referenced interface.

69. What is package?

Ans. A package is a group of similar type of classes interfaces and subpackages. It provides access protection and removes naming collision.

70. Do I need to import java.lang package any time? Why ?

Ans. No. It is by default loaded internally by the JVM.

71. Can I import same package/class twice? Will the JVM load the package twice at runtime?

Ans. One can import the same package or same class multiple times. Neither compiler nor JVM complains about it. But the JVM will internally load the class only once, no matter how many times you import the same class.

72. What is static import ?

Ans. By static import, we can access the static members of a class directly, there is nothing to qualify it with the class name.

73. What is Exception Handling?

Ans. Exception Handling is a mechanism to handle runtime errors. It is mainly used to handle checked exceptions.

74. What is the difference between Checked Exception and Unchecked Exception?

(1) Checked Exception:

The classes that extend Throwable class except RuntimeException and Error are known as checked exceptions, e.g. IOException, SQLException, etc. Checked exceptions are checked at compile-time.

(2) Unchecked Exception:

The classes that extend RuntimeException are known as unchecked exceptions, e.g. ArithmeticException, NullPointerException, etc. Unchecked exceptions are not checked at compile-time.

75. What is the base class for Error and Exception?

Ans. Throwable.

76. Is it necessary that each try block must be followed by a catch block?

Ans. It is not necessary that each try block must be followed by a catch block. It should be followed by either a catch block OR a finally block. And whatever exceptions are likely to be thrown should be declared in the throws clause of the method.

77. What is finally block?

Ans. Finally block is a block that is always executed.more details...

78. Can finally block be used without catch?

Ans. Yes, by try block. finally must be followed by either try or catch.more details...

79. Is there any case when finally will not be executed?

Ans. Finally block will not be executed if program exits (either by calling System.exit() or by causing a fatal error that causes the process to abort).more details...

80. What is difference between throw and throws?

(1) Throw is used to explicitly throw an exception. Throws is used to declare an exception.

(2) Checked exceptions cannot be propagated with throw only. Checked exception can be propagated with throws.

(3) Throw is followed by an instance. Throws is followed by class.

(4) Throw is used within the method. Throws is used with the method signature.

(5) You cannot throw multiple exception. You can declare multiple exception, e.g. public void method()throws IOException, SQLException.

81. Can an exception be rethrown?

Ans. Yes.

82. Can subclass overriding method declare an exception if parent class method doesn't throw an exception ?

Ans. Yes, but only unchecked exception not checked.

83. What is exception propagation ?

Ans. Forwarding the exception object to the invoking method is known as exception propagation.

84. What is the meaning of immutable in terms of String?

Ans. The simple meaning of immutable is unmodifiable or unchangeable. Once string object has been created, its value can't be changed.

85. Why string objects are immutable in Java?

Ans. Because Java uses the concept of string literal. Suppose there are 5 reference variables, all refers to one object "sachin". If one reference variable changes the value of the object, it will be affected to all the reference variables. That is why string objects are immutable in Java.

86. How many ways we can create the string object?

Ans. There are two ways to create the string object, by string literal and by new keyword.

87 How many objects will be created in the following code?

1. String s1="Welcome";
2. String s2="Welcome";
3. String s3="Welcome";

Only one object.

88. Why Java uses the concept of string literal?

Ans. To make Java more memory efficient (because no new object is created if it exists already in string constant pool).

89. How many objects will be created in the following code?

Ans. 1. String s=new String("Welcome");

Two objects, one in string constant pool and other in non-pool(heap).

90. What is the basic difference between string and stringbuffer object?

Ans. String is an immutable object. StringBuffer is a mutable object.

91. What is the difference between StringBuffer and StringBuilder ?

Ans. StringBuffer is synchronized, whereas StringBuilder is not synchronized.

92. How can we create immutable class in Java?

Ans. We can create immutable class as the String class by defining final class and more details...

93. What is the purpose of toString() method in Java ?

Ans. The toString() method returns the string representation of any object. If you print any object, java compiler internally invokes the toString() method on the object. So overriding the toString() method, returns the desired output, it can be the state of an object, etc. depends on your implementation. More details...

94. What is nested class?

Ans. A class which is declared inside another class is known as nested class. There are four types of nested class member inner class, local inner class, annonymous inner class and static nested class.

95. Is there any difference between nested classes and inner classes?

Ans. Yes, of course! Inner classes are non-static nested classes, i.e. inner classes are the part of nested classes.

96. Can we access the non-final local variable, inside the local inner class?

Ans. No, local variable must be constant if you want to access it in local inner class.

97. What is nested interface ?

Ans. Any interface, i.e. declared inside the interface or class, is known as nested interface. It is static by default.

98. Can a class have an interface?

Ans. Yes, it is known as nested interface.

99. Can an interface have a class?

Ans. Yes, they are static implicitly.

100. What is multithreading?

Ans. Multithreading is a process of executing multiple threads simultaneously. Its main advantage is:

- Threads share the same address space.
- Thread is lightweight.
- Cost of communication between process is low.

101. What is thread?

Ans. A thread is a lightweight subprocess. It is a separate path of execution. It is called separate path of execution because each thread runs in a separate stack frame.

102. What is the difference between preemptive scheduling and time slicing?

Ans. Under preemptive scheduling, the highest priority task executes until it enters the waiting or dead states or a higher priority task comes into existence. Under time slicing, a task executes for a predefined slice of time and then reenters the pool of ready tasks. The scheduler then determines which task should execute next, based on priority and other factors.

103. What does join() method?

Ans. The join() method waits for a thread to die. In other words, it causes the currently running threads to stop executing until the thread it joins with completes its task.more details...

104. What is the difference between wait() and sleep() method?

(1) The wait() method is defined in Object class. The sleep() method is defined in Thread class.

(2) wait() method releases the lock. The sleep() method doesn't releases the lock.

105. Is it possible to start a thread twice?

Ans. No, there is no possibility to start a thread twice. If we does, it throws an exception. More details...

106. Can we call the run() method instead of start()?

Ans. Yes, but it will not work as a thread rather it will work as a normal object so there will not be context-switching between the threads.more details...

107. What about the daemon threads?

Ans. The daemon threads are basically the low priority threads that provides the background support to the user threads. It provides services to the user threads. more details...

108. Can we make the user thread as daemon thread if thread is started?

Ans. No, if you do so, it will throw IllegalThreadStateException.more details...

109. What is shutdown hook?

Ans. The shutdown hook is basically a thread, i.e. invoked implicitly before JVM shuts down. So we can use it to perform clean up resource. more details...

110. When should we interrupt a thread?

Ans. We should interrupt a thread if we want to break out the sleep or wait state of a thread.more details...

111. What is synchronization?

Ans. Synchronization is the capability of control the access of multiple threads to any shared resource. It is used:

1. To prevent thread interference.
2. To prevent consistency problem.

112. What is the purpose of Synchronized block?

- Synchronized block is used to lock an object for any shared resource.
- Scope of synchronized block is smaller than the method.

113. Can Java object be locked down for exclusive use by a given thread?

Ans. Yes. You can lock an object by putting it in a "synchronized" block. The locked object is inaccessible to any thread other than the one that explicitly claimed it.

114. What is static synchronization?

Ans. If you make any static method as synchronized, the lock will be on the class and not on object.

115. What is the difference between notify() and notifyAll()?

Ans. notify() is used to unblock one waiting thread, whereas notifyAll() method is used to unblock all the threads in waiting state.

116. What is deadlock?

Ans. Deadlock is a situation when two threads are waiting on each other to release a resource. Each thread waiting for a resource which is held by the other waiting thread.

117. What is Garbage Collection?

Ans. Garbage collection is a process of reclaiming the runtime unused objects. It is performed for memory management.

118. What is gc()?

Ans. gc() is a daemon thread.gc() method is defined in System class that is used to send request to JVM to perform garbage collection.

119. What is the purpose of finalize() method?

Ans. Finalize() method is invoked just before the object is garbage collected. It is used to perform cleanup processing.

120. Can an unreferenced objects be referenced again?

Ans. Yes.

121. What kind of thread is the Garbage collector thread?

Ans. Daemon thread.

122 What is the difference between final, finally and finalize?

Ans. final: Final is a keyword, final can be variable, method or class. You can't change the value of final variable, can't override final method, can't inherit final class.

finally: Finally block is used in exception handling. finally block is always executed.

finalize():finalize() method is used in garbage collection. finalize() method is invoked just before the object is garbage collected. The finalize() method can be used to perform any cleanup processing.

123. What is the purpose of the runtime class?

Ans. The purpose of the runtime class is to provide access to the Java runtime system.

124. How will you invoke any external process in Java?

Ans. By Runtime.getRuntime().exec(?) method.

125. What is the difference between the Reader/Writer class hierarchy and the InputStream/OutputStream class hierarchy?

Ans. The Reader/Writer class hierarchy is character-oriented, and the InputStream/OutputStream class hierarchy is byte-oriented.

126. What an I/O filter?

Ans. An I/O filter is an object that reads from one stream and writes to another, usually altering the data in some way as it is passed from one stream to another.

127. What is serialization?

Ans. Serialization is a process of writing the state of an object into a byte stream. It is mainly used to travel object's state on the network.

128. What is deserialization?

Ans. Deserialization is the process of reconstructing the object from the serialized state. It is the reverse operation of serialization.

129. What is transient keyword?

Ans. If you define any data member as transient, it will not be serialized.

130. What is Externalizable?

Ans. Externalizable interface is used to write the state of an object into a byte stream in compressed format. It is not a marker interface.

131. What is the difference between Serializable and Externalizable interface?

Ans. Serializable is a marker interface but Externalizable is not a marker interface. When you use Serializable interface, your class is serialized automatically by default. But you can override writeObject() and readObject() two methods to control more complex object serialization process. When you use Externalizable interface, you have a complete control over your class's serialization process.

132. How do I convert a numeric IP address like 192.18.97.39 into a hostname like java.sun.com?

Ans. By InetAddress.getByName ("192.18.97.39"). getHostName() where 192.18.97.39 is the IP address.

133. What is reflection?

Ans. Reflection is the process of examining or modifying the runtime behaviour of a class at runtime. It is used in:

- IDE (Integreted Development Environment), e.g. Eclipse, MyEclipse, NetBeans.
- Debugger
- Test Tools, etc.

134. Can you access the private method from outside the class?

Ans. Yes, by changing the runtime behavior of a class if the class is not secured.

Collection Interview Questions

In Java, collection interview questions are mostly asked by the interviewers. Here is the list of mostly asked collection interview questions with answers.

135. What is the difference between ArrayList and Vector?

Ans. ArrayList

(1) ArrayList is not synchronized.
(2) ArrayList is not a legacy class.
(3) ArrayList increases its size by 50% of the array size.

Vector

(1) Vector is synchronized.
(2) Vector is a legacy class.
(3) Vector increases its size by doubling the array size.

136. What is the difference between ArrayList and LinkedList?

Ans. ArrayList

(1) ArrayList uses a dynamic array.
(2) ArrayList is not efficient for manipulation because a lot of shifting is required.

LinkedList

(1) LinkedList uses doubly linked list.

(2) LinkedList is efficient for manipulation.

137. What is the difference between HashMap and Hashtable?

Ans. HashMap

(1) HashMap is not synchronized.

(2) HashMap can contain one null key and multiple null values.

Hashtable

(1) Hashtable is synchronized.

(2) Hashtable neither contain any null key nor value.

138. What is hash-collision in Hashtable and how is it handled in Java?

Ans. Two different keys with the same hash value. Two different entries will be kept in a single hash bucket to avoid the collision.

139. What is the difference between HashSet and HashMap?

Ans. HashSet contains only values, whereas HashMap contains entry(key,value).

140. What is the difference between HashMap and TreeMap?

Ans. HashMap

(1) HashMap can contain one null key.

(2) HashMap maintains no order.

TreeMap

(1) TreeMap cannot contain any null key.

(2) TreeMap maintains ascending order.

141. What is the difference between HashSet and TreeSet?

Ans. HashSet maintains no order, whereas TreeSet maintains ascending order.

142. What is the difference between List and Set?

Ans. List can contain duplicate elements, whereas Set contains only unique elements.

143. What is the difference between Iterator and ListIterator?

Ans. Iterator traverses the elements in forward direction only, whereas ListIterator traverses the elements in forward and backward direction.

144. Can you make List, Set and Map elements synchronized?

Ans. Yes, Collections class provides methods to make List, Set or Map elements as synchronized:

public static List synchronizedList(List l){}
public static Set synchronizedSet(Set s){}
public static SortedSet synchronizedSortedSet(SortedSet s){}
public static Map synchronizedMap(Map m){}
public static SortedMap synchronizedSortedMap(SortedMap m){}

145. What is the difference between Iterator and Enumeration?

Ans. Iterator

(1) Iterator can traverse legacy and non-legacy elements.

(2) Iterator is fail-fast.

(3) Iterator is slower than Enumeration.

Enumeration

(1) Enumeration can traverse only legacy elements.

(2) Enumeration is not fail-fast.

(3) Enumeration is faster than Iterator.

146. What is the difference between Comparable and Comparator?

Comparable

(1) Comparable provides only one sort of sequence.

(2) It provides one method named compareTo().

(3) It is found in java.lang package.

(4) If we implement Comparable interface, actual class is modified.

Comparator

(1) Comparator provides multiple sort of sequences.

(2) It provides one method named compare().

(3) It is found in java.util package.

(4) Actual class is not modified.

147. What is the Dictionary class?

Ans. The Dictionary class provides the capability to store key-value pairs.

148. What are wrapper classes?

Ans. Wrapper classes are classes that allow primitive types to be accessed as objects.

149. What is a native method?

Ans. A native method is a method that is implemented in a language other than Java.

150. What is the purpose of the System class?

Ans. The purpose of the System class is to provide access to system resources.

151. What comes to mind when someone mentions a shallow copy in Java?

Ans. Object cloning.

152. What is singleton class?

Ans. Singleton class means that any given time only one instance of the class is present, in one JVM.

153. Which containers use a border layout as their default layout?

Ans. The Window, Frame and Dialog classes use a border layout as their default layout.

154. Which containers use a FlowLayout as their default layout?

Ans. The Panel and Applet classes use the FlowLayout as their default layout.

155. What are peerless components?

Ans. The peerless components are called light-weight components.

156. What is the difference between a Scrollbar and a ScrollPane?

Ans. A Scrollbar is a Component, but not a Container. A ScrollPane is a Container. A ScrollPane handles its own events and performs its own scrolling.

157. What is a lightweight component?

Ans. Lightweight components are the one which doesn't go with the native call to obtain the graphical units. They share their parent component graphical units to render them. For example, Swing components.

158. What is a heavyweight component?

Ans. For every paint call, there will be a native call to get the graphical units. For example, AWT.

159. What is an applet?

Ans. An applet is a small java program that runs inside the browser and generates dynamic contents.

160. Can you write a Java class that could be used both as an applet as well as an application?

Ans. Yes. Add a main() method to the applet.

161. What is Locale?

Ans. A Locale object represents a specific geographical, political, or cultural region.

162. How will you load a specific locale?

Ans. By ResourceBundle.getBundle(?) method.

163. What is a JavaBean?

Ans. JavaBean are reusable software components written in the Java programming language, designed to be manipulated visually by a software develpoment environment, like JBuilder or VisualAge for Java.

164. Can RMI and Corba based applications interact?

Ans. Yes they can. RMI is available with IIOP as the transport protocol instead of JRMP.

165. What is JDBC?

Ans. JDBC is a Java API that is used to connect and execute query to the database. JDBC API uses JDBC drivers to connects to the database.

166. What is JDBC Driver?

Ans. JDBC Driver is a software component that enables Java application to interact with the database.There are 4 types of JDBC drivers:

1. JDBC-ODBC bridge driver
2. Native-API driver (partially Java driver)
3. Network Protocol driver (fully Java driver)
4. Thin driver (fully Java driver)

167. What are the steps to connect to database in Java?

- Register the driver class
- Creating connection
- Creating statement
- Executing queries
- Closing connection

168. What is the difference between Statement and PreparedStatement interface?

Ans. In case of Statement, query is complied each time, whereas in case of PreparedStatement, query is complied only once. So performance of PreparedStatement is better than Statement.

169. How can we execute stored procedures and functions?

Ans. By using Callable statement interface, we can execute procedures and functions.

170. How can we store and retrieve images from the database?

Ans. By using PreparedStatement interface, we can store and retrieve images. To see examples click here...

Glossary

AccessException	An AccessException is thrown by a certain method of the java.rmi.Maming class to indicate that the caller does not have permission to perform the action requested by the method call.
AlreadyBoundException	An AlreadyBoundException is thrown if an attempt is made to bind an object in the registry to a name that already has an associated binding.
Bootstrap registry	Other name for RMI Registry.
CGI	CGI stands for Common Gateway Interface. It is used to connect dissimilar networks.
COM	COM is used to develop distributed application in a stand alone system. The application may be within the same address space or in different address spaces in a single machine.
ConnectException	A ConnectException is thrown if a connection is refused to the remote host for a remote method call.
ConnectIOException	A ConnectIOException is thrown if an IOException occurs while making a connection to the remote host for a remote method call.
DataInput	DataInput interface includes methods for the input of primitive types.
DataOutput	DataOutput interface includes methods for the output of primitive types.
DCOM	Distributed Component Object Model is an extension of Component Object Model (COM). It is used to develop distributed applications, but within Windows environment.
DGC	The DGC abstraction is used for the server side of the distributed garbage collection algorithm.
Distributed systems	Using distributed system, computations running in different address spaces, potentially on different machines, will be able to communicate.
DLL	Dynamic Link Library is a module in executable form. We can call these modules in our application whenever we need.
ExportException	TheExportException exception extends the RemoteException exception. An ExportException is

	a RemoteException thrown if an attempt to export a remote object fails.
Externalizable object	An Externalizable object is used to handle the serialization of the entire object graph manually. An Externalizable object must extend the Externalizable interface.
Garbage collection	Garbage collection is the process of removing objects from memory when they are no longer used.
HTTP protocol	The HTTP protocol is used by the RMI protocol to invoke remote methods through a firewall.
Java.rmi.dge	java.rmi.dge package provides classes and interfaces for the RMI Distributed Garbage-Collection (DGC).
Java.rmi.server	The java.rmi.server package provides classes and interfaces for supporting the server side of RMI.
Java Object Serialization protocol	The Java Object Serialization protocol is used by the RMI protocol to format the call and return data in RMI calls.
Lease	The Lease class extends the Object class and implements the Serializable interface. A lease contains a unique VM identifier and lease duration. A lease object is used to request and grant leases to remote object references.
LocateRegistry	The LocateRegistry class extends the Object class and is used to obtain a reference to a bootstrap remote object registry on a particular host (including the local host), or to create a remote object registry that accepts calls on a specific port.
LoaderHandler	The LoaderHandler is a deprecated interface used internally by the RMI runtime in previous implementation versions. It should never be accessed by application code.
LogStream	The LogStream is a deprecated class that extends the PrintStream class. The LogStream class provides a mechanism for logging errors that are useful for monitoring a system.
list( )	The list( ) method is present in the Registry interface. This method return an array of names bound in the registry.
MarshalException	A MarshalException is thrown if a java.io.IOException occurs while marshalling the remote call header, arguments or return value for a remote method call, or, if the receiver does not support the protocol version of the sender.

Marshalling	Marshalling is the process of converting input parameters to a format that can be transmitted across the network.
MarshalledObject	The MarshalledObject class extends the Object class and implements the Serializable interface. It facilitates passing objects in RMI calls that are not automatically deserialized immediately by the remote peer.
Multiplexing protocol	The multiplexing protocol provides a model for two endpoints to open multiple full duplex connections. RMI uses this simple multiplexing protocol to allow a client to connect to an RMI server object in some situations where that is otherwise not possible.
Naming	The Naming class extends the Object class and provides methods for storing and obtaining reference to remote objects in the remote object registry.
Naming.lookup( )	The naming.lookup ( ) is a static method that searches for the server using the server name registered with the RMI registry. It takes in a single parameter of String type representing the URL-formatted name for the remote object and returns a reference for the remote object associated with the specified name.
NoSuchObjectException	A NoSuchObjectException is thrown by the method java.rmi.server.RemoteObject.toStub and by the unexportObject method of java.rmi.server. Unicast RemoteObject, or, if we attempt to invoke a method on a remote object that does not exist.
NotBoundException	The NotBoundException class extends the Exception class. A NotBoundException is thrown if an attempt is made to lookup( ) or unbind( ) a name that has no associated binding in the registry.
Object graph	Object graph represents an object and the object references that it contains.
ObjectInput	ObjectInput interface extends the DataInput interface to include objects, arrays and strings. Stream objects implement either this interface or the ObjectOutput interface.
ObjectOutput	ObjectOutput extends the DataIOutput interface to include objects, arrays and strings. Stream objects implement either this interface or the ObjectInput interface.
Object serialization	Object serialization is used to duplicate an object by writing the values of its fields to a stream and recreating a copy of the object from these field values.

Object stream	An object stream transmits a representation of its structure and related data. An object stream contains some header information that is used to rebuild objects that it contains, along with the data for these objects.
ObjID	The ObjID class extends the Object class and implements the Serializable interface. An ObjID is used to identify remote objects uniquely in a VM over time.
Operation	The Operation is a deprecated class that extends the Object class. An Operation contains a description of a Java method.
ORB	An ORB is a software component used for communication between objects in a distributed environment.
RMI	RMI stands for Remote Method Invocation. Java RMI is used for executing distributed applications developed in Java.gf
RMI Registry	Registry is a simple server that keeps track of the addresses of remote objects that are being exported by an application.
readObject( )	readObject( ) method is used in serialization to get data from the object stream.
Recreation	Recreation of an object is a copy of the object.
Reference count algorithm	Reference count algorithm keeps a count of the number of references to a particular object. It is used by the DGC to remove the object from memory whenever this count becomes zero.
Registry Interface	The Registry Interface extends the Remote class. The remote object Registry interfaces helps to obtain reference to remote objects.
RegistryHandler	RegistryHandler is a deprecated interface used internally by the RMI runtime in previous implementation versions.
Remote	The Remote interface serves to identify interfaces whose methods may be invoked from a remote virtual machine. Remote objects must directly or indirectly implement this interface.
RemoteCall	The RemoteCall is a deprecated interface that is an abstraction used solely by the RMI runtime to carry out a call to a remote object.
RemoteException	The RemoteException class extends the IOException class. This exception is the common superclass for a

	number of a remote method call. Each method of a remote interface must list RemoteException in its throws clause.
RemoteObject	The RemoteObject class extends the Object class and implements the two interfaces Remote and Serializable. It provides the remote semantics of Object by implementing methods for hashCode, equals, and toString.
RemoteRef	The RemoteRef interface extends the Externalizable interface. RemoteRef represents the handle for a remote object.
Remote Interface	Remote Interface reference to the remote object. It is used to access the remote object and its methods.
RemoteServer	The RemoteServer class extends the RemoteObject class and is the common superclass to server implementations. It provides the framework to support a wide range of remote reference semantics.
RemoteStub	The RemoteStub class extends the RemoteObject class and is the common superclass to client stubs. It provides the framework to support a wide range of remote reference semantics.
RMIC	RMIC Stands for Remote Method Invocation Compiler. It is used to generate Stub and Skeleton classes.
RMIClassLoader	The RMIClassLoader class that extends the Objects class provides static methods for loading classes from a network location and obtaining the location from which an existing class can be loaded.
RMIClientSocketFactory	An instance of the RMIClientSocketFactory interface is used by the RMI runtime to acquireclient sockets for RMI calls.
RMIFailureHandler	A RMIFailureHandler interface can be registered using the setFailureHandler method of the RMISocketFactory class. Its failure( ) method returns a boolean indicating whether or not the runtime should attempt to re-create the ServerSocket.
RMISecurityException	The RMISecurityException class is a deprecated class that extends the SecurityException class. A RMISecurityException signals that a security exception has occurred during the execution of one of the methods in java.rmi. RMISecurityManager.
RMISecurityManager	The RMISecurityManager extends the java.lang.SecurityManager. It is used to set the security policies to govern the behavior of the stub and skeleton classes.

RMIServerSocketFactory	An instance of the RMIServerSocketFactory interface is used by the RMI runtime in order to obtain server sockets for RMI calls.
RMISocketFactory	The RMISocketFactory class extends the Object class and implements the RMIClientSocketFactoryand RMIServerSocketFactory interfaces. An instance of the RMIServerSocketFactory class is used by the RMI runtime in order to obtain client and server sockets for RMI calls.
RPC	Remote Procedure Call (RPC) provides a function-oriented interface to socket level communications.
Skeleton	The Skeleton is a deprecated interface used solely by the RMI implementation. Every skeleton class generated by the rmic stub compiler in the version 1.1 implements this interface.
SkeletonMismatch Exception	The skeletonMismatchException is a deprecated exception that extends the RemoteException. This exception is thrown when a call is received that does not match the available skeleton.
SkeletonNotFound Exception	The SkeletonNotFoundException is a deprecated exception that extends the RemoteException. It is thrown if the skeleton corresponding to the remote object being exported is not found.
ServerCloneException	The ServerCloneException extends the CloneNotSupportedException. A ServerClone Exception is thrown if a remote exception occurs during the cloning of a UnicastRemoteObject.
ServerError	A ServerError is thrown as a result of a remote method call if the execution of the remote method on the server machine throws a java.lang.Error.
ServerException	A ServerException is thrown as a result of a remote method call if the execution of the remote method on the server machine throws a RemoteException.
ServerNotActive Exception	The ServerNotActiveException extends the Exception. It is thrown if the getClientHost( ) method of the RemoteServer thrown if the getClientHost( ) method of the RemoteServer class is called outside of servicing a remote method call.
ServerRef	The ServerRef interface extends the RemoteRef interface and represents the server handle for a remote object implementation.
ServerRuntimeException	A ServerRuntimeException is a deprecated exception that is thrown as a result of a remote method call if the execution of the remote method on the server machine throws a java.lang. RuntimeException.

Sockets	Socket is a channel through which applications can connect with each other and communicate.
SocketSecurityException	The SocketSecurityException extends the ExportException. It is thrown during a remote object export if the code exporting the remote object does not have permission to create an instance of the ServerSocket class of the java.net package on the port number specified.
Stream	A stream is a series of zeros and ones.
Stub	The Stub is a client side proxy of the remote object.
StubNotFoundException	A StubNotFoundException is thrown if a valid stub class could not be found for a remote object when it is exported or when a remote object is passed in a remote method call as a parameter or return value.
WriteObject( )	WriteObject( ) method is used in serialization to send the data to the object stream.
UID	The UID class extends the Object class and implements the Serializable interface. The UID class is an abstraction for creating identifiers that are unique with respect to the host on which it is generated.
UnexpectedException	An UnexpectedException is thrown if the client of a remote method call receives a checked exception that is not among the checked exception types declared in the throws clause of the method in the remote interface.
UnicastRemoteObject	The UnicastRemoteObject extends the RemoteServer class. The UnicastRemoteObject class defines a non-replicated remote object whose references are valid only while the server process is alive.
UnknownHostException	An UnknownHostException is thrown if a java.net. UnknownHostException occurs while creating a connection to the remote host for a remote method call.
UnmarshalException	An UnmarshalException is thrown while unmarshalling the parameters or results of a remote method call if the protocol for the return value is invalid, an exception occurs while unmarshalling the call header, etc.
Unmarshaling	Unmarshalling is the reverse process of marshalling in which the data is converted from a format portable across the network to the format used by methods.
Unreferenced	A remote object implementation should implement the unreferenced interface in order to receive notification when there are no more client that reference that remote object.
VMID	A Virtual Machine ID (VMID) is an identifier that is unique across all Java virtual machines. The VMID class extends the Object class and implements the Serializable interface.

Index